Soil and Environmental Analysis

BOOKS IN SOILS, PLANTS, AND THE ENVIRONMENT

Soil Biochemistry, Volume 1, edited by A. D. McLaren and G. H. Peterson
Soil Biochemistry, Volume 2, edited by A. D. McLaren and J. Skujiņš
Soil Biochemistry, Volume 3, edited by E. A. Paul and A. D. McLaren
Soil Biochemistry, Volume 4, edited by E. A. Paul and A. D. McLaren
Soil Biochemistry, Volume 5, edited by E. A. Paul and J. N. Ladd
Soil Biochemistry, Volume 6, edited by Jean-Marc Bollag and G. Stotzky
Soil Biochemistry, Volume 7, edited by G. Stotzky and Jean-Marc Bollag
Soil Biochemistry, Volume 8, edited by Jean-Marc Bollag and G. Stotzky
Soil Biochemistry, Volume 9, edited by G. Stotzky and Jean-Marc Bollag
Soil Biochemistry, Volume 10, edited by Jean-Marc Bollag and G. Stotzky

Organic Chemicals in the Soil Environment, Volumes 1 and 2, edited by C. A. I. Goring and J. W. Hamaker
Humic Substances in the Environment, M. Schnitzer and S. U. Khan
Microbial Life in the Soil: An Introduction, T. Hattori
Principles of Soil Chemistry, Kim H. Tan
Soil Analysis: Instrumental Techniques and Related Procedures, edited by Keith A. Smith
Soil Reclamation Processes: Microbiological Analyses and Applications, edited by Robert L. Tate III and Donald A. Klein
Symbiotic Nitrogen Fixation Technology, edited by Gerald H. Elkan
Soil–Water Interactions: Mechanisms and Applications, Shingo Iwata and Toshio Tabuchi with Benno P. Warkentin
Soil Analysis: Modern Instrumental Techniques, Second Edition, edited by Keith A. Smith
Soil Analysis: Physical Methods, edited by Keith A. Smith and Chris E. Mullins
Growth and Mineral Nutrition of Field Crops, N. K. Fageria, V. C. Baligar, and Charles Allan Jones

Semiarid Lands and Deserts: Soil Resource and Reclamation, edited by J. Skujiņš

Plant Roots: The Hidden Half, edited by Yoav Waisel, Amram Eshel, and Uzi Kafkafi

Plant Biochemical Regulators, edited by Harold W. Gausman

Maximizing Crop Yields, N. K. Fageria

Transgenic Plants: Fundamentals and Applications, edited by Andrew Hiatt

Soil Microbial Ecology: Applications in Agricultural and Environmental Management, edited by F. Blaine Metting, Jr.

Principles of Soil Chemistry: Second Edition, Kim H. Tan

Water Flow in Soils, edited by Tsuyoshi Miyazaki

Handbook of Plant and Crop Stress, edited by Mohammad Pessarakli

Genetic Improvement of Field Crops, edited by Gustavo A. Slafer

Agricultural Field Experiments: Design and Analysis, Roger G. Petersen

Environmental Soil Science, Kim H. Tan

Mechanisms of Plant Growth and Improved Productivity: Modern Approaches, edited by Amarjit S. Basra

Selenium in the Environment, edited by W. T. Frankenberger, Jr., and Sally Benson

Plant–Environment Interactions, edited by Robert E. Wilkinson

Handbook of Plant and Crop Physiology, edited by Mohammad Pessarakli

Handbook of Phytoalexin Metabolism and Action, edited by M. Daniel and R. P. Purkayastha

Soil–Water Interactions: Mechanisms and Applications, Second Edition, Revised and Expanded, Shingo Iwata, Toshio Tabuchi, and Benno P. Warkentin

Stored-Grain Ecosystems, edited by Digvir S. Jayas, Noel D. G. White, and William E. Muir

Agrochemicals from Natural Products, edited by C. R. A. Godfrey

Seed Development and Germination, edited by Jaime Kigel and Gad Galili

Nitrogen Fertilization in the Environment, edited by Peter Edward Bacon

Phytohormones in Soils: Microbial Production and Function, William T. Frankenberger, Jr., and Muhammad Arshad

Handbook of Weed Management Systems, edited by Albert E. Smith

Soil Sampling, Preparation, and Analysis, Kim H. Tan

Soil Erosion, Conservation, and Rehabilitation, edited by Menachem Agassi

Plant Roots: The Hidden Half, Second Edition, Revised and Expanded, edited by Yoav Waisel, Amram Eshel, and Uzi Kafkafi

Photoassimilate Distribution in Plants and Crops: Source–Sink Relationships, edited by Eli Zamski and Arthur A. Schaffer

Mass Spectrometry of Soils, edited by Thomas W. Boutton and Shinichi Yamasaki

Handbook of Photosynthesis, edited by Mohammad Pessarakli

Chemical and Isotopic Groundwater Hydrology: The Applied Approach, Second Edition, Revised and Expanded, Emanuel Mazor

Fauna in Soil Ecosystems: Recycling Processes, Nutrient Fluxes, and Agricultural Production, edited by Gero Benckiser

Soil and Plant Analysis in Sustainable Agriculture and Environment, edited by Teresa Hood and J. Benton Jones, Jr.
Seeds Handbook: Biology, Production, Processing, and Storage, B. B. Desai, P. M. Kotecha, and D. K. Salunkhe
Modern Soil Microbiology, edited by J. D. van Elsas, J. T. Trevors, and E. M. H. Wellington
Growth and Mineral Nutrition of Field Crops: Second Edition, N. K. Fageria, V. C. Baligar, and Charles Allan Jones
Fungal Pathogenesis in Plants and Crops: Molecular Biology and Host Defense Mechanisms, P. Vidhyasekaran
Plant Pathogen Detection and Disease Diagnosis, P. Narayanasamy
Agricultural Systems Modeling and Simulation, edited by Robert M. Peart and R. Bruce Curry
Agricultural Biotechnology, edited by Arie Altman
Plant–Microbe Interactions and Biological Control, edited by Greg J. Boland and L. David Kuykendall
Handbook of Soil Conditioners: Substances That Enhance the Physical Properties of Soil, edited by Arthur Wallace and Richard E. Terry
Environmental Chemistry of Selenium, edited by William T. Frankenberger, Jr., and Richard A. Engberg
Principles of Soil Chemistry: Third Edition, Revised and Expanded, Kim H. Tan
Sulfur in the Environment, edited by Douglas G. Maynard
Soil–Machine Interactions: A Finite Element Perspective, edited by Jie Shen and Radhey Lal Kushwaha
Mycotoxins in Agriculture and Food Safety, edited by Kaushal K. Sinha and Deepak Bhatnagar
Plant Amino Acids: Biochemistry and Biotechnology, edited by Bijay K. Singh
Handbook of Functional Plant Ecology, edited by Francisco I. Pugnaire and Fernando Valladares
Handbook of Plant and Crop Stress: Second Edition, Revised and Expanded, edited by Mohammad Pessarakli
Plant Responses to Environmental Stresses: From Phytohormones to Genome Reorganization, edited by H. R. Lerner
Handbook of Pest Management, edited by John R. Ruberson
Environmental Soil Science: Second Edition, Revised and Expanded, Kim H. Tan
Microbial Endophytes, edited by Charles W. Bacon and James F. White, Jr.
Plant–Environment Interactions: Second Edition, edited by Robert E. Wilkinson
Microbial Pest Control, Sushil K. Khetan
Soil and Environmental Analysis: Physical Methods, Second Edition, Revised and Expanded, edited by Keith A. Smith and Chris E. Mullins
The Rhizosphere: Biochemistry and Organic Substances at the Soil–Plant Interface, Roberto Pinton, Zeno Varanini, and Paolo Nannipieri

Additional Volumes in Preparation

Woody Plants and Woody Plant Management: Ecology, Safety, and Environmental Impact, Rodney W. Bovey

Handbook of Postharvest Technology, A. Chakraverty, Arun S. Mujumdar, and G. S. V. Raghavan

Metals in the Environment, M. N. V. Prasad

Soil and Environmental Analysis

Physical Methods

Second Edition
Revised and Expanded

edited by

Keith A. Smith
University of Edinburgh
Edinburgh, Scotland

Chris E. Mullins
University of Aberdeen
Aberdeen, Scotland

CRC Press
Taylor & Francis Group
Boca Raton London New York

CRC Press is an imprint of the
Taylor & Francis Group, an **informa** business

CRC Press
Taylor & Francis Group
6000 Broken Sound Parkway NW, Suite 300
Boca Raton, FL 33487-2742

First issued in paperback 2019

ISBN-13: 978-0-8247-0414-8 (hbk)
ISBN-13: 978-0-367-39804-0 (pbk)

Library of Congress Cataloging-in-Publication Data

Soil and environmental analysis : physical methods/edited by Keith A. Smith, Chris E. Mullins. —2nd ed., rev. and expanded
p. cm. — (Books in soils, plants, and the environment)
Rev. ed. of: Soil analysis. 1991.
ISBN 0-8247-0414-2 (alk. paper)
1. Soil physics—Methodology. 2. Soils—Environmental aspects. I. Smith, Keith A., II. Mullins, Chris E. III. Soil analysis. IV. Series.

S592.3 .S66 2000
631.4′3—dc21 00-060207

Visit the Taylor & Francis Web site at
http://www.taylorandfrancis.com

and the CRC Press Web site at
http://www.crcpress.com

Preface

This second edition retains all of the topics covered in the first edition. Each chapter has been revised, to take account of new developments. The two separate contributions relating to penetrometer measurements have been combined into one chapter, and others have been somewhat shortened, in order to include new material on the measurement of infiltration, the measurement of soil strength and friability, and field methods of assessment of soil physical conditions. The chapter on gas movement and air-filled porosity now covers soil–atmosphere exchange of environmentally important gases, including radon and greenhouse gases.

While some topics have undergone relatively little change in terms of available methods or instrumentation in the period since the first edition appeared, some have changed considerably. The measurement of soil water, which has such an important role in soil physics and which underwent such a change when the neutron probe was developed, can now be undertaken with other sophisticated instruments. For example, time domain reflectrometry (TDR) and frequency domain systems, which share with the neutron method the desirable feature of allowing nondestructive measurements at the same site to study temporal variations, now provide a reliable alternative to the neutron probe, while avoiding the problems of radiation protection. The widespread availability and use of data loggers has also transformed our approach to many measurements, particularly water content, matric potential, penetrometry, and soil thermal properties, and placed a greater emphasis on those instruments that can be logged.

Like the previous edition, this book is aimed at the researcher or agricultural or environmental adviser working in environmental science, soil science, or a related field. It should also be useful to teachers and students in postgraduate courses in soil science, soil analysis, and environmental science. One of the significant trends of the past few years has been the development of interdisciplinary activities, in the attempt to improve understanding of complex phenomena in the life and environmental sciences. This places new emphasis on the concurrent measurement of physical, chemical and biological parameters. One typical example of this is the study of losses of nitrogen from soils into waters and the atmosphere, where information may be needed on soil water infiltration, saturated and unsaturated flow, and water-filled pore space—all of which require physical measurements—as well as on soil mineral nitrogen analysis and plant growth. Researchers who may have trained in chemistry or biological sciences now need to become informed about physical techniques as well. In this book we attempt to provide an introduction to each type of measurement, with enough theory to teach the principles behind the methods, and to help in the selection of methods appropriate to the task at hand.

Keith A. Smith
Chris E. Mullins

Contents

Contributors

David Atkinson Scottish Agricultural College, Edinburgh, Scotland

Bruce C. Ball Land Management Department, Scottish Agricultural College, Edinburgh, Scotland

Tom Batey Department of Plant and Soil Science, University of Aberdeen, Aberdeen, Scotland

A. Glyn Bengough Soil–Plant Dynamics Unit, Scottish Crop Research Institute, Dundee, Scotland

Ken Blyth Department of Bio-Physical Processes, Centre for Ecology and Hydrology, Wallingford, Oxfordshire, England

Graeme D. Buchan Soil and Physical Sciences Group, Lincoln University, Canterbury, New Zealand

Donald J. Campbell Land Management Department, Scottish Agricultural College, Edinburgh, Scotland

Andrée Carter Agricultural Development Advisory Service, Rosemaund, Preston Wynne, Hereford, England

Brent E. Clothier HortResearch, Palmerston North, New Zealand

J. David Cooper Instrument Section, Centre for Ecology and Hydrology, Wallingford, Oxfordshire, England

Lorna Anne Dawson Plant Science Group, Macaulay Land Use Research Institute, Aberdeen, Scotland

A. R. Dexter Department of Soil Physics, Institute of Soil Science and Plant Cultivation, Pulawy, Poland

Christiaan Dirksen Department of Water Resources, Wageningen University, Wageningen, The Netherlands

Catriona M. K. Gardner Jesus College, University of Oxford, Oxford, England

J. Kenneth Henshall Land Management Department, Scottish Agricultural College, Edinburgh, Scotland

Peter J. Loveland Soil Survey and Land Research Centre, Cranfield University, Silsoe, Bedfordshire, England

Chris E. Mullins Plant and Soil Science Department, University of Aberdeen, Aberdeen, Scotland

Michael F. O'Sullivan Engineering Resources Group, Scottish Agricultural College, Edinburgh, Scotland

Malcolm J. Reeve Land Research Associates, Derby, England

David Robinson Centre for Ecology and Hydrology, Wallingford, Oxfordshire, England

Keith A. Smith Institute of Ecology and Resource Management, University of Edinburgh, Edinburgh, Scotland

John Townend Plant and Soil Science Department, University of Aberdeen, Aberdeen, Scotland

Chris W. Watts Department of Soil Science, Silsoe Research Institute, Silsoe, Bedfordshire, England

W. Richard Whalley Department of Soil Science, Silsoe Research Institute, Silsoe, Bedfordshire, England

Edward G. Youngs Institute of Water and Environment, Cranfield University, Silsoe, Bedfordshire, England

Soil and Environmental Analysis

1
Soil Water Content

Catriona M. K. Gardner
Jesus College, University of Oxford, Oxford, England
David Robinson, Ken Blyth, and J. David Cooper
Centre for Ecology and Hydrology, Wallingford, Oxfordshire, England

I. INTRODUCTION

Measurement of the water content of soil and the unsaturated zone is fundamental to many investigations in agriculture, horticulture, forestry, ecology, hydrology, civil engineering, waste management, and other environmental fields. While other factors related to soil water are important, probably the single most useful piece of information about soil water is knowing how much is present, either in a complete profile or within a well-defined volume.

The diverse range of applications means that there is a wide range of demands on the measurements. Some objectives require a single measurement of total soil water content in a field profile, whereas others demand repeated measurements of the spatial distribution of water content to track changes over time. The time scales may vary from minutes to months. Measurements may be undertaken in the laboratory, on loose or repacked samples, on undisturbed cores, in plant containers or lysimeters, or as part of field experiments, trials or larger, catchment scale, studies. The measurement precision and accuracy demanded varies widely and hence so does the sophistication of the methodology which must be employed. As a result of this wide range of demands, no one method can satisfy all requirements. However, three methods are used for the vast majority of determinations today: the thermogravimetric method, neutron thermalization, and a group of techniques based on measurement of soil dielectric properties.

The oldest established and the only truly direct method is the thermogravimetric method, which requires samples for oven-drying. The other two techniques rely on measurement of physical properties of the soil that depend on its water content. The neutron method was adopted for routine use in the 1960s and has been popular ever since, although the radiation hazard and cost preclude semipermanent installation and hence automation. The development of dielectric methods since 1980 has introduced opportunities for rapid collection of soil water content data at short time intervals, five minutes or less if required, and permitted automation and logging of measurements. The ability to log soil water content automatically is opening up ways of soil water monitoring and soil hydrological research that have hitherto been impossible.

In this chapter, the concept of soil water content, definitions of the water content of a block of soil, and the terminology and units used are described briefly. The relative merits of direct and indirect measurements and the spatial and temporal resolution that can be achieved by various methods are considered. The principles and practice of the three methods are then discussed in detail and applications of the neutron and dielectric methods are described. A summary of the more common alternatives to the three major ground-based methods for soil water content measurement, referred to above, is provided in Table 1. A review of techniques for remote sensing of soil water, which complement ground-based techniques, is also provided.

II. SOIL WATER CONTENT

A. Definition

The term "soil water content" is widely accepted as referring to the water that may be evaporated from a soil by heating to between 100 and 110°C, but usually at 105°C, until there is no further weight loss. This is the basis of the thermogravimetric method. It is important to be aware of the arbitrary nature of this definition, which is the standard reference against which other techniques are normally calibrated. As Gardner (1986) stated, "the choice of this particular temperature range appears not to have been based upon scientific consideration of the drying characteristics of soil." Its origin probably has more to do with the notion of ensuring evaporation of liquid or "free" water and the relative ease with which determinations can be made by oven-drying samples.

Water is present in soil as water vapor and liquid. In addition, water molecules are adsorbed in layers on the surfaces of colloidal materials, particularly clays, and molecules are incorporated with hydroxyl groups within clay lattice structures. The distinctions between thin films of water retained by surface tension and water that is adsorbed (bound water), and between bound and structural water, are less precise than this categorization suggests. Water vapor and structural water are disregarded in the conventional definition of soil water content. Structural water is immobile and is generally released only upon heating to temperatures

Table 1 Alternative Methods of Soil Water Content Measurement

Method	Use	Principle	Application	References/ Comments
Calcium carbide method	Lab/field, on samples	Calcium carbide mixed with soil in pressure chamber produces acetylene gas; gas pressure depends on soil water content	Civil engineering purposes as well as agricultural	Morrison (1983)
Sulphuric acid method	Lab, on samples	Concentrated sulphuric acid mixed with soil raises temperature; maximum temperature depends on soil water content	Mainly agricultural	Gupta and Gupta (1981)
Soil matric potential	Lab/field in situ	Soil matric potential measurements are translated into water content using the water release characteristic. As the matric potential–water content relationship is hysteretic, precise determination of water content is not possible. Assumes the soil water release characteristic is known	Field or lab where soil matric potential measurement required but water content measurement precluded, e.g., irrigation	See Chapter 2 for details of measuring soil matric potential
Gamma ray attenuation	Lab/field in situ	When soil is irradiated with gamma rays, the scattering and absorption which occur are primarily a function of soil density. In nonshrink–swell soils, temporal variation in total bulk density is due to water content change and therefore gamma ray attenuation or backscatter can be used to monitor water content	Experimental conditions only due to cost and radioactive hazard. Used in lab and field	See Chapter 7 for details of gamma ray methodology. Wood and Collis George (1980); Morrison (1983)
Nuclear magnetic resonance spectroscopy (NMR)	Lab samples, field in situ	Atomic nuclei change their energy levels when subjected to oscillating electromagnetic fields; different frequencies affect different nuclei, but hydrogen nuclei give the strongest response. Electronic detection of either the energy absorption or nuclear dipole excitation gives the NMR signal. NMR measurement of hydrogen concentration is related to water content by calibration	Geophysical use in boreholes. Experimental NMR equipment for field measurements on samples and of surface water content has been described	Paetzold and Matzkanin (1984); Paetzold et al. (1985)
Thermal conductivity	Lab/field in situ	An electrical heating element and a temperature sensor are placed in soil either directly, or encased in a porous block. The time for a given temperature to be achieved after heat is applied is measured. The rate of heat dissipation is a function of soil thermal diffusivity, which depends on soil water content	Mainly agricultural. Direct contact probes require good contact with soil; blocks respond to soil water potential—see above. Usable in very saline soils	Fritton et al. (1974); Sophocleus (1979)

between 400 and 800°C; an exception is gypsum, from which structural water is lost at only 80°C. Bound water does have a degree of mobility which becomes important at very low water contents and may be exploited by drought-resistant plants. Heating to 105°C is not normally sufficient to remove bound water; most is eliminated from clay surfaces at temperatures between 110 and 160°C.

The conventional definition of soil water content is not a limitation in most work because the quantities of bound and structural water are small relative to the "free" water content and can be assumed to be constant for most purposes. In practice it is usually changes of soil water content with time that are of interest (e.g., seasonal changes in field soils or change in response to irrigation). Alternatively, the quantity of water retained between specific thresholds may be required (e.g., between the liquid and plastic limits or between "field capacity" and "wilting point"). Several methods of water content determination, including the neutron probe and dielectric methods, are sensitive to all the water molecules present in a soil, although this information is effectively lost as they are calibrated against thermogravimetric determinations. Dielectric methods have the potential to discriminate between liquid, bound, and structural water, but this has yet to be exploited.

B. Units

Soil water content may be expressed on either a mass or a volumetric basis, that is, as a mass ratio, kg kg^{-1} (kg water per kg dry soil), or a volume fraction, $m^3\ m^{-3}$ (m^3 water per m^3 of bulk soil volume), respectively. In either case the value is a dimensionless fraction and can be multiplied by 100 to express it as a percentage. One can be obtained from the other if the dry bulk density of the soil, and the density of water, are known:

$$\theta = \frac{w\rho_b}{\rho_w} \tag{1}$$

where θ is volumetric soil water content (volume fraction), w is water content as a mass fraction, ρ_b is the dry bulk density of the soil (kg m^{-3}), and ρ_w is the density of free water (usually approximated as 1000 kg m^{-3}). For most purposes, expression as a volume fraction is more useful, since multiplying θ by the soil depth gives the "depth" of water in that depth of soil, a figure with the same (length) dimensions used to express rainfall, evaporation, transpiration, drainage, and irrigation.

Because the thermogravimetric method is used as a standard for calibration, soil bulk density as well as water content measurements are required to calibrate techniques that measure volumetric water content, unless undisturbed samples of known volume are obtained for oven-drying. This introduces an additional source of error into the calibration. Since a technique can be no more accurate than the procedure used to calibrate it, particular care is required in both the water content and the density determinations when undertaking calibrations.

If soil water content is monitored at several depths in a core or a soil profile, the depth interval z_i to which a measurement θ_i refers is normally taken as the vertical distance separating the two midpoints between the measurement depth and the depths of the measurements immediately above and below it. The water content of the soil profile, P, to a depth z, is obtained by summation of the water contents of each depth interval:

$$P = \sum_{0}^{z} \theta_i z_i \tag{2}$$

The effect of this integration of a step function of the water content is equivalent to trapezoidal integration; although little used, Simpson's rule can reduce the errors involved (Haverkamp et al., 1984).

C. Direct *Versus* Indirect Measurements

Direct measurements involve removal of soil water by evaporation, leaching, or a chemical process, and subsequent determination of the amount of water removed; the thermogravimetric method is the principal example. Direct measurements are beset with problems primarily due to the need for destructive sampling. Thus replicate samples must be taken to determine the variance of measurements made on a given occasion and whether they differ significantly from measurements made on other occasions. This replication can result in the handling of very large numbers of samples. Practical difficulties are compounded if determinations deep in the profile are required. Furthermore, repeated sampling within the same area may cause unacceptable damage to the soil or vegetation present. Provision must also be made for bulk density determinations if volumetric water content data are required. However, taking undisturbed cores of known volume to determine both water content and bulk density avoids this.

Indirect methods depend on monitoring a soil property that is a function of water content (e.g., the basis of the neutron method is detection of hydrogen nuclei in soil, most of which are present in water molecules). Indirect methods usually involve instrumentation placed in or on the soil, or remote techniques involving sensors mounted on a platform over the soil or on aircraft or satellites. Although indirect measurements require calibration, most have the considerable advantage that measurements on the soil in situ are possible and these can be repeated at the same place through time.

Another significant advantage is that change in soil water content is determined directly. The standard error of estimation of change of water content obtained from repeated measurements on the same n samples is simply

$$s.e.(\Delta\theta) = \sqrt{\frac{\mathrm{var}(\Delta\theta)}{n(n-1)}} \quad (3)$$

whereas the standard error associated with a change in water content based on direct measurements made on two sets of n_1 and n_2 independent samples, depends on the variances attached to both sets of samples:

$$s.e.(\overline{\theta_2} - \overline{\theta_1}) = \sqrt{\frac{\mathrm{var}(\theta_1)}{n_1(n_1-1)} + \frac{\mathrm{var}(\theta_2)}{n_2(n_2-1)}} \quad (4)$$

In the latter case, the variation in the water content on each measurement occasion is superimposed on the spatial variation of the change in water content. Therefore, if changes of water content are the focus of interest, rather than absolute water contents, indirect in situ measurements are preferable to direct measurement that involves removing samples.

D. Spatial Resolution of Measurements

The thermogravimetric, neutron, dielectric, and remote sensing methods between them cover various measurement scales in three dimensions (Fig. 1). Most measurements integrate over a volume around a position in the soil, the size of which depends on the technique used, or may be defined by the size of a sample or core taken to the laboratory. Oven-drying of a soil sample produces an integrated water content measurement for that sample. Most instruments integrate the water content unevenly over a volume of soil, with the largest contribution coming from the region close to the sensor. The size of the volume measured is frequently dependent on the water content of the soil. The neutron depth probe measures a sphere of soil, 0.2 to 0.5 m in diameter, centered approximately on the source. Many dielectric instruments have parallel rod type sensors that are usually most influenced by the soil between and immediately around the rods and so measure a roughly cylindrical volume, the length of which is determined by the length of the rods; the measurement integrates the water content along the sensor. Rod spacing in most equipment implies a cylinder of 50 to 100 mm diameter, and rod lengths range from 50 mm to 1 m. In deciding which measurement method to employ, it is important to consider the volume of soil that the measurements will represent and how water content or other gradients within that volume (e.g., wetting fronts, density, or mineralogical differences) may influence the measurements made.

Many techniques make what are referred to as "point measurements." In practice this is actually a measurement of soil water content within a finite volume which is small compared with the overall scale of the area and/or depth range

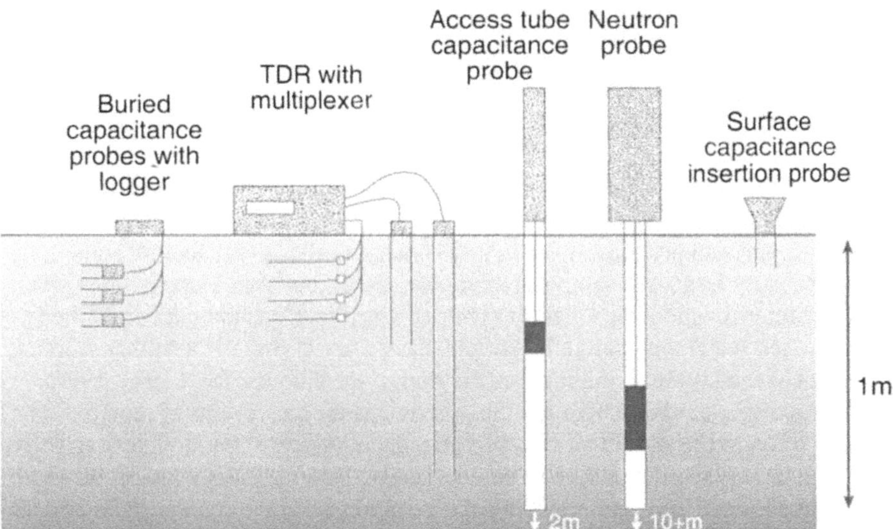

Fig. 1 Spatial arrangement of soil water sensors for in situ measurement. Sensors for dielectric methods (capacitance and time domain reflectometry, TDR) can be installed semipermanently and operated automatically. Installation of access tubes permits manual use of neutron or capacitance depth probes. Capacitance and TDR instruments can also be used for one-off readings at the soil surface.

under study. Water content information is often required over large areas, but research is only now addressing how to make the leap from "point" to areal measurements. Remote sensing techniques are potentially very useful in this respect; although they only allow measurement of water content in the surface soil, the combination of this with point measurements at greater depth, and/or modeling of changes with depth, has considerable potential that has yet to be fully realized.

III. THE THERMOGRAVIMETRIC METHOD

The thermogravimetric method is straightforward. A soil sample is placed in a heat-proof container of known weight, weighed, dried in an oven set at a constant temperature of 105°C, removed and allowed to cool in a desiccator, then reweighed. This procedure is repeated until the sample attains a constant mass (ISO, 1993). The water content, w, of the sample is the mass of water per unit mass of dry soil:

$$w = \frac{\text{Mass of wet soil} - \text{Mass of dry soil}}{\text{Mass of dry soil}} \tag{5}$$

If a sample of known volume obtained by coring is used, the volumetric water content can be obtained directly:

$$\theta = \frac{\text{Mass of wet soil} - \text{Mass of dry soil}}{\text{Soil volume}} \tag{6}$$

(ISO, 1997). An oven temperature of $105 \pm 5°\text{C}$ and a 24 hour drying period are widely adopted. Drying time is influenced by the oven's efficiency and the condition, size, and number of samples in it. 24 hours may be insufficient for some soils and especially large wet samples (Reynolds, 1970), but unnecessarily long when making determinations on small or air-dried samples. Constant mass is defined as that reached when the change in sample mass, after drying for a further 4 hours, does not exceed 0.1% of the mass at the start of the 4 hours (ISO, 1993, 1997).

An oven ventilated by a fan that distributes the heat evenly is required. The drying temperature should be checked periodically using a thermocouple in a dry soil sample. Oven efficiency can be checked by loading it with subsamples of a well mixed moist soil and checking the variation in water content measured. A balance capable of weighing to better than 0.1% of the mass of the dried samples is required. Analyses of the random errors accompanying gravimetric water content determination due to varying degrees of weighing precision and accuracy were provided by Gardner (1986).

Recommended sample sizes range from 10 to 100 g (Australian Water Resources Council, 1974), but 50 to 100 g is preferable for moist samples. If volumetric water content is to be obtained, undisturbed cores of at least 20 cm^3 should be collected and dried. For stony soils, larger samples are necessary; recommendations according to the dimensions of the aggregates and stones in the moist soil are available (ASTM, 1981). Variation of the proportion of stone in samples may cause problems, in which case the water content of the < 2 mm fraction, $\theta_{<2}$, and the volume of the stone (> 2 mm) fraction, S, are determined (Reinhart, 1961). The water content of the whole soil is

$$\theta = \theta_{<2}(1 - S) \tag{7}$$

The water content of the stone fraction, θ_s, is often considered to be negligible (Hanson and Blevins, 1979) but may not be, in which case it should be determined by oven-drying as for the soil and included in the calculation of θ.

When dealing with organic soils, some inaccuracy in water content determination may occur due to the oxidation and decomposition of organic matter at 105°C, causing weight loss other than that due to water evaporation. In certain soils, volatilization of substances other than water may occur at temperatures below 105°C, causing similar problems. Lower drying temperatures may be considered when working with soils where this occurs but can lead to determination of significantly lower water contents.

Because of its simplicity, the oven-drying method is easily abused. In particular, oven temperatures may not be checked and neither they nor the drying time are usually reported. Common problems include drying of the soil during transit before weighing, loss of soil in transit, water uptake from the atmosphere during cooling because no desiccator was used, and poor determination of the volume of the core or the dry bulk density. The use of thermogravimetric determinations as a reference against which to calibrate and investigate the accuracy of other methods of water content measurement requires special care in its application. The advantages of this method are its simplicity, reliability, and low cost in terms of equipment requirements. The disadvantages are the need for destructive sampling, the time required for drying, and the staff time needed to deal with large numbers of determinations.

Drying time may be reduced to ≤ 20 min with the use of microwave ovens, but there are two problems inherent in this approach: drying time increases with initial water content; and, if a dry sample is left in a microwave oven, its temperature will continue to rise beyond 105°C which may cause weight changes other than those due to evaporation of water. Consequently, drying times must be estimated initially. Microwave drying can give water content determinations within 0.5 to 1.0% of those obtained using conventional oven-drying, if trials are conducted to determine appropriate drying times (Gee and Dodson, 1981; Tan, 1992). For some purposes the method may be suitable, but for best accuracy the use of a conventional oven is recommended (Standards Association of Australia, 1986).

IV. THE NEUTRON METHOD

The neutron method uses the ability of hydrogen to slow down fast neutrons much more efficiently than other substances. In any soil, most of the hydrogen is present in water molecules, and therefore changes in hydrogen concentration occur mainly due to changes in water content. A radioactive source, continually emitting fast neutrons, and a slow neutron detector, are housed within a probe that is lowered into the soil down an access tube. The fast neutrons are slowed as they move through the soil. The number of slow neutrons detected is recorded as a count rate and is converted to volumetric water content using a calibration relationship. For depth measurements in soil, an access tube is installed semipermanently and readings are made at successive depths by lowering the probe within the tube. Measurements can be made with ease to depths of 5 m or more in many soils, once the effort of access tube installation has been completed. Neutron meters of different design for use at the soil surface are also available.

The neutron method was first proposed in the 1940s (Brummer and Mardock, 1945; Pieper, 1949) and field tests soon followed (Belcher, 1950). By the mid-1950s, portable instruments for field use had been developed in North

America (Underwood et al., 1954; Stone et al., 1955) and Australia (Holmes, 1956). Equipment soon became available commercially. Instruments available today are considerable refinements of the early designs. Technological developments have permitted use of less hazardous neutron sources, reducing the amount of shielding required and allowing smaller, lighter yet safer designs. The electronics are more reliable and data can now be stored and processed on board.

The emphasis here is on neutron depth probes; surface meters are only considered briefly. Dual-purpose depth probes that measure soil bulk density by gamma ray attenuation (see Chapter 8), and water content by the neutron method, are also available. Three reports, although published some years ago, still represent the most comprehensive accounts of the theoretical and practical aspects of using neutron depth probes (IAEA, 1972; Greacen, 1981; Bell, 1987) and are recommended for further detail. Use of neutron depth probes is now well established, and standard procedures have been agreed upon (ISO, 1996).

A. Neutrons and Neutron Moderation

Neutrons are uncharged particles of mass very slightly greater than a proton. They are classified according to their kinetic energy measured in electron volts (eV). Fast neutrons have kinetic energies exceeding 1 keV. Thermal neutrons have energies of 0.025 to 0.5 eV and are close to thermal equilibrium with the molecules of the surrounding medium; their movement through the medium is controlled by the gas diffusion laws.

Because they have no charge, neutrons are not influenced by electric fields. They are therefore able to penetrate through the electron cloud of an atom to reach the nucleus. When a neutron comes close to, or collides with, a nucleus, a variety of interactions may occur depending on the energy of the neutron and the characteristics of the nucleus. The probability that collisions resulting in a given interaction will occur when a substance is irradiated with neutrons of a given energy is defined by the interaction cross-section of the isotope, measured in units of area called barns; 1 barn is 10^{-28} m^2. The greater the cross-section, the greater is the probability of interactions. The macroscopic interaction cross-section of a unit volume of soil is calculated as the weighted sum of the values for the individual elements present. There are two types of neutron–nucleus interaction: neutron scattering and neutron capture.

1. *Neutron Scattering*

Scattering occurs when the collision of a fast neutron with a nucleus causes the neutron's direction of travel to change and its velocity, and so kinetic energy, to reduce. Such collisions may be elastic, i.e., kinetic energy and momentum are

Table 2 The Effect on Fast Neutrons of Collisions with Nuclei of the Commonest Elements in Soils

Nucleus	% energy lost in head-on collision	Average number of collisions to slow 2 MeV neutron to <0.5 eV
Hydrogen	100	18
Oxygen	22.1	152
Silicon	13.8	252
Aluminum	13.3	279
Iron	6.8	519

Source: Hodnett, 1986.

conserved, or inelastic, i.e., some of the neutron's energy is transferred to the nucleus, resulting in the emission of gamma radiation. Inelastic scattering is unimportant in the present context. The elastic scattering cross-section of most elements is small, less than 5 barns, and relatively constant at neutron energies between 2 eV and 2 MeV.

The loss of energy by a neutron in the course of elastic scattering is inversely related to the mass of the nucleus with which it collides. When a head-on collision takes place with a hydrogen nucleus, the neutron loses all of its energy. In practice, collisions occur at all angles, and many are required to slow a fast neutron (Table 2). Heavier nuclei are most likely to deflect a neutron through a greater angle from its original path without significant loss of energy. Collisions with heavy nuclei therefore reduce the distance that fast neutrons move from a source before they are slowed to thermal energies.

2. *Neutron Capture*

Some collisions between a neutron and a nucleus result in the neutron being absorbed (captured) by the nucleus. The capture cross-section depends on both the type of nucleus and the energy of the neutron. For most elements, it is negligible for neutron energies greater than 1 eV, so only slow neutrons are likely to be captured. The capture cross-section for most soil constituents is between 0.1 and 1 barn, but some elements have much larger values. Important examples are gadolinium (46,000 barn), cadmium (2,450 barn), and boron (755 barn). A trace of one of these in soil will greatly increase the soil's macroscopic capture cross-section and so reduce the slow neutron count rate markedly, thus affecting the calibration curve. Other more common elements, such as manganese (33 barn), chlorine (33 barn), and iron (2.6 barn), may have a significant effect if present in sufficient quantity. Capture reactions with certain elements result in emission of alpha particles or protons, and this is the basis on which slow neutron detectors operate.

B. Neutron Sources, Detectors, and Instrument Design

Fast neutron sources usually contain two elements: one emits alpha particles in the course of radioactive decay; the other is beryllium, which absorbs the alpha particles and in the process emits fast neutrons. The reaction is

$$^{9}_{4}Be + ^{4}_{2}He \rightarrow ^{1}_{0}n + ^{12}_{6}C + 5.74 \text{ MeV}$$

The neutron emitted gains some of the reaction energy plus some of the alpha particle's energy. Most probes use sources with an isotope of americium, ^{241}Am, as the alpha emitter. It has a half-life of 458 years. Source activity in modern probes is usually 1.85 GBq (50 mCi) or less. The sources are constructed to strict safety standards: finely powdered beryllium and sintered americium oxide are contained within a double-walled capsule of stainless steel that is cylindrical or annular in shape. Their working life is at least 20 years, but regular tests for leakage should be conducted (Lorch, 1980).

Improvements in the detection efficiency of thermal neutron detectors have enabled use of lower activity sources in probes. The isotopes ^{10}B, ^{3}He, and ^{6}Li have very high capture cross-sections for neutrons of energy less than 1 eV and are relatively insensitive to high-energy neutrons. Boron trifluoride and helium-3 filled metal tube detectors are most common. Both require a stable 1 to 2 kV supply to operate. Lithium-enriched glass scintillation counters can give 100% detection efficiency but are more complex and delicate than gas counters. They can monitor gamma radiation separately from thermal neutrons and so are useful in dual-purpose probes. The efficiency of a detector declines slowly with time but the useful life is at least 15 years.

The arrangement of the source and detector within the probe contributes to its sensitivity to water content change. Certain geometries result in a linear calibration for the range of water contents commonly encountered. Ideally both source and detector would be placed at the same point, to give a symmetrical distribution of thermal neutrons about the detector. Some designs use an annular source fitted around the midpoint of the detector to achieve a symmetrical arrangement. If the detector is remote from the center of the neutron cloud, a nonlinear calibration results, and the influence of interfaces in the soil and at the surface is exacerbated.

Most neutron depth probes comprise six parts: the probe (containing the source and detector), which is connected by cable to the counting unit; the cable; the counting unit; the power supply; the probe carrier; and a system for lowering the probe into an access tube and locating it at given depths (Fig. 2). The counter unit measures the electronic pulses transmitted from the detector and displays the result. Most instruments count for a preset time, typically between 4 and 64 seconds. Longer count times can be selected on some instruments for high-precision

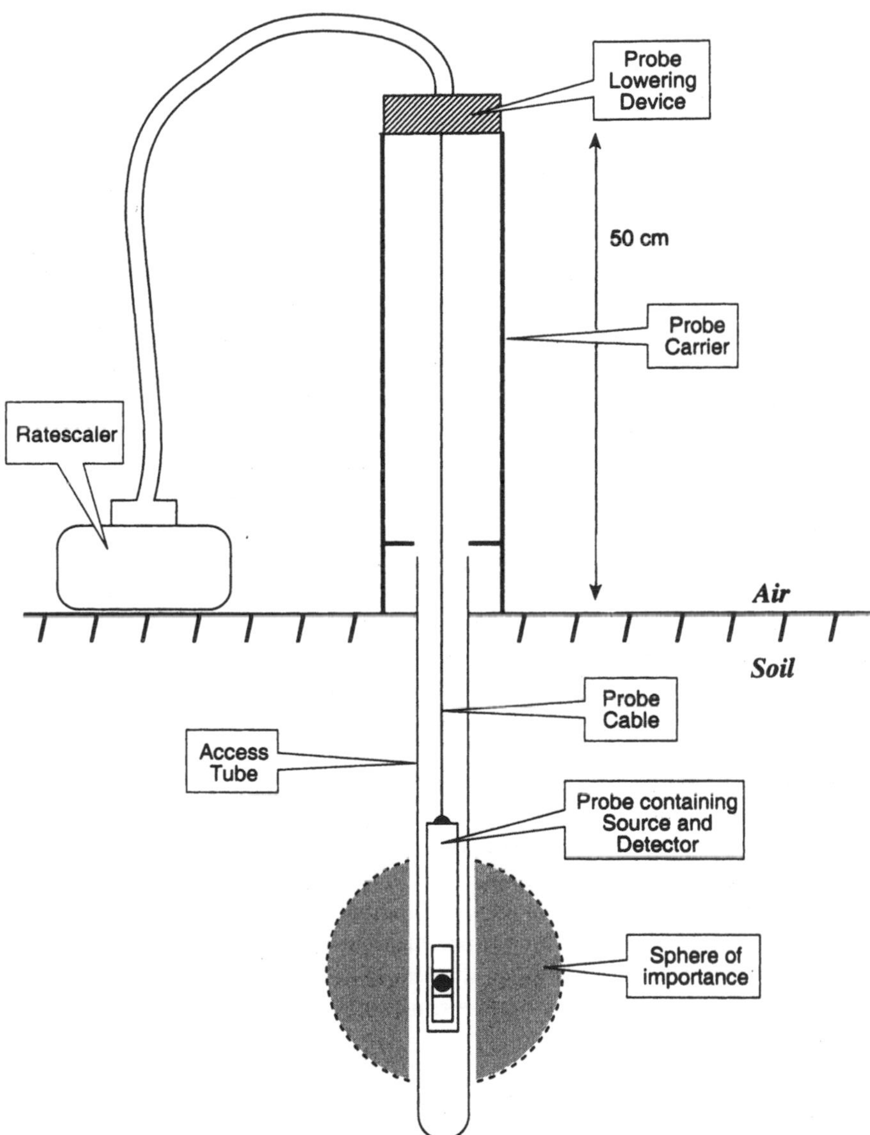

Fig. 2 Principal components of the neutron depth probe. The sphere of importance designates the volume of soil that contributes to the reading.

measurements. Nicolls et al. (1981) provide a useful account of instrument design in relation to sensitivity, accuracy, precision, and convenience of use.

C. Standard Neutron Count Rates

As indicated above, neutron depth probes of different design have different calibrations. However, the sensitivity of instruments of the same design is not identical either, due to differences in source strength and detector efficiency. To ensure data compatibility if slow aging of components occurs, if a component is replaced or a probe is otherwise repaired, or if more than one probe of the same type is in use, neutron count rates in a standard medium should be made at regular intervals. Calibrations should be made in terms of count rate ratio R/R_s, where R is the count rate in soil and R_s the standard neutron count rate. Data from probes of different designs cannot be normalized in this way, but intercalibration is possible (Nakayama and Reginato, 1982). Weekly standard counts are recommended, but if a probe is used less frequently, a standard count should be made before or after each reading occasion. A count time of 1 h minimizes the random error of the standard count, and so of water content measurements obtained with that count.

The use of a water standard is preferred to that of other hydrogen-rich media, such as plastics, because the count rate is almost independent of temperature and there is no possibility of water absorption from the atmosphere (Hodnett and Bell, 1990). A water standard can be cheaply constructed by fixing a water-tight access tube axially in the center of a drum that is then filled with water. The drum should be at least 0.6 m deep and 0.5 m in diameter to ensure that the water surrounding the source, when it is lowered into the access tube, effectively represents an infinite volume.

Some manufacturers suggest taking standard counts in the probe transport shield. This is not advisable, because the shield is not large enough to represent an infinite medium and therefore the counts are easily influenced by surrounding neutron moderators. In addition, temperature and humidity also affect the count rate. Precautions to overcome these shortcomings have been described (Hauser, 1984) but serve more to emphasize the simplicity and reliability of using a water standard.

D. Neutron Movement in Soil—The "Sphere of Importance"

A neutron emitted from the source of a probe travels outward into the soil until it collides with an atomic nucleus. Some energy is lost in the collision and the direction of travel altered. The neutron continues in the new direction until another collision occurs. Most neutrons migrate away from the source, but a proportion return, having been slowed in the process. The further a neutron gets from the source, the smaller its chance of returning; this is particularly so once thermal energies have been attained, as the probability of absorption is then greatly increased. The soil closest to the probe therefore has the greatest influence on the count rate measured. For working purposes a "sphere of importance" can be defined. The center of the sphere of importance lies between the source and the center of the detector. If the source is placed at the center of the detector, these are coincident. The sphere of importance is defined as that which, if the soil and water surrounding the sphere were removed, would result in a thermal neutron count that was a given fraction, usually 0.95, of the count if the medium were infinite in extent (IAEA, 1972).

The size of the sphere of importance depends on

1. The energy spectrum of the neutrons emitted from the source (the type of radionuclide in the source but *not* the source strength)
2. The neutron scattering and capture cross-sections of the soil and its bulk density
3. Soil water content

While the effects of 1 and 2 are constant for a given probe and soil, the influence of soil water content changes with time. The sphere's radius decreases as water content increases, because the greater hydrogen content causes more neutron scattering close to the probe, restricting movement of neutrons away from it. The radius of the sphere of importance of most depth probes with americium–beryllium sources is about 0.15 m in wet soil, increasing to about 0.5 m in very dry soil.

Since water content measurements are thus made on a sizeable volume of soil, there is little advantage to be gained from making readings at depth intervals of less than 0.1 m. When measurements are made through an interface between wet and dry soil, the measurements in the wet soil close to the interface will indicate that the soil is drier than is actually the case. Conversely, the water content of the dry soil near the interface is overestimated, but to a lesser degree than the underestimation for the wet soil (Hodnett, 1986). This effect increases with the difference in water content between the layers. The shape of the measured water content profile is smoothed, and so neutron probes are not suitable if measurements with good depth resolution are required. The slight underestimation of the total soil profile water content is usually disregarded. However, Van Vuuren (1984) found that the bias so introduced can be significant and advocated use of field calibrations to allow for site-specific properties such as the presence of a water table. Wilson (1988a) analyzed the phenomenon and demonstrated theoretically that it would be unwise to rely on measurements closer than about 0.25 m to a marked discontinuity such as a water table.

E. Random Counting Errors

Both radioactive decay and thermal neutron counting are random processes. When repeated neutron counts are made using the same time interval, the number of counts recorded varies. This is an important source of random error in the measurement. (Other errors may arise from changes in the placement depth, calibration uncertainties, thermal effects on the electronics, and warm up.) Repeated counts fit a Poisson distribution. For this distribution, if N is the total number of counts recorded over a time, t, the standard deviation of the mean value of N is $\sqrt{N}$. It is usual to work with a count rate, R, where $R = N/t$, and so the standard deviation of R is

$$\sigma_{\mathrm{R}} = \left(\frac{R}{t}\right)^{0.5} \tag{8}$$

Therefore, if the time taken to obtain a count is increased, the standard deviation of the mean decreases. The absolute error accompanying greater count rates obtained in wet soils is always greater than in dry soils, because if counts are made over a fixed time interval, R is greater, whereas if N is fixed, t is reduced.

The standard deviation of a standard count determination is $(R_s/t_s)^{0.5}$, and that of a water content determination is

$$\sigma_\theta = a\frac{R}{R_s}\left(\frac{1}{Rt} + \frac{1}{R_s t_s}\right)^{0.5} \tag{9}$$

where a is the slope of the calibration curve, R_s is the standard count rate (s^{-1}), and t_s is the standard count time (s). Since the standard count itself introduces a small error, long standard count times of an hour or so should be used, if possible, to minimize that source of error. The depth of water in a layer of soil is obtained by multiplying θ by the layer depth. Similarly σ_θ is multiplied by the layer depth to give the standard error of the layer value. The error associated with the profile water content value is the square root of the sum of squares of the errors attached to the individual layer values.

For field measurement purposes, the advantages of the greater precision obtained at one location associated with longer count times (Fig. 3) needs to be balanced against uncertainties arising from spatial variability of soil water content. Because of the latter, it is usually preferable to conduct measurements in many tubes using a short count time. This provides a better estimate of both the mean water content and its variability than more precise data from fewer tubes.

F. Field Measurements

Before measurements can be made with a depth probe at a new site, access tubes must be installed, measurement depths must be selected, and decisions regarding soil calibration and how to deal with measurements close to the soil surface are necessary. Measurement intervals of 0.1 or 0.2 m, perhaps increasing to 0.3 m at greater depth, are generally appropriate. Once a set of measurement depths has

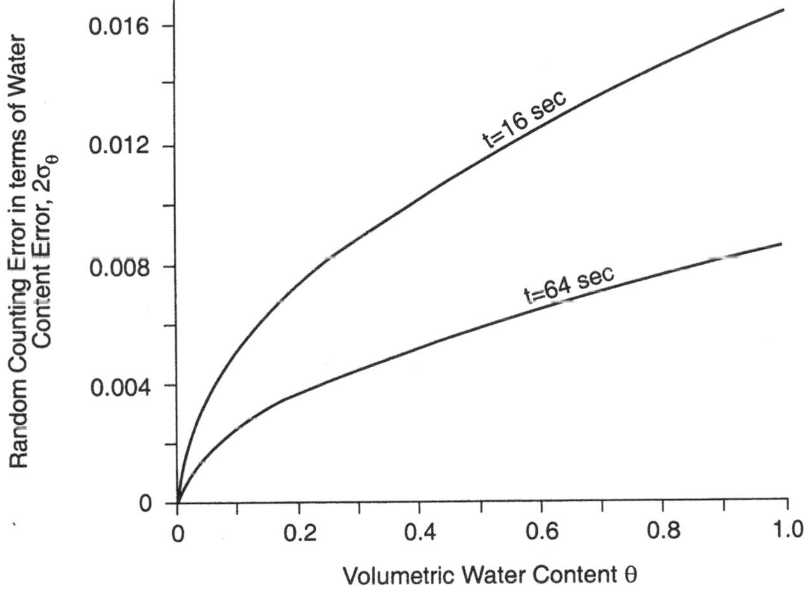

Fig. 3 Relationships between water content error, $2\sigma_\theta$, resulting from the random counting error, and water count, for counting periods of 16 and 64 seconds. (After Bell, 1987.)

been established, it is important to adhere to it. If the depths are changed, the two sets of data will not be strictly comparable because different parts of the soil have been measured. For the same reason, it is important that the chosen measurement depths are accurately maintained on every measuring occasion.

1. Access Tubes

The factors to consider in selecting material for access tubes are transparency to neutrons, mechanical strength and resistance to corrosion in the soils to be investigated, as well as cost and availability. Aluminum, aluminum alloy, stainless steel and some plastics are all suitable; their relative merits are given in Table 3. Aluminum alloy tubing is usually preferred. Polyvinyl chloride (PVC) is not recommended because the chlorine content considerably reduces the neutron count. The iron content of stainless steel has a similar, though less serious, effect, but for some applications the strength is required.

The internal diameter should be sufficient to allow free movement of the probe; a difference of 2 to 4 mm between the outside diameter of the probe and the inner diameter of the tubing normally ensures this. A tubing wall thickness of 1.5 to 5 mm is appropriate. Most equipment is designed for use with 44.5 mm (1.75 inch) or 50.8 mm (2 inch) outer diameter tubing, and the probe carrier fits

Table 3 Relative Advantages of Different Types of Access Tubing

Material	Effect on neutron count	Strength	Resistance to corrosion	Cost
Aluminum	Transparent	Weak	Poor	Expensive
Aluminum alloy	Transparent	Moderate	Poor	Moderate
Stainless steel	Lowers count by 10–15%	Strong	Good	Expensive
Plastic	Increases (PVC decreases)	Moderate	Good	Cheap

on to the top of the access tube while the probe is lowered within it. If tubing of appropriate diameter is not available, an adaptor can be made to allow the probe carrier to be fitted on to the top of larger tubing. Suitable tubing can normally be obtained from stock from suppliers, as can rubber stoppers to close the exposed end. A stopper may be used to close the bottom end, but a turned or cast end-piece of the same material as the tubing, sealed with waterproof adhesive into the end of the access tube, is preferable.

Whichever tubing is selected, it is important that all calibration work and all standard counts are made using tubing of the same material and diameter as used in the field.

2. Access Tube Installation

During installation, disturbance to the soil, the soil surface, and vegetation in the vicinity must be minimized to ensure that subsequent measurements are representative of the surrounding area. The access tube must fit tightly into the soil. Biased measurements will be obtained if there are voids adjacent to the tube or if preferential movement of water occurs beside it (Amoozegar et al., 1989). If there is doubt as to how well a tube has been installed, it is best to re-site it nearby. The extra effort is preferable to collecting suspect data over a long period.

Plenty of time should be allowed for installation work. Two people working in favorable conditions can be expected to install only three or four 2 m access tubes per day, using the method given below. Longer tubes or difficult soils may only permit complete installation of one per day. Installation in heavy clay soils is often difficult both when the soil is wet (due to soil sticking to equipment) and when it is dry (because of hardness). Dry sand makes augering difficult and the sides of the reamed hole may collapse.

The installation method described here has been used successfully to install tubes to 3 m and greater depth in many different soils developed on clays, chalk, silts and sandstones, without resort to power-driven implements. A hole is made for the access tube by using a steel guide tube of the same outer diameter as the access tube. The lower end of the guide tube is sharpened by an internal bevel to give a cutting edge of the same diameter as the external diameter of the access tube. A screw auger that moves easily within the guide tube is used through it to drill out soil to about 0.1 m below the cutting end; the guide tube is then hammered in 0.1 m using a sliding hammer. If this procedure is followed, the guide tube will not be hammered down until a hole of slightly smaller diameter has been augered below it, thus disturbance to the soil surrounding the tube is minimized. The process is repeated until the required depth is reached. The guide tube is then withdrawn and the access tube slid into the reamed hole; gentle tamping may be necessary to drive it fully home. The access tube should then be cut off so that the desired length protrudes from the ground. It should be fitted with a stopper so that the tube remains dry and clean.

If access tubes are to be installed to more than 1 m depth, a series of guide tubes 1.15, 2.15, 3.15 and even 4.15 and 5.15 m in length is used successively with an auger having an extendable shaft. Alternatively, an extendable guide tube with 1 m extensions which can be screwed on to the first tube of 1.15 m length can be employed. A removable collar is necessary to protect the top of the screw thread while hammering. A sharpener, and a file to remove any buckling of the cutting edge caused by stones, should be part of the installation kit.

The top of the guide tube should not be driven in too far, in case it is necessary to fit a clasp if mechanical means are required to extract it. Automobile jacks can be used, and powerful rod-pullers are available from drilling equipment suppliers. It is essential that the pull be exerted along the axis of the tube both to reduce effort and to avoid deforming the hole during extraction. Use of a base plate with a central hole for the guide tube is recommended unless it is likely to damage the crop. This presents a firm base when using tube extractors and minimizes surface soil compaction and enlargement of the neck of the hole.

This installation method can be adapted for use in situations where the soil is unstable, or saturated due to a shallow water table, by using the access tube itself to ream the hole, so avoiding the need to withdraw the tube. The greater strength of a stainless steel access tube may be required, however. Sealing the bottom end of a tube installed in this fashion, particularly below a water table, is not easy; bungs and adhesive, bentonite and other materials have been used (Prebble et al., 1981). This installation method may also be preferred in heavy clay soils if considerable effort is required to extract the guide tube, leading to over-enlargement of the hole near the surface. The timing of installation in swelling clays may affect subsequent cracking adjacent to access tubes and should be considered when planning installation in such soils (Jarvis and Leeds-Harrison, 1987).

A power-driven hammer may be used to drive tubes into very dense or stony soils. The power device should only be used to drive the tube down about 0.1 m after augering. Several attempts at installation may be necessary in stony soils.

Unfortunately there is a tendency for greater success in less stony places, which may result in measurements that are not representative of the soil as a whole. Prebble et al. (1981) addressed this problem and described a variety of installation methods that may be required in other situations. Once installation is complete, precautions should be taken to prevent damage to the surrounding soil and vegetation in the course of making measurements.

3. Measurements Near the Soil Surface

The most satisfactory method of overcoming the influence of the soil–air interface on near-surface measurements is to conduct calibrations specifically for the surface soil layers. Many approaches to deal with the effect (some very elaborate!) have been devised, including use of neutron reflectors placed on the soil surface, use of soil-filled trays placed on the surface to extend the soil medium artificially, correction methods, and use of the probe horizontally on the soil surface. Chanasyk and Naeth (1996) provide a comprehensive review of these. However, a calibration or calibrations for the upper 0.2 to 0.3 m are simple to obtain, as core sampling to such shallow depths is straightforward, and provide the most accurate means of determining water content from neutron counts at shallow depth. Accurate depth placement of the probe for measurements close to the soil surface is particularly important, as Fig. 4 illustrates.

G. Calibration

There are three techniques for calibrating soil water content against count rate ratio: theoretical calibrations, drum calibrations, and field calibrations. A linear relationship between count rate ratio and soil volumetric water content is obtained with most neutron depth probes:

$$\theta = a\frac{R}{R_s} + b \tag{10}$$

where R is the count rate (s^{-1}) in soil and R_s is the standard count rate (s^{-1}). Calibrations are specific to the design of neutron probe used. As described in Sec. C, the use of standard counts to normalize count rate measurements results in a soil-specific calibration that can be used with any probe of the same design. However, it is important to use the same type of access tubing for routine field measurements and calibration purposes because of its influence on count rates. The calibration depends on the soil's neutron scattering and capture cross-sections and bulk density. It is important to be aware of particularly high concentrations of neutron absorbers such as iron and of the presence of any very strong absorbers such as gadolinium and cadmium. For instance, the effect on calibrations of gadolinium concentrations of only 1 to 36 mg kg^{-1} in Tasmanian soils is considerable (Nicolls et al., 1977).

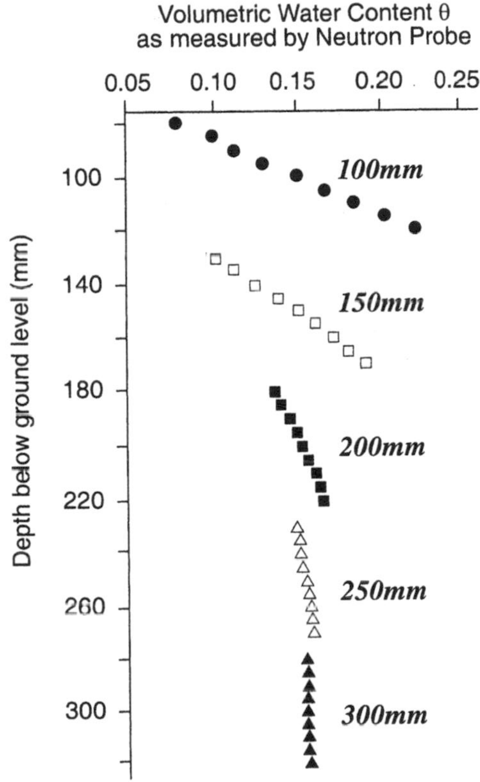

Fig. 4 Effect of depth location on water content measurement at shallow depth. Calibrations for measurements with the probe located at 100, 150, 200, 250, and 300 mm depths were prepared, and measurements precisely at these depths show that the water content of the upper 300 mm of the profile is uniform at 0.15. However, even a small error in the depth location of the probe can cause a significant error in the measured water content. (After Karsten and Van der Vyver, 1979.)

The neutron count rate is influenced by all the hydrogen present in the soil, including that in free and bound water, as well as in other compounds. The hydrogen in adsorbed and structural water and the nonwater hydrogen has the same influence on neutron thermalization as that in free water. Its presence can be expressed in terms of an equivalent water content. Since it does not change with time and is not removed during oven-drying, its effect is incorporated into the intercept term, b, of the calibration equation. Greacen (1981) advocated calibration in terms of total water content (i.e., the sum of the free and equivalent (θ_e) water content); both a laboratory method for determining θ_e and a means of estimating it from clay content are described. For some soils, this permits use of a single calibration for different soil layers, providing θ_e has been determined for each one individually.

An increase in bulk density causes an increase in the number of nuclei close to the source, resulting in more neutron scattering close to it and so an increase in the number of slow neutrons detected. This increase in count rate with increase in density is reinforced if the equivalent water content of the dry soil is large, because of the greater concentration of hydrogen close to the source. However, the concentration of neutron absorbers in the vicinity of the source is also increased, and this counteracts the tendency towards a higher count rate. There is disagreement as to the net effect of bulk density on neutron count rates (Greacen and Schrale, 1976; Rahi and Shih, 1981). If soil-specific field calibrations are used, they will incorporate the effect of bulk density. Otherwise it is important to measure field soil bulk density, ρ, and adjust calibrations to this using

$$R = R_i \left(\frac{\rho}{\rho_i}\right)^{0.5} \tag{11}$$

where R_i is the count rate in soil of density ρ_i, and R is the adjusted count rate (Greacen and Schrale, 1976).

1. Theoretical Calibrations

Theoretical models based on diffusion theory have been developed to simulate neutron flux in soils for which the neutron interaction cross-sections are known. The interaction cross-section of a soil may be determined by direct measurement or by detailed chemical analysis and use of published cross-sections (Mughapghab et al., 1981). Assumptions about soil density are made in the theoretical calibration, which is then adjusted to allow for field soil bulk density.

Determination of soil neutron interaction cross-section by chemical analysis requires detailed analysis of the concentration of at least 20 elements in representative samples of the soil (Olgaard, 1965). Omission of the analysis of a crucial neutron absorber such as gadolinium or boron would have a substantial effect on the resulting calibration. Because of a tendency for overestimation of the neutron absorption effect, the procedure is most satisfactory for light-textured soils with low neutron capture cross-sections, <0.004 barn (Greacen and Schrale, 1976). Wilson (1988b) found that the likely minimum error to be achieved in practice with this calibration method ranged from about $\pm 1.6\%$ to $\pm 3.5\%$ volume fraction, with larger errors occurring in drier soils.

Direct measurement of neutron interaction cross-sections requires access to appropriate specialized equipment, a large neutron source, or even a reactor (Couchat et al., 1975, McCulloch and Wall, 1976). A comparison of calibrations obtained by Couchat et al. (1975), who used a large source in a graphite block, with those determined by the conventional field method for sand, chalk, silt, and chalky clay soils, found good agreement (Vachaud et al., 1977). The method was particularly recommended for use in heavy soils, where obtaining samples over a full range of water content is difficult, and for soils with marked layering, as it enables isolation of the layers from one another for calibration purposes.

2. Drum Calibration

This requires the uniform packing of soil of known water content into a large drum of about 1.5 m diameter and 1.2 m depth. An access tube is installed so that neutron counts can be made within the soil-filled drum. The process is repeated with the soil at a different water content. In principle, as the relationship between soil water content and neutron count is known to be very nearly linear, only two points are required, but it is preferable to obtain several over a range of water contents and bulk densities. The method is very laborious, requiring collection of large quantities of soil from the field and care in wetting up and packing to ensure uniformity in the drum. Use of the bulk density correction (Eq. 11) removes the need to pack the soil to the field bulk density. With care, good calibrations with high correlation coefficients can be obtained for a wide variety of soils (Greacen, 1981).

3. Field Calibration

In this method, a calibration is derived by simultaneous measurement of the neutron count rate and sampling of soil for determination of the volumetric water content of each layer on several occasions, so as to cover the range of hydrological conditions characteristic of the site. The theoretical and drum calibration methods assume a homogeneous soil, whereas field calibrations allow for the presence of site-specific features such as textural boundaries or the fluctuations of a shallow water table. Field calibrations usually result in greater scatter in the calibration points due to soil heterogeneity and sampling errors, but if conducted with care may represent the absolute water content of soil at a site better than the alternative methods.

There are two approaches. Simultaneous neutron counts and samples for volumetric water content determination may be achieved by installation of a temporary access tube in the area used for monitoring the soil of interest. Neutron counts are recorded in the temporary tube at the required depths and then five or six undisturbed samples are taken from immediately around it at each depth by coring and, if necessary, excavating around the tube. The temporary access tube is then removed to be used later. The process is repeated for different depths and times of the year to obtain a calibration over the range of water contents found at the site for each soil layer. Alternatively, neutron counts may be recorded in the access tube used for monitoring and samples collected by coring close to (within 2 m of) the tube. This is suitable in soils where samples can be readily collected by coring; otherwise damage to the vegetation and soil around the access tube may render subsequent measurements in it unrepresentative of the wider area. Again, the process is repeated on several occasions. Irrigation of the area, or encouraging drying with a shelter to keep off rainfall, is acceptable to extend the range of hydrological conditions covered by calibration. It is important to avoid times when a wetting front is moving rapidly through the soil (i.e., immediately after rainfall or irrigation).

The first approach is particularly useful where many access tubes are used to monitor a fairly well defined soil (e.g., in the course of field trials or experiments). The second is appropriate where access tubes are located in differing soils, as in a catchment experiment, and a calibration for the soil at each tube is required. However, if obtaining volumetric samples by coring is difficult, use of a temporary access tube at greater distance from the semipermanent tube will be preferable.

The volumetric water content of the samples is determined by oven-drying; then the paired neutron count and water content data are used to determine a calibration for each soil layer by linear regression. The count rate ratio is considered as the independent variable (x) and the water content as the dependent variable (y). The data from different depths should be analyzed separately, even if the soil appears homogeneous, until the calibrations can be reviewed. Pooling data to reduce the number of calibrations may then be appropriate.

Stones can present a problem in deriving calibrations but cannot be ignored. Stocker (1984) described a method using an access tube and sand to measure the volume of soil samples collected from around the temporary access tube in stony soils.

An alternative procedure for in situ calibration, which is applicable in dry, homogeneous, light-textured soils with a high infiltration rate, is described by Carneiro and De Jong (1985). Known amounts of water are allowed to infiltrate the soil between recording neutron counts. The method assumes that there is no loss of water by evaporation or drainage from the profile during the calibration process.

H. Surface Neutron Meters

Surface neutron meters are used widely in civil engineering and soil mechanics for monitoring the water content of earthworks but have other applications where measurements at a smooth, bare soil surface are required. Ahuja and Williams (1985) used a surface gamma-neutron meter to characterize surface soil properties. Measurements represent a layer about 0.35 m deep in dry soil but only 0.15 m deep in wet soil. Farah et al. (1984) showed that only two calibrations were necessary to represent satisfactorily all or part of the layers 0–0.10 and 0–0.30 m deep. However, if a shallow wetting front is present, measurements are difficult to interpret.

I. Radiological Safety

The acquisition, use, transport, storage and eventual disposal of neutron probes is subject to regulation because of the potential hazard to human health and the environment posed by the neutron source. Most governments have legally enforceable radiological safety regulations that must be followed when using neutron probes. The recommendations of the International Atomic Energy Agency (IAEA, 1972, 1990) and the International Commission on Radiological Protection (ICRP, 1990) should be consulted in the absence of specific regulations.

With sensible usage, the radiation hazard to a trained neutron probe operator is only a little greater than that permitted for members of the public. Precautions such as maximizing one's distance from the source when carrying a probe, or transporting one in a vehicle, are straightforward. A probe should never be left unattended except when locked in its designated storage place.

Regular tests to check for leakage from the source are advisable and mandatory in some countries (e.g., in the U.K., tests must be conducted once every two years). Americium–beryllium sources have a half-life of 458 years, much longer than the useful life of the probe, and longer than the time over which the integrity of the source capsule can be expected to be maintained (up to 30 years). When a source is no longer required it must be disposed of at a designated repository for radiological waste and this cost can add significantly to the lifetime cost of the probe.

V. DIELECTRIC METHODS

Dielectric methods for soil water content measurement exploit the strong dependence of soil dielectric properties on water content. These dielectric properties affect the velocity of an electromagnetic wave (used in TDR), the characteristic impedance of a transmission line (used in the Theta probe), and the capacitance of two electrodes embedded in the soil (used in capacitance techniques).

Smith-Rose (1935) explored the electrical properties of soil as a function of water content, and Thomas (1966) used capacitance instruments, but developments were limited by the lack of an accurate method of measuring high-frequency capacitance. TDR was first applied to dielectric measurement by Fellner-Feldegg (1969) and was soon used to investigate the dielectric properties of soils (Hoekstra and Delaney, 1974; Topp et al., 1980). TDR equipment is now available commercially (Table 4). Interest in capacitance techniques revived in the mid-1980s when developments in electronics meant that capacitance in the 100 MHz frequency range could be measured much more readily, and the method is used in a wide variety of applications.

Early work by Topp et al. (1980) suggested that, for most purposes, a

Table 4 Equipment Manufacturers/Suppliers

Equipment name	Address	Principle
TDR Soil Moisture Measurement System (based around the Tektronix 1502C)	Campbell Scientific Ltd., 815W 1800N Logan, UT 84321-1784, USA	TDR
CS615 Water Content Reflectometer	Campbell Scientific Ltd., 815W 1800N Logan, UT 84321-1784, USA	TDR
Easy Test	Easy Test Ltd., Solarza 8b, 20-815 Lublin 56, PO Box 24, Poland	TDR
Moisture Point	Environmental Sensors, Inc. 100-4243 Glanford Ave, Victoria, BC, Canada V8Z 4B9	TDR (with shorting diodes)
HP 54120	Hewlett-Packard Company, 5161 Lankershim Blvd, No. Hollywood, CA 91601, USA	TDR
Trime	IMKO GmbH, Im Stock 2, D-76275 Ettlingen Germany	TDR
Tektronix 1502B/C	Tektronix, PO Box 1197, 625 S.E. Salmon Street, Redmond, OR 97756-0227, USA	TDR
TRASE	Soil Moisture Equipment Corp., PO Box 30025, Santa Barbara, CA 93105, USA	TDR
Theta Probe	Delta-T Devices Ltd., Burwell, Cambridge, UK	Impedance
EnviroSCAN	Sentek Pty Ltd., 69 King William Street, Kent Town, S. Australia 5067, Australia	Capacitance
IH Capacitance probe	Soil Moisture Equipment Corp., PO Box 30025, Santa Barbara, CA 93105, USA	Capacitance
Humicap 9000	SDEC France, 19 rue E. Vaillant, 37000 Tours, France	Capacitance
Troxler Sentry 200 AP	Troxler Electronic Laboratories, Inc., 3008 Cornwallis Road, PO Box 12057, Research Triangle Park, NC 27709, USA	Capacitance

This list is not exhaustive. Sources are given for the convenience of the reader only, and imply no endorsement on the part of the authors.

universal relationship between dielectric measurements and θ would be applicable to the majority of soils, and so calibration would often be unnecessary. However, further studies have shown that the dependence of soil dielectric properties on water content is more complex and that calibration for individual soils is necessary. Much effort has gone into defining precisely the relationship between water content and soil dielectric properties, using physically based models. Progress is being made, but assessment of results is complicated by the fact that various groups are working with different soils and equipment. At the same time, others are attempting to validate the performance of new designs of equipment. The focus in this chapter is on the practical use of dielectric methods, but a brief explanation of dielectric theory and soil dielectric properties is appropriate. The principles and practice of TDR are described in detail. One impedance technique is described. The theory of capacitance measurements is explained, but as different measurement techniques can be used, only one instrument system is discussed in any detail. The principles governing installation and calibration are the same for all of these instruments and are considered together.

A. Dielectrics

A dielectric is an electrical insulator. When a dielectric is placed in an electrical field, the positive and negative charges within it are pulled in opposite directions, producing a polarization of the dielectric and storing energy in it. Every dielectric is capable of storing electrical energy in this way; this is described by the material's permittivity, ϵ, and is measured in picofarads per meter (pF m^{-1}). As the permittivity of any dielectric is always greater than that of a vacuum, ϵ_0 (8.854 pF m^{-1}), it is convenient to work with the relative permittivity, ϵ_r, which is the ratio of the permittivity of the material to that of a vacuum, ϵ/ϵ_0. (ϵ_0 is also known as the electric constant.) ϵ_r is often called the dielectric constant, but the term relative permittivity is preferred, since ϵ_r varies between materials and depends on temperature and pressure and the frequency of the applied field.

Some substances have individual molecules that possess a permanent electrical dipole. They can therefore store greater amounts of energy than other materials and consequently have high relative permittivities. Water is a prime example of such a polar dielectric. When a molecule with a permanent dipole is placed in an electric field, it will attempt to orientate itself with the field. If the electric field is alternating, the molecule will attempt to rotate with the field, but its rotation will be constrained by the presence of adjacent molecules and by collisions with other molecules.

Whether a substance is polar or nonpolar, when the applied electric field is removed, it takes a short time for the molecules to "relax" to random positions and orientations and the polarization to decay. The time required for this relaxation is characteristic of the material. The same relaxation time governs the response to any change in field strength, so that as the field frequency increases, a point is reached where the polarization cannot change direction as fast as the field. Consequently the permittivity of the substance decreases; the frequency threshold at which this occurs is characteristic for any given substance and is known as the relaxation frequency.

In practice, most substances are imperfect dielectrics and exhibit electrical conduction over a wide range of frequencies. This is often because the substance possesses some ionic conductivity. Soil is such a medium, the soil solution providing an electrically conducting pathway. Soils which have high salinity, contain a lot of clay, or receive regular fertilizer applications exhibit the greatest conductivity. The effect of this conduction may be described in the form of a complex relative permittivity, ϵ_r^*, which has a "real" part, ϵ', describing energy storage and an "imaginary" part, ϵ'', describing energy losses:

$$\epsilon_r^* = \epsilon' - j\epsilon'' \tag{12}$$

where

$$\epsilon'' = \frac{\sigma}{\epsilon_0 \omega} + \text{any other loss mechanisms} \tag{13}$$

σ is the low-frequency electrical conductivity, ϵ_0 is the permittivity of free space, ω is angular frequency ($= 2\pi F$, where F is the ordinary frequency), and j is $\sqrt{-1}$. The effect of this conductivity on relative permittivity measurements depends on which measurement method is used. The aim of most soil water content measuring devices is to measure the real permittivity, ϵ', which is related to volumetric water content, without interference caused by losses due to soil electrical conductivity. Additional measurement of the imaginary part of the permittivity can be used to estimate soil solution conductivity and hence to infer the solute content.

B. Dielectric Properties of Water and Soil

At frequencies below 10 GHz the relative permittivity of pure water at 25°C is 78.38 and increases by ca. 0.36°C^{-1} (0–50°C) as temperature falls. When water freezes, the permittivity falls to about 4 (Fig. 5). Within soil, water molecules in the proximity of colloidal surfaces are influenced by the electrical charge on the surface and lose some of their rotational freedom. The permittivity of bound water in soils is therefore less than that of free water. Research has indicated that values of 4 to 40 for bound water are appropriate at frequencies greater than about 100 kHz (Sposito, 1984). The value varies since the dielectric behavior and relaxation frequency of bound water is influenced by the chemistry of the soil solution and the character of the surface. The other constituents of soil have much lower permittivities than free water; the value for air is 1 and that of most soil solids is usually less than 6.

To make progress in deriving calibration equations to relate permittivity to soil water content, a conceptual framework is required. Much theoretical work has been directed at producing models of the permittivity of mixtures for ordered and

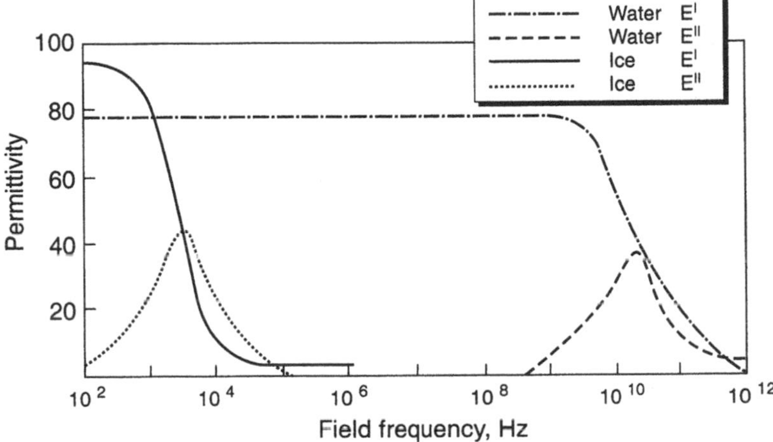

Fig. 5 Change in the real and imaginary permittivity of water and ice, with field frequency.

disordered systems. No real soil conforms to all the assumptions used in deriving these, and indeed, the arrangement of the components in one soil is often quite different from that in another. It is probable that the relationship between permittivity and the concentration of different soil components is similar to that predicted by the models, but the exact values of constants in any one model are unlikely to be realized.

The manner in which soil components contribute to bulk soil permittivity can be illustrated using a straightforward mixing model. Bulk soil is considered as a mixture of four phases: air, solids, free water, and bound water, thus

$$\epsilon^{\alpha} = \epsilon_{a}^{\alpha} f_{a} + \epsilon_{s}^{\alpha} f_{s} + \epsilon_{w}^{\alpha} f_{w} + \epsilon_{bw}^{\alpha} f_{bw} \tag{14}$$

where ϵ_{a}^{α}, ϵ_{s}^{α}, ϵ_{w}^{α}, and ϵ_{bw}^{α} are the permittivities of air, soil solids, free water, and bound water, respectively, and f_a, f_s, f_w, and f_{bw} are their volume fractions. The total water content, θ, is the sum of f_w and f_{bw}. The bound water is often ignored, however. Experimental and theoretical work have shown that a value of about 0.5 for α (Birchak, 1974; Roth et al., 1990; Whalley, 1993; Jacobsen and Schonning, 1994) is appropriate for many soils. Since

$$f_a + f_s + \theta = 1 \tag{15}$$

and

$$f_s = \frac{\rho}{\rho_p} \tag{16}$$

where ρ is soil bulk density and ρ_p particle density, Eq. 14 can be expressed in terms of dry bulk density and particle density:

$$\epsilon^{\alpha} = \epsilon_a^{\alpha}\left(1 - \frac{\rho}{\rho_p} - \theta\right) + \epsilon_s^{\alpha}\frac{\rho}{\rho_p} + \epsilon_w^{\alpha}\theta - (\epsilon_w^{\alpha} - \epsilon_{bw}^{\alpha})f_{bw} \tag{17}$$

If the volume fraction of the bound water, f_{bw}, is assumed to be so small that it can be ignored, then, assuming that α equals 0.5, the permittivity of air is 1, and that of water is 81, Eq. 17 becomes

$$\begin{aligned}\sqrt{\epsilon} &= 1 + \frac{(\sqrt{\epsilon_s} - 1)\rho}{\rho_p} + (\sqrt{81} - 1)\theta \\ &= 1 + \frac{(\sqrt{\epsilon_s} - 1)\rho}{\rho_p} + 8\theta\end{aligned} \tag{18}$$

It is clear that θ makes a very big contribution to the bulk soil permittivity due to the large permittivity of free water. However, it is also notable that dry bulk density has a role, and that its influence will be greater at greater water contents (solving Eq. 18 for ϵ rather than $\sqrt{\epsilon}$ results in $\theta\rho$ terms). More complex dielectric mixing models are available in the literature (e.g., de Loor, 1968) and have been applied to soils (e.g., Dobson et al., 1985).

C. Time Domain Reflectometry

The principle behind TDR is that a fast rise-time electromagnetic pulse is fed into the soil between two or more metal rods, which act as a waveguide. The soil acts as a dielectric between the conductors of this transmission line. The velocity of propagation of the pulse depends only on the permittivity of the soil between the rods. The applied pulse will be reflected either from the end of the transmission line or from impedance mismatches along it (e.g., connectors). The time interval between the incident and reflected pulses is measured. Cable testers use this principle to locate faults and breaks in cables. The cable tester measures the travel time of the pulse to and from any discontinuity and so the distance to it can be determined easily.

The propagation velocity, v, of a transverse electromagnetic (TEM) wave is related to the permittivity of the material by

$$v = \frac{c}{\sqrt{\epsilon_r}} \tag{19}$$

where c is the velocity of light (3×10^8 m s^{-1}). The time, t, taken for a wave to propagate down the transmission line and return is

$$t = \frac{2L}{v} = \frac{2L\sqrt{\epsilon_r}}{c} \tag{20}$$

where L is the length of the line. Topp et al. (1980) used the term apparent relative permittivity of the soil (K_a) in place of ϵ_r to indicate that other factors, principally the imaginary part of the permittivity, influence the measurement. The effect is negligible except when the imaginary part of the permittivity is very large, as in strongly conducting soils.

Because the square root of permittivity is almost linearly related to water content (Eq. 18), the time taken for the pulse to propagate along the line (Eq. 20) is proportional to the square root of permittivity. Thus, the propagation time varies linearly with total water content along the line, even when there are water content variations along it. This makes TDR a good method for estimating total water storage over an extended depth range.

1. TDR Systems

Figure 6 is a block diagram of a TDR instrument. A timer provides synchronizing information to a pulse generator and a receiver. The pulse generator supplies a voltage step with a very fast rise time, effectively feeding a train of high-frequency (predominantly in the range 100 MHz to 1 GHz) electromagnetic waves with a

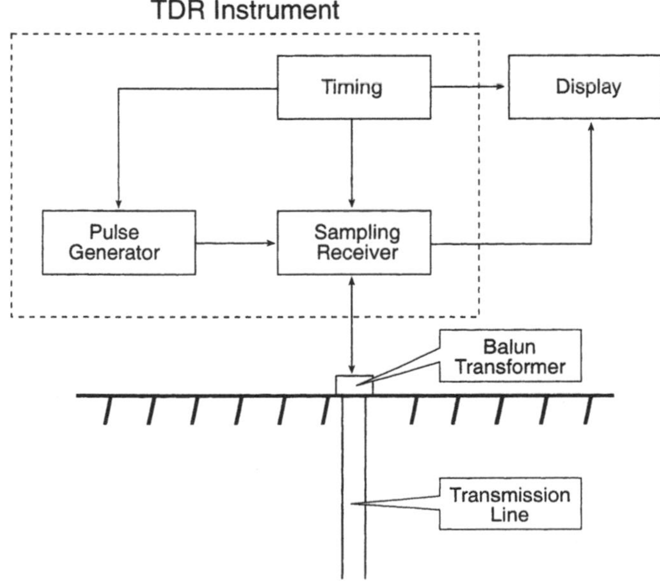

Fig. 6 Block diagram of a TDR instrument.

wide frequency distribution into the sample. The detector circuit measures the sum of the input voltage and the reflected pulse. Because the times involved are very short, a few nanoseconds, the time dependence of the output voltage is determined by sampling the voltage at a series of times after the initial pulse. Pulses are sent repeatedly, every millisecond or so, and one voltage sample is measured after each pulse cycle. Thus a voltage–time curve (the waveform) can be reconstructed from these measurements and used to determine t. It is important to realize that the resultant waveform is the sum of a step input and the reflected voltage.

It is possible to assemble a TDR system for soil water content measurement quite easily, if a cable tester is available. Topp et al. (1980) used a Tektronix 1502B cable tester, which can be linked to a PC using a RS-232 interface. This instrument, or the 1502C model, is commonly used in TDR research because of its adaptability. A number of companies provide systems incorporating Tektronix cable testers, with their own waveguides and software. However, such setups are less convenient than the off-the-shelf systems now available (Table 4). For example, the TRASE system (Fig. 7) incorporates a TDR plus a data logger and

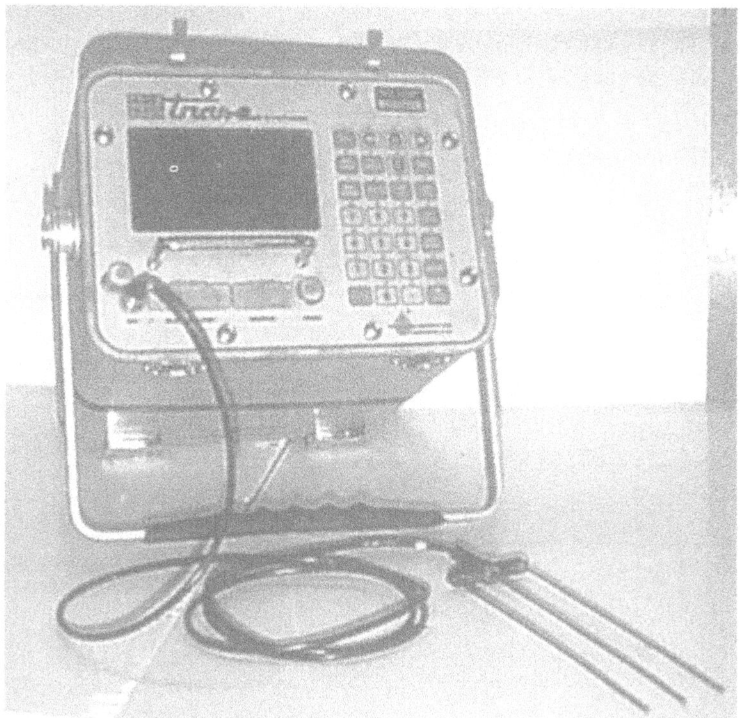

Fig. 7 A TRASE TDR system.

interpretation software. Waveguides are available for TRASE that can be used for measurements at the surface or buried for continuous monitoring. Stored data is easily downloaded into a PC via a RS-232 connection. For routine measurement of soil water content, it is a well integrated user-friendly system.

Commercial TDR systems are supplied with in-built software that analyzes each waveform. Such software works well with waveforms produced in homogeneous media. However, dielectric discontinuities along the waveguide may create reflections other than from the end, and if the soil is particularly conductive, the waveform may be attenuated. Automatic analysis of the waveform may then be unreliable. More specialized software can recognize difficult waveforms and tag them so that the user can examine the waveform to determine the end point reflection manually (Heimovaara and de Water, 1993).

A major advantage of TDR is that readings can be logged automatically, and several waveguides can be attached to a multiplexer, which switches between channels to make a measurement on each (Baker and Allmaras, 1990; Heimovaara and Bouten, 1990; Herkelrath et al., 1991). Up to 70 locations in the soil may be monitored, but as channels cannot be read simultaneously, the reading cycle takes longer the more waveguides are monitored; cycles may take 10 to 15 minutes for a lot of sensors.

2. *Waveguides*

The waveguide is the TDR sensor that is inserted into the soil. Waveguides are also referred to as "guides," "probes," "rods," or "wires." Several designs are illustrated in Fig. 8. There has been much discussion about the design of waveguides, in particular their length, width, and number of electrodes (Heimovaara, 1993; Whalley, 1993; Selker et al., 1993; Baumgartner et al., 1994; Noborio et al., 1996). The minimum requirement is two electrodes for each waveguide, one attached to the central conductor of the coaxial cable and one or more attached to the sheath.

TDR provides a measurement of the integrated water content along the full length of the waveguide. Waveguides of up to about 1 m length can be used in favorable conditions. Use of short waveguides installed horizontally from the walls of a pit may be preferable to vertical installation of long waveguides, if measurements at discrete depths are required. Alternatively, vertically installed waveguides of different lengths may be used to derive water content in different depth ranges by difference.

The Easy Test TDR system differs from others in having very small waveguides (rods <6 mm length, <2 mm diameter and separated by <2 mm) (Malicki et al., 1992). For field use, these are attached to a cylindrical body and so can be installed vertically at the base of a preaugered hole, in a manner similar to that used for tensiometer installation. Their short length means that a needle voltage

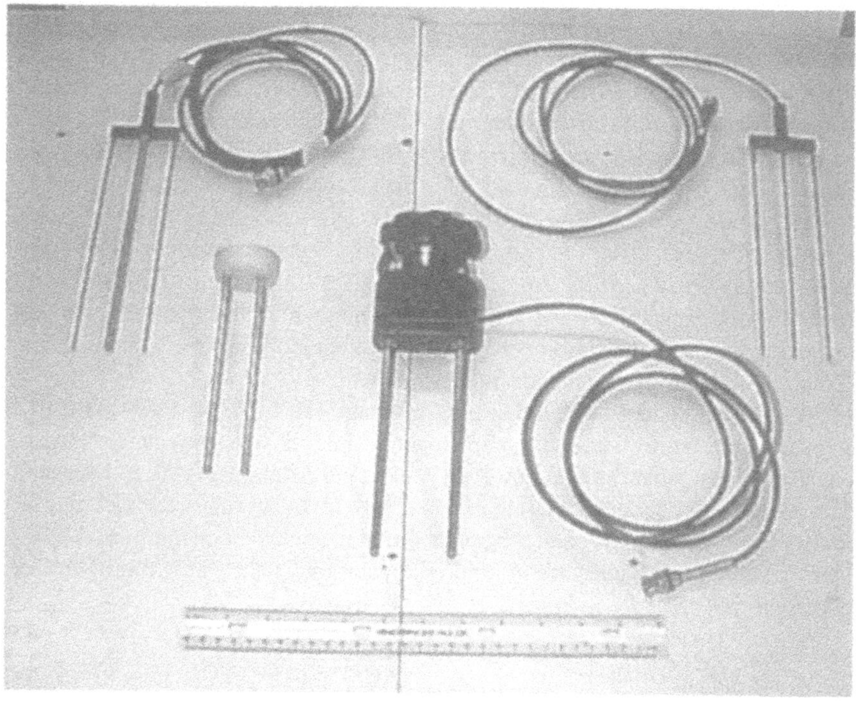

Fig. 8 Different designs of TDR waveguides.

pulse with a very short duration (200 ps) is required, rather than a single step voltage.

Attachment of a coaxial cable to a waveguide results in some reflection of the applied pulse. This is used to identify the position corresponding to the start of the waveguide on the TDR trace. However, too large an impedance mismatch causes only a small proportion of the applied voltage pulse to enter the waveguide, with consequent small signal levels and multiple reflections, making interpretation of the trace difficult (Spaans and Baker, 1993). Two-wire probes normally use a "balun" (an impedance matching transformer) to reduce this problem. Three- and four-wire guides do not normally require the use of a balun. If resistance is also to be measured, a balun cannot be used.

a. Waveguide Sampling Volume

De Clerk (1985) showed that for a waveguide with a rod spacing of 25 mm, 94% of the energy was contained within a cylinder of 50 mm diameter; thus a 20 cm long waveguide has a sampling volume of some 98 cm^3. Whalley (1993) demonstrated that TDR is most sensitive to the soil close to the rod connected to the central conductor of the transmission line. Thus the sampling volume is more concentrated around the central rod of 3- and 4-wire waveguides than around the conductors of a two wire sensor. In addition, the smaller the diameter of the conductors, the smaller the volume of soil to which the measurement is most sensitive. For detailed discussion of waveguide sampling volume see Knight (1992, 1995).

b. Constraints on Waveguide and Cable Length

The length of waveguide used will be dictated by two main factors: the volume of soil to be measured and its electrical characteristics. 15 cm is recommended as the minimum waveguide length for routine field work with most systems. The error in measurement increases as the sensors become shorter, because the accuracy with which the returning pulse can be timed is fixed, and so the proportional accuracy increases as the length of the waveguide increases. However, the shorter the waveguide, and the greater the distance between the electrodes, the smaller the influence of electrical conductivity. In soils with a high electrical conductivity, the length of waveguide that can be used effectively is limited to 50 cm or less. Thus before deciding on a field installation, it is advisable to assess the soil's attenuation characteristics. This can be as simple as taking the TDR to the field site, wetting the soil, and installing a waveguide to see if an interpretable waveform is generated. The effect of attenuation due to conductivity can be reduced using rods coated with heat shrink Teflon to ensure the return of a strong reflection (Kelly et al., 1995). An epoxy-coated waveguide is offered by Soil Moisture Equipment Corp. for use with the TRASE system and has a similar effect.

Cable length also influences the magnitude of the returning step pulse; the longer the cable, the greater is the attenuation of the signal (Heimovaara, 1993). Herkelrath et al. (1991) recommended that coaxial cable runs should be no longer than 30 m. Use of low-loss cable will increase the working distance from the TDR pulser.

3. Waveforms

The output from TDR equipment is a waveform, a graph of voltage versus time. Figure 9 illustrates how the shape of the waveform is made up of voltages from successive reflections at the junctions between the coaxial connector and the waveguide and at the end of the waveguide. The time measured to determine permittivity using Eq. 20 is that between points A and B in Fig. 9a. Figure 9b illustrates the waveforms produced when measuring the permittivity of air and tap water. The travel time for the pulse along a 20 cm waveguide in air is 0.67 ns and 5.97 ns in water; the time increases proportionally with longer waveguides.

Locating the end point, B, of the waveform is fundamental to the measurement of permittivity. In Fig. 10a the position of the reflection from the end of the waveguide is readily distinguished. However, it is not sharp but distributed over

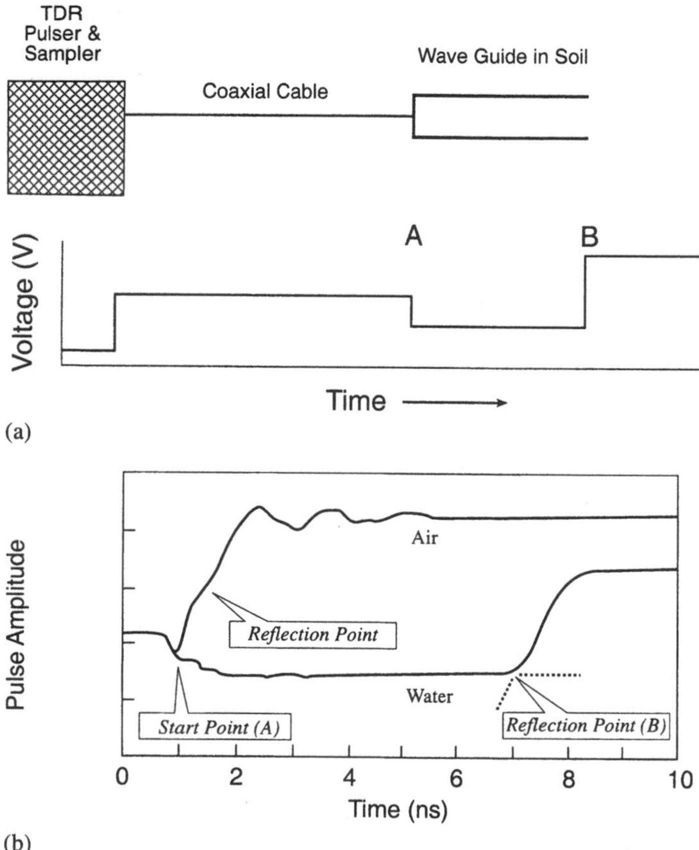

Fig. 9 (a) Relationship between the waveform shown on the TDR screen and the TDR/waveguide setup. Usually only the right-hand part of the waveform is displayed, i.e., from just before A to after B. (b) TDR waveforms produced with a 20 cm waveguide in air and in water.

a range of times. This is due to some dispersion of the pulse (i.e., some frequencies of the wave propagating at slightly different speeds), greater attenuation of some frequency components than others, and penetration of part of the pulse beyond the end of the waveguide. The position of the reflection point can be reliably estimated from the intersection of two tangents to the line (Fig. 10a) and enables estimation of the time of propagation to within 80 ps (Topp et al., 1980). This or similar approaches are used in software for analyzing TDR waveforms. However, in the case of a 20 cm waveguide, the 80 ps results in an uncertainty in water content of about 0.013 by volume.

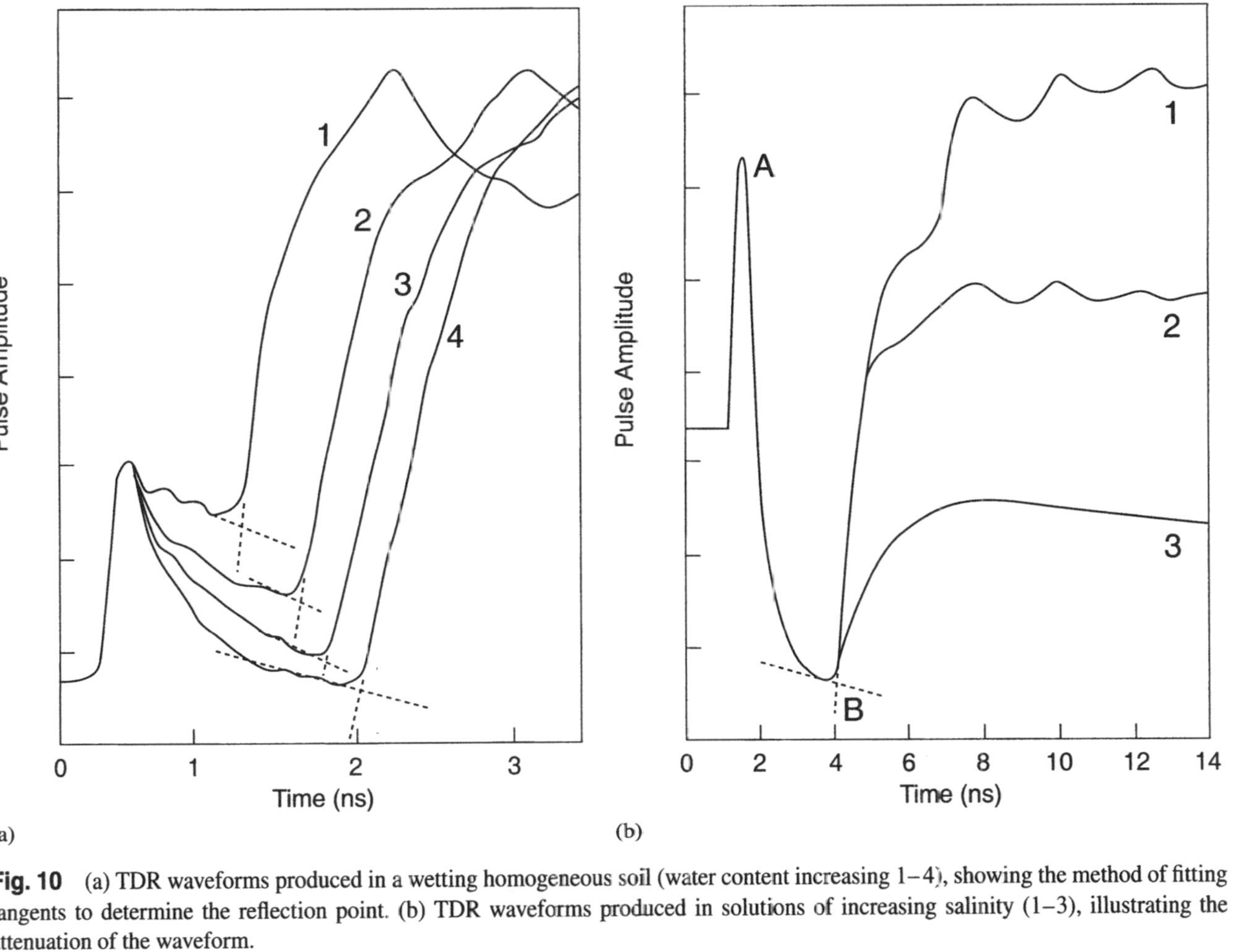

Fig. 10 (a) TDR waveforms produced in a wetting homogeneous soil (water content increasing 1–4), showing the method of fitting tangents to determine the reflection point. (b) TDR waveforms produced in solutions of increasing salinity (1–3), illustrating the attenuation of the waveform.

a. Waveforms in Electrically Conducting, Lossy, Dielectrics

TEM waves travelling through electrically conducting media are liable to attenuation. The higher frequency components of the waveform are usually lost first. As a result, the amplitude of the reflected portion of the pulse is reduced (Fig. 10b). Locating the reflection becomes more difficult, and the errors in the measurement of the travel time increase. In very conductive media, the waveguide is effectively short circuited and permittivity cannot be measured. Advantage can, however, be taken of the attenuation effect and the waveform analyzed to give the low frequency resistance and hence the bulk soil electrical conductivity of the medium through which it has travelled (Dalton et al., 1984; Topp et al., 1988; Dalton, 1992; Kachanoski et al., 1992; Heimovaara et al., 1995).

b. Waveforms From Soils

The waveforms obtained depend on the soil and the manner of installation of the waveguide: horizontal or vertical. Horizontally installed waveguides provide easier traces to work with because they are not usually influenced by water content gradients or other soil changes along the length of the guide. A vertically installed waveguide is more likely to pass through soil density boundaries and wetting or drying fronts that may cause additional reflections, resulting in waveforms that are difficult to interpret (Fig. 11) and that challenge the ability of software to locate the correct end point. If the reflection point can be located, the resulting measurement will represent the integrated water content over the length of the waveguide. Hook et al. (1992) designed TDR waveguides with shorting diodes that make waveform analysis easier for a vertically installed sensor.

D. Impedance Technique

Another property of transmission lines, their impedance, is used in the Theta Probe, developed at the Macaulay Land Use Research Institute (Aberdeen, U.K.). The instrument measures impedance at a fixed frequency of 100 MHz. The technique compares the impedance of a section of fixed transmission line with that of a set of stainless steel electrodes embedded in the soil, whose impedance varies with soil water content (Gaskin and Miller, 1996). The compact buriable sensor produces a voltage output and so can be interrogated with a voltmeter or connected to any logger that takes a dc input. The voltage output can be calibrated directly against water content, or alternatively calibrated to obtain relative permittivity, from which water content can be determined. The suppliers provide two calibration equations, one for mineral and one for organic soils. The volume measured by the probe is much the same as that of the corresponding configuration of TDR probe, where the sampling volume is strongly biased towards the central conductor. The sampling volume of the instrument is ca. 50 cm^3 and gives good

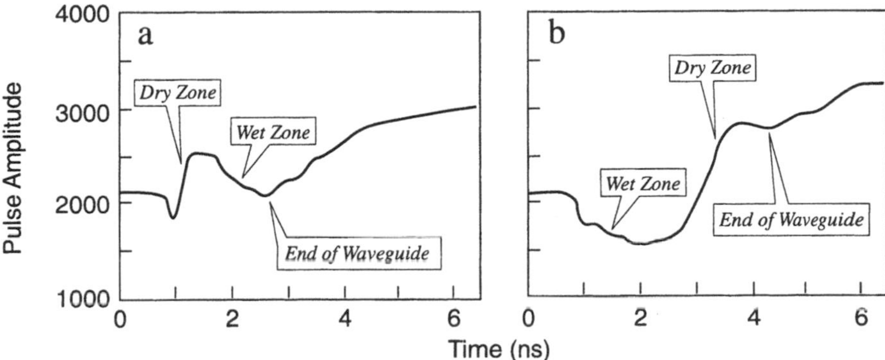

Fig. 11 TDR waveforms produced with waveguides installed vertically in soil with (a) a dry zone overlying a wet layer; (b) a wet zone over a dry layer.

averaging along the 60 mm rod length. Sensors cost about $600 each, so the system is attractive for portable and laboratory use and setups requiring several sensors.

E. Capacitance Techniques

Soil capacitance sensors measure the capacitance between two electrodes whose dielectric is partly or completely the soil to be measured. Capacitance is defined as

$$C = \epsilon_r \epsilon_0 g \tag{21}$$

where g is a geometric constant dependent on the size and arrangement of the electrodes. This measurement is difficult at low frequency unless the material is pure. Impurities lead to complications such as electrical conduction in the material and polarization of colloidal material or at interfaces. As a result, the measured capacitance is different from that of the pure material, and the calculated permittivity is incorrect. To overcome these problems, measurement at frequencies greater than 50 MHz is necessary. High-frequency capacitance can be measured in various ways, and several contrasting soil water sensors are available (Table 4).

It is important to be aware that capacitance sensors may be influenced by soil electrical conductivity, particularly those operating at <50 MHz. However, Gaudu et al. (1993) found that the effects of electrical conductivity were negligible with their system, which operates at about 40 MHz. Eller and Denoth (1996) reported a similar result with an instrument operating at about 32 MHz, except in wet organic soil, when slightly reduced accuracy, due to electrical conductivity, was evident. The IMAG DLO probe, designed to be buried or used for point measurements at the soil surface, operates at 20 MHz and measures the real (capacitive) and imaginary (conductive) parts of the permittivity independently (Hilhorst et al., 1993).

1. IH Capacitance Instruments

The IH capacitance systems, designed at the Institute of Hydrology (Wallingford, U.K.), give an instantaneous measurement of frequency which is a function of the electrode capacitance, from which soil permittivity can be calculated. Several instruments have been developed using the same sensor electronics (Fig. 12). A sensor that can be inserted into the soil via a plastic access tube, much as a neutron probe, is available (Dean et al., 1987). An insertion probe with two rod-shaped electrodes has been developed that can be used at the soil surface or buried (Dean, 1994), and a tine arrangement that can be towed behind a tractor has been tested by Whalley et al. (1992). The principle of operation is to use the capacitor formed by the electrodes in the soil as part of an oscillator circuit comprising capacitors, an inductor, and a driver transistor. The frequency of oscillation (F) of such a circuit is

$$F = \frac{1}{2\pi\sqrt{LC}} \tag{22}$$

where L is the circuit inductance and C its capacitance. The circuit capacitance, C, is determined mainly by the capacitance of the electrodes, which is the only variable element in the circuit. Calibration of the sensor is necessary to relate oscillation frequency to permittivity (Robinson et al., 1998). A frequency of ~150 MHz is obtained in air and ~75 MHz in water for all electrode configurations.

The design of the instrument gives the electrical field good penetrability into the material under test. The depth probe has a sampling volume of about 800 cm^3 with the field penetrating ~7 cm from the sensor body (Dean et al., 1987). The insertion probe has a sampling volume of about 500 cm^3 for 10 cm rods and 250 cm^3 for 5 cm rods and shows good averaging along the length of the rods (Dean, 1994). In soil, the frequency of oscillation is determined by a combination of the capacitance and the parallel conductance caused by electrical conduction. Ionic conductivity of the soil reduces the frequency of oscillation, but the effect is relatively small for bulk soil electrical conductivity of less than 0.05 S m^{-1} (Robinson, 1998). For higher conductivities the effect can often be compensated for (Robinson et al., 1998).

Several studies using IH sensors have related the instrument frequency reading directly to field soil water content. Robinson and Dean (1993), using the surface probe for measurements to 0.1 m depth, developed an inverse square root model to relate water content to oscillation frequency in a clay soil. Bell et al.

Surface Capitance Insertion Probe

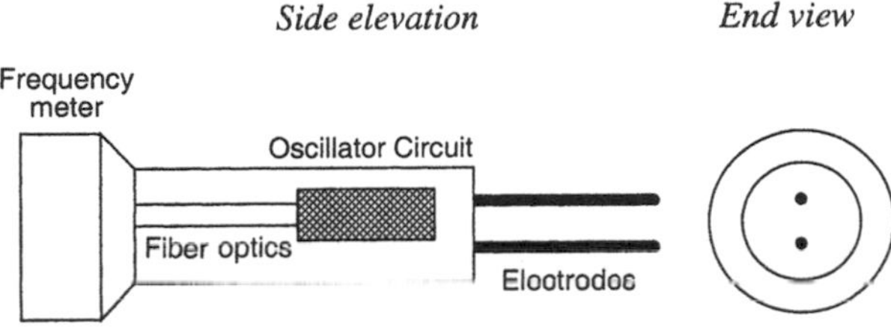

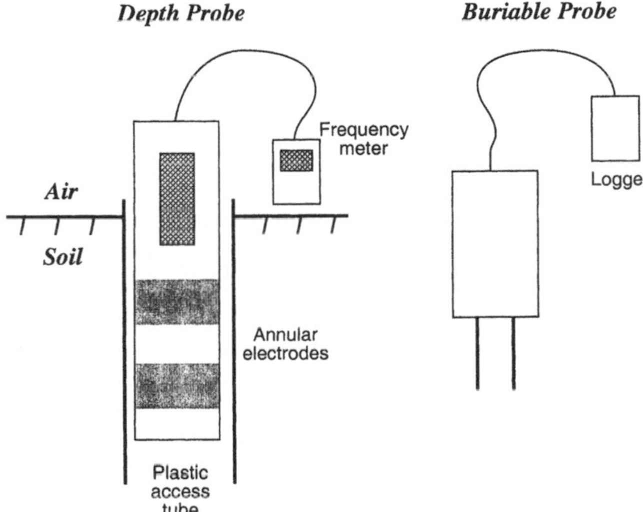

Fig. 12 The Institute of Hydrology surface, depth, and buriable capacitance probes.

(1987) found that linear calibrations satisfactorily represented the water content–frequency relationship measured with the depth probe in four soils, over the normal range of soil water content. Evett and Steiner (1995), using a capacitance depth probe of similar design, also found linear calibrations to be most satisfactory, but Tomer and Anderson (1995), with the same type of equipment, preferred a second order polynomial to represent water content in a fine sand soil. These calibrations are all specific to both the soil and the particular instrument used. Initial calibration of the instruments, using liquids of known permittivity, allows permittivity to be determined from the frequency measurement. This allows more flexibility, permitting soil water content calibration in terms of permittivity; it also enables comparison with other dielectric methods and soil dielectric models. Laboratory trials with the surface probe have shown that well-defined relationships relating water content and permittivity are obtained for individual soils (Gardner et al., 1998). Differences between soils could be described by the parameters of a three-phase mixing model that included a bulk density term and gave results comparable to those obtained by TDR.

F. Field Installation of Dielectric Equipment

As with neutron probe access tubes, the aim during installation must be to minimize disturbance to the surrounding soil and vegetation, so that the water content measurements made are representative of the hydrology of the soil as a whole. The rod-shaped electrodes of most capacitance sensors can be treated similarly to short TDR waveguides and buried at the required depth, from the side of a pit if necessary. The access tube version of the IH capacitance probe requires installation of plastic access tubing, which can be achieved using methods similar to those used for neutron probe access tubing (Bell et al., 1987). However, the volume measured by the depth capacitance probe is smaller than that for the neutron probe, and so the effect of cavities around the tube is more serious.

The physical nature of the soil and its water content at the time of installation are important factors to be taken into account when installing both TDR and capacitance sensors. It is preferable to install sensors into wetted soil if they are to be left for any considerable length of time. Stony soils prevent the use of long TDR waveguides and make installation of depth capacitance access tubes difficult. Very stony soils may preclude any form of installation without completely disturbing the soil around the sensor.

TDR waveguides may be installed horizontally or vertically; the choice depends on the data required. Vertical installation from the surface creates the minimum soil disturbance. Probes of increasing length can be used to give soil profile water contents by subtracting the volumetric water content measured by the shorter sensors, from that measured by the longer ones. Sometimes waveguides may pass through soil horizons and/or density boundaries, giving rise to waveforms that are difficult to interpret and presenting calibration difficulties. The sensors may also act as a focal point for infiltrating water, hence giving unrepresentative field data. Horizontal installation is advantageous for measuring the water content of specific horizons and avoids the problem of channeling water down the waveguide. However, installation requires the digging of a pit, causing major soil disturbance. Hokett et al. (1992) examined the influence of soil cracks next to waveguides and found that an air-filled crack between the rods in an otherwise saturated soil could reduce the measurement of water content by as much as 46%, but water- and air-filled cracks in dry soils had little influence. The evidence suggests that in soils prone to shrinkage, where the rods may act as a focus for cracking, horizontal rather than vertical installation will give more representative results.

G. Calibration

TDR does not require calibration to measure soil permittivity if the length of the waveguide is known accurately, since electromagnetic theory relates the two as in Eq. 20. The calibration of other dielectric sensors in terms of relative permittivity can be achieved using fluids of known permittivity. Tables of the permittivity and temperature coefficients of a large range of fluids are given by Lide (1992). It is important to choose only liquids whose relaxation frequency is much greater than the operating frequency of the equipment.

Soil is inherently a complex material, and yet calibrations between soil permittivity and volumetric water content have been remarkably consistent. The initial suggestion that the relationship between permittivity and soil water content was "universal," so that once established it could be applied to all soils, is too simplistic. However, the Topp et al. (1980) calibration for TDR (Table 5) has been found to be valid for many soils and serves as a good benchmark for comparisons between TDR calibrations and those of other instruments. Different instruments operate at different frequencies, making direct comparisons between calibrations difficult. As the frequency rises, so more components such as bound water will attain their relaxation frequency, resulting in a lowered soil permittivity. In practice this means that instruments such as the IMAG-DLO capacitance probe, operating at 20 MHz, are likely to give greater permittivity measurements for the

Table 5 Empirical Calibration Equations for Obtaining θ from TDR-measured $\in_r$

Soils	Empirical formulae derived for TDR	Source
4 mineral soils	$\theta = (A + B \times \epsilon_r + C \times \epsilon_r^2 + D \times \epsilon_r^3) \times 10^{-4}$ A = −530, B = 292, C = −5.5, D = 0.043	Topp et al. (1980)
Organic soil	$\theta = (A + B \times \epsilon_r + C \times \epsilon_r^2 + D \times \epsilon_r^3) \times 10^{-4}$ A = −252, B = 415, C = −14.4, D = 0.22	Topp et al. (1980)
Loam	$\theta = 0.1138\sqrt{\epsilon_r} - 3.38\rho_b - 0.1529$	Ledieu et al. (1986)
10 mineral soils	$\theta = (A + B \times \epsilon_r + C \times \epsilon_r^2 + D \times \epsilon_r^3 - 370\rho_b + 7.36 \times \%\ clay + 47.7 \times \%\ org.mat.) \times 10^{-4}$ A = −341, B = 345, C = −11.4, D = 0.171	Jacobsen and Schjonning (1994)
62 mineral/organic soils and porous media	$\theta = \dfrac{\sqrt{\epsilon_r} - 0.819 - 0.168\rho_b - 0.159\rho_b^2}{7.17 + 1.18\rho_b}$	Malicki et al. (1996)

same water content than the Topp et al. (1980) calibrations determined using TDR (~ 200 MHz), as found by Perdok et al. (1996). However, although the calibrations may differ, the influential soil factors will, for the most part, be the same.

The number of published calibration models is growing as more measurements are taken, but most apply to TDR. The applicability of any model should be verified where possible by conducting at least a limited calibration for the soil concerned. Calibrations for systems other than TDR are limited, so these instruments will normally require calibration. There is as yet no standard method for calibrating dielectric instruments in terms of soil water content. Some calibration methods are more representative of field conditions than others, but the choice of method will also be based on other factors, including time available and the range of water content required.

1. *Field Calibration*

The principle of field calibration is the same as for deriving calibrations for the neutron method. Measurements are made, and immediately undisturbed soil samples of known volume are collected from the measurement point, for water content determination by oven-drying. Depending on the type of equipment, and the depth of the soil, it may be possible to sample the volume of soil where the instrument measurement was made. Such an approach assumes temporary installation of equipment and is destructive. Sampling at a greater distance from a permanent equipment installation may be preferred. Alternatively, for the depth capacitance probe, samples can be taken from the access tube at the time of installation (Bell et al., 1987). Covers and irrigation may be used to extend the range of water content involved.

2. *Laboratory Calibration*

Laboratory methods offer the advantage of being in a controlled environment. The most rapid method is to wet air-dried sieved soil with deionized water using a mist spray while mixing continuously (Malicki et al., 1996). The soil is then packed into a known volume and weighed; the electrodes or waveguides are inserted and measurements taken immediately. A small sample of the soil, ~50 g, is then removed for oven-drying and water content determination as a mass ratio. Volumetric water content is calculated knowing the weight and volume of the packed soil. The soil can be packed to different bulk densities and measurements for a wide range of water content achieved by gradual wetting. Perdok et al. (1996) used a triaxial soil press to provide soil cores with different bulk densities in which to calibrate the IMAG DLO capacitance probe. A complete calibration curve can be derived in two days, allowing overnight drying of the samples for water content determination.

Undisturbed cores from the field can be used (Heimovaara et al., 1994) so that the complete range of soil water content can be achieved on cores that are as close to their field condition as possible. For most equipment, a core of about 10 cm diameter and 15 to 20 cm length is large enough. Cores need to be encased and a perforated base should be fitted, so that in the laboratory they can be wetted from the base upwards, preventing air entrapment. Cores are saturated using deionized water, and then the electrodes/waveguides are inserted and measurements begun. On each measurement occasion the core is also weighed. The cores will dry out in the laboratory from the open top and through the perforated bottom. Drying can take up to two or three months. Finally, the soil core is removed for oven-drying, and the water content on each measurement occasion is calculated from the corresponding weights. At least two cores must be taken for comparison, as natural inhomogeneities such as stones may cause unrepresentative calibrations. Shrink/swell soils are difficult to deal with in this manner. An alternative approach along similar lines is to sieve soil and pack a core and then to treat the core as above. This homogenizes the soil and eliminates the possibility of large stones, cracks, or pores influencing the calibration.

H. Influence of Soil Properties on Calibrations

1. *Soil Temperature*

The relative permittivity of water decreases almost linearly by 0.36 per °C as temperature rises between 5 and 50°C (Lide, 1992). The permittivity of the solid components is likely to change very little with temperature, and so the average change in soil permittivity with temperature will be less than that for pure water. Experiments by Topp et al. (1980) demonstrated that, for the soils used in their experiment, there was a negligible temperature effect in the range of 10–36°C. Halbertsma et al. (1995) showed that the incorporation of temperature compensation for the permittivity of water into a mixing formula replicated data for sand, but in a clay soil no noticeable change of permittivity occurred with an increase in temperature, and so application of the model overestimated the soil water content. For most purposes, with temperature-stable equipment, it is likely that the effect of temperature on permittivity will be small compared with the other errors in the calibration process.

2. *Bulk Density and Soil Mineralogy*

Bulk density, directly or indirectly, has a significant influence on the calibration of dielectric techniques. Topp et al. (1980), using a limited number of soils, found that bulk density was not an important factor in the calibration they produced. Subsequent work on a wider range of soils found that incorporation of bulk density into calibrations improved results (Ledieu et al., 1986; Jacobsen and Schjonning, 1994). The semiphysical mixing model presented by Whalley (1993) gives a physical explanation of the effect of bulk density. The linear model (Eq. 18) shows that the intercept is a function of the permittivity of the solid and its dry bulk density. This approach has proved useful for exploring the dielectric properties of soil in a physical rather than an empirical way (Robinson, 1998). Work with capacitance instruments has also found that bulk density should be incorporated into calibrations (Perdok et al., 1996; Gardner et al., 1998).

The most likely effect of an increase in soil bulk density is to increase the permittivity of the soil. Jacobsen and Schjonning (1994) suggested that the effect of change in bulk density was more than could be accounted for by a change in the ratio of solids to voids and their respective permittivities. As the effect is most noticeable in certain heavier textured soils, it is likely that this is associated with the clay content. As a clay soil becomes more dense, the quantity of bound water increases, and therefore one might expect a decrease in soil permittivity at the same water content, as bulk density increases. The four-phase mixing formula, Eq. 17, gives, using $\epsilon_a = 1$:

$$\theta = \frac{\epsilon^\alpha + (\epsilon_w^\alpha - \epsilon_{bw}^\alpha) f_{bw} - (\epsilon_s^\alpha - 1)(\rho/\rho_p) - 1}{\epsilon_w^\alpha - 1} \tag{23}$$

where $\theta > f_{bw}$. Typically, $\epsilon_s = 3.5$, $\epsilon_{fw} = 81.0$, $\epsilon_{bw} = 3.2$, $\rho_p = 2.56$, and values for α range from 0.46 to 0.70 (Dirksen and Dasberg, 1993; Roth et al., 1990). This equation combines the effect of both bulk density and surface area changes (Fig. 13). However, changes in bulk density produce a proportionate change in surface area per unit volume and hence in the amount of bound water, which may be a large fraction of the total water in a clay soil. Peplinski et al. (1995) suggested a refinement of the methodology by incorporating the known surface properties of specific clay minerals into the calibration relationship.

Certain minerals may influence soil dielectric properties and thus calibrations because the solid itself has a high permittivity (Roth et al. 1992; Dirksen and Dasberg, 1993; Robinson et al., 1994; Peplinski et al., 1995). Robinson et al. (1995) demonstrated that iron minerals such as haematite and magnetite had higher permittivities than the values of 4 to 6 normally found in common soil minerals. Some titanium and aluminum hydroxides may also fall into this category and might influence calibrations performed in tropical soils.

3. *Organic Soils*

Topp et al. (1980) demonstrated, using TDR, that the calibration relationship for an organic soil with a bulk density of 0.422 Mg m^{-3} was significantly different from the calibration found for mineral soils. This finding was supported by Stein and Kane (1983), Pepin et al. (1992), and Roth et al. (1992) for peat soils with

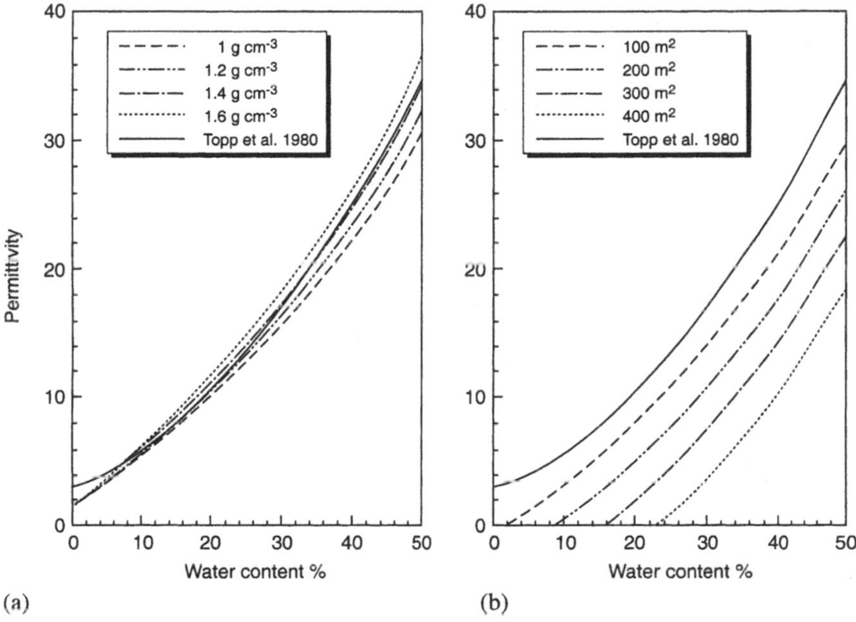

Fig. 13 The effect on the permittivity/water content relationship of (a) increasing bulk density; (b) increasing surface area per unit of soil. (After Dirksen and Dasberg, 1993.)

bulk densities ranging from 0.06 to 0.25 Mg m^{-3}. A calibration derived from measurements in several peat substrates was found to be similar to that of Pepin et al. (1992) by Paquet et al. (1993).

VI. APPLICATIONS OF NEUTRON AND DIELECTRIC METHODS

The examples reviewed briefly in this section illustrate how the neutron and dielectric measurement methods have been used in practical applications. Because neutron probes have been available for so much longer, there are many more reports in the literature of their use. Examples of the application of dielectric methods, particularly capacitance methods, rather than publications on the calibration or evaluation of sensors, are as yet less usual.

Neutron probes have been used most often to measure water content change to depth in the field at weekly, or sometimes more frequent, intervals. Water content distribution has been measured beneath crops (e.g., Bautista et al., 1985), and the soil water regime of different soils and vegetation types, varying from arid rangelands (Nash et al., 1991) to equatorial forest and cleared areas (Hodnett et al.,

1996), has been characterized. Soil water content data are frequently collected to measure crop or soil water balances, where the focus of interest may be the soil evaporation and/or plant transpiration components, or the subsurface and deep drainage (recharge) components. McGowan and Williams (1980) used the depth of the drying front, measured by neutron probe, to define the depth above which water content loss was due to evaporation and transpiration, and below which water content change could be ascribed to drainage, and hence derived a catchment water balance (McGowan et al., 1980). Often additional measurements, particularly of soil matric potential, are made to enable partitioning of water content change in the profile into evaporation (including transpiration) and drainage (e.g., Sophocleus and Perry, 1985; Cooper et al., 1990). Neutron probe measurements have been particularly useful in the study of the hydraulic properties of the unsaturated zone of deep aquifers such as the English Chalk and sandstones (Gardner et al., 1990; Cooper et al., 1990) because it is possible to make measurements to depths of 4 m or more.

In many cases, dielectric monitoring methods could have been used to obtain much the same information, with the advantage that more frequent and automated monitoring, if required, would have been feasible. However, measurements at depths greater than about 1 m using TDR or buried capacitance sensors would have necessitated excavation of pits from which to install equipment, entailing some disturbance to the soil's hydrology. The essential difference between the neutron probe and dielectric methods is that neutron probes permit measurement at many depths (to ≥5 m) infrequently, whereas most dielectric methods permit measurement at relatively few depths (due to cost), but with high temporal frequency. TDR has been used successfully in various field studies to obtain frequent measurements of water content, though generally not to depths much below 0.5 m. The aim of these studies has varied from characterizing soil water regimes in time and space (Van Wesenbeeck and Kachanoski, 1988; Herkelrath et al., 1991; Nyberg, 1996) to determining soil evaporation and transpiration rates (Zegelin et al., 1992; Plauborg, 1995). These studies used vertically installed waveguides of different length to monitor water content distribution by layer in the soil profile, but others have used horizontal installations in similar work. Nielsen et al. (1995) set out to study the immediate surface soil and used horizontally installed waveguides for measurements at just 25 mm depth. Measurement at shallower depth, 13 mm, proved unreliable, however.

Other examples of in situ use of TDR include work in peats, including very low density ones (Pepin et al., 1992). Parkin et al. (1995) measured unsaturated hydraulic conductivity using TDR to 0.4 m depth in field plots irrigated using a rainfall simulator. Temporal variations in soil water composition have been investigated by Heimovaara et al. (1995), both in the field and in laboratory cores, using TDR to monitor both water content and bulk soil electrical conductivity, in combination with soil solution sampling.

The neutron method is much less versatile than dielectric methods for container, glasshouse, and laboratory work, but equipment to permit such experimental work has been designed, e.g., Klenke and Flint (1991) described a neutron collimator for use with a CPN 503 probe. The good space and time resolution of TDR measurements has been used effectively in container studies of water uptake by roots (e.g., Wraith and Baker, 1991; Heimovaara et al., 1993). Topp et al. (1996) were able to record the diurnal uptake of water from, and its release to, relatively dry soil in which maize roots were growing. The Easy Test miniprobe, because of its small size, lends itself to this type of study and has been used, with minitensiometers, to obtain soil water release and hydraulic conductivity functions in undisturbed soil cores 100 mm high and 55 mm in diameter, as the cores dried from saturation (Malicki et al., 1992).

Neutron probes are being used increasingly in work associated with potential environmental pollution due to leakage from landfills and accidental spillage of contaminants. Prospective landfill and hazardous waste sites have been characterized for their suitability prior to use and monitored thereafter (Unruh et al., 1990). For example, Kramer et al. (1995) used a 670 m access tube installed horizontally beneath the leachate collection system of a municipal landfill to detect leachate leaks. No attempt at calibration was made; changes in neutron count with distance along the tube, and with time, were interpreted in terms of water content.

Provision of irrigation scheduling advice on the basis of both neutron probe and dielectric measurements is a service industry in high-value crop growing areas of several countries. Remote interrogation of TDR or capacitance sensors installed in farmers' fields will permit the same information to be gained more cheaply and open up the possibility of using more sensors to define crop water requirements better. Design of intelligent irrigation systems incorporating dielectric sensors to monitor water content, and hence water need, are well underway (e.g., Miller and Ray, 1985). Connecting TDR or capacitance sensors to systems that measure soil temperature, rainfall, soil matric potential, and any other parameters that may be required opens up the possibility of studying soil hydrology and crop water use to a level of detail not previously feasible. The U.K. Institute of Hydrology has an operational Automatic Soil Water Station that combines these sensors, using buried capacitance probes for the water content measurements. The possible uses for such systems in research and commercial applications are only just being explored. The revolution in soil water content measurement that dielectric methods have sparked is already having an impact in soil and environmental work beyond the dreams of most earlier neutron probe users.

VII. REMOTE SENSING OF SOIL WATER CONTENT

The development of remote sensing, which was given considerable impetus by the Soviet and U.S. space programs in the early 1960s, is now a flourishing subdiscipline with a wide range of applications in the monitoring of many aspects of the environment. In remote sensing, several methods are used to convey data about the object of interest, called the "target," to the sensor. Sensors may be mounted just above ground level (e.g., on a tower or moving vehicle), on an aircraft, or on a satellite. In the last case, data are purchased from the relevant space agency for processing by the user. As an alternative, many commercial organizations provide a service if users do not have adequate processing capabilities or expertise.

Figure 14 shows the electromagnetic spectrum, with the sensing technologies that have been most usefully applied in each portion of the spectrum. Remote sensing studies of soil water have exploited a wide range of wavelengths from gamma rays (<0.003–10 nm) to long-wavelength microwave radiometry and radar (1–800 mm). Both "passive" and "active" remote sensing techniques have been successfully employed. With passive techniques, the sensor measures radiation that either is emitted by the target (as a function of its black-body temperature and emissivity) or is reflected, refracted, or polarized by it, having originated from the sun. Active remote sensing uses an artificial source of radiation. This radiation

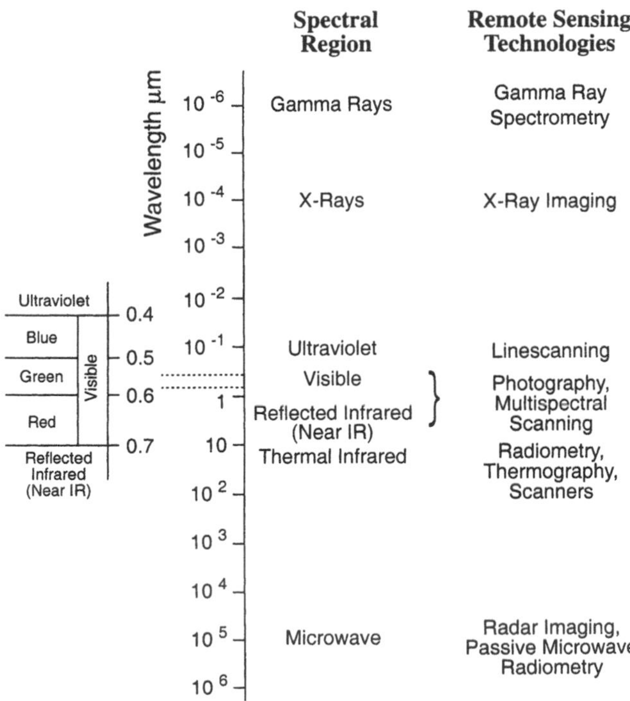

Fig. 14 The electromagnetic spectrum indicating the principal spectral regions exploited in remote sensing and the corresponding technologies. The x-ray and ultraviolet regions are not used in remote sensing of soil water.

is detected after being reflected from the target; sonar, radar, and monochromatic lidar are examples of active systems. A useful technical introduction to remote sensing and data interpretation and analysis in hydrology is provided in Engman and Gurney (1991), while Schmugge (1990) provides a summary specifically in the field of soil water.

Several important factors must be taken into account when using remote sensing for soil water assessment. Sensitivity to soil water content is usually confined to the surface soil layers. Measurements of the average water content to a maximum depth of 0.3 m are possible using gamma-ray spectrometry. At microwave frequencies, penetration (or emission) depth varies with wavelength, soil composition, and water content. In dry, sandy deserts, penetration depths of at least one wavelength may occur (i.e., 200 mm in the L-band), but this reduces to approximately one tenth of a wavelength for wet soils. At visible and infrared frequencies, any interaction with soil water is confined to less than 1 mm from the soil surface. As the measurements are made at a distance from the soil, they are subject to interference from objects between the soil and the sensor. Vegetation and clouds are the most common causes of interference. Wavelengths <25 mm are affected by cloud cover and atmospheric aerosols, while most techniques perform more effectively in the absence of vegetation, particularly the gamma-radiation and polarization techniques. Also, the sensor type and platform must be carefully matched to the measurement requirement. For example, sensors mounted on portable hydraulic arms are commonly employed for detailed process studies to achieve high temporal sampling rates and accurate spatial location.

Passive microwave measurements from satellites may be sensitive to soil water, but, at suitable wavelengths and using current technology, will have a spatial resolution of around 50 km, which will confine their application to very large areas. However, there are many applications for such large-scale areal estimates of soil water, and remote sensing has most potential for these. Practical considerations, such as the cost and availability (or delivery time) of appropriate data (particularly if aircraft or satellite-mounted sensors are used) and difficulties of sensor calibration must be assessed at the project planning stage.

Methods to estimate soil profile water content from remotely sensed surface measurements are being developed. For example, Entekhobi et al. (1994) used a coupled soil water and heat flux model with remotely sensed water content and temperature data to extrapolate the remotely sensed information to greater depths. Progress in the estimation of soil water content aggregated over large areas is also forthcoming. Georgakakos and Baumer (1996) used a technique involving conceptual hydrological models with on-site soil water and discharge measurements. With remotely sensed measurements they were able to produce much improved estimates of aggregated soil water content for large areas, despite the errors associated with the remotely sensed water content. There is great potential for use of remotely sensed and other soil water measurements in understanding land–

atmosphere interactions and global climate, but Nielsen et al. (1996) have also examined the opportunities for soil science studies associated with the increasing amount of information on spatial and temporal variation in surface soil water content. An overview of the use and success of different remote sensing technologies follows. Currently, most work is focussed on passive and active microwave sensing.

A. Techniques Based on Naturally Occurring Gamma Radiation

Natural gamma radiation has been widely used with terrestrial and airborne sensors in mineral prospecting (e.g., Cook et al., 1996). All rocks and soils are inherently radioactive and emit gamma radiation. Since soil water attenuates such radiation, it is possible to deduce changes in soil water content by repeated gamma-ray spectrometry of areas of interest. Average near-surface soil water content in the 0–0.3 m zone can be measured to an accuracy of 10% (Zotimor, 1971; Carroll, 1981). The risk of noise from atmospheric gamma-ray emissions necessitates a very low aircraft altitude, often as low as 100–200 m (Salomonsen, 1983), and consequently the technique can be used only in areas of low relief. Even at such low altitudes, the "ground footprint" of gamma-radiation attenuation techniques is still quite large (approximately twice the aircraft altitude). The most promising future application of gamma-ray spectrometry for soil water assessment probably is in ground-based studies (Loijens, 1980).

B. Reflectance and Polarization Techniques in the Visible and Near-Infrared Regions

Interactions between visible or near-infrared radiation and the ground surface are, in part, a function of soil water content. The spectral reflectance of soil generally decreases at higher water contents (i.e., wet soil is darker in color), and the polarization characteristics of visible light are significantly affected by soil water content. However, soil spectral properties are influenced by a variety of other factors such as soil texture, structure, illumination geometry, and atmospheric conditions (Liang and Townshend, 1996), and care must be taken before ascribing any change in reflectance to water content variation. It has been found that rapid drying of the soil surface provides anomalous indications of underlying conditions that limit the application of bare earth studies to local qualitative comparisons (Evans, 1979). It is not likely that direct-reflectance studies offer an immediately viable method of soil water measurement.

While not a direct measurement, vegetation reflectance may provide a much more practical indication of soil water as it responds to water availability within the whole root zone rather than in a thin surface layer. Vegetation indices based on the red/near infrared reflectance (Steven et al., 1990) are used to express crop vigor and may also provide an indication of water availability, particularly in drier climates. Under controlled drydown conditions, linear relationships have been established between root zone soil water and the normalized difference vegetation index for maize and groundnut crops (Narasimha Rao et al., 1993) but further work is required to determine the effects of different crop types, growth stage, and nutrient application.

C. Techniques Using Thermal Infrared Radiation

Surface soil temperature is influenced by a number of factors, one of which is the water content of the soil below. Wet soil has a higher thermal capacity than dry soil, so it exhibits a smaller diurnal temperature range, appearing cooler during the day and warmer at night. Empirical work established how diurnal variations in observed soil temperature could be related to soil water content at various depths (Idso et al., 1975), and a number of modeling approaches have since been used for both bare soil and vegetated surfaces (Van de Griend et al., 1985). This property has been exploited in ground-based, airborne, and satellite remote sensing studies of soil water, usually employing sensors operating in the 8–14 μm portion of the electromagnetic spectrum, where atmospheric attenuation is at a minimum. Currently the operational orbiting satellites carrying thermal sensors do not provide measurements at the optimal time of day or night for thermal inertia modeling, but attempts have been made to adjust the data acquired by the Advanced Very High Resolution Radiometer (AVHRR) from the NOAA satellite to make this possible (Cracknell and Xue, 1996). For accurate measurement of surface temperature, atmospheric corrections based on profiles of pressure, temperature, and humidity must be applied to both satellite and aircraft-acquired thermal data using some form of radiative transfer model (Price, 1983). For practical application of thermal techniques over different vegetation types and partial vegetation cover, the use of soil-vegetation-atmosphere transfer (SVAT) models are required, and simplified versions have provided sensible results when applied to regional studies and for incorporation into climate models (Saha, 1995; Gillies and Carlson, 1995). The main problem with thermal techniques is that they are ineffective in the presence of clouds, and this severely restricts their application.

D. Passive and Active Microwave Techniques

Microwaves have the advantage of being scarcely affected by atmospheric conditions and, as a result of their longer wavelength, interact with a greater depth of soil than visible and infrared wavelengths. Unlike other techniques, there is a direct physical relationship between soil water and soil dielectric properties (see Sec. V) which determines both microwave emission and reflection. Two distinct types of microwave sensors are used: microwave radiometers, which are passive sensing devices, and microwave radars, which illuminate the target with microwaves and measure the backscattered signal. A useful summary of microwave remote sensing of soil water is given in Engman and Chauhan (1995).

Microwave radiometers measure the natural emission of microwaves from soil as a result of its blackbody temperature and emissivity in the same way as infrared thermometers. The presence of water in the soil and overlying vegetation results in a decrease in emissivity and consequently a reduction in microwave brightness temperature. By contrast, with microwave radar, an increase in soil water (and hence soil permittivity) results in an increase in backscatter caused by the increased number of water dipoles per unit volume of soil; the dipoles oscillate in response to the microwave illumination and reflect more of that energy back to the sensor. Another major difference between active and passive microwave sensors lies in the spatial resolution of the data that can be acquired. From satellite altitudes, a ground resolution of 50 km is typical for microwave radiometers, and advances in antenna technology should provide data at 10 km resolution within the next decade. In comparison, synthetic aperture radar (SAR) typically has a spatial resolution of around 20 m. The latter is thus better suited to local studies, while the former would be more appropriate for regional or global applications.

Currently there are no satellite microwave radiometers designed specifically for soil water measurement, although the Nimbus-SMMR and DMSP-SSM/I satellite–sensor combinations have provided some useful results, particularly in drier and vegetation-sparse environments (Teng et al., 1993). The AgRISTARS Program (Schmugge et al., 1986) was a four-year study that combined field measurements of soil and vegetation parameters with ground-based, aircraft, and satellite microwave data acquisition, both passive and active, at a number of sites throughout the USA. It concluded that the best single channel for radiometric observation of soil water was the L-band (0.21 m wavelength). At this wavelength it should be possible to measure the soil water of the surface layer (0–5 cm) to an accuracy of ±5% absolute about 90% of the time where vegetation permits. The major difficulty was when the soil surface had just been worked and was extremely rough and of low density. The L-band was found to be the least sensitive to the effects of vegetation attenuation and soil roughness variations. It was also felt that the combination of other spectral data (e.g., the use of visible/near-infrared for vegetation estimates and active microwave for roughness estimates) would be more useful than additional microwave radiometer channels. A correction procedure for the effects of surface roughness and crop parameters has since been derived (Paloscia et al., 1993) using a multiband package of ground-based sensors (L, X, and K_a band microwave radiometers plus infrared bands).

The AgRISTARS Project also reported on active microwave applications for soil water and found that the most suitable single-sensor configuration was C-band (wavelength 5 cm) operating within the 10–20° incidence angle range at either HH (horizontal emit, horizontal receive) or HV (horizontal emit, vertical receive) polarization (Dobson and Ulaby, 1986). Since this report, the ERS-1 and ERS-2 satellites have been successfully providing C-band VV (vertical emit, vertical receive) polarization SAR at 23° incidence angle, which is quite close to the optimum configuration. The results of ERS studies, including many relating to the measurement of soil water, have been presented at three ERS Symposia (ESA 1992, 1993, 1997). Some of the most encouraging results have come from ERS Pilot Projects that have supported river basin experiments. In one study, the mean radar backscatter over a river basin in northern France showed clear linear correlation with automatic soil water measurements during autumn, winter, and mid-spring, but the correlation was lost during the end of spring and during the summer, which corresponded to periods of denser vegetation (Cognard et al., 1996). These results were confirmed by a more intensive catchment study in southern England (Stuttard et al., 1998) which derived linear backscatter/soil water relationships for bare earth, crops, and grassland at satellite spatial resolutions of 12.5 m (actual), 150 m, and 1000 m (simulated). Another study used a statistical analysis of the influence of land use and soil type on radar backscatter and incorporated this knowledge into a GIS. Soil water content and matric potential were measured on a single field, and catchment water status was calculated in relation to this point based on the radar backscatter (Mauser et al., 1994).

Aircraft (Chen et al., 1997) and Space Shuttle experiments (Dubois et al., 1996) have also demonstrated the capabilities of multifrequency and fully polarized SAR for determining surface roughness and vegetation type, which is considered to be an essential requirement for the future successful application of satellite soil water monitoring systems.

REFERENCES

Ahuja, L. R., and R. D. Williams. 1985. Use of surface gamma neutron gauge to measure bulk density, field capacity and macro-porosity in the topsoil. In: *Isotope and Radiation Techniques in Soil Physics and Irrigation Studies.* Vienna: IAEA, pp. 469–478.

Amoozegar, A., K. C. Martin, and M. T. Hoover. 1989. Effect of access hole properties on soil water content determination by neutron thermalization. *Soil Sci. Soc. Am. J.* 53: 330–335.

ASTM. 1981. *Annual Book of American Society for Testing and Materials Standards.* Philadelphia: ASTM.

Australian Water Resources Council. 1974. *Soil Moisture Measurement and Assessment.* AWRC Hydrological Series No. 9. Canberra: Australian Department of Environmental Conservation.

Baker, J. M., and R. R. Allmaras. 1990. System for automating and multiplexing soil moisture measurement by time domain reflectometry. *Soil Sci. Soc. Am. J.* 54: 1–6.

Baumgartner, N., G. W. Parkin, and D. E. Elrick. 1994. Soil water content and potential measured by hollow time domain reflectometry probe. *Soil Sci. Soc. Am. J.* 58: 315–318.

Bautista, I., G. Cruz Romero, J. R. Castel, and C. Ramos. 1985. Spatial and time variability of soil moisture in citrus orchards as measured by neutron probe. *Acta Hortic.* 171: 61–73.

Belcher, D. J. 1950. *The Measurement of Soil Moisture by Neutron and Gamma Ray Scattering*. Civil Aeronautics Administration Tech. Development Rept. No. 127, pp. 98–120.

Bell, J. P. 1987. *Neutron Probe Practice*. Wallingford, Oxon, U.K.: Institute of Hydrology Rept. No. 19 (3d ed.).

Bell, J. P., T. J. Dean, and M. G. Hodnett. 1987. Soil moisture measurement by an improved capacitance technique, Part II. Field techniques, evaluation and calibration. *J. Hydrol.* 93:79–90.

Birchak, J. R., C. G. Gardner, J. E. Hipp, and J. M. Victor. 1974. High dielectric constant microwave probes for sensing soil moisture. *Proc. IEEE* 62:93–98.

Brummer, E., and E. S. Mardock. 1945. A neutron method for measuring saturation in laboratory flow measurements. *Proc. Am. Inst. Mining and Metal Engineering*.

Carneiro, C., and E. de Jong. 1985. In situ determination of the slope of the calibration curve of a neutron probe using a volumetric technique. *Soil Sci.* 139:250–254.

Carroll, T. R. (1981). Airborne soil moisture measurement using natural terrestrial gamma radiation. *Soil Sci.* 132:358–366.

Chanasyk, D. S., and M. A. Naeth. 1996. Field measurement of soil moisture using a neutron probe. *Can. J. Soil Sci.* 76:317–323.

Chen, K. S., W. P. Huang, D. S. Tsay, and F. Amar. 1997. Classification of multifrequency polarimentric SAR imagery using dynamic learning neural networks. *IEEE Trans. Geosci. Rem. Sens.* 34:814–820.

Cognard, A.-L., C. Loumagne, M. Normand, P. Olivier, C. Ottle, D. Vidal-Madjar, S. Louahala, and A. Vidal. 1995. Evaluation of the ERS 1 synthetic aperture radar capacity to estimate surface soil moisture: Two-year results over the Naizin watershed. *Wat. Resources Res.* 31:975–982.

Cook, S. E., R. J. Corner, P. R. Groves, and G. J. Grealish. 1996. Use of gamma-radiometric data for soil mapping. *Aust. J. Soil Res.* 34:183–194.

Cooper, J. D., C. M. K. Gardner, and N. MacKenzie. 1990. Soil controls on recharge to aquifers. *J. Soil Sci.* 41:613–630.

Couchat, P., C. Carre, J. Marcesse, and J. le Ho. 1975. The measurement of thermal neutron constants of the soil: Application to the calibration of neutron moisture gauges and to the pedological study of soil. In: *Nuclear Cross-Sections and Technology* (R. A. Schrack and C. D. Bowman, eds.). Washington DC: US Dept. Commerce, pp. 516–579.

Cracknell, A. P., and Y. Xue. 1996. Thermal inertia determination from space–A tutorial review. *Int. J. Rem. Sens.* 17:431–461.

Dalton, F. N. 1992. Development of time domain reflectometry for measuring soil water content and bulk soil electrical conductivity. In: *Advances in Measurement of Soil Physical Properties: Bringing Theory into Practice* (R. Green and G. C. Topp, eds.). Madison, WI: Spec. Publ. No. 30, SSSA, pp. 143–167.

Dalton, F. N., W. N. Herkelrath, D. S. Rawlins, and J. D. Rhoades. 1984. Time domain reflectometry: Simultaneous measurement of soil water content and electrical conductivity with a single probe. *Science* 224:989–990.

Dean, T. J. (1994). *The IH Capacitance Probe for Measurement of Soil Water Content.* Wallingford, Oxon, U.K.: Institute of Hydrology, Rept. No 125.

Dean, T. J., J. P. Bell, and A. J. B. Baty. 1987. Soil moisture measurement by an improved capacitance technique, Part 1. Sensor design and performance. *J. Hydrol.* 93: 67–78.

De Clerk, P. 1985. Mesure de l'évolution de la teneur en eau des sols par voie électromagnétique. *Tech Routière* 3:6–15.

De Loor, G. P. 1968. Dielectric properties of heterogenous mixtures containing water. *J. Microwave Power* 3(2): 67–73.

Dirksen, C., and S. Dasberg. 1993. Improved calibration of time domain reflectometry soil water content measurements. *Soil Sci. Soc. Am. J.* 57:660–667.

Dobson, M. C., and F. T. Ulaby. 1986. Active microwave soil moisture research. *IEEE Trans. Geosci. Rem. Sens.* GE-24:23–36.

Dobson, M. C., F. T. Ulaby, M. T. Hallikainen, and M. A. El-Rayes. 1985. Microwave dielectric behaviour of wet soil, Part II: Dielectric mixing models. *IEEE Trans. Geosci. Rem. Sens.* GE-23:35–46.

Dubois, P. C., J. van Zyl, and E. T. Engman. 1996. Measuring soil moisture with imaging radars (Special issue on SIR-C/X-SAR). *IEEE Trans. Geosci. Rem. Sens.* 33: 915–926.

Eller, H., and A. Denoth. 1996. A capacitive soil moisture sensor. *J. Hydrol.* 185:137–146.

E.S.A. (European Space Agency). 1992, 1993, 1997. *Proc. ERS Symposia: Space at the Service of Our Environment.* ESA Publ. Div., Noordwijk, The Netherlands; 1st Symp. 4–6 Nov. 1992, Cannes, France, ESA Publ. SP-359; 2d Symp. 11–14 Oct. 1993, Hamburg, Germany, ESA Publ. SP-361; 3d Symp. 15–21 Mar. 1997, Florence, Italy, ESA Publ. SP-414.

Engman, E. T., and N. Chauhan. 1995. Status of microwave soil moisture measurements with remote sensing. *Rem. Sens. Environ.* 51:189–198.

Engman, E. T., and R. J. Gurney. 1991. *Remote Sensing in Hydrology.* London: Chapman and Hall, pp. 127–154.

Entekhobi, D., H. Nakmura, and E. G. Njoku. 1994. Solving the inverse problem for soil moisture and temperature profiles by the sequential assimilation of multifrequency remotely sensed observations. *IEEE Trans. Geosci. Rem. Sens.* 32:438–448.

Evans, R. 1979. Air photos for soil survey in lowland England: Factors affecting the photographic images of bare soil and their relevance to assessing soil moisture content and discrimination of soils by remote sensing. *Rem. Sens. Environ.* 8:39–63.

Evett, S. R., and J. L. Steiner. 1995. Precision of neutron scattering and capacitance type soil water content gauges from field calibration. *Soil Sci. Soc. Am. J.* 59:961–968.

Farah, S. M., R. J. Reginato, and F. S. Nakayama. 1984. Calibration of a soil surface neutron meter. *Soil Sci.* 138:235–239.

Fellner-Feldegg, H. 1969. The measurement of dielectrics in the time domain. *J. Phys. Chem.* 73:616–623.

Fritton, D., W. Busscher, and J. Alpert. 1974. An inexpensive but durable thermal conductivity probe for field use. *Soil Sci. Soc. Am. Proc.* 38:854–855.

Gardner, C. M. K., J. D. Cooper, S. R. Wellings, J. P. Bell, M. G. Hodnett, S. A. Boyle, and M. J. Howard. 1990. Hydrology of the unsaturated zone of the chalk of southeast England. In: *Chalk* (J. B. Burland et al., eds.). London: Thomas Telford, pp. 611–618.

Gardner, C. M. K., T. J. Dean, and J. D. Cooper. 1998. Soil water content measurement with a high frequency capacitance sensor. *J. Agric. Eng. Res.* 71:395–403.

Gardner, W. H. 1986. Water content. In: *Methods of Soil Analysis, Part 1.* 2d ed. (A. Klute, ed.). Madison, WI: Am. Soc. Agron., pp. 493–544.

Gaskin, G. J., and J. D. Miller. 1996. Measurement of soil water content using a simplified impedance measuring technique. *J. Agric. Eng. Res.* 63:163–160.

Gaudu, J. C., J. M. Mathieu, J. C. Fumanal, L. Bruckler, A. Chanzy, P. Bertuzzi, P. Stengel, and R. Guennelon. 1993. Mesure de l'humidité des sols par une méthode capacitive: Analyse des facteurs influençant la mesure. *Agronomie* 13:57–73.

Gee, G. W., and M. E. Dodson. 1981. Soil water content by microwave drying: A routine procedure. *Soil Sci. Soc. Am. J.* 45:1234–1237.

Georgakakos, K. P., and O. W. Baumer. 1996. Measurement and utilisation of on-site soil moisture data. *J. Hydrol.* 184:131–152.

Gillies, R. R., and T. N. Carlson. 1995. Thermal remote sensing of surface soil water content with partial vegetation cover for incorporation into climate models. *J. Appl. Meteorol.* 34:745–756.

Greacen, E. L., ed. 1981. *Soil Water Assessment by the Neutron Method.* East Melbourne, Victoria, Australia: CSIRO.

Greacen, E. L., and G. Schrale. 1976. The effect of bulk density on the neutron meter calibration. *Aust. J. Soil Res.* 14:159–169.

Gupta, U. S., and R. K. Gupta. 1981. Evaluation of a sulphuric acid procedure for the determination of soil moisture. *J. Indian Soc. Soil Sci.* 29:156–159.

Halbertsma, J., E. van den Elsen, H. Bohl, and W. Skierucha. 1995. Temperature effects on TDR determined soil water content. In: *TDR Applications in Soil Science* (Proc. TDR Symposium). Foulum, Denmark: Danish Institute for Plant and Soil Science, Rept. No. 11.

Hanson, C. T., and R. L. Blevins. 1979. Soil water in coarse fragments. *Soil Sci. Soc. Am. J.* 43:819–820.

Hauser, V. L. 1984. Neutron meter calibration and error control. *Trans. Am. Soc. Agric. Eng.* 27:722–728.

Haverkamp, R., M. Vauclin, and G. Vachaud. 1984. Error analysis in estimating soil water content from neutron probe measurements: 1. Local standpoint. *Soil Sci.* 137: 78–90.

Heimovaara, T. J. 1993. Design of triple wire time domain reflectometry probes in practice and theory. *Soil Sci. Soc. Am. J.* 57:1410–1417.

Heimovaara, T. J., and W. Bouten. 1990. A computer controlled 36-channel time-domain reflectometry system for monitoring soil water contents. *Wat. Resources Res.* 26: 2311–2316.

Heimovaara, T. J., and E. de Water. 1993. *A Computer Controlled TDR System for Measuring Water Content and Bulk Electrical Conductivity of Soils.* Lab. Physical Geog. and Soil Sci., Univ. of Amsterdam, Rept. No. 41.

Heimovaara, T. J., J. I. Freijer, and W. Bouten. 1993. The application of TDR in laboratory column experiments. *Soil Technol.* 6:261–272.

Heimovaara, T. J., W. Bouten, and J. M. Verstraten. 1994. Frequency domain analysis of time domain reflectometry waveforms 2. A four component complex dielectric mixing model for soils. *Wat. Resources Res.* 30:201–209.

Heimovaara, T. J., A. G. Focke, W. Bouten, and J. M. Verstraten. 1995. Assessing temporal variations in soil water composition with time domain reflectometry. *Soil Sci. Soc. Am. J.* 59:689–698.

Herkelrath, W. N., S. P. Hamburg, and F. Murphy. 1991. Automatic real-time monitoring of soil moisture in a remote field area with time domain reflectometry. *Wat. Resources Res.* 27:857–864.

Hilhorst, M. A., J. Balendonck, and F. W. H. Kampers. 1993. A broad-bandwidth mixed analog/digital integrated circuit for the measurement of complex impedances. *IEEE J. Solid State Circuits* 28:764–768.

Hodnett, M. G. 1986. The neutron probe for soil moisture measurement. In: *Advances in Agricultural Instrumentation* (W. G. Gensler, ed.). Dordrecht, The Netherlands: Martinus Nijhoff, pp. 148–192.

Hodnett, M. G., and J. P. Bell. 1990. Neutron probe standards—Transport shields or a large drum of water? *Soil Sci.* 151:113–120.

Hodnett, M. G., L. P. Dasilava, H. R. Darocha, and R. C. Senna. 1996. Seasonal soil water storage changes beneath Central Amazonian rainforest and pasture. *J. Hydrol.* 170: 233–254.

Hoekstra, P., and A. Delaney. 1974. Dielectric properties of soils at UHF and microwave frequencies. *J. Geophys. Res.* 79:1699–1708.

Hokett, S. L., J. B. Chapman, and S. D. Cloud. 1992. Time domain reflectometry response to lateral soil water content heterogeneities. *Soil Sci. Soc. Am. J.* 56:313–316.

Holmes, J. W. 1956. Calibration and field use of the neutron scattering method of measuring soil water content. *Aust. J. Appl. Sci.* 7:45–58.

Hook, W. R., N. J. Livingston, Z. J. Sun, and P. B. Hook. 1992. Remote diode shorting improves measurement of soil water by time domain reflectometry. *Soil Sci. Soc. Am. J.* 56:1384–1391.

IAEA (International Atomic Energy Agency). 1972. *Neutron Moisture Gauges.* IAEA Tech. Rept. Ser., No. 112. Vienna: IAEA.

IAEA (International Atomic Energy Agency). 1990. *Regulations for the Safe Transport of Radioactive Material, 1985 Edition (Amended 1990).* IAEA Safety Ser., No. 6. Vienna: IAEA.

ICRP (International Commission on Radiological Protection). 1990. *1990 Recommendations of the ICRP.* ICRP Publ. No. 60. Oxford, U.K.: Pergamon Press.

Idso, S. B., T. J. Schmugge, R. D. Jackson, and R. J. Reginato. 1975. The utility of surface temperature measurement for the remote sensing of soil water status. *J. Geophys. Res.* 80:3044 -3049.

ISO (International Standards Organisation). 1993. *Soil Quality—Determination of Dry Matter and Water Content on a Mass Basis—Gravimetric Method.* Doc. ISO 11465. Geneva: ISO.

ISO (International Standards Organisation). 1996. *Soil Quality—Determination of Water Content in the Unsaturated Zone—Neutron Depth Probe Method.* Doc. ISO 10573. Geneva: ISO.

ISO (International Standards Organisation). 1997. *Soil Quality—Determination of Water Content on a Volume Basis—Gravimetric Method.* Doc. ISO 11461. Geneva: ISO.

Jacobsen, O.H., and P. Schjonning. 1994. A laboratory calibration of time domain reflectometry for soil water measurement including effects of bulk density and texture. *J. Hydrol.* 151:147–157.

Jarvis, N., and P. Leeds-Harrison. 1987. Some problems associated with the use of the neutron probe in swelling/shrinking clay soils. *J. Soil Sci.* 38:149–156.

Kachanoski, R. G., E. Pringle, and A. Ward. 1992. Field measurement of solute travel times using time domain reflectometry. *Soil Sci. Soc. Am. J.* 56:47–52.

Karsten, J. H. M., and C. J. Van der Vyver. 1979. The use of a neutron moisture meter near the surface. *Agrochemophysica* 11:45–49.

Kelly, S. F., J. S. Selker, and J. L. Green. 1995. Using short soil moisture probes with high bandwidth time domain reflectometry instruments. *Soil Sci. Soc. Am. J.* 59:97–102.

Klenke, J. M., and A. L. Flint. 1991. Collimated neutron probe for soil water content measurements. *Soil Sci. Soc. Am. J.* 55:916–923.

Knight, J. H. 1992. Sensitivity of time domain reflectometry measurements to lateral variations in soil water content. *Wat. Resources Res.* 28:2345–2352.

Knight, J. H. 1995. Sampling volume of TDR probes used for water content monitoring: Theoretical investigation. Proc. TDR Symposium, *TDR Applications in Soil Science*. Foulum, Denmark: Danish Institute for Plant and Soil Science. Rept. No. 11.

Kramer, J. H., S. Cullen, and L. G. Everett. 1995. Vadose zone monitoring with the neutron probe. In: *Handbook of Vadose Zone Characterization and Monitoring* (L. G. Wilson, L. Everett, and S. J. Cullen, eds.). Boca Raton, FL: Lewis Publishers, pp. 291–309.

Leidieu, J., P. De Ridder, P. De Clerck, and S. Dautrebande. 1986. A method of measuring soil moisture by time domain reflectometry. *J. Hydrol.* 88:319–328.

Liang, S., and J. R. G. Townshend. 1996. A parametric soil BRDF model: A four stream approximation for multiple scattering. *Int. J. Rem. Sens.* 17:1303–1315.

Lide, D. R. 1992. *Handbook of Chemistry and Physics*. 73rd ed. London: CRC Press.

Loijens, H. S. 1980. Determination of soil water content from terrestrial gamma radiation measurements. *Wat. Resources Res.* 16:565–573.

Lorch, E. A. 1980. The concept of the recommended working life applied to radiation sources. *Radiol. Protection Bull.* 34:20–22.

Malicki, M. A., R. Plagge, M. Renger, and R. T. Walzak. 1992. Application of time-domain reflectometry (TDR) soil moisture miniprobe for the determination of unsaturated soil water characteristics from undisturbed soil cores. *Irrig. Sci.* 13:65–72.

Malicki, M. A., R. Plagge, and C. H. Roth. 1996. Improving the calibration of dielectric TDR soil moisture determination taking into account the solid soil. *Eur. J. Soil Sci.* 47:357–366.

Mauser, W., M. Rombach, H. Bach, R. Stolz, A. Demerican, and J. Kellndorfer. 1994. The use of ERS-1 data for spatial surface-moisture determination. *Proc. First ERS-1 Pilot Project Workshop, Toledo, Spain, 22–24 June 1994*. Publ. No. SP-365. Noordwijk, The Netherlands: European Space Agency, pp. 61–73.

McCulloch, D. B., and T. Wall. 1976. A method of measuring neutron absorption cross-sections of soil samples for calibration of the neutron moisture meter. *Nucl. Instrum. Meth.* 137:516–579.

McGowan, M., and J. B. Williams. 1980. The water balance of an agricultural catchment: 1. Estimation of evaporation from soil water records. *J. Soil Sci.* 31:217–230.

McGowan, M., J. B. Williams, and J. L. Monteith. 1980. The water balance of an agricultural catchment: 3. The water balance. *J. Soil Sci.* 31:245–262.

Miller, R. N., and K. B. Ray. 1985. System for optimum irrigating and fertilizing. US Patent No. 4545396. Washington DC: U.S. Patent Office.

Morrison, R. D. 1983. *Groundwater Monitoring Technology: Procedures, Equipment and Applications*. Prairie du Sac, WI: Timco Mfg. Inc.

Mughapghab, S. F., M. Divadeenam, and N. E. Holden. 1981. *Neutron Cross Sections: Vol. 1*. New York: Academic Press.

Nakayama, F. S., and R. J. Reginato. 1982. Simplifying neutron moisture meter calibration. *Soil Sci.* 133:48–52.

Narasimha Rao, P. V., L. Venkataratnam, P. V. Krishna Rao, K. V. Ramana, and M. N. Singarao. 1993. Relation between root zone soil moisture and normalised difference vegetation index of vegetated fields. *Int. J. Rem. Sens.* 14:441–449.

Nash, M. S., P. J. Wierenga, and A. Gutjahr. 1991. Time series analysis of soil moisture and rainfall along a line transect in arid rangeland. *Soil Sci.* 152:189–198.

Nicolls, K. D., J. T. Hutton, and J. L Honeysett. 1977. Gadolinium in soils and its effect on the count rate of the neutron moisture meter. *Aust. J. Soil Res.* 14:159–169.

Nicolls, K.D., J. L Honeysett, and M. W. Hughes. 1981. Instrument design. In: *Soil Water Assessment by the Neutron Method* (E. L. Greacen, ed.). East Melbourne, Victoria, Australia: CSIRO, pp. 24–34.

Nielsen, D., H. J. Lagae, and R. L. Anderson. 1995. Time-domain reflectometry measurements of surface soil water content. *Soil Sci. Soc. Am. J.* 59:103–105.

Nielsen, D., M. Kutilek, and M. B. Parlange. 1996. Surface soil water content regimes: Opportunities in soil science. *J. Hydrol.* 184:35–55.

Noborio K., K. J. McInnes, and J. L. Heilman. 1996. Measurements of soil water content, heat capacity and thermal conductivity with a single TDR probe. *Soil Sci.* 161: 22–28.

Nyberg, L. 1996. Spatial variability of soil water content in the covered catchment at Gardsjon, Sweden. *Hydrol. Processes* 10:89–103.

Olgaard, P. L. 1965. *On the Theory of the Neutronic Method for Measuring the Water Content in Soil*. Riso Report No. 97. Danish Atomic Energy Commission.

Paetzold, R. F., and G. A. Matzkanin. 1984. NMR measurement of water in clay. ILRI Publication No. 37. Beltsville, MD: USDA Conservation Service, pp. 316–319.

Paetzold, R. F., G. A. Matzkanin, and A. De los Santos. 1985. Surface soil water content measurement using pulsed nuclear magnetic resonance techniques. *Soil Sci. Soc. Am. J.* 49:537–540.

Paloscia, S., P. Pampaloni, L. Chiarantini, P. Coppo, S. Gagliani, and G. Luzi. 1993. Multifrequency passive microwave remote sensing of soil moisture and roughness. *Int. J. Rem. Sens.* 14:467–483.

Paquet, J. M., J. Caron, and O. Banton. 1993. In situ determination of the water desorption characteristics of peat substrates. *Can. J. Soil Sci.* 73:329–339.

Parkin, G. W., R. G. Kachanoski, D. Elrick, and R. Gibson. 1995. Unsaturated hydraulic conductivity measured by time domain reflectometry under a rainfall simulator. *Wat. Resources Res.* 31:447–454.

Pepin S., A. P. Plamondon, and J. Stein. 1992. Peat water content measurement using time domain reflectometry. *Can. J. For. Res.* 22:534–540.

Peplinski, N. R., F. T. Ulaby, and M. C. Dobson. 1995. Dielectric properties of soils in the 0.3–1.3 GHz range. *IEEE Trans Geosci. Rem. Sens.* 33:803–807.

Perdok, U. D., B. Kroesbergen, and M. A. Hilhorst. 1996. Influence of gravimetric water content and bulk density on the dielectric properties of soils. *Eur. J. Soil Sci.* 47: 367–371.

Pieper, G. F. 1949. The measurement of soil moisture by the slowing down of neutrons. Ph.D. Thesis, Cornell Univ., Ithaca, NY.

Plauborg, F. 1995. Evaporation from bare soil in a temperate humid climate—measurement using micro-lysimeters and time domain reflectometry. *Agric. For. Meteorol.* 76:1–17.

Prebble, R. E., J. A. Forest, J. L. Honeysett, M. W. Hughes, D. S. McIntyre, and G. Schrale. 1981. Field installation and maintenance. In: *Soil Water Assessment by the Neutron Method* (E. L. Greacen, ed.). East Melbourne, Victoria, Australia: CSIRO, pp. 82–98.

Price, J. C. 1983. Estimating surface temperatures from satellite thermal infrared data—A simple formulation for the atmospheric effect. *Rem. Sens. Environ.* 13:353–361.

Rahi, G. S., and S. F. Shih. 1981. Effect of bulk density on calibration of neutron moisture probes for organic soils. *Trans. Am. Soc. Agric. Eng.* 24:1230–1233.

Reinhart, K. G. 1961. The problem of stones in soil moisture measurement. *Soil Sci. Soc. Am. Proc.* 25:268–270.

Reynolds, S. G. 1970. The gravimetric method of soil moisture determination: 1. A study of equipment and methodological problems. *J. Hydrol.* 11:258–273.

Robinson, D. A. 1998. Soil water content estimates based on the measurement of soil relative permittivity: Use of capacitance and time domain reflectometry. PhD Thesis, Univ. of Ulster, Coleraine, Northern Ireland.

Robinson, D. A., J. P. Bell, and C. H. Batchelor. 1994. The influence of iron minerals on the determination of soil water content using dielectric techniques. *J. Hydrol.* 161: 169–180.

Robinson, D. A., J. P. Bell, and C. H. Batchelor. 1995. Influence of iron and titanium minerals on water content determination by TDR. Proc. TDR Symposium. *TDR Applications in Soil Science*. Rept. No. 11. Foulum, Denmark: Danish Institute for Plant and Soil Science.

Robinson, D. A., C. M. K. Gardner, J. Evans, J. D. Cooper, M. G. Hodnett, and J. P. Bell. 1998. The dielectric calibration of capacitance probes for soil hydrology using an oscillating frequency response model. *Hydrol. Earth Systems Sci.* 2:111–120.

Robinson, M., and T. J. Dean. 1993. Measurement of near surface soil water content using a capacitance probe. *Hydrol. Processes* 7:77–86.

Roth, K., R. Schulin, H. Fluhler, and W. Attinger. 1990. Calibration of time domain reflectometry for water content measurement using a composite dielectric approach. *Wat. Resources Res.* 26:2267–2273.

Roth, C. H., M. A. Malicki, and R. Plagge. 1992. Empirical evaluation of the relationship between soil dielectric constant and volumetric water content and the basis for calibrating soil moisture measurements by TDR. *J. Soil Sci.* 43:1–13.

Saha, S. K. 1995. Assessment of regional soil moisture conditions by coupling satellite sensor data with soil-plant system heat and moisture balance model. *Int. J. Rem. Sens.* 16:973–980.

Salomonsen, V. V. 1983. Water resources assessment. In: *Manual of Remote Sensing, Vol. 2* (R. N. Colwell, D. S. Simonett, and J. E. Estes, eds.). Falls Church, VA: Am. Soc. Photogrammetry, pp. 1497–1570.

Schmugge, T. J. 1990. Measurements of surface soil moisture and temperature. In: *Remote Sensing of Biosphere Functioning* (R. J. Hobbs and H. A. Mooney, eds.). New York: Springer-Verlag, pp. 31–63.

Schmugge, T. J., P. E. O'Neill, and J. R. Wang. 1986. Passive microwave soil moisture research. *IEEE Trans. Geosci. Rem. Sens.* GE-24:13–22.

Selker, J. S., L. Graff, and T. Steenhuis. 1993. Non-invasive time domain reflectometry moisture measurement probe. *Soil Sci. Soc. Am. J.* 57:934–936.

Smith-Rose, R. L. 1935. The electrical properties of soil at frequencies up to 100 megacycles per second; with a note on the resistivity of ground in the United Kingdom. *Proc. Phys. Soc. London* 47:923–931.

Sophocleus, M. 1979. A thermal conductivity probe designed for easy installation and recovery from shallow depths. *Soil Sci. Soc. Am. J.* 43:1056–1058.

Sophocleus, M., and C. A. Perry. 1985. Experimental studies in natural groundwater recharge dynamics: The analysis of observed recharge events. *J. Hydrol.* 81:297–332.

Spaans, E. J. A., and J. M. Baker. 1993. Simple baluns in parallel probes for time domain reflectometry. *Soil Sci. Soc. Am. J.* 57:668–673.

Sposito, G. 1984. *The Surface Chemistry of Soils*. New York: Oxford Univ. Press.

Standards Association of Australia. 1986. *Determination of the Moisture Content of a Soil: Microwave Oven Drying Method.* AS 1289.B1.4–1986.

Stein, J., and D. L. Kane. 1983. Monitoring the unfrozen water content of soil and snow using time domain reflectometry. *Wat. Resources Res.* 19:1573–1584.

Steven, M. D., T. J. Malthus, T. H. Demetriades-Shah, F. M. Danson, and J. A. Clark. 1990. High spectral resolution indices for crop stress. In: *Application of Remote Sensing in Agriculture* (J. A. Clark and M. D. Steven, eds.). London: Butterworths, pp. 209–228.

Stocker, R. V. 1984. Calibration of neutron moisture meters on stony soils. *J. Hydrol. N.Z.* 23:34–46.

Stone, J. F., D. Kirkham, and A. A. Read. 1955. Soil moisture determination by a portable neutron scattering moisture meter. *Soil Sci. Soc. Am. Proc.* 19:419–423.

Stuttard, M., K. Blyth, A. Zmuda, P. Bird, and D. Corr. 1998. Study of spatial and radiometric resolution of space-borne SAR data for hydrological applications. ESTEC Establishment Contractor Rept. No. 11978/96/NL/CN, Apr. 1998, European Space Agency, Noordwijk, The Netherlands.

Tan, C. S. 1992. Effect of different water content, sample number and soil type on determination of soil water using a home microwave oven. *Soil Sci. Plant Nutr.* 38: 381–384.

Teng, W. L., J. R. Wang, and P. C. Doraiswamy. 1993. Relationships between satellite microwave radiometric data, antecedent precipitation index and regional soil moisture. *Int. J. Rem. Sens.* 14:2483–2500.

Thomas, A. M. 1966. In-situ measurement of moisture in soil and similar substances by 'fringe' capacitance. *J. Sci. Instr.* 43:21–27.

Tomer, M. D., and J. L. Anderson. 1995. Field evaluation of a soil water capacitance probe in a fine sand. *Soil Sci.* 159:90–98.

Topp, G. C., J. L. Davies, and A. P. Annan. 1980. Electromagnetic determination of soil water content: Measurements in coaxial transmission lines. *Wat. Resources Res.* 16: 574–582.

Topp, G. C., M. Yanuka, W. D. Zebchuk, and S. J. Zegelin. 1988. Determination of electrical conductivity using Time Domain Reflectometry: Soil and water experiments in coaxial lines. *Wat. Resources Res.* 24:945–952.

Topp, G. C., M. Watt, and H. N. Hayhoe. 1996. Point specific measurement and monitoring of soil water content with an emphasis on TDR. *Can. J. Soil Sci.* 76:307–316.

Underwood, N., C. H. M. van Bavel, and R. W. Swanson. 1954. A portable slow neutron flux meter for measuring soil moisture. *Soil Sci.* 77:339–340.

Unruh, M. E., C. Corey, and J. M. Robertson. 1990. Vadose zone monitoring by fast neutron thermalisation (neutron probe): A 2-year case study. In: *Groundwater Management, No. 2*. Dublin, OH: NWWA, pp. 431–444.

Vachaud, G., J. M. Royer, and J. D. Cooper. 1977. Comparison of methods of calibration of a neutron probe by gravimetry or neutron capture model. *J. Hydrol.* 34:343–356.

Vaitekunas, D., G. S. V. Raghavan, and F. R. Van de Voort. 1989. Drying characteristics of a soil in a microwave environment. *Can. Agric. Eng.* 31:117–123.

Van de Griend, C. J., P. J. Camillo, and R. J. Gurney. 1985. Discrimination of soil physical parameters, thermal inertia and soil moisture from diurnal surface temperature fluctuations. *Wat. Resources Res.* 21:997-1009.

Van Vuuren, W. E. 1984. Problems involved in soil moisture determination by means of a neutron depth probe. In: *Recent Investigations in the Zone of Aeration.* Proc. Int. Symposium, Munich, pp. 281–293.

Van Wesenbeeck, I. J., and R. G. Kachanoski. 1988. Spatial and temporal distribution of soil water in the tilled layer under a corn crop. *Soil Sci. Soc. Am. J.* 52:363–368.

Whalley, W. R. 1993. Considerations on the use of time domain reflectometry (TDR) for measuring soil water content. *J. Soil Sci.* 44:1–9.

Whalley, W. R., T. J. Dean, and P. Izzard. 1992. Evaluation of the capacitance technique as a method for dynamically measuring soil water content. *J. Agric. Eng. Res.* 52: 147–155.

Wilson, D. J. 1988a. Uncertainties in the measurement of soil water content, caused by abrupt layer changes, when using a neutron probe. *Aust. J. Soil Res.* 26:8796.

Wilson, D. J. 1988b Neutron moisture meters: The minimum error in the derived water density. *Aust. J. Soil Res.* 26:97–104.

Wood, B., and N. Collis-George. 1980. Moisture content and bulk density measurements using dual-energy beam gamma radiation. *Soil Sci. Soc. Am. J.* 44:662–663.

Wraith, J. M., and J. M. Baker. 1991. High-resolution measurement of root water uptake using automated time-domain reflectometry. *Soil. Sci. Soc. Am. J.* 55:928–932.

Zegelin, S. J., I. White, and G. F. Russell. 1992. A critique of the time domain reflectometry technique for determining soil water content. In: *Advances in Measurement of Soil Physical Properties: Bringing Theory into Practice* (G. C. Topp et al., eds.). Spec. Publ. No. 30. Madison, WI: Soil Sci. Soc. Am., pp. 187–208.

Zotimor, N. V. 1971. Use of the gamma field of the earth to determine the water content of the soil. *Sov. Hydrol.* 4:313–320.

2
Matric Potential

Chris E. Mullins
University of Aberdeen, Aberdeen, Scotland

I. INTRODUCTION

The total potential ψ_t of soil water refers to the potential energy of water in the soil with respect to a defined reference state. Various components of this potential control water flow in the soil (Chaps. 4, 5, and 6), from the soil into roots, and through plants. Matric potential refers to the tenacity with which water is held by the soil matrix (Marshall, 1959). In the absence of high concentrations of solutes, it is the major factor that determines the availability of water to plants. After allowing for differences in elevation, differences in matric potential between different parts of the soil drive the unsaturated flow of soil water (Chap. 5).

A. Definition

The soil physics terminology committee of the ISSS provided agreed-upon definitions for total potential and its various components (Aslyng, 1963), which were slightly modified in 1976 (Bolt, 1976). A brief summary is given here. More detailed discussions of the meaning and significance of these definitions are given in soil physics books such as those of Marshall et al. (1996) and Hillel (1998).

Total potential of soil water can be divided into three components:

$$\psi_t = \psi_p + \psi_g + \psi_o \tag{1}$$

The pressure potential ψ_p is defined as "the amount of useful work that must be done per unit quantity of pure water to transfer reversibly and isothermally to the soil water an infinitesimal quantity of water from a pool at standard atmospheric pressure that contains a solution identical in composition to the soil water and is at the elevation of the point under consideration" (Marshall et al., 1996). Similar definitions have been given for gravitational potential, ψ_g, and osmotic potential, ψ_o, which refer to the effects of elevation (i.e., position in earth's gravitational field) and of solutes on the energy status of soil water. The sum of gravitational and pressure potential is called the hydraulic potential ψ_h. Differences between the hydraulic potential at different places in the soil drive the movement of soil water. Matric potential ψ_m is a subcomponent of pressure potential and is defined as the value of ψ_p where there is no difference between the gas pressure on the water in the reference state and that of gas in the soil.

The above definition of pressure potential includes (1) the positive hydrostatic pressure that exists below a water table, (2) the potential difference experienced by soil that is under a gas pressure different from that of the water in the reference state, and (3) the negative pressure (i.e., suction) experienced by soil water as a result of its affinity for the soil matrix. In the past, some authors (Taylor and Ashcroft, 1972; Hanks and Ashcroft, 1980) have used the term "pressure potential" to refer only to subcomponents 1 and 2. However, all authors use equivalent definitions for matric potential, which is subcomponent 3. Matric potential can have only a zero or negative value. As water becomes more tightly held by the soil its matric potential decreases (becomes more negative). Matric or soil water suction or tension refers to the same property but takes the opposite sign to matric potential. In a swelling soil, overburden pressure can cause a slight error in applications where it is intended to relate matric potential to soil water content (Towner, 1981).

The sum of matric and osmotic potential is called the water potential ψ_w and is directly related to the relative humidity of water vapor in equilibrium with the liquid phase in soils and plants. ψ_w is an important indicator of plant water status and is also important in saline soils, where the osmotic potential of the soil solution is sufficient to influence plant water uptake.

B. Units

Since potentials are defined as energy per unit mass, they have units of joules per kilogram. However, it is also possible to define potentials as energy per unit volume or per unit weight. Thus, since the dimensions of energy per unit volume are identical to those of pressure, the appropriate unit is the pascal (1 bar = 100 kPa). Similarly, the dimensions of energy per unit weight are identical to those of length, so the appropriate unit is the meter. Because it is common to refer to the pressure due to a height *h* of a column of water as a pressure head (or simply head) *h*, this term is often used to describe the potential energy per unit weight. The relation

$$\psi\,(\mathrm{J\ kg^{-1}}) - \gamma\psi\,(\mathrm{Pa}) = \frac{\psi\,(\mathrm{m})}{g} \tag{2}$$

where γ is the density of water and g is the acceleration due to gravity ($\sim 1000\ \text{kg m}^{-3}$ and $9.81\ \text{m s}^{-2}$, respectively), is used to convert potentials from one set of dimensions to another. A logarithmic (pF) scale (Schofield, 1935), where

$$\text{pF} = \log_{10}\ (\text{negative pressure head in cm of water}) \tag{3}$$

has also been used.

II. AN OVERVIEW OF METHODS FOR MEASURING MATRIC POTENTIAL

The main features of methods for measuring matric potential and the addresses of some manufacturers and suppliers are given in Table 1. The web sites for many of the manufacturers list their suppliers in many countries. In considering the cost of instruments, it is important to decide whether a data logger is required, and to consider the cost of the logger or meter as well as the cost of the sensor, since some sensors are more easily logged than others and some are available with cheap loggers. Consequently Table 1 should be treated only as an initial guide to purchase, because of the pace of development in the choice of loggers and meters. There are many earlier reviews of the design and use of such methods (Marshall, 1959; Rawlins, 1976; Cassell and Klute, 1986; Rawlins and Campbell, 1986). Methods have been classified according to the measurement principle involved and are discussed in detail in the following sections. Tensiometers (Sec. III) consist of a porous vessel attached via a liquid-filled column to a manometer. Porous material sensors (Sec. IV) consist of a porous material whose water content varies with matric potential in a reproducible manner; a physical property of the material that varies with its water content is measured and related to matric potential using a calibration curve. Psychrometers (Sec. V) measure the relative humidity of water vapor in equilibrium with the soil solution. Because they measure the sum of matric and osmotic potentials, they are also readily applicable for measurements in various parts of plants.

There have been large improvements in the performance and availability of data loggers over the past ten years, some improvements in methods for measuring potential, and a growing use and awareness of the importance of measurements of potential. Despite this, there is still a need for a single sensor that can log matric potential to a field accuracy that is sufficient for understanding water movement and soil aeration under wet conditions (e.g. 0 to -100 ± 0.2 kPa) while being able to measure to a reasonable accuracy (say $\pm$ 5%) down to < -1.5 MPa. This is a tall order, but it explains the continuing interest in the osmotic tensiometer and improved porous material sensors.

III. TENSIOMETERS

A tensiometer consists of a porous vessel connected to a manometer, with all parts of the system water filled (Fig. 1). When the cup is in contact with the soil, films of water make a hydraulic connection between soil water and the water within the cup via the pores in its walls. Water then moves into or out of the cup until the (negative) pressure inside the cup equals the matric potential of the soil water.

The following equations are used to obtain matric and hydraulic potential from the mercury manometer readings shown in Fig. 1.

$$\psi_m = \frac{h - 12.6b - c}{g}$$

$$\psi_h = \frac{-(12.6b + c)}{g} \tag{4}$$

The factor of 12.6 is the difference between the relative densities of mercury and water. c is a factor to correct for the capillary depression that occurs at the mercury–water interface. If g is omitted from these two equations, they will give the potentials in head units.

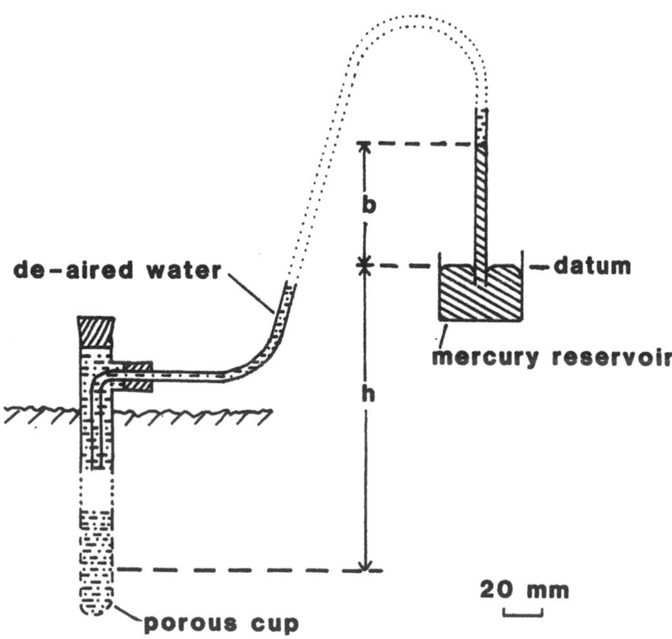

Fig. 1 Mercury manometer tensiometer.

Tensiometers are also available with Bourdon vacuum gauges, with pressure transducers (for data logging), and for portable use. Cassell and Klute (1986) provide a good discussion of methods for installing and maintaining tensiometers. I have discussed limitations common to most designs before considering each type of tensiometer.

A. Design Limitations

1. *Trapped Air*

All water-filled tensiometers have a lower measuring limit of about −85 kPa because, at more negative potentials, there is a tendency for air bubbles to nucleate at microscopic irregularities within the instrument. At such a low pressure relative to atmospheric pressure these bubbles expand, augmented by dissolved air coming out of solution, and can eventually block the tubing, making further readings unreliable. Filling with deaired water, which has had some of its dissolved air removed by boiling or by leaving it for some hours under a vacuum, is done to counteract this effect. Despite this, because dissolved air tends to move into the porous cup and come out of solution, tensiometers often incorporate an air trap that allows air to collect without blocking the instrument (Fig. 1). However, since this air causes the reponse time to increase (become slower), it is usual to "purge" tensiometers at regular intervals (ca. weekly or less often under cool wet conditions) by replacing the trapped air with deaired water (Cassell and Klute, 1986). The temporary release of suction during purging allows some water to pass into the surrounding soil so that readings are not reliable for some time after purging.

2. *Response Time*

Because any change in matric potential will cause a change in the volume of liquid in the tensiometer, time is required for this water to move into or out of the instrument and hence for it to respond. The conductance of the porous cup and the unsaturated hydraulic conductivity of the soil control the response time as well as the amount of water movement required for a given change in potential (the "gauge" sensitivity). Mercury manometers and Bourdon vacuum gauges are much less sensitive than pressure transducers. However, since most tensiometers operate with some trapped air within them, and since their tubing is not completely rigid, differences in response time between pressure transducers and other tensiometer types are much less than would be expected from the sensitivity of the gauges.

A tensiometer is said to be tensiometer limited if its response time is not influenced by soil properties, but only by the cup conductance and gauge sensitivity; otherwise it is soil limited. Tensiometer-limited response time is inversely proportional to cup conductance and gauge sensitivity (Richards, 1949), and cups with 100 times greater conductivity than normal cups are available for specialized applications. It is not difficult to obtain tensiometer-limited conditions, although in some soils tensiometers may be soil limited in drier soils (Towner, 1980).

Tensiometer-limited conditions are advantageous because instrument behavior is reproducible and not dependent on variable soil conditions (Klute and Gardner, 1962). This is particularly important when the potential is changing fast. However, obtaining a tensiometer-limited response is not the main consideration when tensiometers are used to monitor field conditions over periods of weeks or months and are read at infrequent intervals. Furthermore, too high a sensitivity can cause problems if the tensiometer is then too sensitive to other factors that can cause a change in the liquid-filled volume such as temperature changes (Watson and Jackson, 1967) and bending of the tubing. In field use, all tensiometer tubing should be shaded from direct sunlight where possible. Otherwise, sudden exposure to the sun can cause the tubing (and any air it contains) to expand and temporarily perturb the readings. High sensitivity/fast response tensiometers require careful handling and operate better under laboratory conditions.

Porous cups are usually made of a ceramic and must have pores that are small enough to prevent air from entering the cup when it is saturated. The cup must also have a reasonably high conductance. Ceramic tensiometer cups for field use have a conductance of about $3 \cdot 10^{-9}\ m^2\ s^{-1}$, and even a mercury-manometer tensiometer with such a cup will have a (tensiometer-limited) response time of about one minute in the absence of trapped air (Cassell and Klute, 1986), more than adequate for most field use.

B. Mercury Manometer and Bourdon Gauge Tensiometers

A manometer scale can easily be read to the nearest millimeter, so that mercury tensiometers have a scale resolution of $\pm$ 0.1 kPa. However, with the smallest (1.7 mm diameter) nylon tubing commonly used for the manometer, there is a significant capillary correction ($\sim$ 0.8 kPa) and hysteresis, caused by the mercury meniscus sticking to the walls of the tube. If the tube is agitated, to cause a small fluctuation in the mercury level, an accuracy of $\pm$ 0.25 kPa can be achieved; otherwise much larger errors can occur (Mullins et al., 1986). Bourdon vacuum gauges are less accurate, typically with a scale division of 2 kPa, but friction in the gauge mechanism and the difficulty of setting an accurate zero further limit their accuracy. Mercury tensiometers suffer from the environmental hazard of mercury and require a 1 m manometer post but are preferable if high accuracy is required (e.g., when measuring vertical gradients in hydraulic potential).

Mercury tensiometers can be constructed very cheaply, without the need for workshop facilities (Webster, 1966; Cassell and Klute, 1986). Where several tensiometers are used in the same vicinity, it is common to share a single mercury reservoir among 6–30 tensiometers. Because the mercury withdrawn from the reservoir will cause a slight drop in its level, for high accuracy, the level should be measured each time a reading is taken, or the reservoir should have a cross-section many times greater than the sum of the cross-sections of the tubes that dip into it. It is also advisable to check each tensiometer for air leaks before installation. This is done by soaking the cup in water, then applying an air pressure of 100 kPa to the inside of the tensiometer while it is immersed in water (Cassell and Klute, 1986). To minimize thermal effects, the manometer tubing should be shielded from direct sunlight (e.g., by facing the manometer post away from the midday sun). With prolonged outside use, some plasticizer may come out of the nylon tubing and collect as a white deposit, which can eventually block the tube. We have not found this to be a problem over a single season, but 1.7 mm tubing may need to be occasionally replaced over longer periods.

C. Pressure Transducer and Automatic Logging Systems

Because pressure transducers have a high gauge sensitivity, they are particularly useful when a short response time is important. They can also be used with data loggers. Transducers (e.g., piezoresistive silicon types) that are not temperature sensitive and have a precision of ± 0.2 kPa can be bought for ~ $140. Types that are vented to the atmosphere should be used so that changes in atmospheric pressure have no effect.

In the unusual case that matric potentials are required at a considerable depth (say 10 m), a pressure transducer located close to the measuring depth is essential because a hanging water column will break once the tension in it approaches 100 kPa.

1. Automatic Logging Systems

Automatic logging systems are required at remote sites, when measurements are required more often than the site can be visited, and to study laboratory or field situations in which many measurements are required over a period of hours or days (e.g., drainage studies). In the former case a provision for automatic purging may also be necessary if weekly visits (or less frequently in wet conditions) are not possible. Systems that use a motor-driven fluid-scanning switch allow a number of tensiometers to be connected each in turn to a single pressure transducer (Anderson and Burt, 1977; Lee-Williams, 1978; Blackwell and Elsworth, 1980).

It is necessary to have a transducer attached to each tensiometer if very short measurement intervals are required because re-equilibration, when a transducer is switched between tensiometers at different potentials, can take 2 minutes (Blackwell and Elsworth, 1980) or more (Rice, 1969). The effect of temperature

Table 1 Methods, Range, Accuracy, Typical Cost, and Suppliers for Measuring Matric (ψ_m) or (Where Indicated) Water (ψ_m) Potential

Method, range, and accuracy[a]	Unit cost (U.S.$)	Manufacturers/suppliers and References
Tensiometers (0 to −85 kPa)		
Bourdon gauge, ± 2 kPa	150	C, D, F[b]
Mercury manometer, ≤ ± 0.25 kPa	30 + post & Hg	Homemade with commercial cups (Webster, 1966; Cassell and Klute, 1986)
Ceramic cups for tensiometers	15	E, F
Pressure transducer: normal, miniature,[c] ± 0.2 kPa	250, 450	B, G, H
Portable Bourdon gauge, ± 2 kPa, but see text	1,000	C, D, F (Mullins et al., 1986)
Puncture tensiometer, ≥ + 0.7 kPa (systematic) + portable readout	40 each + 1,000	G, H
Filter paper (ψ_m/ψ_w) (−1 kPa to −100 MPa), 0 to −50 kPa ± 150%, −50 kPa to −2.5 MPa ± 180%	1	All suppliers of Whatman filter paper (Deka et al., 1995)
Electrical resistance,[c] **Watermark** (−10 to −400 kPa) ± 10%, **Gypsum block** (−50 to −1500 kPa)	50, 25	F, G, H, I
Heat dissipation[c] (−10 kPa to −100 MPa) ± 10%	200 + 2,500	A
Equitensiometer[c] (0 to −100 kPa) ± 5 kPa (−100 to −1000 kPa) ± 5% + portable δ meter	800 + 500	B
Psychrometers (ψ_w), all for disturbed samples except the Spanner psychrometer		
Isopiestic (0 to < −40 MPa) ± 10 kPa	15,000	(see text) (Boyer, 1995)
Dew point (0 to −40 MPa) ± 100 kPa	4,500	A
Richards (0 to −300 MPa) ± 5–10% + meter	2,500 + 2,500	A (but may no longer be available)
Spanner (0 to −7 MPa) ± 5–10% + meter	40 + 2,600	I (field/in situ measurement)

[a] Accuracy represents the best reliable reported values or manufacturers' figures, but see text for details, since accuracy can be limited by a number of factors.

[b] Key (many web sites list local suppliers): A, Decagon Devices Inc., U.S.A. (http://www.decagon.com). B, Delta T, U.K. (http://www.delta-t.co.uk). C, Eijkelkamp, The Netherlands (http://www.eijkelkamp.com). D, ELE International Ltd., U.K. (http://www.eleint.co.uk). E, Fairey Industrial Ceramics Ltd., Filleybrook, Stone, Staffs., ST15 0PU, U.K. F, Soilmoisture Equipment Corp., U.S.A. (http://www.soilmoisture.com). G, Skye Instruments Ltd. (http://www.skyeinstruments.com). H, UMS GmbH, Germany (http://www.ums-muc.de). I, Wescor Inc., U.S.A. (http://www.wescor.com).

[c] Can be used with data loggers ($1000–3000).

fluctuations on readings, which is most notable where nylon tubing is exposed above ground (Watson and Jackson, 1967; Rice, 1969), is also minimized with the transducer attached directly to the tensiometer. Such tensiometers and loggers are commercially available (Table 1).

2. *Systems with Portable Transducers (Puncture Tensiometers)*

A puncture tensiometer consists of a portable pressure transducer attached to a hypodermic needle that can be used to puncture a septum at the top of a permanently installed tensiometer and hence measure the pressure inside it (Fig. 2) (Marthaler et al., 1983; Frede et al., 1984). In this way, one transducer and readout unit can be used to measure the pressure in a large number of tensiometers. Each tensiometer simply consists of a porous cup attached to the base of a water-filled tube topped by a rubber or plastic septum that reseals each time the needle is removed. A small air pocket is deliberately left at the top of each tensiometer to reduce any thermal effects on the reading and the small pressure change caused

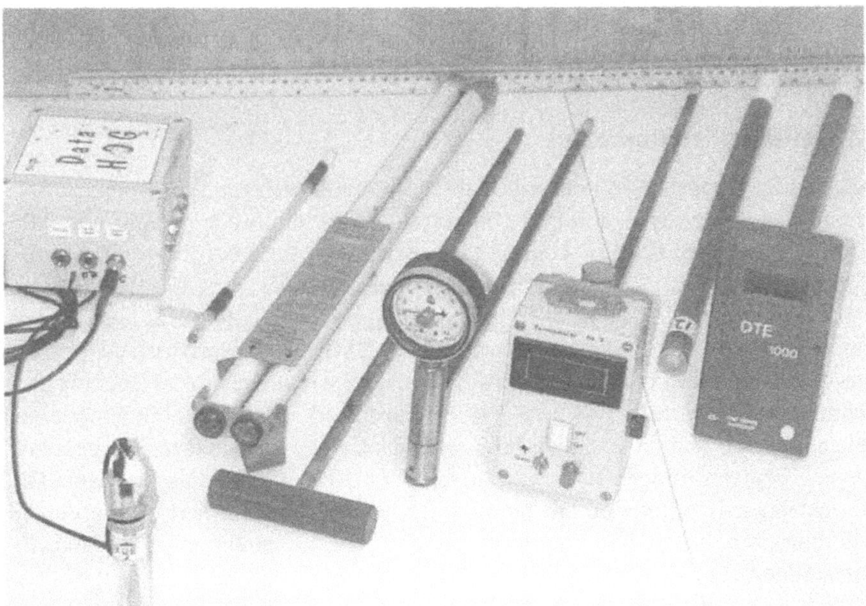

Fig. 2 Various tensiometers. From left to right: data logger attached to a pressure transducer tensiometer (only the top part with cover removed to reveal transducer); Webster (1966) type mercury manometer tensiometer; "quick draw" portable tensiometer (case, auger, and tensiometer); portable tensiometer with a pressure transducer and readout; puncture tensiometer without, and with, portable meter attached.

by the introduction of the needle. The needle and sensor are designed to have a very small dead volume to minimize this. However, Marthaler et al. reported systematic errors of $\sim$ 0.7 kPa in potentials close to zero (-2 to -3.6 kPa) but a good overall relation between mercury manometer and puncture tensiometer readings. Eventually the septum needs to be replaced, and careful insertion is required to ensure that there is no leak into the system. Consequently, these devices are not as accurate as systems with an in situ manometer or pressure sensor.

D. Portable Tensiometers

Portable tensiometers with Bourdon vacuum gauges (Table 1) and ones with a pressure transducer (available from UMS, Table 1) that can be read to $\pm$ 0.1 kPa are commercially available. These can be stored with their tips in water when not in use so that there is little accumulation of air within them, and they rarely need to be refilled. They can be used when single or occasional measurements are required. However, they cannot usually give a reliable reading quickly after insertion because of the effect of soil deformation during insertion. Mullins et al. (1986) found that re-equilibration of the disturbed soil with that surrounding it took only a few minutes in soil at > -5 kPa but > 2 h in soil at < -30 kPa (irrespective of the use of the null-point device supplied on one model).

E. Osmotic Tensiometers

Peck and Rabbidge (1969) described the design and performance of an osmotic tensiometer. It consists of a cell containing a high molecular weight (20,000) polyethylene glycol solution confined between a pressure transducer and a semipermeable membrane supported behind a porous ceramic. The cell is pressurized so that it registers 1.5 MPa when immersed in pure water, allowing the tensiometer to measure matric potentials between 0 and -1.5 MPa. However, there were problems due to polymer leakage and sensitivity to temperature changes (Bocking and Fredlund, 1979). Biesheuvel et al. (1999) have used an improved membrane to prevent leakage and have shown how readings can be corrected for temperature effects. Their tensiometer had an accuracy of $< 10\%$ at potentials < -100 kPa. The technique is promising but requires further development and testing in soil to demonstrate that it has long-term stability and acceptable accuracy and response time.

IV. POROUS MATERIAL SENSORS

These sensors are made of a porous material whose water content varies with matric potential in a reproducible manner. A physical property of the material that varies with water content is measured and related to matric potential, using a calibration curve. Sensors based on the measurement of the water content of filter paper, electrical conductivity, heat dissipation, and dielectric constant are discussed.

Irrespective of the method used to measure the water content of the porous material, its physical properties determine the range of matric potentials over which the sensor will be sensitive and accurate. Sensitivity depends on the rate of change of water content with matric potential, and hence on the pore size distribution of the porous material. A major limitation to accuracy is the amount of hysteresis that the material displays, and special materials have been developed to have low hysteresis and good sensitivity for recently developed sensors. The porous material is calibrated by equilibrating it at a set of known matric potentials. The reliability of published calibration curves or those supplied by manufacturers depends on how closely the water characteristic of the sensor resembles that of the sensor used in the original calibration. For greater accuracy, users should calibrate all, or a representative sample, of their sensors in the range of interest. Apart from the filter-paper technique, which is used on disturbed samples, the other sensors described here are nondestructive and can be logged. Because their response time will depend on the amount of water that has to flow out of the sensor for any given change in potential, there will be a lag in response, especially at low potentials. Sensitivity and accuracy also vary along the sensing range. Since the accuracy figures quoted by manufacturers normally refer to optimal conditions (laboratory equilibration at constant temperature and the most accurate portion of the sensing range using calibrated sensors), these should be treated with considerable caution. Finally, when left in the soil the sensors are likely to accumulate fine material, including microbial debris that can progressively clog the pores, so that it is desirable to recheck the calibration after prolonged field use. Although electrical resistance sensors are becoming much less popular due to the availability of better techniques, the sections on the sensor material, response time, hysteresis, and calibration of these sensors are of relevance to all porous material sensors.

A. Filter Paper Method

The filter paper method, originally used by Gardner (1937) as a simple means for obtaining the soil water release characteristic, is a cheap and simple method for measuring matric potential that is only beginning to receive the use it deserves. The method consists of placing a filter paper in contact with a soil sample ($>$ 100 g) in a sealed container at constant temperature until equilibrium is reached. The gravimetric water content of the filter paper is then determined, and this is converted to matric potential using a calibration curve. Apart from calibrated filter papers, this technique requires only a homemade lagged sample equilibration box, an oven set at 105°C, and a balance accurate to ±1 mg. Deka et al. (1995) give a full description of how to perform the technique.

The water retention characteristic of a filter paper (which is its calibration curve) can usefully cover a wide range of potentials from −1 kPa to −100 MPa (Fawcett and Collis-George, 1967). At the wetter end of this range, equilibration occurs by liquid water flow between soil and the filter paper. It is therefore important that the soil sample makes good contact with the paper and fully covers it. It is best to sandwich the paper between two halves of a core or two layers of soil. Vapor equilibrium becomes increasingly important in dryer soil, so that the paper responds to the water potential. Vapor equilibration is a slower process. Although equilibration times from 3 to 7 days have been used (Fawcett and Collis-George, 1967; McQueen and Miller, 1968; Hamblin, 1981), Deka et al. (1995) have shown that at least 6 d was required for full equilibration, even at −50 kPa, although this was still sufficient at −2.5 MPa. Small temperature fluctuations during equilibration can disturb the process and may even cause distillation (i.e., condensation of water on the walls of the container) (Al-Khafaf and Hanks, 1974). To avoid these problems, the sealed containers should be kept thermally insulated in Styrofoam (expanded polystyrene) containers, out of direct sunlight, and in a room or cupboard that does not have a large diurnal temperature variation (Campbell and Gee, 1986).

Since the potential of a sample can be altered by deformation, it is important to use an undisturbed soil core or soil that has been removed with minimal disturbance, to transport it with a minimum of vibration, or to equilibrate it in situ (Hamblin, 1981). Hamblin has also used the technique in situ by introducing papers into slits cut with a spatula in field soils.

Many authors have found it necessary to impregnate their filter papers to avoid fungal degradation during equilibration. Both 0.005% $HgCl_2$ and 3% pentachlorophenol in ethanol have been successfully used by moistening the filters, which are then allowed to dry before use. This has not been found to affect the calibration curve (Fawcett and Collis-George, 1967; McQueen and Miller, 1968). We have not found that a fungicide was necessary for equilibration times of up to 7 d, but this probably depends on soil type. Various methods have been proposed to cope with the soil that can stick to the equilibrated filter paper. Often it can be detached by a combination of flicking the paper with a fingernail and using a fine brush. Gardner (1937) corrected for the mass of soil adhering to the paper by determining its oven-dry mass (when it was brushed off the dry paper) and then back-calculating what its moist mass would have been from a knowledge of the water content of the soil sample. It is also possible to use a stack of three papers and only use the central one for measurement (Fawcett and Collis-George, 1967). However, we have found that this is often less accurate than using a single paper and that the central paper does not always reach equilibrium.

1. *Calibration and Accuracy*

Because filter papers have a measurable hysteresis (Fawcett and Collis-George, 1967; McQueen and Miller, 1968; Deka et al., 1995) it is necessary to bring them to equilibrium in the same way during calibration as when they are used. Thus, since the filter papers are dry before use, they should be calibrated on their wetting curve (Fawcett and Collis-George, 1967; Hamblin, 1981). Calibrations can be performed using a tension table, pressure plate, psychrometer, and/or vapor equilibration to cover different parts of the calibration (Campbell and Gee, 1986; Deka et al., 1995).

Deka et al. (1995) have critically reviewed the literature on calibration. They have shown that the calibrations for Whatman No. 42 filter paper determined by most authors are quite similar and give the following average calibration equations:

$$\begin{aligned} \log_{10}(-\psi_m) &= 5.144 - 6.699M \qquad \text{for } \psi_m < -51.6 \text{ kPa} \\ \log_{10}(-\psi_m) &= 2.383 - 1.309M \qquad \text{for } \psi_m > -51.6 \text{ kPa} \end{aligned} \tag{5}$$

where ψ_m is in kPa and M is the water content of the filter paper in g g^{-1}. The "broken stick" shape of the calibration curve is the result of water release from within the cellulose fibers at low potentials and from between the fibers at high potentials.

With calibrated batches of filter papers, accuracies of $\pm 150\%$ and $\pm 180\%$ can be expected between 0 and -50 kPa, and -50 kPa and -2.5 MPa, respectively (Deka et al., 1995). Where less accuracy is acceptable, the above equation can be used with uncalibrated papers. Because accuracy is mainly limited by the variability in the properties of individual filter papers, the accuracy obtainable from calibrated batches can be improved by replicating measurements. This is shown by the very good agreement between the mean value obtained from replicate filter papers and tensiometer measurements (Deka et al., 1995).

B. Electrical Resistance

Electrical resistance sensors consist of two electrodes enclosed or embedded within a porous material and have been used since the 1940s. At equilibrium, the matric potential of the solution within the sensor is equal to that of the surrounding soil. Commercial sensors can be purchased cheaply (Table 1), and it is also not difficult to construct large numbers of sensors at very little cost. However, the method is subject to a series of limitations that restrict the accuracy that can be obtained.

The potential of the sensor is obtained by measuring the electrical resistance between the two electrodes, which is a function of the water content of the porous material, and hence of its matric potential. Unfortunately, the resistance is also a function of temperature and of the concentration of solutes in the soil solution. Empirical equations to correct the resistance of gypsum sensors for temperature effects are available (Aitchison et al., 1951; Campbell and Gee, 1986) and have been reviewed by Aggelides and Paraskevi (1998). However, sensors cannot be used in saline soils unless the electrical conductivity of the soil solution is also known or can be compensated for. Scholl (1978) has described the construction and use of a combined salinity–matric potential sensor designed to overcome this limitation. More commonly, the sensor is cast from, or contains, gypsum, which slowly dissolves and maintains a saturated solution of calcium sulfate within itself. At 20°C, the solubility of calcium sulfate is about 1 g/dm^3, which should be more than ten times greater than the soil solution concentration in nonsaline soils, rendering gypsum sensors insensitive to the electrical conductivity of the soil solution in such soils.

1. Sensor Materials and Measurement Range

Many authors have given construction details for gypsum sensors (Pereira, 1951; Cannell and Asbell, 1964; Fourt and Hinton, 1970). Other types of sensor material have been tried, including fiberglass and nylon encased in gypsum (Perrier and Marsh, 1958) and fired mixtures of ground charcoal and clay (Scholl, 1978). The geometry of the electrodes depends on the material used but must aim to minimize electrical conduction through the soil (e.g., by using concentric electrodes), which would bias the reading. In practice, there are only two commercial sensors that are widely available: the Watermark sensor and the gypsum block (Table 1). The Watermark sensor is 76 mm long and 20 mm in diameter, contains a proprietary porous material held behind a synthetic membrane, and includes an internal gypsum tablet to neutralize solution conductivity effects. Its range is from −10 to −400 kPa ± 10%, although the distributors claim that an accuracy of ± 1% is possible in the range −10 to −200 kPa with individually calibrated sensors (Wescor web site). The gypsum block sensor is 32 mm long and 22 mm in diameter and covers the range −50 to −1500 kPa.

Gypsum sensors have a limited lifetime because they slowly dissolve in the soil, and their calibration will consequently change with time (Bouyoucos, 1953; Wellings et al., 1985). Bouyoucos (1953) suggested that gypsum sensors may last more than 10 years in dry soil but that their useful life in very wet (or acid) soil may not exceed 1 year. Aitchison et al. (1951) reported that gypsum sensors degenerate much faster in saline soils. Both the durability and the calibration of gypsum sensors depend on the source of the plaster of Paris used in their construction and the ratio of plaster to water used in casting (Aitchison et al., 1951; Perrier and Marsh, 1958).

Irrespective of the sensor material, it seems likely that the calibration curve may change significantly, well before the sensor shows obvious signs of wear. Thus the only guarantee of consistent behavior is to recheck at regular intervals (< 1 year) the calibration of a sample set of sensors taken from the whole range of soil conditions in which the sensors are installed.

2. Response Time

It is not possible to generalize about sensor response time because this can depend on the unsaturated hydraulic conductivity of the soil and the goodness of the soil–sensor contact as well as the potential towards which the sensor is equilibrating and the physical properties of the sensor. Gypsum sensors require about 1 week to equilibrate fully on a pressure plate at potentials between -0.1 and -1.5 kPa, but most of the equilibration has occurred within the first 48 h (Haise and Kelly, 1946; Wellings et al., 1985). Thus such sensors cannot be expected to respond any faster in the soil. In practice, fast changes in potential in the field are associated with rewetting events to which sensors are found to respond quickly (Goltz et al., 1981), whereas it is unlikely that sensors will lag much behind the rate at which soils dry out, except near to the soil surface.

3. Hysteresis and Uniformity

Tanner et al. (1948) found that vacuum saturation of gypsum sensors gave a lower resistance than saturation by immersion, while capillary saturation gave an intermediate value. They suggested that vacuum wetting is the most appropriate wetting method for testing a set of sensors for uniformity, since other wetting methods gave greater variability in the resistances of a set of saturated sensors. These effects are due to trapped air. Capillary saturation, in which each sensor is allowed to wet slowly from one end, was suggested as the most appropriate procedure before field installation, since this is closest to how they might become rewetted in the field.

The effect of rewetting is one aspect of the hysteresis in resistance exhibited by sensors, whereby the resistance of a sensor on a drying curve is less than that on a wetting curve. Since sensors are calibrated by desaturation and since they are often installed at the start of a growing season into a wet soil that subsequently dries out, it has often been argued that hysteresis problems may not be serious. However, in nearly all applications there are likely to be transient rewetting events (rain or irrigation) that result in partial rewetting of the soil profile, so that some inaccuracy due to hysteresis is unavoidable. Laboratory measurements of the hysteresis of gypsum sensors (Tanner and Hanks, 1952; Bourget et al., 1958) show that, in the range -30 to -1000 kPa, calibration based on a drying curve can typically result in a 100% overestimation of the matric potential measured during rewetting.

4. Calibration

Detailed methods have been given for the calibration of gypsum sensors using a pressure membrane (Haise and Kelly, 1946) or pressure plate (Wellings et al., 1985). Care is required to ensure good hydraulic contact between the sensors, which are initially saturated, and the membrane or plate. This can be achieved by attaching sensors to the membrane with plaster of Paris or embedding them into a paste of ground chalk on top of a pressure plate. Electrical connection to the sensors through the wall or lid of the pressure chamber is made via metal-through-glass or metal-through-ceramic insulated connectors (commercially available with some chambers), and the leads within the chamber must be sleeved to avoid condensation providing an additional electrical pathway. Each sensor requires a separate pair of lead-through connections to avoid current flow from adjacent sensors, and sealing the wires with silicone rubber at the connector is recommended (Wellings et al., 1985).

5. Meters

To avoid polarization effects, sensor resistance must be measured with an alternating current. Low frequency (~1 kHz) ac bridge circuits were used to measure this resistance, but because the sensor also has a capacitance that varies with its water content, this also had to be balanced in order to obtain a satisfactory null reading. Modern circuits operate on a different principle, in which a voltage output is produced that is proportional to the sensor's resistance (Wellings et al., 1985) and can be directly read from a meter or logged.

C. Heat Dissipation

This technique involves sensing the heat dissipation in a porous material sensor, to the center of which a short (150 s) heat pulse has been applied. The thermal diffusivity of the sensor, which determines its rate of heat dissipation, is related to the water content and hence matric potential of the sensor. Heat dissipation is measured as the difference between the temperature at the center of the sensor before and after the heat pulse has been applied. Performance is unaffected by the thermal properties of the surrounding soil because the sensor is large enough to contain the heat pulse. The original sensors were made of a germanium junction diode used to measure temperature, around which was wrapped a heating coil, and both were then encased in a cylinder of plaster of Paris or of a ceramic material. Unlike electrical resistance sensors, they are not responsive to the salinity of the soil solution.

The sensor is calibrated by equilibrating it at a range of matric potentials as described for electrical resistance sensors (Sec. IV.B.4). Theory, design, and constructional details are given by Phene et al. (1971a), who have also compared the performance of these sensors against that of psychrometers (1971b).

Sensor performance depends on the porous material that is used. Phene et al. (1971b) report a calibration accuracy of ± 20 kPa for matric potentials from 0 to −300 kPa and ± 100 kPa from −300 to −600 kPa for homemade ground ceramic/Castone sensors. Campbell and Gee (1986) estimated a precision of ± 10 kPa in the range 0 to −100 kPa for commercially available sensors (which are 50 mm long and 14 mm in diameter). As with electrical resistance sensors, accuracy will be further restricted by hysteresis of the porous material. Although the sensors can be used with data loggers, they cannot be read too frequently because each heat pulse requires time to dissipate fully before the next reading can be taken (Campbell and Gee, 1986).

D. Equitensiometers

This is the commercial name for a sensor (first produced in 1997) that is based on measurement of the water content of a proprietary porous material using a high-frequency capacitance-sensing technique (the theta probe, see Chap. 1). The porous sensor is claimed to have minimal hysteresis but is comparatively large (40 mm in diameter and ~60 mm long), so that it is not appropriate for use in small containers. The sensor covers the range 0 to −1 MPa and is most sensitive and accurate in the range 0 to −100 kPa. Because of its principle of operation, it should not be sensitive to soil salinity. Other authors have reported on the use of commercially available TDR water content sensors (Chap. 1) embedded in a ceramic disk (Or and Wraith, 1999a) or dental plaster (gypsum) (Noborio et al., 1999) to measure matric potential. Noborio's probe is sensitive to potentials between −30 and −1000 kPa and simultaneously measures water content using a separate part of the probe.

In addition to the limitations of all porous material sensors, all of these probes share two further problems. Firstly, the method of sensing water content means that the probes have to be comparatively large, and this in turn means that the time to approach equilibrium after a change in potential can be large. Noborio et al., for example, show that their probe takes over 2 weeks to reach equilibrium after a step change in potential from 0 to −100 kPa. Secondly, there is some evidence of temperature effects on the dielectric properties of material with fine pores (Or and Wraith, 1999b). It seems clear that laboratory tests and field comparisons with other sensors are now needed, to establish how accurate these type of probes can be expected to be in field use and to study response time and long-term stability.

E. Summary

In the past, gypsum sensors which can cover a range of potentials down to about −1.5 MPa (the approximate limit for water extraction by roots) offered a useful complement to the use of tensiometers to cover the full range of water availability to plants in applications where limited accuracy is acceptable. However, because of their temperature dependence, limited life in soil, and the change in calibration with time, the heat dissipation sensors, which are of comparable dimensions, are a better alternative. Techniques based on the TDR or theta probe (the so-called equitensiometer) are promising, but they have larger sensors, and their suitability is yet to be fully demonstrated.

V. PSYCHROMETERS

Psychrometers sense the relative humidity of vapor in equilibrium with the liquid phase in soils or plants. They can measure water potential in a range that overlaps the lower limit of tensiometer response (~ -80 kPa) and extends well beyond the limits of available water (< -1.5 MPa). They are widely used to measure plant water status (Boyer, 1995), and equipment has been commercially available for over 20 years (Table 1).

Psychrometers cover a range of potentials in which there is a lack of measurement techniques whose absolute accuracy can be theoretically guaranteed. Laboratory psychrometers are therefore used as a standard against which to compare and calibrate other methods.

A. Modes of Operation and Accuracy

The principle of measurement using psychrometry falls into three categories: isopiestic, dew point, and nonequilibrium (Spanner/Peltier and Richards). Boyer (1995) provides a readable review and description of these techniques from the viewpoint of plant measurements.

1. Isopiestic Psychrometers

Isopiestic psychrometers work by placing a solution of known water potential into a wire loop containing a thermocouple junction and enclosing this in a thermally insulated container just above the sample. (A thermocouple is made by joining two dissimilar metals. If this junction is at a different temperature from the temperature at which both metals are joined to another metal, such as the terminals of a voltmeter, a small voltage is generated that can be related to this temperature difference.) Any tendency for water to evaporate or condense onto the solution is registered by the thermocouple as a change in temperature. By repeating this procedure with solutions with known potentials that are close to that of the sample, the potential of a solution that would give the same reading as a dry thermocouple can be determined. This will be the same as the water potential of the sample. Consequently no calibration is required, and an absolute accuracy of ± 10 kPa can be achieved (Boyer, 1995).

2. *Dew Point Hygrometers*

In these devices, the sample is kept in an enclosed, thermally insulated container with a thermocouple that is maintained at the dew point (Neumann and Thurtell, 1972). This is the temperature at which vapor just starts to condense on the thermocouple junction and is related to the water potential of the sample. The sensing chamber is similar in construction to other psychrometers but is called a hygrometer because of its mode of operation. The sensing junction is cooled by passing a current through it in the reverse direction, which results in cooling (the Peltier effect). The sensing junction is alternately connected to a nanovoltmeter, to measure its temperature difference from the surroundings, and to a cooling current. The temperature of the sensing junction is controlled by an electronic feedback mechanism that switches the cooling current on for just the correct proportion of time to hold the junction at the dew point. Dew point hygrometers operate close to equilibrium but have to be calibrated with a range of solutions of known water potential. Commercial laboratory units that can accommodate small samples of soil or plant material have an accuracy of ± 100 kPa. The most recent versions use a chilled mirror dew point technique (www.decagon.com) in which the temperature of a small mirror is controlled by Peltier cooling and the (dew point) temperature at which condensation first occurs on the mirror is detected by a photocell from the change in reflectance of the mirror (Table 1). Such instruments still take 5 minutes to obtain each reading because of the time taken for equilibrium conditions to be approached in the measuring cell.

3. *Nonequilibrium Psychrometers*

Nonequilibrium Richards (Richards and Ogata, 1958) and Spanner (1951) psychrometers work by measuring the temperature drop caused by a water droplet evaporating from the tip of a fine thermocouple suspended in an enclosed insulated container over the sample. Water evaporates from the droplet at a rate controlled by its temperature and the relative humidity of the surrounding air. Within a few seconds, a steady rate of evaporation is reached when the junction has a constant temperature difference ΔT from its surroundings, such that the heat loss by evaporation is balanced by the heat gained in various ways (radiation, conduction along the thermocouple wires, etc.). ΔT is measured by having two thermocouple junctions, one consisting of the sensing junction and the other a reference junction attached to some thermal ballast (e.g., a piece of metal whose mass is much greater than that of the sensing junction and which is in good contact with the soil and the surroundings).

In commercial versions of the Richards psychrometer, the sensing junction is coated with a porous ceramic to form a bead that is wetted by immersion in water just before measurement. In the Spanner psychrometer, the Peltier effect is used to condense water onto the junction, and consequently this psychrometer can also be operated in the dew point mode. Irrespective of their mode of operation, Spanner psychrometers are limited to a range of potentials > -7 MPa because a larger cooling current is necessary to cool the sensing junction sufficiently at lower potentials, and this results in Joule heating of the thermocouple wires. In both nonequilibrium psychrometers, the way in which vapor diffuses from the thermocouple to the sample affects the measurements, causing a systematic error that is usually 5 to 10% for plant material but can be greater (Boyer, 1995). Savage and Cass (1984) also indicated that such psychrometers have a reproducibility of about $\pm$ 150 kPa for plant tissues and soils, although Rawlins and Campbell (1986) reported a much better precision under near-ideal laboratory conditions.

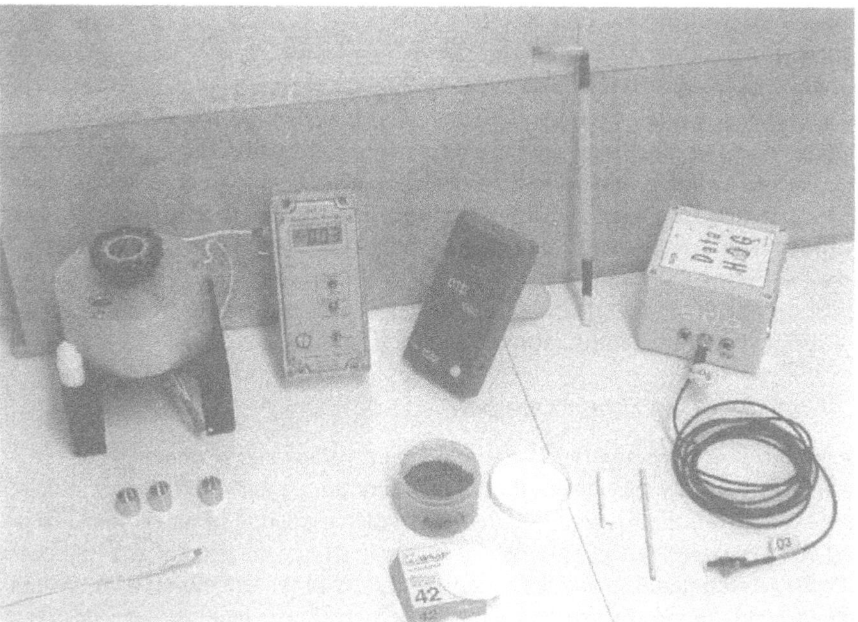

Fig. 3 From left to right: Richards laboratory psychrometer with three sample cups shown and nanovoltmeter attached; bottom left, field psychrometer sensor; portable meter for puncture tensiometer; Webster (1966) tensiometer sensor; data logger with pressure transducer. A porous ceramic tube and cup that can be attached to the transducer are shown to the left; bottom center, filter paper ready to be placed on the soil in the plastic sample container and covered with more soil.

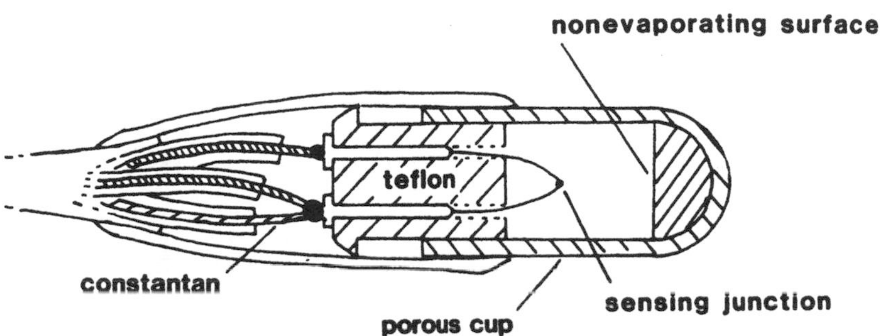

Fig. 4 Three-wire Spanner psychrometer (adapted from Rawlins and Campbell, 1986). A stainless steel screen can be used in place of the porous cup.

The discussion of methods so far has only considered designs that have been used on disturbed soil samples in the laboratory. However, Spanner psychrometers suitable for insertion into the soil for field or laboratory logging of water potential are commercially available (Table 1, Figs. 3 and 4) and can be used in the dew point or nonequilibrium mode. Psychrometers using all three principles of operation are commercially available for use in the laboratory with small (2–15 cm^3) samples, although the nonequilibrium psychrometers may no longer be available. Nanovoltmeters and automatic dew point control systems, made for use with psychrometers, and systems that can automatically log a number of field psychrometers, are also commercially available (Table 1). Wiebe et al. (1971) gave instructions for the construction of homemade psychrometers.

B. Limitations on Accuracy

All psychrometers are limited at the wet end of the range by the smallest temperature difference that can be meaningfully detected. Modern portable nanovoltmeters have a readability of $\pm$ 10 nV, corresponding to a potential of $\pm$ 2 kPa. However, the problems associated with measuring such small temperature differences (~0.0002°C) probably limit the useful range of current field psychrometers to potentials below -100 kPa. The major factors that influence the accuracy of psychrometer results and can cause large systematic errors are mainly associated with temperature and diffusive error (Boyer, 1995). Temperature errors and how to cope with them are shown in Table 2. A detailed review of the factors in this table is given by Rawlins and Campbell (1986).

Precautions to minimize temperature gradients for laboratory bench psychrometers include use in a room where temperature changes are not rapid and there is little air movement, minimizing hand contact with the sample changer, and encasing the sample changer in polyurethane foam or other thermal

Table 2 Factors That Can Introduce Systematic Errors in Soil Psychrometer Readings[a]

Factor and source	Effect	Remedy
1. Temperature gradients (variations in temperature of surroundings, electrical heating of thermocouple wires, absorption of external radiation)	Temperature difference between reference and sensing junction	i. (L) Use thermal insulation and/or a water bath to avoid gradients, allow $\frac{1}{2}$ h for samples to equilibrate in sample holder ii. (Ps, Pd) If reference junction is isolated from sample, measure temperature difference before Peltier cooling and subtract it from the reading iii. (F) Align psychrometer, with reference and sensing junctions parallel to isotherms (i.e., insert parallel to soil surface) iv. (F) Use a thermally shielded psychrometer with shield attached to reference junction
2. Temperature fluctuations with time	As for 1 above	As for 1 above
3. Variation in temperature of surroundings	Variation in relative humidity within chamber	Arrange sample to surround the sensing junction as nearly as possible
4. Vapor pressure gradient (L) only (extraneous sources or sinks of water vapor, especially where samples are warmer than the chamber, and water condenses on chamber walls)	Relative humidity in chamber is not controlled by the sample and reading is erroneous	As for 3 above. Ensure that sample and holder have reached the same temperature before moving under the sensing junction; do not insert samples that are warmer than the holder into it
5. Contamination of sensing junction or chamber walls	Unreliable readings	Clean junction and chamber and recalibrate
6. Zero offset	Nonzero output when calibrated over water	Subtract offset reading before converting it to a potential
7. Temperature correction (calibration temperature was not the same as measurement temperature)	Not important for Pd; incorrect readings for Pr and Ps	Calibrate at more than one temperature and interpolate to measurement temperature or use a theoretical correction procedure
8. Insufficient equilibration (L)	Incorrect reading	Plot psychrometer reading versus time to gain familiarity with its performance and use an adequate time. Equilibration time reduced by remedy in 3 above

[a] Key: L, laboratory sample changer arrangement; F, field psychrometer; Pr, Richards psychrometer; Ps, Spanner psychrometer; Pd, dew point mode.

insulation. For samples with a high relative humidity (e.g. $\psi_w < -6$ MPa), samples should be transferred to and loaded into the sample changer in a humid atmosphere (e.g., a box lined with wetted paper towels and with limited access, ideally a glove box). Before measurement, samples should be kept in the same room for at least 30 minutes to reach a similar temperature to the sample changer and can require between 4 and 30 minutes within the sample changer for conditions to approach vapor equilibrium (or steady state in a nonequilibrium psychrometer). Suggested times are given in the manufacturer's manuals and depend on the apparatus and the magnitude of the potential being measured.

Use of laboratory apparatus on samples that have been taken from the field, transported in sealed and thermally insulated containers, and then subsampled to fill the sample holder, will depend on factors such as water loss by distillation onto the container walls, variation of sample potential with temperature, and the effects of mechanical disturbance on the measured potential.

C. Calibration and Solutions of Known Potential

Isopiestic psychrometers do not require calibration but do require solutions of known potentials. Other psychrometers are usually calibrated by placing the sensing junction over a range of salt solutions of known potentials in a constant-temperature enclosure. Field psychrometers, for example, can be enclosed with the solution in a sealed container in a water bath. There are published values of the water potential of solutions of KCl (Campbell and Gardner, 1971), NaCl (Lang, 1967), and sucrose (Boyer, 1995) at a range of temperatures. Details of calibration of laboratory psychrometers are given in the manufacturer's instructions. Merrill and Rawlins (1972) described calibration of field psychrometers, and, for both laboratory and field psychrometers, recommended calibration procedures were given by Rawlins and Campbell (1986). If the sample temperature is not the same as the temperature at which calibration was performed, and the psychrometer is used in the nonequilibrium mode, it is necessary to make a temperature correction. This can be done either by calibrating at a series of temperatures and interpolation of the correct calibration curve or by a theoretical correction procedure (Merrill and Rawlins, 1972; Rawlins and Campbell, 1986).

D. Psychrometers for Insertion into the Soil

Only Spanner type psychrometers, which may be used in the dew point or nonequilibrium mode, are available for field use. Figures 3 and 4 show a three-wire psychrometer that includes a thermocouple to sense soil temperature. These are particularly important for use in the nonequilibrium mode where temperature correction is required for accurate results (Merrill and Rawlins, 1972). Diurnal soil temperature variations depend on climate. Their amplitude is considerably reduced by vegetation cover and decays exponentially with depth. They can impose a serious limitation to the accuracy of psychrometer readings taken near to the soil surface (< 0.25 m). Merrill and Rawlins (1972) have discussed the installation and calibration stability of soil psychrometers. They observed errors of 50% for Wescor ceramic-enclosed psychrometers installed vertically at a depth of 0.25 m in soil with a bare surface. Diurnal temperature variation at this depth was ± 1.3°C, and when the psychrometers were installed horizontally to minimize the influence of temperature gradients, the variation in readings was reduced to ~ 10%. Improved design can further reduce sensitivity to temperature gradients (Bruini and Thurtell, 1982). In addition to horizontal placement, Merrill and Rawlins (1972) recommended that 50–100 mm of the lead adjacent to the psychrometer be horizontally oriented. They also observed a 5.3% median change in calibration sensitivity of 33 Wescor ceramic psychrometers after 8 months of field use; only one psychrometer changed by > 15%. They considered that field psychrometers were able to distinguish day-to-day changes in water potential to within ± 50 kPa.

There are two psychrometer versions that are commercially available, one encased in a ceramic cup and one encased in a wire screen–shielded case (Fig. 3). The ceramic cup excludes contamination by fungal hyphae and prevents flooding of the chamber if it is below the water table for short periods. The screen-shielded version should be more suitable in soils that are likely to shrink away from the sensor during drying and may be less sensitive to temperature gradients (Merrill and Rawlins, 1972).

E. Summary

For laboratory use, particularly as a standard against which to compare other techniques, the isopiestic psychrometer is the most accurate but the most expensive option, and a cheaper dew point hygrometer may have acceptable accuracy. Results obtained with a nonequilibrium psychrometer in optimal laboratory conditions may also be useful where diffusive error can be minimized.

Field psychrometers are cheap and small but are limited in many situations to use at > 0.25 m depth due to sensitivity to thermal gradients and are most appropriate where measurement of low matric potentials (say < −300 kPa) are required.

VI. APPLICATIONS

Measurement of soil matric, hydraulic, and water potentials are so fundamental for studying water movement, germination, plant growth, and soil strength that the literature is full of examples of the use of these measurements. Examples of some of the major applications are given here.

Irrigation scheduling can be based on data from tensiometers (Hagan et al., 1967; Cassell and Klute, 1986), electrical resistance (Goltz et al., 1981), or heat dissipation sensors (Phene and Beale, 1976), all of which can be adapted to continuous logging and automatic irrigation control. Tensiometers, with their greater accuracy but restricted lower limit, are most suitable for applications such as the irrigation of vegetables and glasshouse crops, where it is intended to keep the soil permanently at a high potential and where fairly accurate control is required to avoid overwatering. Small portable tensiometers can be used for testing the suitability of conditions for germination and establishment in seedbeds, peat blocks, and other media used to raise plants (Goodman, 1983).

For monitoring the potential in the root zone under nonirrigated conditions, the best accuracy will be obtained with a combination of tensiometers and either psychrometers or heat dissipation sensors. If there is little recharge of the soil profile during the growing season, it is possible to identify a zero flux plane, where there is zero hydraulic potential gradient. This plane represents an imaginary watershed above which water moves upward to plant roots and below which drainage may occur (McGowan, 1974; Arya et al., 1975; Cooper, 1980). By following the movement of the zero flux plane down the profile during the growing season, it is possible to follow changes in the maximum depth of root water extraction and to obtain improved estimates of the soil water balance. Psychrometers designed for attachment to leaves or stems (McBurney and Costigan, 1987) can be used in combination with soil sensors to provide detailed information on the diurnal pattern of the plant water regime (Bruini and Thurtell, 1982).

For measuring matric and hydraulic potential under wet conditions, there is still no substitute for the accuracy of tensiometers, especially as they will function equally well below the water table. Tensiometers can be used to study the water regime in relation to restrictions on soil aeration and root growth (King et al., 1986; Nisbet et al., 1989) and to follow the pattern of water flow that determines the water regime on hillsides and in hollows (Anderson and Burt, 1977). Under wet ($\Psi_m > -10$ kPa) conditions, portable tensiometers can be used to study spatial variation of matric potential and hence the effectiveness of field drainage systems (Mullins et al., 1986).

Where data logging systems are too costly or impractical, the filter paper technique has proved to be useful for studying temporal and spatial variations of matric potential at remote sites, for example across gaps in the rainforest (Veenendaal et al., 1995). It is also useful for studying near-surface conditions such as in seedbeds (Townend et al., 1996), where sensor size, response time, and temperature fluctuations limit the use of other techniques.

In addition to spatial variations resulting from plant water uptake, the soil water regime may be heterogeneous in structured soils. Sensors that connect with cracks or biopores, which form preferred pathways for infiltration, may then give readings that differ from those installed within structural units. In such cases there is no single representative value, and the positioning of sensors must be related to the aim of the particular investigation. Superimposed on such structure-related variability there is also likely to be longer range variability in the soil water regime. Greminger et al. (1985) observed significant spatial variability between tensiometer readings at a separation of > 10 m.

Use of matric potential sensors for in situ determination of the water release characteristic (Greminger et al., 1985) and for determination of unsaturated hydraulic conductivity is discussed in Chaps. 3 and 6, respectively.

REFERENCES

Al-Khafaf, S., and R. J. Hanks. 1974. Evaluation of the filter paper method for estimating soil water potential. *Soil Sci.* 117:194–199.

Aggelides, S. M., and A. Paraskevi. 1998. Comparison of empirical equations for temperature correction of gypsum sensors. *Agron. J.* 90:441–443.

Aitchison, G. D., P. F. Butler, and C. G. Gurr. 1951. Techniques associated with the use of gypsum block soil moisture meters. *Aust. J. Appl. Sci.* 2:56–75.

Anderson, M. G., and T. P. Burt. 1977. Automatic monitoring of soil moisture conditions in a hillslope spur and hollow. *J. Hydrol.* 33:27–36.

Arya, L. M., D. A. Farrell, and G. R. Blake. 1975. A field study of soil water depletion patterns in presence of growing soybean roots: I. Determination of hydraulic properties of the soil. *Soil Sci. Soc. Am. Proc.* 39:424–430.

Aslyng, H. C. 1963. Soil physics terminology. *Int. Soc. Soil Sci. Bull.* 23:7–10.

Biesheuval, P. M., R. Raangs, and H. Verweij. 1999. Response of the osmotic tensiometer to varying temperatures: Modeling and experimental validation. *Soil Sci. Soc. Am. J.* 63:1571–1579.

Blackwell, P. S., and M. J. Elsworth. 1980. A system for automatically measuring and recording soil water potential and rainfall. *Agric. Water Manage.* 3:135–141.

Bocking, K. A., and D. G. Fredlund. 1979. Use of the osmotic tensiometer to measure negative pore water pressure. *Geotech. Test J.* 2:3–10.

Bolt, G. H. 1976. Soil physics terminology. *Int. Soc. Soil Sci. Bull.* 49:16–22.

Bourget, S. J., D. E. Elrick, and C. B. Tanner. 1958. Electrical resistance units for moisture measurements: Their moisture hysteresis, uniformity and sensitivity. *Soil Sci.* 86: 298–304.

Bouyoucos, G. J. 1953. More durable plaster of Paris moisture blocks. *Soil Sci.* 76: 447–451.

Boyer, J. S. 1995. *Measuring the Water Status of Plants and Soil.* San Diego, CA: Academic Press.

Bruini, O., and G. W. Thurtell. 1982. An improved thermocouple hygrometer for in situ measurements of soil water potential. *Soil Sci. Soc. Am. J.* 46:900–904.

Campbell, G. S., and W. H. Gardner. 1971. Psychrometric measurement of soil water potential: Temperature and bulk density effects. *Soil Sci. Soc. Am. Proc.* 35:8–12.

Campbell, G. S., and G. W. Gee. 1986. Water potential: Miscellaneous methods. In: *Meth-*

ods of Soil Analysis, Part 1 (A. Klute, ed.). Madison, WI: Am. Soc. Agron., pp. 619–633.

Cassell, D. K., and A. Klute. 1986. Water potential: Tensiometry. In: *Methods of Soil Analysis, Part 1* (A. Klute, ed.). Madison, WI: Am. Soc. Agron., pp. 563–596.

Cannell, G. H., and C. E. Asbell. 1964. Prefabrication of mould and construction of cylindrical electrode-type resistance units. *Soil Sci.* 97:108–112.

Cooper, J. D. 1980. Measurement of moisture fluxes in unsaturated soil in Thetford Forest. Report No. 66. Wallingford, Oxfordshire, U.K.: Inst. Hydrol.

Deka, R. N., M. Wairiu, P. W. Mtakwa, C. E. Mullins, E. M. Veenendaal, and J. Townend. 1995. Use and accuracy of the filter paper technique for measurement of soil matric potential. *Eur. J. Soil Sci.* 46:233–238.

Fawcett, R. G., and N. Collis-George. 1967. A filter-paper method for determining the moisture characteristics of soils. *Aust. J. Exp. Agric. Animal Husb.* 7:162–167.

Fourt, D. F., and W. H. Hinton. 1970. Water relations of tree crops. A comparison between Corsican pine and Douglas fir in south-east England. *J. Appl. Ecol.* 7:295–309.

Frede, H. G., W. Weinzerl, and B. Meyer. 1984. A portable electronic puncture tensiometer. *Z. Planzenernaehr. Bodenk.* 147:131–134.

Gardner, R. 1937. A method of measuring the capillary tension of soil moisture over a wide moisture range. *Soil Sci.* 43:277–293.

Goltz, S. M., G. Benoit, and H. Schimmelpfennig. 1981. New circuitry for measuring soil water matric potential with moisture blocks. *Agric. Meteorol.* 24:75–82.

Goodman, D. 1983. A portable tensiometer for the measurement of water tension in peat blocks. *J. Agric. Eng. Res.* 28:179–182.

Greminger, P. J., Y. K. Sud, and D. R. Neilsen. 1985. Spatial variability of field-measured soil-water characteristics. *Soil Sci. Soc. Am. J.* 49:1075–1082.

Hagan, R. M., H. R. Haise, and T. W. Edminster, eds. 1967. *Irrigation of Agricultural Lands.* Madison, WI: Am. Soc. Agron.

Haise, H. R., and O. J. Kelly. 1946. Relation of moisture tension and electrical resistance in plaster of Paris blocks. *Soil Sci.* 61:411–422.

Hamblin, A. P. 1981. Filter-paper method for routine measurement of field water potential. *J. Hydrol.* 53:355–360.

Hanks, R. J., and G. L. Ashcroft. 1980. *Applied Soil Physics.* Berlin: Springer-Verlag.

Hillel, D. 1998. *Environmental Soil Physics.* New York: Academic Press.

King, J. A., K. A. Smith, and D. G. Pyatt. 1986. Water and oxygen regimes under conifer plantations and native vegetation on upland peaty gley soil and deep peat soils. *J. Soil Sci.* 37:485–497.

Klute, A., and W. R. Gardner. 1962. Tensiometer response time. *Soil Sci.* 93:204–207.

Lang, A. R. G. 1967. Psychrometric measurement of soil water potential in situ under cotton plants. *Soil Sci.* 106:460–468.

Lee-Williams, T. H. 1978. An automatic scanning and recording tensiometer system. *J. Hydrol.* 39:175–183.

Marshall, T. J. 1959. Relations between water and soil. Tech. Commun. No. 50. Harpenden, U.K.: Commonwealth Bureau Soils.

Marshall, T. J., J. W. Holmes, and C. W. Rose. 1996. *Soil Physics*, 3d ed. Cambridge, U.K.: Cambridge Univ. Press.

Marthaler, H. P., W. Vogelsanger, F. Richard, and P. J. Wierenga. 1983. A pressure transducer for field tensiometers. *Soil Sci. Soc. Am. J.* 47:624–627.

McBurney, T., and P. A. Costigan. 1987. Plant water potential measured continuously in the field. *Plant Soil* 97:145–149.

McGowan, M. 1974. Depths of water extraction by roots: Applications to soil-water balance studies. In: *Isotopes and Radiation Techniques in Soil Physics and Irrigation Studies.* Vienna: IAEA, pp. 435–445.

McQueen, I. S., and R. F. Miller. 1968. Calibration and evaluation of a wide-range gravimetric method for measuring stress. *Soil Sci.* 106:225–231.

Merrill, S. D., and S. L. Rawlins. 1972. Field measurement of soil water potential with thermocouple psychrometers. *Soil Sci.* 113:102–109.

Mullins, C. E., O. T. Mandiringana, T. R. Nisbet, and M. N. Aitken. 1986. The design, limitations, and use of a portable tensiometer. *J. Soil Sci.* 37:691–700.

Neumann, H. H., and G. W. Thurtell. 1972. A Peltier cooled thermocouple dewpoint hygrometer for in situ measurement of water potentials. In: *Psychrometry in Water Relations Research* (R. W. Brown and B. P. van Haveren, eds.). Logan, UT: Utah State University, pp. 103–112.

Nisbet, T. R., C. E. Mullins, and D. A. MacLeod. 1989. The variation of soil water regime, oxygen status and rooting pattern with soil type under Sitka spruce. *J. Soil Sci.* 40:183–197.

Noborio, K., R. Horton, and C. S. Tan. 1999. Time domain reflectometry probe for simultaneous measurement of soil matric potential and water content. *Soil Sci. Soc. Am. J.* 63:1500–1505.

Or, D., and J. M. Wraith. 1999a. A new soil matric potential sensor based on time domain reflectometry. *Water Resour. Res.* 35:3399–3408.

Or, D., and J. M. Wraith. 1999b. Temperature effects on soil bulk dielectric permittivity measured by time domain reflectometry: A physical model. *Water Resour. Res.* 35:371–383.

Peck, A. J., and R. M. Rabbidge. 1969. Design and performance of an osmotic tensiometer for measuring capillary potential. *Soil Sci. Soc. Am. Proc.* 33:196–202.

Pereira, H. C. 1951. A cylindrical gypsum block for moisture studies in deep soils. *J. Soil Sci.* 2:212–223.

Perrier, E. R., and A. W. Marsh. 1958. Performance characteristics of various electrical resistance units and gypsum materials. *Soil Sci.* 86:140–147.

Phene, C. J., and D. W. Beale. 1976. High-frequency irrigation for water nutrient management in humid regions. *Soil Sci. Soc. Am. J.* 40:430–436.

Phene, C. J., G. J. Hoffman, and S. L. Rawlins. 1971a. Measuring soil matric potential in situ by sensing heat dissipation within a porous body. I. Theory and sensor construction. *Soil Sci. Soc. Am. Proc.* 35:27–33.

Phene, C. J., S. L. Rawlins, and G. J. Hoffman. 1971b. Measuring soil matric potential in situ by sensing heat dissipation within a porous body. II. Experimental results. *Soil Sci. Soc. Am. Proc.* 35:225–229.

Rawlins, S. L. 1976. Measurement of water content and the state of water in soils. In: *Water Deficits and Plant Growth*, Vol. 4 (T. T. Kozlowski, ed.). New York: Academic Press, pp. 1–55.

Rawlins, S. L., and G. S. Campbell. 1986. Water potential: Thermocouple psychrometry. In: *Methods of Soil Analysis, Part 1* (A. Klute, ed.). Madison, WI: Am. Soc. Agron., pp. 597–618.

Richards, L. A. 1949. Methods of measuring soil moisture tension. *Soil Sci.* 68:95–112.

Richards, L. A., and G. Ogata. 1958. Thermocouple for vapor-pressure measurement in biological and soil systems at high humidity. *Science* 128:1089–1090.

Rice, R. 1969. A fast response, field tensiometer system. *Trans. Am. Soc. Agric. Eng.* 12: 48–50.

Savage, M. J., and A. Cass. 1984. Measurement of water potential using in situ thermocouple hygrometers. *Adv. Agron.* 37:73–126.

Schofield, R. K. 1935. The pF of the water in soil. *Trans. 3rd Int. Congr. Soil Sci.*, Vol. 2, pp. 37–48.

Scholl, D. G. 1978. A two-element ceramic sensor for matric potential and salinity measurements. *Soil Sci. Soc. Am. J.* 42:429–432.

Spanner, D. C. 1951. The Peltier effect and its use in the measurement of suction pressure. *J. Exp. Bot.* 11:134–168.

Tanner, C. B., and R. J. Hanks. 1952. Moisture hysteresis in gypsum moisture blocks. *Soil Sci. Soc. Am. Proc.* 16:48–51.

Tanner, C. B., E. Abrams, and J. C. Zubriski. 1948. Gypsum moisture-block calibration based on electrical conductivity in distilled water. *Soil Sci. Soc. Am. Proc.* 13: 62–65.

Taylor, S. A., and G. L. Ashcroft. 1972. *Physical Edaphology.* San Francisco: Freeman.

Townend, J., P. W. Mtakwa, C. E. Mullins, and L. E. Simmonds. 1996. Factors limiting successful establishment of sorghum and cowpea in two contrasting soil types in the semi-arid tropics. *Soil Till. Res.* 40:89–106.

Towner, G. D. 1980. Theory of time response of tensiometers. *J. Soil Sci.* 31:607–621.

Towner, G. D. 1981. The correction of in situ tensiometer readings for overburden pressures in swelling soils. *J. Soil Sci.* 32:499–504.

Veenendaal, E. M., M. D. Swaine, V. K. Agyeman, D. Blay, I. Abebrese, and C. E. Mullins. 1995. Differences in plant and soil water relations in and around a forest gap in West Africa during the dry season may influence seedling establishment and survival. *J. Ecol.* 83:83–90.

Watson, K. K., and R. D. Jackson. 1967. Temperature effects in a tensiometer-pressure transducer system. *Soil Sci. Soc. Am. Proc.* 31:156–160.

Webster, R. 1966. The measurement of soil water tension in the field. *New Phytol.* 65: 249–258.

Wellings, S. R., J. P. Bell, and R. J. Raynor. 1985. The use of gypsum resistance blocks for measuring soil water potential in the field. Report No. 92. Wallingford, Oxfordshire, U.K.: Inst. Hydrol.

Wiebe, H. H., G. S. Campbell, W. H. Gardner, S. L. Rawlins, J. W. Cary, and R. W. Brown. 1971. *Measurement of Plant and Soil Water Status.* Bull. No. 484. Logan, UT: Utah State University.

3
Water Release Characteristic

John Townend
University of Aberdeen, Aberdeen, Scotland

Malcolm J. Reeve
Land Research Associates, Derby, England

Andrée Carter
Agricultural Development Advisory Service, Rosemaund, Preston Wynne, Hereford, England

I. INTRODUCTION

The water release characteristic is the relationship between water content (usually volumetric water content) and matric potential (or matric suction) in a drying soil. The water release characteristic is one of the most important measurements for characterizing soil physical properties, since it can (1) indicate the ability of the soil to store water that will be available to growing plants, (2) indicate the aeration status of a drained soil, and (3) be interpreted in nonswelling soils as a measure of pore size distribution.

There are a range of methods used for measurement of the water release characteristics of soils. This chapter describes the physical properties that determine the release characteristic, outlines the most common methods used to measure it and their suitability for a range of analytical environments, and briefly illustrates the ways in which the results can be presented and applied.

II. THE SOIL WATER RELEASE CHARACTERISTIC

A. Energy of Soil Water

Soil water that is in equilibrium with free water is by definition at zero matric potential. Water is removed from soil by gravity, evaporation, and uptake by plant roots. As the soil dries, water is held within pores by capillary attraction between the water and the soil particles. The energy required to remove further water at any stage is called the matric potential of the soil (more negative values indicate more energy is required to remove further water). The term matric suction is also used. This represents the same quantity but is given as a positive value (e.g., a matric potential of −1 kPa is the same as a matric suction of 1 kPa). The units used to express the energy of soil water are diverse, and Table 1 provides a conversion for some of those more commonly used. The kilopascal is the most commonly applied SI unit. Schofield (1935) proposed the pF scale, which is the logarithm of the soil water suction expressed in cm of water. The scale is analogous to the pH scale and is designed to avoid the use of very large numbers, but it has not been universally adopted.

As the soil dries the largest pores empty readily of water. More energy is required to remove water from small pores, so progressive drying results in decreasing (more negative) values of matric potential. Not only is water removed from soil pores, but the films of water held around soil particles are reduced in thickness. Therefore there is a relationship between the water content of a soil and its matric potential. Laboratory or field measurements of these two parameters can be made and the relationship plotted as a curve, called the soil moisture characteristic by Childs (1940). Soil water retention characteristic, soil moisture characteristic curve, pF curve, and soil water release characteristic have also been used as synonymous terms.

B. Hysteresis

The term "water *release* characteristic" implies a measurement made by desorption (drying) from saturation or a low suction. However, this curve is different

Table 1 Conversion Factors for Energy of Soil Water

$$
\begin{aligned}
-1\ \text{kPa} &= -1\ \text{J kg}^{-1} \\
&= -0.01\ \text{bar} \\
&= -10\ \text{hPa} \\
&= -10.2\ \text{cm } H_2O \text{ at } 20°\text{C} = -0.75\ \text{cm Hg} \\
\text{pF} &= \log_{10}(-\text{cm } H_2O \text{ at } 20°\text{C}) \qquad (\text{e.g., } -10.2\ \text{cm} = \text{pF } 1.01)
\end{aligned}
$$

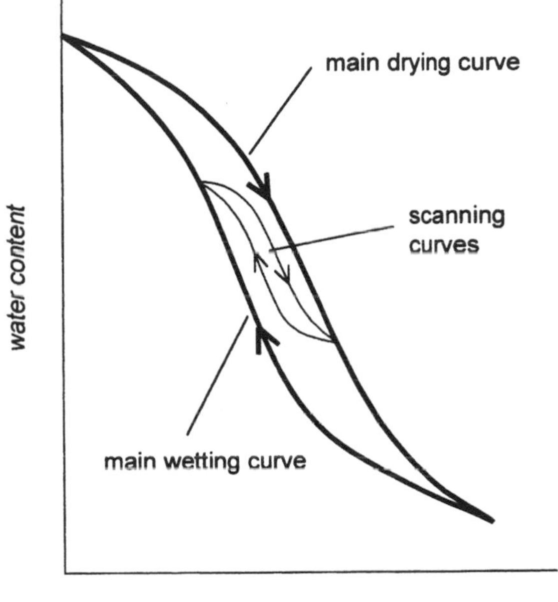

Fig. 1 The hysteresis loop. Scanning curves occur when a partially dried soil is rewetted or a wetting soil is redried.

from the sorption (wetting) curve, obtained by gradually rewetting a dry sample. Both curves are continuous, but they are not identical and form a hysteresis loop (Fig. 1). Partial drying followed by rewetting, or partial wetting followed by drying, can result in intermediate curves known as scanning curves, which lie within the hysteresis loop. The phenomenon of hysteresis (Haines, 1930) has been frequently documented, more recently by Poulovassilis (1974) and Shcherbakov (1985).

The main reasons for hysteresis, described in detail by Hillel (1971), are

1. Pore irregularity. Pores are generally irregularly shaped voids interconnected by smaller passages. This results in the "inkbottle" effect, illustrated in Fig. 2.

2. Contact angle. The angle of contact between water and the solid walls of pores tends to be greater for an advancing meniscus than for a receding one. A given water content will tend therefore to exhibit greater suction in desorption than in sorption.

3. Entrapped air. This can decrease the water content of newly wetted soil.

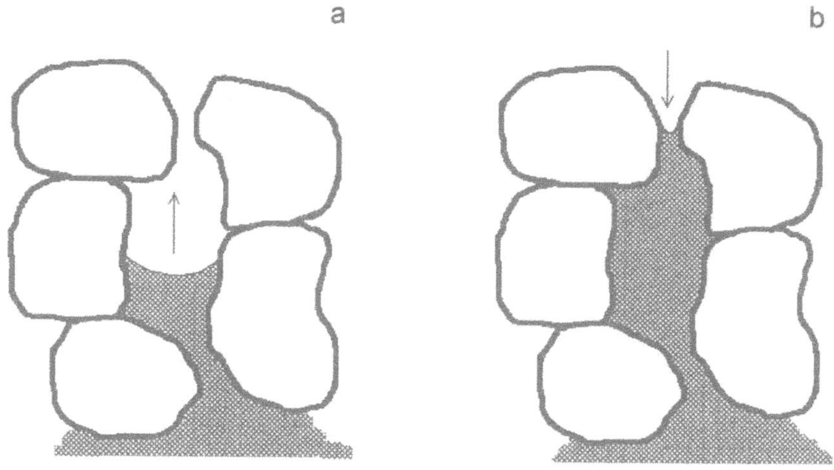

Fig. 2 The "inkbottle" effect. The pore does not fill until the suction is quite low due to its large diameter (a). Once full, this pore does not reempty until a high suction is applied because of the small diameter of the pore neck (b).

4. Swelling and shrinking. Volume changes cause changes of soil fabric, structure, and pore size distribution, with the result that interparticle contacts differ on wetting and drying.

Poulovassilis (1974) added that the rate of wetting or drying may also affect hysteresis.

For accurate work a knowledge of the wetting and drying history of a soil is therefore essential to interpret results. However, for most practical applications the drying curve only is measured and the effect of hysteresis ignored. Although an understanding of hysteresis is central to any explanation of soil water release characteristics, the overriding influence on the shape of the water release curve is soil composition.

C. Effect of Soil Properties

The amount of water retained at low suctions (0–100 kPa) is strongly dependent on the capillary effect and hence, in nonshrinking soils, on pore size distribution. Sandy soils contain large pores, and most of the water is released at low suctions, whereas clay soils release small amounts of water at low suctions and retain a large proportion of their water even at high suctions, where retention is attributable to adsorption (Fig. 3). Clay mineralogy is also important, smectitic clays with high cation-exchange capacity and specific surface area having greater adsorption than kaolinitic clays (Lambooy, 1984). Organic matter increases the amount of water retained, especially at low suctions, but at higher suctions soils rich in organic materials release water rapidly. The presence of free iron oxides and calcium carbonate has also been shown to affect the release characteristic (Stakman and Bishay, 1976; Williams et al., 1983), though the effect of free iron is difficult to separate from the effect of the high clay contents and good structural conditions with which it is often associated (Prebble and Stirk, 1959).

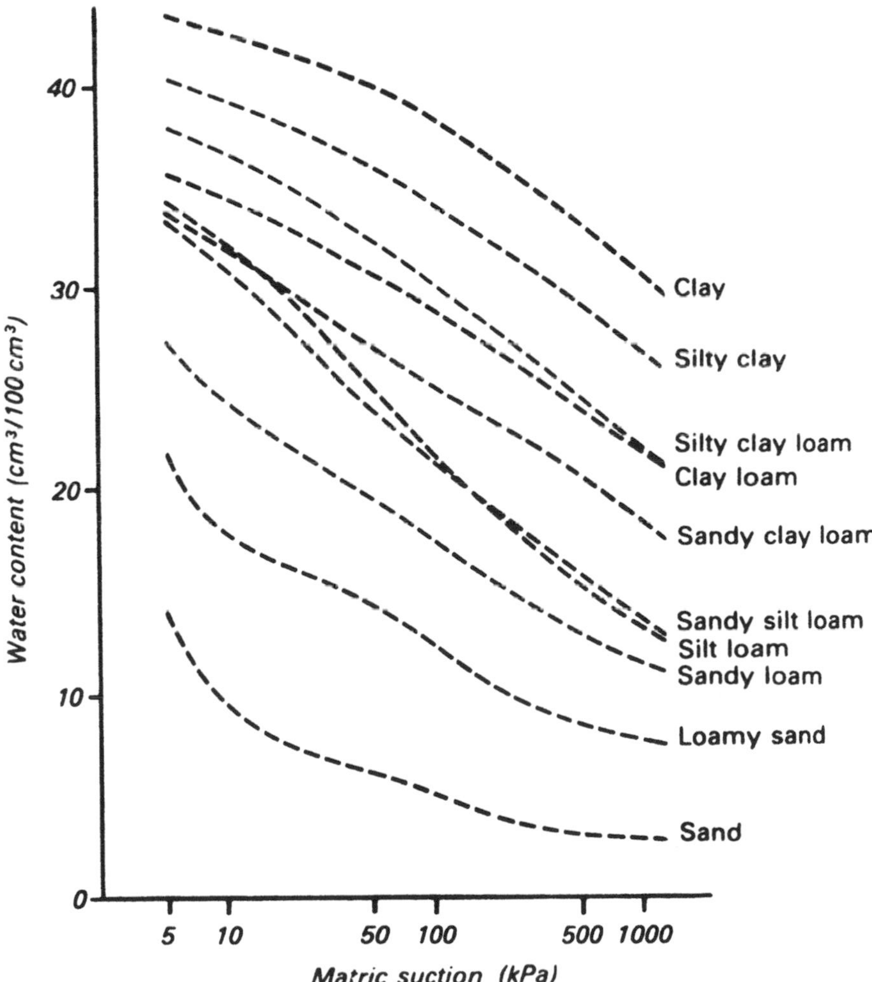

Fig. 3 Water release characteristics for subsoils of different texture. (After Hall et al., 1977.)

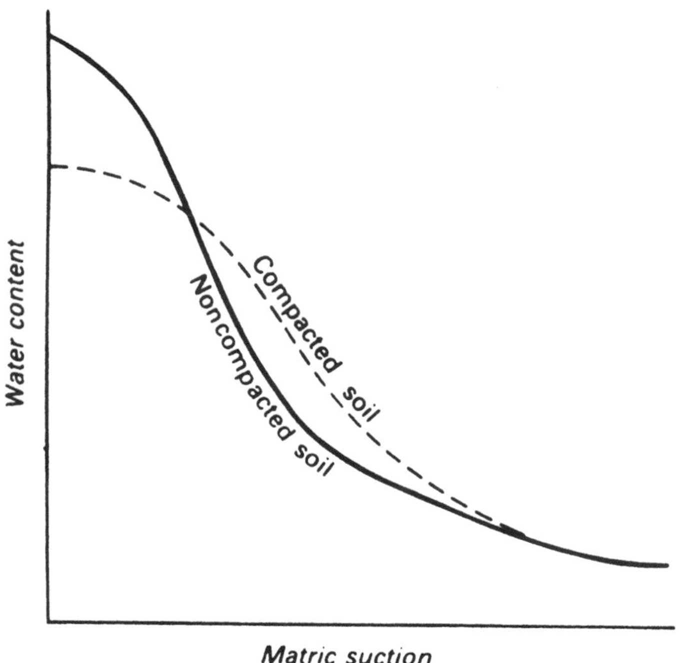

Fig. 4 The effect of compaction on the water release characteristic of an aggregated soil.

Soil structure and density have significant effects. For example, compaction decreases the total pore space of a soil (Archer and Smith, 1972), mainly by reducing the volume occupied by large pores, which retain water at low suctions (Fig. 4). Whereas the volume of fine pores remains largely unchanged, that occupied by pores of intermediate size is sometimes increased, and this can increase the amount of water retained between specific matric suctions of agronomic importance (Archer and Smith, 1972).

D. Suction and Pore Size

In a simple situation of a rigid soil containing uniform cylindrical pores, the applied suction is related to the size of the largest water-filled pores by the equation

$$d = \frac{4\sigma}{\rho g h} \tag{1}$$

where d is the diameter of pores, σ is the surface tension, ρ is the density of water, h is the soil water suction, and g is the acceleration due to gravity. At 20°C Eq. 1 gives $d = 306/h$, where h is in kilopascals and d is in micrometers. Pores larger than diameter d will be drained by a suction h.

The volume of water released by an increase in matric suction from h_1 to h_2 therefore equals the volume of pores having an effective diameter between d_1 and d_2, where d and h are related by Eq. 1. This simple relationship will operate only in nonshrinking soils and where the pore space consists of broadly circular pores with few "blind ends" or random restrictions (necks). Real soils can contain planar voids, pores with blind ends, and/or restrictions. If a void of 200 μm diameter has a neck exit of only 30 μm, water in the void will be released only when the suction exceeds 10 kPa. Thus the water release characteristic is at best only a general indicator of the effective pore size distribution.

The size distribution of pores in a soil can be used as a means of quantifying soil structure (Hall et al., 1977) or to give a general indication of saturated hydraulic conductivity, the value of which is largely determined by the volume of larger pores. Aeration is also largely a function of larger pores. Whereas larger pores may be defined as macropores and related to the water released at an arbitrary low suction, other pore sizes may be termed meso- or micropores (Beven, 1981), the latter being related to the water release characteristic at higher suctions. Conversely, the water release characteristic of soil can also be used to estimate the distribution of the size of the pores that make up its pore space. In clay soils, however, this is complicated by the fact that shrinkage results in pores reducing in size as water is withdrawn.

III. MEASUREMENT METHODS

There are three distinct ways to obtain a release characteristic. The usual procedure is to equilibrate samples at a chosen range of potentials and then determine their moisture contents. Suction tables, pressure plates, and vacuum desiccators are examples of this approach. In the second procedure, samples are allowed to dry out progressively and their potential and moisture content are both directly measured. A third option is to produce a theoretical model of the water release characteristic, based on other parameters measured from the soil such as the particle size distribution, or fractal dimensions obtained from image analysis of resin-impregnated samples of the soil.

A. Methods for Equilibrating Soils at Known Matric Potentials

1. *Main Laboratory Methods for Potentials of 0 to −1500 kPa*

Diverse methodologies for the determination of water release characteristics have evolved since Buckingham (1907) introduced the concept of using energy relations to characterize soil water phenomena. The most important techniques of measuring water release characteristics in the laboratory and the ranges of suction for which each method can be used are shown in Table 2.

Table 2 Methods of Determining Soil Water Release Characteristics in the Laboratory

Method	Approximate range (kPa, suction)	Type of potential measured	Early reference to method
Büchner funnel	0–20	Matric	Haines, 1930
Porous suction plate	0–70	Matric	Loveday, 1974
Sand suction table	0–10	Matric	Stakman et al., 1969
Sand–kaolin suction table	10–50	Matric	Stakman et al., 1969
Porous pressure plate (including Tempe cell)	0–1500	Matric	Richards, 1948 Reginato and van Bavel, 1962
Pressure membrane	10–10,000	Matric	Richards, 1941 Richards, 1949
Centrifuge	10–3000	Matric	Russell and Richards, 1938
Osmosis	30–2500	Matric and osmotic	Zur, 1966 Pritchard, 1969
Consolidation	1–1000	Matric	Croney et al., 1952
Vapor pressure (vacuum desiccator)	3000–1,000,000	Matric and osmotic	Croney et al., 1952
Sorption balance	3000–1,000,000	Matric and osmotic	Wadsworth, 1944
Filter paper	0–10,000	Matric	McQueen and Miller, 1968

a. Vacuum or Suction Methods for Measurement at High Potentials (< 100 kPa suction)

The basis of these methods is that soil is placed in hydraulic contact with a medium whose pores are so small that they remain in a saturated state up to the highest suction to be measured. The suction can be applied by using either a hanging water column or a pump and suction regulator. The soil in contact with the medium loses or gains water depending on whether the applied suction is greater or less than the initial value of soil water suction. Because it is more common to carry out such measurements on the desorption segment of the hysteresis curve, we are usually concerned with the loss of water. Attainment of equilibrium with the applied suction can be determined by regularly weighing the soil sample or by measuring the outflow of water until either the weight loss or outflow ceases or becomes minimal. The main restriction to such methods is the bubbling pressure of the medium used. The bubbling pressure (which is negative) is the suction applied to the medium that empties the largest pores, thus allowing air to pass through the pores and causing a breakdown in the applied suction. Various experimental arrangements to apply the suction are discussed in the following sections.

Büchner Funnel. In the simplest application of the suction principle, a Büchner funnel and a filter paper support the soil. The apparatus, introduced by Bouyoucos (1929) and later adapted by Haines (1930) to demonstrate hysteresis effects, is still occasionally referred to as the Haines apparatus, even in installations where the funnel is fitted out with a porous ceramic plate (Russell, 1941; Burke et al., 1986; Danielson and Sutherland, 1986).

One type of installation is illustrated in Fig. 5. One end of a flexible PVC tube is connected to the base of a funnel and the other end to an open burette. The tubing should be flexible but resistant to collapse, which can result in measurement errors. The tubing and funnel are filled with deaerated water and the burette adjusted until the water is level with the ceramic plate or filter paper. Air bubbles trapped within the funnel can be expelled upward by tapping the funnel while applying a gentle air pressure through the end of the burette. If a porous ceramic plate is used, as in Fig. 5, deaerated water will need to be drawn through the plate by applying a vacuum to the open end of the burette while the funnel is inverted in the water. Once the system is air-free, a prewetted soil sample (normally a soil core) is placed in contact with the filter paper or ceramic plate. The water level is maintained level with the base of the sample until it is saturated, whereupon the volume in the burette is recorded. A suction, h cm of water, can then be applied by adjusting the burette so that the water level in it is h cm below the midpoint of the sample. Water that flows out of the sample in response to the applied suction can be measured by the increase in volume of the water in the burette after the water level has stopped rising.

No detectable change in burette water level within 6 hours is suggested as a satisfactory definition of equilibrium (Vomocil, 1965), but a shorter period without change might be acceptable. Small evaporative losses through the open end of the burette can be suppressed by adding a few drops of liquid paraffin to the water in it. Evaporative losses from the sample can be minimized by covering the open top of the funnel or creating a closed system as in Fig. 5. If the final level in the burette is h', then the final suction applied is h', rather than h. However, by altering the level of the free water surface to h at each inspection, the desired suction can be maintained. By repeating the exercise at successively increasing suctions, a soil moisture characteristic curve can be plotted by calculating back from the final moisture content of the soil sample (determined gravimetrically) using the volumes of water extracted between successive applied suctions.

Using a filter paper, the maximum suction that can be applied is only 50–70 cm of water before air entry occurs around the sides of the paper; but using a porous ceramic plate, the maximum suction attainable is much higher, depending

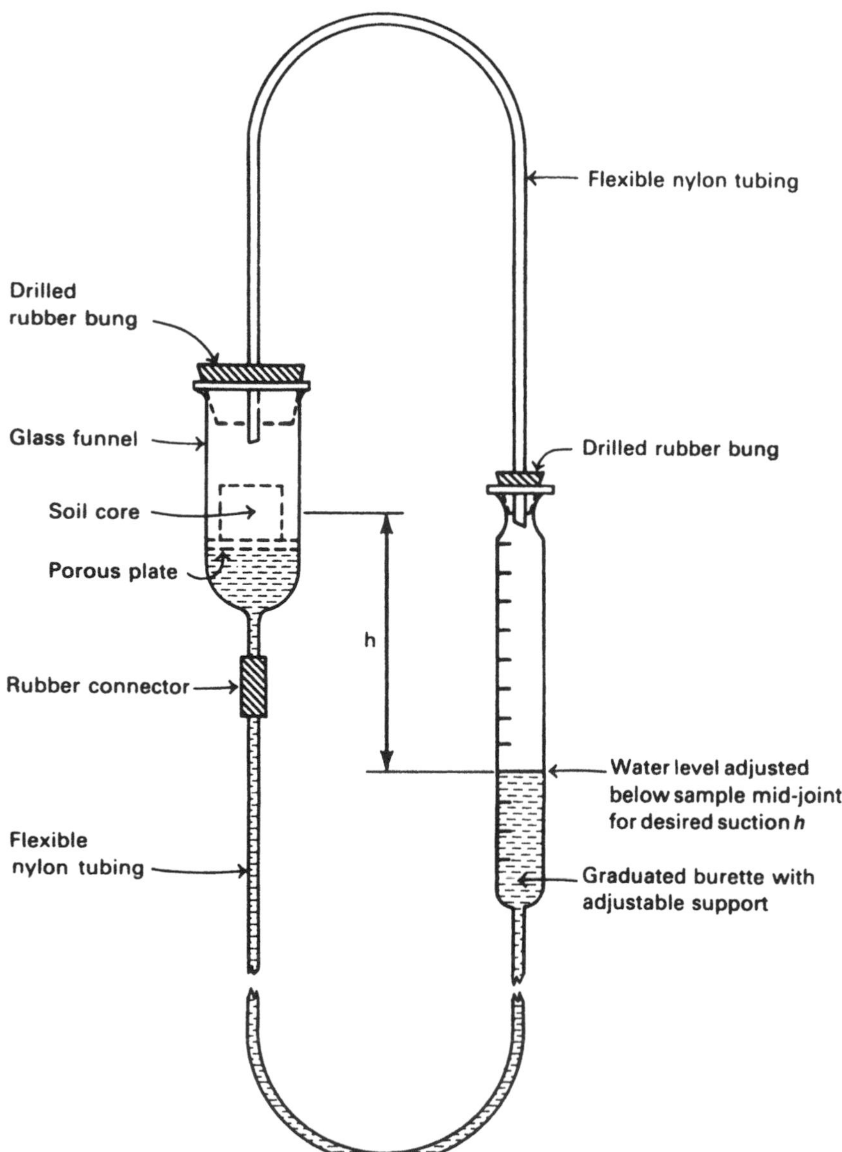

Fig. 5 Büchner funnel or Haines apparatus tension method.

on the air-entry (bubbling) pressure of the plate. In practice, the maximum suction applied using a ceramic insert is restricted by the distance to which the levelling burette can be lowered below the funnel (i.e., typically < 200 cm of water).

The Büchner funnel technique is not only very suitable as a teaching method, it is also trouble free. Even with the limitations of using filter paper, a curve can be obtained that can be used to interpret the soil pore size distribution in a range important for soil drainage. The volume of water extracted from some soils between successive suctions might be small and difficult to measure accurately in the burette. An alternative, possible only if a ceramic plate is used in the Büchner funnel, is to determine the water content of the soil sample gravimetrically after each successive equilibrium is reached (Burke et al., 1986). Because the Büchner funnel method requires a separate piece of apparatus for each soil sample, it lends itself to small research and/or teaching laboratories, where large numbers of samples are not normally analyzed. However, the method should not be disregarded for other situations, as accuracy is claimed to be good and material costs are low (Burke et al., 1986).

Porous Suction Plate. The Büchner funnel method has been adapted in a variety of ways (Jamison, 1942; Croney et al., 1952), but most assemblies retain the common property of accommodating only one sample at a time. Czeratzki (1958) described the construction and use of a ceramic suction plate 500 mm by 350 mm, capable of taking several samples, and several European institutions were reported as using the method (de Boodt, 1967). Loveday (1974) described three designs of ceramic suction plate extractor, although noting that only one was commercially available in Australia. One design consists of a large ceramic plate sealed onto a clear, water-filled acrylic container with outlet. The space between the plate and container is kept water filled, and air bubbles trapped below the plate can be readily seen and removed. A cover to the whole assembly reduces evaporative losses and, depending on the size of the plate, several soil cores can be brought to equilibrium at one time. The suction can be applied either by using a hanging water column (as for the Büchner funnel) attached to a levelling bottle or burette, or by a vacuum pump and regulator. A design using 330 mm diameter ceramic plates is shown in Fig. 6. If several contrasting soils are being analyzed at the same time, some might reach equilibrium much more quickly than others. Then, if water outflow were used as a criterion of equilibrium, the samples could not be removed until the last sample had reached equilibrium. Because the water extracted from each sample cannot be measured by the outflow and must be determined from the equilibrium weight, it is easier to determine equilibrium of each individual sample by regular weighing, as for sand suction tables (see next section). Regaining hydraulic contact between samples and plate after weighing can be a problem. This can be overcome by setting a layer of fine plaster of Paris in the bottom of the sample to provide a flat base that can repeatedly make good

Fig. 6 Ceramic suction plates. The suction is controlled by the height of the bottle on the left. A cover is placed over the apparatus when in use to reduce evaporation.

hydraulic contact with the plate, or using a fine layer of silt on the plate, but care must then be taken to remove silt adhering to the sample before it is weighed.

The requirement for regular weighing means that porous suction plates must be maintained at working height, thus limiting the height available below the plate for a suspended water column (unless in multifloor buildings it can be extended into an underlying storey). For suctions in excess of 10 kPa, a complex sequence of bubbling towers (Loveday, 1974) or an accurately controlled mechanical vacuum system (Croney et al., 1952) is then required, and this has probably limited the widespread adoption of the porous suction plate.

Sand Suction Tables. The use of sand suction tables is fully described by Stakman et al. (1969), who refer to them as the sandbox apparatus. Instead of applying a suction to a ceramic plate or filter paper, suction is applied to saturated coarse silt or very fine sand held in a rigid container, and core samples are then put into contact with it. The maximum suction that can be applied before air entry occurs is related to the pore size distribution of the packed fine sand or coarse silt and is thus related to its particle size distribution. The original design has been adapted, sometimes with minor modifications, elsewhere (Fig. 7). They are available commercially, but one of the attractions of sand suction tables is that they can be constructed easily and cheaply from readily available materials, although care

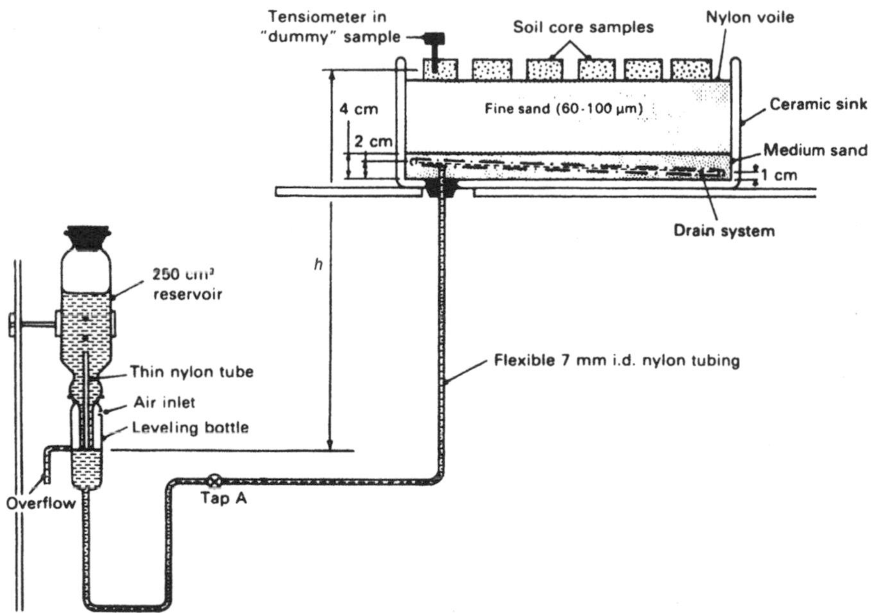

Fig. 7 Components of a sand suction table. The suction is equivalent to the difference in height h. (After Hall, et al., 1977.)

must be taken during assembly. They are thus well suited to laboratories in locations where supplies of more sophisticated equipment are available only at great cost as imports, or not at all. The container need not be a ceramic sink, though such receptacles are very suitable. Any rigid, watertight, nonrusting container, with a cover to prevent evaporative losses, will suffice, and slightly flexible plastic stacking storage bins can be used successfully, provided the sides cannot flex away from the sand to allow air entry. Industrial sands with a narrow particle size distribution are most suitable because they contain few fines; the particle size distribution of some suitable grades available commercially in Britain is given in Table 3. In practice, local sources of sediments, such as from rivers, estuaries, coastal flats (Stakman et al., 1969), or the washing lagoons of aggregate plants, can often provide a suitable particle size distribution. Fine glass beads and aluminum oxide powder have been shown to have adequately high air-entry values and hydraulic conductivities for use as tension media (Topp and Zebchuk, 1979), but these materials cost considerably more than sand. Ball and Hunter (1980) reported a shallower design of suction table, which utilizes a strengthened Perspex tray with integral drainage channels overlain by glass microfiber paper and a thin layer of commercially available silica flour with particles mainly of 10–50 μm.

Table 3 Industrial Sands and Silica Flour for Suction Tables[a]

		Typical particle size distribution (μm)					
Type	Use	>500	250–500	125–250	63–125	20–63	<20
Congleton CN HST 60	Base of suction tables	2	33	62	3	0	0
Redhill 110	Surface of suction tables (< 50 cm suction)	0	1	45	51	3	0
Redhill HH	Surface of suction tables (< 110 cm suction)	0	0	6	43	46	5
Oakamoor HPF2	Surface of suction tables (< 210 cm suction)	0	0	0	1	43	56

[a] All samples available in U.K. from Hepworth Minerals and Chemicals Ltd., Brookside Hall, Sandbach, Cheshire, CW11 0TR.

It follows that sand suction tables can be of a variety of designs and sizes. Typically though, each should hold 30–50 undisturbed presaturated soil cores. The upper face of the core is kept covered by a lid, while the lower face is covered by a piece of nylon voile secured with an elastic band. Vomocil (1965) considered that the voile interferes with hydraulic contact only if a suction of more than 15 kPa is applied. By placing tensiometers beneath the surface of the sand and in the samples, we have confirmed that hydraulic contact is maintained to suction of at least 10 kPa. Sand baths up to 10 kPa suction are fairly reliable and maintenance free. The applied suction can be monitored by a tensiometer embedded in a "dummy" sample and connected to a mercury manometer (Hall et al., 1977) or by a standard nondegradable porous sample weighed at regular intervals. The occasional air locks that do occur can be cured by temporarily flooding the bath with deaerated water and drawing it through under vacuum.

For full characterization of the water release at high potentials, samples on sand baths need to be brought to equilibrium at a series of increasing suctions (Stakman et al., 1969). Regular alteration of the tension applied to a single suction table can result in more frequent air locks, and furthermore, all samples must reach equilibrium before the tension can be changed. A more practical solution is to wait until samples have reached equilibrium and then transfer them to tables set at progressively higher suctions (Hall et al., 1977).

The attainment of equilibrium at a given suction is determined by weighing the samples at 2–3 day intervals. If the decline in weight does not follow the general shape of the curves in Fig. 8 but continues at the same magnitude, hydraulic contact is likely to have been lost. Weight loss criteria for equilibrium

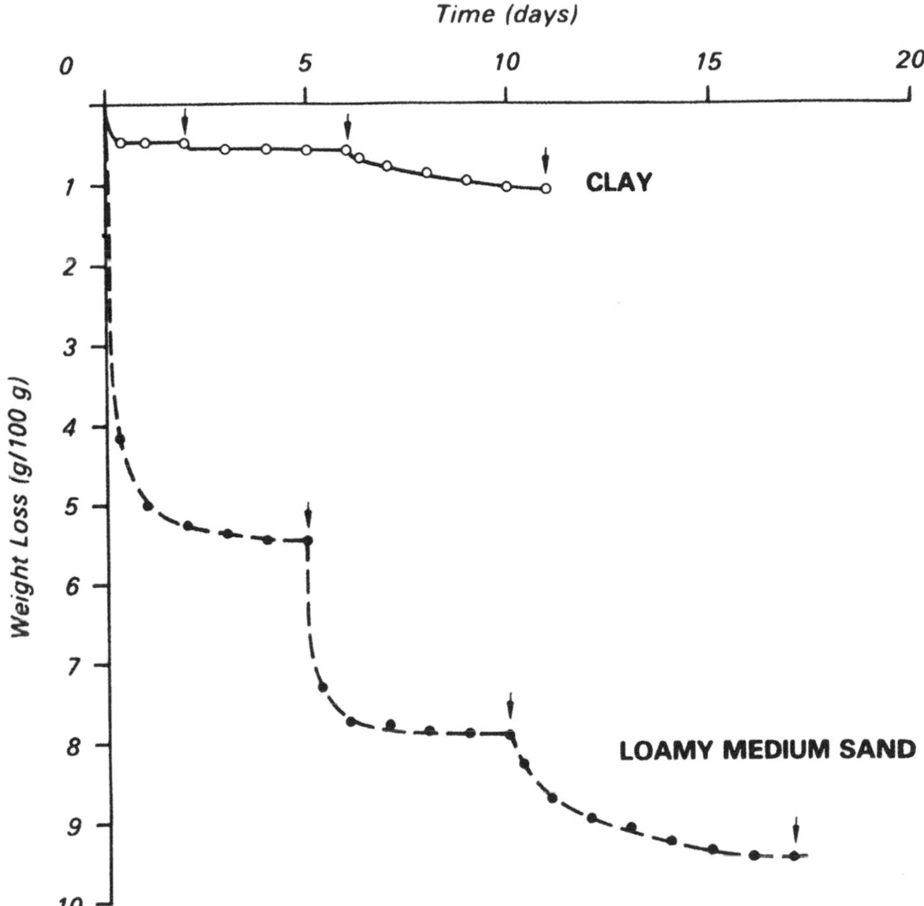

Fig. 8 Outflow curve for two soils equilibrated from natural saturation at three successive suctions (2.5, 5, and 10 kPa) on sand suction tables.

depend on sample size and accuracy required, and thus quoted equilibration times (Czeratzki, 1958; Ball and Hunter, 1980) may not be appropriate in some situations. By recording the equilibrium weight, the moisture content at any given suction can later be calculated after the sample has been oven dried. The time taken to reach equilibrium depends on sample height, the particle size distribution of the sample, its organic matter content, and the suction being applied. For example, equilibration times for sandy soils are often longer than those for clayey soils (Fig. 8). This is because a loamy sand that has the same unsaturated hydraulic conductivity as a clay loam at 1 kPa suction has an unsaturated hydraulic conductivity of only around one tenth that of the clay loam at 10 kPa (Carter and Thomasson, 1989).

The air-entry value of fine sand precludes the use of sand suction tables at suctions above about 10 kPa. Stakman et al. (1969) extended the range of the sand suction table by first applying layers of a sand–kaolin mixture and then pure kaolin to the top of a sand suction table. The required suction was maintained by a vacuum pump. The kaolin–sand suction table has been reported to be in use elsewhere (Hall et al., 1977), but it is more difficult to construct than a sand suction table. It also suffers from problems of entrapped air (Topp and Zebchuk, 1979) and capillary breakdown and thus requires more maintenance than a sand suction table. However, versions are available commercially. The kaolin used has a low hydraulic conductivity; hence samples require a long time to reach equilibrium. Ball and Hunter (1980) reported achieving suctions of 20 kPa with their silica flour assembly but did not report an air-entry value for it. Such a medium might be usable up to 33 kPa and might result in fewer problems than the sand–kaolin combination.

Because sand or silt suction tables provide an excellent low-cost method of measuring the soil water characteristic for a large number of samples at high potentials, they have been adopted by many researchers (see, e.g., Hall et al., 1977; Stakman and Bishay, 1976). Their main limitation is capillary breakdown as larger suctions are applied, and for this reason, pressure methods are more commonly adopted for suctions in excess of 10 kPa.

b. Gas Pressure Methods (0 to −1500 kPa potential)

As with the vacuum or suction methods, soils are placed on a porous medium, but they are brought into equilibrium at a given matric potential by applying a positive gas pressure (e.g., applying a pressure of 100 kPa brings the sample to equilibrium at a matric potential of −100 kPa, a matric suction of 100 kPa). To maintain this pressure, the porous medium and samples are contained within a pressure chamber while the underside of the porous medium is maintained at atmospheric pressure. Various designs of pressure chamber have been reported (Hall et al., 1977; Loveday, 1974) since Richards (1941; 1948) developed the original designs. All use either a porous plate or a cellulose acetate membrane as the porous medium. The pressure is supplied via regulators and gauges, by bottled nitrogen, or by a mechanical air compressor. Most designs of pressure chamber can take soils in a variety of physical states, but as equilibration times in pressure cells depend on the height of the soil sample, core samples in excess of 5 cm high are undesirable. At −1500 kPa, a sample height of 1 cm is convenient. Because the water in samples equilibrated at low potentials is held in small pores, it is acceptable to use disturbed samples, provided the soil is not compressed or remolded.

Pressure Plate Extractor. With the development of porous ceramics, pressure plate extractors have become available to cover a range of potentials down to −1500 kPa (Fig. 9) and have been widely used (Gradwell, 1971; Lal, 1979; Datta and Singh, 1981; Kumar et al., 1984; Lambooy, 1984; Puckett et al., 1985) for measurement of the water release characteristic, although some research (Madsen et al., 1986) casts some doubt over their accuracy. Most are designed to accommodate several samples contained within soil sample retaining rings in contact with the porous plate. Once the extractor has been sealed, a gas pressure is applied to the air space above the samples, and water moves downward from the samples through the plate, for collection in a burette or measuring cylinder. Equilibrium is judged to have been attained when outflow of water ceases. The samples can then be removed and their moisture content determined gravimetrically. Since samples are usually disturbed and the sample volume may not be known accurately for pressure plate measurements, the equivalent volumetric water content in the undisturbed state can be obtained by multiplying the gravimetric water content by the dry bulk density of the soil in its undisturbed state, and dividing by the density of water (usually taken as 1 g cm^{-3}). Burke et al. (1986) report that 2–14 days is necessary to establish equilibrium. Precision of the method is good, a coefficient of variation of 1–2% being attainable (Richards, 1965). However, clogging of the

Fig. 9 Two designs of pressure plate extractors with pressure control manifold.

ceramic plates by soil particles or algal growth can occur after repeated use, reducing the efficiency of the extractor. Furthermore, Chahal and Yong (1965) discovered that because of air bubbles trapped or nucleated in the water-filled pores, the soil water characteristic curve obtained with the pressure plate apparatus at high potentials (low suction) is higher than that obtained using the suction method of Haines. Thus pressure plate extractors are best suited to suctions of 33 kPa or greater.

Pressure Membrane Apparatus. In contrast to pressure plate extractors, in the pressure membrane apparatus the soil sample sits in contact with a semipermeable cellulose acetate (Visking) membrane. This allows passage of water from the sample but retains the air pressure applied to the upper surface of the membrane. Since the first pressure membrane cell was developed (Richards, 1941), designs have varied, and the technique has been used in many parts of the world (Heinonen, 1961; Gradwell, 1971; Stackman and Bishay, 1976; Hall et al., 1977; Kuznetsova and Vinogradova, 1982). Larger cells take several small disturbed samples contained in retaining rings, and some designs incorporate in the lid a diaphragm that expands during use to hold the soil samples in firm contact with the cellulose membrane. As with pressure plate extractors, outflow from large cells is measured in a single container, and thus all samples must have reached equilibrium before any can be removed for gravimetric determination of moisture content. Because gas diffuses slowly through the membrane and is replaced by drier gas from the pressure source, samples that reach equilibrium several days before others may start to dry by evaporation (Collis-George, 1952) and give erroneous results. This is likely to be a more serious problem with systems powered by bottled dry nitrogen gas than with those using humid laboratory or outdoor air compressed mechanically. Evaporation is also less likely to be a problem with smaller cells, designed to take only one sample (Hall et al., 1977) from which the outflow is monitored by a single collection device. With these, the sample can be removed as soon as equilibrium is reached. Texture-related equilibrium times for pressure membrane analysis were given by Stakman and van der Harst (1969). The pressure membrane apparatus gives moisture contents comparable to those from pressure plate extractors at the same applied pressure (Waters, 1980) but is found by some authors (Richards, 1965; Waters, 1980) to be prone to membrane leaks due to microbial action, iron rust from the chamber, or sand grains trapped near the gasket seals. These problems are a greater nuisance with a large cell containing many samples, and we find that such problems are rare when we use brass or stainless steel pressure cells and two membranes for high pressures (> 1000 kPa), and exercise care in operation.

Tempe Cells. Most pressure membrane and pressure plate extractors have been designed to extract moisture from small disturbed soil samples and are thus not suitable for characterizing the low suction range, where soil structure is all-important. Because of this, an individual cell, similar to the individual pressure membrane cells described by Hall et al. (1977) but of lightweight construction, has been developed for measurement on undisturbed soil cores using pressures of 0–100 kPa. The commercially available design is a development of that described by Reginato and van Bavel (1962), and equilibrium at a given gas pressure can be determined by periodically weighing the complete assembly including soil core. A submersible variant of the Tempe cell has been developed (Constantz and Herkelrath, 1984) to overcome problems due to air bubbles, which can result in inaccuracies in volumetric water content measurements and porous plate failure. Tempe cells are a useful addition to installations equipped only with large pressure plate and pressure membrane extractors. They are typically used at potentials between 0 and −100 kPa (Puckett et al., 1985); for potentials in the 0 to −20 kPa range sand suction tables are cheaper and easier to use.

c. *Centrifugation*

The use of a centrifuge to extract water from soils was introduced by Briggs and McLane (1907). These investigators centrifuged saturated soils in perforated containers at a speed that exerted a force of 1000 times gravity and termed the resulting moisture content the "moisture equivalent."

Russell and Richards (1938) improved on the technique, and it has since been reported to be in fairly wide use (Croney et al., 1952; Odén, 1975/76; Kyuma et al., 1977; Scullion et al., 1986) for measuring moisture retained at a variety of applied suctions. The soil sample is commonly supported on a porous medium in a cup containing a water table at the opposite end from the soil. The force exerted by the centrifuge during spinning is related to the angular velocity and the distances of the water table and sample from the center of rotation, given by

$$\log_{10} h = \log_{10}\left(\frac{r_2^2 - r_1^2}{2} \cdot \frac{w^2}{g}\right) \tag{2}$$

where h is the suction in centimeters of water, r_1 and r_2 are the distances (cm) between the center of rotation and the midpoint of the sample and of the water table, respectively, w is the angular velocity, and g is the acceleration due to gravity.

Thus, by varying the angular velocity, different suctions can be applied to the soil sample. Odén (1975/76) recommended centrifugation times ranging between 5 and 60 min for equilibrating saturated soils 3 cm high and with a volume of 50 cm^3 to matric suctions between 1 and 2500 kPa, though the precise time will depend also on the sample composition. The advantage of centrifugation as a method is, therefore, that it can quickly produce a soil water release curve. However, as Childs (1969) pointed out, the suction actually varies over the thickness of the sample, and other methods give better accuracy. While the centrifuge stops spinning and before the sample can be removed for weighing, the sample might reabsorb some moisture from the porous medium on which it sits. Furthermore, in saturated compressible samples thicker than 0.5 cm, consolidation during centrifugation can introduce further errors (Croney et al., 1952).

2. *Main Laboratory Methods for Potentials of Less than −1500 kPa*

Although it is uncommon to measure the water release characteristic to a matric suction greater than 1500 kPa, several methods are available to extend the curve to greater suctions. Some methods, such as the pressure membrane apparatus, can be considered direct, while others are indirect (vapor pressure and sorption balance), involving the thermodynamic relationships between the suction of retained water and freezing point or vapor pressure depression.

a. *Pressure Membrane*

By using strengthened assemblies, the usefulness of the pressure membrane apparatus can be extended to extract water held at potentials less than −1500 kPa. Richards (1949) measured moisture retention in soils to −10,000 kPa potential, while the apparatus of Coleman and Marsh (1961) can accept pressures of almost 150,000 kPa. Even though pressure membranes measure matric potential, while a sorption balance (see below) measures water potential (the sum of matric and osmotic potentials), Coleman and Marsh (1961) found good agreement between results from the two methods applied to a clay soil at around −10,000 kPa.

b. *Vapor Pressure*

The relationship between relative humidity at 20°C and soil water suction h (cm H_2O) is expressed by

$$\log_{10} h = 6.502 + \log_{10}(2 - \log_{10} H) \qquad (3)$$

where H is the relative humidity in percent (Schofield, 1935). This relationship can be used in two ways to determine the water release characteristic at high suctions.

Vacuum Desiccator. By placing soil that has been broken into small aggregates (passed through a 2 mm sieve) on a petri dish, into constant-humidity atmospheres in a vacuum desiccator or other sealed container, soil can be equilibrated at a chosen water potential before its moisture content is determined gravimetrically. Aqueous sulfuric acid solutions have been used, but Loveday (1974) recommends the use of several easily available neutral or acid salts to achieve a range of vapor pressures (Table 4). Although equilibrium times are long (5–15 days), the accuracy of the method is claimed to be good (Burke et al., 1986). To minimize errors due to temperature fluctuations, however, it is essential that the vapor pressure method be used only in an environment (room or insulated container) with temperature control to better than 1°C, especially for potentials higher than −10,000 kPa (Coleman and Marsh, 1961).

Table 4 Saturated Salt Solutions and Vapor Pressures at 20°C

Salt	Relative humidity (%)	Potential (kPa)
$CaSO_4 \cdot 5H_2O$	98	−2730
$Na_2SO_3 \cdot 7H_2O$	95	−6935
$ZnSO_4 \cdot 7H_2O$	90	−14245
NaCl	75	−38893
$Ca(NO_3)_2 \cdot 4H_2O$	56	−78389
$CaCl_2 \cdot 6H_2O$	32	−154047

Source: Loveday, 1974.

Sorption Balance. The sorption balance also uses the relationship between the soil water potential and the vapor pressure of the atmosphere with which the soil is in equilibrium. In the sorption balance, water from the sample is allowed to evaporate into a previously evacuated chamber, and the potential is deduced from measurements of the vapor pressure (Croney et al., 1952). The sample is weighed continuously by a sensitive balance as the vapor pressure is changed. It is important to maintain a constant temperature, but Coleman and Marsh (1961) found the sorption balance less prone than the vacuum desiccator to temperature-induced errors.

3. *Other Laboratory Methods*

a. *Osmosis*

Zur (1966) was the first to present a method of analysis based on the osmotic pressure of different solutions. A polyethylene glycol solution is separated from a soil–water system by a membrane that is permeable to water and small ions but impermeable to certain larger solute ions and soil particles. The water in the solution has a lower partial free energy than that of the water in the soil, and this tends to move water from the soil to the glycol solution until equilibrium is established. Since the membranes are permeable to most of the ions found in soil solution, the osmotic system actually controls the soil matric potential only. By using solutions of different concentrations, calibrated to apply given matric potentials, a water release characteristic can be determined. Pritchard (1969) developed the apparatus and extended the method to cover a range of potentials from −30 to −1500 kPa but encountered problems with microbial breakdown of membranes. Although there is fairly good agreement between water release characteristics obtained by the osmotic method and those by pressure membrane (Zur, 1966), the osmotic method has not been applied widely because of long sample equilibration times (Klute, 1986).

b. Consolidation

Measurement of the water release characteristic by applying a direct load to the soil was described by Croney et al. (1952). A saturated soil sample, laterally confined and sandwiched top and bottom between two porous disks, is loaded with successive weights on a consolidation frame (oedometer) (Head, 1982). The excess pore water pressure induced by each load is dissipated through the porous disks at a rate dependent on the hydraulic conductivity of the soil, and the soil compresses to a new state of equilibrium in which the load is equated by the matric potential of the new soil–water system. When compression ceases for any given load, the equilibrium moisture content can be calculated from reduction in sample thickness (measured by micrometer) and plotted against applied pressure. The method is applicable only to compressible soils such as shrinking clays and only over the primary consolidation phase (Head, 1982). Croney et al. (1952) pointed out that the friction between the sample and the containing ring can affect accuracy at low suctions. However, our research on disturbed clays indicates that the method gives a water release characteristic for clays comparable to that obtained by a combination of sand suction tables and pressure membrane apparatus (Fig. 10). The consolidation method is also faster than most others (the curves in Fig. 10 were obtained in 6 days), but it is mainly likely to find application in laboratories with an interest in the engineering application of soil physical data and already possessing the necessary equipment.

B. Methods for Measuring the Matric Potential for Soils Dried to a Range of Water Contents

1. Filter Paper

The filter paper method is based on the assumption that the matric potential of moist soil and the potential of filter paper in contact with it will be the same at equilibrium; it is described in Chap. 2. To plot the water release characteristic, however, soil samples uniformly dried to a range of moisture contents are required. These are best obtained by successive sampling of field soils as they dry out, though the climate and the season will then determine the scope of the water release characteristic obtained. One of the main interests in the filter paper method is for measurements of soil water potential, which, in fine-grained soils, controls soil strength (Chandler and Gutierrez, 1986). Deka et al. (1995) carried out trials to quantify the accuracy of the method and found it to be sufficient for many types of field experiments. They also gave a detailed sampling and handling procedure that could be used for determination of matric potential in the laboratory or field. The technique has the advantages of being cheap and not requiring specialized equipment. The water content of the soil sample can readily be determined by

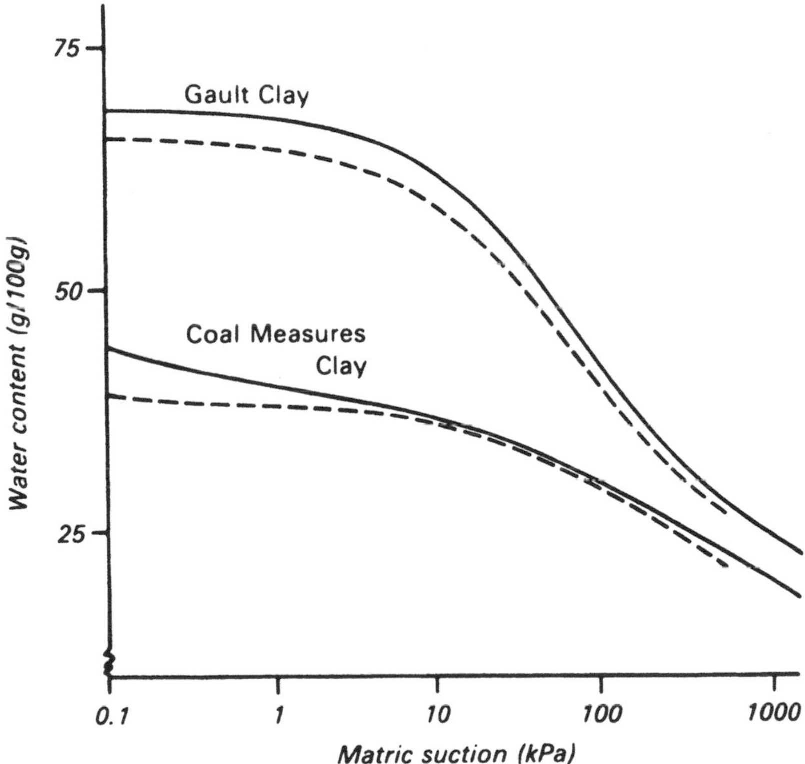

Fig. 10 Comparison of water release characteristics obtained by consolidation (---) and by sand suction table-pressure membrane apparatus (—) for two sieved and rewetted subsoil clays.

oven drying after removal of the filter paper, and hence a water release characteristic can be built up.

2. *Psychrometry*

The application of, and equipment for, thermocouple psychrometry is described in Chap. 2. Provided that samples uniformly dried to a suitable range of moisture contents are available, laboratory psychrometers such as those described by Rawlins and Campbell (1986) can also be used to determine the water release characteristic (Fig. 11). However, psychrometers are mainly suited to the drier end of the water release curve (< -100 kPa).

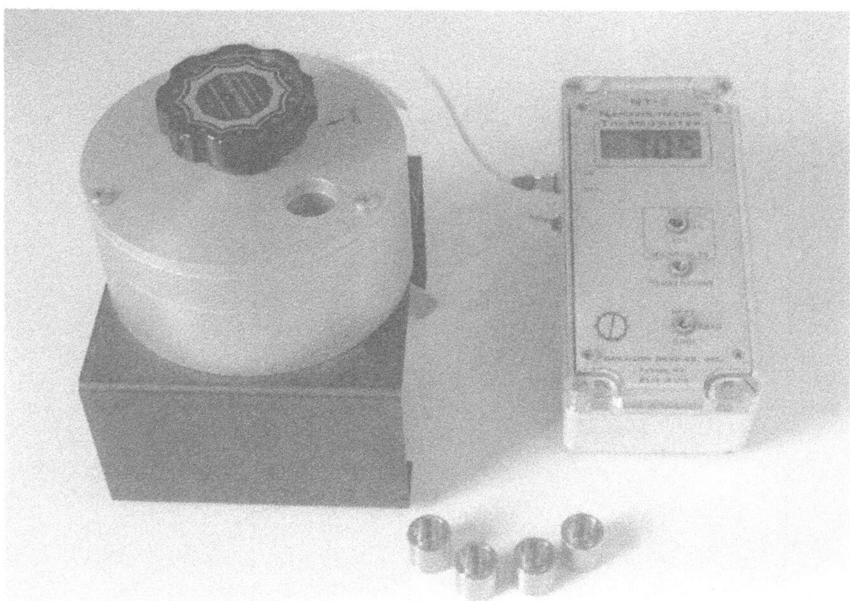

Fig. 11 Richards' psychrometer for laboratory determination of matric potential in dry soils. Samples are placed in the small stainless steel cups and then inserted into the device. Readings may be taken in a few minutes.

3. Field Methods

It is relevant briefly to discuss field methods of determining the soil water release characteristic, as these are done in situ and consequently are more representative than laboratory measurements. Laboratory measurements often deviate significantly from the field-measured water release curve, especially in fine-grained compressible soils where there is the influence of overburden load in the field (Yong and Warkentin, 1975). Thus Ratliff et al. (1983) recommended that if absolute accuracy is required (e.g., in soil water balance calculations), field-measured curves should be taken. By installing tensiometers at different depths in the field, readings of potential can be related to water content determined either gravimetrically (hence destructively) or by a neutron probe (Greminger et al., 1985; Burke et al., 1986). The method is limited by the range of tensiometers (0 to −80 kPa), and although use of electric resistance sensors (Campbell and Gee, 1986) or thermocouple psychrometers can extend this range, there can be calibration problems, and a long time is needed before a soil water characteristic curve can be obtained. If the soil rewets between readings, hysteresis can be a problem, and fluctuations in soil temperature cause further complications through their effect on the viscosity of soil water. For these reasons, field methods are less commonly used than laboratory methods. Spaans and Baker (1996) suggested that the dry end of the water release curve can effectively be derived from the soil freezing characteristic (the relationship between quantity and energy status of liquid water in frozen soil), which can be measured in the field during freezing weather in soils that experience suitably low temperatures. Bruce and Luxmoore (1986) provided a useful summary of references describing measurement of the release characteristic in the field.

C. Methods Based on Modeling

Attempts to model the water release curve from a few point measurements on the curve, or measurements of other parameters, date back over 30 years and have largely been restricted to academic studies. However, this field of research has attracted renewed interest in recent years with the advent of computers able to perform the extensive calculations required, making the methods potentially of practical value.

1. Pedotransfer Functions

Estimation methods that describe the soil water release characteristic based on other soil characteristics have been referred to as pedotransfer functions by Tietje and Tapkenhinrichs (1993), who divided them into three categories:

a. Point Regression Methods

Water contents are measured for a range of matric potentials and in each case regressed on a range of soil parameters such as silt and clay content, organic matter content, and dry bulk density, using a range of soils. The regression equations can then be used for estimation of water content at these matric potentials, given the relevant parameters, for other soils.

b. Physical Model Methods

The water release curve is estimated from theory starting with a given particle size distribution. Assumptions must be made about the shape of particles, packing arrangements, and the capillary attraction of water in pores of different sizes.

c. Functional Parameter Regression Methods

A form of equation describing the water release curve is decided upon, and the parameters of the curve for a particular soil are derived using regression analysis with measured values on a water release curve.

An early attempt at the parameter regression method was that of Brooks and Corey (1964). Their model, usually in the slightly revised form below (Buchan and Grewel, 1990), has been used as the basis of many models since (e.g., Campbell, 1985):

$$\frac{\psi}{\psi_e} = \left(\frac{\theta}{\theta_s}\right)^b \tag{4}$$

where ψ is the matric potential, ψ_e is the air entry potential (the potential needed to drain the largest pores in the soil), θ is the water content, θ_s is the saturated water content, and b is a constant. Gregson et al. (1987) and Gregson (1990) argued that a single-parameter model is satisfactory in many situations, other parameters in their model being fairly constant for a wide range of soils. This raises the possibility of estimating a water release characteristic for a soil from a single point measurement such as the field capacity. Conversely, others have suggested that a greater number of parameters are required for the curve to fit near to saturation, where the Brooks, Corey relationship has been shown not to hold. Van Genuchten's five-parameter sigmoidal model (van Genuchten, 1980) has been widely used. Some models also attempt to account for hysteresis (Haverkamp and Parlange, 1986; Tietje and Tapkenhinrichs, 1993; Viaene et al., 1994).

There have been many independent attempts to compare pedotransfer functions with each other and/or with measured data, often using a combination of the above methods (Haverkamp and Parlange, 1986; Vereecken et al. 1989; Felton and Nieber, 1991; Tietje and Tapkenhinrichs, 1993; Danalatos et al., 1994; Viaene et al., 1994; Nandagiri and Prasad, 1997). The van Genuchten (1980) model appears to produce accurate results in many of these studies but has the disadvantage of requiring at least five measurements to fit it. The ability to describe the water release curve for a soil as an equation is required for most soil water transport models.

2. *Fractal Models of Soil Structure*

Although these fall within the definition of a pedotransfer function used by Tietje and Tapkenhinrichs (1993), they represent a new and distinct development. Recently it has been argued by Crawford et al. (1995) that the parameters of the water release curve for a soil using a model such as the Brooks, Corey model are related to the fractal dimensions if soil structure is simulated by a fractal model. These authors measured fractal dimensions of soils using image analysis of thin sections prepared by impregnating the soils with resin, and compared these with fractal dimensions derived from a model fitted to the measured water release curves. Perfect et al. (1996) suggest that three fractal dimensions are required to produce accurate models of the water release curve. The limitations of these methods are discussed further by Bird et al. (1996), Bird (1997), and Crawford and Young (1997) and include the problems of considering the "inkbottle effect" (see Sec. II.B), pore connectivity in fractal models, and the fact that a fractal relationship (similarity of structure at different scales) only holds over a limited range of scales. Potential developments are also discussed, and the subject remains an active area of research at this time.

3. *Other Models of Soil Structure*

Recent advances in computing power open up the possibility of creating models of soil structure as three-dimensional arrays of pixels representing solid, air, or water. The models can be built up to simulate different soil structures using a range of possible methods including fractal dimensions (Crawford et al., 1995), Boolean models of overlapping spheres (Horgan and Ball, 1994), or from a measured particle size distribution. Image analysis of sections of resin-impregnated soils may be used to determine the parameters to model a particular soil structure (Glasbey et al., 1991; Crawford et al., 1995; Anderson et al., 1996; Bruand et al., 1996; Ringrose-Voase, 1996; Vogel and Kretzschmar, 1996). The water release characteristic, and other hydraulic data, can then be calculated by modeling the movement of water into, through, or out of each individual pore in the structure under varying hydraulic potentials. The method has the advantages of including hysteresis effects, pore connectivity, and irregularly shaped pores. We have found close agreement between modeled and measured water release curves over the range -10 to -100 kPa for a range of structureless soils. However, such models are limited in their ability to model water release at high suctions by the resolution of the array used to represent the soil, and at low suctions by the overall size of the array. These restrictions are likely to diminish with improvements in computing power. The practical usefulness of this approach has, therefore, yet to be proved.

D. Choice of Method

Having reviewed the various methods available to measure the soil water release characteristic, it is pertinent to consider external factors that might influence the choice of method in any particular situation.

1. *Analysis Time*

Most methods of measuring the water release characteristic involve leaving samples until their potential reaches equilibrium with an applied suction or pressure. Because of this, the time taken for "full characterization" can be considerable when compared, for example, with many methods of soil chemical analysis. Samples can take 4 to 12 days to reach each successive equilibrium on sand suction tables and in pressure cells (Ball and Hunter, 1980). Thus determination of five or six equilibrium points using one sample can result in a total analysis time of 3 to 4 months, once peripheral laboratory tasks such as oven-drying and data collection have been taken into account. This time scale might not be a problem for a laboratory servicing a large strategic soil resource survey, but it is totally unacceptable for short-term, customer-oriented projects. Analysis time in such situations can be shortened by careful division of samples so that different equilibrium points can be determined simultaneously on subsamples or by taking a large number of replicate undisturbed samples. Any requirement for more rapid analysis is likely to be met only by methods such as those using a centrifuge and will entail any inaccuracies inherent in such methods.

2. Equipment Availability and Price

Perhaps the major influence on methods adopted in soil physics laboratories around the world is the availability of an extensive range of soil moisture extractors manufactured by the Soilmoisture Equipment Corporation (Santa Barbara, CA). Smaller ranges of similar equipment are available in the United Kingdom, Australia, and the Netherlands, but they are not in wide use outside their country of origin. A list of suppliers is given in Table 5. In many developing countries, however, acquisition of imported equipment is strongly discouraged by fiscal policies. Thus although a range of suitable equipment may be available, it is not easily obtainable, and alternative supplies or methodologies may need to be adopted. Under such circumstances, it might be pertinent to consider adopting methods that are less capital intensive, or manufacturing equipment locally. It must be remembered though that whereas a commercially available system such as a pressure plate extractor and peripherals comes well documented with a complete set of instructions, a proven methodology for measurement, and a single source of replacement parts, self-designed installations require staff with the necessary aptitude for construction and maintenance and often necessitate considerable effort in locating and obtaining component parts. Whatever the degree of sophistication of the equipment used, the usefulness of the data will be affected by many other factors including the quality of available staff. Maintenance of a near-constant temperature for laboratory measurement is also important because of the effect of temperature changes on the viscosity of water (Hopmans and Dane, 1985, 1986).

3. Safety and Statutory Requirements

The most common techniques used to characterize the low-potential part of the water release characteristic employ high air pressures. Thus it is essential that the equipment used and the peripheral supply lines be designed not only to withstand the pressure range applied but also to do so within an acceptable safety margin. This is an important consideration not only for equipment made locally according to laboratory specifications but also for internationally available standard pieces of equipment. Different countries interpret safety criteria differently and apply different safety margins. In the United Kingdom, for example, the design, operation, and maintenance of air receivers come under the control of the Factories Act

Table 5 Some Equipment Suppliers and Typical Prices

Equipment	Typical unit cost 1998 (US $)	Suppliers	Remarks
Büchner funnel	35–80	D	Available from general laboratory suppliers
Suction plate	150	G	(e.g., 330 mm diameter × 13 mm)
Sand suction tables	2500–3150	B, C	Can be handmade
Sand–kaolin tables	5600–6350	B, C	Can be handmade
Pressure plate extractor, 500 kPa	1500–2500	A, B	
Pressure plate extractor, 1500 kPa	1800–5600	A, B	
Pressure membrane extractor, 1500 kPa	1900–3200	A, C	
Pressure membrane extractor, 10 MPa	4225	A	
Tempe cell, 100 kPa	170–280	A	
Centrifuge	1300–4300	B, D	
Laboratory psychrometers	660–4200	E, F	
Sample corers and corer sets	150–1500	A, B, C	
Lab compressor, 1500 kPa	2850–5600	A, B	
Pressure control manifold	3200–4500	A, B	

Key: A. Soilmoisture Equipment Corp., PO Box 30025, Santa Barbara, CA 93105, USA. B. ELE International Ltd., Eastman Way, Hemel Hempstead, Hertfordshire, HP2 7HB, UK. C. Eijkelkamp, Nijverheidsstraat 14, 6987 EM Giesbeek, The Netherlands. D. Fisher Scientific UK, Bishop Meadow Road, Loughborough, Leicestershire, LE11 5RG, UK. E. Decagon Devices Inc., 950 NE Nelson Court, PO Box 835, Pullman, Washington 99163, USA. F. Wescor Inc., 459, South Main Street, Logan, Utah 84321, USA. G. Fairey Industrial Ceramics Ltd., Filleybrook, Stone, Staffs., ST15 0PU, UK.

of 1961. This act is normally interpreted as including pressure plate and pressure membrane extractors. These devices are subject to initial inspections and pressure tests to ensure that their design incorporates a sufficient safety margin against failure, and then to regular (26-month) inspections to ensure that they are maintained in a safe condition. The same rules apply to the air receivers of compressors, which may be used to pressurize the extractors.

The application of these stringent safety regulations in the early 1980s prevented many U.K. laboratories from using the pressure plate extractors with which they were already equipped, thus disrupting research programs and incurring considerable costs for re-equipping. Thus it is advisable to be aware of the statutory or local constraints on the use of pressure apparatus before equipping a laboratory for measurement of the soil water characteristic.

4. Standardization

In certain allied disciplines, such as in soil analysis for engineering purposes, there are well-documented standard methods (British Standards Institution, 1975) using equipment of standard design. There have been attempts at some degree of standardization for methods of determining the water release characteristic, e.g., by Burke et al. (1986). However, a variety of analytical methods are still in use worldwide and will continue to be used as long as individual requirements differ. Given the wide variety of physical states in which samples are tested, any attempt at standardization should start with sampling procedure and sample preparation. These are major factors in analytical differences, and a correct choice of sample state and sample size will largely decide the analytical technique used.

IV. SAMPLING METHODOLOGY AND PRETREATMENT FOR ANALYSIS

A. Field Sampling

Soil samples taken for water release analysis should be isolated with minimal disturbance so that they are closely representative of the in situ soil property. McKeague (1978) stated that the quality of samples depends on the judgment and ingenuity of the sampler, and the reliability of the physical data depends on the original soil sample more than any other factor. Burke et al. (1986) list the following as important factors that should be carefully considered to obtain a representative sample: the method to be used, the sample dimension, the sampling location within the field and within the soil profile, the number of replicates, and the time of sampling. Loveday (1974) provided a comprehensive discussion on sampling technique and sampler design.

1. Location

If soil samples are to be taken to represent an area of land such as a field or soil mapping unit, they should be taken from several soil pits located at random within the area, to characterize the natural variability. Areas that contain different site or soil types should first be divided into smaller, relatively homogeneous areas, and a number of sampling positions located at random within each of these. Soil survey information may help in determining suitable boundaries (Burke et al., 1986). Greminger et al. (1985) present field-measured water release data for 100 locations, demonstrating variability attributable to soil changes along a 100 m line.

2. Sampling from a Soil Profile

Samples should be taken from representative locations within a freshly dug soil profile (e.g., the midpoints of discrete soil layers or horizons), taking special note of such management-induced boundaries as plough pans, deep loosening, and drainage treatments. Where obvious differences occur within a soil horizon or layer, each discrete area should be sampled. Detailed profile descriptions, whether in soil science (U.S. Department of Agriculture, 1951; Hodgson, 1976) or geotechnical (Carter, 1983) terminology, and particle size analysis are important aids to the interpretation of analytical results.

3. Time of Sampling

To standardize procedures and to minimize the effect of hysteresis, water release samples should ideally be taken when the soil is fully wetted. This is most important where clay soils with shrink–swell properties are being investigated. In this case, it is preferable to sample a few months after the soil has returned to and remained close to field capacity, to ensure that maximum soil expansion has occurred.

4. Sample Type and Dimensions

Disturbed Versus Undisturbed. As discussed in Secs. II.C and D, the shape of the water release curve at high potentials is largely dependent on soil structure and the associated pore size distribution. Thus if a sample is disturbed or sieved it cannot reflect the true properties of a relatively undisturbed field soil, because its pore size distribution will have been greatly altered. Figure 12 shows the effect of sample disturbance on the water release curves of a loamy medium sand. Unger (1975), who made comparative water retention analyses using core and sieved samples, found that disturbance generally decreased water storage in coarse-textured soils but increased it in fine-textured ones, although organic matter content and structural development in the undisturbed soil affected this general trend. Similar results have been recorded by others (Elrick and Tanner, 1955; Young and Dixon, 1966).

Disturbed samples, provided they have not been crushed, compressed, or in any other way remolded, may however be acceptable for measurements at matric suctions greater than 100 kPa, and remolded samples might be used for certain geotechnical applications.

Sample Size. The minimum sample volume required to represent a given soil layer without producing unacceptable variation is termed the representative elementary volume (Burke et al., 1986). For each soil type this is largely dependent on soil structure, being smaller for sandy soil with a single grain structure

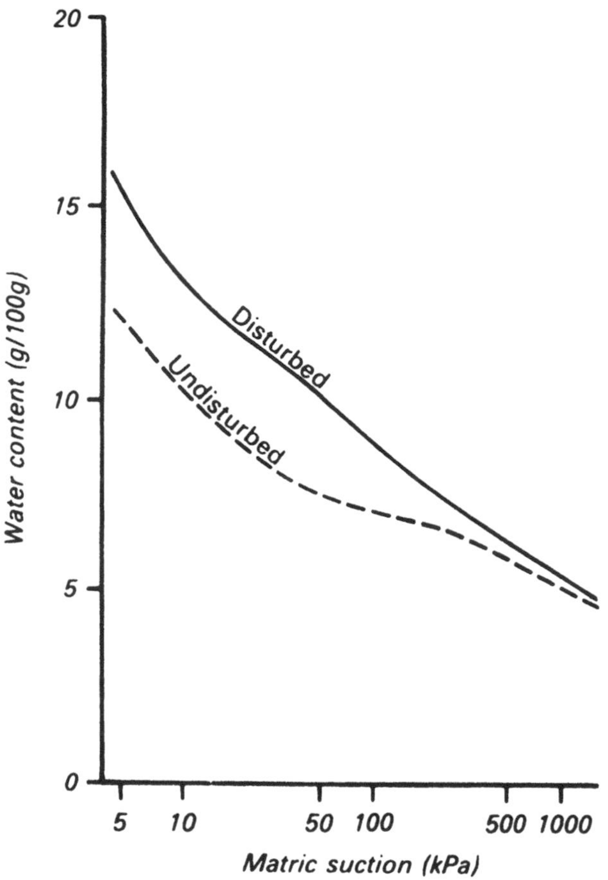

Fig. 12 The effect of sample disturbance on the water release characteristic of a loamy sand subsoil.

than for a clay soil with larger natural aggregates or peds. Although samples of different size should be taken for different representative elementary volumes, many workers use a standard sample size because of the fixed dimensions of the sampler and increase the volume sampled by replication. In practice, the number of replicates is often limited by the time and expense of fieldwork and laboratory analysis. Generally cores with diameters of at least 5 cm but preferably 10 cm are the most practical for measurements at potentials in the 0 to −100 kPa range. A core length between 2 and 7 cm is usually used, since longer cores would take a long time to equilibrate, and to limit the difference in suction between the top and bottom of the cores when they are being equilibrated on suction tables.

Coring Devices. The core method normally uses a cylindrical metal sampler that is pressed or driven into the soil to the desired depth and is carefully removed to preserve a known volume of soil as it existed in situ. Dagg and Hosegood (1962) devised a sampler incorporating several existing designs, which, with further slight modifications, is used on a routine basis in England and Wales (Hall et al., 1977). A tin-plated sleeve 7.5 cm diameter and 5.0 cm high is placed in a machined steel barrel and a cutting ring attached. The coring device is driven carefully, using an integral 3.5 kg sliding hammer, into a flat, horizontal surface prepared in the relevant soil layer. Compaction of the sample is avoided by not coring beyond a level marked on the barrel of the corer. The corer is dug out with a trowel and the core ejected by means of a spring-loaded plunger. Various other designs are available internationally, and the suppliers of some of these are listed in Table 5.

Stony Soils. Many soils are difficult to sample because of stones, and although specially designed corers have been recommended (McLintock, 1959; Jurgensen et al., 1977), sample disturbance is unavoidable in many soils. Rimmer (1982), working on reclaimed colliery spoil heaps with large stone contents, filled cans with disturbed material. Alternatively, water release data can be derived from sieved soil repacked to field density and the results corrected using a stone content measured in the field (Hodgson, 1976). Where it is not possible to obtain core samples, or expansion or excessive shrinking of a sample is expected, a clod sample can be taken. Loveday (1974) described a method in which natural clods are immersed in Saran resin; after initial measurements of the sample volume, the Saran coating is removed from one flat face and the clods can be equilibrated at various potentials.

B. Sample Preparation

In the field, the soil core should be trimmed roughly with a knife before being fitted with lids at each end and labeled clearly. Samples should be wrapped in plastic bags to prevent drying and if necessary packed in foam or polystyrene to avoid damage in transit. Cores taken to the laboratory should be stored in a refrigerator at 1–2°C if they are to be stored for long periods before use, to reduce evaporation and suppress biological activity. Biotic activity in soil cores can make the determination of equilibrium conditions difficult, and where activity is evident, samples should be treated with an inhibitor, such as a 0.05% solution of copper sulfate or copper chloride. Freezing of samples is to be avoided at all costs, because it is likely to alter the pore size distribution and hence the release curve. Preparation for water retention measurements varies between laboratories. Hall et al. (1977) described in detail a procedure in which the ends of the core were trimmed flush with the sleeve, and then one end was covered with nylon mesh or voile and secured with an elastic band. The lid for the other end was sprayed with a dry film lubricant to ease removal, as the tins could become corroded after a few weeks in the moist atmosphere during equilibration. When trimming the cores, small projecting stones sometimes had to be carefully removed and the cavity filled with surplus soil or a smaller stone. Samples with large projecting stones were discarded. The samples were wetted by standing on a sheet of saturated foam rubber to ensure that they were brought to a suction of less than 0.05 bar (the first equilibration point). The time required for wetting varied with particle size class, being a day or two for sands and as much as two weeks for clayey soils. It was recommended that sandy soils should not be left wetting for too long, since they may slake. Low-density subsoil sands without the stabilizing influence of organic matter or roots are the most susceptible to this problem.

Klute (1986) suggested that a wetting solution of deaerated 0.005 M $CaSO_4$ was preferable to either deionized or tap water. Deionized water promotes dispersion of clays in the sample, and dissolved air in tap water can come out of solution, affecting the water content at a given potential.

Fast wetting such as by submergence is not recommended for swelling soils or those with a fragile structure. Klute (1986) pointed out that wetting in the fashion described by Hall et al. (1977) brings the sample to natural saturation rather than total saturation because of the presence of trapped air. The water release characteristic will then follow a different curve initially from that from total saturation. It will be representative of field situations but, for detailed studies of pore size distribution, vacuum saturation may be necessary. Too great a vacuum should be avoided, as the water can boil under the reduced pressure and disrupt the sample.

A final point concerns the representativeness of measurements on unconfined swelling clays. In situ they are subjected to an overburden load. To mimic this situation, a similar external load should be applied in the laboratory before wetting and subsequent measurement, but routine techniques for this have not been developed.

V. APPLICATIONS OF WATER RELEASE MEASUREMENTS

Knowledge of the amount of water held at various matric potentials is used in agronomic, engineering, and environmental applications. In agronomic applications a number of soil moisture constants are regularly used as these relate to the availability of water to crops. These are discussed below.

A. Soil Moisture Constants

1. Field Capacity (FC)

Field capacity is defined as the water content of soil that has been allowed to drain freely for two days from saturation with negligible loss due to evaporation. Initially the hydraulic conductivity is close to the saturated value, so drainage is relatively fast. As water is lost from the soil, the matric potential decreases and the hydraulic conductivity begins to drop rapidly as the soil dries. By the time the matric potential has reached −5 kPa, drainage is extremely slow from most soils. This point is typically reached after about 2 days, and the water content of the soil is then termed the field capacity for that soil. Since the water that has drained from the soil has done so too quickly to be useful to plants, the field capacity is often considered to be the upper limit of the amount of water that can be stored in any particular soil after rainfall or irrigation.

Many problems arise with the assumption of a single value for field capacity. The redistribution of draining water in a soil profile is a continuous process, which may be influenced by many factors (Hillel, 1982; Beukes, 1984; Cassel and Nielsen, 1986), including antecedent moisture conditions, depth of wetting, soil texture, type of clay present, organic matter, presence of slowly permeable horizons, and the rate of evapotranspiration. Consequently the matric potential can be different in deep horizons of less permeable soils than in an overlying topsoil. The field capacity concept is most acceptable for coarser and loamy textured soils, where a static state is more easily defined because of the sharp decrease in unsaturated hydraulic conductivity with a comparatively small drop in matric potential.

Values ranging from −3 to −8 kPa have been reported for the matric potential at field capacity of a range of freely draining soils (Webster and Beckett, 1972; Dent and Scammell, 1981; U.S. Department of Agriculture, 1983; Cassel and Nielsen, 1986). Ideally, field capacity should be determined in the field by monitoring soil water content. However, this is time-consuming, so in most applications a value for field capacity is estimated by equilibrating soil cores at published values of matric potentials that are thought to approximate to field capacity. Such values vary from −5 to −50 kPa (Cassel and Nielsen, 1986), but the water content at −5 kPa or −10 kPa is widely used to represent the field capacity for any soil.

The amount of water lost readily by the soil after heavy rain (i.e., the difference between saturation and FC) is also significant in designing drainage (Scullion et al., 1986) and irrigation systems (Reeve, 1986).

2. *Permanent Wilting Point (PWP)*

The permanent wilting point is defined as the soil water content at which the leaves of a growing plant first reach a stage of wilting from which they do not recover. Different plants wilt at different values of soil matric potential, with values between −800 and −3000 kPa being reported (Loveday, 1974). Early research on plant response to low soil moisture contents (Richards and Weaver, 1943; Veihmeyer and Hendrickson, 1949) indicated that sunflowers wilt permanently at a suction of about 1500 kPa (15 bar) and, since the change in moisture with matric suction is so small in this range for most soils, the water content at a potential of −1500 kPa is generally taken to be an approximation of the permanent wilting point. Water remaining in the soil at this point or drier is, therefore, considered to be unavailable to plants.

3. *Available Water Capacity (AWC) and Profile Available Water Capacity (PAWC)*

The difference between FC and PWP represents the amount of water held in a soil after heavy rain or irrigation that is available for plant growth, and is therefore termed the available water capacity. The concept is widely used, although it is subject to many limitations (Hillel, 1982). The amount of water actually available to a crop will be reduced by evaporation (Cassel and Nielsen, 1986). The soil water release curve provides a means of obtaining the volumetric available water capacity (θ_A) for any soil horizon:

$$\theta_A = \theta(5) - \theta(1500) \tag{5}$$

where $\theta(5)$ is the volumetric water content at a potential of -5 kPa (FC) and $\theta(1500)$ is the volumetric water content at a potential of -1500 kPa (PWP).

Available water for the soil horizon is then the product of the horizon thickness and θ_A, while that for the whole profile (profile available water capacity) is the sum of such values down to a specified depth or a barrier to rooting.

4. *Air Capacity*

Air capacity (or coarse porosity) is obtained as the difference between the total porosity and the volumetric water content at field capacity. Such pores are normally air filled except during short periods following heavy rainfall. Because air capacity is a measure of the fractional volume of large pores in the soil, it also provides a reasonable indication of saturated hydraulic conductivity, where the large pores are continuous (Ahuja et al., 1984).

B. Diagrammatic Presentation of Data

The relationship between soil air, soil water, and the soil solids can be obtained from the water release characteristic and can be presented diagrammatically for a complete soil profile (Fig. 13). The horizontal axis is divided into unavailable water, available water (at stated suctions), air capacity, fine earth (< 2 mm), and stones, all on a percentage volume basis. The vertical axis represents depth below the soil surface, and mean results for each sampling depth are plotted. The points for each sampling depth are then connected by a line added solely for diagrammatic clarity and having no analytical basis. Soil horizons or a change to bedrock can be shown where appropriate. Particle size distribution can be presented in a similar format for easy comparison. The Newport series profile in Fig. 13 is a haplumbrept with a large amount of fine sand (60–200 μm) in all

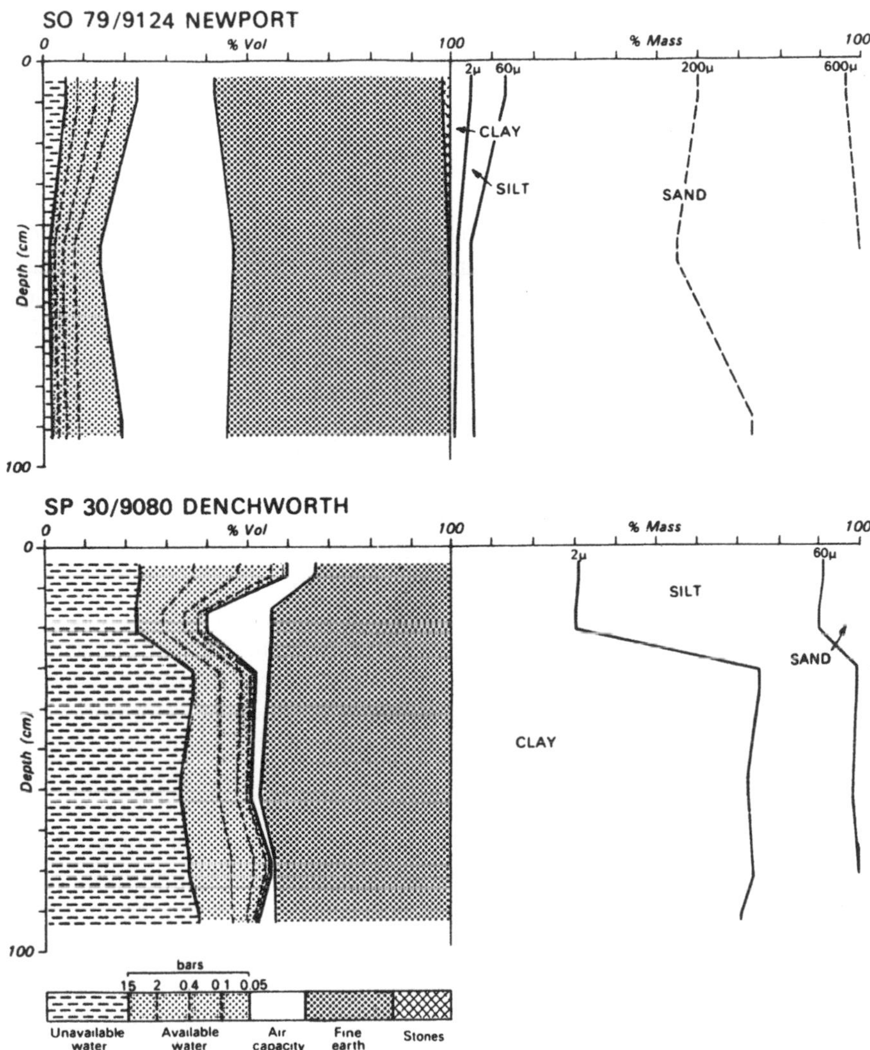

Fig. 13 Water release profiles of two contrasting soils. (After Hall et al., 1977.)

horizons. The Denchworth series profile is a haplaquept formed on Mesozoic clay shales.

Advantages of this style of representation are that data for a soil profile can be presented concisely and that changes in air–water–solid relationships down the profile can be seen at a glance. Careful study of the diagrams can give information about potential problems of drainage, water storage, stoniness, and poor aeration at different depths in the profile.

C. Agronomic Applications

1. Crop Water Supply

For annual crops, the amount of available water that is genuinely accessible varies with crop and soil. However, various approaches have been taken to assess long-term moisture limitations to optimum crop production. On the broad scale, one can classify profile available water according to climatic moisture regime (U.S. Department of Agriculture, 1983). At a more detailed level, the available water range can be split into easily and less easily available portions, and empirical models can be set up to obtain crop adjusted profile available water values. These can then be used with data on potential soil moisture deficit to assess soil droughtiness in a given area (Thomasson, 1979).

At a field scale, water retention data are important when considering a soil for irrigation requirements. A full water release curve is required for each soil type to assess available water capacity, critical deficits, and optimum frequency and volume of water applications (Dent and Scammell, 1981). Reeve (1986) has explained the relevance of water retention measurements to irrigation planning in New Zealand in terms of the ability of a soil to sustain crop transpiration during drought or between irrigation events, the ability of soil to absorb irrigation water when dry, potential losses of irrigation water by drainage, the possibility of waterlogging caused by slowly permeable subsoils, and the existence of dense or compact layers that may restrict rooting. The slope of the release characteristic, termed the differential or specific water capacity, is also an important function in calculating soil water diffusivity (Chap. 5) used in modeling water use by crops.

2. Porosity and Structure

Values of air capacity have been used as a guide to the recognition of impermeable horizons (Avery, 1980), and values integrated down to the top of an impermeable horizon have been used to represent the storage capacity of soils for irrigation water (Reeve, 1986) and for rainwater in flood response studies.

In addition, the water release characteristic can be used as a measure of soil structure in an undisturbed situation (Hall et al., 1977), or to record the recovery of land after damage (Bullock et al., 1985).

D. Other Applications

A knowledge of the water release characteristic is useful in various engineering applications such as off-road trafficability and stability of earthworks formed from clay. In the latter situation the shape of the curve can be important. From Fig. 10, a small water loss over the middle section of the characteristic represents a much larger strength increase in the Mesozoic (Gault) Clay than in the Paleozoic (Coal Measures) Clay. Further applications are in relating the soil water release curve to other physical parameters such as bearing capacity (Mullins and Fraser, 1980) and soil shrinkage (Reeve et al., 1980), both of which are important in construction and in agriculture. The water release characteristic can also be used to estimate unsaturated hydraulic conductivity as a function of water content, providing that a single value of unsaturated conductivity at a known water content is available (see Chap. 5, Sec. X).

Many physically based models depend on the use of water release data. These models include assessments of soil suitability for restoring damaged land and accepting municipal sewage sludge (U.S. Department of Agriculture, 1983), predictions of nitrate leaching (Addiscott, 1977), aquifer vulnerability measurements (Carter et al., 1987), and descriptions of the residual behavior of pesticides (Nicholls, 1989) in the profile. Substances such as nitrate and certain pesticides are readily soluble in water, and their movement in the profile is largely controlled by the water release characteristic of that soil.

Regional simulations of moisture availability and soil water fluxes often incorporate soil water release data. Predictions of the effect of groundwater lowering on crop production may require water release data and hydraulic conductivities for all soil horizons (Wosten et al., 1985; Bouma et al., 1986).

Many of these applications require a large amount of data, which may present a formidable barrier to progress. In these cases, rapid measurement methods (e.g., Wosten et al., 1985) or estimations may be necessary. Estimations can be based on tables relating soil moisture constants to texture classes and horizon types (McKeague et al., 1984) or on multiple regression equations (Peterson et al., 1968; Hall et al., 1977; Gupta and Larson, 1979; Rawls et al., 1982).

REFERENCES

Addiscott, T. M. 1977. A simple computer model for leaching in structured soils. *J. Soil Sci.* 28:554–563.

Ahuja, L. R., Naney, J. W., Green, R. E., and Nielsen, D. R. 1984. Macroporosity to characterise spatial variability of hydraulic conductivity and effects of land management. *Soil Sci. Soc. Am. J.* 48:699–702.

Anderson, S. H., McBratney, A. B., and FitzPatrick, E. A. 1996. Soil mass, surface, and spectral fractal dimensions estimated from thin section photographs. *Soil Sci. Soc. Am. J.* 60:962–969.

Archer, J. R., and Smith, P. D. 1972. The relation between bulk density, available water capacity, and air capacity of soils. *J. Soil Sci.* 23:475–480.

Avery, B. W. 1980. *Soil Classification for England and Wales*. Soil Survey Tech. Monogr. No. 14. Harpenden, U.K.

Ball, B. C., and Hunter, R. 1980. *Improvements in the Routine Determination of Soil Pore Size Distribution from Water Release Measurements on Tension Tables and Pressure Plates*. Dept. Note No. SIN/314. Penicuik, U.K.: Scottish Inst. Agric. Eng.

Beukes, D. J. 1984. The effect of certain soil properties on matric potential at, and time duration to, field capacity. *S. Afr. J. Plant Soil* 1 : 126–131.

Beven, K. 1981. Micro-, meso-, macroporosity and channeling flow phenomena in soils. *Soil Sci. Soc. Am. J.* 45 : 1245.

Bird, N. R. A. 1997. Comments on 'The relationship between the moisture release curve and the structure of soil' by J. W. Crawford, N. Matsui and I. M. Young. *Eur. J. Soil Sci.* 48 : 188–189.

Bird, N. R. A., Bartoli, F., and Dexter, A. R. 1996. Water retention models for fractal soil structures. *Eur. J. Soil Sci.* 47 : 1–6.

Bouma, J., van Lanen, H. A. J., Breeusma, A., Wosten, H. J. M., and Kooistra, M. J. 1986. Soil survey data needs when studying modern land use problems. *Soil Use Management* 2 : 125–130.

Bouyoucos, G. J. 1929. A new, simple and rapid method for determining the moisture equivalent of soils, and the role of soil colloids on this moisture equivalent. *Soil Sci.* 27 : 233–241.

Briggs, L. J., and McLane, J. W. 1907. The moisture equivalent of soils. *Bur. Soils Bull.* No. 45. U.S. Department of Agriculture.

British Standards Institution. 1975. *Soils for Civil Engineering Purposes, BS1377.* London.

Brooks, R. H., and Corey, A. T. 1964. Hydraulic properties of porous media. Hydrol. Pap. 3. Fort Collins: Colorado State Univ.

Bruand, A., Cousin, I., Nicoullard, B., Duval, O., and Bergon, J. C. 1996. Backscattered electron scanning images of soil porosity for analysing soil compaction around roots. *Soil Sci. Soc. Am. J.* 60 : 895–901.

Bruce, R. R., and Luxmoore, R. J. 1986. Water retention: Field methods. In: *Methods of Soil Analysis, Part 1* (A. Klute, ed.). Madison, WI: Am. Soc. Agron., pp. 663–686.

Buchan, G. D., and Grewel, K. S. 1990. The power-function model for the soil moisture characteristic. *J. Soil Sci.* 41 : 111–117.

Buckingham, E. 1907. Studies on the movement of soil moisture. U.S. Bur. Soils, Bull. No. 38. Washington DC: U.S. Department of Agriculture.

Bullock, P., Newman, A. C. D., and Thomasson, A. J. 1985. Porosity aspects of the regeneration of soil structure after compaction. *Soil Till. Res.* 5 : 325–341.

Burke, W., Gabriels, D., and Bouma, J., eds. 1986. *Soil Structure Assessment*. Rotterdam: A. A. Balkema.

Campbell, G. S. 1985. *Soil physics with BASIC, Transport models for soil–plant systems.* Amsterdam: Elsevier.

Campbell, G. S., and Gee, G. W. 1986. Water potential: Miscellaneous methods. In: *Methods of Soil Analysis, Part 1* (A. Klute, ed.). Madison, WI: Am. Soc. Agron., pp. 619–633.

Carter, M. 1983. *Geotechnical Engineering Handbook*. London: Pentech Press.

Carter, A. D., and Thomasson, A. J. 1989. Data to feed and calibrate land evaluation models. In: *Application of Computerized EC Soil Map and Climate Data* (H. A. J. van Lanen and A. K. Bregt, eds.). Wageningen, The Netherlands: Soil Survey Institute, pp. 35–43.

Carter, A. D., Palmer, R. C., and Monkhouse, R. A. 1987. Mapping the vulnerability of groundwater to pollution from agricultural practice, particularly with respect to nitrate. In: *Vulnerability of Soil and Groundwater to Pollutants* (W. van Duijvenbooden and H. G. van Waegeningh, eds.). The Hague, The Netherlands: CHO-TNO, pp. 333–342.

Cassel, D. K., and Nielsen, D. R. 1986. Field capacity and available water capacity. In: *Methods of Soil Analysis, Part 1* (A. Klute, ed.). Madison, WI: Am. Soc. Agron., pp. 901–926.

Chahal, R. S., and Yong, R. N. 1965. Validity of the soil water characteristics determined with the pressurized apparatus. *Soil Sci.* 99:98–103.

Chandler, R. J., and Gutierrez, C. I. 1986.The filter-paper method of suction-measurement. *Geotechnique* 36:265–268.

Childs, E. C. 1940. The use of soil moisture characteristics in soil studies. *Soil Sci.* 50: 239–252.

Childs, E. C. 1969. *An Introduction to the Physical Basis of Soil Water Phenomena.* London: John Wiley.

Coleman, J. D., and Marsh, A. D. 1961. An investigation of the pressure-membrane method for measuring the suction properties of soil. *J. Soil Sci.* 12:343–362.

Collis-George, N. 1952. A note on the pressure plate-membrane apparatus. *Soil Sci.* 74: 315–322.

Constantz, J., and Herkelrath, W. N. 1984. Submersible pressure outflow cell for measurement of soil water retention and diffusivity from 5 to 95°C. *Soil Sci. Soc. Am. J.* 48:7–10.

Crawford, J. W., Matsui, N., and Young, I. M. 1995. The relation between the moisture-release curve and the structure of soil. *Eur. J. Soil Sci.* 46:369–375.

Crawford, J. W., and Young, I. M. 1997. Reply to: Comments on 'The relation between the moisture release curve and the structure of soil' by N. R. A. Bird. *Eur. J. Soil Sci.* 48:189–191.

Croney, D., Coleman, J. D., and Bridge, P. M. 1952. The suction of moisture held in soil and other porous materials. Road Research Tech. Paper No. 24. Road Research Laboratory. London: HMSO.

Czeratzki, W. 1958. Eine keramische Platte zur Serienmässigen Untersuchung von Porengrössen in Boden in Spannungsbereich bis ca. −1 Atm. *Z. Pflanzenernähr. Düng. Bodenk.* 81:50–56.

Dagg, M., and Hosegood, P. H. 1962. Details of hand sampling tools for taking undisturbed soil cores. *E. Afr. Agric. For. J.* Special Issue: 129–131.

Danalatos, N. G., Kosmas, C. S., Driessen, P. M., and Yassoglou, N. 1994. Estimation of the draining soil moisture characteristic from standard data as recorded in routine soil surveys. *Geoderma* 64:155–165.

Danielson, R. E., and Sutherland, P. L. 1986. Porosity. In: *Methods of Soil Analysis, Part 1* (A. Klute, ed.). Madison, WI: Am. Soc. Agron., pp. 443–461.

Datta, B., and Singh, O. P. 1981. Comparative study of moisture release behaviour of soils, soil clays and pure clays. *Aust. J. Soil Res.* 19:79–82.

de Boodt, M., ed. 1967. *West European Methods for Soil Structure Determination*. State Faculty of Agricultural Sciences, Univ. of Ghent, Belgium.

Deka, R. N., Wairiu, M., Mtakwa, P. W., Mullins, C. E., Veenendaal, E. M., and Townend, J. 1995. Use and accuracy of the filter-paper technique for the measurement of soil matric potential. *Eur. J. Soil Sci.* 46:233–238.

Dent, D. L., and Scammell, R. P. 1981. Assessment of long-term irrigation need by integration of data for soil and crop characteristics and climate. *Soil Survey Land Eval.* 1:51–57.

Elrick, D. E., and Tanner, C. B. 1955. Influence of sample pretreatment on soil moisture retention. *Soil Sci. Soc. Am. Proc.* 19:279–282.

Felton, G. K., and Nieber, J. L. 1991. Four soil moisture characteristic curve functions evaluated for numerical modeling of sand. *Trans. Am. Soc. Agric. Eng.* 34: 417–422.

Glasbey, C. A., Horgan, G. W., and Darbyshire, J. F. 1991. Image analysis and three-dimensional modelling of pores in soil aggregates. *J. Soil Sci.* 42:479–486.

Gradwell, M. W. 1971. The available water capacities of North Auckland soils. *N.Z. J. Agric. Res.* 14:253–287.

Gregson, K. 1990. Modelling the soil moisture characteristic. *J. Soil Sci.* 41:677.

Gregson, K., Hector, D. J., and McGowan, M. 1987. A one-parameter model for the soil water characteristic. *J. Soil Sci.* 38:483–486.

Greminger, P. J., Sud, Y. K., and Nielsen, D. R. 1985. Spatial variability of field-measured soil-water characteristics. *Soil Sci. Soc. Am. J.* 49:1075–1092.

Gupta, S. C., and Larson, W. E. 1979. Estimating soil water retention characteristics from particle size distribution, organic matter percent and bulk density. *Water Resour. Res.* 15:1633–1635.

Haines, W. B. 1930. Studies in the physical properties of soils. *J. Agric. Sci.* 20:97–116.

Hall, D. G. M., Reeve, M. J., Thomasson, A. J., and Wright, V. F. 1977. *Water Retention, Porosity and Density of Field Soils*. Soil Survey Tech. Monogr. No. 9. Harpenden, U.K.

Haverkamp, R., and Parlange, J.-Y. 1986. Predicting the water-retention curve from particle-size distribution: 1. Sandy soils without organic matter. *Soil Sci.* 142: 325–339.

Head, K. H. 1982. *Manual of Soil Laboratory Testing*. Vol. 2. London: Pentech Press.

Heinonen, R. 1961. On the pre-treatment of samples of heavy clay soil for determinations by the pressure membrane apparatus. *J. Sci. Agric. Soc. Finland* 33:153–158.

Hillel, D. 1971. *Soil and Water*. New York: Academic Press.

Hillel, D. 1982. *Introduction to Soil Physics*. New York: Academic Press.

Hodgson, J. M., ed. 1976. *Soil Survey Field Handbook*. Tech. Monogr. No. 5. Harpenden, U.K.

Hopmans, J. W., and Dane, J. H. 1985. Effect of temperature-dependent hydraulic properties on soil water movement. *Soil Sci. Soc. Am. J.* 49:51–58.

Hopmans, J. W., and Dane, J. H. 1986. Combined effect of hysteresis and temperature on soil-water movement. *J. Hydrol.* 83:161–171.

Horgan, G. W., and Ball, B. C. 1994. Simulating diffusion in a Boolean model of soil pores. *Eur. J. Soil Sci.* 45:483–491.

Jamison, V. C. 1942. Structure of a Dunkirk silty clay loss in relation to pF moisture measurements. *J. Am. Soc. Agron.* 34:307–321.

Jurgensen, M. F., Larsen, M. J., and Harvey, A. E. 1977. *A Soil Sampler for Steep, Rocky Sites.* U.S. Dept. Agric. Forest Service Research Note No. INT-217. Washington, DC: U.S. Government Printing Office.

Klute, A. 1986. Water retention: Laboratory methods. In: *Methods of Soil Analysis, Part 1* (A. Klute, ed.). Madison, WI: Am. Soc. Agron., pp. 635–662.

Kumar, S., Malik, R. S., and Dahiya, I. S. 1984. Water retention, transmission and contact characteristics of Ludas Sand as influenced by farmyard manure. *Aust. J. Soil Res.* 22:253–259.

Kuznetsova, I. V., and Vinogradova, G. B. 1982. Wilting moisture of plants in compacted soil horizons. *Pochvovedeniye* 5:58–64.

Kyuma, K., Suh, Y.-S., and Kawaguchi, K. 1977. A method of capability evaluation for upland soils: I. Assessment of available water retention capacity. *Soil Sci. Plant Nutr.* 23:135–149.

Lal, R. 1979. Physical properties and moisture retention characteristics of some Nigerian soils. *Geoderma* 21:209–223.

Lambooy, A. M. 1984. Relationship between cation exchange capacity, clay content and water retention of Highveld soils. *S. Afr. J. Plant Soil* 1:33–38.

Loveday, J., ed. 1974. *Methods for Analysis of Irrigated Soils.* Tech. Commun. No. 54, Commonwealth Bureau of Soils, Farnham Royal, U.K.

Madsen, H. B., Jensen, C. R., and Boysen, T. 1986. A comparison of the thermocouple psychrometer and the pressure plate methods for determination of soil water characteristic curves. *J. Soil Sci.* 37:357–362.

McKeague, J. A., ed. 1978. *Manual on Soil Sampling and Method of Analysis.* Ottawa, Ontario: Can. Soc. Soil Sci.

McKeague, J. A., Eilers, R. G., Thomasson, A. J., Reeve, M. J., Bouma, J., Grossman, R. B., Favrot, J. C., Renger, M., and Strebel, O. 1984. Tentative assessment of soil survey approaches to the characterization and interpretation of air-water properties of soils. *Geoderma* 34:69–100.

McLintock, T. F. 1959. A method for obtaining soil-sample volume in stony soils. *J. For.* 57:832–834.

McQueen, I. S., and Miller, R. F. 1968. Calibration and evaluation of a wide-range gravimetric method for measuring moisture stress. *Soil Sci.* 106:225–231.

Mullins, C. E., and Fraser, A. 1980. Use of the drop-cone penetrometer on undisturbed and remoulded soils at a range of soil-water tensions. *J. Soil Sci.* 31:25–32.

Nandagiri, L., and Prasad, R. 1997. Relative performances of textural models in estimating soil moisture characteristic. *J. Irrig. Drainage Eng.* 123:211–214.

Nicholls, P. H. 1989. Predicting the availability of soil-applied pesticides. *Aspects Appl. Biol.* 21:173–184.

Odén, S. 1975/76. An integral method for the determination of moisture retention curves by centrifugation. *Grundförbättring* 27:137–143.

Perfect, E., McLaughlin, N. B., Kay, B. D., and Topp, G. C. 1996. An improved fractal equation for the soil water retention curve. *Water Resources Res.* 32:281–287.

Petersen, G. W., Cunningham, R. L., and Matelski, R. P. 1968. Moisture characteristics of Pennsylvania soils: I. Moisture retention as related to texture. *Soil Sci. Soc. Am. Proc.* 32:271–275.

Poulovassilis, A. 1974. The uniqueness of the moisture characteristics. *J. Soil Sci.* 25: 27–33.

Prebble, R. E., and Stirk, G. B. 1959. Effect of free iron oxide on range of available water in soils. *Soil Sci.* 88:213–217.

Pritchard, D. T. 1969. An osmotic method for studying the suction/moisture content relationships of porous materials. *J. Soil Sci.* 20:374–383.

Puckett, W. E., Dane, J. H., and Hajek, B. F. 1985. Physical and mineralogical data to determine soil hydraulic properties. *Soil Sci. Soc. Am. J.* 49:831–836.

Ratliff, L. F., Ritchie, J. T., and Cassel, D. K. 1983. Field-measured limits of soil water availability as related to laboratory measured properties. *Soil Sci. Soc. Am. J.* 47: 770–775.

Rawlins, S. L., and Campbell, G. S. 1986. Water potential: Thermocouple psychrometry. In: *Methods of Soil Analysis, Part 1* (A. Klute, ed.). Madison, WI: Am. Soc. Agron., pp. 597–618.

Rawls, W. J., Brakensiek, D. L., and Saxton, K. E. 1982. Estimation of soil water properties. *Trans. Am. Soc. Agric. Eng.* 25:1316–1328.

Reeve, M. J. 1986. Water retention, porosity and composition interrelationships of alluvial soils in mid-Hawke's Bay and their relevance in irrigation planning. *N.Z. J. Agric. Res.* 29:457–468.

Reeve, M. J., Hall, D. G. M., and Bullock, P. 1980. The effect of soil composition and environmental factors on the shrinkage of some clayey British soils. *J. Soil Sci.* 31: 429–442.

Reginato, R. J., and van Bavel, C. H. M. 1962. Pressure cell for soil cores. *Soil Sci. Soc. Am. Proc.* 26:1–3.

Richards, L. A. 1941. A pressure-membrane extraction apparatus for soil solution. *Soil Sci.* 51:377–386.

Richards, L. A. 1948. Porous plate apparatus for measuring moisture retention and transmission by soil. *Soil Sci.* 66:105–110.

Richards, L. A. 1949. Methods of measuring soil moisture tension. *Soil Sci.* 68:95–112.

Richards, L. A. 1965. Physical condition of water in soil. In: *Methods of Soil Analysis, Part 1* (C. A. Black, ed.). Madison, WI: Am. Soc. Agron., pp. 128–152.

Richards, L. A., and Weaver, L. R. 1943. Fifteen-atmosphere percentages as related to the permanent wilting percentage. *Soil Sci.* 56:331–339.

Rimmer, D. L. 1982. Soil physical conditions on reclaimed colliery spoil heaps. *J. Soil Sci.* 33:567–579.

Ringrose-Voase, A. J. 1996. Measurement of macropore geometry by image analysis of sections through impregnated soil. *Plant Soil* 183:27–47.

Russell, M. B. 1941. Pore size distribution as a measure of soil structure. *Soil Sci. Soc. Am. Proc.* 6:108–112.

Russell, M. B., and Richards, L. A. 1938. The determination of soil moisture energy relations by centrifugation. *Soil Sci. Soc. Am. Proc.* 3:65–69.

Schofield, R. K. 1935. The pF of the water in soil. *Trans. 3rd Int. Congr. Soil Sci.* 2: 37–48.

Scullion, J. A., Mohammed, R. A., and Ramshaw, G. A. 1986. Statistical evaluation of drainage treatments in simple field trials with special reference to former opencast coal mining land. *J. Agric. Sci. Camb.* 107:515–520.

Shcherbakov, R. A. 1985. Model of the hysteresis of water retention by soils. *Pochvovedeniye* 8:54–60.

Spaans, E. J. A., and Baker, J. M. 1996. The soil freezing characteristic: Its measurement and similarity to the moisture characteristic. *Soil Sci. Soc. Am. J.* 60:13–19.

Stakman, W. P., and Bishay, B. G. 1976. Moisture retention and plasticity of highly calcareous soils in Egypt. *Neth. J. Agric. Sci.* 24:43–57.

Stakman, W. P., and van der Harst, G. G. 1969. *Pressure Membrane Apparatus Range pF 3.0 to 4.2.* Wageningen, The Netherlands: Institute for Land and Water Management Research, P.O. Box 35.

Stakman, W. P., Valk, G. A., and van der Harst, G. G. 1969. *Determination of Soil Moisture Retention Curves: I. Sand-Box Apparatus—Range pF 0 to 2.7.* 3d rev. ed. Wageningen, The Netherlands: Institute for Land and Water Management Research.

Thomasson, A. J. 1979. Assessment of soil droughtiness. In: *Soil Survey Applications* (M. G. Jarvis and D. Mackney, eds.) Tech. Monogr. No. 13. Harpenden, U.K., pp. 43–50.

Tietje, O., and Tapkenhinrichs, M. 1993. Evaluation of pedo-transfer functions. *Soil Sci. Soc. Am. J.* 57:1088–1095.

Topp, G. C. and Zebchuk, W. 1979. The determination of soil-water desorption curves for soil cores. *Can. J. Soil Sci.* 59:19–26.

Unger, P. W. 1975. Water retention by core and sieved soil samples. *Soil Sci. Soc. Am. Proc.* 39:1197–1200.

U.S. Department of Agriculture. 1951. *Soil Survey Staff, Soil Survey Manual.* Handbook No. 18. Washington, DC: U.S. Government Printing Office.

U.S. Department of Agriculture. 1983. *National Soils Handbook.* Washington, DC.

van Genuchten, M. Th. 1980. A closed-form equation for predicting the hydraulic conductivity of unsaturated soils. *Soil Sci. Soc. Am. J.* 44:892–898.

Viaene, P., Vareecken, H., Diels, J., and Feyen, J. 1994. A statistical analysis of six hysteresis models for the moisture retention characteristic. *Soil Sci.* 157:345–355.

Veihmeyer, F. J., and Hendrickson, A. H. 1949. Methods of measuring field capacity and permanent wilting percentage of soils. *Soil Sci.* 68:75–94.

Vereecken, H., Maes, J., Feyen, J., and Darius, P. 1989. Estimating the soil moisture retention characteristic from texture, bulk density, and carbon content. *Soil Sci.* 148: 389–403.

Vogel, H. J., and Kretzschmar, A. 1996. Topological characterisation of pore space in soil-sample preparation and digital image-processing. *Geoderma* 73:23–38.

Vomocil, J. A. 1965. Porosity. In: *Methods of Soil Analysis, Part 1* (C. A. Black, ed.). Madison, WI: Am. Soc. Agron., pp. 299–314.

Wadsworth, H. A. 1944. An interpretation of the moisture content-surface force curve for soils. *Soil Sci.* 58:225–242.

Waters, P. 1980. Comparison of the ceramic plate and the pressure membrane to determine the 15 bar water content of soils. *J. Soil Sci.* 31:443–446.

Webster, R., and Beekett, P. H. T. 1972. Matric suctions to which soils in South Central England drain. *J. Agric. Sci. Camb.* 78:379–387.

Williams, J., Prebble, R. E., Williams, W. T., and Hignett, C. T. 1983. The influence of texture, structure and clay mineralogy on the soil moisture characteristic. *Aust. J. Soil Res.* 21:25–32.

Wosten, J. H. M., Bouma, J., and Stoffelsen, G. H. 1985. The use of soil survey data for regional water simulation. *Soil Sci. Soc. Am. J.* 49:1238–1245.

Yong, R. N., and Warkentin, B. P. 1975. *Soil Properties and Behaviour.* Amsterdam: Elsevier.

Young, K. K., and Dixon, J. D. 1966. Overestimation of water content at field capacity from sieved sample data. *Soil Sci.* 101:104–107.

Zur, B. 1966. Osmotic control of the matric soil-water potential: I. Soil-water system. *Soil Sci.* 102:394–398.

4
Hydraulic Conductivity of Saturated Soils

Edward G. Youngs
Cranfield University, Silsoe, Bedfordshire, England

I. INTRODUCTION

The physical law describing water movement through saturated porous materials in general and soils in particular was proposed by Darcy (1856) in his work concerned with the water supplies for the town of Dijon. He established the law from the results of experiments with water flowing down columns of sands in an experimental arrangement shown schematically in Fig. 1. Darcy found that the volume of water Q flowing per unit time was directly proportional to the cross-sectional area A of the column and to the difference Δh in hydraulic head causing the flow as measured by the level of water in manometers, and inversely proportional to the length L of the column. Thus

$$Q = \frac{KA\,\Delta h}{L} \tag{1}$$

where the proportionality constant K is now known as the *hydraulic conductivity* of the porous material. The dimensions of K are those of a velocity, LT^{-1}. Typical values of K for soils of different textures are given in Table 1. Conversion factors relating various units are given in Table 2. Since the hydraulic conductivity of a soil is inversely proportional to the viscous drag of the water flowing between the soil particles, its value increases as the viscosity of water decreases with increasing temperature, by about 3% per °C.

The hydraulic head is the sum of the soil water pressure head (the pressure potential discussed in Chap. 2 expressed in units of energy per unit weight) and the elevation from a given datum level. It is measured directly by the level of water in the manometers above a datum in Darcy's experiment and is the water potential

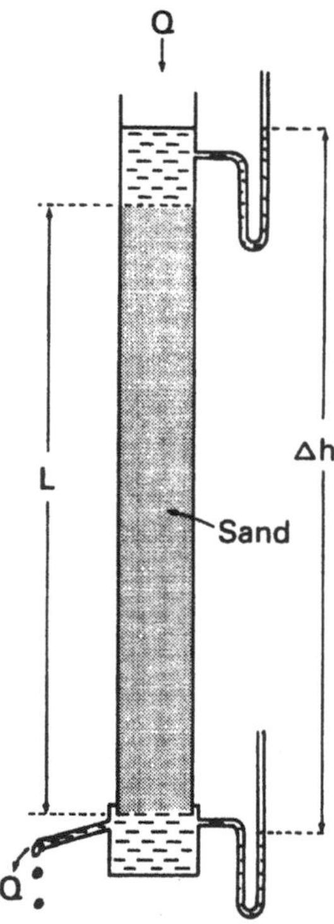

Fig. 1 Darcy's experimental arrangement.

expressed as the work done per unit weight of water in transferring it from a reference source at the datum level. The potential may also be defined as the work done per unit volume of water, in which case the potential difference causing the flow would be $\rho g \Delta h$, where ρ is the density of water and g is the acceleration due to gravity; Darcy's law using potentials defined in this way would give K in units with dimensions $M^{-1}L^3T$. Here we will adopt the usual convention of defining the potential as the work done per unit weight, that is as a head of water, so that K is simply expressed in units of a velocity. This is very convenient when computing water flows in soils, but it has the disadvantage that the value of the hydraulic conductivity of a porous material depends on g. This means that the hydraulic conductivity of a given porous material depends on altitude and is smaller at the top of a mountain than at sea level, but this is of little importance in most practical problems concerned with groundwater movement.

Equation 1 describes the flow of water in porous materials at low velocities when viscous forces opposing the flow are much greater than the inertial forces.

Table 1 Hydraulic Conductivity Values of Saturated Soils

Soil	Hydraulic conductivity ($mm\ d^{-1}$)
Fine-textured soils	< 10
Soils with well-defined structure	10–1000
Coarse-textured soils	> 1000

Table 2 Conversion Factors for Units of Hydraulic Conductivity*

$m\ d^{-1}$	$cm\ h^{-1}$	$cm\ min^{-1}$	$mm\ s^{-1}$
1	4.17	0.0694	0.0116
0.24	1	0.0167	0.00278
14.4	60	1	0.167
86.4	360	6	1

* *Example:* To convert x cm min^{-1} to meters per day, find 1 in the cm min^{-1} column. Numbers on the same horizontal row are values in other units equivalent to 1 cm min^{-1}, so that 1 cm min^{-1} = 14.4 m d^{-1} and x cm min^{-1} = $14.4x$ m d^{-1}.

The ratio of the inertial forces to the viscous forces is represented by the Reynolds number (Muskat, 1937; Childs, 1969) which may be defined as

$$\mathrm{Re} = \frac{vd\rho}{\eta} \tag{2}$$

where v is the mean flow velocity, d a characteristic length (for example, the mean pore diameter), ρ the density of water as before, and η the viscosity of water. When Re exceeds a value of about 1.0, Darcy's law no longer describes the flow of water through porous materials. Under field conditions this is unlikely to occur except in some situations of flow in gravels and in structural fissures and worm holes.

Darcy's work was concerned with one-dimensional flow. However, flows in soil are most often two- or three-dimensional, so Eq. 1 has to be extended to take into account multidimensional flow. Slichter (1899) argued that the flow of water in soil described by Darcy's law is analogous to the flow of electricity and heat in conductors, and so generally Darcy's law may be written in vectorial notation as

$$v = -K \operatorname{grad} h \tag{3}$$

where v is the flow velocity and h is the hydraulic potential of the soil water expressed as the hydraulic head as in Eq. 1, with the flow normal to the equipotentials. If the water is considered to be incompressible and the soil does not shrink or swell, the equation of continuity is

$$\mathrm{div}\, v = 0 \tag{4}$$

so that h is described by Laplace's equation

$$\nabla^2 h = 0 \tag{5}$$

Thus it is only a matter of solving Eq. 5 for the hydraulic head h with the given boundary conditions in order to obtain a complete solution to a given flow problem in saturated soil in one, two, or three dimensions. With h known throughout the flow region from Eq. 5, flows can be found from Eq. 3 if K is known. Conversely, if flows and hydraulic heads are measured in the flow region, the hydraulic conductivity can be deduced. Measurement techniques for the determination of hydraulic conductivities of porous materials in general, including soils, make use of solutions of Laplace's equation with the prescribed boundary conditions imposed by the particular method.

The concept of hydraulic conductivity is derived from experiments on uniform porous materials. Methods of measuring hydraulic conductivity assume implicitly that the flow in the soil region concerned is given by Darcy's law with the head distribution described by Laplace's equation (Eq. 5); that is, among other factors they presuppose that the soil is uniform. As discussed in Sec. II, soils can be far from uniform because of heterogeneities at various scales, and measurements need to be made on some representative volume of the whole flow region. Thus although values of "hydraulic conductivity" for a soil in a given region can always be obtained using any method, such values will be of little relevance in the context of predicting flows if the volume of soil sampled by the method is unrepresentative of the soil region as a whole.

In the above discussion it has been tacitly assumed that the hydraulic conductivity of the soil is the same in all directions. However, anisotropy in soil properties can occur because of structural development and laminations, giving different hydraulic conductivity values in different directions. Darcy's law then has to be expressed in tensor form (Childs, 1969). In anisotropic soils the streamlines of flow are orthogonal to the equipotential surfaces only when the flow is in the direction of one of the three principal directions. The theory of flow in anisotropic soils (Muskat, 1937; Maasland, 1957; Childs, 1969) shows that Laplace's equation can still be used to obtain solutions to flow problems if a transformation incorporating the components of hydraulic conductivity in the principal directions is applied to the spatial coordinates. If the soil is anisotropic, the two- and three-dimensional flows usually used in hydraulic conductivity measurement techniques in the field require analysis using this theory to obtain values of the hydraulic conductivity in the principal directions.

II. FUNDAMENTAL CONSIDERATIONS OF FLOW THROUGH SOILS

A. Soil Considered as a Continuum

The movement of water through soils takes place in the tortuous channels between the soil particles with velocities varying from point to point and described by the Stokes–Navier equations (Childs, 1969). Darcy's law does not consider this microscopic flow pattern between the particles but instead assumes the water movement to take place in a continuum with a uniform flow averaged over space. It therefore describes the flow of water macroscopically in volumes of soil much larger than the size of the pores. It can thus only be used to describe the macroscopic flow of water through soil regions of volume greater than some *representative elementary volume* that encompasses many soil particles.

The concept of representative elementary volume of a porous material is most easily illustrated by considering the measurement of the water content of a sample of unstructured "uniform" saturated soil, starting with a very small volume and then increasing the sample size. For volumes smaller than the size of the soil particles the sample volume would include only solid matter if located wholly within a soil particle, giving zero soil water content, but would contain only water if located wholly in a pore, giving a soil water content of one. All values between zero and one are possible when the sample is located partly within a soil particle and partly within the pore. As the volume is increased with the sample having to contain both pore volume and solid particle, the lower limit of measured water content increases while the upper limit decreases, as shown in Fig. 2a. When the size of sample is sufficiently large, repeated measurements on random samples of the soil give the same value of soil water content. The smallest sample volume that produces a consistent value is the representative elementary volume. Measurements of hydraulic conductivity and other soil properties need to be made on volumes larger than this volume. While additive soil properties, such as the water content, can be obtained by averaging a large number of measurements made on smaller volumes within the representative elementary volume, the hydraulic conductivity cannot be obtained in this way because of the interdependent complex pattern of flows in between soil particles that this property embraces.

Figure 2a illustrates the variability of a soil physical property that exists in all porous materials at a small enough scale because of their particulate nature. Variability can also be present in soils at larger scales. For example, in aggregated and structured soils where a distribution of macropores between the aggregates or

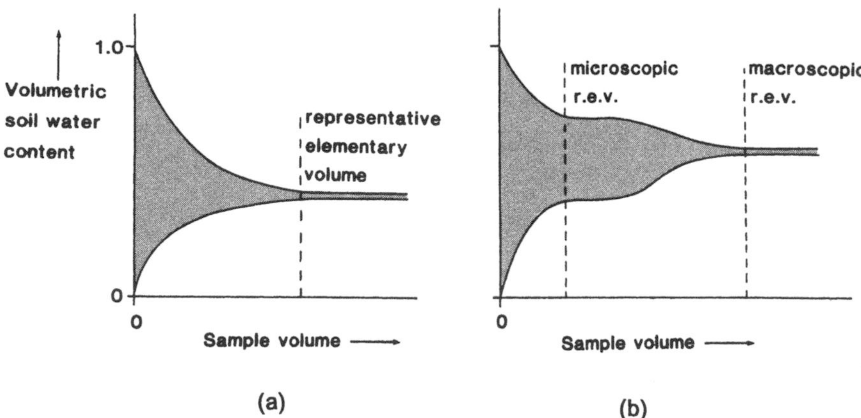

Fig. 2 Measurement of soil water content (a) of a saturated "uniform" soil and (b) of a saturated soil with superimposed macrostructure (r.e.v. = representative elementary volume).

peds is superimposed on the interparticle micropore space, the soil water content would vary with sample size as shown in Fig. 2b; only when the sample size encompasses a representative sample of macropore space do we have a representative volume. This volume will be characteristic of the soil's structure that determines the hydraulic conductivity of the bulk soil.

It is only in materials that show behavior similar to that depicted in Fig. 2a that continuum physics, such as that implied by Darcy's law, can be applied macroscopically without difficulty to soil water flow problems. In materials such as that illustrated in Fig. 2b, boundary conditions at the surfaces of the aggregates and fissures affect the flow patterns throughout the soil region. However, for saturated conditions, so long as sufficiently large volumes are considered, continuum physics can still be applied to water flows at this larger scale using an appropriate value of hydraulic conductivity measured on the bulk soil.

B. Heterogeneity

Because of the complex geometry of the pore system of soils, there is an inherent heterogeneity at pore size dimensions that is not observed when measurements are made on volumes containing a large number of pores. Soil heterogeneity usually implies variations of soil properties between soil volumes containing such a large number of pores. Such heterogeneity occurs at many scales in the following progression:

Particle → aggregate → pedal/fissure → field → regional

The objective in making measurements of hydraulic conductivity is to enable quantitative predictions of soil water flows under given conditions. In a soil showing heterogeneity at various scales, different values of hydraulic conductivity apply at different spatial scales and need to be obtained by appropriate measurement techniques. For example, the calculation of water movement to roots requires measurements at the scale of the soil aggregates, whereas the calculation of the flow to land drains in the same soil requires measurements at a much larger scale that takes into account the flow through fissures. For hydrological purposes measurements need to be made at an even larger scale in order to consider flows at the field or regional scale.

The discussion so far has considered soil heterogeneity as stochastic so that measurements of physical properties can be made on a sample larger than some representative elementary volume. However, changes in soil occur often abruptly or as a trend, that is, in a deterministic manner. One particularly important aspect of soil variability occurs with the variation of the soil with depth. This has a profound effect on field soil water regimes. There is often a gradual change of soil properties with depth that makes it impossible to define a representative elementary volume as previously described. In such cases it is assumed that Eq. 1 defines the hydraulic conductivity; hence with vertical flow in soils with a hydraulic conductivity $K(z)$ varying with the height z, we have

$$K(z) = \frac{v}{dh/dz} \tag{6}$$

where v is the vertical flow velocity; that is, we assume the soil to be a continuum with properties varying with depth.

C. Equivalent Hydraulic Conductivity

As noted in Sec. I, the measurement of the flow that occurs with imposed boundary conditions in a uniform soil allows the determination of the hydraulic conductivity. For a nonuniform soil the measurement gives an equivalent hydraulic conductivity value for the flow region with the given imposed boundary conditions; that is, a value of hydraulic conductivity that would give the measured flow under the same conditions if the soil were uniform.

If the hydraulic conductivity varies spatially so that $K = K(x, y, z)$, the arithmetic and harmonic mean values K_a and K_h of a unit cube of soil are given by

$$K_a = \int_0^1 \int_0^1 \int_0^1 K(x, y, z)\, dx\, dy\, dz \tag{7}$$

and

$$K_h = \frac{1}{\int_0^1 \int_0^1 \int_0^1 1/K(x, y, z)\, dx\, dy\, dz} \tag{8}$$

It can be shown that (Youngs, 1983a)

$$K_a > K_e > K_h \tag{9}$$

where K_e is the equivalent hydraulic conductivity that would actually be measured in any given direction. Since

$$K_a > K_g > K_h \tag{10}$$

where K_g is the geometric mean value, this result is in keeping with the fact that the geometric mean is often taken as the equivalent hydraulic conductivity value for groundwater flow computations. For an isotropic soil it can be argued (Youngs, 1983a) that

$$K_e = \sqrt[3]{K_a^2 K_h} \tag{11}$$

The measurement of hydraulic conductivity by any method gives an equivalent value for the particular flow pattern produced in a uniform soil by the boundary conditions used in the measurement. The value will be different for different boundary conditions if the soil varies spatially. For example, strata of less permeable soil at right angles to the direction of flow, that is strata coinciding approximately with the equipotentials, reduce the value significantly, whilst more permeable strata have little effect. When, however, such strata are in the direction of flow, the reverse is the case. The dependence of the equivalent hydraulic conductivity value on the boundary conditions of the flow region has been further demonstrated in calculations of flow through an earth bank with a complex spatial variation of hydraulic conductivity (Youngs, 1986).

Hydraulic conductivities obtained by methods employing any boundary conditions will give correct predictions when used in computations of groundwater flows in uniform soils. However, the accuracy of predictions in a nonuniform soil will be dependent on the relevance of the measured equivalent hydraulic conductivity. If the measurement imposes boundary conditions that produce flow patterns very different from those of the flows to be calculated, then the predictions will lack accuracy. For accurate predictions the pattern of flow in the measurement must approximate as near as possible to that of the problem, since local variations of hydraulic conductivity can distort flows profoundly.

Thus the measurement of hydraulic conductivity is not a simple matter when the soil is nonuniform. Methods used to make measurement in such soils must be conditioned by the purpose for which they are made. Otherwise values obtained are of little relevance. Unless otherwise stated, the methods described in this chapter, as in other reviews of methods (Reeve and Luthin, 1957; Childs, 1969; Bouwer and Jackson, 1974; Kessler and Oosterbaan, 1974; Amoozegar and Warrick, 1986), assume that the soil is uniform and isotropic; that is, it is assumed that the measurements are on flow regions made up of several representative elementary volumes with no preferential direction.

III. LABORATORY MEASUREMENTS

A. General Principles

Many laboratory measurements of hydraulic conductivity on saturated samples of soils essentially repeat Darcy's original experiments described in Sec. I. The principles that apply for soil samples taken from the field are the same as those for the sands used by Darcy. The soil is removed from the field, hopefully undisturbed, so as to form a column on which measurements can be made, with the sides enclosed by impermeable walls. With the column of soil standing on a permeable base, the soil is saturated and the surface ponded so that water percolates through the soil. The soil water pressure head in the soil is measured at positions down the column, and the rate of flow of water through the soil is measured. The hydraulic conductivity is the rate of flow per unit cross-sectional area per unit hydraulic head gradient. An arrangement used for measuring hydraulic conductivity is known as a *permeameter.* While gravity is the usual driving force for flow in permeameters, use can be made of centrifugal forces to increase the hydraulic head gradients when measuring the hydraulic conductivity of saturated low permeability soils (Nimmo and Mellow, 1991).

In addition to methods that involve measurements on a completely saturated material, there are other methods that involve wetting up an unsaturated sample from a surface maintained saturated at zero soil water pressure. These methods utilize infiltration theory (described in Chap. 6) in order to obtain the hydraulic conductivity of the saturated soil from measurements on the rate of uptake of water by the soil.

B. Collection and Preparation of Soil Samples

For loosely bound soil materials such as sands and sieved soils that are often used in various tests, care has to be taken to obtain uniform packing of columns on which measurements are to be made. If the material is not packed uniformly as the column is filled, separation of different-sized particles can occur, resulting in a column with spatially variable hydraulic conductivity; even columns of coarse sand can pack to give a two-fold variation of hydraulic conductivity down the column (Youngs and Marei, 1987). In filling columns it is useful to attach a short extension length to the top of the column and fill above the top, pouring continuously but slowly while tamping to obtain a uniform density. The material in the top extension is then removed, leaving the bottom part for the measurement. For granulated materials with particles passing through a 2 mm sieve, the representative elementary volume is small enough to allow columns of small diameter, 100 mm or less, to be used.

The taking of field soil samples requires great care so as to obtain samples as near representative of the field soil as possible. The size of sample required cannot easily be inferred from visual inspection because fine cracks in soils, that contribute largely to the hydraulic conductivity of a soil, may not be noticed. In poorly structured soils small samples of cross-sectional area 0.01 m^2 or less can be representative for such purposes as groundwater-flow calculations. In highly structured soils the size of a sample that is representative for a measurement will depend on the purpose for which the measurement is required. Small samples of the size of those suitable for poorly structured soils might suffice for some purposes, for example for studies on water movement in the soil matrix between cracks in a fissured soil, but for groundwater-movement predictions generally a much larger sample that includes the highly conducting cracks and fissures is required. Cylindrical samples 0.4 m in diameter and 0.6 m high have been used (Leeds-Harrison and Shipway, 1985; Leeds-Harrison et al., 1986). For special purposes larger "undisturbed" samples can be obtained as for lysimeter studies (Belford, 1979; Youngs, 1983a), typically 0.8 m in diameter.

Soil samples can be collected in large-diameter PVC or glass fiber cylinders. A steel cutting edge is first attached to one end and the sample taken by jacking the cylinder into the soil hydraulically. While samples are usually taken vertically, horizontal samples can also be taken. As the sampling cylinder is forced into the soil, the surrounding soil is removed to lessen resistance to passage. When the required sample is contained in the cylinder, the surrounding soil is dug away to a greater depth to allow a cutting plate to be jacked underneath, separating the sample from the soil beneath. The sample is then removed to the laboratory, covered by plastic sheeting in order to retain moisture. In the laboratory the upper and lower faces are carefully prepared by removing any smeared or damaged surfaces before saturating the samples for the hydraulic conductivity measurements by infiltrating water through the base to minimize air entrapment.

While taking and removing the sample, soil disturbance or shrinkage may occur, notably with the soil coming detached from the side of the sampling cylinder. A seal can be made by pouring liquid bentonite down the edge. The wetting of the sample will swell the soil and make the seal watertight.

An alternative method of preparing a sample for hydraulic conductivity measurements has been devised by Bouma (1977). A cylindrical column of soil is sculptured in situ so that the column is left in the middle of a trench. Plaster of Paris is then poured over it to seal the sides. The column can then either be cut from the base and removed to the laboratory for measurements of hydraulic conductivity, both in saturated and unsaturated conditions, or alternatively left in place for measurements to be made in the field. A cube of soil is sometimes cut (Bouma and Dekker, 1981) so that flow measurements can be made in different directions after the removal of the plaster from the appropriate faces, allowing the components of hydraulic conductivity in the different directions to be obtained in anisotropic soils. In a modification of the method (Bouma et al., 1982) a cube of soil is carved around a tile drain so that measurements of hydraulic conductivity can be made in this sensitive region in drained lands.

C. Constant Head Permeameter

The constant head permeameter uses exactly the same arrangement as Darcy used in 1856 as illustrated in Fig. 1. The soil column is supported on a permeable base such as a wire gauze or filter, or sometimes a sand table. Water flows through the column from a constant head of water on the soil surface and is collected for measurement from an outlet chamber attached to the base. Slichter (1899) recommended that soil water pressures be measured within the soil column since he noted that "there appears sudden reduction in pressure as the liquid enters the soil." The error arising from not accounting for this reduction is considered to be of no great importance today because of the recognition of the true degree of accuracy that can be expected for hydraulic conductivity values due to inhomogeneities in most soils. The hydraulic conductivity is given from the measurements by

$$K = \frac{QL}{A\,\Delta h} \tag{12}$$

where Q is the flow rate, L the length of the column, A its cross-sectional area, and Δh the head difference causing the flow. In Eq. 12, as with all formulae for K in this chapter, the units of K are the same as the units used for length and time for the quantities on the right hand side of the equation. The measurements made using a constant head permeameter are interpreted as hydraulic conductivity values assuming the soil to be uniform; that is, equivalent hydraulic conductivity values are inferred from measurements of the hydraulic conductance between the levels at which the measurements of head are made.

Errors often occur because of preferential boundary wall flow between the soil and the sides of the permeameter. This can be reduced by separately collecting and measuring the throughput from the central area of the sample (McNeal and Roland, 1964).

Youngs (1982) has described an alternative technique to measure the hydraulic conductivity in saturated soil columns with piezometers that are usually used to measure the soil water pressure head down the column, acting as interceptor drains, as illustrated in Fig. 3. With only one of the piezometers at a height Z above the base acting as a drain and removing water at a rate Q_Z, and with no flow through the base, the hydraulic conductance C_{LZ} between the top of the column at height L and the height Z is given by

$$C_{LZ} = \frac{Q_Z}{h_L - h_0} \tag{13}$$

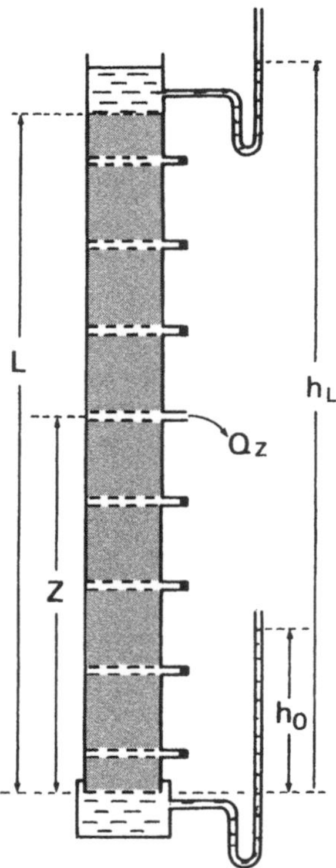

Fig. 3 Measurement of hydraulic conductivity profiles down soil monoliths using interceptor drains.

where h_L is the measured head of the ponded water on the surface and h_0 is that measured at the base of the column. When the conductance profile is obtained by making measurements of flows from successive piezometers down the column, the hydraulic conductivity profile is given by

$$K(Z) = \left[A \frac{d}{dZ}\left(\frac{1}{C_{LZ}}\right)\right]^{-1} \tag{14}$$

where $K(Z)$ is the hydraulic conductivity at height Z. This technique therefore can be used (Youngs, 1982) to obtain the variation of hydraulic conductivity with depth on a soil monolith contained in a lysimeter.

D. Falling Head Permeameter

The falling head permeameter is similar to the constant head permeameter except that, instead of maintaining a constant head of water on the surface of the soil sample, no water is added after a head is applied initially to the soil surface, and the changing level of the head is observed as the water percolates through the sample. Such an arrangement is shown in Fig. 4. Magnification of the rate of fall of the standing head is achieved by containing it in a tube of smaller cross-sectional area A' than the cross-sectional area A of the soil sample. With the height of the water level h_0 (measured from the level of water in a manometer measuring the head at the base of the column) at time t_0 falling to h_1 at time t_1, the hydraulic conductivity is given by

$$K = \frac{A'L \ln(h_0/h_1)}{A(T_1 - t_0)} \tag{15}$$

E. Oscillating Permeameter

A drawback of the constant head and falling head permeameters is that a fairly large volume of water percolates through the soil sample during the course of a measurement of hydraulic conductivity. If the material is surface active, structural changes may occur during the test because of changes in chemical constitution, thus producing changes in the hydraulic conductivity of the soil sample.

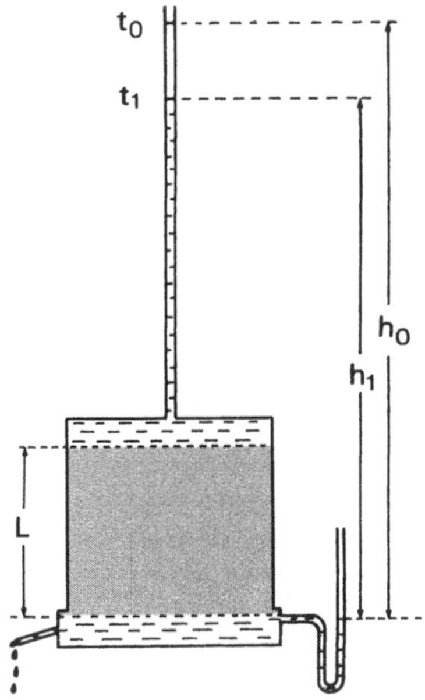

Fig. 4 Falling head permeameter.

A variation of the falling head permeameter is the oscillating permeameter (Childs and Poulovassilis, 1960). This utilizes the passage of water to and fro through the soil sample contained in a limited volume of water, very little in excess of that required to saturate the pore space. Such a small quantity of water quickly comes to chemical equilibrium with the soil without affecting greatly its chemical composition, therefore remaining in equilibrium throughout the test, however long its duration. Water flows through the saturated soil sample contained in a tube under a head of water at the base of the column sinusoidally varying about a mean position. This and the head of water standing on the surface of the soil sample are recorded with time, for example with pressure transducers. After a few cycles, the two heads oscillate out of phase and with different amplitudes. If the amplitude of the forcing head is H_0 and that on the surface of the soil sample is h_0, the phase angle β is given by

$$\tan \beta = \sqrt{\frac{H_0^2}{h_0^2} - 1} \tag{16}$$

and the hydraulic conductivity of the sample is given by

$$K = \frac{2\pi A' L}{AT \tan \beta} \tag{17}$$

where A is the cross-sectional area of the sample of length L, A' is that of the tube containing the water imposing the forcing head, and T is the period of one cycle. The hydraulic conductivity can thus be found from the phase angle obtained either by direct measurement or from measurements of the amplitudes of the heads and the use of Eq. 16.

F. Infiltration Method

Infiltration theory shows that the infiltration rate from a ponded surface into a long vertical column of uniform porous material eventually approaches a constant rate, equal to the hydraulic conductivity of the saturated material. The approximate Green and Ampt (1911) theory of infiltration gives the infiltration rate di/dt when the wetting front has advanced to a depth Z as

$$\frac{di}{dt} = K\left(\frac{h_f}{Z} + 1\right) \tag{18}$$

where $-h_f$ is the soil water pressure head at the wetting front. Thus a plot of di/dt against $1/Z$ gives an intercept K on the di/dt axis, as sketched in Fig. 5. The hydraulic conductivity of saturated uniform porous materials can thus be obtained by observing the position of the wetting front while measuring the infiltration rate

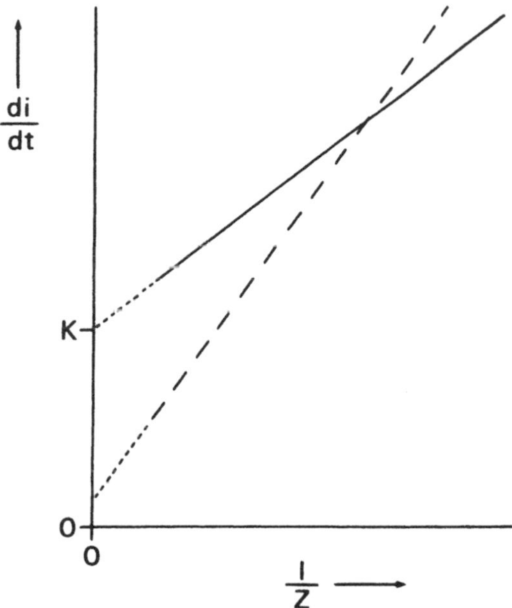

Fig. 5 Plot of the rate of infiltration *di/dt* against the reciprocal of the depth of wetting front 1/Z. Solid line: uniform soil; broken line: soil with hydraulic conductivity decreasing with depth.

from a ponded surface. However, the fact that a linear plot is found when plotting *di/dt* against 1/Z should not be taken as proof that the column is uniform, since it has been found (Childs, 1967; Childs and Bybordi, 1969; Youngs, 1983b) that such a linear plot is obtained in certain situations when there is a decrease in hydraulic conductivity with depth. The intercept in this case is less than if the soil were uniform, and it can even become negative. The method is therefore only reliable if the soil profile is known to be uniform within the wetted depth, and this may be difficult to ascertain.

G. Varying Moment Permeameter

The varying moment permeameter (Youngs, 1968a), although originally used to measure the hydraulic conductivity of unsaturated soils, provides a quick method of measuring the hydraulic conductivity of soil samples that are initially unsaturated. Water is infiltrated horizontally at a positive pressure head into columns of the unsaturated soil, and the rate of change of moment of the advancing water profile about the plane through which infiltration takes place is measured. It can be shown that this rate of change of the moment is equal to the integral of the hydraulic conductivity with respect to the soil water pressure along the column multiplied by the cross-sectional area A of the column. Thus

$$\frac{dM}{dt} = A\left[\int_{p_i}^{p_0} \rho g K' \, dp\right] = A\left[\int_{p_i}^{0} \rho g K' \, dp + \rho g K p_0\right] \tag{19}$$

where M is the moment of the advancing soil water profile at time t, p is the soil water pressure head with the subscripts 0 and i referring to that at the infiltration surface and that in the soil not yet reached by the advancing water front, respectively, and $K'(p)$ is the hydraulic conductivity of the soil that is a function of the soil water pressure head p in unsaturated soils but equal to K for saturated soils. By measuring dM/dt for different pressure heads p_0 of infiltrating water, the hydraulic conductivity of the saturated soil can be obtained using Eq. 19 from the slope of the plot of dM/dt against p_0.

IV. FIELD MEASUREMENTS BELOW A WATER TABLE

A. General Principles

In situ measurements of hydraulic conductivity below the water table provide the most reliable values for use in estimating groundwater flows, especially when they sample large volumes of soil. Techniques usually employ unlined or lined wells sunk below the water table and involve measurements of flow into or out of the wells when the water levels in them are perturbed from the equilibrium. The hydraulic conductivity values are calculated from the solution of the potential problem for the flow region with the imposed boundary conditions. If no analytical solution is available, recourse can be made to electric analogs or numerical methods to obtain solutions. The various well techniques for measuring the hydraulic conductivity of soils when the water table is near the soil surface are given particular attention in books on land drainage (Reeve and Luthin, 1957; Bouwer and Jackson, 1974) where values are required for design purposes. Since all gave satisfactory results in a comparison of well methods in a hydraulic sand tank (Smiles and Youngs, 1965), it would appear that the choice of method depends largely on site conditions, resources available, and individual preference. However, in some methods the flow is predominantly horizontal while in others it is vertical, so that if the soil is suspected of being anisotropic, the method to be employed must take into consideration the direction of flow in the region under investigation.

For satisfactory measurements, wells must be large enough to allow a representative volume of soil to be sampled. However, it is not easy to deduce the volume of soil sampled in a given measurement. Some indication of this volume might be obtained from the volume traced out by 90% (say) of the streamtubes for a 90% (say) reduction in head. It obviously increases with the size of well used. It will also depend on other geometrical factors of the flow system; for example, the area of the well walls through which water can flow, and the spacing of wells in a multiwell system.

Well radii of 50 mm or more are typically used. The wells are best made with post augers,* and special tools can be used to form the holes into an exact cylindrical shape. Some difficulties may be encountered doing this (Childs et al., 1953). First, there is the common problem of making holes when the soil is stony; stones may have to be cut with chisels during the operation. Secondly, there is the problem of unstable soils slumping below the water table; permeable liners can be used to alleviate this problem. And thirdly, in clay soils there is the problem of smearing of the sides of the walls of the wells, thus creating surfaces of low conductance that restrict flow; to lessen this effect the wells are first emptied to allow inflowing water to unblock the pores before measurements are made.

While the use of wells gives a practical and convenient method of providing an arrangement of groundwater flows that can be analyzed to give hydraulic conductivity values, any arrangement of sinks and/or sources that produce flows that can be analyzed may be used for the purpose. For example, land drains, which sample much larger regions of soil than can be sampled with wells, can be used as permeameters (Hoffman and Schwab, 1964; Youngs, 1976).

B. Auger-Hole Method

In the auger-hole method of determining the hydraulic conductivity of a soil, an unlined cylindrical hole is made below the water table (Fig. 6). The position of the water table is found by allowing the water in the hole to return to its equilibrium water level. The water level in the hole is then lowered by removing water by pumping or bailing, and its rate of rise is observed as it returns to equilibrium. Alternatively, the water level can be raised by adding water, and measurements made on the falling level. This is useful when the equilibrium depth of water in the hole is small. The hydraulic conductivity is calculated from measurements taken during the early stage of return before there is appreciable water table drawdown around the hole, using the formula

$$K = C\frac{dy}{dt} \tag{20}$$

where y is the depth of the water level in the hole below the water table at time t and C is a factor that depends on the radius r of the hole, the depth s of a stratum

* A comprehensive range of augers are given in the catalogue of Eijkelkamp Agrisearch Equipment bv, P.O. Box 4, 6987 ZG Gesbeek, The Netherlands.

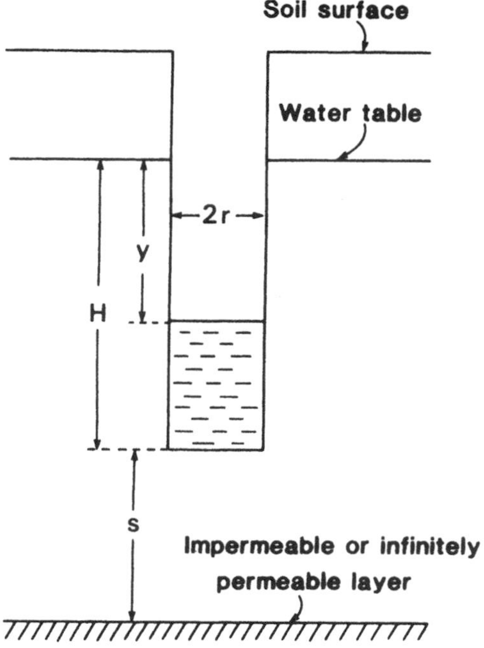

Fig. 6 Geometry of the auger-hole method.

of different hydraulic conductivity below the bottom of the hole, and the depth y, all expressed as a fraction of the depth H of the water in the hole when in equilibrium with the water table; thus we can write $C = C(r/H, s/H, y/H)$.

Formulae for obtaining the factor C in Eq. 20 have been given by Diserens (1934), Hooghoudt (1936), Kirkham and van Bavel (1949), and Ernst (1950). An exact mathematical solution in the form of an infinite series was obtained for C by Boast and Kirkham (1971). Their results are presented in Table 3. Ernst's formulae may be written:

$$K = \frac{4.63}{(20 + H/r)(2 - y/H)} \frac{r}{y} \frac{dy}{dt} \qquad \text{for } s > 0.5H \tag{21}$$

and

$$K = \frac{4.17}{(10 + H/r)(2 - y/H)} \frac{r}{y} \frac{dy}{dt} \qquad \text{for } s = 0 \tag{22}$$

and can be used when the hole is in soil that is effectively infinitely deep and when the hole extends down to an impermeable layer, respectively. These formulae provide a simple means of calculating the shape factor with sufficient accuracy for

Table 3 Values of the Shape Factor $C \times 10^3$ for Auger Holes

		Impermeable layer at s/H =									Infinitely permeable layer at s/H =			
H/r	y/H	0	0.05	0.1	0.2	0.5	1	2	5	∞	5	2	1	0.5
1	1	518	490	468	435	375	331	306	296	295	292	280	247	193
	0.75	544	522	503	473	418	376	351	339	338	335	322	287	230
	0.5	643	623	605	576	521	477	448	441	440	437	416	376	306
2	1	215	204	193	178	155	143	137	135	133	133	131	123	106
	0.75	227	216	208	195	172	160	154	152	152	151	148	140	123
	0.5	271	261	252	240	218	203	196	194	194	193	190	181	161
5	1	60.2	56.3	53.6	49.6	44.9	42.8	41.9		41.5		41.2	40.1	37.6
	0.75	63.6	60.3	57.9	54.3		47.6	46.6		46.4		45.9	44.8	42.1
	0.5	76.7	73.5	71.1	67.4	62.5	60.2	59.1		58.8		58.3	57.1	54.1
10	1	21.0	19.6	18.7	17.5	16.4	15.8	15.5		15.5		15.4	15.2	14.6
	0.75	22.2	21.0	20.2	19.1	18.0	17.4	17.2		17.2		17.1	16.8	16.2
	0.5	27.0	25.9	24.9	23.9	22.6	22.0	21.8		21.7		21.6	21.3	20.6
20	1	6.86	6.41	6.15	5.87	5.58	5.45	5.4		5.38		5.36	5.31	5.17
	0.75	7.27	6.89	6.65	6.38	6.09	5.97	5.9		5.89		5.88	5.82	5.67
	0.5	8.90	8.51	8.26	7.98	7.66	7.52	9.5		7.44		7.41	7.35	7.16
50	1	1.45	1.37	1.32	1.29	1.24	1.22			1.21			1.19	1.18
	0.75	1.54	1.47	1.42	1.39	1.35	1.32			1.31			1.30	1.28
	0.5	1.90	1.82	1.79	1.74	1.69	1.67			1.66			1.65	1.61
100	1	0.43	0.41	0.39	0.39	0.38	0.37			0.37			0.37	0.36
	0.75	0.46	0.44	0.42	0.42	0.41	0.41			0.41			0.39	0.39
	0.5	0.57	0.54	0.53	0.52	0.51	0.51			0.51			0.50	0.50

Source: After Boast and Kirkham (1971).

most purposes; however, Ploeg and van der Howe (1988) pointed out that values using these formulae can differ from Boast and Kirkham's values by as much as 25%. Equations 21 and 22 give the hydraulic conductivity K in the same units as those for the rate of rise of the water level dy/dt, as are the values of C given in Table 3; published presentations for the shape factor usually require dy/dt values to have units cm s^{-1} to give K in units m d^{-1}, and this can give rise to confusion. Measurements are sometimes made using seepage into large holes below the water table, a method sometimes referred to as the "pit-bailing" method. Then shape factors are required for $r > H$, a situation not encountered with the normal use of auger holes. These have been given by Boast and Langebartel (1984).

The flow into auger holes is primarily horizontal, so that in anisotropic soils the results obtained approximate to the horizontal component of the hydraulic conductivity. Although the method has been developed, as have most other methods, for use in uniform soils, it can be used in layered soils to estimate the hydraulic conductivity in the different layers (Hooghoudt, 1936; Ernst, 1950; Kessler and Oosterbaan, 1974).

C. Piezometer Method

A *piezometer* is an open-ended pipe driven into the soil that measures the groundwater pressure below the water table. The piezometer method uses pipes or lined wells with diameters usually much larger than for those used for groundwater pressure measurements, sunk below the water table, with or without a cavity at the bottom, as illustrated in Fig. 7. The cavity is usually cylindrical in shape, although other shapes, for example hemispherical, can be used. As in the auger-hole method, after the water level in the well has come into equilibrium with the water table, it is depressed by pumping or bailing and its rate of rise observed as it returns to equilibrium. The hydraulic conductivity is then given by

$$K = \frac{\pi r^2 \ln(y_0/y)}{A(t - t_0)} \tag{23}$$

where y_0 and y are the depths of the water level in the well below the equilibrium level at time t_0 and at time t, respectively, and A is a shape factor that depends on the depth d of water in the well at equilibrium, the length w of the cavity at the bottom of the well, and the depth s of soil to a stratum of different hydraulic conductivity, all expressed as a fraction of the radius r of the well; that is, $A = A(d/r, w/r, s/r)$.

Shape factors obtained with an electric analog were given by Frevert and Kirkham (1948). More accurate values were presented by Smiles and Youngs (1965), and a comprehensive table of accurate values, reproduced in Table 4, was given by Youngs (1968b). As shown by these values, so long as the cavity is not

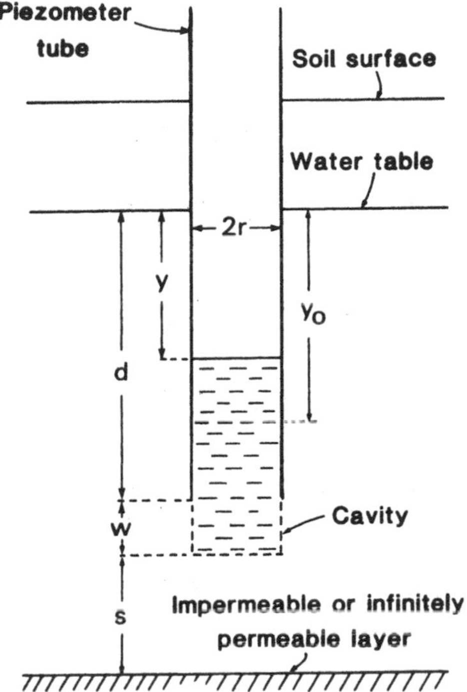

Fig. 7 Geometry of the piezometer method.

less than about a radius from an impermeable or permeable stratum, the results are very nearly the same as for an infinitely deep soil and so are unaffected by changes of hydraulic conductivity at this distance away. Thus accurate determinations of hydraulic conductivity can be made with this method in layered soils, so long as measurements are made in the different layers with the cavity properly located at least one radius above the change in soil. With cavities of small length, the flow is mainly vertical, so that values reflect the vertical component of hydraulic conductivity in anisotropic soils.

Piezometers installed for soil water pressure measurements may also be used to measure hydraulic conductivity. For example, Goss and Youngs (1983) used an existing installation of piezometers inserted horizontally from the walls of an inspection pit. Such piezometers may not have cavities that conform to those for which shape factors are available, so that shape factors for the particular piezometers have to be determined with an electric analog. An arrangement of piezometers located at intervals down the soil profile allows the hydraulic conductivity variation with depth to be determined; and when the installation is from an

Table 4 Values of the Shape Factor A (Expressed as A/r) for Piezometers with Cylindrical Cavities

		A/r, impermeable layer at $s/r =$							Infinitely permeable layer at $s/r =$						
w/r	d/r	∞	8.0	4.0	2.0	1.0	0.5	0	∞	8.0	4.0	2.0	1.0	0.5	0
0	20	5.6	5.5	5.3	5.0	4.4	3.6	0	5.6	5.6	5.8	6.3	7.4	10.2	∞
	16	5.6	5.5	5.3	5.0	4.4	3.6	0	5.6	5.6	5.8	6.4	7.5	10.3	∞
	12	5.6	5.5	5.4	5.1	4.5	3.7	0	5.6	5.7	5.9	6.5	7.6	10.4	∞
	8	5.7	5.6	5.5	5.2	4.6	3.8	0	5.7	5.7	5.9	6.6	7.7	10.5	∞
	4	5.8	5.7	5.6	5.4	4.8	3.9	0	5.8	5.8	6.0	6.7	7.9	10.7	∞
0.5	20	8.7	8.6	8.3	7.7	7.0	6.2	4.8	8.7	8.9	9.4	10.3	12.	15.2	∞
	16	8.8	8.7	8.4	7.8	7.0	6.2	4.8	8.8	9.0	9.4	10.3	12.2	15.2	∞
	12	8.9	8.8	8.5	8.0	7.1	6.3	4.8	8.9	9.1	9.5	10.4	12.2	15.3	∞
	8	9.0	9.0	8.7	8.2	7.2	6.4	4.9	9.0	9.2	9.6	10.5	12.3	15.3	∞
	4	9.5	9.4	9.0	8.6	7.5	6.5	5.0	9.5	9.6	9.8	10.6	12.4	15.4	∞
1.0	20	10.6	10.4	10.0	9.3	8.4	7.6	6.3	10.6	11.0	11.6	12.8	14.9	19.0	∞
	16	10.7	10.5	10.1	9.4	8.5	7.7	6.4	10.7	11.0	11.6	12.8	14.9	19.0	∞
	12	10.8	10.6	10.2	9.5	8.6	7.8	6.5	10.8	11.1	11.7	12.8	14.9	19.0	∞
	8	11.0	10.9	10.5	9.8	8.9	8.0	6.7	11.0	11.2	11.8	12.9	14.9	19.0	∞
	4	11.5	11.4	11.2	10.5	9.7	8.8	7.3	11.5	11.6	12.1	13.1	15.0	19.0	∞

2.0	20	13.8	13.5	12.8	11.9	10.9	10.1	9.1	13.8	14.1	15.0	16.5	19.0	23.0	∞
	16	13.9	13.6	13.0	12.1	11.0	10.2	9.2	13.9	14.3	15.1	16.6	19.1	23.1	∞
	12	14.0	13.7	13.2	12.3	11.2	10.4	9.4	14.0	14.4	15.2	16.7	19.2	23.2	∞
	8	14.3	14.1	13.6	12.7	11.5	10.7	9.6	14.2	14.8	15.5	17.0	19.4	23.3	∞
	4	15.0	14.9	14.5	13.7	12.6	11.7	10.5	15.0	15.4	16.0	17.6	20.1	23.8	∞
4.0	20	18.6	18.0	17.3	16.3	15.3	14.6	13.6	18.6	19.8	20.8	22.7	25.5	29.9	∞
	16	19.0	18.4	17.6	16.6	15.6	14.8	13.8	19.0	20.0	20.9	22.8	25.6	29.9	∞
	12	19.4	18.8	18.0	17.1	16.0	15.1	14.1	19.4	20.3	21.2	23.0	25.8	30.0	∞
	8	19.8	19.4	18.7	17.6	16.4	15.5	14.5	19.8	20.6	21.4	23.3	26.0	30.2	∞
	4	21.0	20.5	20.0	19.1	17.8	17.0	15.8	21.0	21.5	22.2	24.1	26.8	31.5	∞
8.0	20	26.9	26.3	25.5	24.0	23.0	22.2	21.4	26.9	29.6	30.6	32.9	36.1	40.6	∞
	16	27.4	26.6	25.8	24.4	23.4	22.7	21.9	27.4	29.8	30.8	33.1	36.2	40.7	∞
	12	28.3	27.2	26.4	25.1	24.1	23.4	22.6	28.3	30.0	31.0	33.3	36.4	40.8	∞
	8	29.1	28.2	27.4	26.1	25.1	24.4	23.4	29.1	30.3	31.2	33.8	36.9	41.0	∞
	4	30.8	30.2	29.6	28.0	26.9	25.7	24.5	30.8	31.5	32.8	35.0	38.4	42.0	∞

Source: Youngs (1968). by Williams and Wilkins, MD.

inspection pit, measurements can be made from one year to another in a soil that remains undisturbed at depth, with normal cultivation practices being carried out above.

D. Two-Well Method

The two-well method of Childs (Childs, 1952; Childs et al., 1953, 1957; Smiles and Youngs, 1965) uses two unlined wells sunk to the same depth below the water table, as illustrated in Fig. 8. Water is pumped at a constant rate from one well into the other, thus depressing the level in one and raising it in the other. When a steady state ensues, the hydraulic conductivity of the soil is given by

$$K = \frac{Q}{\pi \, \Delta H(L + L_{\mathrm{f}})} \cosh^{-1}\left(\frac{b}{2r}\right) \tag{24}$$

where Q is the steady flow rate, L the length of the wells below the water table, L_{f} an end correction to be added to take into account flow in the capillary fringe together with the flow beneath the wells if they do not reach to an impermeable floor, b the distance between centers of the wells, r the radius of the wells, and ΔH the difference in water level in the two wells. The hydraulic conductivity profile may be obtained when there is a soil variation with depth by making measurements on wells sunk successively deeper. Alternatively, the seepage analysis of

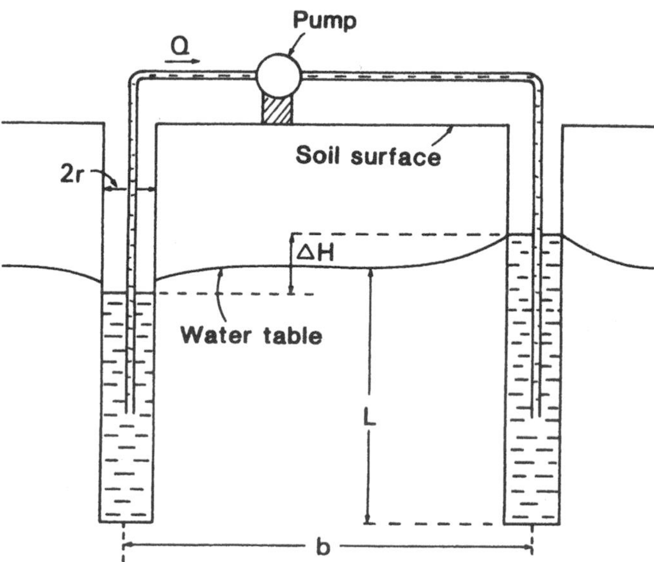

Fig. 8 Geometry of the two-well method.

Youngs (1965, 1980) can be used to measure this variation with depth by making measurements using a range of drawdowns in the pumped well.

Childs' two-well method may be extended to a radial symmetrical array of wells (Smiles and Youngs, 1963), alternate ones discharging and receiving the same rate of flow. The formula for obtaining K for this case is

$$K = \frac{2Q}{n\pi \, \Delta H(L + L_{\mathrm{f}})} \ln \left(\frac{4a}{nr} \right) \tag{25}$$

where n is the even number of wells of radius r, arranged symmetrically on the circumference of a circle of radius a and sunk to a depth L below the water table, L_{f} is an end correction as in the two-well method, and Q is now the total rate of water being pumped from the wells in the system when there is a head difference of ΔH between the levels of water in the pumped and receiving wells.

In uniform soils the depression of the water level in the pumped well is equal to the elevation in the receiving well. However, in field soils this is rarely found to be the case because of soil variation. Some indication of the variability of the soil is given by the differences between the elevations and depressions in the wells (Childs et al., 1957; Smiles and Youngs, 1963).

A modification of the two-well method (Kirkham, 1955) employs two inspection wells symmetrically installed between the two wells to measure the heads in the flow system at these locations. This arrangement overcomes difficulties associated with clogging of pores in the return well. The formula for calculating K is

$$K = \frac{BQ}{\Delta H \, L} \tag{26}$$

where B is a factor, given by a set of graphs, that depends on the geometry of the system, and ΔH is now the difference in level in the two inspection wells (Snell and van Schilfgaarde, 1964).

The flow produced in the unlined two-well and multiple-well methods is mainly horizontal, so that values obtained with these methods in anisotropic soils approximate to the horizontal component of the hydraulic conductivity. The methods can be used in conjunction with Kirkham's piezometer method at the same site to obtain both the vertical and horizontal components of hydraulic conductivity (Childs, 1952).

E. Pumped Wells

Pumped wells discharging at a constant rate are used extensively to measure aquifer characteristics for groundwater supplies. They may be employed to determine the hydraulic conductivity of the soil by measuring the drawdown of the water table at some distance from the pumped wells as a function of time. The transmissivity T, which is the product of the hydraulic conductivity and the depth of the aquifer, is given by Theis' (1935) formula

$$z = -\frac{Q}{4\pi T}\mathrm{Ei}\left(-\frac{r^2S}{4Tt}\right) \tag{27}$$

where z is the drawdown at time t at a radial distance r from the well pumped at a constant rate Q, and S is the storage coefficient of the aquifer. Ei is the exponential integral of the expression within brackets (Reeve and Luthin, 1957; Abramowitz and Stegun, 1972). T and S are found by making a log–log plot of the experimental results of z and r^2/t, and overlaying it on top of a plot of the function Ei(x) against x on identical scales, matching experimental points with the curve while keeping the axes on each plot parallel. Values of $Q/(4\pi T)$ and $4T/S$ are the values of the coordinates z and r^2/t, respectively, which superimpose values of 1.0 on the type curve. Some difficulties in matching may arise because of delayed yield with the value of S varying with the time of pumping.

F. Land Drains Used as Permeameters

Drainage equations that give the relationship between water table height and drain discharge for a particular drainage installation provide a means whereby land drains can be used as large permeameters to give equivalent hydraulic conductivity values of soils for the flows to the drains. Land-drainage theory (van Schilfgaarde et al., 1957; Youngs, 1983c) shows that for steady-state conditions with parallel drain lines, drainage equations are of the form

$$\frac{q}{K} = \mathrm{f}\left(\frac{H_\mathrm{m}}{D}\right) \tag{28}$$

where q is the flux through the water table derived from a uniform steady rainfall on the soil surface and hence given by the drain discharge rate per unit area of drained land, and $\mathrm{f}(H_\mathrm{m}/D)$ is a function of the ratio of the maximum water table height H_m midway between the drains to the half-drain spacing D (Fig. 9). The hydraulic conductivity K is thus given by:

$$K = \frac{q}{\mathrm{f}(H_\mathrm{m}/D)} \tag{29}$$

so that from measurements of q, H_m, and D, and knowing the form of $\mathrm{f}(H_\mathrm{m}/D)$, K can be determined.

The difficulty in using this method of determining values of hydraulic conductivity from measurements on drained lands is in making a correct choice of drainage equation from the many available. These equations involve physical and mathematical assumptions in their derivation, and Lovell and Youngs (1984)

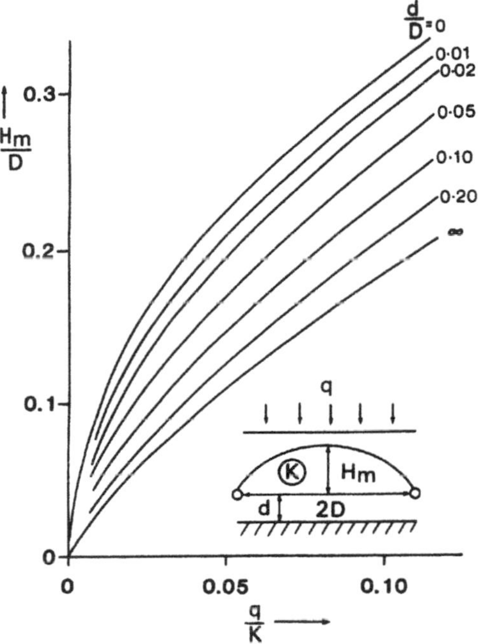

Fig. 9 Water flow to land drains: relationship between the maximum water table height H_m and the uniform rainfall rate q for various depths to the impermeable floor d, shown as plots of H_m/D against q/K for different values of d/D.

showed, in comparing ten commonly used equations, that these assumptions lead often to large errors. However, one empirical equation that approximates well to the correct relationship when the drain is larger than the optimum size, and so does not affect the water table height H_m midway between drains, is the power-law relationship

$$\frac{q}{K} = \left(\frac{H_m}{D}\right)^{\alpha} \tag{30}$$

where $\alpha = 2(d/D)^{d/D}$ for $0 < d/D < 0.35$ and $\alpha = 1.36$ for $d/D > 0.35$, and where d is the depth of an impermeable layer below the drains (Youngs, 1985a).

Equation 30 is particularly useful in analyzing drain hydrographs in moving water table situations and has been used to predict water table drawdowns (Youngs, 1985a). However, this involves the specific yield, a knowledge of which is therefore required in order to obtain hydraulic conductivity values from water table recessions in drained land. Nevertheless, while it may not be possible to estimate hydraulic conductivity values directly from these drain hydrographs if the specific yield is not known, a drain installation's characteristics, once determined from a recession, allows future drain performances to be predicted without the need of actual hydraulic conductivity values and instead using a parameter that involves the drain spacing and the soil's specific yield as well as the hydraulic conductivity (Youngs, 1985b).

The drainage inequality obtained from seepage analysis (Youngs, 1965; 1980) can be used to interpret field results of drainage performance in terms of the depth-dependent hydraulic conductivity (Youngs, 1976). For parallel drains that lie on top of an impermeable layer, the depth-dependent hydraulic conductivity $K(z)$ is given approximately by

$$K(z) = A \frac{d^2 q}{dH_{\mathrm{m}}^2} \tag{31}$$

at $z = H_{\mathrm{m}}$, where the factor A depends on the shape and dimensions of the drainage installation and for parallel ditch drains with ditches dug to an impermeable base, equals $D^2/2$. Thus the dependence of hydraulic conductivity with depth can be obtained by determining the relationship between the water table height and drain discharge on a given drainage installation. However, the precision of $K(z)$ is poor because of the second differential in Eq. 31.

V. FIELD MEASUREMENTS IN THE ABSENCE OF A WATER TABLE

A. General Principles

Values of hydraulic conductivity of saturated soils are sometimes required when there is no water table at the time of measurement, in order to plan and design works for the future when the groundwater level is expected to rise. Techniques have been developed that allow measurements to be made in such circumstances. These measure the water uptake by the unsaturated soil from a saturated surface as in laboratory infiltration methods (see Secs. III.F and III.G) and so rely for their interpretation on infiltration theory. The measured flow depends not only on the hydraulic conductivity of the saturated soil but also on the capillary absorptive properties of the unsaturated soil, represented by the negative soil water pressure head at the wetting front as in the Green and Ampt (1911) analysis of infiltration or by the sorptivity in more exact analyses of the infiltration process (Philip, 1957). Hydraulic conductivity values are often obtained from formulae derived using theory with assumed hydraulic conductivity functions, so that their reliability is sometimes difficult to establish.

In the wetting-up process, entrapped bubbles of air may be left behind the advancing wetting front, so that the soil is not completely saturated and there is a reduction of pore space for water conduction. Values of hydraulic conductivity obtained using infiltration methods have been found to be smaller than those made with techniques that involve measurements below a water table, typically by as much as 50% (Youngs, 1972). Caution should be exercised therefore in using values obtained in this way for computing groundwater flows.

B. Borehole Permeameter

One of the oldest techniques for measuring the hydraulic conductivity of soils in the absence of a water table is the borehole permeameter, which uses water seeping into the soil from a vertical cylindrical hole made in the unsaturated soil to the depth at which the measurement is required. Hydraulic conductivity values of the saturated soil are obtained from the steady-state seepage from the borehole that occurs after some time when the depth of water in the hole is maintained at some constant level, often using a Mariotte bottle arrangement (see Fig. 10) (Talsma

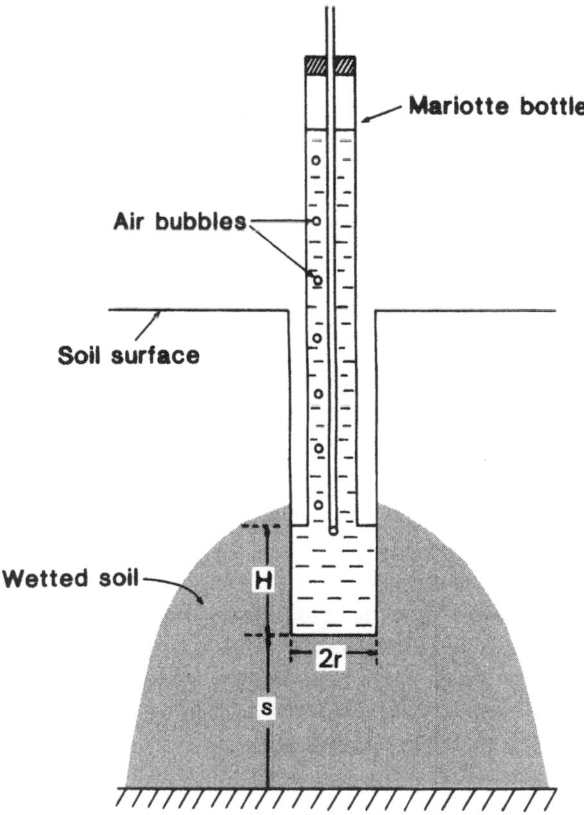

Fig. 10 Borehole permeameter.

and Hallam, 1980; Reynolds et al., 1983; Nash et al., 1986). The hydraulic conductivity is calculated from formulae, cited in many reviews of the method (see, for example, that by Stephens and Neuman, 1982), that have been derived from an approximate consideration of the physical situation.

For deep water tables Glover's (1953) formula is commonly used, giving K in the form

$$K = \frac{CQ}{2\pi H^2} \tag{32}$$

with $C = \sinh^{-1}(H/r) - 1$ for $H \gg r$, or more accurately according to Reynolds et al. (1983) by an expression that for $H \gg r$ reduces to

$$C = 2\left[\sinh^{-1}\left(\frac{H}{2r}\right) - 1\right] \tag{33}$$

where Q is the steady seepage rate, H the depth of water in the borehole, and r the radius of the borehole.

When an impermeable layer is at a relatively small depth s below the borehole ($s < 2H$), K is given by (Jones, 1951; Bouwer and Jackson, 1974)

$$K = \frac{3Q}{\pi H(3H + 2s)} \ln\left(\frac{H}{r}\right) \tag{34}$$

These formulae overestimate values of hydraulic conductivity (Reynolds and Elrick, 1985); better values can be obtained using an extension of theory that takes into account the effect of flow in the unsaturated soil (Reynolds et al., 1985). Although the borehole method has been considered to have great potential for field measurements (Reynolds et al., 1983), some doubt has been expressed (Philip, 1985) concerning the utility of the method because of the difficulties in the theoretical interpretation of the field data. Nevertheless, the method has been used in the Guelph Permeameter* (Reynolds and Elrick, 1985) and in Amoozegar's (1989) compact constant head permeameter.

C. Auger-Hole Method

A simple borehole method uses an auger hole made to a given depth in the soil in the absence of a water table (Kessler and Oosterbaan, 1974); it is sometimes referred to incongruously as the "inversed" auger-hole method. Water is added to fill the hole to a given level, and then the fall of the water level is observed with time. The hydraulic conductivity is given approximately by

---

* The Guelph Permeameter is sold by ELE International Ltd., Eastman Way, Hemel Hempstead, Hertfordshire, HP2 7HB, U.K.; cost ca. $2,500.

$$K = \frac{r}{2(t - t_0)} \ln \left(\frac{1 + 2H_0/r}{1 + 2H/r} \right) \tag{35}$$

where H_0 and H are the depths of water in the hole at time t_0, when measurements are begun, and time t, respectively, and r is the radius of the hole.

In the derivation of Eq. 35, a unit hydraulic head gradient is assumed for the flow through the bottom and side of the hole. Because of this crude assumption, the use of the method can only be expected to give a very approximate indication of the actual hydraulic conductivity value.

D. Air-Entry Permeameter

With the air-entry permeameter (Bouwer, 1966; Bouwer and Jackson, 1974) a column of soil is contained within an infiltration cylinder driven into the soil. Water under a pressure head is infiltrated into the soil, and the rate is measured after the wetting front has penetrated some distance down the isolated column of soil. The hydraulic conductivity is determined using the Green and Ampt (1911) analysis. This method and its limitations are described in Chapter 6.

E. Ring Infiltrometer Method

Since the infiltration capacity (that is, the steady infiltration rate that is approached at large times when water infiltrates over the whole land surface) is identified with the hydraulic conductivity of the saturated soil, infiltration measurements into dry soil provide a means of obtaining hydraulic conductivity values. Such measurements are usually made using infiltration rings.

As discussed in Chapter 6, flow from a surface pond, as presented by an infiltration ring, has a lateral component of flow due to capillarity. The flow approaches a steady rate after some time, and for infiltration from a circular pond into a deep uniform soil this rate is described by Wooding's (1968) formula that can be written (White et al., 1992) as

$$\frac{Q}{\pi R^2} = K \left(1 + \frac{4bS^2}{\pi RK\,\Delta\theta} \right) \tag{36}$$

where Q is the steady flow rate that is approached after long time, R the radius of the ring, S the sorptivity of the soil, $\Delta\theta$ the difference between the saturated and initial soil water contents, and b a parameter that depends on the shape of the soil water diffusivity function. b is in the range $0.5 < b < \pi/4$, and a "typical" value of a soil is 0.55. Alternatively, the Wooding equation can be put in the form (Youngs, 1991)

$$\frac{Q}{\pi R^2} = K \left(1 + \frac{4h_f}{\pi R} \right) \tag{37}$$

where $-h_f$ is the soil water pressure head at the wetting front as in the Green and Ampt analysis of infiltration. The steady rate is approached quickly, more so as the radius of the ring becomes smaller (Youngs, 1987). It follows therefore that the use of small rings, for which the steady rate occurs when wetting of soil has occurred only to a small depth, allows the hydraulic conductivity of soils very close to the surface to be estimated.

In practice the rings have to be pressed into the soil to give a seal against leaks around the edge when a small head of water is maintained on the soil surface within the ring. Alternatively, earth bunds can be formed to seal round large infiltration areas. The cumulative infiltration is measured with time, usually by observing the time the ponded water on the surface takes to fall a small distance when a measured amount of water is applied to bring the height back to its original height. The steady rate, from which the hydraulic conductivity is obtained, can take less than an hour for a small ring on sandy soil or many days in the case of a large area on a compacted clay soil.

There are several ways of obtaining the hydraulic conductivity from the infiltration data. The type curve shown in Fig. 11 may be used (Youngs, 1972). This shows a log–log plot of $Q/(\pi KR^2)$ against R/h_f, where Q is the steady rate of water infiltrating into the soil after large times, R is the radius of the ring, and h_f is the negative pressure head at the wetting front of the saturated zone that is assumed to advance into the soil. By obtaining values of $Q/(\pi R^2)$ with rings of different radii R, and plotting these against one another on identical log–log scales to those used for the type curve of $Q/(\pi KR^2)$ plotted against R/h_f, the data can be superimposed on top of the type curve. Values of K and h_f are the values of the coordinates $Q/\pi R^2$ and R, respectively, that superimpose values of 1.0 on the type curve when they are matched.

Alternatively, the hydraulic conductivity can be obtained from infiltrometer results at early times using the semiempirical equation (Youngs, 1987)

$$K = \frac{\rho g \eta R^4 (\Delta\theta)^2}{\sigma^2 t^2} \left[-0.365 + \sqrt{0.133 + \frac{I}{R^3 \Delta\theta}} \right]^4 \tag{38}$$

where I is the total volume of infiltration up to time t, R the radius of the infiltration ring, $\Delta\theta$ the difference between the saturated and initial water contents of the soil, g the acceleration due to gravity, and ρ, η, and σ the density, viscosity, and surface tension, respectively, of water. Equation 38 was obtained by curve fitting laboratory experimental results, scaled according to similar media theory (Miller and Miller, 1956), incorporating a microscopic characteristic length defined in terms of the hydraulic conductivity of the porous material. This equation can only be used during the early stage of the infiltration when $I < R^3\,\Delta\theta$. If the unit of length is the meter and the unit of time is the day, $\rho g \eta/\sigma^2 = 0.0216\ \mathrm{m}^{-3}\ \mathrm{d}$ to give the units of K in m d^{-1}.

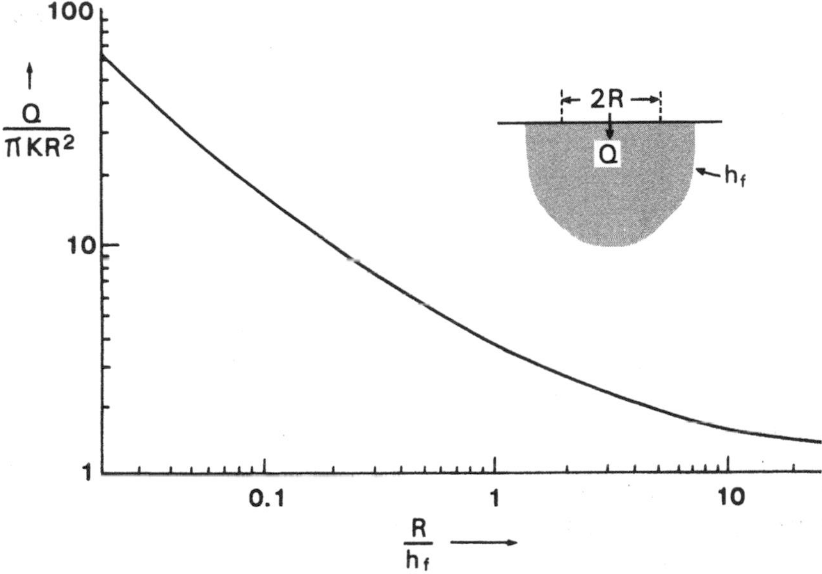

Fig. 11 Type curve of $Q/\pi R^2 K$ against R/h_f for steady flow from infiltrometer rings.

Another way of interpreting the steady-state infiltrometer rate is to determine the sorptivity from the infiltration results at the beginning of the test when

$$S = \lim_{t \to 0} \left[\frac{dI}{d\sqrt{t}} \right] \tag{39}$$

and using Eq. 36 to obtain the hydraulic conductivity value when a steady state infiltration rate occurs.

As noted earlier, the infiltrometer method can give results that can be analyzed after only a short time of infiltration, allowing hydraulic conductivity values to be measured near the soil surface. It thus provides a means of monitoring structural changes of the soil. The method is very sensitive to worm and root holes as well as structural fissures (Bouwer, 1966; Youngs, 1983a), and care must be taken to use rings large enough to sample a representative area.

In order to overcome the complications of taking into account the lateral flow component in analyzing infiltrometer results, two concentric rings can be used and measurements of flow made only on the inner ring where it is considered that the flow is mainly vertical and hence the steady rate after a long time is the hydraulic conductivity.

The determination of hydraulic conductivity values using infiltrometers depends on measurements being taken with infiltration taking place with the wetting front advancing into uniform soil at a uniform water content. Variations with depth of both the soil and water content affect the infiltration process, and care must be taken in analyzing results. This was demonstrated in tests on a silt loam soil overlying a very permeable terrace under an artesian head (Youngs et al., 1996). After an initial steady state infiltration period into uniform unsaturated soil, the infiltration rate abruptly changed to a lower rate when the advancing wetting front met the capillary fringe.

F. Dripper Method

An alternative to using an infiltration ring is to supply water from an irrigation dripper at known rates and observe the ultimate extent of the surface ponding (Shani et al., 1987) at several measured rates. With water supplied as a point source on the surface, the circular ponded area increases during the early stages of infiltration but approaches a constant maximum radius after some time. Then it is supposed that the infiltration proceeds in the same way as for infiltration from a ponded ring after a long time, so Wooding's equation can be applied. Thus, if measurements of the maximum wetted radius R_{max} are made for a range of dripper rates Q, from Eq. 36 or 37 the hydraulic conductivity is the intercept on the $Q/\pi R^2_{max}$ axis of a plot of $Q/\pi R^2_{max}$ against $1/R_{max}$.

G. Sorptivity Measurement Method

The measurement of the steady state infiltration rate from small surface sources at pressure heads less than atmospheric that maintain the soil surface saturated although under tension, can be used to obtain values of the hydraulic conductivity of small volumes of soil material, such as that of soil aggregates (Leeds-Harrison and Youngs, 1997). With the hydraulic conductivity equal to that of the saturated soil over a range of negative soil water pressure heads, the steady state infiltration rate Q given by Eq. 36 at a pressure head p can be shown to be given by

$$Q = \frac{4bRS^2}{\Delta\theta} + 4RKp \tag{40}$$

for a small circular infiltration area of radius R. Thus by measuring Q over a range of p, K can be found. In the apparatus described, contact with the soil surface was obtained through the use of a small sponge and the water uptake measured using the observations on the meniscus in a small capillary tube that supplied the infiltration water.

H. Pressure Infiltrometer

The pressure infiltrometer was developed especially for the measurement of the hydraulic conductivity of low permeability soils (Fallow et al., 1993; Youngs et al., 1995). It employs a stainless steel ring that is driven into the soil to a depth of about one radius. Water is supplied to the soil surface at a head through the sealed top lid from a small capillary tube that also acts as a measuring device. The ring has to be anchored or weighted down because of the upthrust on the sealed lid. The steady state flow Q that occurs after a relatively short time with a head H is given by

$$Q = \pi R^2 K + \frac{R}{G}(KH + \phi_{\mathrm{m}}) \tag{41}$$

where ϕ_{m} is the matric flux potential and G is a factor depending on the depth d of penetration of the ring, given by

$$G = 0.316\frac{d}{R} + 0.184 \tag{42}$$

When used on very wet soils, as is often the case, the situation is analogous to that of the piezometer method of measuring the hydraulic conductivity in the presence of a water table. Youngs et al. (1995) provided shape factors to be used in this situation.

I. Bouwer's Double Ring Method

The Bouwer's (1961) double ring method is an infiltration method performed at the bottom of an auger hole. The rates of flow in a central ring and in a peripheral ring are measured when the heads feeding the water in each section are maintained at the same height and also when no water is fed to maintain the head of the central ring so this head falls. A flow of water is thus induced between the inner and outer rings. The hydraulic conductivity is obtained from sets of graphs that have been obtained with an electric analog. The method is sensitive to the hydraulic conductivity of the soil in the vicinity of the inner ring, where soil disturbance is likely to occur during installation, and thus results may not give the soil's undisturbed hydraulic conductivity.

VI. SUMMARY AND DISCUSSION

Hydraulic conductivity measurements are needed for various purposes. Methods used generally depend on the application. For example, the auger-hole method is used commonly in land-drainage investigations (Bouwer and Jackson, 1974), while pumping tests are used as the standard for aquifer investigations in water resource engineering (Kruseman and de Ridder, 1990); other special techniques are required for investigating the low-permeability compacted clay soils used for lining landfill sites (Daniel, 1989). This chapter, while attempting to provide an

Table 5 Summary of Methods for Measuring the Hydraulic Conductivity of Saturated Soils

Method	Comments
Constant head permeameter (LS)	Used on small soil cores and packed soil columns. (SE)
Falling head permeameter (LS)	Used on small soil cores and packed soil columns. (SE)
Oscillating permeameter (LS)	Used on small soil cores and packed soil columns. Only small quantity of added water needed. (SA)
Infiltration method (LU)	Used on long uniform soil columns.(SE)
Varying moment permeameter (LU)	Used on short uniform soil columns. (SA)
Auger-hole method (FW)	Samples soil over depth of hole below water table. (SE)
Piezometer method (FW)	Samples soil in vicinity of open base. (SE)
Two-well method (FW)	Samples soil between wells. (SE)
Pumped wells (FW)	Used in aquifer tests at depth. Well boring equipment required.
Land drains (FW)	Samples soil between drain lines. (SE)
Borehole permemeater (FA)	Samples soil in vicinity of wetted surface. (SE)
"Inversed" auger hole method (FA)	Samples soil in vicinity of wetted surface. (SE)
Air-entry permeameter (FA)	Samples soil within isolated tube. (SA)
Ring infiltrometer method (FA)	Samples soil near soil surface. (SE)
Dripper method (FA)	Samples soil near soil surface. (SE)
Sorptivity method (LU/FA)	Samples small volumes. (SA)
Pressure infiltrometer method (FW/FA)	Used on low permebility soils. (SA)
Double ring infiltrometer method (FA)	Samples soil near soil surface. (SE)

LS = laboratory method on saturated soil; LU = laboratory method on unsaturated soil; FW = field method below water table; FA = field method in the absence of a water table; SE = simple equipment usually found in the soil laboratory or easily fabricated. Field methods usually require soil augers; SA = special apparatus requiring workshop facilities for assembly.

overview of techniques, has concentrated on those methods that are used in determining the hydraulic conductivity near the soil surface, which is the concern of soil scientists and soil hydrologists. These are summarized in Table 5. Many methods require simple equipment that is readily available or easily constructed in most soil laboratories. Some methods, however, require special apparatus that has to be constructed in a workshop or purchased from specialist manufacturers.

Implicit in making measurements of hydraulic conductivity and their use in calculating water flow in soils is that Darcy's law describes the flow of water both in the soil sample used in the measurement and in the flow region as a whole. Thus it is assumed that the soil is "uniform" and that the same "uniformity" is "seen" in the measurement as in the soil region at large. A hydraulic conductivity measurement must therefore use a flow region at least the size of a representative volume of the soil. Techniques should allow, if possible, an assessment of any spatial variability by replicating measurements, preferably with different flow geometries at different scales. In all cases, in selecting the method and considering the size of sample, attention has to be paid to any natural macropore development (Bouma, 1983) and the possibility of heterogeneity.

REFERENCES

Abramowitz, M., and Stegun, L. A. 1972. *Handbook of Mathematical Functions*. Applied Mathematics Series No. 55. Washington, DC: National Bureau of Standards.

Amoozegar, A., and Warrick, A. W. 1986. Hydraulic conductivity of saturated soils: field methods. In: *Methods of Soil Analysis. Part 1. Physical and Mineralogical Methods* (A. Klute, ed.) Madison, WI: Am. Soc. Agron., pp. 735–770.

Amoozegar, A. 1989. A compact constant-head permeameter for measuring saturated hydraulic conductivity of the vadose zone. *Soil. Sci. Soc. Am. J.* 53:1356–1361.

Belford, R. K. 1979. Collection and evaluation of large soil monoliths for soil and crop studies. *J. Soil Sci.* 30:363–373.

Boast, C. W., and Kirkham, D. 1971. Auger hole seepage theory. *Soil Sci. Soc. Am. Proc.* 35:365–374.

Boast, C. W., and Langebartel, R. G. 1984. Shape factors for seepage into pits. *Soil Sci. Soc. Am. J.* 48:10–15.

Bouma, J. 1977. *Soil Survey and the Study of Water in Unsaturated Soil.* Soil Survey Papers, No. 13. Wageningen: Netherlands Soil Survey Institute, pp. 1–107.

Bouma, J., and Dekker, L. W. 1981. A method for measuring the vertical and horizontal K_{sat} of clay soil with macropores. *Soil Sci. Soc. Am. J.* 45:662–663.

Bouma, J., van Hoorn, J. H., and Stoffelsen, G. H. 1982. Measuring the hydraulic conductivity of soil adjacent to tile drains in a heavy clay soil in the Netherlands. *J. Hydrol.* 50:371–381.

Bouma, J. 1983. Use of soil survey data to select measurement techniques for hydraulic conductivity. *Agric. Water Manage.* 6:177–190.

Bouwer, H. 1961. A double tube method for measuring hydraulic conductivity of soil in situ above a water table. *Soil Sci. Soc. Am. Proc.* 25:334–339.

Bouwer, H. 1966. Rapid field measurement of air-entry value and hydraulic conductivity of soil as significant parameters in flow system analysis. *Water Resour. Res.* 2:729–738.

Bouwer, H., and Jackson, R. D. 1974. Determining soil properties. In: *Drainage for Agriculture* (J. van Schilfgaarde, ed.). Madison, WI: Am. Soc. Agron., pp. 611–672.

Childs, E. C. 1952. The measurement of the hydraulic permeability of saturated soil in situ. I. Principles of a proposed method. *Proc. Roy. Soc. London* A215:525–535.

Childs, E. C. 1967. Soil moisture theory. *Adv. Hydrosci.* 4:73–117.

Childs, E. C. 1969. *An Introduction to the Physical Basis of Soil Water Phenomena.* London: John Wiley.

Childs, E. C., and Bybordi, M. 1969. The vertical movement of water in stratified porous material. 1. Infiltration. *Water Resour. Res.* 5:446–459.

Childs, E. C., Cole, A. H., and Edwards, D. H. 1953. The measurement of the hydraulic permeability of saturated soil in situ. II. *Proc. Roy. Soc. London* A216:72–89.

Childs, E. C., Collis-George, N., and Holmes, J. W. 1957. Permeability measurements in the field as an assessment of anisotropy and structure development. *J. Soil Sci.* 8: 27–41.

Childs, E. C., and Poulovassilis, A. 1960. An oscillating permeameter. *Soil Sci.* 90: 326–328.

Daniel, D. E. 1989. In situ hydraulic conductivity tests in compacted clay. *J. Geotech. Eng.* 115:1205–1226.

Darcy, H. 1856. *Les Fontaines publiques de la ville de Dijon.* Paris: Dalmont.

Diserens, E. 1934. Beitrag zur Bestimmung der Durchlässigkeit des Bodens in natürlicher Bodenlagerung. *Schweiz. Landw. Monatsh.* 12:188–198, 204–212.

Ernst, L. F. 1950. *Een nieuwe formule voor de berekening van de doorlaatfactor met de boorgatenmethode.* Groningen, The Netherlands: Rap. Landbouwproefsta. en Bodemkundig Inst. T.N.O.

Fallow, D. J., Elrick, D. E., Reynolds, W. D., Baumgartner, N., and Parkin, G. W. 1993. Field measurement of hydraulic conductivity in slowly permeable materials using early-time infiltration measurements in unsaturated soils. In: *Hydraulic Conductivity and Water Contaminant Transport in Soils* (D. E. Daniel and S. J. Trautwein, eds.). Am. Soc. for Testing Materials ASTM SJP 1142, pp. 375–389.

Frevert, R. K., and Kirkham, D. 1948. A field method for measuring the permeability of soil below a water table. *Proc. Highw. Res. Board* 28:433–442.

Glover, R. E. 1953. Flow from a test-hole located above groundwater level. In: *Theory and Problems of Water Percolation* (C. N. Zangar, ed.). Eng. Monogr. No. 8. Washington, DC: US Bur. Reclam., pp. 69–71.

Goss, M. J., and Youngs, E. G. 1983. The use of horizontal piezometers for in situ measurements of hydraulic conductivity below the water table. *J. Soil Sci.* 34:659–664.

Green, G. H., and Ampt, G. A. 1911. Studies on soil physics. 1. Flow of air and water through soils. *J. Agric. Sci. Camb.* 4:1–24.

Hoffman, G. J., and Schwab, G. O. 1964. The spacing prediction based on drain outflow. *Am. Soc. Agrc. Eng. Trans.* 7:444–447.

Hooghoudt, S. B. 1936. Bijdragen tot de kennis van eenige natuurkunige grootheden van den grond: 4. *Versl. Landb. Ond.* 42(13)B:449–541.

Jones, C. W. 1951. *Comparison of Seepage Based on Well Permeameter and Ponding Tests, Earth Materials.* Lab. Rep., No. EM-264. Denver, CO: US Bur. Reclam.

Kessler, J., and Oosterbaan, R. J. 1974. Determining hydraulic conductivity of soils. *Drainage principles and applications: III. Surveys and investigations.* Publ. 16. Wageningen, The Netherlands: ILRI, pp. 253–296.

Kirkham, D. 1955. Measurement of the hydraulic conductivity of soil in place. In: *Proc. Symp. on Permeability of Soils.* ASTM Spec. Tech. Publ. No. 163, pp. 80–97.

Kirkham, D., and van Bavel, C. H. M. 1949. Theory of seepage into auger holes. *Soil Sci. Soc. Am. Proc.* 13:75–82.

Kruseman, G. P., and de Ridder, N. A. 1990. *Analysis and Evaluation of Pumping Test Data.* 2d ed. Publ. 47. Wageningen, The Netherlands: ILRI.

Leeds-Harrison, P. B., and Shipway, C. J. P. 1985. Variations in hydraulic conductivity under different wetting regimes. In: *Proc. I.S.S.S. Symp. on Water and Solute Movement in Heavy Clay Soils* (J. Bouma and P. A. C. Raats, eds.). Publ. 37. Wageningen, The Netherlands: ILRA, pp. 67–70.

Leeds-Harrison, P. B., Shipway, C. J. P., Jarvis, N. J., and Youngs, E. G. 1986. The influence of soil macroporosity on water retention, transmission and drainage in a clay soil. *Soil Use Manage.* 2:47–50.

Leeds-Harrison, P. B., and Youngs, E. G. 1997. Estimating the hydraulic conductivity of soil aggregates conditioned by different tillage treatments from sorption measurements. *Soil Tillage Res.* 41:141–147.

Lovell, C. J., and Youngs, E. G. 1984. A comparison of steady-state land-drainage equations. *Agric. Water Manage.* 9:1–21.

Maasland, M. 1957. Soil anisotropy and land drainage. *Drainage of Agricultural Lands* (J. N. Luthin, ed.). Madison, WI: Am. Soc. Agron., pp. 216–285.

McNeal, L., and Roland, C. 1964. Elimination of boundary-flow errors in laboratory hydraulic conductivity measurements. *Soil Sci. Soc. Am. Proc.* 28:713–714 .

Miller, E. E., and Miller, R. D. 1956. Physical theory for capillary flow phenomena. *J. Appl. Phys.* 27:324–332.

Muskat, M. 1937 *The Flow of Homogeneous Fluids Through Porous Media.* New York: McGraw-Hill.

Nash, D. M., Willatt, S. T., and Uren, N. C. 1986. The Talsma-Hallam well permeameter—Modifications. *Aust. J. Soil Res.* 24:317–320.

Nimmo, J. R., and Mello, K. A. 1991. Centrifugal techniques for measuring saturated hydraulic conductivity. *Water Resour. Res.* 27:1263–1269.

Philip, J. R. 1957. The theory of infiltration:4. Sorptivity and algebraic infiltration equations. *Soil Sci.* 84:257–264.

Philip, J. R. 1985. Approximate analysis of the borehole permeameter in unsaturated soil. *Water Resour. Res.* 21:1025–1033.

Ploeg, R. R., and van der Huwe, B. 1988. Einige Bemerkungen zur Bestimmung der Wasserleitfähigkeit mit der Bohrlochmethode, *Z. Pflanz. Boden.* 151:251–253.

Reeve, R. C., and Luthin, J. N. 1957. Drainage investigation methods: I. Methods of measuring soil permeability. In: *Drainage of Agricultural Lands* (J. N. Luthin, ed.). Madison, WI: Am. Soc. Agron., pp. 395–413.

Reynolds, W. D., and Elrick, D. E. 1985. In situ measurement of field-saturated hydraulic conductivity, sorptivity, and the A-parameter using the Guelph permeameter. *Soil Sci.* 140:292–302.

Reynolds, W. D., Elrick, D. E., and Clothier, B. E. 1985. The constant head well permeameter: Effect of unsaturated flow. *Soil Sci.* 139:172–180.

Reynolds, W. D., Elrick, D. E., and Topp, G. C. 1983. A re-examination of the constant head well permeameter method for measuring saturated hydraulic conductivity above the water table. *Soil Sci.* 136:250–268.

Shani, U., Hanks, R. J., Bresler, E., and Oliveira, C. A. S. 1987. Field method for estimating hydraulic conductivity and matric potential—Water content relations. *Soil Sci. Soc. Am. J.* 51:298–302.

Slichter, C. S. 1899. *Theoretical Investigation of the Motion of Ground Waters.* U.S. Geol. Surv. 19th. Ann. Rep. Part 2, pp. 295–384.

Smiles, D. E., and Youngs, E. G. 1963. A multiple-well method for determining the hydraulic conductivity of a soil in situ. *J. Hydrol.* 1:279–287.

Smiles, D. E., and Youngs, E. G. 1965. Hydraulic conductivity determinations by several field methods in a sand tank. *Soil Sci.* 99:83–87.

Snell, A. W., and van Schilfgaarde, J. 1964. Four-well method of measuring hydraulic conductivity in saturated soils. *Am. Soc. Agric. Eng. Trans.* 7:83–87, 91.

Stephens, D. B., and Neuman, S. P. 1982. *Vadose Zone Permeability Tests.* Hydrol. Div. ASCE, 108 HY5), Proc. Paper 17058, 623–639.

Talsma, T., and Hallam, P. M. 1980. Hydraulic conductivity measurement of forest catchments. *Aust. J. Soil Res.* 30:139–148.

Theis, C. V. 1935. The relation between the lowering of the piezometric surface and the rate of duration of discharge of a well using ground water storage. *Trans. Am. Geophys. Union* 16:519–524.

Topp, G. C., and Binns, M. R. 1976. Field measurement of hydraulic conductivity with a modified air-entry permeameter. *Can. J. Soil Sci.* 56:139–147.

van Schilfgaarde, J., Engelund, F., Kirkham, D., Peterson, D. F., and Maasland, M. 1957. Theory of land drainage. In: *Drainage of Agricultural Lands* (J. N. Luthin, ed.). Madison, WI: Am. Soc. Agron., pp. 79–285.

White, I., Sully, M. J., and Perroux, K. M. 1992. Measurement of surface-soil hydraulic properties: Disk permeameters, tension infiltrometers, and other techniques. In: *Advances in Measurement of Soil Physical Properties: Bringing Theory into Practice* (G. C. Topp, W. D. Reynolds, and R. E. Green, eds.). Madison, WI: SSSA Special Publ. No. 30, Soil Sci. Soc. Am., pp. 69–103.

Wooding, R. A. 1968. Steady infiltration from shallow circular ponds.*Water Resour. Res.* 4:1259–1273.

Youngs, E. G. 1965. Horizontal seepage through unconfined aquifers with hydraulic conductivity varying with depth. *J. Hydrol.* 3:283–296.

Youngs, E. G. 1968a. An estimation of sorptivity for infiltration studies from moisture moment considerations. *Soil Sci.* 106:157–163.

Youngs, E. G. 1968b. Shape factors for Kirkham's piezometer method for soils overlying an impermeable floor or infinitely permeable stratum. *Soil Sci.* 106:235–237.

Youngs, E. G. 1972. Two- and three-dimensional infiltration: Seepage from irrigation channels and infiltrometer rings. *J. Hydrol.* 15:301–315.

Youngs, E. G. 1976. Determination of the variation of hydraulic conductivity with depth in drained lands and the design of drainage installations. *Agric. Water Manage.* 1: 57–66.

Youngs, E. G. 1980. The analysis of groundwater seepage in heterogeneous aquifers. *Hydrol. Sci. Bull.* 25:155–165.

Youngs, E. G. 1982. The measurement of the variation with depth of the hydraulic conductivity of saturated soil monoliths. *J. Soil Sci.* 33:3–12.

Youngs, E. G. 1983a. Soil physical theory and heterogeneity. *Agric. Water Manage.* 6: 145–159.

Youngs, E. G. 1983b. Soil physics and the water management of spatially variable soils. In: *Proc. FAO/IAEA Symp. on Isotope and Radiation Techniques in Soil Physics and Irrigation Studies.* Aix-en-Provence, pp. 3–22.

Youngs, E. G. 1983c. The contribution of physics to land drainage. *J. Soil Sci.* 34:1–21.

Youngs, E. G. 1985a. A simple drainage equation for predicting water-table drawdowns. *J. Agric. Eng. Res.* 31:321–328.

Youngs, E. G. 1985b. Characterization of hydrograph recessions of land drains. *J. Hydrol.* 82:17–25.

Youngs, E. G. 1986. The analysis of groundwater flows in unconfined aquifers with non-uniform hydraulic conductivity. *Transport in Porous Media.* 1:399–417.

Youngs, E. G. 1987. Estimating hydraulic conductivity values from ring infiltrometer measurements. *J. Soil Sci.* 38:623–632.

Youngs, E. G. 1991. Infiltration measurements—A review. *Hydrol. Processes* 5:309–320.

Youngs, E. G., Leeds-Harrison, P. B., and Elrick, D. E. 1995. The hydraulic conductivity of low permeability wet soils used as landfill lining and capping material: Analysis of pressure infiltrometer measurements. *Soil Technol.* 8:153–160.

Youngs, E. G., and Marei, S. M. 1987. The influence of air access on the water movement down soil profiles with impeding layers. In: *Proc. Int. Conf. on Infiltration Development and Application* (Y.-S. Fok, ed.). Honolulu, pp. 50–58.

Youngs, E. G., Spoor, G., and Goodall, G. R. 1996. Infiltration from surface ponds into soils overlying a very permeable substratum. *J. Hydrol.* 186:327–334.

5
Unsaturated Hydraulic Conductivity

Christiaan Dirksen
Wageningen University, Wageningen, The Netherlands

I. INTRODUCTION

The unsaturated zone plays an important role in the hydrological cycle. It forms the link between surface water and ground water and has a dominant influence on the partitioning of water between them. The hydraulic properties of the unsaturated zone determine how much of the water that arrives at the soil surface will infiltrate into the soil, and how much will run off and may cause floods and erosion. In many areas of the world, most of the water that infiltrates into the ground is transpired by plants or evaporated directly into the atmosphere, leaving only a small proportion to percolate deeper and join the ground water. Surface runoff and deep percolation may carry pollutants with them. Then it is important to know how long it will take for this water to reach surface or ground water resources.

Besides providing water for plants to transpire, the unsaturated zone also provides oxygen and nutrients to plant roots, thus having a dominant influence on food and fiber production. Water content also determines soil strength, which affects anchoring of plants, root penetration, compaction by cattle and machinery, and tillage operations. To mention just one other role of the unsaturated zone, its water content has a great influence on the heat balance at the soil surface. This is well illustrated by the large diurnal temperature variations in deserts.

To understand and describe these and other processes, the hydraulic properties that govern water transport in the soil must be quantified. Of these, the unsaturated hydraulic conductivity is, if not the most important, certainly the most difficult to measure accurately. It varies over many orders of magnitude not only between different soils but also for the same soil as a function of water content. Much has been published on the determination and/or measurement of the unsaturated hydraulic conductivity, including reviews (Klute and Dirksen, 1986; Green et al., 1986; Mualem, 1986a; Kool et al., 1987; Dirksen, 1991; Van Genuchten et al., 1992, 1999). There is no single method that is suitable for all soils and circumstances. Methods that require taking "undisturbed" samples are not well suited for soils with many stones or with a highly developed, loose structure. It is better to select an in situ method for such soils. Hydraulic conductivity for relatively dry conditions cannot be measured in situ when the soil in its natural situation is always wet. It is then necessary to take samples and dry them first. The latter process presents problems if the soil shrinks excessively on drying. These and other factors that influence the choice between laboratory and field methods are discussed separately in Sec. IV.

Selection of the most suitable method for a given set of conditions is a major task. The literature is so extensive that it is neither necessary nor possible to give a complete review and evaluation of all available methods. Instead, I have focused on what I think should be the selection criteria (Sec. III) and described the most familiar types of methods (in Secs. VI to IX) with these criteria in mind. This includes some very recent work. The need for and selection of a standard method is discussed separately in Sec. V. Since some of the methods used to study infiltration are also used to determine unsaturated hydraulic conductivity, reference is made to the appropriate section in Chap. 6 where relevant.

There are two soil water transport functions which, under restricting conditions, can be used instead of hydraulic conductivity, namely hydraulic diffusivity and matric flux potential. Diffusivity can be measured directly in a number of ways that are easier and faster than the methods available for hydraulic conductivity. Moreover, the latter can also be derived from the former. The same is true for yet another transport function, the sorptivity, which can also be measured more easily than the hydraulic conductivity. At the outset I have summarized the theory and transport coefficients used to describe water transport in the unsaturated zone (Sec. II). Theoretical concepts and equations associated with specific methods are given with the discussion of the individual methods. Readers who have little knowledge of the physical principles involved in unsaturated flow and its measurement can find these discussed at a more detailed and elementary level in soil physics textbooks (Hillel, 1980; Koorevaar et al., 1983; Hanks, 1992; Kutilek and Nielsen, 1994) and would be advised to consult one of these before attempting this chapter.

Apparatus for determining unsaturated hydraulic conductivity is not usually commercially available as such. However, many of the methods involve the measurement of water content, hydraulic head and/or the soil water characteristic, and methods and commercial supplies of equipment to determine these properties are given in Chaps. 1, 2, and 3, respectively. Where specialized or specially constructed equipment is required, this is indicated with the discussion of individual methods.

In general, it is difficult if not impossible to measure the soil hydraulic transport functions quickly and/or accurately. Therefore it is not surprising that attempts have been made to derive them indirectly. The derivation of the hydraulic transport properties from other, more easily measured soil properties is discussed in Sec. X, and the inverse approach of parameter optimization in Sec. XI.

II. TRANSPORT COEFFICIENTS

A. Hydraulic Conductivity

In general, water transport in soil occurs as a result of gradients in the hydraulic potential (Koorevaar et al., 1983):

$$H = h + z \tag{1}$$

where H is the hydraulic head, h is the pressure head, and z is the gravitational head or height above a reference level. These symbols are generally reserved for potentials on a weight basis, having the dimension J/N = m. Although h is called a pressure head, in unsaturated flow it will have a negative value with respect to atmospheric pressure and can be referred to as a suction or tension. In rigid soils there exists a relationship between volumetric water content or volume fraction of water, $\theta(\mathrm{m^3\,m^{-3}})$, and pressure head, called the soil water retention characteristic, $\theta[h]$ (see Chap. 3). Here, and throughout this chapter, square brackets are used to indicate that a variable is a function of the quantity within the brackets. The function $\theta[h]$ often depends on the history of wetting and drying; this phenomenon is called hysteresis. Water transport in soils obeys Darcy's law, which for one-dimensional vertical flow in the z-direction, positive upward, can be written as

$$q = -k[\theta]\frac{\partial H}{\partial z} = -k[\theta]\frac{\partial h}{\partial z} - k[\theta] \tag{2}$$

where q is the water flux density ($\mathrm{m^3\,m^{-2}\,s = m\,s^{-1}}$) and $k[\theta]$ is the hydraulic conductivity function ($\mathrm{m\,s^{-1}}$). k is a function of θ, since water content determines the fraction of the sample cross-sectional area available for water transport. Indirectly, k is also a function of the pressure head. $k[h]$ is hysteretic to the extent that $\theta[h]$ is hysteretic. Hysteresis in $k[\theta]$ is second order and is generally negligible. Determinations of k usually consist of measuring corresponding values of flux density and hydraulic potential gradient, and calculating k with Eq. 2. This is straightforward and can be considered as a standard for other, indirect measurements.

B. Hydraulic Diffusivity

For homogeneous soils in which hysteresis can be neglected or in which only monotonically wetting or drying flow processes are considered, $h[\theta]$ is a single-valued function. Then, for horizontal flow in the x-direction, or when gravity can be neglected, Eq. 2 yields

$$q = D[\theta]\frac{\partial \theta}{\partial x} \qquad \text{for} \qquad D[\theta] = k[\theta]\left(\frac{dh}{d\theta}\right) \tag{3}$$

where $D[\theta]$ is the hydraulic diffusivity function (m s^{-2}). Thus under the above stated conditions, the water content gradient can be thought of as the driving force for water transport, analogous to a diffusion process. Of course, the real driving force remains the pressure head gradient. Therefore, $D[\theta]$ is different for wetting and drying. There are many methods to determine $D[\theta]$, some of which are described later. They usually require a special theoretical framework with simplifying assumptions. Once $D[\theta]$ and $h[\theta]$ are known, the hydraulic conductivity function can be calculated according to

$$k[\theta] = D[\theta]\left(\frac{d\theta}{dh}\right)[\theta] \tag{4}$$

Because of hysteresis, one should combine only diffusivities and derivatives of soil water retention characteristics that are both obtained either by wetting or by drying. Since $k[\theta]$ is basically nonhysteretic, the $k[\theta]$ functions obtained in the two ways should agree closely.

C. Matric Flux Potential

Water transport in soils in response to pressure potential gradients can also be described in terms of the matric flux potential (Raats and Gardner, 1971):

$$\phi = \int_{-\infty}^{h} k[h]\,dh = \int_{0}^{\theta} D[\theta]d\theta \tag{5}$$

Equation 3 then becomes

$$q = \frac{\partial \phi}{\partial z} \tag{6}$$

The matric flux potential (m^2 s^{-1}) integrates the transport coefficient and the driving force. In homogeneous soil without hysteresis, the horizontal water flux density is simply equal to the gradient of ϕ. This formulation of the water transport process offers distinct advantages in certain situations, especially in the simulation of water transport under steep potential gradients (Ten Berge et al., 1987). It also allows one to obtain analytical solutions for steady-state multidimensional flow problems, including gravity, where the hydraulic conductivity is expressed as an exponential function of pressure head (Warrick, 1974; Raats, 1977). Like k and D, ϕ is a soil property that characterizes unsaturated water transport and is a direct function of θ and only indirectly of h. A method for measuring ϕ directly is described in Sec. VI.E.

D. Sorptivity

Sorptivity is an integral soil water property that contains information on the soil hydraulic properties $k[\theta]$ and $D[\theta]$, which can be derived from it mathematically (Philip, 1969). Generally, sorptivities can be measured more accurately and/or more easily than $k[\theta]$ and $D[\theta]$, so it is worth considering whether to determine the latter in this indirect way (Dirksen, 1979; White and Perroux, 1987). One-dimensional absorption (gravity negligible), initiated at time $t = 0$ by a step-function increase of water content from θ_0 to θ_1 at the soil surface, $x = 0$, is described by

$$I = S[\theta_1, \theta_0]t^{1/2} \tag{7}$$

where I is the cumulative amount of absorbed water (m) at any given time t, and sorptivity S (m $s^{-1/2}$) is a soil property that depends on the initial and final water content, usually saturation. Saturated sorptivity characterizes ponding infiltration at small times, as it is the first term in the infiltration equation of Philip (1969) and equal to the amount of water absorbed during the first time unit. With the flux-controlled sorptivity method (Sec. VIII.F), the dependence of S on θ_1 at constant θ_0 is determined experimentally. From this, $D[\theta]$ can be derived algebraically (see Eq. 20, below). The $t^{1/2}$-relationship of Eq. 7 has also been used for scaling soils and estimating hydraulic conductivity and diffusivity of similar soils (Sec. X.D).

III. SELECTION FRAMEWORK

A. Types of Methods

There are many published methods for determining soil water transport properties. No single method is best suited for all circumstances. Therefore it is necessary to select the method most suited to any given situation. Time spent on this selection is time well spent. Table 1 lists various types of methods that have been proposed and evaluates them on a scale of 1 to 5 using the selection criteria listed in Table 2. These tables form the nucleus of this chapter. In subsequent sections, the various methods are reviewed in varying detail. In general, the theoretical framework and/or main working equations are described, and other pertinent information is added to help substantiate the scores given in Table 1. For the more familiar methods, mostly only evaluating remarks are made; some experimental details are given also for the less familiar and newest methods. The scores are a reflection of my own insight and experience and are not based solely on the information provided. Further information is given in the references quoted.

Table 1 Evaluation of Methods to Measure Soil Water Transport Properties According to Criteria and Gradations in Table 2

Method	Criteria											
	A	B	C	D	E	F	G	H	I	J	K	L
STEADY STATE												
Laboratory												
Head-controlled	5	5	5	3(5)	5	3	2(1)	3(2)	3(2)	4	4	4
Flux-controlled	5	5	5	3(5)	5	3(4)	3(2)	3(1)	3(2)	4(3)	(4)2	4
Steady-rate (long column)	5	4	4	4	5	2	1	3	3	5	4	4
Regulated evaporation	5	2	2	3	3	2	2	3	3	4	2	4
Matric flux potential	3	3	3	5	3	3	3	4	4	5	5	4
Field												
Sprinkling infiltrometer	5	4	3	2	5	3(4)	2(1)	1	2	1	1	3
Isolated column (crust)	5	4	3	3	2	2	3	3	3	2	2	3
Spherical cavity	5	4	3	3	3	4	2	4	2	3	4	3
Tension disk infiltrometer	5, 3	2	3	5	3	2	4	2	4	3	2	3
TRANSIENT												
Laboratory												
Pressure plate outflow	4	2	4	5	3	2	2	3	4	3	4	3
One-step outflow	4	2	4	5	3	2	3	3	4	3	4	3
Boltzmann, fixed time	4	4	5	2	1	5	4	3	4	5	3	3
Boltzmann, fixed position	4	4	5	2	1	5	5	1	2	4	2	2
Hot air	4	4	1	4	1	5	4	4	4	4	3	2
Flux-controlled sorptivity	4	4, 2	5	4	3	5	4	3(1)	3	3	2	4
Instantaneous profile	5	5	5	2	2	3	2	2	2	2	2	2
Wind evaporation	5	3	5	3	4	4	2	2	3	3	4	4
Field												
Instantaneous profile	5	5	3	2	2	3	2	2	2	2	2	2
Unit gradient, prescribed	5	2	3	2	2	3	3	4	2	2	4	2
Unit gradient, simple	5, 4	1	1	4	2	3	2	4	3	3	4	2
Sprinkling infiltrometer	5	4	3	2	2	3	2	1	1	1	1	2

Table 2 Selection Criteria and Gradations for Methods to Measure Soil Water Transport Properties

A. Determined parameter
   5. Hydraulic conductivity
   4. Hydraulic diffusivity
   3. Matric flux potential
   2. Sorptivity
   1. Any other transport property

B. Theoretical basis
   5. Simple Darcy law or rigorously exact
   4. Exact, or minor simplifying assumptions
   3. Quasi-exact, simplifying assumptions
   2. Major simplifying assumptions
   1. Minimal theoretical basis

C. Control of initial or boundary conditions
   5. Exact, no requirements
   4. Indirect and accurate
   3. Approximate
   2. Approximate part of the time
   1. Little control, if any

D. Accuracy of measurements
   5. Weight, water volume, time
   4. Water content measurements, direct
   3. Pressure head measurements
   2. Indirect calibrated measurements
   1. Approximate uncalibrated measurements

E. Error propagation in data analysis
   5. Simple quotient (Darcy law)
   4. Accurate operations on accurate data
   3. Inaccurate operations on accurate data
   2. Accurate operations on inaccurate data
   1. Inaccurate operations on inaccurate data

F. Range of application
   5. Saturation to wilting point ($h > -160$ m)
   4. Tensiometer range ($h > -8.5$ m)
   3. Hydrological range ($k > 0.1$ mm/d)
   2. Wet range ($h > -0.5$ m)
   1. Psychrometer range ($-10 > h > -160$ m)

G. Duration of method
   5. 1 hour
   4. 1 day
   3. 1 week
   2. 1 month
   1. More than 1 month

H. Equipment
   5. Standard for soil laboratory
   4. General-purpose, off-the-shelf
   3. Easily made in average machine shop
   2. Special-purpose, off-the-shelf
   1. Special-purpose, custom-made

I. Operator skill
   5. No special skill required
   4. Some practice required
   3. General measuring experience adequate
   2. Special training of experimentalist
   1. Highest degree of specialization needed

J. Operator time
   5. Few simple and fast operations
   4. Few elaborate operations
   3. Repeated simple and fast operations
   2. Repeated elaborate operations
   1. Operator required continuously

K. Simultaneous measurements
   5. No limit
   4. Large number, at significant cost
   3. Small number, at little cost
   2. Small number, at substantial cost
   1. No potential

L. Check on measurements
   5. Continuous monitoring of all parameters
   4. Easy verification at all times
   3. Each verification requires effort
   2. Single check is major effort
   1. Check not possible

A major division is made between *steady-state* and *transient* measurements. In the first category, all parameters are constant in time. For this reason, steady-state measurements are almost always more accurate than transient measurements, usually even with less sophisticated equipment. Their main disadvantage is that they take much more time, often prohibitively so. Therefore, the choice between these two categories usually involves balancing costs, time available, and the required accuracy. For one-dimensional infiltration in a long soil column and for three-dimensional infiltration in general, the infiltration rate after some time becomes steady, but the flow system as a whole is transient due to the progressing wetting front. These flow processes, therefore, form an intermediate category that will be characterized as *steady-rate*.

The methods are divided further into *field* and *laboratory* methods, the choice of which is discussed in Sec. IV. Methods for measuring soil water transport coefficients can also be divided into those that measure hydraulic conductivity directly and all other methods (column A). From what follows it should become clear that one should measure hydraulic conductivity as a function of volumetric water content, whenever possible. When the hydraulic diffusivity is measured or the hydraulic conductivity as a function of pressure head, it is important to make a distinction between wetting and drying flow regimes in view of the hysteretic character of soil water retention.

B. Selection Criteria

The methods listed in Table 1 are evaluated on the basis of the criteria in Table 2, which include the following: the degree of exactness of the theoretical basis (B), the experimental control of the required initial and boundary conditions (C), the inherent accuracy of the measurements (D), the propagation of errors in the experimental data during the calculation of the final results (E), the range of application (F), the time (duration) required to obtain the particular transport coefficient function over the indicated range of application (G), the necessary investment in workshop time and/or money (H), the skill required by the operator (I), the operator time required while the measurements are in progress (J), the potential for measurements to be made simultaneously on many soil samples (K), and the possibility for checking during and/or after the measurements (L).

Depending on the particular situation, only a few or all of these criteria must be taken into account to make a proper choice. For example, accuracy will be a prime consideration for detailed studies of water transport processes at a particular site, whereas for a study of spatial variability the ability to make a large number of measurements in a reasonably short time is mandatory. These often do not have to be very accurate. If the absolute accuracy of a newly developed method must be established, the most accurate method already available should be selected, since there is no "standard" material with known properties available with which the method can be tested. The need and selection of a "standard method" for this purpose is discussed in Sec. V. When facilities for routine measurements must be set up, the last four criteria are particularly pertinent. Finally, there may be particular (difficult) conditions under which one method is more suitable than others, and these conditions may dominate the choice of method. Such criteria are not covered by Table 1 but are mentioned with the description of individual methods when appropriate.

The selection criteria used (Table 2) are mostly self-explanatory and will become clearer with the discussion of the individual methods. At this stage only a few general remarks are made about accuracy (relating to criteria B–E) and the range of application (G), which, out of practical considerations, is associated with pressure heads. For examples, reference is made to methods that are described later in more detail.

C. Accuracy

Direct measurements of weight, volume of water, and time, made in connection with the determination of soil hydraulic properties, are simple and very accurate (maximum score 5). An exception is measuring very small volumes of water while maintaining a particular experimental setup, for example a small hydraulic head gradient. Although the mass and water content of a soil sample can usually be measured accurately, the water content may not conform to what it should be according to the theoretically assumed flow system. For example, for Boltzmann transform methods a water content profile must be determined after an exact time period of wetting or drying. Gravimetric determinations cannot be performed instantaneously; during the destructive sampling water contents will change due to redistribution and evaporation of water and due to manipulation of the soil. Indirect water content measurements can be made nondestructively and repeatedly during a flow process. For high accuracy, these measurements normally require extensive calibration under identical conditions; usually this is not possible or takes too much time.

Derivation of hydraulic properties from other measured parameters introduces two kinds of errors. Firstly, the theoretical basis of the method may not be exact, either because it involves simplifying assumptions or because the theoretical analysis of the water flow process yields only an approximation of the transport property. Secondly, errors in the primary experimental data are propagated in the calculations required to obtain the final results. Mathematical manipulations each have their own inherent inaccuracies, a good example being differentiation. Another common source of error is that the theoretically required initial and/or boundary conditions cannot be attained experimentally. For example, it is impossible to impose the step-function decrease of the hydraulic potential at the soil surface under isothermal conditions, as is assumed with the hot air method.

Hydraulic potential measurements are relatively difficult and can be very inaccurate. Water pressure inside tensiometers in equilibrium with the soil water around the porous cup can in principle be measured to any desired accuracy with pressure transducers, but temperature variations can render such measurements very inaccurate. Mercury manometers are probably the least sensitive to large errors, but their accuracy is at best about ± 2 cm (Chap. 2). Near saturation, water manometers should respond quickly to changing pressure heads with an accuracy of about ± 1 mm. Beyond the tensiometer range, soil matric potentials are mostly determined indirectly from soil water characteristics or by measuring the electrical conductivity, heat diffusivity, or other properties of probes in equilibrium with soil water, with all the inaccuracies associated with indirect measurements. Direct measurements can be made with psychrometers (which also measure the osmotic component of the soil water potential) but these can be used only by workers experienced with sophisticated equipment and are at best accurate to about ± 500 cm. However, for many studies, such as that of the soil-water-plant-atmosphere continuum, such accuracies are acceptable, because hydraulic conductivities in this dry range are so low that hydraulic head gradients must be very large to obtain significant flux densities.

D. Range of Application

The range of application of a particular method depends to a large extent on whether, and if so how, soil water potentials are to be measured. For convenience and based on practical experience, therefore, the range of application is characterized in somewhat vague terms, which are identified further by approximate ranges of pressure head or flux density. Tensiometers can theoretically be used down to pressure heads of about −8.5 m, but in practice air intrusion usually causes problems at much higher values. Fortunately, hydraulic transport properties need not be known in the drier range, except where water transport over small distances is concerned (e.g., evaporation at the soil surface, and water transport to individual plant roots). Water transport over large distances occurs mostly in the saturated zone (or as surface water), for which the saturated hydraulic conductivity must be known. However, there are some exceptions, such as saline seeps, which are caused by unsaturated water transport over large distances during many years. Although unsaturated water transport normally occurs over short distances, it plays a key role in hydrology, as mentioned in the introduction. The unsteady, mostly vertical water transport in soil profiles is only significant when the hydraulic conductivity is in the range from the maximum value at saturation to values down to about 0.1 mm d^{-1}, since precipitation, transpiration, and evaporation can generally not be measured to that accuracy. This "hydrological" range ($k >$ 0.1 mm d^{-1}) corresponds to a pressure head range between 0 and −1.0 to −2.0 m, depending on the soil type.

The pressure head range over which hydraulic transport properties must be known should be carefully considered and be a major consideration in the selection process. It makes no sense, for instance, to determine hydraulic conductivities with the hot air method (which yields very inaccurate results over the entire pressure head range) when the results are only required for use in the hydrological range, for which much better methods are available. Conversely, it is dangerous to select an attractive method suitable only in the wetter range and to extrapolate the results to a drier range. In practice, the range of application of a particular method depends also on the time required to attain appropriate measurement conditions. Criteria F and G are interdependent: the time needed to measure the soil water property function often increases exponentially as the range of potentials is extended to lower values.

E. Alternative Approaches

Because measurements of the soil water transport properties leave much to be desired in terms of their accuracy, cost, applicability, and time, it is not surprising that other ways to obtain these soil properties have been investigated. The most extreme of these approaches is not to make any water transport measurements, but to derive the water transport functions from other, more easily measured soil properties (e.g., particle size distribution and soil water retention characteristic). These procedures are usually based on a theoretical model of the relationship, but they can also be of a purely statistical nature, in which case one should be cautious in applying the results to soil types outside the range used to derive the relationship. An intermediate approach forms the so-called inverse or "parameter optimization" techniques, which have recently received renewed attention. To be able to decide how the hydraulic transport functions can best be determined in a given situation, the possibilities and limitations of these alternative approaches should also be considered. They are briefly described in Secs. X and XI.

IV. LABORATORY VERSUS FIELD METHODS

A. Working Conditions

A major division between available methods is that of laboratory versus field methods. Laboratory measurements have many advantages over field measurements. In the laboratory, facilities such as electricity, gas, water, and vacuum are available, and temperature variations are usually modest and controllable. Standard equipment (e.g., balances and ovens) is also more readily available than in the field. Expensive and delicate equipment can often not be used in the field because of weather conditions, theft, vandalism, etc. One can usually save much time by working in the laboratory. Samples from many different locations can then first be collected and measurements carried out consecutively or in series. Considering all these advantages, it would seem good practice to carry out measurements in the laboratory, unless there are overriding reasons to perform them in situ. This may be necessary for experiments involving plants, but in situ hydraulic conductivity measurements are normally only needed to determine the hydraulic properties of a strongly layered soil profile as a whole or when heterogeneity and instability of soil structure make it very difficult if not impossible to obtain large enough, undisturbed soil samples and transport them to the laboratory.

B. Sampling Techniques

Because the hydraulic conductivity of soil is very sensitive to changes in soil structure due to sampling and/or preparation procedures, these operations should be carried out with utmost care. Fractures formed during sampling that are oriented in the direction of flow are disastrous for saturated hydraulic conductivity determinations but have very little influence on unsaturated hydraulic conductivities. Fractures perpendicular to the direction of flow have the very opposite effect on both types of measurements.

To obtain as nearly "undisturbed" soil samples as possible, soil columns have been isolated in situ by carefully excavating the surrounding soil and shaving off the top soil to the desired depth. Usually, a plaster of Paris jacket is cast around the soil column to facilitate applying water from an airtight space above the soil surface (needed, e.g., for the crust method), installing tensiometers, etc. The jacket also allows saturated measurements (it is not necessary to seal the soil column for unsaturated measurements) and protects the soil column in the field and during transport to a laboratory. Somewhat more disturbed soil columns from entire soil profiles can be obtained by driving a cylinder, supplied with a sharp, hardened steel cutting edge, into the soil with a hydraulic press. If the stroke of this press is smaller than the height of the sample, care should be taken to maintain exactly the same alignment for each stroke. We have been able to accomplish this easily and satisfactorily by pushing a sample holder hydraulically against a horizontal crossbar anchored firmly by four widely spaced tie lines (Fig. 1). To reduce compaction of the soil inside the cylinder due to the friction between the cylinder wall and the soil, the diameter of the cylinder should be kept large and/or a sampling tool with a moving sleeve should be used (Begemann, 1988). Driving cylinders into the ground by repeated striking with a hammer should not be tolerated for quantitative work, not even for short samples, because of the lateral forces that are likely to be applied. A compromise between a hammer and a hydraulic press is a cylindrical weight that, sliding along a steady vertical rod, is dropped repeatedly onto a sampleholder. For measurements of hydraulic conductivity of packed soil columns, it is essential that the packing be done systematically to attain the best possible reproducibility and uniformity. At the moment this appears to be more an art than a science.

Fig. 1 Hydraulic apparatus for obtaining short (left) and long (right) "undisturbed" soil columns. The apparatus is stabilized by a crossbar and four widely anchored tie lines.

C. Sample Representativeness

Other important aspects of soil sampling are the size and number of samples required to be representative in view of soil heterogeneity and spatial variability. The development and size of the natural structural units (peds) dictate the size of the sample needed for a particular measurement. If a soil property were measured repeatedly on soil samples of increasing size, the variance of the results would normally decrease until it reached a constant value, the variance of the method alone. The smallest sample for which a constant variance of a specific soil property is obtained is called the representative elementary volume (REV) for that property (Peck, 1980). Assuming that a soil sample should contain at least 20 peds to be representative, Verlinden and Bouma (1983) estimated REVs for various combinations of texture and structure. These varied from the commonly used 50-mm-diameter (100 cm^3) samples to characterize the hydraulic properties of field soils with little structure, to 10^5 cm^3 soil samples for heavy clays with very large peds or soils with strongly developed layering. The desirable length of (homogeneous) soil samples depends on the particular measurement method that is used.

Considering the number of soil samples needed, Warrick and Nielsen (1980) listed the unsaturated hydraulic conductivity under the category of soil properties with the highest coefficient of variation. They reported that about 1300 independent samples from a normally distributed population (field) were needed to estimate mean hydraulic conductivity values with less than a 10% error at the 0.05 significance level. The theory of regionalized variables or geostatistics (Journel and Huibregts, 1978) provides insight into the minimum number and spatial distribution of soil samples required to obtain results with a certain accuracy and probability. Of course, the same applies to the required number and locations of sites for in situ measurements.

V. STANDARD METHOD

A major problem associated with the determination of soil hydraulic transport properties is the lack of uniform soils or other porous materials with constant, known transport properties, which could serve as standard reference materials with which to establish the absolute accuracy of any method. It is impossible to pack granular material absolutely reproducibly, and consolidated porous materials (e.g., sandstone) are not suitable for most of the methods used on soils. Also, repeated wetting or drying of a soil sample to the same overall water content does not lead to the same water content distribution and hydraulic conductivity. Given these insuperable difficulties, hydraulic transport properties are almost always presented without any indication of their accuracy. Only the method used to determine them is described and sometimes, for good measure, a comparison between the results of two methods is given. Agreement between two methods is still not a guarantee that both are correct. Often the results of two methods are said to correspond well when in fact they differ by as much as an order of magnitude. There is no way to decide which is the more accurate. The only recourse is to evaluate the potential accuracy of the required measurements, possibility of experimentally attaining the theoretically required initial and boundary conditions, and error propagation in the required calculations. In this way, instead of a standard material with accurately known properties, a "standard method" can be selected for reference. While searching for such a standard method, a number of features that enhance the accuracy should be kept in mind.

Since hydraulic conductivity is defined by *Darcy's law* (Eq. 2), its determination as the quotient of simultaneously and directly measured water flux density and hydraulic head gradient is most accurate. Determinations according to other equations, such as those of the Boltzmann transform methods (see Eq. 13), or derivations from other measured parameters, such as flux density derived from measured water contents for the instantaneous profile method, introduce (additional) errors in the measurements that are propagated in the more complex algebraic operations. Water flux densities and hydraulic head gradients can be measured most accurately when they do not change in time. Attainment of such *steady*

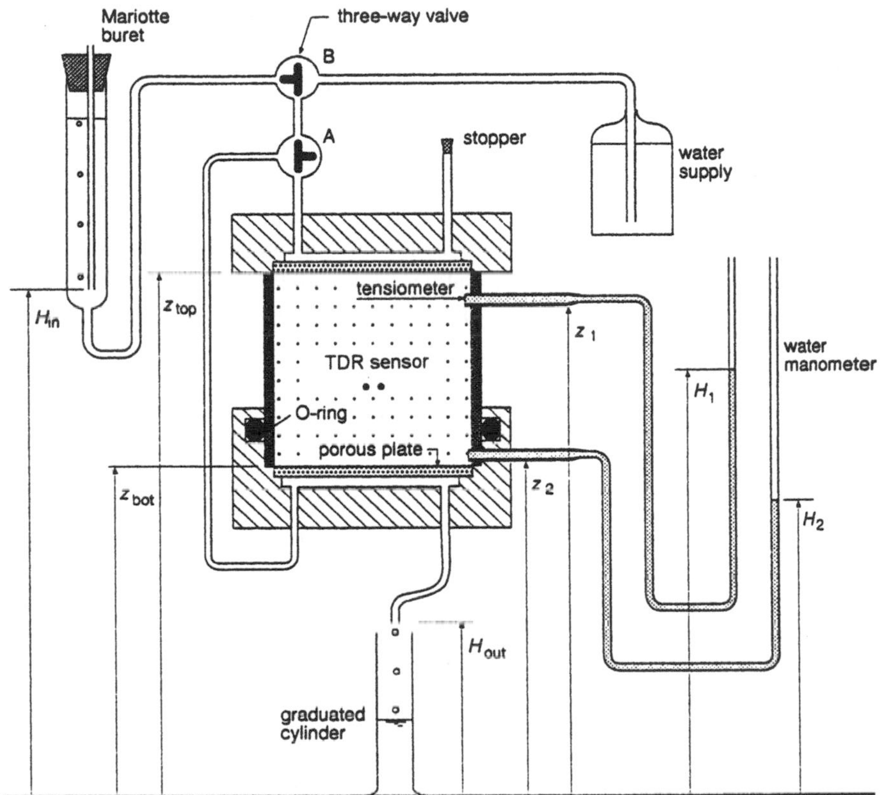

Fig. 2 Schematic experimental apparatus for head-controlled hydraulic conductivity measurements, illustrating the accuracy-enhancing features.

flow in a soil column can be checked by verifying that the measured influx and outflux are equal (Fig. 2). This also increases the accuracy of the water flux density determination. Because resistances of tubing and at the contact between the soil and porous plates are often too large and unpredictable to permit reliance on measurement of an externally applied hydraulic gradient, the *hydraulic head gradient within the soil* should be measured with sensitive and accurate tensiometers (Fig. 2).

Unless measured hydraulic conductivities are associated with an identifying parameter, they are, literally, useless. Hydraulic conductivity depends on the distribution of water in the pore space, usually adequately characterized by the volume fraction of water. A relationship with pressure head is valid only for the specific conditions of the measurements. It can be converted to a water content relationship only if the soil column was homogeneous, hysteresis was negligible, and the soil water retention characteristic is known accurately. Since it is virtually impossible to carry out hydraulic conductivity measurements so that all parts of a soil column have only been consistently wetting or drying, measured hydraulic conductivities should be related to *simultaneously measured water contents*. When the water content in the soil column is not uniform, there is a question about which water content should be associated with the obtained hydraulic conductivity. When water flows vertically downward in a soil column under unit hydraulic head gradient, gravity is the only driving force. The pressure head is then everywhere the same and, without hysteresis, the water content will be as uniform as possible. Under monotonically attained *gravitational flow conditions*, therefore, the indicated ambiguity hardly exists.

The features described above approach most closely the requirements for a "standard method" for measuring soil hydraulic conductivity. A soil hydraulic conductivity function $k[\theta]$ can be determined most accurately by performing these measurements on a series of such steady flow systems, preferably all in one soil column and changing the water content monotonically to minimize errors due to hysteresis. This requires nondestructive water content measurements. These can be made conveniently by time-domain reflectometry (Chap. 1) or improved dielectric measurements in the frequency domain (Dirksen and Hilhorst, 1994). This leaves the application and measurement of small, uniform water flux densities to soil columns often for extended time periods as the major experimental hurdle to this approach. If the system is flux controlled, such as the atomized spray system described in Sec. VI, the hydraulic conductivity that will be measured is predictable. Head-controlled flow through a porous plate, crust, etc. often is unsteady and yields unpredictable hydraulic gradients and conductivities. Very small water fluxes can be measured accurately by weighing and by observing the movement of air bubbles in thin glass capillaries.

Theoretically, these measurements are limited to pressure heads in the tensiometer range, approximately 0 to -8.5 m water. Before this "dry" limit is reached, however, the time needed to reach a steady state becomes prohibitively long, either due to practical considerations or because long term effects (e.g., microbial activity, loss of water through tubing walls) reduce the overall accuracy to an unacceptable level. Therefore, the practical range probably does not extend much below a pressure head of -2.0 m. This is sufficient for characterization of water transport over the relatively large distances of a soil profile. However, for analyses of water transport to plant roots, and of evaporation near the soil surface, hydraulic conductivities for much lower pressure heads and water contents are needed. These can be determined only with other, usually indirect methods. Selection of a standard method for this higher tension range does not yet seem to be possible. For field measurements, steady infiltration over a large surface area (with tensiometer measurements in the center) with a sprinkling infiltrometer approaches most closely to the requirements for a "standard method" (Sec. VI).

VI. STEADY-STATE LABORATORY METHODS

A. Head-Controlled

The classical head-controlled method used by Darcy is featured in most soil physics textbooks. It involves steady-state measurements on a soil column in which water flows under a hydraulic gradient controlled by means of a porous plate at both ends. Principles, apparatus, procedures, required calculations, and general comments are given in great detail by Klute and Dirksen (1986).

The head-controlled setup of Fig. 2 shows all the accuracy-enhancing features discussed in Sec. V. Soil water contents can be measured nondestructively with sensors for dielectric measurements in the time or frequency domain (see Chap. 1), making this setup suitable as a standard method. This is reflected in the maximum scores in Table 1 for theoretical basis (B), control of initial and boundary conditions (C), and error propagation in data analysis (E). As the flux density decreases, the ease and accuracy with which it can be measured also decreases, whereas the time to attain steady state increases. Therefore while theoretically the entire tensiometer range of pressure heads can be covered, the practical limit of this method is probably -2.0 m (F). When used as standard, water contents and hydraulic heads can be measured with greater than normal accuracy and the application can be extended beyond the practical range by using more expensive equipment and spending more time, as indicated by the additional score within parentheses for criteria D, F. G, H, and I.

Indirect determinations of hydraulic conductivity (see Sec. X) call for one measured hydraulic conductivity value as a correction (matching) factor. Usually the saturated hydraulic conductivity is used for this, but it is a poor choice because of the dominating influence of macropores on these measurements. At slightly negative pressure heads ($< h = -10$ cm), these macropores are empty, and the hydraulic conductivity is then a much truer reflection of the soil matrix. The head-controlled setup of Fig. 2 presents few problems, and one measurement takes little time for all but the least permeable soils. For these reasons and the inherent accuracy of the measurements, I recommend that the type of setup shown in Fig. 2 be used as the standard method.

B. Flux-Controlled

Hydraulic conductivities can also be measured at steady state by controlling the flux density rather than the hydraulic head at the input end of a vertical soil column (Klute and Dirksen, 1986). The major experimental hurdle of flux-controlled measurements is a device that can deliver small, uniform, steady water flux densities for extended time periods (Wesseling and Wit, 1966; Kleijn et al., 1979). To determine $k[\theta]$ functions, it is desirable that rates can be changed easily to predictable values that can be measured accurately. This was true for the reservoir with hypodermic needles and pulse pump described in the first edition of this book (Dirksen, 1991). When this apparatus proved still less than satisfactory, Dirksen and Matula (1994) developed an automated *atomized water spray system* (Fig. 3) capable of delivering steady average fluxes down to about 0.1 mm d^{-1}, which was considered the minimum flux density needed for hydrological applications (criterion F3).

In this system, water and air are mixed in a nozzle assembly to produce an atomized water spray. By decreasing the water pressure and increasing the air pressure, a minimum continuous uniform water spray of about 200 mm d^{-1} has been obtained. The average water application rate can be reduced further by spraying intermittently under control of a timer with independent ON and OFF periods. Figure 3 shows the spray system in the laboratory set up for 20-cm diameter soil columns. The soil columns are placed on very fine sand that can be maintained at

Fig. 3 Laboratory setup of atomized water spray system for 20-cm diameter soil columns, with very fine sand box and hanging water column, and tensiometry and TDR equipment.

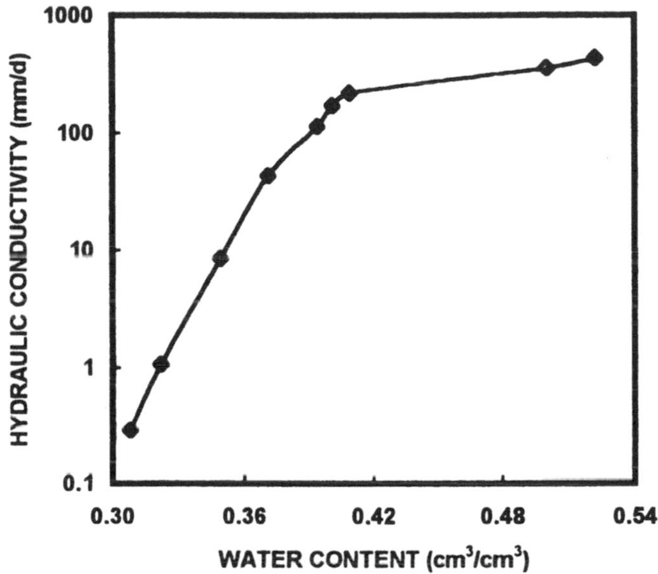

Fig. 4 Hydraulic conductivity as a function of volumetric water content, for a Typic Hapludoll measured with the setup shown in Fig. 3.

constant pressure heads of minimally −120 cm water by means of a hanging water column with overflow. With proper protection of the exposed sand surface, the discharge from the overflow is a measure of the flux density out of the soil column. Hydraulic heads are measured with a sensitivity of 1 mm water at 5 cm depth intervals. Water contents are measured with 3-rod TDR sensors installed halfway between and perpendicular to the tensiometers. Thus all the accuracy-enhancing features are present.

Figure 4 shows the hydraulic conductivities as function of water content measured in a (Typic Hapludoll) soil column. The water flux density was easily varied over more than three orders of magnitude from virtual saturation ($h = -0.9$ cm) to an average flux density of 0.22 mm d^{-1}, attained with 0.1% actual spraying time. After this lowest application rate was discontinued, hydraulic heads changed within two days to essentially hydrostatic equilibrium with the sand, indicating that this low water application rate had indeed produced steady downward flow. In the intermediate range, the discharge from the sand agreed exactly with the applied water flux densities. The time needed to attain steady state varied from about one hour at the highest water application rate to about four days at the lowest rate.

The atomized water spray setup has been tested successfully under field conditions, using a gasoline-powered 220 VAC electric generator. If 12 VDC solenoid valves and a compressed-air cylinder are used, measurements could be made in situ without an electric generator. After months of inoperation, the assembly can be started up almost instantaneously without problems of clogging. It has proven to be a reliable, versatile apparatus for measuring quickly and accurately any soil hydraulic conductivity from that near saturation to about 0.1 mm d^{-1}. The flux densities, and thus the hydraulic conductivities, are predictable. These features make it very attractive to incorporate this flux-controlled system into a standard method.

C. Steady Rate

An early flux-controlled variant is the so-called *Long Column Infiltration* method. By applying a constant flux density to the soil surface of a long, vertical (dry) soil column (Childs and Collis-George, 1950; Wesseling and Wit, 1966; Childs, 1969), the potentials on both ends of the flow system approach constant values, while the distance between them increases with time. If the pressure head gradient becomes negligible with respect to the constant gravitational potential gradient before the wetting front reaches the bottom of the column, a "quasi-steady" state will be attained in which the infiltration rate approaches a steady value. During this "steady-rate" condition, the upper part of the column automatically approaches the water content at which the hydraulic conductivity is equal to the externally imposed, known flux density. Thus if that water content is measured, tensiometers are not needed, and the method can theoretically be used beyond the tensiometer range. As long as there is still dry soil in the bottom of the column, porous plates are not needed, and problems with plate and contact resistances are eliminated. When the wetting front reaches the bottom of the soil column, water can exit only after it reaches zero suction (water table). This limits the range of pressure heads and water contents that can be covered, unless there is a (negative) head-controlled boundary at the bottom of the column. Youngs (1964) applied water directly at constant pressure head to a long soil column.

D. Regulated Evaporation

Steady state can also be attained when water from a water table or a supply at constant negative pressure head is evaporated at the soil surface at a constant rate. Under these conditions of regulated evaporation, there is no measuring zone with a uniform pressure head and water content. The water content, and thus the hydraulic conductivity, decreases towards the surface. Since at steady state the flux density is everywhere the same, the hydraulic gradient is inversely proportional to the hydraulic conductivity and thus will become larger and more difficult to measure accurately towards the soil surface. The hydraulic conductivity obtained will be some kind of average for the range of water contents, and the correct water content to which it should be assigned will be uncertain.

A slightly different experimental arrangement was used by Gardner and Miklich (1962). Their soil column was closed at one end, which makes it theoretically impossible ever to reach a steady state. Nevertheless, they claimed that various constant fluxes could be attained by regulating the evaporation from the column by the size and number of perforations in a cover plate. This would seem to require a lot of manipulation. The rates of water loss were determined by weighing the column. The hydraulic gradient was measured with two tensiometers. By assuming k and θ were constant between the tensiometers for each evaporation rate, they derived an approximate equation for the hydraulic conductivity. The rather severe assumptions limit the applicability of the method and it has not been frequently used.

E. Matric Flux Potential

A controlled evaporative flux from a short soil column in which the pressure head at the other end is controlled (previous section) was used by Ten Berge et al. (1987) in a steady-state method for measuring the matric flux potential as function of water content. They assumed that the matric flux potential function has the form

$$\phi[\theta] = -\frac{A}{x + B} \qquad \text{for} \qquad x = 1 - \frac{\theta}{\theta_0} \tag{8}$$

where A is a scale factor ($m^2\ s^{-1}$) and B is a dimensionless shape factor, both typical for a given soil, and θ_0 is a reference water content, experimentally controlled at the bottom of the soil column. Whereas these authors used the diffusivity function proposed by Knight and Philip (1974),

$$D[\theta] = a(b - \theta)^{-2} \tag{9}$$

where a and b are constants, the method can be used with any set of two-parameter functions of $\phi[\theta]$ and $D[\theta]$.

After a small soil column is brought to a uniform water content (pressure head) and weighed, it is exposed to artificially enhanced evaporation at the top, while the bottom is kept at the original condition with a Mariotte-type water supply. When the flow process has reached steady state, the flux density is measured, as well as the wet and oven dry weights of the soil column. From these simple, accurate experimental data the parameters A and B, and thus $\phi[\theta]$ and $D[\theta]$, can be evaluated by assuming that gravity can be neglected. In this case the matric flux potential at steady state decreases linearly with height so that this method does not suffer from any ambiguity (generally associated with upward flow) in the assignment of appropriate values of water content and pressure head to the calculated values of the water transport parameter.

It is better not to start from saturation, but at a small negative pressure head, to reduce the influence of gravity and to be able to meet the theoretically required upper boundary condition ($\theta = 0$). The method is rather slow and covers a limited range of θ and h, but the measurements require little attention while in progress. The major source of errors appears to be that the theoretically prescribed initial and boundary conditions are hard to obtain experimentally. Furthermore, the theoretical basis involves a number of assumptions. However, direct measurement of $\phi[\theta]$ is likely to be more accurate than methods involving separate measurements of $h[\theta]$ and $D[\theta]$ for flow processes involving steep gradients such as thin, brittle soil layers. For an analysis of the propagation of errors, see Ten Berge et al. (1987).

VII. STEADY-STATE FIELD METHODS

A. Sprinkling Infiltrometer

Analogous to the measurements in long laboratory soil columns (Sec. VI.C), hydraulic conductivities can be measured in the field under steady-rate conditions delivered by a sprinkling infiltrometer (Hillel and Benyamini, 1974; Green et al., 1986). It is the counterpart to the flux-controlled atomized spray laboratory setup (Sec. VI.B) and appears to be the best candidate for "standard field method." In such applications, elaborate sprinkling equipment, which must normally be attended whenever in operation, is justified. Measurements may extend over days or even weeks, depending on the range of water contents to be covered. This range is technically limited by the ability to reduce the sprinkling rate while retaining uniformity. This can be done best by intercepting an increasing proportion of the artificial rain, rather than reducing the discharge from a nozzle (Amerman et al., 1970; Rawitz et al., 1972; Kleijn et al., 1979). Green et al. (1986) give 1 mm h^{-1} as a practical lower limit for the flux density. To prevent hysteresis, the flux density of the applied water should be increased monotonically with time. Because soil profiles are frequently inhomogeneous, and because of possible lateral flow, the hydraulic gradient cannot be assumed to be unity, and it should be measured when a high accuracy is required. Sprinkling infiltrometers are used frequently for soil erodibility studies. In such applications, the impact energy of the water drops emitted by the sprinkling infiltrometer should be as nearly equal to that of natural rain drops as possible (Petersen and Bubenzer, 1986), since changes of the soil physical properties due to structural breakdown (e.g., crust formation) have a great effect on the erosion process (Baver et al., 1972; Lal and Greenland, 1979). For hydraulic conductivity measurements, in contrast, the soil surface should be protected against crust formation as much as possible (e.g., by covering the soil surface with straw).

Field measurements of hydraulic conductivity with a sprinkling infiltrometer may take a long time, during which large temperature variations may occur. Temperature changes and gradients may have a significant influence on the water transport process, especially for small water flux densities and/or hydraulic head gradients near the soil surface. Therefore it is good practice to ensure that all field measurements minimize temperature changes as much as possible (e.g., by shielding the soil surface from direct sunlight).

B. Isolated Soil Column with Crust

Instead of applying water over a large soil surface and concentrating the measurements in the center of the wetted area to approach a one-dimensional flow system (preceding Sec.), true one-dimensionality can be obtained in situ by carefully excavating the soil around a soil column (Green et al., 1986; Dirksen, 1999, Fig. 8.1). Although not strictly necessary for unsaturated conditions, a plaster of Paris jacket is usually cast around the "isolated" soil column assembly for protection or for saturated conductivity measurements. Use of such truly undisturbed soil columns is especially suitable for soils with a well-developed structure, since large-scale "undisturbed" samples, which are easily damaged during transport, would otherwise be required. The isolated soil column in its jacket may also be broken off its pedestal and transported to the laboratory for (additional) measurements.

Water has been applied to such soil columns via crusts of different hydraulic resistance, usually made of mixtures of hydraulic cement and sand (Bouma et al., 1971; Bouma and Denning, 1972). If the space above the crust is sealed off airtight, water can be applied to the soil column at constant pressure head regulated by a Mariotte device. Initially, it was commonly assumed that the crust soon causes the flux density to become steady at unit hydraulic gradient (Hillel and Gardner, 1969), so that a single tensiometer just below the crust could provide the pressure head to be associated with the hydraulic conductivity obtained. However, the hydraulic head gradient generally does not attain unity and should be measured with at least two tensiometers. By using different values of the controlled pressure head and/or crust resistance, a number of points on the $k[h]$ function can be obtained. In practice, the minimum pressure head that can thus be attained appears to be about -50 cm.

In comparison with ponding infiltration, the claim that crusts enhance the attainment of a steady flux is correct, but I suspect that often the final measurements are made before a steady-rate condition has been reached. If measurements are made at a range of pressure heads, one should proceed from dry to progressively wetter conditions (by replacing more resistant crusts with progressively less resistant ones), since a wetter wetting front will quickly overtake a preceding dryer one. Letting the soil dry before applying a smaller flux density takes much time and introduces hysteresis into the measurements. The latter is unacceptable if the obtained hydraulic conductivities are related only to the pressure head. Crust resistances have proved to be quite unpredictable, often nonuniform, and unstable in time. Making and replacing good crusts is tedious work, and curing takes at least 24 hours. Crusts may also add to the soil solution chemicals that alter the hydraulic conductivity. I advocate, therefore, that the "crust method" no longer be used.

C. Spherical Cavity

In one dimension, steady state can be achieved under two types of steady boundaries, either potentials or flux densities. In the field, it is not too difficult to force the flow to be one-dimensional by isolating a small cylindrical soil column (previous Sec.) or a large rectangular soil block. The latter can be done easily by excavating (preferably with a mechanical digger) narrow vertical trenches, covering the inside vertical walls with plastic sheets and refilling the trenches with soil. However, a major experimental effort is required to impose a steady boundary condition at the bottom of a flow system in the field. The practical alternative of a constant-shape wetting front moving downward at a steady rate in the center of a large wetted area (Sec. VI.C) can be attained only in a uniform soil profile that is deep enough for the pressure head gradient to become negligible compared to gravity.

In three-dimensional flow, the influence of gravity is much smaller than in one- or two-dimensional flow. As a result, three-dimensional infiltration from a point source reaches a large-time steady-rate condition irrespective of the influence of gravity (Philip, 1969). Without gravity, three-dimensional infiltration from a point source is spherically symmetric. Raats and Gardner (1971) showed that the hydraulic conductivity can be derived from a series of such steady-rate conditions in which the pressure heads also approach steady values. This presents a very attractive set of conditions for measuring hydraulic conductivity, especially in situ, because (1) only one controlled boundary is required, (2) the influence of gravity, which must be neglected, is especially small, and (3) steady-rate and steady tensiometer measurements are inherently accurate. For these reasons, I have explored the possibilities of this "spherical cavity" method and have analyzed the influence of gravity (Dirksen, 1974).

Water is supplied to the soil (which needs to be initially at uniform pressure head) through the porous walls of a spherical cavity maintained at a constant pressure head until both the flux Q and the pressure head h_a, at the radical distance $r = a$ from the center of the spherical cavity, have become constant. This is re-

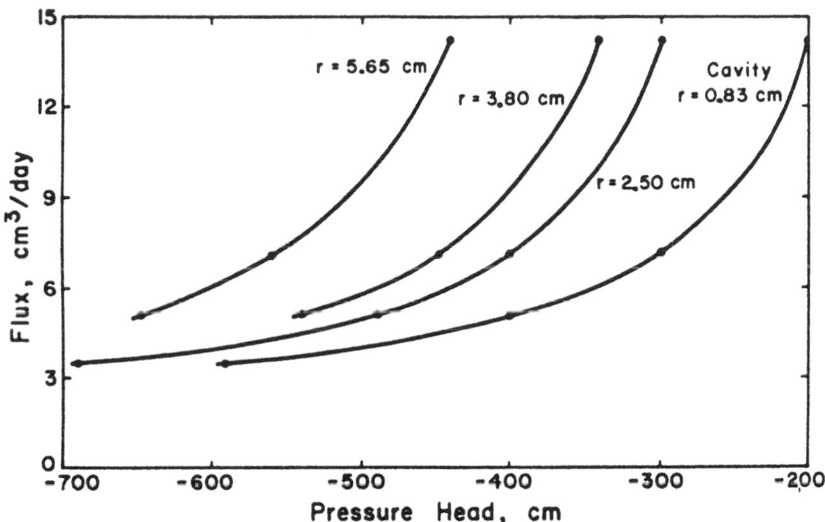

Fig. 5 Steady fluxes from a spherical cavity versus steady pressure heads in the cavity and in three tensiometers at the radial distances indicated. (From Dirksen, 1974.)

peated for progressively larger (less negative) controlled pressure heads in the cavity. Hydraulic conductivity can then be calculated according to

$$k[h_r] = \frac{1}{r}\frac{dQ}{dh_r} \tag{10}$$

which is simply the slope of the graphs in Fig. 5 at any desired pressure head, divided by the radial distance of the particular measuring point. In this way hydraulic conductivities down to $h = -700$ cm were obtained in about 2 weeks, with each tensiometer and the cavity yielding its own result. This overlap provides an internal check. Note that the pressure head range can be expanded downward easily by increasing the radial distance of the measuring point. Of course, the time required to attain a constant pressure head increases with radial distance. It is possible to use the regulated pressure head in the cavity as the only "tensiometer" data. This reduces the experimental duration and operations to a minimum. The resistance between the water supply and the soil (porous walls and soil–ceramic interface) must then be negligible. The effect of gravity is minimized when tensiometers, if used, are placed directly below the cavity. The method has been demonstrated only in the laboratory, although there have been some exploratory measurements in the field. Because of its very attractive features, especially as an in situ method, the approach is worthy of further investigation. If tensiometer measurements can be omitted, placement of the spherical cavity without undue contact resistance with, and disturbance of, the soil presents the only great experimental challenge. This would be reduced even further if the spherical cavity could be placed at the soil surface. Then the measuring system is essentially reduced to that for the tension disc infiltrometers described in the next section. These are operated, however, only at rather low tensions ($h > -30$ cm).

D. Tension Disk Infiltrometer

Perroux and White (1988) developed disk infiltrometers that are very attractive for use in the field. A circular disk provides water at constant pressure head to the surface of homogeneous soil without confinement. Initially, the flow is one-dimensional and the effect of gravity is negligible, so that the sorptivity can be determined. From the steady flow rate, generally attained within a few hours (Philip, 1969), the hydraulic conductivity can be determined (for more details, see Chap. 6).

Tension disk infiltrometers are very user-friendly. They are quickly filled with water, the regulated tension is varied easily, and only the soil surface needs to be prepared. The data analysis is relatively simple but is based on many simplifying assumptions. Not infrequently, negative hydraulic conductivity values are obtained which, of course, is physical nonsense. Apart from measurement errors, this may be due to the simplifying assumptions, to the wetting front reaching soil that is different from that at the surface, etc. There is no way to distinguish between the sources of error. This makes more elaborate measurements and derivations questionable (e.g., measurements made with one disk at different pressure heads (Ankeny, 1992) and with disks of different radii (Smettem and Clothier, 1989; Thony et al., 1991). It also applies to measurements made at saturation, for which the results are extrapolated to negative pressure heads (Scotter et al., 1982; Shani et al., 1987), that were extensively discussed in the first edition (Dirksen, 1991). Clothier et al. (1992) determined the volume fractions of mobile and immobile water by introducing successively reactive and nonreactive tracers during steady flow and afterwards sampling the soil underneath the disk for tracer concentrations. Surprisingly, these authors found that the steady rate of infiltration quickly attained its original value after the necessary interruptions that generally lasted less than two minutes. Ankeny et al. (1988) increased the measuring precision nearly tenfold by using two pressure transducers to measure the infiltration rate. Quadri et al. (1994) developed a numerical model of the axisymmetric water and solute transport system. Tension disk infiltrometers have been used also to monitor changes in soil structure after soil tillage operations. However, if the plow layer is very loose, the weight of the water-filled apparatus may compact the soil, and good contact with the rough surface may be difficult to obtain.

VIII. TRANSIENT LABORATORY METHODS

A. Pressure Plate Outflow

In contrast to the steady-state methods, most transient laboratory methods yield in the first place hydraulic diffusivities. A good example is the pressure-plate outflow method (Gardner, 1956). A near-saturated soil column at hydraulic equilibrium on a porous plate is subjected to a step decrease in the pressure head at the porous plate (e.g., by a hanging water column) or a step increase in the air pressure. The resulting outflow of water is measured with time. The step decrease or increase must be so small that it can be assumed that the hydraulic conductivity is constant and that the water content is a linear function of pressure head. The experimental water outflow as a function of time is matched with an analytical solution, yielding after many approximations

$$\ln(Q_0 - Q) = \ln\left(\frac{8Q_0}{\pi^2}\right) - \left(\frac{\pi}{2L}\right)Dt \qquad (11)$$

where Q is the cumulative outflow at time t, Q_0 is the total outflow, and L is the length of the soil sample. According to Eq. 11, the diffusivity D, for the mean pressure head, can be derived from the slope of a plot of $\ln(Q_0 - Q)$ versus t. This is repeated for other step increases in pressure, which must only be initiated after a new state of hydraulic equilibrium has first been reached. The pressure increments must be small enough for the assumptions to be valid, but large enough to allow accurate measurement of water outflow, while the more steps there are, the more time it takes to cover the desired range of water content. This method was initially widely used, but it generally failed to yield satisfactory results. Much effort was spent to improve it, especially with respect to the correction for the resistance of the porous plate or membrane, but without much success. Applications such as those by Ahuja and El-Swaify (1976) and Scotter and Clothier (1983) have been outdated more recently by the use of outflow experiments as a basis for the inverse approach of parameter optimization discussed in Sec. XI (Van Dam et al., 1994; Eching et al., 1994).

B. One-Step Outflow

Doering (1965) proposed the one-step variant of the previous method, which is much faster and not very sensitive to the resistance of the plate or membrane. If uniform water content in the soil column is assumed at every instant, diffusivities can be calculated from instantaneous rates of outflow and average water content

$$D[\theta] = \frac{-4L^2}{\pi^2(\theta - \theta_f)}\frac{\partial\theta}{\partial t} \qquad (12)$$

where L is the length of the soil sample, θ is the average water content when the outflow rate is $\partial\theta/\partial t$, and θ_f is the final water content. This can be determined by measuring the cumulative outflow and the final weight. Doering found the results as reliable as those obtained with the original version (Sec. VIII.A), and there were large time savings.

Gupta et al. (1974) showed that the analysis of one-step outflow data according to Gardner (1956) and used by Doering can be in error by a factor of 3. They improved the analysis by first estimating a weighted mean diffusivity. This does not require the assumption of a constant diffusivity over the pressure increment, nor over the length of the soil sample, and it also reduces the effect of membrane impedance. Passioura (1976) obtained about the same improvement in accuracy with a much less complicated calculation procedure (given in detail) by assuming that the rate of change of water content at any time is uniform throughout the entire soil sample. He also estimated that a 60-mm long soil sample will take about 5 weeks to run and a 30-mm sample about 1 week. Measurements have been automated by Chung et al. (1988) for up to 16 samples.

Ahuja and El-Swaify (1976) determined the soil hydraulic properties by measuring one-step cumulative inflow or outflow from short soil cores through high-resistance plates at one end and measuring the pressure head at the other end. They obtained good results for pressure heads down to -150 cm. Scotter and Clothier (1983) claimed, without referring to the previous authors, that it is better to analyze the results of a series of small pressure head changes than of one large change, because the former approach does not involve the difficult task of measuring small flow rates. The accuracy relies mainly on the time delay of the outflow, not on the shape of the outflow curve. Eching et al. (1994) also used tensiometer measurements.

The one-step outflow method is attractive for its experimental simplicity; the theoretical analysis of the data remains its weakest point. Since this limitation does not apply to the simulation of the flow process, it is not surprising that recently the same measurements were selected as basis for the parameter optimization approach (Sec. XI).

C. Boltzmann Transform

The theory of the so-called Boltzmann transform methods is well known and can be found in soil physics textbooks (Kirkham and Powers, 1972; Koorevaar et al., 1983). If gravity is neglected, the general flow equation can be written in terms of the diffusivity (Eq. 3). For a step-function increase or decrease of the water content at the adsorption or desorption interface of an effectively semi-infinite uniform soil column, this partial differential equation can be transformed into an ordinary differential equation using the Boltzmann variable $\tau = xt^{-1/2}$, where x is the distance from the sample surface and t is time. Integration of this equation for the also transformed initial and boundary conditions yields the diffusivity

$$D[\theta'] = \frac{1}{2}\left(\frac{d\tau}{d\theta}\right)_{\theta'} \int_{\theta_0}^{\theta'} \tau[\theta]d\theta \tag{13}$$

where θ_0 is the initial water content, and θ' is the water content at which D is evaluated. By measuring the function $\tau[\theta]$ experimentally, the diffusivity at any water content can be calculated as half the product of the slope and area indicated in Fig. 6, which can be determined graphically.

The function $\tau[\theta]$ can be determined experimentally in two ways; by measuring either the water content distribution in a soil column at a fixed time (Bruce and Klute, 1956) or the change of water content with time at a fixed position (Whisler et al., 1968). The first is often done gravimetrically; the latter needs to be done nondestructively (see Chap. 1). A major drawback for both methods is the sensitivity of the calculated diffusivities to irregularities and/or errors in the bulk density and water contents in the soil column and the propagation of these errors in the subsequent calculations. Gravimetric measurements are subject to redistribution and evaporation of water during sampling and must therefore be done as quickly as possible. The fixed-position method is free from these prob-

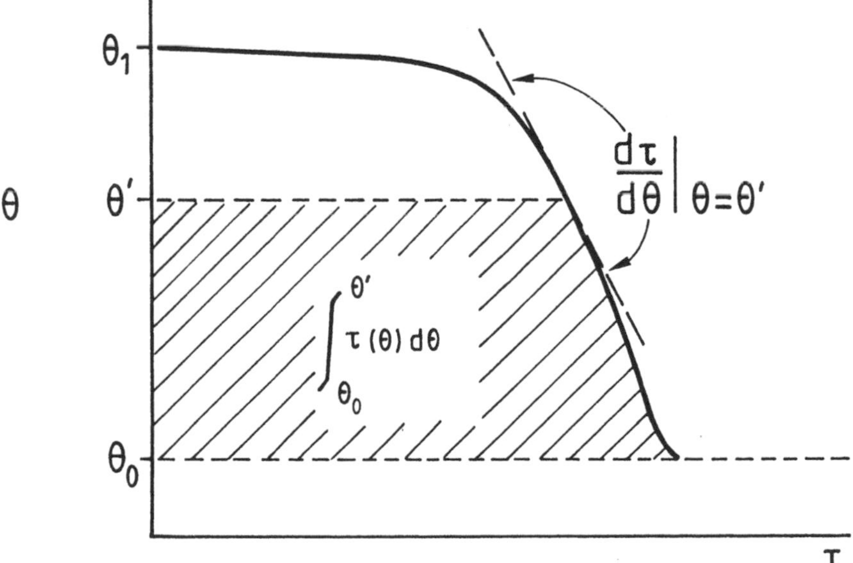

Fig. 6 Graphical solution of Boltzmann transform equation (Eq. 13).

lems. A comparative study of the two variants (Selim et al., 1970) yielded similar errors. With the introduction of dielectric water content measurements, especially in the frequency domain (Dirksen, 1999), the fixed-position variant appears to deserve renewed attention.

Derivation of a $D[\theta]$ function from experimental $\tau[\theta]$ data according to Eq. 13 involves differentiating experimental data with scatter, which is inherently inaccurate and yields poor results, especially near saturation where the water content profile is quite flat (Jackson, 1963; Clothier et al., 1983). The latter authors showed that it is much better to find a value for a parameter p by fitting the experimental $\tau[\theta]$ data to the function

$$\tau[\theta] = \epsilon(1 - \Theta)^p \qquad \text{for } p > 0 \tag{14}$$

where ϵ is a parameter that can be derived from p and the sorptivity, and Θ is the dimensionless soil water content

$$\Theta = \frac{(\theta - \theta_0)}{(\theta_1 - \theta_0)} \tag{15}$$

where θ_1 is the final water content at the adsorption/desorption interface and θ_0 is the initial water content. The corresponding equation for the diffusivity is then

$$D[\theta] = p(p + 1)S^2 \frac{(1 - \theta)^{p-1} - (1 - \theta)^{2p}}{2(\theta_s - \theta_0)^2} \tag{16}$$

This analysis of the experimental data ensures correct integral properties of the $D[\theta]$ function, because it is fitted to the primary data set $\tau[\theta]$ and the measured value of the sorptivity. Moreover, it never leads to physically nonsensical $D[\theta]$ functions that decrease with increasing θ, as least-squares fitting of $\tau[\theta]$ can do. Instead, it yields S-shaped diffusivity curves with infinite diffusivity at saturation (Fig. 7), as observed for many soils (Reichardt and Libardi, 1974).

De Veaux and Steele (1989) proposed another improvement for the analysis of experimental $\tau[\theta]$ data, which yields an estimate for $D[\theta]$ according to Eq. 13 that is guaranteed to be smooth and monotonic, exhibits correct behavior near saturation, and is genuinely guided by the data and not by a preassumed parametric form of the function. Although this method requires specialized knowledge of statistics, it deserves attention, since many smoothing methods lead to virtually useless estimates of $d\tau/d\theta$. With exploratory use of the so-called alternating conditional expectation (ACE) algorithm and the bulge rule, they search for those power transformations $F[\theta]$ and $G[\tau]$ that yield the greatest linear association according to

$$F[\theta] = a + bG[\tau] \tag{17}$$

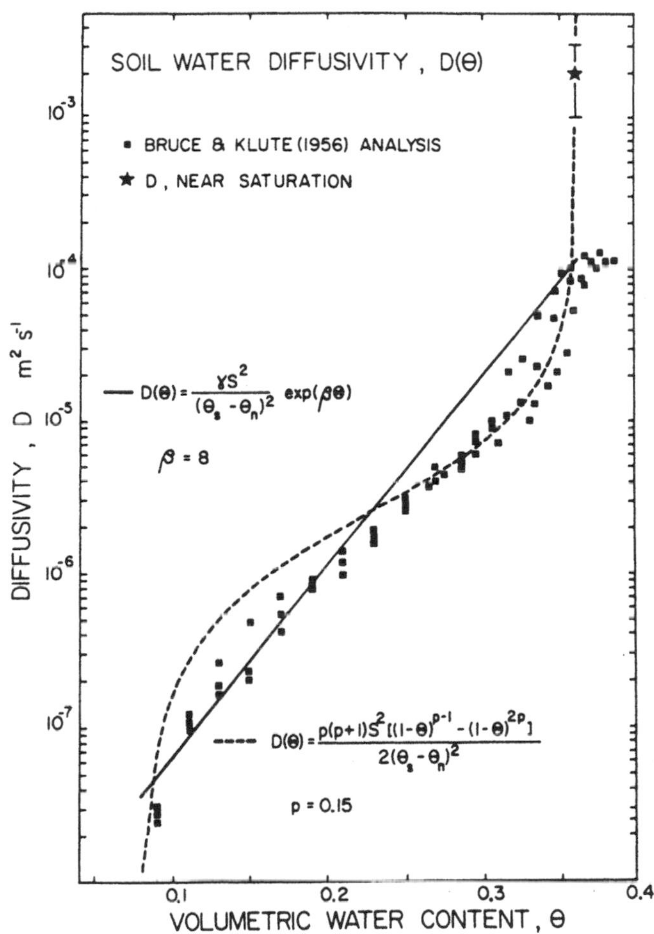

Fig. 7 Diffusivity function derived graphically according to Fig. 6 and derived from fit to Eq. 16, for $p = 0.15$, and diffusivity measured near saturation. (From Clothier et al., 1983.)

De Veaux and Steele (1989) demonstrated the procedure using data for a Manawatu sandy loam (Clothier and Scotter, 1982) and found $F[\theta] = \theta^3$ and $G[\tau] = e^\tau$, $a = 4.48 \times 10^{-2}$ and $b = -1.20 \times 10^{-4}$. The slope indicated in Fig. 6 can then be calculated according to

$$\frac{d\tau}{d\theta} = \frac{F'[\theta]}{bG'[\tau]} = 3\theta^2(\theta^3 - a)^{-1} \tag{18}$$

and the area can be obtained by analytical integration of

$$\tau[\theta] = \log\left(\frac{\theta^3 - a}{b}\right) \tag{19}$$

More details on these improved data analyses are given by the authors.

D. Hot Air

A third variant of the Boltzmann transform method is the "hot air" method (Arya et al., 1975). It has become quite popular in some areas due to the simplicity and speed of the required measurements, and the large range of θ over which $D[\theta]$ values are obtained. It is the drying counterpart of the Bruce and Klute (1956) variant. However, it has not only all the disadvantages of this variant, but also many others. Whereas the required boundary condition of a step-function change in potential (water content) can be attained easily in the case of wetting, a drying step-function is nearly impossible experimentally. It is imposed by a stream of hot air directed at the soil surface, while the rest of the soil column (usually 10 cm long and 5 cm diameter) is shielded from it as much as possible. Air temperatures of up to 240°C have been required for sandy soils. Even then it takes normally several minutes to dry the soil surface, while the total evaporation period normally lasts from 10 to 15 minutes. Whereas temperatures in excess of 90°C have been measured in the soil (Van Grinsven et al., 1985), the data can be analyzed only by assuming isothermal conditions. The effects of temperature on variables (viscosity, surface tension, etc.) and of any water transport due to the thermal gradient are significant but are ignored. Because the soil is hot, there is significant water loss due to evaporation during sampling. The method has been performed on initially saturated, vertically oriented soil columns. Ensuing errors due to gravity, and loss of water as a result of compaction at the wet end during sampling, can be reduced by equilibrating the soil column first at a moderate negative pressure head (around −30 cm).

Often the hot air method appeared to yield useful results, but this is likely to be accidental; several sources of errors tend to cancel each other (Van Grinsven et al., 1985). Even if the obtained $D[\theta]$ function is kept within the theoretically acceptable framework by analyzing the $\tau[\theta]$ data with specially devised software (Van den Berg and Louters, 1986) or using the improved data analyses mentioned above, the result is still based on very dubious experimental measurements. I feel, therefore, that the hot air method should be abandoned. It may be possible to find a way to impose the boundary condition by using hygroscopic agents, eliminating the temperature effects, but in view of all the other objections this does not seem worth the effort. In this connection, it should be pointed out that it is not necessary to dry the soil instantaneously at the surface; only a constant water content or pressure head must be imposed. This does not need to go beyond the range over which the diffusivity or conductivity function is required.

E. Flux-Controlled Sorptivity

This method entails the determination of the sorptivity S as function of the water content at the absorption interface, θ_1, for constant initial water content, θ_0 (Dirksen, 1975, 1979; Klute and Dirksen, 1986). This can be accomplished by means of a series of one-dimensional absorption runs, each yielding one set of (S, θ_1) values. The wetting hydraulic diffusivity function can then be calculated from this experimentally determined $S[\theta_1, \theta_0 = \text{constant}]$ relationship according to

$$D[\theta_1]_w = \frac{\pi S^2}{4(\theta_1 - \theta_0)^2}\left(\frac{\theta_1 - \theta_0}{(1 + \gamma)\log e}\frac{\partial}{\partial\theta_1}(\log S^2) - \frac{1 - \gamma}{1 + \gamma}\right) \tag{20}$$

The value of the weighting parameter γ can be varied between 0.50 and 0.67 without significant effect ($\gamma = 0.62$ is recommended).

Sorptivity measurements require only one controlled boundary. Many experimental problems encountered with a potential-controlled boundary could be eliminated by using a flux-controlled boundary. Sorptivities are imposed by driving a syringe pump so that the cumulative volume of water delivered is proportional to the square root of the elapsed time (Eq. 7). Problems in doing this with shaped rotating disks have now been solved by driving the syringe with a fine-threaded rod rotated by a stepping motor (Dirksen, 1999). One electrical pulse advances the rotor only 1/400th of a revolution. A PC calculates and generates the number of electrical pulses required as a function of time to produce the sorptivity, specified as the value of $\log[S^2/(\text{mm}^2\ \text{s}^{-1})]$. For most soils, this value varies between -0.5 and -5.0 from saturation to wilting point.

For each run, a flat (dry) soil surface must be carefully prepared. After each run, only a thin slice of soil from the top is needed to determine θ_1 gravimetrically. With a specially designed soil column apparatus, the soil surface preparation is facilitated, the porous plate can be brought in contact with the soil at exactly the same time as the pump is started, and the one soil sample at the end of each run can be obtained in less than 10 seconds (Dirksen, 1999). This virtually eliminates errors due to evaporation and redistribution during sampling. Moreover, near the soil surface θ changes neither with time, nor with position, limiting experimental errors even further. The differentiation required in Eq. 20 is performed algebraically on a polynomial regression of $\log S^2$ in terms of θ_1. All this keeps the effect of error propagation in the calculation of $D[\theta_1]_w$ to a minimum.

The sorptivity method is especially attractive because it combines the speed of transient methods with the experimental simplicity and accuracy of stationary measurements. Depending on the desired accuracy, a diffusivity function can be obtained from 1 to 3 soil columns of 10 cm length. By first air-drying these columns, the required uniform initial water content is easily obtained, and a maximum water content range can be covered. The effect of nonuniformity of soil samples on the final results still requires further investigation. The theoretical basis of Eq. 20, although not rigorously exact, appears to be accurate (Dirksen, 1975; Brutsaert, 1976; White and Perroux, 1987). Although water is applied through porous plates, diffusivities well beyond the "tensiometer range" have been obtained. Individual runs need to be continued for only a few minutes near saturation to a few hours when the final water content is very low. This means that a hydraulic diffusivity function can be explored in about 1 day and measured accurately in a few days. For accurate results, the method requires special-purpose, custom-made apparatus.

The hydraulic conductivity function can be calculated according to Eq. 4. This must be done with the wetting soil water retention characteristic, $\theta[h]_w$, which usually is not available. It has been obtained by measuring during the sorptivity runs the pressure head at the soil surface, h_1, with a small tensiometer, mounted slightly protruding in the center of the porous plate, and a sensitive pressure transducer. The line in Fig. 8 indicated by "sorptivity method" was obtained by such simultaneous measurements; only 7 sorptivity runs each lasting from 6 to 12 minutes yielded k values for water contents less than $\theta = 0.10$ (Dirksen, 1979). The results with the instantaneous profile method, obtained on cores of the same packed soil, required several weeks and still yielded k values only for water contents larger than 0.20.

F. Instantaneous Profile

The instantaneous profile method, in its many variants, is probably the most used method to determine the hydraulic conductivity function $k[\theta]$ of laboratory soil columns nondestructively under transient conditions. Quite sophisticated, automated equipment for measuring soil water content and hydraulic heads can allow more complete and/or accurate determination of $k[\theta]$ than is normally the case. This is reflected in the higher scores for this method as a laboratory method, in comparison with the scores as a field method in Table 1. Since this method is especially suited for use in situ, it is discussed in more detail in the next section.

G. Wind Evaporation Method

Wind (1966) proposed a simplified instantaneous profile method to measure simultaneously the water retention characteristic and the hydraulic conductivity of the same soil sample. An initially saturated and homogeneous sample is allowed to evaporate at the top. The total weight and the pressure heads at at least two depths are recorded. From these data the water retention characteristic is calcu-

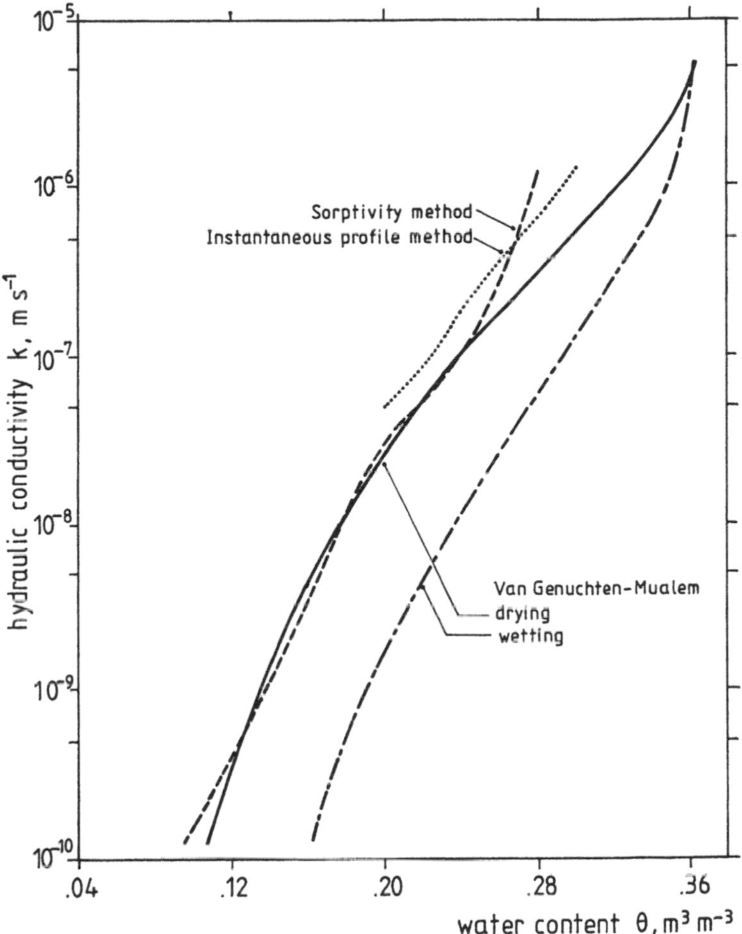

Fig. 8 Hydraulic conductivity functions of Pachappa sandy loam measured with the flux-controlled sorptivity method (and simultaneously measured pressure heads), and with the instantaneous profile method. The Van Genuchten–Mualem functions (Eq. 28) are based on the fitted soil water retention characteristics in Fig. 9.

lated by an iterative method. With this and the measured pressure heads, the water contents per compartment around the tensiometers can be calculated. Then, from the known flux densities at the bottom (zero) and the top (measured evaporation rate), flux densities in between and the hydraulic conductivities can be determined similar to the instantaneous profile method. Boels et al. (1978) designed an automatic recording system for these measurements on many soil samples. They also proposed a direct calculation method by approximating the soil water retention characteristic by a polygon. Tamari et al. (1993) and Wendroth et al. (1993) found that results obtained with this modification compared well with computer simulations, except at water contents near saturation. An error analysis of this method was presented recently by Mohrath et al. (1997). Since its initiation, this Wind evaporation method has been modified and improved so that it is now the major method at the institute where it was developed; the measurements are fully automated, and all calculations can be made with customized computer programs. The data from these experiments can also be used for the inverse parameter optimization approach (Sec. XI).

IX. TRANSIENT FIELD METHODS

A. Instantaneous Profile

The relative merits of laboratory and field measurements were discussed in Sec. IV. It was argued that only special circumstances, such as many thin soil layers or large, unstable structural elements, warrant in situ determinations of the unsaturated hydraulic conductivity function. The instantaneous profile method, in particular the unsteady drainage flux variant (Watson, 1966; Klute, 1972; Hillel et al., 1972; Green et al., 1986), is well suited for this. It requires measuring of water contents and hydraulic potentials as function of time and depth during drainage of an initially saturated, bare soil profile. When the water flux density q is known for all time t at one depth z_0, the flux density at any depth and time can be calculated from the water contents according to

$$q[z, t] = q[z_0, t] - \int_{z_0}^{z} \frac{\partial\theta[z, t]}{\partial t} \partial z \tag{21}$$

This equation assumes vertical transport only, without root uptake. The boundary condition $q[z_0, t]$ is usually set as a zero flux at the soil surface obtained by covering the surface to prevent evaporation. Hydraulic conductivity at any time and depth can then be determined by combining the flux density according to Eq. 21 and the measured hydraulic potential gradients (if needed after smoothing and interpolation) according to

$$k[\theta, z_i, t_j] = -\frac{q[z_i, t_j]}{(\partial H/\partial z)\,[z_i, t_j]} \tag{22}$$

Hydraulic conductivities can thus be obtained for any soil layer between two tensiometers. Also, a soil water retention characteristic for any position can be compiled from corresponding measured θ and h values.

The range of water contents that can be covered is limited at the wet end by the degree of saturation that can be attained by ponding water on the soil surface. This is often no more than 90% of the available poor volume because air tends to be entrapped by the wetting front. At the drier end, the water content range is limited by the drainage characteristics of the particular soil in its hydrological setting. At first, near saturation, θ and H should be measured as frequently as possible, because they vary so quickly that it is hard to obtain accurate results without automated data collection. After the first few days, further accurately measurable differences in water contents will take days or weeks (cf. field capacity), and even then will yield k values only for pressure heads that usually do not go below -200 cm. Thus the main disadvantage is the limited range of θ and h over which $k[\theta]$ can be determined.

The error propagation analysis of Flühler et al. (1976) is not very encouraging; especially toward the dry end, errors can be very large. At small times tensiometer errors predominate, while later water content measurements introduce the largest errors. To reduce errors in fine-textured soils, water content measurements should be intensified; in coarse-textured soils it is better to increase the number and/or frequency of tensiometer measurements. When the draining surface area is large, water contents could be determined gravimetrically by taking soil samples with an auger; otherwise, indirect nondestructive measurements by neutron scattering, TDR, etc. must be taken. Hydraulic potentials should be measured directly with tensiometers with good depth resolution and accurate pressure measuring devices. The h-range can be expanded by allowing evaporation from the soil surface and determining the zero-flux plane from the tensiometer data (Richards et al., 1956). However, the overall results will be even less accurate. The same is true if only either water contents or hydraulic potentials are measured and the others are derived from an independently determined soil water retention characteristic.

B. Unit Gradient with Prescribed *k*-Function

With the present emphasis on studying the spatial variability of soil hydraulic properties, there is a need for simple in situ measurements. Tensiometric measurements are much less convenient for this purpose than indirect water content measurements. A simplified version of the instantaneous profile method involving only water content measurements was used by Jones and Wagenet (1984). They installed 100 neutron access tubes in a 50×100 m fallow field and wetted the soil around them by ponding water in rings 37 cm in diameter, inserted 15 cm into the soil. When water contents were steady down to 120 cm, the access tube sites were covered and redistribution was followed for 10 days. At the end, gravimetric samples were taken to back up the neutron measurements. The results were analyzed in five somewhat different ways, all assuming the hydraulic gradient to be unity at all times, and exponential hydraulic conductivity functions

$$k[\theta] = k_0 \exp(\beta(\theta - \theta_0)) \tag{23}$$

where k_0 and θ_0 are values measured during steady ponded infiltration, sometimes called "satiation." All five analyses yielded values of the constants k_0 and β, with their mean and variance, for selected depths. The difference between the analyses mostly concerned further assumptions on the water content distributions. Jones and Wagenet concluded that the five approximate analyses will be most useful in developing relatively rapid preliminary estimates of soil water properties over large areas, but not as useful when k_0 and β at a particular location need to be known precisely.

C. Simple Unit Gradient

In an even more simplified version, uniform water content and pressure head (unit hydraulic gradient) are assumed throughout the draining profile (Green et al., 1986). This implies that the increase of k with depth, needed to accommodate the increasing flux density with depth, is assumed to occur with a negligible increase of θ. The hydraulic conductivity is then

$$k[\theta^*] = L\frac{d\theta^*}{dt} \tag{24}$$

where θ^* is the average water content of the profile above depth L. With a single tensiometer at depth L and making the same assumptions, the diffusivity can be determined analogously (Gardner, 1970):

$$D[h] = L\frac{dh}{dt} \tag{25}$$

Unless the soil profile is highly uniform, it is doubtful that these versions can yield results better than an educated guess.

D. Sprinkling Infiltrometer

If hydraulic properties must be known for wetting conditions, the instantaneous profile analysis may be used on transient data obtained with a sprinkling infiltrometer. This equipment is unlikely to be used much for this purpose, however, since it is quite elaborate and normally must be attended whenever it is in operation (Sec. VII.A).

E. Sorptivity Measurements

Sorptivity is the first term in the Philip infiltration equation (Philip, 1969) and is a function of θ_1 and θ_0 (see Secs. II.D and VIII.F). This function contains composite information on other soil hydraulic transport properties (Brutsaert, 1976; White and Perroux, 1987), which can be obtained mathematically. Sorptivity can be measured rather easily in the field (Talsma, 1969; Dirksen, 1975; Clothier and White, 1981), including during the first few minutes of tension disk infiltrometer measurements (Perroux and White, 1988) as discussed in Chap. 6. Measuring at very small negative pressure heads prevents macropores from dominating "saturated" sorptivity measurements.

X. DERIVATION FROM OTHER SOIL PROPERTIES

Physical measurements of soil hydraulic conductivities are time-consuming and tedious, and therefore expensive. Moreover, despite considerable effort, the accuracy most often is very poor. With the tremendous variability in hydraulic conductivity, both in space and in time, the practical value of such measurements is difficult to estimate. It is worthwhile, therefore, to consider the possibility of deriving hydraulic conductivity from more easily measured soil properties. In particular, soil water retention characteristics and soil textural data have been used to derive so-called pedotransfer functions. Scaling relationships can also be used for this purpose. More details on these and other indirect methods for estimating the hydraulic properties of unsaturated soils can be found in Van Genuchten et al. (1992, 1999).

A. Soil Water Retention Characteristic

The pressure difference across an air–water interface is inversely proportional to the equivalent diameter of that interface. Thus in the range of water contents where capillary binding of water is predominant, the soil water retention characteristic reflects the pore size distribution. The water content at any given pressure head is equal to the porosity contributed by the pores that are smaller than the equivalent diameter corresponding to that pressure head. To derive the hydraulic conductivity, Childs and Collis-George (1950) converted the soil water retention characteristic into an equivalent pore size distribution, distinguishing a number of pore size classes. Then they assumed that (1) if two imaginary cross-sections of a soil were to be brought into contact with each other, the hydraulic conductivity of the assembly depends on the number and sizes of pores on each side that connect up with each other; (2) the pores are randomly distributed and thus the chances of pores of two sizes connecting is proportional to the product of the relative contributions of their respective pore size classes to the total cross-sectional area; (3) since the contribution of a pore to the hydraulic conductivity is proportional to the square of its radius, the flow through two matching pores is determined by the smallest of the two. The hydraulic conductivity function can then be calculated by carrying out the calculations for each water content up to the pore radius that is still just water filled. Jackson (1972) reviewed various versions of this calculation procedure (e.g., Marshall, 1958) and proposed a simpler calculation procedure without making basic changes. For an example calculation, see Hillel (1980, p. 223). One measured value of hydraulic conductivity is used to correct the calculated curve. Experimental tests of this approach (Green and Corey, 1971; Jackson et al., 1965; Jackson, 1972) found that the correction factor based on measured saturated hydraulic conductivities was unpredictable and varied between 2.0 and 0.004. The shape of the theoretical and experimental hydraulic conductivity functions also differed, sometimes substantially.

Another approach to calculating soil hydraulic conductivities from soil water retention characteristics originated in petroleum engineering and is based on the generalized Kozeny equation. It was introduced into the soil literature by Brooks and Corey (1964); a good summary of this theory and the final working equations can be found in Laliberte et al. (1968). The determinations of the pore size distribution index, air-entry value of pressure head, and residual saturation, required for the Brooks and Corey equations, are also not always straightforward. Brooks, Corey, and their coworkers invariably tested these equations with the hydrocarbon fluid "Soltrol," which has altogether different soil wetting properties than water. There is, therefore, some doubt whether these equations are valid for soil–water systems. Van Schaik (1970) found large internal discrepancies, even for studies that have been claimed to yield the best results for the Brooks and Corey equations. For these reasons, I would caution against the use of these equations.

B. Van Genuchten–Mualem Equations

Mualem (1986b) introduced a few basic changes to the theory of Childs and Collis-George. For instance, he calculated the contribution to the hydraulic conductivity of a larger pore (r_1) following a smaller one (r_2), by assuming that the length of a pore is equal to its diameter and defining an equivalent radius of the two pores as $(r_1 r_2)^{1/2}$. Combining his theory with elements of that of Brooks and Corey (1964) and of Burdine (1953), Mualem derived a simple dimensionless relationship for the relative hydraulic conductivity, k_r (ratio of the value to that at saturation) and found quite good agreement with experimental data for 45 soils. Van Genuchten (1980) combined this relationship with a newly proposed approximation for the soil water retention characteristic to yield the following set of equations.

$$\Theta = \frac{(\theta - \theta_r)}{(\theta_s - \theta_r)} \tag{26}$$

$$\Theta = \frac{1}{(1 + \alpha|h|^n)^m}, \; m = 1 - \frac{1}{n} \tag{27}$$

$$k = k_{ref}\Theta^{\lambda}(1 - (1 - \Theta^{1/m})^m)^2 \tag{28}$$

where Θ is the dimensionless water content, θ_r is the residual water content at which the hydraulic conductivity becomes negligibly small, θ_s is the saturated water content, and α, λ, n, and m are fitting parameters. Fitting of soil water retention data by Eq. 27 and substituting the parameter values obtained into Eq. 28 yields a relative hydraulic conductivity function $k_r = k/k_{ref}$. To obtain absolute hydraulic conductivities, the value of k_{ref} must be determined. According to Eq. 28, this is the hydraulic conductivity at $\Theta = 1$. It is common practice, therefore, to use measured saturated hydraulic conductivities to match calculated and measured values. In general, this is about the worst choice for k_{ref}. The standard deviation of such measurements is normally very large, since they can be totally dominated by wormholes, old root channels, fractures resulting from poor sampling procedures, etc. More importantly, such features have no relation with the pore size distribution of the soil matrix. At small negative pressure heads, all large spaces not associated with the soil matrix are empty and do not conduct water. Therefore, I recommend that k_{ref} be derived from a measurement of k (and θ) at a small tension. This can be done accurately and fast with a head-controlled setup as in Fig. 2 (Dirksen, 1999).

Figure 9 shows the fits of Eq. 27 to experimental wetting and drying soil water retention characteristics of Pachappa fine sandy loam. The corresponding absolute hydraulic conductivity functions according to Eq. 28 were given in Fig. 8. The reference hydraulic conductivity was derived from measurements at "satiation" ($\theta = 0.36$). The comparison with the experimental hydraulic conductivity data is very good for the drying optimized parameter values, especially in the drier range, but very poor for the wetting values. The reason for this is not clear, nor whether this result can be expected generally. For extensive reviews of this and other models to calculate hydraulic conductivities, see Van Genuchten and Nielsen (1985) and Mualem (1986a).

Van Genuchten and his colleagues (Leij et al., 1992; Yates et al., 1992) have developed a program, RETC, that optimizes part or all of the parameters in Eq. 26 to Eq. 28: n, m, α, λ, θ_r, θ_s, and k_s. The optimization can be performed on differently weighted experimental data of $h[\theta]$ as well as $k[\theta]$. The relationship between n and m, given with Eq. 27, is optional. The exponent λ of Θ is usually fixed at the value of $\frac{1}{2}$.

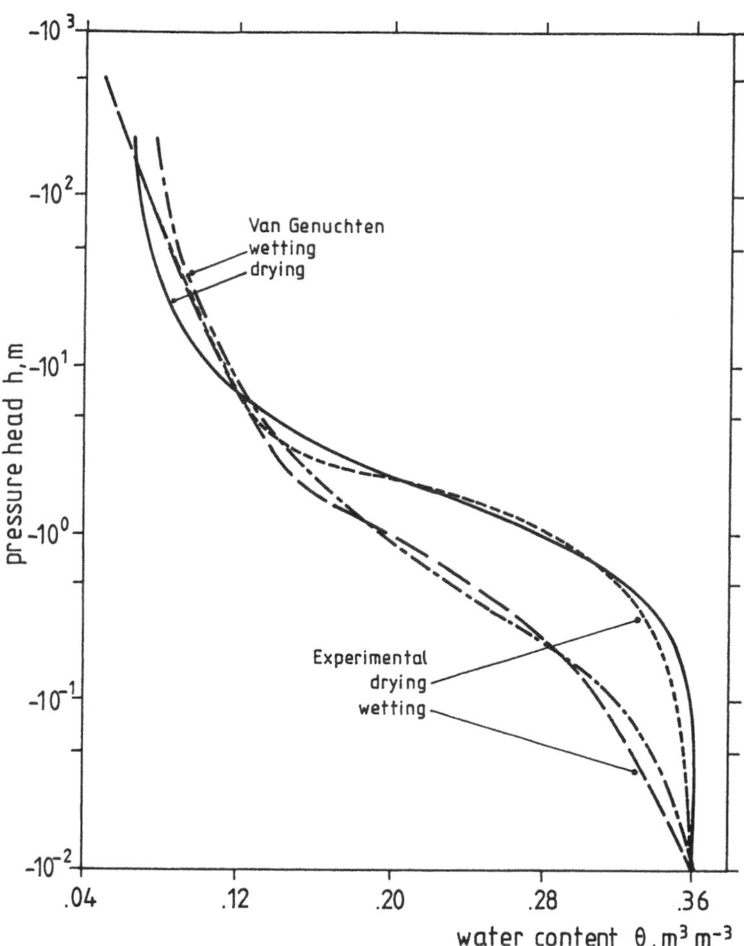

Fig. 9 Soil water retention characteristics of Pachappa sandy loam composed of various experimental data, and the fits of these to Eq. 27. The corresponding hydraulic conductivity functions according to Van Genuchten–Mualem are shown in Fig. 8.

C. Soil Texture

Soil water retention characteristics and hydraulic conductivities have been correlated with soil textural data (Bloemen, 1980; Schuh and Bauder, 1986; Wösten and Van Genuchten, 1988; Vereecken et al., 1990). These so-called pedotransfer functions lack a direct physical basis and must be regarded as statistics. To obtain them still requires many direct measurements, while it remains uncertain whether they can be extrapolated to other soils. Espino et al. (1995) evaluated the use of pedotransfer functions for estimating soil hydraulic properties and sounded many cautionary notes.

D. Scaling

If the scaling relationships of Miller and Miller (1956; see also Miller, 1980) are assumed, soil hydraulic properties can often be determined with much less work than otherwise required. For example, Reichardt et al. (1972) measured hydraulic diffusivities of 12 different soils with the fixed-time Boltzmann method (Bruce and Klute, 1956) and converted these to hydraulic conductivities according to Eq. 4. When these hydraulic conductivities were scaled according to the square of a characteristic microscopic length λ, the data coalesced nicely into one relationship (Fig. 10). For k in cm/s, the solid line in Fig. 10 can be described by (Reichardt et al., 1975)

$$k[\theta] = 1.942 \times 10^{-12} m^4 \exp(-12.235\theta^2 + 28.061\theta) \tag{29}$$

λ was assumed proportional to the square of the slope m of the linear relationship between advance of wetting front and the square root of time during horizontal infiltration (see Eq. 7) and is listed for each soil in Fig. 10 as a ratio to the value of the standard soil. If a soil belongs to the group for which this assumed scaling relationship is valid (which normally will not be known beforehand and must be verified), the hydraulic conductivity function can be obtained with Eq. 29 and just one simple, short infiltration run to measure m, θ_1, and θ_0.

Miller and Bresler (1977) showed that the experimental data of Reichardt et al. (1972), on which Eq. 29 is based, can be transformed to what they suggest is a "universal" equation for the diffusivity:

$$D[\theta] = \alpha m^2 \exp(\beta\theta) \tag{30}$$

with $\alpha = 10^{-3}$ and $\beta = 8$.

Bresler et al. (1978) derived a relationship for the hydraulic conductivity from the same experimental data:

$$k[\theta] = 0.27\ m^4\ \theta^{7.2} \tag{31}$$

XI. PARAMETER OPTIMIZATION

About 30 years ago, the so-called inverse approach for determining the soil hydraulic properties was proposed (Whisler and Watson, 1968; Skaggs et al., 1971), but it found little acceptance due to limitations in mathematical and computational facilities. Recently, the inverse approach has received renewed attention as a parameter optimization technique. It calls for the performance of a relatively simple

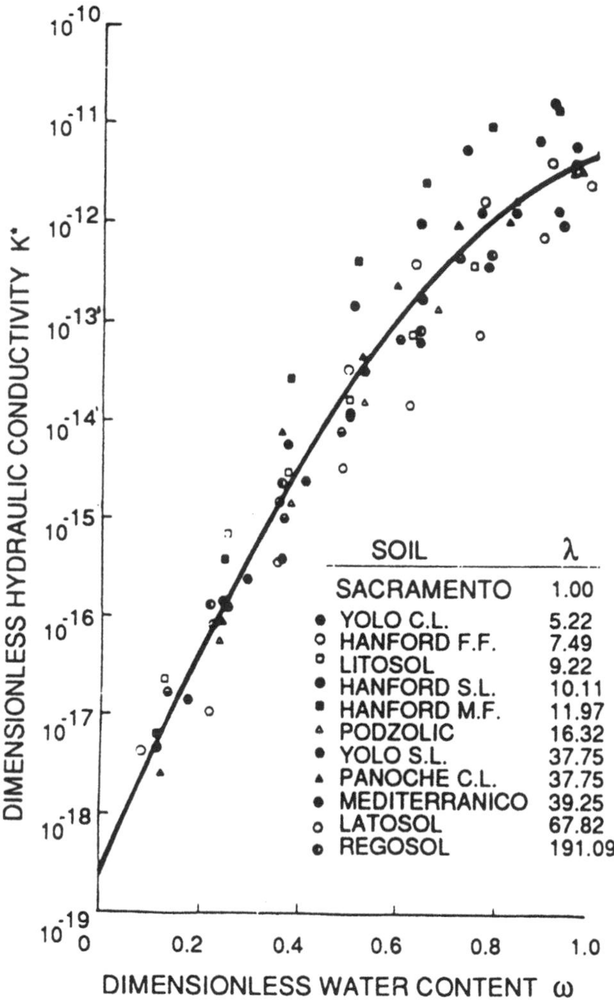

Fig. 10 Hydraulic conductivities of 12 soils scaled according to λ^2 (or m^4) versus dimensionless water content. (From Reichardt et al., 1975.)

experiment with inherently accurate measurements. Assuming algebraic forms of the hydraulic property functions, such as Eqs. 26 to 28, the water transport process is then simulated on a computer, starting with guessed values of the parameters in the transport functions and repeated with newly estimated values until the simulated results agree with the experimental results to within the desired degree of accuracy. Thus the problem is reduced to optimizing the parameters in the hydraulic property functions. Optimization is a specialized mathematical technique which is still being improved. With the progress in computer capabilities and the development of adapted programs, it has become attractive for determining the soil hydraulic property functions indirectly. More details can be found in the review by Kool et al. (1987) and in Van Genuchten et al. (1992). The merits of this inverse approach should be evaluated in a decision how to determine the hydraulic transport functions in a given situation.

Whereas in principle many flow systems with different initial and/or boundary conditions can be used for the parameter optimization, the one-step outflow method appears especially suitable (Kool et al., 1985a; Parker et al., 1985; Van Dam et al., 1992). It only requires inherently accurate measurements of the cumulative (external) outflow as function of time from an initially saturated short soil column in a pressure cell as a result of an applied step-increase of the pneumatic pressure. It allows a large water content range to be covered in a reasonably short time. The influence of the resistance of the porous plate on the outflow, which complicates the traditional analysis of the experimental results, is easily accounted for in the simulation. The program ONE-STEP (Kool et al., 1985b) and its modifications (e.g., Van Dam et al., 1992) have been used by many investigators for the parameter optimization. Lately, the multistep variant is advocated as being even more suitable (Van Dam et al.,1994; Eching et al., 1994). One dimensional infiltration (Sir et al., 1988) and drainage (Zachman et al., 1981; Dane and Hruska, 1985) have also been used for optimization, but these are less attractive flow processes.

A major aspect of the parameter optimization technique is *convergence*. The first guess of the parameter values may be so far off from the actual values that the optimization procedure cannot yield the correct values or can do this only after a prohibitively long computing time. For the optimization of the parameters in Eqs. 26 to 28 based on experimental one-step outflow data, Parker et al. (1985) suggest as first guess for medium textured soils the values of $\alpha = 2.50\ m^{-1}$, $n = 1.75$ and $\theta_r = 0.150$, with suitable adjustments for differently textured soils. Nielsen and Luckner (1992) discussed theoretical aspects for estimating initial parameter values. Convergence also may be a problem when too little information is contained in the input data. Therefore, the input data should cover as large a range of water contents, time, etc. as is practical. If the solution fails to converge after a specified maximum number of function evaluations, a new solution can be started with different initial parameter values.

Another aspect of the inverse approach is *uniqueness*: there may be more than one solution to the problem as stated, and the solution obtained may not be the correct one. This is not expected to be a serious problem with the one-step outflow measurements, if the pressure step and the time period are kept relatively large. Eching et al. (1994) found that additional soil water pressure head values yield unique parameter values. Solutions obtained should be verified and, in case of doubt, the optimization process should be repeated with different initial estimates of the parameters.

The *accuracy* of the optimized parameters depends on the accuracy of the experimental data used as input in the optimization procedure. The sensitivity for this source of errors is different for each combination of flow process and parametric function and deserves further study. Of course, if the preselected algebraic functions are incapable of describing the actual soil hydraulic properties accurately, even a perfect optimization process will not yield an accurate result.

XII. SUMMARY AND CONCLUSIONS

Water transport in soils that are not fully saturated plays an important role in hydrology, water uptake by plant roots, irrigation management, transport of pollutants through the environment, and other areas. This transport is to a large extent characterized by the dependence of the hydraulic conductivity k, diffusivity D, matric flux potential ϕ, and sorptivity S, on the volume fraction of water θ. For a given soil, these soil water transport functions vary over several orders of magnitude and can differ by orders of magnitude between soils. Measuring these functions is a difficult task, which continues to absorb much time and effort. Many methods have been proposed, but no single approach is suitable for all conditions and/or purposes. Most methods lack accuracy, take a prohibitively long time, and/or are costly. In general, steady-state methods are more accurate than transient methods, but they take a lot more time and are therefore more expensive. One also must choose between laboratory and field measurements. The former may have many advantages, but they require the acquisition of undisturbed soil samples and the transport of these to the laboratory.

The absolute accuracy of any given method cannot be established by using it on a "standard" porous medium with very accurately known hydraulic properties. As a result, it is standard practice to compare the results obtained by two (or more) different methods, without knowing the accuracy of either of them separately. It is necessary, therefore, to evaluate the available methods on the basis of their inherent features and potential accuracy. Methods of various types were described and evaluated in Table 1 with respect to a number of criteria given in Table 2. Where the highest accuracy is required, methods should be selected according to soundness of theoretical basis (criterion B), control of initial and boundary conditions (C), inherent accuracy of the required measurements (D), and error propagation (E). On these criteria, steady head-controlled (Sec. VI.A) and flux-controlled (Sec. VI.B) measurements on laboratory soil columns both score higher than any other method. It is proposed, therefore, in view of the lack of a "standard" material, to elevate these methods to the status of "standard method," against which other available methods could and should be evaluated.

Flux-controlled conditions offer additional advantages over head-controlled conditions, especially in the dryer range. The hydraulic conductivity to be measured is predictable and thus it will take less time (G), and the practical range of application is likely to be larger (F). This is at the expense of the need for more special purpose equipment (H). Both methods can be used conveniently only over a pressure head range from saturation to about -2.0 m, or a minimum flux density of about 0.1 mm d^{-1} (F). This is normally more than sufficient for hydrological studies. With special effort (parentheses in Table 1) a larger application range can be covered at the expense of more time (G) and better equipment (H). This is justified when a "standard" measurement is needed.

As for the other laboratory methods, the Wind evaporation method scores quite highly on criteria B–F and deserves to be used more widely. The flux-controlled sorptivity method scores highly on most criteria and is particularly attractive for its speed and rather large range of application. Its weak points appear to be the differentiation in the data analysis (E) and the need for special equipment. The requirement of a long, uniform soil column makes the steady-rate infiltration approach impractical and little used.

A major disadvantage of all field methods is that the boundary and initial conditions generally can be controlled only approximately (C). The instantaneous profile method is attractive but has a very limited pressure head range over which it can yield results, even after rather long time periods. The error analysis of Flühler et al. (1976) shows that even with directly measured pressure heads and using only Darcy's law, the accuracy of the final results can be very poor. Use of the sprinkling infiltrometer under steady-state conditions at least eliminates large errors introduced when fluxes are calculated from indirectly measured water contents. Therefore, the sprinkling infiltrometer appears to be the strongest candidate for a standard field method. Operation of this equipment is very cumbersome and time-consuming. However, if accuracy is of overriding importance, criteria of required time (G), investments (H), skill (I), and operator time (J) should play a secondary role.

When accuracy is not as important as speed and minimizing cost, criteria G–J, as well as the potential for simultaneous measurements (K), become dominant. When many simultaneous measurements are made, it is also important (especially when these are carried out by unskilled workers) to provide for some check on the quality of the work (L). The matric flux potential method scores quite high on these criteria and warrants more consideration than it has received. Also the hot air method is very attractive with respect to these criteria. However, the theoretical basis, control of boundary conditions, error propagation, and limitations on measurement accuracy are in my opinion so totally unacceptable that the hot air method should no longer be used. The wetting-type Boltzmann methods do not have the disadvantage of poor boundary control and nonisothermal conditions, but the inaccuracy of the measurements and the unreliability of the analysis thereof are serious disadvantages. The spherical cavity method has a number of attractive features that appear to deserve further investigation. The pressure plate outflow method in its one-step variant is not good as a direct method, due to the approximate nature of the analysis of the experimental data. As a basis for the inverse approach of parameter optimization, however, the simple, accurate measurements involved make this method very attractive. This appears to be true even more for the multistep variant.

The application of very small uniform flux densities to soil surfaces over extended periods of times presents the largest experimental challenge with direct hydraulic conductivity measurements. Given the unpredictability and nonuniformity of the conductivity of the crusts, as they have been made for the "crust method," the potential accuracy of this approach is questionable. Moreover, the pressure head range is very small. The crust method is too cumbersome and too time-consuming to be suitable for routine measurements at many sites. The hypodermic needles with a pulsating pump introduced easy control and predictability of the flux density, while eliminating or improving most of the limiting factors of crusts. This small, simplified version of the sprinkling infiltrometer, however, still has limitations in uniformity and minimum magnitude of the flux density, and it also requires frequent replacement of needles due to clogging. The atomized water spray system described in Sec. VI.B offers significant improvements on these points. Based on my experience, I encourage others to consider using this equipment.

Derivation of the water transport functions from other soil properties may be a good alternative to direct measurements, particularly when absolute accuracy is not of primary importance but many results are required (e.g., studies of spatial or temporal variability as such). Often, the required input data are already available. The Van Genuchten-Mualem model appears to have an edge on other alternatives. It has an adequate theoretical basis, is generally available in user-friendly PC programs (and is, therefore, widely used), and has given good results for many studies. The same model is also used for the parameter optimization technique. This "inverse" approach seeks the values of the parameters of the model that give the best agreement between measured and numerically simulated quantities. It would seem that as the mathematical procedure is further improved in terms of convergence, uniqueness, and accuracy, this approach should be used more and more. This will be true, particularly, if the selected experimental flow system can be tailored to the actual situation and conditions in which the results will be used.

REFERENCES

Ahuja, L. R., and S. A. El-Swaify. 1976. Determining both water characteristics and hydraulic conductivity of a soil core at high water contents from a transient flow experiment. *Soil Sci.* 121:198–204.

Amerman, C. R., D. Hillel, and A. E. Petersen. 1970. A variable-intensity sprinkling infiltrometer. *Soil Sci. Soc. Am. Proc.* 34:830–832.

Ankeny, M. D. 1992. Methods and theory for unconfined infiltration measurement. In: *Advances in Measurement of Soil Physical Properties: Bringing Theory into Practice* (G. C. Topp, W. D. Reynolds, and R. E. Green, eds.). Spec. Publ. No. 30. Madison, WI: SSSA, pp. 123–141.

Ankeny, M. D., T. C. Kaspar, and R. Horton. 1988. Design for an automated tension infiltrometer. *Soil Sci. Soc. Am. J.* 52:893–896.

Arya, L. M., D. A. Farrell, and G. R. Blake. 1975. A field study of soil water depletion patterns in presence of growing soybean roots. I. Determination of hydraulic properties of the soil. *Soil Sci. Soc. Am. J.* 39:424–430.

Baver, L. D., W. H. Gardner, and W. R. Gardner. 1972. *Soil Physics*. 4th ed. New York: John Wiley.

Begemann, H. K. S. Ph. 1988. *The 66 mm Continuous Sampling Apparatus*. Delft, The Netherlands: Delft Soil Mechanics Laboratory.

Bloemen, G. W. 1980. Calculation of hydraulic conductivities of soils from texture and organic matter content. *Z. Pflanzenernähr. Bodenkd.* 143:581–605.

Boels, D., J. B. H. M. Van Gils, G. J. Veerman, and K. E. Wit. 1978. Theory and system of automatic determination of soil moisture characteristics and unsaturated hydraulic conductivities. *Soil Sci.* 126:191–199.

Bouma, J., and J. L. Denning, 1972. Field measurement of unsaturated hydraulic conductivity by infiltration through gypsum crusts. *Soil Sci. Soc. Am. Proc.* 36:846–847.

Bouma, J., D. Hillel, F. D. Hole, and C. R. Amerman. 1971. Field measurement of unsaturated hydraulic conductivity by infiltration through artificial crusts. *Soil Sci. Soc. Am. Proc.* 35:362–364.

Bresler, E., D. Russo, and R. D. Miller. 1978. Rapid estimate of hydraulic conductivity function. *Soil Sci. Soc. Am. J.* 42:170–172.

Brooks, R. H., and A. T. Corey. 1964. *Hydraulic properties of porous media*. Colorado State Univ. Hydrology Paper No. 3.

Bruce, R. R., and A. Klute. 1956. The measurement of soil-moisture diffusivity. *Soil Sci. Soc. Am. Proc.* 20:458–462.

Brutsaert, W. H. 1976. The concise formulation of diffusive sorption of water in a dry soil. *Water Resour. Res.* 12:1118–1124.

Burdine, N. T. 1953. Relative permeability calculations from pore-size distribution data. *Trans. AIME* 198:71–78.

Childs, E. C. 1969. *Introduction to the Physical Basis of Soil Water Phenomena*. New York: John Wiley.

Childs, E. C., and N. Collis-George. 1950. The permeability of porous materials, *Proc. Roy. Soc. Aust.* 201:392–405.

Chung, C. L., S. H. Anderson, C. J. Ganzer, and Z. Haque. 1988. Automated one-step outflow method for measurement of unsaturated hydraulic conductivity. *Agron. Abstr.* 181.

Clothier, B. E., and D. R. Scotter. 1982. Constant-flux infiltration from a hemi-spherical cavity. *Soil Sci. Soc. Am. J.* 46:696–700.

Clothier, B. E., and I. White. 1981. Measurement of sorptivity and soil water diffusivity in the field. *Soil Sci. Soc. Am. J.* 45:241–245.

Clothier, B. E., D. R. Scotter, and A. E. Green. 1983. Diffusivity and one-dimensional absorption experiments. *Soil Sci. Soc. Am. J.* 47:641–644.

Clothier, B. E., M. B. Kirkham, and J. E. McLean. 1992. In situ measurement of the effective transport volume for solute moving through soil. *Soil Sci. Soc. Am. J.* 56: 733–736.

Dane, J. H., and S. Hruska. 1985. In-situ determination of soil hydraulic properties during drainage. *Soil Sci. Soc. Am. J.* 47:619–624.

De Veaux, R. D., and J. M. Steele. 1989. ACE guided-transformation method for estimation of the coefficient of soil-water diffusivity. *Technometrics* 31:91–98.

Dirksen, C. 1974. Measurement of hydraulic conductivity by means of steady, spherically symmetric flows. *Soil Sci. Soc. Am. Proc.* 38:3–8.

Dirksen, C. 1975. Determination of soil water diffusivity by sorptivity measurements. *Soil Sci. Soc. Am. Proc.* 39:22–27; 39:1012–1013.

Dirksen, C. 1979. Flux-controlled sorptivity measurements to determine soil hydraulic property functions. *Soil Sci. Soc. Am. J.* 43:827–834.

Dirksen, C. 1991. Unsaturated hydraulic conductivity. In: *Soil Analysis, Physical Methods* (K. A. Smith and C. E. Mullins, eds.). New York: Marcel Dekker, pp. 209–269.

Dirksen, C. 1999. *Soil Physics Measurements.* Reiskirchen Germany: Catena Verlag.

Dirksen, C., and M. A. Hilhorst. 1994. Calibration of a new frequency domain sensor for soil water content and bulk electrical conductivity. In: *Time Domain Reflectometry in Environmental, Infrastructure, and Mining Applications.* Proc. Symp. and Workshop, Evanston, Illinois, U.S. Bureau of Mines, Special Publ. No. SP 19–94, pp. 143–153.

Dirksen, C., and S. Matula. 1994. Automatic atomized water spray system for soil hydraulic conductivity measurements. *Soil Sci. Soc. Am. J.* 58:319–325.

Doering, E. J. 1965. Soil-water diffusivity by the one-step method. *Soil Sci.* 99:322–326.

Eching, S. O., J. W. Hopmans, and O. Wendroth. 1994. Unsaturated hydraulic conductivity from multistep outflow and soil water pressure data. *Soil Sci. Soc. Am. J.* 58: 687–695.

Espino, A., D. Mallants, M. Vanclooster, and J. Feijen. 1995. Cautionary notes on the use of pedo-transfer functions for estimating soil hydraulic properties. *Agri. Water Mgmt.* 29:235–253.

Flühler, H., M. S. Ardakani, and L. H. Stolzy. 1976. Error propagation in determining hydraulic conductivities from successive water content and pressure head profiles. *Soil Sci. Soc. Am. J.* 40:830–836.

Gardner, W. R., 1956. Calculation of capillary conductivity from pressure plate outflow data. *Soil Sci. Soc. Am. Proc.* 20:317–320.

Gardner, W. R. 1970. Field measurement of soil water diffusivity. *Soil Sci. Soc. Am. Proc.* 34:832–833.

Gardner, W. R., and F. J. Miklich. 1962. Unsaturated conductivity and diffusivity measurements by a constant flux method. *Soil Sci.* 93:271–274.

Green, R. E., and J. C. Corey. 1971. Calculation of hydraulic conductivity: A further evaluation of some predictive methods. *Soil Sci. Soc. Am. Proc.* 35:3–8.

Green, R. E., L. R. Ahuja, and S. K. Chong. 1986. Hydraulic conductivity, diffusivity, and sorptivity of unsaturated soils: Field methods. In: *Methods of Soil Analysis, Part I,*

Physical and Mineralogical Methods. 2d ed. (A. Klute, ed.). Agronomy Monograph No. 9. Madison, WI: Am. Soc. Agron., pp. 771–798.

Gupta, S. C., D. A. Farrell, and W. E. Larson. 1974. Determining effective soil water diffusivities from one-step outflow experiments. *Soil Sci. Soc. Am. Proc.* 38:710–716.

Hanks, R. J. 1992. *Applied Soil Physics, Soil Water and Temperature Applications*. 2d ed. New York: Springer-Verlag.

Hillel, D. 1980. *Fundamentals of Soil Physics*. New York: Academic Press.

Hillel, D., and Y. Benyamini. 1974. Experimental comparison of infiltration and drainage methods for determining unsaturated hydraulic conductivity of a soil profile in situ. In: *Isotope and Radiation Techniques in Soil Physics and Irrigation Studies*. Vienna: IAEA, pp. 271–275.

Hillel, D., and W. R. Gardner. 1969. Steady infiltration into crust-topped profiles. *Soil Sci.* 108:137–142.

Hillel, D., V. D. Krentos, and Y. Stylianou. 1972. Procedure and test of an internal drainage method for measuring soil hydraulic characteristics in-situ. *Soil Sci.* 114:395–400.

Jackson, R. D. 1963. Porosity and soil-water diffusivity relations. *Soil Sci. Soc. Am. Proc.* 27:123–126.

Jackson, R. D. 1972. On the calculation of hydraulic conductivity. *Soil Sci. Soc. Am. Proc.* 36:380–382.

Jackson, R. D., R. J. Reginato, and C. H. M. Van Bavel. 1965. Comparison of measured and calculated hydraulic conductivities of unsaturated soils. *Water Resour. Res.* 1: 375–380.

Jones, A. J., and R. J. Wagenet. 1984. In situ estimation of hydraulic conductivity using simplified methods. *Water Resour. Res.* 20:1620–1626.

Journel, A., and C. Huibregts. 1978. *Mining Geostatistics*. New York: Academic Press.

Kirkham, D., and W. L. Powers. 1972. *Advanced Soil Physics*. New York: John Wiley.

Kleijn, W. B., J. D. Oster, and N. Cook. 1979. A rainfall simulator with nonrepetitious movement of drop outlets. *Soil Sci. Soc. Am. J.* 43:1248–1251.

Klute, A. 1972. The determination of the hydraulic conductivity and diffusivity of unsaturated soils. *Soil Sci.* 113:264–276.

Klute, A., and C. Dirksen. 1986. Hydraulic conductivity and diffusivity: Laboratory methods. In: *Methods of soil Analysis, Part I, Physical and Mineralogical Methods*. 2d ed. (A. Klute, ed.). Agronomy Monograph No. 9. Madison, WI: Am. Soc. Agron., pp. 687–734.

Knight, J. H., and J. R. Philip. 1974. Exact solutions of non-linear diffusion. *J. Eng. Math.* 8:219–227.

Kool, J. B., J. C. Parker, and M. Th. Van Genuchten. 1985a. Determining soil hydraulic properties from one-step outflow experiments by parameter estimation: I. Theory and numerical studies. *Soil Sci. Soc. Am. J.* 49:1348–1354.

Kool, J. B., J. C. Parker, and M. Th. Van Genuchten. 1985b. ONE-STEP: A nonlinear parameter estimation program for evaluating soil hydraulic properties from one-step outflow experiments. Virginia Agric. Exp. Stat. Bull. 85–3.

Kool, J. B., J. C. Parker, and M. Th. Van Genuchten. 1987. Parameter estimation for unsaturated flow and transport models—A review. *J. Hydrol.* 91:255–293.

Koorevaar, P., G. Menelik, and C. Dirksen. 1983. *Elements of Soil Physics.* Amsterdam: Elsevier.

Kutilek, M., and D. R. Nielsen. 1994. *Soil Hydrology.* Reiskirchen, Germany: Catena Verlag.

Lal, R. and D. J. Greenland. 1979. *Soil Physical properties and Crop Production in the Tropics, Part 8. Soil and Water Conservation.* New York: John Wiley.

Laliberte, G. E., R. H. Brooks, and A. T. Corey. 1968. Permeability calculated from desaturated data. *J. Irrig. Drainage Div. ASCE* 94:57–71.

Leij, F. J., M. Th. Van Genuchten, S. R. Yates, W. B. Russell, and F. Kaveh. 1992. RETC: A computer program for analyzing soil water retention and hydraulic conductivity data. In: *Indirect Methods for Estimating the Hydraulic Properties of Unsaturated Soils* (M. Th. Van Genuchten, F. J. Leij, and L. J. Lund, eds.). Riverside: Univ. of California, pp. 263–272.

Marshall, T. J. 1958. A relation between permeability and size distribution of pores. *J. Soil Sci.* 9:1–8.

Miller, E. E. 1980. Similitude and scaling of soil water phenomena. In: *Applications of Soil Physics* (D. Hillel, ed.). New York: Academic Press, pp. 300–318.

Miller, E. E., and R. D. Miller. 1956. Physical theory for capillary flow phenomena. *J. Appl. Phys.* 27:324–332.

Miller, R. D., and E. Bresler. 1977. A quick method for estimating soil water diffusivity functions. *Soil Sci. Soc. Am. J.* 41:1022–1022.

Mohrath, D., L. Bruckler, P. Bertuzzi, J. C. Gaudu, and M. Bourlet. 1997. Error analysis of an evaporation method for determining hydrodynamic properties in unsaturated soil. *Soil Sci. Soc. Am. J.* 61:725–735.

Mualem, Y. 1986a. Hydraulic conductivity of unsaturated soils: Prediction and formulas. In: *Methods of Soil Analysis, Part I, Physical and Mineralogical Methods.* 2d ed. (A. Klute, ed.). Agronomy Monograph No. 9. Madison, WI: Am. Soc. Agron., pp. 799–823.

Mualem, Y. 1986b. A new model for predicting the hydraulic conductivity of unsaturated porous media. *Water Resour. Res.* 12:513–522.

Nielsen, D. R., and L. Luckner. 1992. Theoretical aspects to estimate reasonable initial parameters and range limits in identification procedures for soil hydraulic properties. In: *Indirect Methods for Estimating the Hydraulic Properties of Unsaturated Soils* (M. Th. Van Genuchten, F. J. Leij, and L. J. Lund, eds.). Riverside: Univ. of California, pp. 147–160.

Parker, J. C., J. B. Kool, and M. Th. Van Genuchten. 1985. Determining soil hydraulic properties from one-step outflow experiments by parameter estimation: II. Experimental studies. *Soil Sci. Soc. Am. J.* 49:1354–1359.

Passioura, J. B. 1976. Determining soil water diffusivities from one-step outflow experiments. *Aust. J. Soil Res.* 15:1–8.

Peck, A. J. 1980. Field variability of soil physical properties. In: *Advances in Irrigation,* Vol. 2. New York: Academic Press, pp. 189–221.

Perroux, K. M., and I. White. 1988. Designs for disc infiltrometers. *Soil Sci. Soc. Am. J.* 52:1205–1215.

Petersen, A. E., and G. D. Bubenzer. 1986. Intake rate: Sprinkler infiltrometer. In: *Methods of Soil Analysis, Part I, Physical and Mineralogical Methods*. 2d ed. (A. Klute, ed.). Agronomy Monograph No. 9. Madison, WI: Am. Soc. Agron., pp. 845–870.

Philip, J. R. 1969. Theory of infiltration. *Hydroscience* 5:215–296.

Quadri, M. B., B. E. Clothier, R. Angulo-Jaramillo, M. Vauclin, and S. R. Green. 1994. Axisymmetric transport of water and solute underneath a disc permeameter: Experiments and numerical model. *Soil Sci. Soc. Am. J.* 58:696–703.

Raats, P. A. C. 1977. Laterally confined, steady flows of water from sources and to sinks in unsaturated soils. *Soil Sci. Soc. Am. J.* 41:294–304.

Raats, P. A. C., and W. R. Gardner. 1971. Comparison of empirical relationships between pressure head and hydraulic conductivity and some observations on radially symmetric flow. *Water Resour. Res.* 7:921–928.

Rawitz, E., M. Margolin, and D. Hillel. 1972. An improved variable intensity sprinkling infiltrometer. *Soil Sci. Soc. Am. Proc.* 36:533–535.

Reichardt, K., and P. L. Libardi. 1974. A new equation for the estimation of soil water diffusivity. In: *Isotopes and Radiation Techniques in Studies of Soil Physics, Irrigation and Drainage in Relation to Crop Production*. Vienna: IAEA, pp. 45–51.

Reichardt, K., D. R. Nielsen, and J. W. Biggar. 1972. Scaling of horizontal infiltration into homogeneous soils. *Soil Sci. Soc. Am. Proc.* 36:241–245.

Reichardt, K., P. L. Libardi, and D. R. Nielsen. 1975. Unsaturated hydraulic conductivity determination by a scaling technique. *Soil Sci.* 120:165–168.

Richards, L. A., W. R. Gardner, and G. Ogata. 1956. Physical processes determining water loss from soil. *Soil Sci. Soc. Am. Proc.* 20:310–314.

Schuh, W. M., and J. W. Bauder. 1986. Effect of soil properties on hydraulic conductivity–moisture relationships. *Soil Sci. Soc. J.* 50:848–854.

Scotter, D. R., and B. E. Clothier. 1983. A transient method for measuring soil water diffusivity and unsaturated hydraulic conductivity. *Soil Sci. Soc. Am. J.* 47:1069–1072.

Scotter, D. R., B. E. Clothier, and E. R. Harper. 1982. Measuring saturated hydraulic conductivity and sorptivity using twin rings. *Aust. J. Soil Res.* 20:295–304.

Selim, H. M., D. Kirkham, and M. Amemiya. 1970. A comparison of two methods for determining soil water diffusivity. *Soil Sci. Soc. Am. Proc.* 34:14–18.

Shani, U., R. J. Hanks, E. Bresler, and C. A. S. Oliveira. 1987. Field method for estimating hydraulic conductivity and matric potential-water content relations. *Soil Sci. Soc. Am. J.* 51:298–302.

Sir, M., M. Kutilek, V. Kuraz, M. Krejca, and F. Kubik. 1988. Field estimation of the soil hydraulic characteristics. *Soil Technol.* 1:63–75.

Skaggs, W. R., E. J. Monk, and L. F. Huggins. 1971. An approximate method for defining the hydraulic conductivity–pressure potential relationship for soils. *Trans. ASAE* 14:130–133.

Smettem, K. R. J., and B. E. Clothier. 1989. Measuring unsaturated sorptivity and hydraulic conductivity using multiple disc permeameters. *J. Soil Sci.* 40:563–568.

Talsma, T. 1969. In situ measurement of sorptivity. *Aust. J. Soil Res.* 7:269–276.

Tamari, S., L. Bruckler, J. Halbertsma, and J. Chadoeuf. 1993. A simple method for determining soil hydraulic properties in the laboratory. *Soil Sci. Soc. Am. J.* 57:642–651.

Ten Berge, H. F. M., K. Metselaar, and L. Stroosnijder. 1987. Measurement of matric flux potential: A simple procedure for the hydraulic characterisation of soils. *Neth. J. Agric. Sci.* 35:371–384.

Thony, J. L., G. Vachaud, B. E. Clothier, and R. Angulo-Jaramillo. 1991. Field measurement of the hydraulic properties of soil. *Soil Techn.* 4:111–123.

Van Dam, J. C., J. N. M. Stricker, and P. Droogers. 1992. Inverse method for determining soil hydraulic functions from one-step outflow experiments. *Soil Sci. Soc. Am. J.* 56:1042–1050.

Van Dam, J. C., J. N. M. Stricker, and P. Droogers. 1994. Inverse method to determine soil hydraulic functions from multistep outflow experiments. *Soil Sci. Soc. Am. J.* 58: 647–652.

Van den Berg, J. A., and T. Louters, 1986. An algorithm for computing the relationship between diffusivity and soil moisture content from the hot air method. *J. Hydrol.* 83:149–159.

Van Genuchten, M. Th. 1980. A closed-form equation for predicting the hydraulic conductivity of unsaturated soils. *Soil Sci. Soc. Am. J.* 44:892–898.

Van Genuchten, M. Th., and D. R. Nielsen. 1985. On describing and predicting the hydraulic properties of unsaturated soils. *Ann. Geophys.* 3:615–628.

Van Genuchten, M. Th., F. J. Leij, and L. J. Lund (eds.). 1992. *Indirect Methods for Estimating the Hydraulic Properties of Unsaturated Soils.* Riverside: Univ. of California.

Van Genuchten, M. Th., F. J. Leij, and L. Wu, eds. 1999. Characterization and Measurement of the Hydraulic Properties of Saturated Porous Media. Riverside: University of California.

Van Grinsven, J. J. M., C. Dirksen, and W. Bouten. 1985. Evaluation of the hot air method for measuring soil water diffusivity. *Soil Sci. Soc. Am. J.* 49:1093–1099.

Van Schaik, J. C. 1970. Soil hydraulic properties determined with water and with a hydrocarbon liquid. *Can. J. Soil Sci.* 50:79–84.

Vereecken, H., J. Maes, and J. Feijen. 1990. Estimating unsaturated hydraulic conductivity from easily measured soil properties. *Soil Sci.* 149:1–11.

Verlinden, H. L., and J. Bouma. 1983. *Fysische Bodemonderzoekmethoden voor de Onverzadigde Zone*, VROM-Rapport BO 22, The Netherlands.

Warrick, A. W. 1974. Time-dependent linearized infiltration. I. Point sources. *Soil Sci. Soc. Am. J.* 38:383–386.

Warrick, A. W., and D. R. Nielsen. 1980. Spatial variability of soil physical properties in the field. In: *Applications of Soil Physics* (D. Hillel, ed.). New York: Academic Press, pp. 319–344.

Watson, K. K. 1966. An instantaneous profile method for determining the hydraulic conductivity of unsaturated porous materials. *Water Resour. Res.* 2:709–715.

Wendroth, O., W. Ehlers, J. W. Hopmans, H. Kage, J. Halbertsma, and J. H. M. Wösten. 1993. Reevaluation of the evaporation method for determining hydraulic functions in unsaturated soil. *Soil Sci. Soc. Am. J.* 57:1436–1443.

Wesseling, J., and K. E. Wit. 1966. An infiltration method for the determination of the capillary conductivity of undisturbed soil cores. In: *Proc. UNESCO/IASH Symp. Water in the Unsaturated Zone*. Wageningen, The Netherlands, pp. 223–234.

Whisler, F. D., and K. K. Watson. 1968. One-dimensional gravity drainage of uniform columns of porous materials. *J. Hydrol.* 6:277–296.

Whisler, F. D., A. Klute, and D. B. Peters. 1968. Soil water diffusivity from horizontal infiltration. *Soil Sci. Soc. Am. Proc.* 32:6–11.

White, I., and K. M. Perroux. 1987. Use of sorptivity to determine field soil hydraulic properties. *Soil Sci. Soc. Am. J.* 51:1093–1101.

Wind, G. P. 1966. Capillary conductivity data estimated by a simple method. In: *Proc. UNESCO/IASH Symp. Water in the Unsaturated Zone.* Wageningen, The Netherlands, pp. 181–191.

Wösten, J. H. M., and M. Th. Van Genuchten. 1988. Using texture and other soil properties to predict the unsaturated soil hydraulic functions. *Soil Sci. Soc. Am. J.* 52: 1762–1770.

Yates, S. R., M. Th. Van Genuchten, and F. J. Leij. 1992. Analysis of predicted hydraulic conductivities using RETC data. In: *Indirect Methods for Estimating the Hydraulic Properties of Unsaturated Soils* (M. Th. Van Genuchten, F. J. Leij, and L. J. Lund, eds.). Riverside: Univ. of California, pp. 273–283.

Youngs, E. G. 1984. An infiltration method of measuring the hydraulic conductivity of unsaturated porous material. *Soil Sci.* 97:307–311.

Zachmann, D. W., P. C. Du Chateau, and A. Klute. 1981. The calibration of the Richards flow equation for a draining column by parameter identification. *Soil Sci. Soc. Am. J.* 45:1012–1015.

6
Infiltration

Brent E. Clothier
HortResearch, Palmerston North, New Zealand

1. INTRODUCTION

A water droplet incident at the soil surface has just two options: it can infiltrate the soil or it can run off. This partitioning process is critical. Infiltration, and its complement runoff, are of interest to hydrologists who study runoff generation, river flow, and groundwater recharge. The entry of water through the surface concerns soil scientists, for infiltration replenishes the soil's store of water. The partitioning process is critically dependent of the physical state of the surface. Furthermore, infiltrating water acts as the vehicle for both nutrients and chemical contaminants.

Infiltration, because it is both a key soil process and an important hydrological mechanism, has been twice studied: once by soil physicists and again by hydrologists. Historically, their approaches have been quite different. In the former case, infiltration was the prime focus of detailed study of small-scale soil processes, and in the latter, infiltration was just one mechanism in a complicated cascade of processes operating across the scale of a catchment. Latterly, access to powerful computers has meant that hydrologists have been able to incorporate the soil physicists' detailed mechanistic descriptions of infiltration into their hydrological models of watershed functioning. This has increased the need to measure the parameters that control infiltration.

To the memory of John Philip (1927–1999), for without his endeavors this would have been a very short chapter.

In this chapter, I first describe the development of one-dimensional ponded infiltration theory, discussing both analytical and quasi-analytical solutions. In passing, I mention empirical descriptions of infiltration before discussing the key development of a simple algebraic expression for infiltration that employs physically based parameters. Emphasis is placed on theoretical approaches, for they can predict infiltration through having parameters capable of field measurement. The preeminent roles of the physical state of the soil surface and the nature of the upper boundary condition are stressed. Infiltration of water into soil can occur as a result of there being a pond of free water on the soil surface, so that the soil controls the amount infiltrated, or water can be supplied to the surface at a given rate, say by rainfall, so the soil only controls the profile of wetting, not the amount infiltrated.

Next, I show how measurement of infiltration can be used, in an inverse sense, to determine the soil's hydraulic properties. In this way, it is possible to predict infiltration into the soil, and general prediction of water movement through soil can also be made using these measured properties. Hydraulic interpretation of the theoretical parameters in the governing equations is outlined, as is the impact of infiltration on solute transport through soil. A list is presented of the various devices that have been developed to measure, in the field, the soil's capillary and conductive properties that control infiltration. An outline of their respective merits is presented, as is a comparative ranking of utility. Finally, I conclude with a presentation of some illustrative results and identify some issues that still remain problematic.

Elsewhere in this book, there are complementary chapters on measurement of the soil's saturated conductivity (Chap. 4) and the unsaturated hydraulic conductivity function (Chap. 5). Here the emphasis is on devices capable of in situ observation of infiltration, and the measurement in the field of those saturated and unsaturated properties that control the time course, and quantity, of infiltration.

II. THEORY

A. One-Dimensional Ponded Infiltration

Significant theoretical description of water flow through a porous medium began in 1856 with Henry Darcy's observations of saturated flow through a filter bed of sand in Dijon, France (Philip, 1995). Darcy found that the rate of flow of water, J (m s^{-1}), through his saturated column of sand of length L (m), was proportional to the difference in the hydrostatic head, H (m), between the upper water surface and the outlet:

$$J = K\left(\frac{\Delta H}{L}\right) \tag{1}$$

in which Darcy called K "un coefficient dépendent du degré de perméabilité du sable." We now call this the saturated hydraulic conductivity K_s (m s^{-1}) (Chap. 4). In 1907, Edgar Buckingham of the USDA Bureau of Soils established the theoretical basis of unsaturated soil water flow. He noted that the capillary conductance of water through soil, now known as the unsaturated hydraulic conductivity, was a function of the soil's water content, θ (m^3 m^{-3}), or the capillary pressure head of water in the soil, h (m). The characteristic relationship between h and θ (Chap. 3) was also noted by Buckingham (1907), so that he could write $K = K(h)$, or if so desired, $K = K(\theta)$. The total head of water at any point in the soil, H, is the sum of the gravitational head due to its depth z below some datum, conveniently here taken as the soil surface, and the capillary pressure head of water in the soil, h: $H = h - z$. Here, h is a negative quantity in unsaturated soil, due to the capillary attractiveness of water for the many nooks and crannies of the soil pore system. Thus locally in the soil, Buckingham found that the flow of water could be described by

$$J = -K(h)\left(\frac{dH}{dz}\right) = -K(h)\frac{dH}{dz} + K(h) \tag{2}$$

which identifies the roles played by capillarity, the first term on the right hand side, and gravity, the second term. These two forces combine to move water through unsaturated soil (Chap. 5). In deference to the discoverers of the saturated form, Eq. 1, and the unsaturated form, Eq. 2, the equation describing water flow at any point in the soil is generally referred to as the Darcy–Buckingham equation.

L. A. Richards (1931) combined the mass-balance expression that the temporal change in the water content of the soil at any point is due to the local flux divergence,

$$\left(\frac{\partial\theta}{\partial t}\right)_z = -\left(\frac{\partial J}{\partial z}\right)_t \tag{3}$$

with the Darcy–Buckingham description of the water flux J, to arrive at the general equation of soil water flow,

$$\frac{\partial\theta}{\partial t} = \frac{\partial}{\partial z}\left(K(h)\frac{\partial h}{\partial z}\right) - \frac{dK(h)}{dh}\cdot\frac{\partial h}{\partial z} \tag{4}$$

where t (s) is time. Unfortunately, this formula, known as Richards' equation, does not have a common dependent variable, for θ appears on the left and h on the right. The British physicist E. C. Childs "decided to try some other hypothesis . . . that water movement is decided by the moisture concentration gradient . . . [and] that water moves according to diffusion equations" (Childs, 1936). Childs and Collis-George (1948) noted that the diffusion coefficient for water in soil could be written as $K(\theta)\, dh/d\theta$. From this, in 1952 the American soil physicist Arnold Klute wrote Richards' equation in the diffusion form of

$$\frac{\partial \theta}{\partial t} = \frac{\partial}{\partial z}\left(D(\theta)\frac{\partial \theta}{\partial z}\right) - \frac{dK}{d\theta} \cdot \frac{\partial \theta}{\partial z} \tag{5}$$

This description shows soil water flow to be dependent on both the soil water diffusivity function $D(\theta)$, and the hydraulic conductivity function $K(\theta)$, but this nonlinear partial differential equation is of the Fokker–Planck form, which is notoriously difficult to solve. Klute (1952) developed a similarity solution to the gravity-free form of Eq. 5, subject to ponding of free water at one end of a soil column.

Five years later, the Australian John Philip developed a power-series solution to the full form of Eq. (5), subject to the ponding of water at the surface of a vertical column of soil, initially at some low water content θ_n (Philip, 1957a). This general solution predicts the rate of water entry through the soil surface, $i(t)$ (m s^{-1}), following ponding on the surface. The surface water content is maintained at its saturated value, θ_s. The cumulative amount of water infiltrated is I (m), being the integral of the rate of infiltration since ponding was established. As well, I can be found from the changing water content profile in the soil,

$$I = \int_0^t i(t')\, dt' = \int_0^\infty \left[\theta(z') - \theta_n\right] dz' = \int_{\theta_n}^{\theta_s} z(\theta)\, d\theta \tag{6}$$

Philip's (1957a) series solution for $I(t)$ can be written

$$I(t) = St^{1/2} + At + A_3 t^{3/2} + A_4 t^{4/2} + \cdots \tag{7}$$

where the sorptivity S (m $s^{-1/2}$) and the coefficients A, A_3, A_4, . . . can be iteratively calculated from the diffusivity and conductivity functions, $D(\theta)$ and $K(\theta)$. The form of Eq. 7 indicates that I increases with time, but at an ever-decreasing rate. In other words, the rate of infiltration $i = dI/dt$ is high initially, due to the capillary pull of the dry soil. But with time the rate declines to an asymptote.

Special analytical solutions can be found for cases where certain assumptions are made about the soil's hydraulic properties. When the soil water diffusivity can be considered to be a constant, and K varies linearly with θ, an analytical solution is possible. This is because Eq. 5 becomes linearized (Philip, 1969) and so there is an analytical solution for infiltration into a soil whose hydraulic properties can be considered only weakly dependent on θ. At the other end of the scale of possible behavior, Philip (1969) presented an analytical solution for a soil whose diffusivity could be considered a Dirac δ-function, in which D is zero, except at θ_s where it goes to infinity. For the analytical solution, this so-called delta-soil, or Green and Ampt soil, also needs to have $K = K_s$ at θ_s, and $K = 0$ for all other θ.

Philip and Knight (1974) showed that the Dirac δ-function solution produces a rectangular profile of wetting (shown later in Fig. 4). It was this geometric form of wetting that was used as the physical basis for Green and Ampt's (1911) functional model of infiltration. If a rectangular profile of wetting is assumed, then behind the wetting front located at depth z_f, $\theta(z) = \theta_s$, for $0 < z < z_f$; and beyond the wetting front, $\theta(z) = \theta_n$, $z > z_f$. If the soil has a shallow free-water pond at the surface, and if it is considered that there is a wetting front potential head, h_f, at z_f, then the Darcy–Buckingham law (Eq. 2) predicts the rate of water infiltrating through the surface as

$$J = \frac{K_s(z_f - h_f)}{z_f} \tag{8}$$

The rectangular profile of wetting allows easy evaluation of the mass balance integral of Eq. 6, and its differentiation to provide the rate of infiltration into the soil,

$$i = \frac{dI}{dt} = \frac{\mathrm{d}(z_f(\theta_s - \theta_n))}{dt} = \frac{dz_f}{dt} \cdot (\theta_s - \theta_n) \tag{9}$$

Equating Eqs. 8 and 9 provides a variables-separable ordinary differential equation in z_f,

$$K_s \frac{(z_f - h_f)}{z_f} = \frac{dz_f}{dt}(\theta_s - \theta_n) \tag{10}$$

which can be solved to provide the penetration of the wetting front into the soil with time,

$$t = \frac{(\theta_s - \theta_n)}{K_s}\left[z_f + h_f \ln\left(1 - \frac{z_f}{h_f}\right)\right] \tag{11}$$

Althrough this expression is not explicit, it does allow implicit prediction of $I(t)$ from basic soil properties. By considering flow in the absence of gravity, z_f is eliminated from the numerator of Eq. 8, and an explicit expression for gravity-free infiltration is arrived at that only contains a square-root-of-time term, as would be expected for a diffusion-like process. By comparing coefficients with Eq. 7, it is found that a Green and Ampt soil must have the sorptivity

$$S^2 = -2K_s h_f(\theta_s - \theta_n) \tag{12}$$

So, if the soil is considered to have the characteristics that lead to a rectangular profile of wetting (shown in Fig. 4), and there is a constant pressure-potential head, h_f, always associated with the wetting front, then simple expressions can be derived to predict infiltration into such a soil (Eqs. 11 and 12).

More recently, Parlange (1971) developed a new and general quasi-analytical solution for infiltration into any soil (Eq. 5). This was extended by Philip and

Knight (1974) using a flux–concentration relationship, $F(\Theta)$, that hides much of the nonlinearity in the soil's hydraulic properties of D and K. Here Θ is the normalized water content.

Considering these mathematical solutions to the flow equation for infiltration I, subject to ponding, Childs (1967) commented that "further investigations to throw yet more light on the basic principles of the flow of water. . . tend to be matters of crossing t's and dotting i's . . . serious difficulties remain in the path of practical application of theory . . . [being] held back by the inadequate development of methods of assessment of the relevant parameters."

These analytical or quasi-analytical solutions are seldom used to predict infiltration directly from the soil's hydraulic properties. The theory and its development are presented here, for they identify the underpinning physics of infiltration. Nowadays, however, the current power of computers, coupled with the burgeoning growth of numerical recipes for solving nonlinear partial differential equations, has meant that brute-force numerical solutions to Eq. 5 for infiltration are easily obtained, *provided* that the functional properties of D and K are known. Thus given a knowledge of the soil's hydraulic properties, it is a reasonably straightforward exercise to predict infiltration, either analytically or numerically.

Infiltration measurements hold the key to obtaining in situ characterization of these soil properties. It is possible to use Eq. 7 or the like in an inverse sense, to use infiltration observations to infer the soil's hydraulic character. The time course of water entry into soil, $I(t)$, depends, as Eq. 7 shows, on coefficients that relate to the hydraulic properties of $D(\theta)$ and $K(\theta)$. Infiltration can quite easily be measured in the field. Hence, I will proceed to show how this measurement of I can be used to extract in situ information about the soil's capillary and conductive properties.

1. *Empirical Descriptions*

Before passing to the discussion of the developments that have led to the use of measurements to predict one-dimensional infiltration behavior, I sidetrack a little to review some of the empirical descriptions of the shape of $i(t)$. This digression is simply to complete our historical record of the study of infiltration, for such empirical equations have little merit nowadays. The Kostiakov–Lewis equation, $I = at^b$ (Swartzendruber, 1993), has descriptive merit through its simplicity, yet comparison with Eq. 7 indicates the inadequacy of this power-law form, for in reality b needs to be a function of time. The hydrologist Horton (1940) proposed that the decline in the infiltration rate could be described using $i = i_\infty + (i_o - i_\infty)\exp(-\beta t)$, where the subscripts o and ∞ refer to the initial and final rates. If description is all that is sought, then the three-term expression will perform better due to its greater fitting "flexibility." In neither case do the fitted parameters have physical meaning, so care needs to be exercised in their extrapolation beyond the fitted range.

2. Physically Based Descriptions

However, the two-term algebraic equation of Philip (1957b) is different from other empirical descriptions. It rationally incorporates physical notions. Simply by truncating the power series of Eq. 7, Philip (1957a) arrived at the expression for the infiltration rate of

$$i = \frac{1}{2} S t^{-1/2} + A \tag{13}$$

which will be applicable at short and intermediate times. However at longer times, we know for ponded infiltration that

$$\lim_{t \to \infty} i(t) = K_s \tag{14}$$

The means by which these two expressions can best be joined has worried some soil physicists, with A/K_s having been found to be bracketed between $1 - 2/\pi$ and 2/3, but probably lying nearer the lower limit (Philip, 1988). However, as Philip (1987) noted, relative to practical incertitudes, a two-term algebraic expression often suffices, with both terms having physical meaning, plus correct short- and long-time behavior, viz.

$$i = \frac{1}{2} S t^{-1/2} + K_s \tag{15}$$

The coefficient of the square-root-of-time term, the sorptivity S, integrates the capillary attractiveness of the soil. Mathematically, as we will see later, this can be linked to the soil water diffusivity function $D(\theta)$. The role of capillarity declines with the square root of time. The second term, which is time independent, is the saturated conductivity K_s, which is the maximum value of the conductivity function $K(h)$ that occurs when the soil is saturated, $h = 0$. If the soil is initially saturated ($S = 0$), or if infiltration has been going on for a long time, then gravity will alone be drawing water into the soil at the steady rate of K_s. Eq. 15 is aptly named Philip's equation.

To understand Eq. 15 is to understand the basics of infiltration.

B. Multidimensional Ponded Infiltration

However, a one-dimensional geometric description is not always appropriate. For example, infiltration into soil might be from a buried and leaking pipe, or it might be from a finite surface puddle of water. In these cases, the physics previously

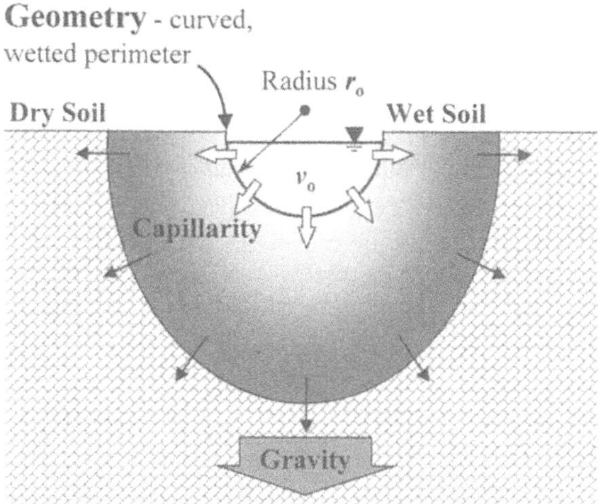

Fig. 1. An idealized multidimensional infiltration source, in which water infiltrates into the soil through a wetter perimeter of radius of curvature r_o. Capillarity and gravity combine to draw water into the dry soil.

described above must now be referenced to the geometry of the source. The respective roles of capillarity and gravity in establishing the rate of multidimensional infiltration, $v_o(t)$ (m s^{-1}), through a surface of radius of curvature r_o (m), are now more complex. Following Philip (1966), let m be the number of spatial dimensions required for geometric description of the flow. The geometry depicted in Fig. 1 might be a transverse section through a cylindrical channel. This would be a 2-D source with $m = 2$. Or it could be a diametric cut through a spherical pond that would be represent a 3-D geometry. So here $m = 3$. The more curved the wetted perimeter of the source, the smaller is r_o, and the greater is the role of the soil's capillarity relative to gravity. In the limit as $r_o \rightarrow \infty$, the geometry becomes one-dimensional ($m = 1$), and the source spreads right across the soil's surface.

As already noted, if the soil is considered to have a constant diffusivity D, and a linear $K(\theta)$, then ananalytical solution can be found for one-dimensional infiltration because the governing equation is linearized. This also applies to multidimensional infiltration, if the flow description of Eq. 5, which has $m = 1$, is written in a form appropriate to a flow geometry with either $m = 2$ or $m = 3$ (Philip, 1966). Philip's (1966) linearized multidimensional infiltration results are illustrative and are presented in Fig. 2. There, the infiltration rate through the wetted perimeter, v_o, is normalized with respect to the saturated conductivity K_s, and the time is normalized by a nondimensional time, $t_{grav} = (S/K_s)^2$. To allow easy comparison, the radius of curvature is also normalized, and given as $R_o = r_o[K_s(\theta_s - \theta_n)^2/\pi S^2]$.

For the one-dimensional case in Fig. 2 ($m = 1$), the infiltration rate can be seen to fall, as the effects of capillarity fade with the square, and higher, roots of time (Eq. 7). At around $t/\pi t_{grav}$, the infiltration rate is virtually the asymptote of $v_o = K_s$. Such behavior is predicted by Eq. 15. Two cases are given for two-dimensional flow from cylindrical channels, $m = 2$. For the tightly curved channel ($R_o = 0.05$), the effect of the source geometry on capillarity is clearly seen, and the infiltration rate is nearly two times K_s at dimensionless time 10. For the less-curved channel ($R_o = 0.25$), the geometry-induced enhancement of capillary effects is correspondingly less. In the three-dimensional case ($m = 3$), for the curved spherical pond with $R_o = 0.05$, the effect of capillarity is so enhanced by the 3-D source geometry that the infiltration rate through the pond walls achieves a steady flux of over $5K_s$ by unit time.

Whereas infiltration in one dimension ($m = 1$) gradually approaches K_s, the source geometry in 2-D and 3-D ($m = 2$ and 3) ensures that the infiltration rate finally arrives at a true steady-state value, v_∞. In Fig. 2, the time taken to realise v_∞

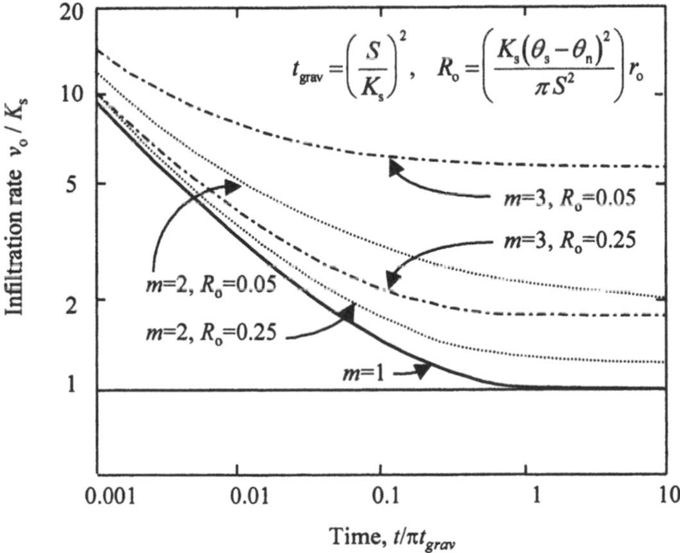

Fig. 2. The normalized temporal decline in the rate of infiltration through the ponded surface into a one-dimensional soil profile ($m = 1$), and from two cylindrical channels ($m = 2$) of contrasting radii of curvature (r_o), as well as from two spherical ponds ($m = 3$) of different radii. To allow comparison of one-, two-, and three-dimensional flows, the infiltration rate, time, and radii of source curvature have all been normalized. (Redrawn from Philip, 1966.)

is more rapid in 3-D than it is for $m = 2$. This achievement of a steady flow rate in 3-D is, as we will see later, a major advantage for certain devices in the field measurement of infiltration.

In this device-context, it is useful to consider in more detail the three-dimensional flow from a shallow, circular pond of water of radius r_o. The history of the study of this problem is given in Clothier et al. (1995), so here we need only concern ourselves with the seminal result of Wooding (1968). The New Zealander Robin Wooding was concerned about the radius requirements for double-ring infiltrometers (shown later in Fig. 5), and he found a complex-series solution for the steady flow from a shallow, circular surface pond of free water. However, he did note that the steady flow could be approximated by a simple equation in which capillary effects were added to the gravitational flow in inverse proportion to the length of the wetted perimeter of the pond,

$$v_\infty = K_s + \frac{4\phi_s}{\pi r_o} \tag{16}$$

Here the sum effect of the soil's capillarity is expressed in terms of the integrals of the hydraulic properties of D and K, the so-called matric flux potential

$$\phi_s = \int_{\theta_n}^{\theta_s} D(\theta)d\theta = \int_h^0 K(h)dh \tag{17}$$

It was necessary for Wooding (1968) to consider that the soil's unsaturated hydraulic conductivity function could be given by the exponential form

$$K(h) = K_s \exp(\alpha h) \tag{18}$$

with the unsaturated slope α (m^{-1}), so that

$$\phi_s = \frac{K_s}{\alpha} \tag{19}$$

This formulation allows Wooding's equation (Eq. 16) for the steady volumetric infiltration from the circular pond, $Q_\infty = \pi r_o^2 v_\infty$ ($m^3\ s^{-1}$), to be written as

$$Q_\infty = K_s\left(\pi r_o^2 + \frac{4r_o}{\alpha}\right) \tag{20}$$

In this way we can see the role of the pond's area in generating the gravitational component of infiltration, and that of the perimeter in creating a capillary contribution. We will return later to this special form of multidimensional ponded infiltration.

C. Boundary Conditions

Thus far, we have only considered the case where water is supplied by a surface pond of free water, namely

$$\theta(0, t) = \theta_s \qquad h(0, t) = 0 \qquad z = 0, t \geq 0 \tag{21}$$

This is termed a constant-concentration boundary condition and known mathematically as a first-type or Dirichlet boundary condition. This is appropriate to cases where water is ponded on the ponded on the soil surface. The soil's hydraulic properties, and source geometry, determine the rate and temporal decline in infiltration (Fig. 2). The water content at the soil's surface is always at its saturated value, θ_s.

However, often water arrives at the soil surface as a flux, as might occur during rainfall, or irrigation. In this case, the upper boundary condition is the applied flux v_o,

$$D(\theta)\frac{\partial\theta}{\partial z} = K(h)\frac{\partial H}{\partial z} = -v_o \qquad z = 0, t \geq 0 \tag{22}$$

This case is mathematically termed a second type or Neumann boundary condition, and the amount and rate of water infiltrating the soil is independent of the soil's hydraulic properties. Rather, it is determined by v_o. Whereas in Eq. 21 the water content at the soil surface is constant, under a flux condition, as the soil wets, the water content at the surface, θ_o, rises: $\theta_o = \theta_o(t)$.

Should the flux of water always be less than K_s, then the water content at the surface will always be less than θ_s. The soil at the surface will remain unsaturated, $h_o < 0$, and all the incident water will enter the soil, with $I = v_o t$.

However, if the rate of water falling on the soil surface exceeds K_s, then eventually at some time t_p, the time to incipient ponding, the soil at the surface will saturate; $h_o = 0$; $\theta_o = \theta_s$, $t \geq t_p$. After this incipient ponding, runoff from the free water pond can occur, and not all the applied water need enter the soil: $I < v_o t$, for $t \geq t_p$. For the case of a constant flux, Perroux et al. (1981) found that a good approximation for the time to ponding was

$$t_p = \frac{S^2}{2v_o(v_o - K_s)} \tag{23}$$

So the greater the flux the quicker the soil surface ponds. Conversely, the drier the soil initially, the greater is the capillarity of the soil, the higher is S, and so the longer can the soil maintain its acceptance of all the applied water.

The presence or absence of a surface pond of free water is critical for infiltration behavior in the macropore-ridden soils of the field. This is shown in Fig. 3. Only free water ($h_o > 0$) can enter surface-vented macropores. Thus during nonponding flux infiltration, $v_o < K_s$, or prior to the time to ponding, $t < t_p$, the soil surface remains unsaturated, $h_o < 0$, so that water does not enter macropores. Rather the water droplets are absorbed right where they land. Hence the pattern of infiltration and soil wetting is quite uniform, as capillarity attempts to even out local heterogeneities. However, following incipient ponding, $t > t_p$, a free-water film develops on the soil surface. This free water can enter surface-vented macro-

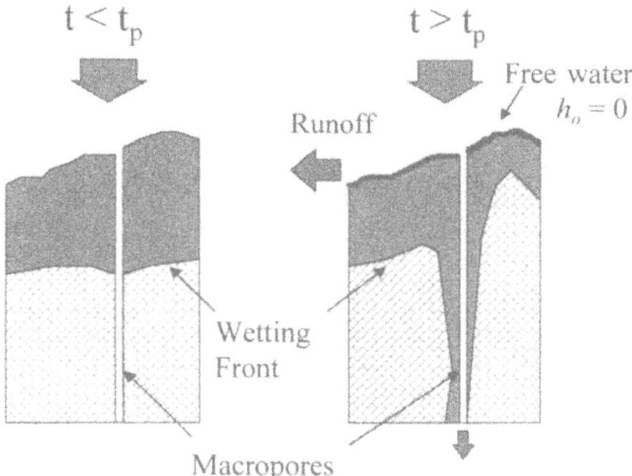

Fig. 3. Infiltration of an applied flux of water into soil. Left: non-ponding infiltration when $v_o < K_s$, or ponding infiltration $v_o > K_s$ prior to the time to ponding t_p. Right: pattern of infiltration after incipient ponding, $t > t_p$, when the possibility of runoff exists, as does the entry of free water into macropores.

pores, creating preferential flow through the soil, and lead to local variability in the pattern of soil wetting. If the infiltration capacity of the soil, both by matrix absorption and macropore flow, is exceeded, there is the possibility of local runoff once the surface storage has been overwhelmed (Dixon and Peterson, 1971).

The magnitude of the flux v_o relative to the soil's K_s is critical in determining infiltration behavior, and during flux infiltration it is critical to know whether the time to ponding has been reached. The value of t_p can be deduced from a knowledge of the soil's sorptivity S, and conductivity K_s, given v_o (Eq. 23). So it is imperative that S and K_s be measured for field soils.

D. Hydraulic Characteristics of Soil

There are three functional properties necessary to describe completely the hydraulic character of the soil: the soil water diffusivity function $D(\theta)$, the unsaturated hydraulic conductivity function $K(h)$, and the soil water characteristic $\theta(h)$. However since $D = K\, dh/d\theta$, only two are sufficient to parameterize Eq. 5. Whereas it is possible to measure these functions in the laboratory, albeit with some difficulty, it is virtually impossible to do so in the field (Chaps. 3 and 5).

Nonetheless, if we were to observe the time course of ponded infiltration in the field, $i(t)$, then by inverse procedures we should be able to use Eq. 15 to infer values of the saturated sorptivity S, and the saturated conductivity K_s. In the first case, we would then have obtained a measurement of something that integrates the soil's capillarity, and in the second case we would know the maximum value of the $K(h)$ function. Because we know in one case an integral measure, and in the other a functional maximum, if we were willing to make some assumptions about functional forms, we could infer the D and K functions from measurements of just S and K_s, and some observations of θ_s and θ_n. Thus observations of infiltration in the field can be used to establish the hydraulic characteristics of field soil.

Formally, the sorptivity can be written as a complicated integral of the soil water diffusivity function

$$S^2(\theta_s, \theta_n) = 2\int_{\theta_n}^{\theta_s} \frac{(\theta - \theta_n)\, D(\theta)}{F(\Theta)}\, d\theta \tag{24}$$

where F is the flux–concentration relation of the quasi-analytical solution of Philip and Knight (1974) (see Sec. II.A). Parlange (1975) independently found some useful and simple algebraic versions of Eq. 24. Eq. 24 is difficult to invert in order that $D(\theta)$ might be deduced from S. However, if we revisit the Kirchhoff transform of Eq. 17, we have the integral of the diffusivity as

$$\phi_s = \int_{\theta_n}^{\theta_s} D(\theta)\, d\theta \tag{25}$$

so that by inspection of Eqs. 24 and 25, we would expect a relationship between ϕ_s and S^2. White and Sully (1987) wrote this as

$$\phi_s = \frac{bS^2}{\theta_s - \theta_n} \tag{26}$$

where it is known theoretically that $\frac{1}{2} < b < \pi/4$. For a wide range of soils they found $b = 0.55$ to be a robust assumption. Thus from a measurement of the sorptivity, we can infer the integral of the diffusivity function ϕ_s. If we were willing to make some assumption about the form of $D(\theta)$, say an exponential with slope 8 (Brutsaert, 1979), then by measuring S, θ_s, and θ_n, and using Eq. 26, we would be able to realize a functional representation of the soil water diffusivity that is capable of parameterization in the field (Clothier and White, 1981). At least, it would be integrally correct.

If we look yet again at Eq. 17, we see that ϕ_s is also the integral of the $K(h)$ function. If an exponential conductivity function (Eq. 18) is assumed, then

$$\phi_s = \int_h^0 K(h)\, dh = \frac{K_s}{\alpha} \tag{27}$$

This can be combined with Eq. 26 to obtain the slope,

$$\alpha = \frac{K_s}{\phi_s} = \frac{K_s(\theta_s - \theta_n)}{bS^2} \tag{28}$$

So by monitoring infiltration to infer both K_s and S (Eq. 15), and by measuring θ_s and θ_n, Eqs. 26 and 28 give us functional descriptions of the soil's $D(\theta)$ and $K(h)$.

These capillary and gravity properties allow us to infer some pore-geometric characteristics of the soil's hydraulic functioning. Philip (1958) defined a macroscopic, mean "capillary length" λ_c, which can be written over the range from h_n to saturation as

$$\lambda_c = \frac{\int_{h_n}^{0} K(h)\, dh}{K_s} = \frac{\int_{\theta_n}^{\theta_s} D(\theta)\, d\theta}{K_s} \tag{29}$$

if the conductivity at h_n is considered to be negligible. This corresponds to the capillary fringe of Myers and van Bavel (1963), and the critical pressure of Bouwer (1964). Note that if the soil's $K(h)$ is exponential (Eq. 18), then Eq. 27 shows us that $\lambda_c = \alpha^{-1}$. Using Eq. 28 gives λ_c in terms of easily measurable quantities,

$$\lambda_c = \frac{bS^2}{K_s(\theta_s - \theta_n)} \tag{30}$$

Using Laplace's capillary-rise formula, Philip (1987) related λ_c to the characteristic mean pore radius, λ_m:

$$\lambda_m = \frac{\sigma}{\rho g} \frac{1}{\lambda_c} \approx \frac{7.4}{\lambda_c} \tag{31}$$

if appropriate values are taken for the surface tension σ and density ρ of water, and for the acceleration due to gravity. White and Sully (1987) called λ_m a "physically plausible estimate of flow-weighted mean pore dimensions." By combining Eqs. 30 and 31 it is possible to use properties measured during infiltration (S and K_s; θ_s and θ_n) to deduce something dynamic about the magnitude of the pore size class involved in drawing water into the soil. Namely,

$$\lambda_m = \frac{13.5(\theta_s - \theta_n)K_s}{S^2} \tag{32}$$

E. Solute Transport During Infiltration

Water is the vehicle for solutes in soil. Here, for simplicity, we consider a soil lying horizontally with water being absorbed in the x direction. During infiltration, water-borne chemicals are transported into the soil. The entry of water into soil is a hydrodynamic phenomenon: the wetting front rides into the soil on "top" of the antecedent water content, θ_n (Fig. 4). For the case of a δ-function soil, that is, one possessing Green and Ampt's (1911) rectangular profile of wetting, Eq. 6 gives the penetration of the wetting front as

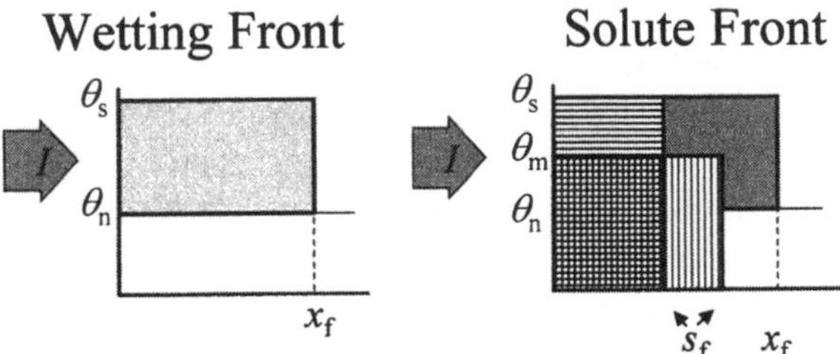

Fig. 4. Left: rectangular profile of wetting that pertains during infiltration into a soil whose diffusivity function is a Dirac δ-function (Green and Ampt, 1911). The position of the wetting front x_f is given by Eq. 33. Right: a dispersion-free invasion front of the solution infiltrating a soil in which all the water is assumed to be mobile, and one in which the mobile water content is just θ_m. The solute fronts for these two cases, s_f, are given by Eq. 34.

$$x_f = \frac{I}{(\theta_s - \theta_n)} \tag{33}$$

The transport of water-borne solute, during this hydrodynamically driven infiltration process, is an invasion mechanism. If all the soil's water is mobile, and if dispersion is ignored, then the invading solute profile will also be rectangular (Fig. 4). In this case, the solute front will be at

$$s_f = \frac{I}{\theta_s} \tag{34}$$

For field soils, due to preferential flow paths, it has been found useful to treat chemical invasion as if not all of the soil's water is mobile. As an approximation, the soil's water can be conveniently partitioned into a mobile phase, θ_m, and a complementary domain that is considered effectively immobile, θ_{im}; $\theta_s = \theta_m + \theta_{im}$ (van Genuchten and Wierenga, 1976). In this mobile-immobile case, the solute front would be further ahead at

$$s_f = \frac{I}{\theta_m} \tag{35}$$

because $\theta_m < \theta_s$. Thus if some inert tracer solution were allowed to infiltrate the soil, then the position of the wetting front, relative to that of the solute front, would be

$$\Re = \frac{x_f}{s_f} = \frac{\theta_m}{\theta_s - \theta_n} \tag{36}$$

So in a fully mobile case $\theta_m = \theta_s$, which is initially dry, $\theta_n = 0$, the wetting front and the invading inert solute front will be coincident; $\Re = 1$. If the soil is not initially dry, then the wetting front will be ahead of the invasion front of the solute, $\Re > 1$. If not all the soil's water is mobile, then the solute will preferentially infiltrate the soil through just the mobile domain, and the solute front may be closer to the wetting front. The simple notions contained in Fig. 4 and Eq. 35 provide a useful means to model chemical transport processes during infiltration. Later, I will discuss how values of θ_m and $\Re$ might be measured and interpreted.

III. DEVICES AND MEASUREMENT

In this section, I now consider eight devices that have been developed to measure infiltration in the field. The relative merits of these devices and instruments are listed in Table 1 and discussed later.

A. Rings

1. *Buffered Rings*

The easiest way to observe ponded infiltration in the field is simply to watch the rate that water disappears from a surface puddle. However, as shown in Fig. 1, two factors control infiltration from a pond, capillarity and gravity. In order to eliminate the perimeter effects of capillarity, buffered rings have been used so that the flow in the inner ring is due only to gravity (Fig. 5). By this arrangement, it is hoped that the steady flux from the inner ring, v_∞, might be the saturated hydraulic conductivity K_s, since capillary effects would be quenched by flow from the buffer ring, v_o^*. To determine what size the radius of the inner ring, r_1, needs to be relative to that of the buffer, r_2, Bouwer (1961) and Youngs (1972) used an electrical-analog approach, whereas Wooding (1968) provided a simple expression based on the properties of the soil (Eq. 16). The ASTM standard double-ring infiltrometer has radii of 150 and 300 mm (Lukens, 1981), although the correct ratio will be soil dependent, and related to the relative sizes of the conductivity K and the sorptivity S (Eq. 16). The flows v_o and v_o^* can be measured using a Mariotte supply system that maintains a constant head within the rings (Constantz, 1983). Or more simply, a nail can be pushed into the soil, and a measuring cylinder used to top-up the water level to it at regular intervals. This approach may require a large amount of water, especially if the soil is dry and has a high S, such that in the buffer ring v_o^* is large. From the measured steady flux it is assumed that $v_\infty = K_s$. The role of the buffer ring is to eliminate capillary effects, so this method provides only the saturated hydraulic conductivity and leaves unresolved any measure of the soil's capillarity.

Table 1 The relative merits of field infiltration devices against a set of criteria where the ranking of 5 implies cheap, easy, or high, and 1 suggests expensive, difficult, or low. Each attribute column contains at least one 5 (top) and at least one 1 (worst). The overall Utility of each device was found as the sum of the first six columns, multiplied by the Information content. A high Utility score indicates usefulness, with the maximum range possible being from 150 down to 6

Device	Cost $5 \approx$ US$100 $1 \approx$ US$10,000	Physical ease of field use	Technical skills required	Site disturbance	Ease of data analysis	Ease of time–space replication	Information content	Utility score
Rings	5	5	5	1	5	4	3	75
Wells, auger hole permeameters	2	3	3	2	3	4	4	68
Pressure infiltrometers	2	2	2	1	3	4	3	42
Closed-top permeameters	2	2	1	1	1	1	2	16
Crust test	3	2	2	3	2	1	3	39
Tension and disc infiltrometers	2	3	2	4	2	4	5	85
Drippers	3	2	2	5	2	5	3	57
Rainfall simulators	1	1	1	5	3	1	1	12

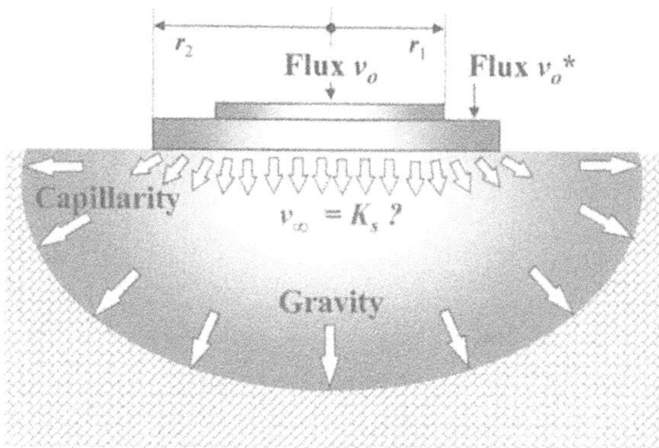

Fig. 5. Infiltration into soil from two concentric rings pressed gently into the soil. The flow in the outer ring of radius r_2, is v_0*, and this is presumed to eliminate perimetric capillary effects so that the steady flux in the inner ring v_∞ can be considered K_s.

2. *Single Ring*

If a single ring were forced into the soil to some depth, *L*, then that ring would confine the flow to the vertical and thereby eliminate the multidimensional confusion created by capillarity. Talsma (1969) developed a method whereby it is possible to measure both the sorptivity and the conductivity. A metal ring of a diameter about 300 mm and length *L* of around 250 mm is pressed into the soil so as to minimize the disturbance of the soil's structure. A free-board of about 50 mm is left, and a graduated scale is laid diametrically across the ring, with one end on the rim and the other on the soil surface. The slope of the scale is measured. A fixed volume of water is then carefully poured into the ring, and the early-time rate of infiltration is obtained from the descent of the water surface along the sloping scale. At very short times, soon after infiltration commences and before gravitational effects intercede, it is reasonable to assume that the integral form of Eq. 15 can be written as

$$\lim_{t \to 0} I = St^{1/2} \tag{37}$$

so that the sorptivity can be found as the slope of $I(\surd t)$. Because gravity's impact grows slowly, it can be difficult to select the length of period within which to fit Eq. 37. Smiles and Knight (1976) found that plotting $(It^{-1/2})$ against $\surd t$ allowed a more robust means of extracting *S* from the cumulative infiltration data.

After the initial wetting, typically after about 10 to 15 minutes, Talsma's method requires that the ring containing the soil be exhumed and placed on a fine-mesh metal grid. A Mariotte device is then used to maintain a small head of water,

h_o, on the surface of the soil. Once water is dripping out the bottom, the steady flow rate J can be measured, and Darcy's law (Eq. 1) used to find the saturated hydraulic conductivity K_s.

This simple and inexpensive method allows measurement of both the soil's capillarity via S and the saturated conductivity of K_s. However, extreme care has to be taken to minimize the disturbance of the soil during insertion. In macroporous soil this will be difficult, and furthermore any macropores that are continuous through the entire core will short-circuit the matrix and result in an erroneously high value of K_s.

3. *Twin Rings*

With the buffered-ring system, capillarity effects are hopefully eliminated. With the single-ring technique, hopefully the effects due to capillarity are measured before those of gravity intervene. But in the twin-ring method of Scotter et al. (1982), two separate rings of different size are used to exploit the dependence of capillarity on the radius of curvature of the wetted source (Fig. 1). The capillary and gravitational influences on infiltration can be separated (Youngs, 1972). Two rings of different diameters are used, and these are simply pressed lightly into the soil surface. A constant head of water is maintained inside both rings, so that the unconfined steady 3-D flow (Figs. 1 and 2) can be measured: v_1 for the smaller ring of radius r_1, and v_2 for the larger ring of r_2. The flux density of flow from the smaller ring will be higher than that of the larger ring by an amount that will reflect the soil's capillarity, namely its sorptivity (Figs. 1 and 2). Substituting r_1 and r_2 into Eq. 16 gives simultaneous equations that can be resolved to find the conductivity as

$$K_s = \frac{v_1 r_1 - v_2 r_2}{r_1 - r_2} \tag{38}$$

and the matric flux potential as

$$\phi_s = \left(\frac{\pi}{4}\right) \cdot \frac{v_1 - v_2}{1/r_1 - 1/r_2} \tag{39}$$

From ϕ_s it is possible to obtain the sorptivity S (Eq. 26), as long as θ_n and θ_s are measured before and after infiltration. In practice, replicates are taken so that the mean values of $\bar{v}_1$ and $\bar{v}_2$ are used in Eqs. 38 and 39. Scotter et al. (1982) showed how the variance in S and K_s can be calculated.

This twin-ring technique allows both the soil's capillarity and its conductivity to be measured, and here the disturbance to the soil's structure is minimal. It is only necessary to press the rings gently into the soil surface, and a mud caulking can be used to seal the ring to the surface. The technique requires, however, that there be a significant difference in the fluxes between rings, and this is dependent upon the relative sizes of the soil's capillarity and conductivity (Figs. 1 and 2). Scotter et al. (1982) showed that these effects are equal when a ring of radius $r_e = 4\phi_s/\pi K_s$ is used. Larger rings are required to obtain an estimate of the K_s of finer-textured soils, and small rings are required to obtain a good estimate of the ϕ_s of coarse-textured soils. Scotter et al. (1982) thought rings of $r = 0.025$ and 0.5 m would be suitable for a wide range of soils. If the difference in the radii is not large enough, or if there are too few replicates to obtain a reliable estimate of the $\bar{v}$'s, erroneous values will result (Cook and Broeren, 1994).

B. Wells and Auger Holes

1. Glover's Solution

It has long been known that water flow from a small surface well soon attains a steady rate, Q ($m^3\ s^{-1}$), and that in some way this flux is related to the soil's hydraulic character, the depth of water in the hole, H, and its radius, a (Fig. 6). If capillarity is ignored, and if it can be considered that the surrounding soil is wet and draining at the rate of K_s, then it is the pressure head H that generates the flow Q. Glover (1953) found that the soil's hydraulic conductivity could thus be found as

$$K_s = \frac{CQ}{2\pi H^2} \tag{40}$$

where the geometric factor C is given by

$$C = \sinh^{-1}\frac{H}{a} - \sqrt{\left(\frac{a}{H}\right)^2 + 1} + \frac{a}{H} \tag{41}$$

Thus simply by creating a small auger hole of radius a, and using a Mariotte device to maintain a constant head H, it is possible to use Q to infer the soil's saturated conductivity, K_s. Holes with $a \cong 20–50$ mm and $H \cong 100–200$ mm have commonly been used. Talsma and Hallam (1980) used this method to measure the hydraulic conductivity for various soils in some forested catchments. The Mariotte device can be simple, and the technique is quite rapid. Measurements are easy to replicate spatially. Especial care must, however, be taken when creating the hole to ensure that no smearing or sealing of the walls occurs. The surface condition of the walls in the well is critical, for it exerts great control on Q. Any smearing will throttle discharge from the well.

Philip (1985) showed that the neglect of capillarity can result in Eq. 40 providing an estimate of K_s that might be an order of magnitude too high, especially in fine-textured soils where ϕ_s is large. Capillarity establishes the size of the saturated bulb around the well and controls in part the flow Q. Its role in the infiltration process needs to be considered.

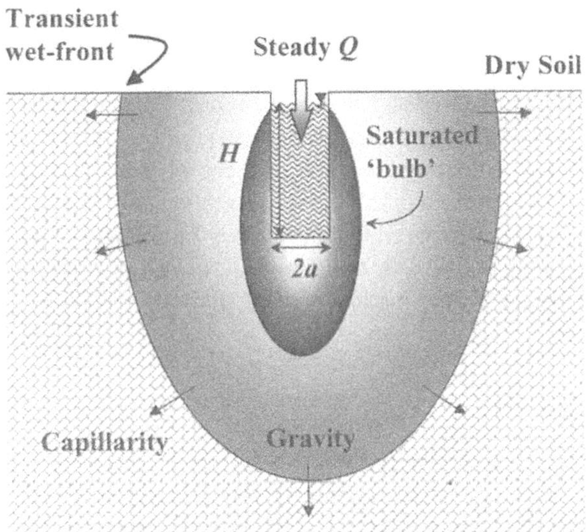

Fig. 6. Diagram to show that after some time, the flow of water Q from a small surface hole becomes steady. This Q in some way reflects the soil's capillarity, gravity, plus the depth of water in the well, H, and the hole's radius a.

2. *Improved Theory and New Devices*

Independently, and via different means, Stephens (1979) and Reynolds et al. (1985) developed new theory of the role of the soil's capillarity in establishing the steady flow Q from a well. Reynolds et al. (1985) proposed that two simultaneous measurements be made using different ponded heights H_1 and H_2 so that an approach similar to Eqs. 38 and 39 might be used. However, the difficulty in obtaining a sufficiently large range in $H_1 - H_2$ weakens the utility of this method.

The approach of Stephens et al. (1987) was to use the shape of the soil water characteristic $\theta(h)$ to correct Q for capillarity. This correction came from results obtained using a numerical solution to the auger-hole problem.

Alternatively, Elrick et al. (1989) simply estimated a value of α (Eq. 18) from an assessment of the soil's texture and structure. For compacted, structureless media they considered α to be about 1 m^{-1}, for fine-textured soils 4 m^{-1}, and structured loams 12 m^{-1}. For coarse-textured or macroporous soils they thought α could be taken as 36 m^{-1}. Given α, the solution of Reynolds and Elrick (1987) gives the value of K_s from Q as

$$K_s = \frac{Q}{\pi a^2 + (H/G)[H + \alpha^{-1}]} \tag{42}$$

where $G = C/2\pi$.

Thus new theories have improved the determination of conductivity from field measurements of infiltration from an auger hole or well. But also, there have been new devices for measuring of Q. The Guelph Permeameter (Norris and Skaling, 1992, and Soilmoisture Corp., Table 2), and the Compact Constant Head Permeameter (Amoozegar, 1992) both permit easy measurement of Q.

Layers in the soil, fractures or macropores that intersect the well, and air entrapped in the soil can all serve to make difficult the interpretation of Q in terms of K_s (Stephens, 1992). Furthermore, it is reiterated that care in the creation of the hole, and the avoidance of smearing and sealing of the walls, are critical to ensure the success of this simple and often effective method of using infiltration measurements to find K_s.

C. Pressure Infiltrometers

The problems of smearing, and of the inability to obtain sufficient separation in the ponded depths H_1 and H_2, without encroaching onto soil of different structure, led Reynolds and Elrick (1990) to develop a variant of the Guelph Permeameter. This instrument maintains a positive pressure head, H, in the water in the headspace of a ring pressed into the soil to some shallow depth, d. Generally d is of the order of 50 mm, and H is less than 250 mm. This device is commercially known as the Guelph Pressure Infiltrometer. Flow from the pressure infiltrometer is therefore confined for $z < d$, while flow beyond the ring, $z > d$, is unconfined so that an equation of the form of Eq. 42 can be employed. Reynolds and Elrick (1990) found that infiltration Q from the pressure infiltrometer could be used to find the soil's conductive and capillary properties from

$$K_s = \frac{Q}{\pi a^2 + (a/G)[H + \alpha^{-1}]} \tag{43}$$

but now $G = 0.316(d/a) + 0.184$. This technique can be used with a single head H, given that α is estimated, or it can be used with multiple heads so that both K_s and α are measured. The advantage in the latter case is that a wide separation in the heads can be achieved, but now infiltration in the different cases proceeds through the same surface. A further advantage of this pressure device is that for slowly permeable soils, or artificial clay-liners, large heads can be used to enhance infiltration so that it can be more easily observed. The device is simple and easy to use (Elrick and Reynolds, 1992). Nonetheless, insertion of the ring, coupled with the high operating water pressure, can create problems due to the creation of preferential flow paths in structured or easily disturbed soils.

D. Closed-Top Permeameters

1. Air-Entry Permeameter

There is a seductive utility in Green and Ampt's Eq. 8, for if we could find both the saturated conductivity K_s and the wetting front potential h_f, we would be able to describe infiltration using Eq. 11. Bouwer (1966) described a device that allowed this, his so-called air entry permeameter. A ring is driven into the soil to a depth of about 150 to 200 mm to constrain infiltration to one dimension. A clear acrylic top with an attached reservoir, air escape valve, and pressure gauge is sealed to the top of the ring. Once the head space is filled with water, the air-escape valve is closed. Infiltration continues until the wetting front has z_f penetrated to about 100 mm. The flow from the reservoir is then stopped, and the changing pressure in the head space monitored. The pressure reaches a minimum before air starts penetrating the soil surface. Bouwer (1966) considered this pressure to be $-2h_f$. By measuring the depth of the wetting front, either by tensiometer or observation at the end, this wetting front pressure head can be used in Darcy's law (Eq. 8) to infer K_s from the measurements of the changing level of water in the reservoir during the infiltration.

Installation of the ring can disturb the soil's structure, especially in the near-saturated range of pore sizes that are especially critical in controlling infiltration. Physically, the device is somewhat cumbersome and quite tiresome to use, so it can be difficult to obtain a large number of replicates. The device is little used nowadays. Anyway, Fallow and Elrick (1996) have recently shown how the wetting front pressure head might be easily measured using a pressure infiltrometer (Sec. III.C), simply via the addition of a tension attachment.

2. Suction Closed-Top Infiltrometers

The dimensions and connectedness of the larger pores are especially important for the determination of water entry into the soil. These pores operate in the near-saturated range of pressure heads, $-150 \text{ mm} \leq h \leq 0$. Closed-top infiltrometers have been developed to operate in this range. To provide measurements to support his views on the role played by the matrix–macropore dichotomy of field soils, Dixon (1975) developed a closed-top device to measure infiltration at pressure heads down to -0.03 m. Topp and Binns (1976) also built a closed-top suction infiltrometer that could be used down to -0.05 m.

By only measuring unsaturated infiltration, the results from these devices might be less affected by any disturbance resulting from insertion. However, the plumbing of these devices still makes their use tedious. Closed-top infiltrometers, either air-entry or suction, tend to be little used nowadays.

E. Crust Test

If there is, on the soil surface, a crust that impedes the transmission of infiltrating free water, then the pressure head at the underlying crust–soil interface, h_o, will be unsaturated; $h_o < 0$. Bouma et al. (1971) developed a crust test by which the soil's unsaturated hydraulic conductivity $K(h_o)$ could be measured in the vicinity of saturation. The procedure is described in Chap. 5 (Sec. VII.B).

The effort required for site preparation, crust installation, and tensiometer measurement makes this a somewhat tedious procedure, and so routine use is not common.

F. Tension Infiltrometers and Disk Permeameters

Infiltration into unsaturated soil reflects the dual influences of the soil's capillarity and of gravity (Fig. 1). The complex and finicky plumbing of the devices reviewed in Secs. III.B to III.E has meant that observation of the effect of the soil's capillarity was overlooked for a long time. Rather, capillarity was eliminated by insertion of rings into the soil, quenched by the addition of a buffer ring, or accounted for by a "guesstimate" of the soil's capillarity.

During the 1930s, in Utah, Willard and Walter Gardner developed a simple, no-moving-parts infiltrometer that could operate at unsaturated heads h_o near saturation. Water can only flow out of the basal porous plate and infiltrate the soil if air can enter the sealed reservoir through a narrow tube in which the capillary rise is h_o. This capillary attraction of water into the air-entry tube means that the soil has to "suck" at h_o to get the water out. The design and operation of this so-called "shower-head" permeameter was never written up, but it was later described in the thesis of Bidlake (1988).

Independently, Clothier and White (1981) developed a device called the sorptivity tube, in which the air entry into the reservoir was via a hypodermic needle and the base plate was sintered glass. A needle of different radius could be used simply to change the operating head. Employing a ring to confine the flow to one dimension, they used the short- and long-time method of Talsma (1969) (Sec. III.A) to determine the sorptivity and the conductivity from measurement of the infiltration rate $i(t)$ at $h_o = -40$ mm.

The disk permeameter of Perroux and White (1988) evolved from the sorptivity tube, but with the pressure head h_o simply controlled by a bubble tower (Fig. 7). This allows the imposed head to be changed more easily. The disk has a basal membrane of 20 to 40 μm nylon mesh, and fine sand is used to ensure a good contact between the soil surface and the permeameter. The permeameter is easy to use, economical on water, and portable, and several can be operated at the same time. Measurement in the field, across a range of heads, minimally disturbs

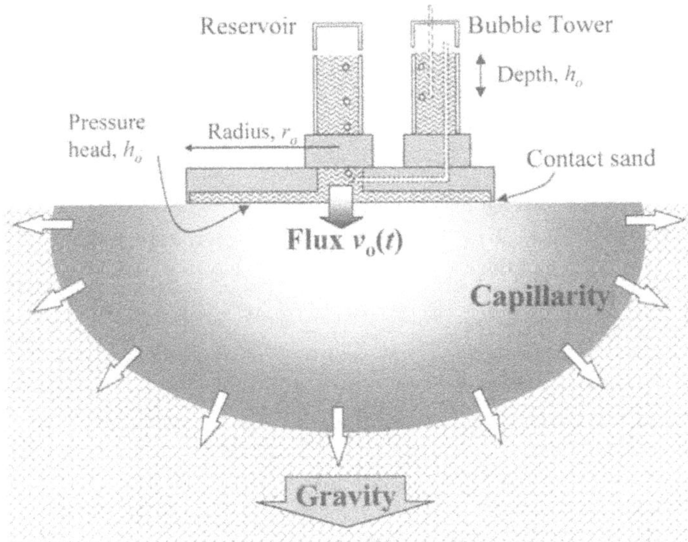

Fig. 7. A transverse section through a disk permeameter of radius r_0. At the imposed unsaturated head of h_0, both capillarity and gravity combine to draw water into the soil at flux density v_0 (m s^{-1}). Contact sand is used to ensure good hydraulic connection between the permeameter and the soil.

the soil. The disk permeameter, or tension infiltrometer as it is sometimes known, has become so popular that several companies now produce the device, and the cost ranges from US$1500 to $3000, depending on accessories (Tables 1 and 2). A variant of the shower-head permeameter, called the Mini Disk Infiltrometer, is now in commercial production (Table 2).

The disk permeameter is set at head h_0 and then placed on the smooth flat surface of contact sand, which has previously been prepared to ensure good contact between the permeameter and the soil. The unconfined infiltration $v_0(t)$ is monitored by observing the drop of water level in the reservoir, or it can be recorded automatically using pressure transducers (Ankeny et al., 1988). There are various means by which this observation can be used to infer the soil's hydraulic character. I discuss three of these below, before outlining the use of the permeameter to infer the chemical transport characteristics of field soil.

1. *Short and Long-Time Observations*

At very short time, just after the disk permeameter is placed on the soil, the flow from the surface disk is not greatly affected by the 3-D geometry, so that $v_0(t)$ is

Table 2 Major Suppliers of Infiltration Measurement Devices, along with Contact Details and the Nature of the Devices Sold

Manufacturer	Address	Devices sold
Decagon Devices Inc.	PO Box 835, Pullman, Washington 99163, USA http://www.decagon.com	Mini tension infiltrometers
Eijkelkamp	Nijverheidsstraat 14, PO Box 4 6987 ZG Giesbeck, The Netherlands Tel +31 313 631 941 Fax +31 313 632 16 http://www.eijkelkamp.com	Rings, auger hole permeameters
Soil Measurement Systems	7090 N Oracle Road, Suite 178, PMB #170, Tucson, Arizona 85704-4383, USA Tel +1 520 742 4471 Fax +1 520 797 0356 http://www.soilmeasurement.com	Tension infiltrometers
Soilmoisture Equipment Corp.	PO Box 30025 801 S Kellogg Avenue, Santa Barbara, California 93105, USA Tel +1 805 964 3525 Fax +1 805 683 2189 http://www.soilmoisture.com	Auger hole permeameters (Guelph) Pressure and tension infiltrometers

akin to the 1-D $i(t)$. During this very early stage of infiltration it can be expected that for some short period the cumulative infiltration I will be a function of the square root of time, at least until gravity effects intercede (Perroux and White, 1988). Thus early observations of infiltration from the disk can, in theory, be used in Eq. 37 to infer the unsaturated sorptivity, $S_o = S(\theta_o)$.

Here the unsaturated sorptivity is given by Eq. 24, with the upper limit of integration being the $\theta_o = \theta(h_o)$ imposed by the permeameter's head of h_o. Given the measurement of θ_n, and final observation of the water content θ_o just under the disk, then Eq. 26 gives the unsaturated matric flux potential $\phi_o = \phi(\theta_o)$. Thus the short-time observations of 3-D infiltration from the disk can be used in a manner similar to that employed in 1-D by Talsma (1969) (Sec. III.A). However, it can be difficult to ensure that only the true square-root-of-time signal is observed in the measured $v_o(t)$. This sorptive period is unfortunately even shorter in 3-D than it is in 1-D, so determination of the true S_o can be difficult. Furthermore, if a significant amount of fine sand is used to ensure good disk–soil contact, the short $I(\sqrt{t})$ period can be obscured by imbibition of water into the contact material.

After this short time period, the permeameter, still at h_o, is left until the flow v_o has become effectively steady at v_∞. This can take anywhere between 15 minutes and several hours. From this final steady-flow observation, Wooding's equation (Eq. 16) can used to find $K_o = K(h_o)$, since ϕ_o has already been found from the short-time analysis for the sorptivity S_o (Eq. 26). Care must be taken that the operator's enthusiasm to conclude the test does not override the requirement to ensure that the flow is effectively steady, rather than still declining, albeit slowly.

So using an approach akin to that of Talsma (1969), the approach of Perroux and White (1988) permits measurement of both the soil's capillarity and its conductivity from observations of 3-D infiltration from a disk permeameter set at h_o.

2. *Twin and Multiple Disks*

To get around the problem of finding the sorptivity from the short-time infiltration curve, the twin ponded ring technique of Scotter et al. (1982) (Eqs. 38 and 39) was used by Smettem and Clothier (1989) with disk permeameters of different radii. Both K_o and ϕ_o could now be found from the steady unsaturated flows leaving permeameters of different radii. Again a sufficiently wide separation of r_1 and r_2 is required so that good estimates of the soil's ϕ_o and K_o are realized. To overcome this, three or more different radii can be used, and a regression of v_∞ on r^{-1} used to resolve K_o as the intercept, and $4\phi_o/\pi$ as the slope (Thony et al. 1991). These twin or multiple disk techniques require that there be sufficient replications to obtain robust measures of the mean study flows $\bar{v}_{1,2,3,\ldots}$. This requirement has the advantage that some indication of the soil's variability is obtained. However, that variation can make difficult the application of Eqs. 38 and 39 because the various measurements are not made on the same infiltration surface.

3. *Multiple Heads*

Rather than use permeameters of different radii, Ankeny et al. (1991) proposed a simultaneous solution of Wooding's equation (Eq. 20) based on a single permeameter and observations of steady infiltration at the two, or more, different heads $h_1, h_2, \ldots$. For simplicity, I present here a two-head version of this approach that assumes that the soil has an exponential conductivity function (Eq. 18), so that

$$K_1 = K_s \exp(\alpha h_1) \qquad K_2 = K_s \exp(\alpha h_2) \tag{44}$$

so from Eq. 20,

$$Q_1 = K_1\left(\pi r_o^2 + \frac{4r_o}{\alpha}\right) \quad Q_2 = K_2\left(\pi r_o^2 + \frac{4r_o}{\alpha}\right) \tag{45}$$

Combination of these gives

$$\frac{Q_1}{Q_2} = \frac{K_1}{K_2} = \exp[\alpha(h_1 - h_2)] \tag{46}$$

which can be rearranged to provide a measure of the soil's capillarity from

$$\alpha = \frac{\ln(Q_1/Q_2)}{h_1 - h_2} \tag{47}$$

This α can be inserted back into Eq. 45 to find the conductivities K_1 and K_2 at these two heads.

This procedure is best started by initial placement of the permeameter on the soil at the lowest head, say -150 mm. Once the flow becomes steady, the head can be easily changed by raising the air-entry tube in the bubble tower (Fig. 7), and a new steady flow observed. This procedure can be repeated several times, in jumps of $\Delta h \approx 20$–50 mm, so that the soil's near-saturated conductivity function can be constructed as a piece-wise exponential. This approach only relies on the conductivity being described as an exponential just over Δh. A real advantage of the technique is that all the infiltration from the disk is through the same surface, so that spatial variability does not pose the problem it can in the multiple-radii case.

Of all the measurement approaches that rely on inverse interpretation of flow from a disk permeameter, that of Ankeny et al. (1991) is probably the most robust means by which to obtain the hydraulic properties of the soil.

4. *Solute Transport*

Infiltrating water is the vehicle for transporting solutes through the soil. However, deeper-than-expected penetration of surface-applied chemicals has lead to the realization that not all of the soil's pore water is actively and equally involved in solute transport. Better description of this transport process can be achieved if only some portion of the wetted pore space is considered mobile (Eq. 35), say θ_m. So field measurement of θ_m is needed if we are to be able to model the movement of chemicals through structured soil.

Clothier et al. (1992) proposed a method for achieving this using a disk permeameter whose reservoir was filled with a tracer solution at some concentration c_m (mol L^{-1}). Inert anionic tracers such as bromide or chloride are suitable for most soils, except of course those variably charged soils that undergo anion exchange. The tracer-laden permeameter can be first used, as described above, to obtain the soil's hydraulic properties. However, at the end of infiltration, the permeameter is lifted and a vertical face cut along the diameter in the soil under the disk (Fig. 8). Samples are taken from this face so that their water content θ_o and

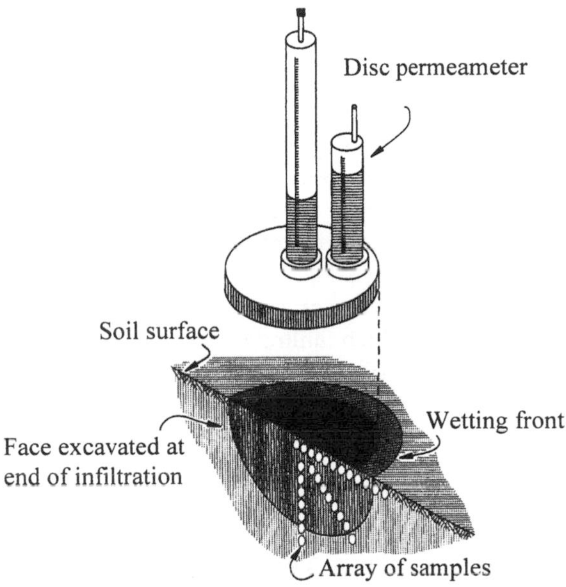

Fig. 8. Disk permeameter, with sampling locations to allow measurement of the resident concentration c^* of tracer in the soil for use in Eq. 49, at the end of infiltration.

tracer concentration c^* can be determined. The tracer solute in the sample will be partitioned between the mobile and immobile domains:

$$\theta_o c^* = \theta_m c_m + \theta_{im} c_{im} \tag{48}$$

If interdomain exchange can be ignored during the period of infiltration, and if there were no tracer initially present in the soil ($c_{im} = 0$), then

$$\theta_m = \theta_o \frac{c^*}{c_m} \tag{49}$$

So if the measured resident concentration of solute in the soil under the disk is not that of the flux concentration of tracer leaving in the reservoir, then some of the soil's antecedent water must have remained immobile during the invasion of the tracer. To be valid, it is necessary that there be a depth of infiltration of $I \cong$ 25 mm so that hydrodynamic dispersive effects have locally dissipated, and that the resident concentration has reached its steady value (van Genuchten and Wierenga, 1976).

Jaynes et al. (1995) have proposed an alternative means of measuring the mobile fraction, which through the use of multiple tracers can provide information on the interdomain exchange coefficient. Clothier et al. (1996) developed a technique whereby the permeameter reservoir contained two tracers, one inert and one reactive. The inert tracer could be used to provide the mobile fraction, and the retardation of the reactive tracer behind the inert tracer front could be used to infer the chemical exchange isotherm (cf. Eq. 36). Vogeler et al. (1996) devised a means by which time domain reflectometry probes could be inserted directly through a permeameter so that measurements under the disk of the changing water content and electrical conductivity could be used to infer the solute transport characteristics during infiltration.

Tension infiltrometers, or disk permeameters as they might be known, have become one of the most popular means by which infiltration can be measured in the field, and these data can be used in an inverse sense to obtain the soil's hydraulic characteristics.

G. Drippers

Whereas all the previous methods use the instrument itself to define or constrain the flow domain, the dripper method of Shani et al. (1987) actually uses the size and character of the infiltration zone around an unconfined dripper to infer the soil's hydraulic properties. Commercial drip-irrigation emitters are used to create a range of discharges Q upon a parcel of soil, and the radius of the wetted pond, r_o, is measured for each. Shani et al. (1987) found that the steady-state radius of the free-water pond under each dripper would be achieved after about 15 min. By plotting these various observations of r_o^{-1} against Q, from Eq. 20, both K_s and ϕ_s can be deduced. For less permeable soils, discharges in the range of 0.5 to 5 L h^{-1} are apt, whereas for more permeable soils it may be that $100 < Q < 700$ L h^{-1} is required to get an appropriately sized pond.

Shani et al. (1987) also considered that a Green and Ampt (1911) rectangular profile of wetting (Fig. 4) could be used to interpret observations of the radial distance between the wetting front and the ponded radius. Care would need to be exercised in this case, for as Philip (1969) showed, such a rectangular profile of wetting is not theoretically possible in 3-D.

The simple procedure of using $Q(r^{-1})$ to infer the soil's hydraulic properties offers a useful means of field measurement, especially because it is possible to obtain easily a large number of replicates.

H. Rainfall Simulators

Many devices have been constructed over the last century to mimic rainfall landing on the soil surface. Generally these quite expensive instruments have been built to investigate the impact of rainfall intensity on the generation of runoff and soil erosion (Grierson and Oades, 1977). However, rainfall simulators can also be used to observe the role of the soil's hydraulic character in controlling infiltration (Amerman, 1979).

In the simplest of arrangements, the simulator can be used to supply rainfall to the soil surface at a rate v_o that will generate runoff, the time rate of which can be monitored. From the difference between the rate of applied rainfall, and variation in the measured rate of runoff, a ponded infiltration curve $i(t)$ can be found. The quality of the data from this differencing is never very high, even though the results do represent an areal integration across a surface area of about 1 m^2.

Alternatively, the simulator can be used in a nonponding mode, $v_o < K_s$. Eventually the surface water content, θ_o, will attain equilibrium with the applied flux v_o, so that one point of the unsaturated conductivity curve is obtained, since functionally $\theta_o = K^{-1}(v_o)$ (Clothier et al., 1981). Cores can be extracted from the soil surface to obtain θ_o. For example, with a Bungedore fine sand they found that when $v_o/K_s = 0.283$, $\theta_o = 0.21$, whereas $\theta_s = 0.335$ and $K_s = 5 \times 10^{-6}$ m s^{-1}. The rate can then be raised, and another value of θ_o obtained. Since the soil is unsaturated, the removal of cores has little influence on infiltration (Fig. 3). Time domain reflectometry (TDR) would make such measurements of θ_o easier. Nonetheless, because of the expense and complexity of rainfall simulators, they are more likely to remain used in studies of soil erosion (Amerman et al., 1979).

I. Summary

To allow an easy intercomparison of the various devices that might be used to study infiltration in the field, Table 1 has been constructed. The eight instruments are rated with regard to their cost, ease of use in the field, technical difficulty, soil disturbance, ease of analysis, and ability to replicate. In Table 1 a low number is unfavorable, whereas a higher number indicates utility. A 5 is the maximum, with 1 being the minimum. Every column has at least one of each. The sum of the values in each row is multiplied by what I consider to be the information content of the results, to give an overall Utility Score.

I consider that the age-old technique of rings still has merit, whereas the newer tension infiltrometers, or disk permeameters, score highest in terms of usefulness in the study of infiltration. Rainfall simulators, since they are expensive and can only provide coarse measures of the soil's hydraulic properties, score worst on my scale of utility. Their merit lies elsewhere. A list of suppliers of commercial devices is presented in Table 2.

Given that nowadays it is possible to measure infiltration in the field with relative ease using new devices, and that modern theory presently allows cogent interpretation of the observations, the following section considers what these studies have told us about infiltration, and what remains to be wary of.

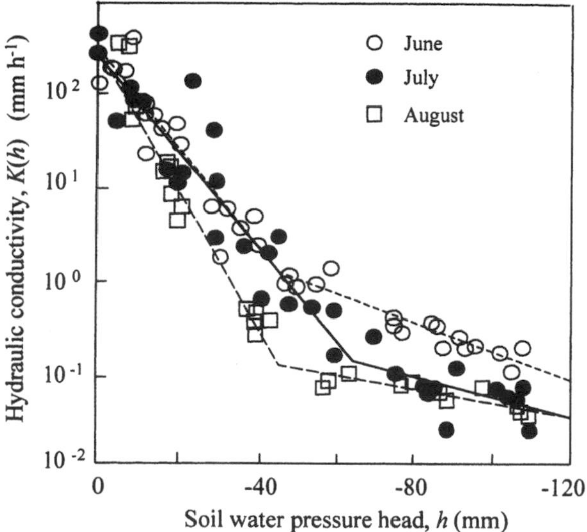

Fig. 9. Disk permeameter measurements of the $K(h)$ of a Swedish clay soil at three times of the year (redrawn from Messing and Jarvis, 1993). They fitted a two-line linear regression model to the $\ln K - h$ data at each sampling, as shown by the lines.

IV. RESULTS AND RESERVATIONS

A. Conductivity

In coarser textured soils, the role of conductivity is paramount in controlling infiltration, especially the conductivity close to saturation. Furthermore, it is the surface-ventedness and connectedness of the mesopores and the macropores that operate at near-saturated heads that play dominant roles in establishing the shape of the soil's near-saturated $K(h)$ (Clothier, 1990). Disk permeameters, which operate in the range of $-150 < h \text{ (mm)} < 0$, are useful tools with which to observe the soil's near-saturated conductivity. Messing and Jarvis (1993) used permeameters, with the multiple-head approach of Ankeny et al. (1991) (Eqs. 45 and 47), to determine the conductivity of a Swedish clay soil at three times during the summer (Fig. 9).

With the disk permeameter a wealth of in situ infiltration information has been obtained with relative ease. The dramatic drop in the conductivity as the head decreases is seen, highlighting the role played by macropores during near-saturated flow. Messing and Jarvis (1993) showed that all their data were better described by a two-line regression than a single linear fit. They considered the breakpoint to be the separation between macropores and mesopores. The other feature of Fig. 9 is the temporal change in the soil's hydraulic character between the three samplings. In the region between heads of −40 and −20 mm, the soil's conductivity dropped by an order of magnitude during the summer. They considered this to be due to structural breakdown by rain impact and surface capping or sealing. Disk permeametry has allowed exploration of the role of the soil's conductivity in generating a change in the way water would infiltrate the soil during the summer period.

B. Capillarity

In finer textured soils, capillarity can be the most important control on infiltration. Thony et al. (1991) found that for a heavy clay soil in Spain the capillary-dominated period of I being a function of $\sqrt{t}$ lasted for 5 hours. This square-root-of-time capillarity only extended to about 8 seconds for their French loam.

Sauer et al. (1990) used disk permeameters to examine the impact of plowing on the capillary properties of a Plainfield sand. The sorptivity S_o in the near-saturated range from 0 down to −90 mm is shown in Fig. 10 either for soil that

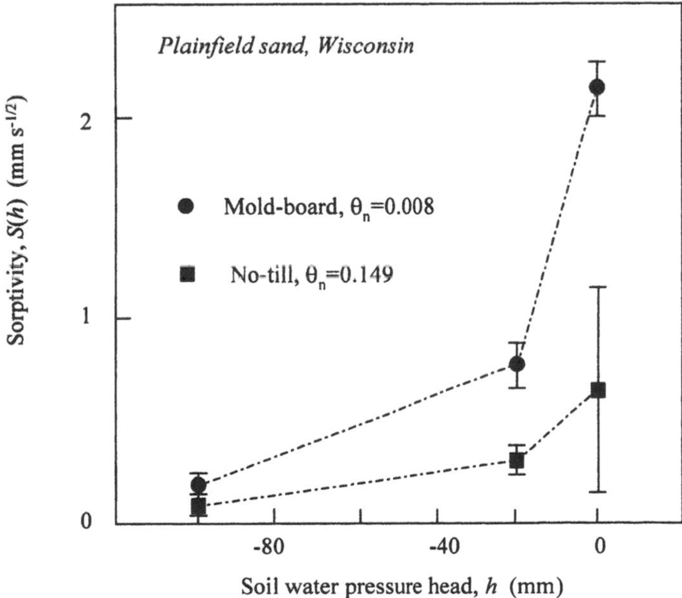

Fig. 10. Disk permeameter measurements of the unsaturated sorptivity S_o of a Plainfield sand that had either been regularly tilled by a mold-board plow (●), or in which maize (*Zea mays* L.) had been direct drilled (■). The error bars are the standard deviations of the replicated measurements. (Redrawn from Sauer et al., 1990.)

had been mold-board plowed every year and planted with maize (*Zea mays* L.) or for the same soil which had been not tilled, but direct drilled.

The soil that had been mold-board plowed was single-grained, and the inter-row space where the measurements were made was exposed to direct sunshine, so that the antecedent water content there was low: $\theta_n = 0.008$. In the no-till case, the trash from last year's crop provided a mulch, such the soil underneath on which the measurements were made was much wetter at $\theta_n = 0.149$.

The sorptivity of the tilled soil was much higher than that of the no-till soil—for two reasons. Sorptivity is the definite integral of the diffusivity function (Eq. 24), and so it is affected by both the upper and the lower limits on the integral: $S = S(\theta_o, \theta_n)$. Simply because the θ_n of the tilled soil was lower than that of the no-till soil, the sorptivity of the tilled soil is higher, irrespective of any changes induced by tillage. However, tillage of this sand destroyed the structure, leaving a single grained medium with a high surface area that generated a high degree of capillarity. In contrast, the no-till soil was riven with macropores so that the surface area for water absorption was less. Also the higher variability in the sorptivity at saturation for the no-till soil reflects the variation due to this macroporosity. In contrast, the mold-board plow had homogenized the tilled soil. In both cases the sorptivity drops off as h_o declines, and θ_o drops, for less of the near-saturated $D(\theta)$ contributes to the integral. Indeed, this drop-off in S_o, measured during infiltration, can be used to infer, in an inverse sense, the soil's diffusivity function $D(\theta)$ (Smiles and Harvey, 1973, Chap. 5).

C. Pore Size Characteristics

The soil's hydraulic properties can be used to obtain a measure of the soil's mean pore size characteristics (Eqs. 30 and 32). White and Perroux (1989) determined the characteristic mean pore size λ_m (Eq. 32) of a Murrumbateman silty clay loam (Fig. 11) using permeameters at heads h_o of -93 mm and -23 mm. At the lower head, this soil was characterized by a mean pore size of about 20 μm, whereas closer to saturation λ_m was over 0.1 mm. Measurements were made just prior to drought-breaking rains, and immediately after. The impact of the rain was negligible in the micro-mesopore range up to the lower head measurement, indicating that the pore size characteristics of the matrix of this soil remained unaffected by the rain. However, the rain affected this characteristic in the macropore range at the higher head. A structural change is evident, with macroporosity collapse, macropore infilling, and surface sealing all causing the mean pore size of around 0.25 mm, prior to rain, to drop to about 100 μm. This drop in macroporosity was matched by a loss in the near-saturated conductivity at $h_o = -23$ mm, with K_o dropping from 1.25 to 0.235 μm s^{-1}.

Because infiltration is strongly influenced by pore size and connectedness, it is very useful to be able to use infiltration to detect changes in the functioning

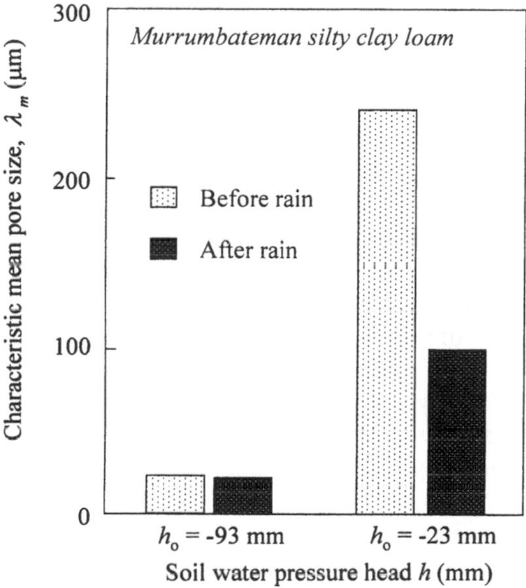

Fig. 11. The flow-weighted mean pore size (Eq. 31) of a Murrumbateman silty clay loam as determined by permeameters set at heads of $h_o = -93$ and -23 mm, both before drought-breaking rains and immediately thereafter. (Redrawn from White and Perroux, 1989.)

of the soil's macroporosity. Methods such as micromorphology or bulk density determination cannot offer such powers of functional discrimination.

D. Fingered Flow and Hydrophobicity

Not always does the infiltrating wetting front move into the soil in a stable manner. Rather, viscous fingering can occur as the front becomes unstable, and certain portions of the front advance more rapidly. Hill and Parlange (1927) and Philip (1975) noted that soil crusts, layering of a finer medium over a coarser underlay, and air entrapment were all conditions that could lead to frontal instability and the generation of fingered flow. However, probably the major cause of fingered flow in field soils is that generated by the widespread phenomenon of water repellency (Ritsema and Dekker, 1996).

Thus far, we have assumed that the soil is hydrophilic, and that the infiltrating water easily wets the soil. However decaying organic matter, plus humic and fulvic acids, induce a degree of water repellency so that the soils can become hydrophobic. This repellency is most pronounced under dry conditions, but it can slowly break down during wetting. Clothier et al. (1996) found for a structured loam that the infiltration rate from a disk permeameter remained low for about 100 minutes, during which time an I of only 5 mm infiltrated. Then the rate rose rapidly to attain a steady flow rate of around 5 μm s^{-1}. Such a time course of infiltration defies description in terms of sorptivity and conductivity (Eq. 15). Tillman et al. (1989) proposed a means by which the soil's water repellency could be characterized using infiltration measurements. Using glass sorptivity tubes (Clothier and White, 1981) filled either with ethanol or water, two measures of the soil's sorptivity can be obtained; one for water and the other for ethanol. It is important that glass be used, for ethanol will crack acrylic reservoirs. For a hydrophilic soil, the sorptivity of water should be 1.95 times that of the ethanol, since S should scale by $(\mu\sigma)^{1/2}$ for different fluids, where μ is the dynamic viscosity (N s m^{-1}) and σ is the surface tension (N m^{-1}). They suggested that the measured ratio of the ethanol S over that of water be used as an index of repellency. Hydrophilic soils would have an index of 0.5, and anything above indicates repellency. In the hydrophobic case described above (Clothier et al., 1996), the ethanol S was 0.6 mm $s^{-1/2}$, whereas the water S was just 0.03 mm $s^{-1/2}$, or a repellency index of 40! Dekker and Ritsema (1994) developed a water-drop penetration time test to provide a measure of the soil's water repellency.

Water repellency by soil, a biologically induced phenomenon, is widespread (Wallis and Horne, 1992), and its consequences can be dramatic. Ritsema and Dekker (1996) found that fingers of wetting had passed a depth of 700 mm in a hydrophobic Dutch soil, some 3 days after just a 24 mm rainstorm. The main infiltration "front," however, had only penetrated to 100 mm.

V. CONCLUSIONS

Up until the 1970s, the focus of infiltration studies was the analytical description of the flow process. Field experiments were carried out in attempts to validate directly these theoretical models. However, since then, a change of direction has occurred. These theories are now being used in an inverse sense to infer the hydraulic characteristics of field soil from observations of infiltration obtained in the field with new devices. These hydraulic and chemical transport properties are then being used in numerical models to predict, in a forward sense, the hydrologic functioning of soil.

Further development of theory would seem unlikely, except perhaps in areas of macropore flow, fingering, and hydrophobicity. However, we can look forward to the further development of new devices and improved techniques for measuring infiltration in the field.

LIST OF SYMBOLS

Roman

a	radius of auger hole	(m)
A	coefficients in Eq. 7	
b	parameter in Eq. 26	
C	Glover's parameter in Eq. 41	
c	solution concentration of chemical	(mol L^{-1})
D	soil-water diffusivity function	($m^2\ s^{-1}$)
d	depth of ring pressed into soil	(m)
F	flux-concentration relation	
G	Guelph permeameter coefficient	
g	acceleration due to gravity	(m s^{-2})
H	total hydraulic pressure head, or ponded height	(m)
h	soil water pressure head	(m)
I	cumulative infiltration	(m)
i	infiltration rate	(m s^{-1})
J	Darcy flux of water	(m s^{-1})
K	hydraulic conductivity function	(m s^{-1})
L	column length	(m)
m	number of spatial dimensions	
Q	volume flux of water	($m^3\ s^{-1}$)
R	normalized radius of curvature	
r	radius of curvature, or ring radius, or disk radius	(m)
$\Re$	retardation of solute front relative to the wetting front	
S	sorptivity	(m $s^{-1/2}$)
s	solute front	(m)
t	time	(s)
v	flux through a surface	(m s^{-1})
x	horizontal distance	(m)
z	depth	(m)

Greek

α	slope of the exponential $K(h)$ function	(m^{-1})
β	coefficient	
δ	Dirac delta function	
ϕ	matric flux potential	($m^2\ s^{-1}$)
λ	capillary length scale	(m)
μ	dynamic viscosity	(N s m^{-2})
Θ	normalized water content	
θ	volumetric soil water content	($m^3\ m^{-3}$)
ρ	density of water	(kg m^{-3})
σ	surface tension	(N m^{-1})

Superscripts and Subscripts

c capillary
f front
grav gravity
im immobile
m mobile, or matrix
n antecedent
o surface, or unsaturated
p ponded
s saturated
* buffer ring
∞ long time, or steady value

REFERENCES

Amerman, C. R. 1979. Rainfall simulation as a research tool in infiltration. In: *Proceedings of the Rainfall Simulator Workshop*, Tucson, Arizona. March 7–9, 1979. Sidney, Montana: USDA-SEA-AR, ARM-W-10/July 1979, pp. 85–90.

Amerman, C. R., et al. 1979. *Proceedings of the Rainfall Simulator Workshop,* Tucson, Arizona, March 7–9, 1979. Sidney, Montana: USDA-SEA-AR, ARM-W-10/July 1979, p. 185.

Amoozegar, A. 1992. Compact constant head permeameter: A convenient device for measuring hydraulic conductivity. In: *Advances in Measurement of Soil Physical Properties: Bringing Theory into Practice* (G. C. Topp, et al., eds.). Soil Sci. Soc. Am. Special Publ. 30, pp. 31–43.

Ankeny, M. D., T. C. Kaspar, and R. Horton. 1988. Design of an automated tension infiltrometer. *Soil Sci. Soc. Am. J.* 52:893–896.

Ankeny, M. D., M. Ahmed, T. C. Kaspar, and R. Horton. 1991. Simple field method for determining unsaturated hydraulic conductivity. *Soil Sci. Soc. Am. J.* 55:467–470.

Bidlake, W. R. 1988. Seed zone temperatures and moisture conditions under conventional-till and no-till systems in Alaska. Ph.D. thesis, Washington State Univ., Pullman (Diss. Abstr. 89-15895).

Bouma, J., D. Hillel, F. D. Hole, and C. R. Amerman. 1971. Field measurement of hydraulic conductivity by infiltration through artificial crusts. *Soil Sci. Soc. Am. J.* 35: 362–364.

Bouwer, H. 1961. A study of final infiltration rates from cylinder infiltrometers and irrigation furrows with an electrical resistance network. Trans. 7th Int. Congr. Soil Sci., Madison, Wisconsin, Vol. 1, pp. 448–456.

Bouwer, H. 1964. Unsaturated flow in groundwater hydraulics. *J. Hydraul. Eng.* 90(HY5): 121–155.

Bouwer, H. 1966. Rapid field measurement of the air-entry value and hydraulic conductivity of soil as significant parameters in flow system analysis. *Water Resour. Res.* 2: 729–738.

Brutsaert, W. 1979. Universal constants for scaling the exponential soil water diffusivity? *Water Resour. Res.* 15:481–483.

Buckingham, E. 1907. *Studies on the Movement of Soil Moisture.* Bull. 38, USDA Bureau of Soils, Washington DC.

Childs, E. C. 1936. The transport of water through heavy clay soils: I. *J. Agr. Sci.* 26: 114–127.

Childs, E. C. 1967. Soil moisture theory. *Adv. Hydrosci.* 4:73–117.

Childs, E. C., and Collis-George, N. 1948. Soil geometry and soil-water equilibria, *Disc. Faraday Soc.* 3:78–85.

Clothier, B. E. 1990. Soil water sorptivity and conductivity. *Remote Sens. Rev.* 5:281–291.

Clothier, B. E., and I. White. 1981. Measurement of sorptivity and soil water diffusivity in the field. *Soil Sci. Soc. Am. J.* 45:241–245.

Clothier, B. E., I. White, and G. J. Hamilton. 1981. Constant-rate rainfall infiltration: Field experiments. *Soil Sci. Soc. Am. J.* 45:245–249.

Clothier, B. E., M. B. Kirkham, and J. E. MacLean. 1992. In situ measurement of the effective transport volume for solute moving through soil. *Soil Sci. Soc. Am. J.* 56: 733–736.

Clothier, B. E., S. R. Green, and H. Katou. 1995. Multi-dimensional infiltration: Points, furrows, basins, wells, and disks. *Soil Sci. Soc. Am. J.* 59:286–292.

Clothier, B. E., G. N. Magesan, L. Heng, and I. Vogeler. 1996. In situ measurement of the solute adsorption isotherm using a disc permeameter. *Water Resour. Res.* 32: 771–778.

Constantz, J. 1983. Adequacy of a compact double-cap infiltrometer compared to the ASTM double ring infiltrometer. In: *Advances in Infiltration.* Proc. Nat. Conf. on Advances in Infiltration, Am. Soc. Agric. Eng., ASAE Publ. 11–83, pp. 226–230.

Cook, F. J. and A. Broeren. 1994. Six methods for determining sorptivity and hydraulic conductivity with disc permeameters. *Soil Sci.* 157:2–11.

Dekker, L. W., and C. J. Ritsema. 1994. How water moves in a water-repellent soil. 1. Potential and actual water repellency. *Water Resour. Res.* 30:2507–2517.

Dixon, R. M. 1975. Design and use of closed-top infiltrometers. *Soil Sci. Soc. Am. J.* 39: 755–763.

Dixon, R. M., and A. E. Peterson. 1971. Water infiltration control: A channel system concept. *Soil Sci. Soc. Am. J.* 35:968–973.

Elrick, D. E., and W. D. Reynolds. 1992. Infiltration from constant-head well permeameters and infiltrometers. In: *Advances in Measurement of Soil Physical Properties: Bringing Theory into Practice* (G. C. Topp, et al., eds.), Soil Sci. Soc. Am. Special Publ. 30, pp. 1–25.

Elrick, D. E., W. D. Reynolds, and K. A. Tan. 1989. Hydraulic conductivity measurements in the unsaturated zone using improved well analyses. *Ground Water Monit. Rev.* 9: 184–193.

Fallow, D. J., and D. E. Elrick. 1996. Field measurement of air-entry and water-entry soil water pressure heads. *Soil Sci. Soc. Am. J.* 60:1036–1039.

Glover, R. E. 1953. Flow from a test-hole located above groundwater level, pp. 69–71. In: *Theory and Problems of Water Percolation.* (C. N. Zanger, ed.). Denver, Colorado: US Bur. Reclam. Eng. Monogr. 8.

Green, W. H., and G. A. Ampt. 1911. Studies on soil physics: Part 1. The flow of air and water through soil. *J. Agric. Sci.* 4:1–24.

Grierson, I. T., and J. M. Oades. 1977. A rainfall simulator for field studies of run-off and soil erosion. *J. Agric. Eng. Res.* 22:37–44.

Hill, D. E. and J.-Y. Parlange. 1972. Wetting front instability in layered soils. *Soil Sci. Soc. Am. J.* 36:697–702.

Horton, R. E. 1940. Approach toward a physical interpretation of infiltration capacity. *Soil Sci. Soc. Am. J.* 5:339–417.

Jaynes, D. B., S. D. Logsdon, and R. Horton. 1995. Field method for measuring mobile/immobile water content and solute transfer rate coefficient. *Soil Sci. Soc. Am. J.* 59: 352–356.

Klute, A. 1952. Some theoretical aspects of the flow of water in unsaturated materials. *Soil Sci. Soc. Am. J.* 16:144–148.

Lukens, R. P., ed. 1981. *Annual Book of ASTM Standards, Part 19: Soil and Rock; Building Stones.* Washington DC: ASTM, pp. 509–514.

Messing, I., and N. J. Jarvis. 1993. Temporal variation in the hydraulic conductivity of a tilled clay soil as measured by tension infiltrometers. *J. Soil Sci.* 44:11–24.

Myers, L. E., and C. H. M. van Bavel. 1963. Measurement and evaluation of watertable elevations, 5th Congress, Intern. Comm. Irrig. Drain. Tokyo, May 1963.

Norris, J. M., and W. Skaling. 1992. Guelph permeameter: Commercial and regulatory demands for acceptance of a new method. In: *Advances in Measurement of Soil Physical Properties: Bringing Theory into Practice* (G. C. Topp et al., eds.). Soil Sci. Soc. Am. Special Publ. 30, pp. 25–31.

Parlange, J.-Y. 1971. Theory of water movement in soils. I. One-dimensional absorption. *Soil Sci.* 111:134–137.

Parlange, J.-Y. 1975. On solving the flow equation in unsaturated soils by optimization: Horizontal infiltration. *Soil Sci. Soc. Am. J.* 39:415–418.

Perroux, K. M., D. E. Smiles, and I. White. 1981. Water movement in uniform soils during constant flux infiltration. *Soil Sci. Soc. Am. J.* 45:237–240.

Perroux, K. M., and I. White. 1988. Designs for disc permeameters. *Soil Sci. Soc. Am. J.* 52:1205–1215.

Philip, J. R., 1957a. The theory of infiltration: 1. The infiltration equation and its solution. *Soil Sci.* 83:345–357.

Philip, J. R. 1957b. The theory of infiltration: 4. Sorptivity and algebraic infiltration equations. *Soil Sci.* 84:257–267.

Philip, J. R. 1966. Absorption and infiltration in two- and three-dimensional systems. In: *Water in the Unsaturated Zone,* Vol. 1 (R. E. Rijtema and H. Wassink, eds.). Paris: UNESCO, pp. 503–525.

Philip, J. R. 1969. Theory of infiltration. *Adv. Hydrosci.* 5:215–296.

Philip, J. R. 1975. The growth of disturbances in unstable infiltration flows. *Soil Sci. Soc. Am. J.* 39:1049–1053.

Philip, J. R. 1985. Reply to "Comments on 'Steady infiltration from spherical cavities.'" Soil Sci. Soc. Am. J. 49:788–789.

Philip, J. R. 1987. The infiltration joining problem. *Water Resour. Res.* 23:2239–2245.

Philip, J. R. 1988. Quasianalytical and analytical approaches to unsaturated flow. In: *Flow and Transport in the Natural Environment: Advances and Applications* (W. L. Steffen and O. T. Denmead, eds.) Berlin: Springer Verlag, pp. 30–48.

Philip, J. R. 1995. Desperately seeking Darcy in Dijon. *Soil Sci. Soc. Am. J.* 59:319–324.

Philip, J. R., and J. H. Knight. 1974. On solving the flow equation: 3. New quasi-analytical technique. *Soil Sci.* 117:1–13.

Reynolds, W. D., and D. E. Elrick. 1987. A laboratory and numerical assessment of the Guelph permeameter. *Soil Sci.* 144:282–299.

Reynolds, W. D., and D. E. Elrick. 1990. Ponded infiltration from a single ring. I. Analysis of steady flow. *Soil Sci. Soc. Am. J.* 54:1233–1241.

Reynolds, W. D., D. E. Elrick, and B. E. Clothier. 1985. The constant head well permeameter: Effect of unsaturated flow. *Soil Sci.* 139:172–180.

Richards, L. A. 1931. Capillary conduction of liquids through porous mediums. *Physics* 1: 318–333.

Ritsema, C. J., and L. W. Dekker. 1996. Water repellency and its role in forming preferred flow paths in soils. *Aust. J. Soil Res.* 34:475–487.

Sauer, T. J., B. E. Clothier, and T. C. Daniel. 1990. Surface measurement of the hydraulic properties of a tilled and untilled soil. *Soil Till. Res.* 15:359–369.

Scotter, D. R., B. E. Clothier, and E. R. Harper. 1982. Measuring saturated hydraulic conductivity and sorptivity using twin rings. *Aust. J. Soil Res.* 20:295–340.

Shani, U., R. J. Hanks, E. Bresler, and C. A. S. Oliveira. 1987. Field method for estimating hydraulic conductivity and matric potential water content relations. *Soil Sci. Soc. Am. J.* 51:298–302.

Smettem, K. R. J., and B. E. Clothier. 1989. Measuring unsaturated sorptivity and hydraulic conductivity using multiple disc permeameters. *J. Soil Sci.* 40:563–568.

Smiles, D. E., and A. G. Harvey. 1973. Measurement of moisture diffusivity in wet swelling systems. *Soil Sci.* 116:391–399.

Smiles, D. E., and J. H. Knight. 1976. A note on the use of the Philip infiltration equation. *Aust. J. Soil Res.* 14:103–108.

Stephens, D. B. 1979. Analysis of constant head borehole infiltration tests in the vadose zone. In: *Report on Natural Resources Systems. Series 35*. Tucson: Univ. of Arizona.

Stephens, D. B. 1992. Application of the borehole permeameter. In: *Advances in Measurement of Soil Physical Properties: Bringing Theory into Practice* (G. C. Topp et al., eds.), Soil Sci. Soc. Am. Special Publ. 30, pp. 43–69.

Stephens, D. B., K. Lambert, and D. Watson. 1987. Regression models for hydraulic conductivity and field test of the borehole permeameter. *Water Resour. Res.* 23: 2207–2214.

Swartzendruber, D. 1993. Revised attribution of the power form of the infiltration equation. *Water Resour. Res.* 29:2455–2456.

Talsma, T. 1969. In situ measurement of sorptivity. *Aust. J. Soil Res.* 7:269–276.

Talsma, T., and P. M. Hallam. 1980. Hydraulic conductivity measurements of forest catchments. *Aust. J. Soil Res.* 30:139–148.

Thony, J.-L., G. Vachaud, B. E. Clothier, and R. Angulo-Jaramillo. 1991. Field measurement of the hydraulic properties of soil. *Soil Technol.* 4:111–123.

Tillman, R. W., D. R. Scotter, M. G. Wallis, and B. E. Clothier. 1989. Water-repellency and its measurement by using intrinsic sorptivity. *Aust. J. Soil Res.* 27:637–644.

Topp, C. J., and M. R. Binns. 1976. Field measurement of hydraulic conductivity with a modified air-entry permeameter. *Can J. Soil Sci.* 56: 139–147.

van Genuchten, M. Th., and P. J. Wierenga. 1976. Mass transfer studies in sorbing porous media. I. Analytical solutions. *Soil Sci. Soc. Am. J.* 40: 473–480.

Vogeler, I., B. E. Clothier, S. R. Green, D. R. Scotter, and R. W. Tillman. 1996. Characterizing water and solute movement by Time Domain Reflectometry and disk permeametry. *Soil Sci. Soc. Am. J.* 60: 5–12.

Wallis, M. G., and D. J. Horne, 1992. Soil water repellency. *Adv. Soil Sci.* 20: 91–146.

White, I., and K. M. Perroux. 1989. Estimation of unsaturated hydraulic conductivity from field sorptivity measurements. *Soil Sci. Soc. Am. J.* 53: 324–329.

White, I., and M. J. Sully. 1987. Macroscopic and microscopic capillary length and time scales from field infiltration. *Water Resour. Res.* 23: 1514–1522.

Wooding, R. A. 1968. Steady infiltration from a shallow circular pond. *Water Resour. Res.* 4: 1259–1273.

Youngs, E. G. 1972. Two- and three-dimensional infiltration: Seepage from irrigation channels and infiltrometer rings. *J. Hydrol.* 15: 301–315.

7
Particle Size Analysis

Peter J. Loveland
Cranfield University, Silsoe, Bedfordshire, England

W. Richard Whalley
Silsoe Research Institute, Silsoe, Bedfordshire, England

I. INTRODUCTION

This chapter is not a laboratory manual. It is more concerned with the principles underlying the concepts of particle, size, and distribution, the relationships between them, and the methods by which they may be measured. There are now some 400 reported techniques for the determination of particle size (Barth and Sun, 1985; Syvitski, 1991), although the large body of measurements amassed by soil scientists has generally been made using simple methods and equipment, principally sieving, gravitational settling, the pipet, and the hydrometer. There is also a large body of experience in interpreting these data. However, there is still a surprising lack of uniformity in these simple procedures, and for that reason we consider them in some detail.

The classification of soils in terms of particle size stems essentially from the work of Atterberg (1916). He built on the work of Ritter von Rittinger (1867) in relation to rationalization of sieve apertures as a function of (spherical) particle volume, and that of Odén (1915), who applied Stokes' law to soil science for the first time. In 1927 the International Society of Soil Science adopted proposals to standardize the method for the "mechanical analysis" of soils by a combination of sieving and pipeting and, equally important, resolved to analyze (at least for agricultural soils) only the fraction passing a round-hole 2 mm sieve—the so-called "fine earth" (ISSS, 1928).

There have been many revisions of the particle size classes promulgated in 1927, and it is now recognized that soil science can make little further headway in the interpretation of particle size distribution in the submicrometer range, because the simple methods are incapable of further resolution. For that reason we have reviewed a number of less common or more recent instrumental techniques, which are capable of extending our understanding of the distribution of particles in this region. We have also quoted much of the older literature, as this gives the physics and mathematics from which more recent developments have arisen.

A large number of standard methods for particle size analysis is available. Many have been published by bodies responsible for national standards*, and others by the ISO* (e.g., AFNOR, 1983c; DIN, 1983, 1996; BSI, 1990, 1998; ISO, 1998). Other key sources are Klute (1986), Head (1992), Carter (1993), USDA (1996), and ASTM (1998b). Readers should consult these publications, especially those by the ISO, for practical details of methods of analysis, as use of them will reduce the divergence of analytical results often found in interlaboratory "ring-tests."

II. BASIC CONCEPTS

A. Particles

A particle is a coherent body bounded by a clearly recognizable surface. Particles may consist of one kind of material with uniform properties, or of smaller particles bonded together, the properties of each being, possibly, very different. Soils are formed under particular conditions, and the particles are to a greater or lesser extent products of those conditions. If the soil is disturbed, the particles may change: for example, salts and cements can dissolve, organic remains can be fragile, bonding ions can hydrolyze, and bonds thus be weakened. Not all these changes may be desirable if the original material is to be fully and properly characterized. AFNOR (1981b) has given a useful vocabulary that defines terms relating to particle size.

Few natural particles are spheres, and often the smaller they are, the greater is the departure from sphericity. One method of size analysis may not be enough, and the methods chosen should reflect the information desired; there may be little point in characterizing as spheres particles that are plates. Allen et al. (1996) listed a number of measures of particle size applicable to powders. In soil analysis, the commonest by far is the volume diameter, which is generally equated with Stokes' diameter.

---

* Throughout this chapter, AFNOR stands for Association Française de Normalisation (Paris); ASTM for American Society for Testing and Materials (Philadelphia); BSI for British Standards Institution (London); DIN for Deutsches Institut für Normung (Berlin); ISO for International Standards Organisation (Geneva).

Sedimentologists often characterize irregular particles in terms of "sphericity" or, more usually, an index to indicate departure from sphericity, although all the methods involve much labor to acquire enough measurements on enough grains to obtain statistically valid data (Griffiths, 1967). The introduction of image-analyzing computers has made the task of size analysis much easier and has extended the techniques beyond the range of the optical microscope (e.g., Ringrose-Voase and Bullock, 1984). Tyler and Wheatcraft (1992) made a useful review of the application of fractal geometry to the characterization of soil particles, and cautioned against the use of simple power law functions for particles as diverse as those found in soils. Barak et al. (1996) went further, and concluded that fractal theory offers no useful description of sand particles in soils and hence doubted the applicability of these methods to soils with large amounts of coarser particles. Grout et al. (1998) came to an almost identical conclusion. However, Hyslip and Vallejo (1997) stated that fractal geometry *can* be used to describe the particle size distribution of well-graded coarser materials. The utility of fractal mathematics in soil particle size analysis is clearly an area likely to develop further.

B. Size and Related Matters

Soils may contain particles from > 1 m in a maximum dimension to < 1 μm, i.e., a size ratio of 1,000,000 : 1 or more. For the larger particles, which can be viewed easily by the naked eye, a crude measure of size is the maximum dimension from one point on the particle to another. In many cases, only a scale for the coarse material is needed—for example, as a guide to the practicalities of plowing land. It is the smaller particles, however, on which most interest focuses, as these have a proportionately greater influence on soil physical and chemical behavior.

Size and shape are indissoluble. The only particle whose dimensions can be specified by one number (viz., its diameter) is the sphere. Other particle shapes can be related to a sphere by means of their volume. For example, a 1 cm cube has the same volume as a sphere of 1.24 cm diameter. This is the concept of equivalent sphere (or spherical) diameter (ESD). Thus the behavior of spheres of differing diameters can be equated to particles of similar behavior to those spheres in terms of their ESD. However, the limitations of the equivalent sphere diameter concept are illustrated by the fact that a sphere of diameter 2 μm has a volume of approximately 4×10^{-12} cm^3, but the same volume is occupied by a particle of 100 nm $\times$ 2 μm $\times$ 20 μm.

Most soil scientists are interested in the proportion (usually the weight percent) of particles within any given size class, as defined by an upper and lower limit (e.g., 63–212 μm). Size classes are usually identified by name, such as clay, silt, or sand, and each class corresponds to a grade (Wentworth, 1922). It is common, particularly among sedimentologists, to describe a deposit in terms of its principal particle size class, for example, of being "sand grade." Soil scientists use a similar system when using the proportions of material in different size fractions to construct so-called texture triangles or particle size class triangles (Figs. 1 and 2). There is considerable variation among countries as to the limits of the different particle size classes (Hodgson, 1978; BSI, 1981; ASTM, 1998d), and hence the meaning of such phrases as "silt loam," "silty clay loam," etc. Rousseva (1997) has proposed functions that allow translation between these various particle size class systems.

The distribution of particles in the different size classes can be used to construct particle size distribution curves, the commonest of which is the cumulative curve, although there are others. Interpolation of intermediate values of particle size from such curves should be undertaken with care. The curves are only as good as the method used to obtain the data and the number of points used to construct them. Serious errors can arise if the latter are inadequate (Walton et al., 1980). Thus curve fitting, especially though software, should only be undertaken with a proper understanding of the underlying mathematics (ISO, 1995a, b; AFNOR, 1997b; ASTM, 1998c).

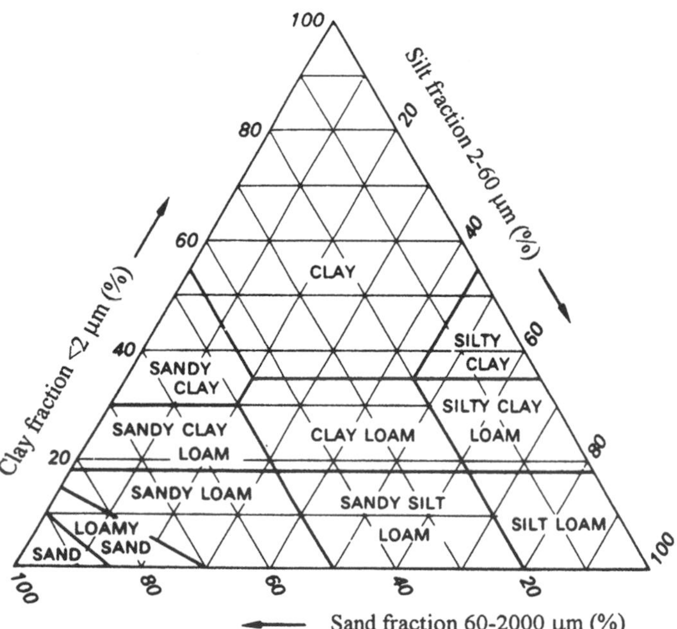

Fig. 1. Triangular diagram relating proportions of sand, silt, and clay to particle size classes as defined in England and Wales.

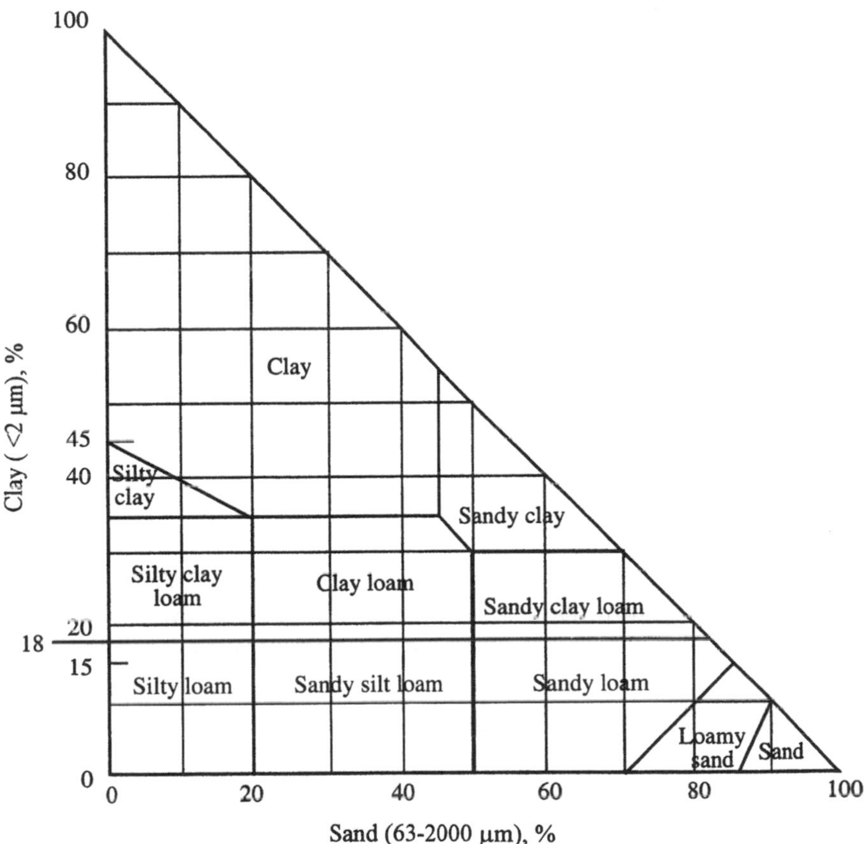

Fig. 2 Particle size classes drawn as an orthogonal diagram using only clay and sand fractions.

C. Sampling and Treatment of Data

Sampling and treatment of data have been discussed exhaustively by many authors (e.g., Klute, 1986; Webster and Oliver, 1990). The cardinal principle is that the sample must be representative of the soil under study; otherwise, the resulting data will be inadequate or misleading, and no amount of statistical massaging will compensate for this. Head (1992) gave recommended minimum quantities of soil to be taken for analysis based on the maximum size of particle forming more than 10% of the soil (Table 1). It is clear that as particle size increases, the problems of representative sampling become formidable.

Ideally, laboratory subsamples should be taken from a moving stream of the bulk material (Allen et al., 1996). A rotary sampler or chute splitter is the best tool

Table 1 Minimum Quantities of Soils for Sieve Analysis

Maximum size of particle forming more than 10% of soil (mm)	Minimum mass of soil for sieve analysis (kg)
63	50
50	35
37.5	15
28	5
20	2
<20[a]	1

[a] It is recommended that the minimum sample mass be 1 kg, however small the particles.
Source: Modified from Head (1992) and ASTM (1998b).

for obtaining relatively small samples of soil of < 2 mm size from a larger bulk sample (Mullins and Hutchinson, 1982), while riffling can be used up to about 10 cm. The only practicable method thereafter is coning and quartering (BSI, 1981).

D. Accuracy, Precision and Reference Materials

The *accuracy* of particle size analysis methods for soils is difficult to establish, as there are no natural soils made up of perfectly spherical particles for use as standards. Further, because of the varied shape of naturally occurring particles, there is no general agreement on how the accuracy, i.e., the approach to an absolute or true value, of this shape should be measured and reported. The *precision* is less difficult to assess. Provided that the technique is followed carefully, then sufficient data can be acquired to perform normal quality control statistics (ISO, 1998), which can be used to express the "repeatability" of a method *for a particular class of materials*. The latter may have to be more specific than just "soils," for a particular method of determination, e.g., soils dominated by sand grains may give different performance criteria from soils dominated by clay particles.

Synthetic reference materials (obtainable as Certified Reference Materials, CRMs), such as glass beads ("ballotini"), latex spheres, and so on, are of limited application in assessing the performance of methods for the particle size analysis of natural materials. They may be useful in certain techniques, e.g., image analysis, electrical sensing zone methods, and methods dependent on the interaction with radiation (Hunt and Woolf, 1969). However, such applications are less common than the need to assess method performance on a routine basis, e.g., in a teaching or commercial laboratory.

Other CRMs, such as powdered quartz, are also available (Table 2), but any particular CRM covers only a limited size range, is relatively expensive (ca. US$2/g at the time of writing), and is available in relatively small amounts, e.g., 100 g lots. Thus any laboratory using these materials to cover a wide range of particle sizes, using the quantities required by many methods of analysis—10 g is not uncommon—may find the expense of including a standard in every analytical batch (often considered to be the minimum requirement of "good laboratory practice") unsustainable.

An alternative is to use in-house reference materials, which can, if prepared and subsampled carefully, be more than adequate to *monitor* the long-term performance of the method of analysis. They have the added advantage that continuity of supply can be ensured by careful selection of the source site(s). Our own experience suggests that ca. 10 kg of each of one material representing fine-textured soils, e.g., a clay or clay loam, and another representing coarse textured soils, e.g., a sandy loam or loamy sand, is adequate for quality control of 25,000 or more routine particle size analyses (ca. 10 g of each reference material for every batch of 30 samples). It should be well within the capabilities of the average soil laboratory to obtain, prepare, and subsample such modest amounts of material.

There is a widespread view that a few percent error either way in the particle size determination of a specific size class is not very important. This seems to stem from the beliefs that soils are inherently variable and that, in most cases, the analytical data are used only to place a soil in a particle size class. However, size classes have exact numerical boundaries, and major decisions can flow from the class in which a soil is placed. Therefore, the class should be decided on the basis of the best possible data that can be obtained.

III. PARTICLE SIZE TECHNIQUES AND APPLICATIONS

A. Introduction

Methods for determining particle size can be divided into the following broad groups:

Direct measurement (ruler, caliper, microscope, etc.)
Sieving
Elutriation
Sedimentation (gravity, centrifugation)
Interaction with radiation (light, laser light, x-rays, neutrons)
Electrical properties
Optical properties
Gas adsorption
Permeability

Table 2 Suppliers of Equipment, Software, and Other Materials[a,b]

Type of equipment	Supplier
General equipment (samplers, sieves, shakers, splitters, crushers, elutriators, etc.)	Amherst Process Instruments Inc., The Pomeroy Building, 7 Pomeroy Lane, Amherst, MA 01002-2905, USA (www.aerosizer.com/)
	Dispersion Technology Inc., Hillside Avenue, Mt. Kisco, NY 10549, USA (www.dispersion.com/)
	Eijkelkamp Agrisearch Equipment, P.O. Box 4, 6987 ZG Giesbeek, The Netherlands (www.diva.nl/eijkelkamp/)
	ELE International (Agronomics), Eastman Way, Hemel Hempstead, Herts. HP2 7HB, UK (www.eleint.co.uk/)
	Endecotts Ltd., 9 Lombard Road, London. SW19 3TZ, UK (www.martex.co.uk/)
	Fritsch Laborgerätebau GmbH, Industriestraße 8, D-55743, Idar-Oberstein, Germany (www.fritsch.de/)
	The Giddings Machine Company, 401 Pine Street, P.O. Drawer 2024, Fort Collins, Colorado 80522, USA (www.soilsample.com/)
	Gilson Company Inc., P.O. Box 677, Worthington, Ohio 43085-0677, USA (www.globalgilson.com/)
	Glen Creston Ltd., 16, Dalston Gardens, Stanmore, Middlesex HA7 1BU, UK (www.labpages.com/)
	Ladal (Scientific Equipment) Ltd., Warlings, Warley Edge, Halifax, Yorks. HX2 7RL, UK (www.members.aol.com/fpsconsult/)
	Pascal Engineering Co. Ltd., Gatwick Road, Crawley, Sussex. RH10 2RD, UK
	Seishin Enterprise Co. Ltd., Nippon Brunswick Buildings, 5-27-7 Sendagaya, Shibuya-ku, Tokyo, Japan (www.betterseishin.co.jp/)
	Wykeham Farrance Engineering Ltd., 812 Weston Road, Slough, Berks. SL1 2HW, UK (www.wfi.co.uk/)
Centrifugal analyzers	Brookhaven Instruments Corp., 750 Blue Point Road, Holtsville NY 11742, USA (www.bic.com/)
	Horiba Ltd., 17671 Armstrong Ave., Irvine, CA 92714, USA (www.horiba.com/)
	Joyce-Loebl Ltd., 390 Princesway, Team Valley, Gateshead, NE11 0TU, UK (www.mjhjl.demon.co.uk/)
Digital density meters	Anton Paar GmbH., Kaerntner Straße 322, A-8054 Graz, Austria (www.anton-paar.com/)
Electrical sensing zone devices	Beckmann Coulter Inc., 4300 N. Harbour Boulevard, PO Box 3100, Fullerton, CA 92834-3100, USA (www.coulter.com/)
	Micromeritics Instrument Corp., One Micromeritics Drive, Norcross, GA 30093-1877, USA (www.micromeritics.com/)
Light-scattering devices/ Photosedimentometers	Brookhaven Instruments Corp., 750 Blue Point Road, Holtsville NY 11742, USA (www.bic.com/)
	Beckmann Coulter Inc., 4300 N. Harbour Boulevard, PO Box 3100, Fullerton, CA 92834-3100, USA (www.coulter.com/)
	Fritsch Laborgerätebau GmbH, Industriestraße 8, D-55743, Idar-Oberstein, Germany (www.fritsch.de/)

Table 2 *Continued*

Type of equipment	Supplier
Light-scattering devices/ Photosedimentometers (*continued*)	High Accuracy Products Corp. (HIAC), 141 Spring Street, Claremont, CA 91711, USA (www.hiac.com/) Honeywell Inc., 16404 N. Black Canyon Road, Phoenix AZ85023, USA (Mictotrac Analyzers) (www.iac.honeywell.com/) LECO Corporation Svenska AB, Lövängsvägen 6, S-194 45 Upplands, Väsby, Sweden (www.lecoswe.se/) Malvern Instruments Ltd., Enigma Business Park, Grovewood Road, Malvern, Worcs. WR14 1XZ, UK (www.malvern.co.uk/) Quantachrome Corp., 1900 Corporate Drive, Boynton Beach, FL 33426, USA (Cilas Analyzers) (www.quantachrome.com/) Sequoia Scientific, Inc., PO Box 592, Mercer Island, WA 98040, USA (www.sequoiasci.com/) (includes submersible instruments)
X-ray sedimentation equipment (Sedigraph)	Micromeritics Instrument Corp., One Micromeritics Drive, Norcross, GA 30093-1877, USA (www.micromeritics.com/)
Software	Most electronic instruments come with built-in software to process, display, or output data. Many earth science and civil engineering departments of universities offer software for aspects of particle size analysis, and the following also supply more general-purpose software: Fritsch Laborgerätebau GmbH, Industriestraße 8, D-55743, Idar-Oberstein, Germany (www.fritsch.de/) (sieve analysis) SPSS Inc., 233 S. Wacker Drive, 11th Floor, Chicago, IL 60606-6307, USA (www.spss.com/) (image analysis) Fine Particle Software, 6 Carlton Drive, Heaton, Bradford, W. Yorkshire, BD9 4DL, UK (www.members.aol.com/lsvarovsky/) (most areas of particle size data manipulation) Texture Autolookup (www.members.xoom.com/drsoil/tal.html) (places particle size analysis data in USDA and UK "texture" classes; see also Christopher & Mokhtaruddin, 1996) Advanced American Biotechnology and Imaging, 116 E. Valencia Drive, #6C, Fullerton, CA 93831, USA. (www.aabi.com/) (image analysis, including shape factors)
Certified Reference Materials (CRMs)	Many National Standards' Organisations (but not ISO) produce, or participate in the production of, Certified Reference Materials for environmental analysis. The following have particularly wide coverage, but a search of the WWW will reveal very many more: Community Bureau of Reference—BCR, Commission of the European Communities, rue de la Loi 200, B-1049 Brussels, Belgium Promochem GmbH, Postfach 101340, 46469 Wesel, Germany

[a] This list is not claimed to be exhaustive. We give manufacturers/suppliers only of items specific to particle size analysis, and generally give the headquarters' address and world wide web site. All addresses were checked at the time of writing, and all quoted web-sites visited to test that they existed and were working. The mention of any company or product is not a recommendation or warranty of any kind, but is given merely for information.

[b] All world wide web site addresses given between brackets are assumed to start with: http://.

Some procedures make use of combinations of these methods. This chapter touches on some of the techniques available. We aim to discuss the principles, origins, and limitations of some standard methods and to point to newer methods that may provide more and/or better information as to how particles in soils can be characterized, and hence how soil behavior can be better predicted. Table 2 gives commercial sources of some of the instrumentation.

B. Direct Measurement

Although soil scientists generally concentrate on the soil fraction passing a 2 mm aperture sieve, many soil classification systems categorize soils according to the amounts of particles greater than a given size (e.g., ASTM 1998d). Engineers faced with moving much soil may find its complete grading to be essential (BSI, 1981). Although even large particles may be sized by sieving, it is often more practical to resort to direct measurement in situ. The very largest particles can be measured with a tape, and those up to some tens of cm in size by wooden or light alloy templates into which are cut holes of differing shapes and dimensions (Billi, 1984). Caroni and Maraga (1983) used an adjustable caliper connected to a tape-punch so that the results could be fed directly to a computer back at the laboratory; nowadays an electronic caliper and data-logger would be possible. Hodgson (1997) gave a method by which the *volume* of particles above a particular sieve size may be estimated by means of plastic balls. Laxton (1980) has used a photographic technique for estimating the grading of the boulder- and cobble-grade material in exposed working faces of quarries. Buchter et al. (1994) found good correlation between the amounts of very coarse material in a rendzina, as measured by volume, conventional particle size analysis, and thin section.

For particles between about 10 cm and 1 mm, there is little *practical* alternative to sieving (Sec. III.C), as the particles are too numerous for the methods outlined above. Between 1 mm and about 20 μm, optical microscopic methods are suitable, while for smaller particles electron microscopy can be used. The advantage of microscopy is that it allows full consideration of shape factors. Microscopy requires careful sampling for the measurement of many individual particles to obtain statistically valid results (Griffiths, 1967; Kiss and Pease, 1982; AFNOR, 1988). The use of automatic image analysis can also speed matters. All microscopic techniques, but especially those for very small particles, require good dispersion of the material. This usually means destruction of organic matter, solvation with a particular cation, commonly sodium, with subsequent removal of excess salt, and/or dissolution of cementing agents (Klute, 1986). The basic techniques for sizing by microscopy were reviewed by Allen et al. (1996). Many Standards give specific procedures for optical microscopy (e.g., AFNOR, 1990; BSI, 1993). Tovey and Smart (1982) covered electron microscopy techniques in detail, while Nadeau (1985) discussed measuring the "thickness" of very small particles and clay mineral platelets by shadowing.

Where particles are roughly equidimensional, microscopy can yield a single or average dimension, relatively easily checked against accurately sized graticules (BSI, 1993). However, soil particles $< 5\ \mu m$ are usually far from equidimensional, and the sizes measured along different particle axes may differ enormously. In such cases, it may be more useful to express size in terms of particle thickness or equal volume diameter, together with the *aspect ratio*, that is, the distance between parallel crystallographic faces divided by thickness, itself often the distance between two other crystallographically related surfaces such as cleavage planes (Nadeau et al., 1984).

With nonspherical, platy, or angular particles, "size" as measured rarely corresponds exactly in geometric terms with the surface resting on the support (Fig. 3). Where the particles are very thin, and the dimensions measured are very large in relation to the vertical dimension, the error is small. When the vertical dimension increases greatly in relation to dimensions in the horizontal plane, however, the error can be much greater (Allen et al., 1996). Dimensions in the plane

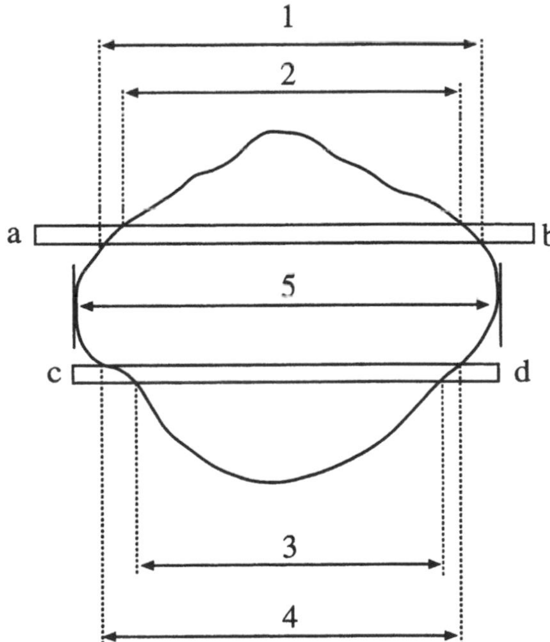

Fig. 3 Side view of two sections, a–b and c–d, through a particle, showing how the dimensions measured can differ depending on the plane in which the measurement is made.

of a sectioned particle can be used to calculate the particle size probabilistically (Kellerhals et al., 1975). However, there will always be uncertainty as to how well the plane of section represents a random pass through the "true" dimensions of the particles. In optical microscopy, it can be difficult to locate particle edges because of diffraction effects. For this reason, it is recommended that optical microscopy not be used for particles smaller than 0.8 μm, and the accuracy obtainable should be qualified below 2.3 μm (BSI, 1993). Shiozawa and Campbell (1991) have described a method of characterizing soils by a mean particle diameter and geometric mean standard deviation, based on the content of sand, silt, and clay fractions.

C. Sieving

Sieves are available with apertures ranging from 125 mm to 5 μm, either in round-hole or square-hole forms, depending on aperture size. Round-hole sieves size material by one dimension only, whereas square-hole sieves size particles by two dimensions: the distance between two parallel faces and the diagonal between corners, respectively. Using a mixture of round-hole and square-hole sieves can cause serious errors in constructing particle size distribution curves of soils, because of which, many standards now preclude the use of round-hole sieves (Tanner and Bourget, 1952). Larger apertures are usually made by punching steel plate. Below 2 mm aperture, square-hole, woven-wire sieves are usual, while electroformed square-hole sieves are increasingly popular below about 37 μm (e.g., ISO, 1988, 1990a–e, 1998). For fibrous materials, e.g., peats, it may be necessary to use special slotted-aperture sieves. Sieve apertures are manufactured to tolerances, not to absolute values; that is, the stated aperture may vary between given limits. For example, the nominal 2 mm aperture of a wire-woven sieve may have an average variation of $\pm 3\%$ (1.94–2.06 mm), with no one aperture being more than 12% larger than the nominal aperture, i.e., 2.24 mm (BSI, 1986).

One still finds sieves described by their *mesh number*, a practice that is to be deplored. The mesh number of a sieve is the number of wires per linear inch, which (in theory) is one more than the number of holes over the same distance. However, without a knowledge of wire diameter, one cannot derive the sieve aperture from the mesh number. While it is perfectly possible to memorize a table of mesh numbers and apertures, there seems to be little point to this exercise when the aperture itself can be stated so simply. The use of mesh numbers is also against the trend to move to SI (Système International) units.

It is very common to round-off sieve apertures when reporting results, e.g., 53 μm will be given as 50 μm. The reason for this widespread practice is obscure. We strongly recommend that it be discouraged, as it degrades hard-won information, and is misleading: sieves of, for example, 50 μm aperture are nowhere used in soil analysis. Most standards organizations nowadays strongly support the

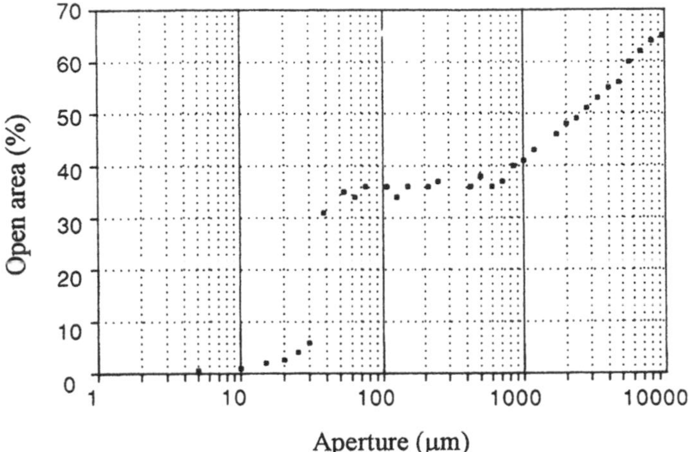

Fig. 4 Relationship between open area of sieve and sieve aperture (for square-hole sieves).

manufacture of sieves in accordance with the "preferred number series" of ISO. The principal series are based on geometric progressions of $n\sqrt{10}$, where n is 5, 10, 20, 40 etc. (ISO, 1973, 1990a). These give the least numerical error in relating one sieve aperture to the next in the *same* series (switching from one series to another to construct a "tower" of sieve apertures is discouraged by ISO and most other standards' bodies).

Mechanical sieve shaking is commonly used in preference to hand sieving, and with careful control it can give very precise results. Most errors arise from worn or damaged sieve screens or variation in sieve loading—especially overloading, variation in shaking time, poor fit between sieves, lids, and receivers, and failure to keep shaking equipment horizontal (Metz, 1985; Head, 1992). Kennedy et al. (1985) commented on the sorting and sizing of particles during sieving, according to their shape.

Sieving becomes increasingly laborious below apertures of approximately 30 μm, because the area of hole drops sharply as a percentage of total sieve area (Fig. 4), and dry sieving is not recommended in this range. If such sieving is attempted, the air-jet technique is both quicker and more reproducible than conventional sieving (AFNOR, 1979). For finer materials that may "ball" (aggregate), wet-sieving equipment is available (AFNOR, 1982).

Sieve apertures tend to block, and are usually brushed clean, which can damage sieves, especially those of smaller aperture, both by stretching and by breaking the weave. Sieves can be cleaned in an ultrasonic bath filled with propan-2-ol, although the frequency of oscillation must be chosen with care to avoid cavitation and hence mesh weakening. It is *always* worth inspecting sieves and their accessories for damage after each shaking, whence fresh-looking, bright, shiny fragments of brass or stainless steel, however small, are an infallible guide to sieve mesh failure.

D. Sedimentation

Methods of particle size determination using a combination of sieving and sedimentation are undoubtedly the commonest in soil science. "Sedimentation" means the settling of particles in a fluid under the influence of gravity or centrifugation. The amount of material above or below a specified size is determined by abstraction of an aliquot of suspension that is then dried and the residue weighed, by measuring the change in the density or opacity of the suspension, or by measuring the amount of sediment that has settled in a suitable vessel after a certain time.

Whichever method of measurement is chosen, *all* assume that the particles in suspension behave according to the Stokes equation (Stokes, 1849), as applied to soil analysis by Odén (1915). This can be written *for spheres* as follows:

$$t = \frac{18\eta h}{(\rho - \rho_0) g d^2} \tag{1}$$

where t is the time in seconds for a particle to fall h cm once terminal velocity has been attained, ρ is the particle density (g cm^{-3}), ρ_0 is the density of the suspending medium (g cm^{-3}), g is the acceleration due to gravity (cm s^{-2}), d is the equivalent sphere particle diameter (cm), and η is the viscosity of the suspending medium (poise, where 1 poise = 0.1 Pa s^{-1}). Because this is not an empirical equation, it is equally valid if SI units are used throughout.

This equation is modified in a centrifugal field (Dewell, 1967) to

$$t = \frac{18\eta}{(\rho - \rho_0)\omega^2 d^2} \ln\left(\frac{R}{S}\right) \tag{2}$$

where ω is the angular velocity of the centrifuge, i.e., the number of revolutions per second $\times$ 2π, S is the distance (cm) of particles from the axis of rotation of the centrifuge at the start of analysis and is measured from the surface of the suspension, and R (cm) is the distance the particle has reached in time t (s).

Stokes' equation for spheres is applicable when the following criteria are met:

1. The particles are rigid and smooth.
2. The particles settle independently of each other.
3. There is no interaction between fluid and particle.
4. There is no "slip" or shear between the particle surface and the fluid.
5. The diameter of the column of suspending fluid is large compared to the diameter of the particle.

6. The particle has reached its terminal velocity.
7. The settling velocity is small.

Stokes' law refers to an equation that describes the drag force on a particle of any shape, and is valid for nonspherical particles if (and *only* if) the concept of equivalent sphere diameter (ESD) is used. Whalley and Mullins (1992) have discussed its application to plate-like particles.

Allen et al. (1996) pointed out that Stokes' equation is valid only under conditions of laminar flow when the Reynolds number (R_e) is ≤ 0.2 (R_e is dimensionless and is a measure of turbulence in fluid flow; if R_e is small, flow is nonturbulent—see Anon., 1997, for a fuller explanation), and that the critical value of the Stokes diameter (d), which sets an upper limit to the use of Stokes' law, is given by

$$d = \frac{3.6\eta^2}{(\rho - \rho_0)\rho_0 g} \tag{3}$$

For quartz particles settling in water, Allen et al. (1996) showed that Stokesian behavior for spherical particles holds only for those less than about 61 μm in diameter. They also considered each of the criteria listed above in considerable detail. For soils and clays their findings may be summarized as follows:

1. Flat, thin plates will settle more slowly than their equivalent spheres; hence the amount of such material may be overestimated. This slowing of the fall rate is partly because the plates trace out a zigzag path as they settle.
2. Below ca. 1 μm ESD, Brownian motion can displace a settling particle by an amount equal to or greater than the settling induced by gravitation. Below this limit gravitational sedimentation becomes increasingly unreliable.
3. Electrical interactions between a dilute electrolyte and soil particles have a negligible effect on settling, as does the time taken for particles to reach terminal velocity.

Particle–particle interaction is more difficult to deal with, as the number of particles in suspensions of different soil can differ enormously. Extensive experience has shown that the maximum concentration of suspended material should be no more than 1% by volume, or about 2.5% by weight. However, suspensions of bentonitic soils may exhibit thixotropy at smaller concentrations of suspended solids. Dilution of the suspension usually overcomes this, but may introduce greater error because of the difficulty of determining very small residue weights, or differences in suspension density or suspension opacity, accurately. It is axiomatic that the soil should be well dispersed in an electrolyte, usually following the destruction of organic matter. Dispersion is almost always in an alkaline solution, most commonly sodium hexametaphosphate buffered to about pH 9.5 with sodium carbonate or ammonia solution (Klute, 1986), although there are many others (see, e.g., AFNOR, 1983b). Dispersion may be aided by ultrasonic treatment (Pritchard, 1974), particularly in volcanic ash soils, for which dispersion in alkaline media is inappropriate due to their, often large, content of positively charged material. For these soils, an acid dispersion routine should be followed (Maeda et al., 1977). Most particle size determinations are carried out on <2 mm air-dried soil, but highly weathered soils, especially those from the tropics, may be difficult to disperse once dried. It may be preferable to analyze them while still wet (ISO, 1998).

1. Pipet Method

For the size fractions < 63 μm obtained after sieve analysis, the pipet method is the officially preferred ISO method (ISO, 1998), and in the U.K. (BSI, 1998), Germany (DIN, 1983), and France (AFNOR, 1983c). It is also the method of choice of the U.S. Soil Conservation Service (USDA, 1996) and Agriculture Canada (Carter, 1993).

Gee and Bauder (1986) have discussed the basic pipet methodology for routine soil analysis. A common complaint is that the method is tedious for the fraction < 2 μm ESD. Coventry and Fett (1979) showed how the efficiency of pipet analysis can be greatly improved by attention to time-saving details at every step of the process. In our Soil Survey laboratory we have much shortened the analysis time by developing a programmable automatic sampling device for taking the silt-plus-clay and clay aliquots. Computerized calculation can give large savings in operator time, and commercial software is now available (Table 2; Christopher and Mokhtaruddin, 1996). Given sufficient care in dispersion and sampling, the pipet method is capable of great precision (Gee and Bauder, 1986). However, the relatively large spread of values found during an interlaboratory comparison shows that there is still room for improvement (Pleijsier, 1986). Burt et al. (1993) described a micropipet method, which compared well with the USDA "macropipet" method. They recommended the micropipet particularly for use in Soil Survey offices where there could be a need to assess the particle size distribution of large numbers of field samples.

2. Density Methods

The density of a suspension is proportional to the amount of solid present, and to the difference between the densities of the suspending liquid and the suspended solid. The density of the liquid is usually fixed by controlling its temperature and electrolyte content, while that of the solid is usually assigned some constant value, commonly 2.65 Mg m^{-3} for soils and clays. However, soil particles, or aggregates behaving as such, can be porous and thus have a smaller density, as can those particles containing much organic material. Conversely, particles composed largely of iron (e.g., hematite, goethite, lepidocrocite, ferrihydrite, maghemite, magnetite), manganese (e.g., pyrolusite, birnessite), or titanium (e.g., ilmenite, titanomagnetite) minerals can have very much higher densities. Further, if the soils under study contain considerable amounts of soluble salts, these can greatly affect the principles on which routine density methods are based.

If the density of a suspension is measured at known depths and time intervals following agitation, it is relatively easy to relate this to the mass of material above or below the Stokes diameter. By far the most widespread procedure is that based on the "Bouyoucos" hydrometer. A detailed ISO procedure for agricultural soils is available (ISO, 1998), as are the precautions for the proper use and calibration of hydrometers (ISO 1977, 1981a, b). Head (1992) has discussed the special problems of soil hydrometers. The greatest source of error in hydrometer methods is the accurate reading of the hydrometer scale, which becomes almost impossible if there is a layer of undecomposed organic matter on the surface of the suspension. Even after suitable oxidation treatment or with purely mineral soils, frothing following agitation can be a problem. This may be controlled by adding a drop or two of a surfactant such as octan-2-ol *after* the suspension has been stirred. [*Warning*: Some authors recommend the use of pentan-1-ol (amyl alcohol) or pentan2-ol (isoamyl alcohol) to control frothing. This is effective, but these alcohols can become addictive. Octan-2-ol is equally effective, but has an unpleasant smell and is less likely to encourage addiction.]

A further difficulty with the hydrometer method relates to the density of the suspension. For accurate determination, this should be significantly different from that of the suspending fluid. Gee and Bauder (1986) recommended 40 g of soil per liter of suspension. This should ensure that even where the soil contains only a few percent of clay or silt, this is enough to give an accurately measurable increase in the suspension density. Should *all* the soil be of clay or silt size, the suspension may contain so many particles that hindered settling occurs, and the determinations may need to be made with less soil. Bentonitic clays will gel at this concentration. Allen et al. (1996) cautioned against the use of hydrometers in suspensions that are not reasonably continuous distributions of sizes, because the relatively large length of the hydrometer bulb may give an average density for two or more zones, with the effect of smoothing out sharp changes in the grading that actually occur.

There have been numerous comparisons between the pipet and hydrometer methods, and it is generally agreed that the former is more precise; see Gee and Bauder (1986) for relevant references. Sur and Kukal (1992) have described modifications of the principles inherent in the hydrometer method, which make its application much more rapid.

Stabinger et al. (1967) were the first to use an ultrasonic technique to measure the density of suspensions. The equipment requires only a small volume of

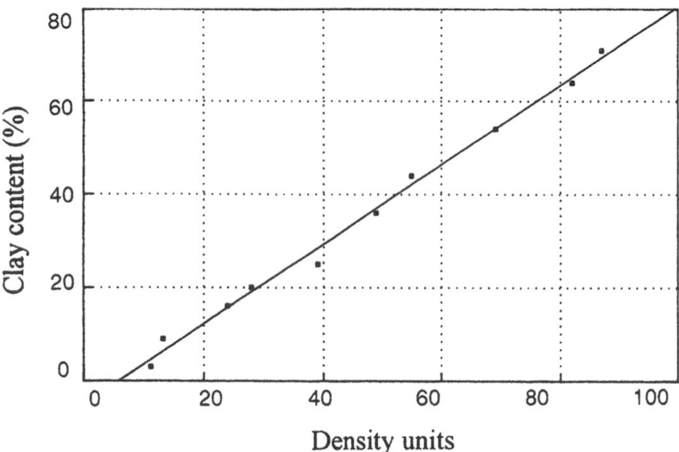

Fig. 5 Relationship between clay content by the pipet method and density units measured by a digital density meter. [Density unit is calculated from (density of suspension minus density of electrolyte) × 10^4.]

suspension, which can be abstracted from a larger volume automatically and with little disturbance. The ultrasonic signal can be processed digitally and hence offers the prospect of automation (Table 2). Work done in the Soil Survey laboratory indicates a reasonable relationship between measured suspension density and clay ($< 2\ \mu m$ ESD) content determined by the pipet method (Fig. 5).

E. Centrifugation

Centrifugation is an extension of sedimentation under gravity, and it offers a means of determining the amounts of particles $< 1\ \mu m$ ESD in suspension, i.e., those whose settling under gravity is seriously affected by Brownian motion. Tanner and Jackson (1947) published comprehensive nomograms for the settling times of particles of different Stokes diameters under centrifugation. This approach was adopted by Avery and Bascomb (1982) and by the U.S. Soil Conservation Service (USDA, 1996) for the determination of particles $< 0.2\ \mu m$ ESD (the so-called "fine clay").

The volumes of suspension involved are usually large, and the design of standard laboratory centrifuges is not suited to controlled sedimentation, because the cylindrical sedimentation vessels are usually long compared with the centrifuge radius. This results in the particles colliding with the vessels' walls during centrifugation. Two designs of modern centrifugal analyzer attempt to overcome this problem. These are defined by the radius of the measurement zone (R) and the radius to the inner surface (S) of the sedimenting column, respectively. In the most common type, S/R tends to zero, and radial sedimentation occurs in a hollow rotating disk. Hence they are known as disk centrifuges. Typically, such disks are no more than a few cm thick and perhaps 20 cm in diameter. In the second type, which are often called long-arm centrifuges, S is large, S/R tends to 1, and the sedimentation paths of particles are assumed to be parallel. The two types can be distinguished by observing whether the concentration of an initially homogeneous suspension is reduced at the sampling point immediately after startup. This happens in a disk centrifuge, due to the dilution effect of radial sedimentation, whereas in long-arm centrifuges the suspension concentration remains constant until the larger size fractions settle out of the measurement zone. The upper Stokes diameter that can be determined, with water as a suspension medium, is about 7 μm ESD, but the range can be extended by the use of more viscous liquids. The lower limit is still controlled by Brownian motion, and is thought to lie between 10 and 50 nm ESD (BSI, 1994b).

Centrifugal particle size analyzers are operated in one of two modes. Either the sedimentation vessel is filled with a homogeneous suspension at the start of analysis, or the vessel is filled with a clear carrier liquid onto which the suspension is floated. These two techniques are known as the *homogeneous-start* and *line-start* techniques, respectively. Pipet sampling is not recommended for use with the line-start technique because the suspension concentration is usually very low (Allen et al., 1996). Examples of common types of centrifugal analyzers are discussed in the following sections.

1. *Pipet-Sampling Centrifuges*

When disk centrifuges are used with the homogeneous-start technique, as is the case with pipet sampling, the reduction in suspension concentration at the sampling point can be attributed to the sedimentation of various size fractions and the diverging radial sedimentation paths of particles which give rise to additional dilution. To calculate particle size distributions, this radial dilution effect must be corrected. The calculation of the exact solution is complicated, but provided sampling is modified so that successive values of Stokes' diameter occur in a ratio of $1:\sqrt{2}$, a much simpler approximate solution can be applied (see, e.g., Allen et al., 1996). However, the use of this approximation may lead to some error when the sample under analysis has a bimodal particle size distribution. It has been suggested that in some cases improved results can be obtained by fitting experimental data to a curve defined by a mean and a standard deviation or other assumed function. A complete mathematical analysis of the required theory was presented by Svarovsky and Svarovska (1975).

2. *X-Ray and Photosedimentation Centrifuges*

The centrifugal x-ray and photosedimentation techniques continually monitor the sedimenting suspension by measuring the transmission of radiation (either visible or x-ray) in a well-defined measurement zone. A centrifugal disk x-ray particle size analyzer operates on principles essentially similar to those of gravitational x-ray sedimentation described below (Sec. III.G). However, since a homogeneous start is used with a disk centrifuge, the data analysis must follow the theory given by Svarovsky and Svarovska (1975) to compensate for the radial dilution effect.

Centrifugal photosedimentation, i.e., using visible radiation, has been widely used for particle size analysis. The use of light is better suited for soils than x-rays, because, as explained below (Sec. III.G), quartz and clay minerals can be translucent to x-rays. However, since clay fractions, i.e., $< 2\ \mu m$ ESD, contain particles both greater and smaller than the wavelength of visible light, photosedimentation data must be corrected for the large variation in light scattering that occurs with change in particle size. The theory and techniques of this correction are described by Whalley et al. (1993). Analysis may be performed with either the line-start or the homogeneous-start technique, and examples of both modes of use are discussed below.

Homogeneous-start sedimentation produces a monotonic relationship between turbidity (the absorption coefficient of the suspension) and Stokes' diameter. The initial suspension concentration has to be adjusted to ensure that the turbidity data obtained from the start of the analysis are within the region in which the Beer–Lambert law is valid, i.e., suspension concentration is proportional to turbidity. When analyzing clays or other very small particles, it is preferable to split the whole analysis into a series of overlapping or contiguous runs, e.g., 20 nm to 0.1 μm, 0.1 to 2 μm, and 1 to 10 μm (Whalley et al., 1993). This is necessary because the smaller particles scatter very little light compared to larger particles, so, to obtain measurable turbidity values, higher suspension concentrations are required for smaller particles. Typically, suspension concentrations of 10 g dm^{-3} are required for the 20 nm to 0.1 μm size range to obtain reliable turbidity data, while concentrations in the 1 to 10 μm size range may have to be as low as 0.3 g dm^{-3} to ensure compliance with the Beer–Lambert law (Whalley, 1988).

At completion of the photosedimentation, the turbidity data can be normalized to a single suspension concentration to give a continuous curve that covers the overlapping runs. After correction for the variation in light scattering with particle size, the results from a long-arm centrifuge, i.e., neglecting radial dilution effects, represent a particle size distribution by area. Some assumption about particle shape is necessary to convert it into a particle size distribution by mass (Whalley et al., 1993). Suitable theories and methods for correcting for both light scattering and absorption effects in clays were given by Whalley (1988).

In line-start centrifugal photosedimentation, the dispersed sample is floated on top of the already spinning disk of liquid, and the sedimentation of the particles out of their narrow start zone is monitored at some fixed distance in the disk fluid by light transmission. It is usual for the disk liquid to be slightly denser than the suspension to prevent irregular streaming of the sample from the narrow start zone; 10% glycerol to 90% water is suitable. Once the relationship between turbidity and Stokes' diameter has been corrected for the variation in light scattering with particle size, it represents a size distribution by mass, in contrast to the distribution by area initially given by homogeneous-start photosedimentation. Correction of disk centrifuge data for light-scattering effects was described by Oppenheimer (1983). Churchman and Tate (1987) used such a disk centrifuge in an investigation of allophanic soils in New Zealand. Whalley and Mullins (1992) found that, in high centrifugal fields, platelike clay particles settle with their minimum dimension in the direction of motion. This phenomenon is in accordance with hydrodynamic theory (Davis, 1947), and excessive force fields should therefore be avoided in all types of centrifugal particle size analysis.

The main criticism of all photosedimentation analysis, particularly with fine clays, is that large corrections to the experimentally obtained data are required to compensate for light-scattering effects. The study of the effect of the saturating cation on aggregate (tactoid) size in dilute bentonite suspensions by Whalley and Mullins (1991) provided an example of the high size resolution of photosedimentation, and the way in which such data can be used to give relative estimates of particle size in a given clay sample subjected to different treatments. AFNOR (1983a), BSI (1994b), and ASTM (1998e) have published Standards for centrifugal photosedimentation.

F. Electrical Sensing Zone Method

The basis of the electrical sensing zone (ESZ) method is commonly known as the Coulter principle, from its discoverer, and commercially available instruments, although not all made by the Beckman-Coulter Corporation, are generally called Coulter counters. Coulter discovered that the resistance measured between two electrodes in an electrolyte, separated by an aperture of known size and hence electrical characteristics, changes in proportion to the volume of a particle passing through the aperture. These changes in resistance can be scaled and counted at the rate of several thousand per second.

In the ESZ method, a measured volume of suspension is drawn through the aperture by automatic operation of a manometer, and the change in resistance between the electrodes caused by the passage of each particle is detected as a voltage pulse. This is scaled, amplified, and assigned electronically to a particular size class or channel. There may be up to 256 such channels to cover the range of the particular aperture in use. With the aid of microprocessors, the instrument output can be expressed directly as "percent oversize," as a cumulative distribution curve, and so on. It is important to remember, however, that the output is a number size distribution, in which the total volume of the particles is deduced (with some assumptions) from the size class itself. The mathematics of conversion to a weight basis were considered by Batch (1964).

It is assumed that the particle resistivity is extremely high, due to very stable electrical double layers or to oxide films, although this may not be true for some of the iron and iron–titanium minerals found in sediments (Walker and Hutka, 1971). The crucial parameter is the relationship between particle and aperture cross-sectional areas, and Lines (1981) recommended a particle-to-aperture ratio $< 40\%$ for routine analysis. Lloyd (1982) investigated the response of the aperture to nonspherical particles using a model system and found no serious deviations, while Atkinson and Wilson (1981) gave details of the underlying principles of calibration.

Two kinds of coincidence counting can occur. In *primary coincidence*, two particles pass through the aperture so closely together that the instrument counts them as one. In *secondary coincidence*, two small particles, which are normally below the detection or "threshold" voltage measurement limit, give rise to a combined signal that is above the limit. The answer to both is to use extremely dilute, effectively dispersed, suspensions to ensure that particles are counted singly and separately.

The size range in soils that can be studied with this technique is from 1.5 mm to 0.5 μm. To cover the entire range, several apertures may be necessary (Allen et al., 1996). Large particles cannot be kept suspended adequately in water, but 10:90 saline/glycerol solution will suspend quartz particles up to 1 mm in diameter (McCave and Jarvis, 1973).

There is a considerable literature that compares the ESZ method to other methods of particle size determination (see, e.g., Syvitski, 1991). However, the most thorough report on the use of the ESZ technique for soils is still that of Walker and Hutka (1971). Although the equipment they used is now outmoded, many of their findings are relevant today:

1. The satisfactory size range is 2–100 μm using apertures of 50, 100, and 200 μm.
2. It is necessary to split soil suspensions at 31.5 μm to avoid blockage of the 50 and 100 μm apertures.
3. Careful attention needs to be given to a choice of electrolyte to ensure that flocculation does not occur. The electrolyte may need to be different for different apertures.
4. The clay fraction (< 2 μm ESD) can be determined with reasonable accuracy by a difference technique based on the measurement of the 0–31.5 μm and the 2–31.5 μm fractions (although this presupposes that one has an acceptable method for splitting the suspension at 2 μm, e.g., by repeated sedimentation and siphoning: laborious at best).
5. Clear relationships exist between ESZ size fraction percentages and sieve weight percentages in the 37.2–88.5 μm range. However, conversion of one to the other requires a different factor for each size fraction.

6. Materials of low resistivity, e.g., magnetite, hematite, ilmenite, are probably not sized properly. (But then, neither are they in conventional sedimentation because of their large specific gravities.)
7. The technique is especially useful where only very small amounts of sample are available, or for already existing very dilute suspensions, e.g., river and marine waters.
8. The ESZ technique compares well with conventional sieving and sedimentation in terms of reproducibility and efficiency for *detailed* size analysis. However, the need to change apertures and electrolytes, and to perform considerable mathematical analysis of the data to achieve results on a mass basis, make the technique difficult to use for rapid routine use. The use of a multiaperture instrument, with all the apertures in operation in the same suspension at the same time, coupled with computerized data processing, could overcome many of these difficulties. However, as far as we know, such an instrument has never been built.

Walker et al. (1974) applied the method to the analysis of very small deposits such as laminae, and to suspended sediment in freshwater streams. Dudley (1976) found the reproducibility over the 2–60 μm range to be extremely good in forensic samples. Duke et al. (1970) also found the method to be highly reproducible for lunar soil between 1 μm and 125 μm ESD, using 200 μm and 50 μm apertures, with good agreement over the same sieve and ESZ equivalent ranges. Sapetti (1963) considered ESZ to be superior to the "Andreasen" pipet and to agree well with results from a sedimentation balance, as did Walker and Hutka (1971). The ESZ method and the hydrometer technique diverge at small particle sizes (Muller and Tisne, 1977). Rybina (1979) showed that the ESZ method oversizes the finer material relative to the pipet method. Furthermore, the ESZ method generally *undersizes* the 10–50 μm fraction, which Walker and Hutka (1971) also reported to be the case for the 44–53 μm fraction of their soils. Pennington and Lewis (1979) found a reasonably linear relationship between silt content (2–53 μm) by both ESZ and pipet methods using 43 soils of different particle size classes and mineralogies. However, inspection of their data suggests that the clay relationship was curvilinear. These authors also noted that background "noise" in ESZ systems can be greatly reduced if all water and electrolytes are filtered at 0.45 μm and 0.22 μm before use. Lewis et al. (1984) used an ESZ instrument to identify loess by constructing very detailed particle size distribution curves between 2 and 50 μm ESD. More recently, McTainsh et al. (1997) have proposed a combined approach, which recommends the pipet (<2 μm), ESZ "Multisizer" (2–75 μm), and sieving (>75 μm) in combination. The ESZ technique is the subject of at least three Standards (BSI, 1994a; AFNOR, 1997a; ASTM, 1998a).

In summary, the ESZ method is probably best used to obtain very detailed particle size distributions over a narrow range of ESD. There is little doubt, however, that ESZ instruments do not always "see" particles in the same way as more conventional methods, such as sedimentation. This, however, is true of all methods and does *not* mean that the electrical sensing zone approach is thereby invalidated. One drawback to the ESZ method is the need to work with more than one aperture to cover a range exceeding 50 μm ESD.

G. Interaction with Electromagnetic Radiation

A particle may absorb, scatter, refract, diffract, or reradiate incident electromagnetic radiation. Such interactions can be used to estimate the mass of material encountered by a beam of radiation, or they can be used directly to yield information about the size of the particles encountered. Generally speaking, modern instruments utilizing these principles fall into two groups, radiation absorbers and radiation scatterers. These two principles, and their applications to particle size analysis, were discussed by Barth (1984).

1. Absorption

The simplest application of absorption involves total light extinction, in which each particle intercepts a collimated beam of light, the obscuring of which is determined electronically. The sample cell causes turbulent flow, so the particles present a constantly changing cross-section to the beam as they pass through, and it is the *maximum* cross-sectional area that is recorded. This principle has been incorporated in the HIAC instrument, which (in theory at least) can cover the range from about 1 to 9000 μm ESD (Barth, 1984). Gibbs (1982) found that floc breakage was a severe problem as material passed through the sensor.

Zaneveld et al. (1982) used optical attenuation in conjunction with photosedimentation, and found good agreement with the ESZ and gravitational settling tube techniques. Coates and Hulse (1985) reappraised photoextinction techniques, and found that, despite good precision, the so-called hydrophotometer gave results very different from those yielded by the pipet and hydrometer methods. Melik and Fogier (1983) examined both the theory and the practice of turbidimetric particle size analysis and concluded that for particles with regular shapes the method is reliable between ~0.1 and 3 μm ESD. AFNOR (1984) gives a standard method for photosedimentation.

The principle by which the mass of material in suspension at a fixed depth is determined from the attenuation of a beam of x-rays was first described by Hendrix and Orr (1971) and is used in the Micromeritics Corporation "Sedigraph" (Table 2). This instrument consists of an x-ray source (tungsten L-line, wavelength 14.76 nm), a cell (~1.25 cm wide, 3.5 cm high, and 0.35 cm thick; volume ~1.65 cm^3) through which the finely collimated x-ray beam passes, an x-ray detector and signal processor, a pump, and a recorder/digital output. The chart is set at 100% with the pump in operation, i.e., with the suspension thoroughly agitated. Once the pump is switched off, the particles begin to settle and a "run" begins. The unique feature of the "Sedigraph" is that the cell is slowly lowered relative to the x-ray beam during measurement, thus greatly reducing the effective settling time. The manufacturers state that the suspension density is measured every 1.88 μm throughout the cell length—a total of more than 13,000 measurements. The instrument is programmed to solve Stokes' equation automatically as modified by the movement of the cell, and it produces the cumulative mass percentage versus ESD.

Olivier et al. (1971) discussed instrument performance and showed that as long as the area irradiated by the beam is small, the errors from irradiation of the cell wall are negligible, and attenuation of the beam is then dependent on the mass absorption coefficients of the suspending liquid and the particles in suspension. This raises two problems:

1. The absorption of x-rays becomes increasingly poor for elements below atomic number 14. This includes aluminum and silicon.
2. The mass absorption coefficients of soil materials cover a range of values, and average values have to be assumed. However, it is unlikely that these values will remain constant over the whole size range being examined in polymineralic mixtures such as soils (Buchan et al. 1993).

Stein (1985) showed that the suspension concentration should be $< 2\%$ v/v to achieve reproducible results for fractions < 63 μm, but that samples with more than 50% montmorillonite in the same size fraction gave unreliable results due to thixotropy.

As for the ESZ technique, there is a large literature for the "Sedigraph." For *soils*, the majority of authors have used it most successfully between 63 and 2 μm ESD. With a cell volume of 1.65 cm^3 and, say, 50 g dm^{-3} of <100 μm soil in the suspension, the cell will contain < 0.1 g of material. This may simply yield too few particles to give reliable values for the larger ones. Because of Brownian motion (Sec. III.D), the determination of the proportion of particles below about 1 μm ESD is unreliable by gravitational sedimentation. Buchan et al. (1993) showed that much better correlations could be obtained between the "Sedigraph" and pipet methods if the results for the former were adjusted for the Fe content of the soils (Fe being a strong x-ray absorber) and gave regression equations for this purpose.

Given these constraints, and the need to bear in mind the mineralogy of the sample, the "Sedigraph" offers a rapid method of determining the size distribution of soil material between about 2 and 60 μm (taking about 20 min per sample). The smaller (<2 μm) fraction may need to be determined by difference. The use of the "Sedigraph" principle is the subject of at least two national Standards (AFNOR, 1981a; ASTM, 1998f).

2. Scattering

Developments in modern electronics, signal processing, and microcomputing ensure that scattering is the most rapidly developing area of particle size measurement. Two problems are, however, inherent in *all* light-scattering devices:

1. The theories on which they are based, and that can readily be evaluated, are available *only* for spheres and other regular shapes such as ellipsoids.
2. There are considerable theoretical and technical problems in obtaining meaningful information for particles whose size is of the order of (or smaller than) the wavelength of the incident radiation.

Size information about smaller particles is yielded by large-angle scattering (commonly 90° to the plane of the incident light), and for larger particles by so-called forward scattering. The former is dealt with by the Mie theory, the latter usually by Fraunhofer diffraction theory (Dahneke, 1983). By careful instrument design, the smaller particle region can be considered to cover the range from about 0.04 to 3 μm, and the larger particle region from about 1 to 2000 μm or more (Barth, 1984).

The submicrometer range can be dealt with by photon (or auto-) correlation spectroscopy (PCS) (ISO, 1996). This relies on the fluctuations in light intensity with time, caused by Brownian movement of particles. Although the theory is well understood for monodisperse systems of spheres, this is not the case for polydisperse systems of particles of differing shapes and refractive indices. A related device, which also depends on the fluctuation of light intensity, is the fiber-optic Döppler anemometer (FODA). In this case, laser light is passed down a fiber into a suspension, and particles passing the end of the fiber reflect light back to a detector. There is a Döppler shift in the wavelength of the reflected light due to the Brownian motion of the particle, which is related to its size (Ross et al., 1978). Kosmas et al. (1986) used this method to obtain size distribution information for synthetic iron oxides, but no comparison was given with more conventional methods. Since there is no sample cell in FODA, the fiber can be dipped into a vessel, and it becomes possible, in theory, to follow the change in particle size inside a reaction vessel, and to make measurements rapidly in a large number of vessels.

There are several Fraunhofer-based and Mie-based light-scattering devices on the market, e.g., Microtrac, Cilas, Malvern Instruments, Quantachrome, and Sequoia (Table 2). All use low-power lasers as light sources. There is considerable variation in the manner in which the signal is detected, and the physical principles were considered in detail by Swithenbank et al. (1977), Plantz (1984), and Cornillault (1986), for the Malvern Instruments, Microtrac, and Cilas machines, respectively.

The application of light-scattering instruments to sedimentological analysis has, to date, been limited. Cooper et al. (1984) reported the use of the Microtrac in soil particle size analysis, over the size range 1.9–176 μm, in comparison with the pipet method. They presented their data as a statistical comparison of the percentage of each size fraction found by each method, with and without removal of organic matter and soluble salts. Their findings were as follows:

1. Removal of organic matter improved the agreement between each particle size range.
2. Agreement was better between separate size fractions than between the complete range studied.
3. Agreement was best between the ranges 31–62 μm and 16–31 μm ($r^2 = 0.92$ in both cases).
4. Statistical agreement for all size ranges improved when the 62–176 μm sieve data were omitted.
5. The greatest differences between the pipet and Microtrac methods were found in the 1.9–3.9 μm fractions.

Differences were also found on the basis of mineralogy. Samples containing a greater proportion of platy minerals such as mica and kaolin, and expansible clays, gave higher contents for the finer fractions than did samples in which such minerals were less abundant. In general, there was no very clear pattern of agreement between the methods for any given sample.

Mohnot (1985) found the Cilas instrument to be useful as a rapid means of checking flocculation phenomena in drilling muds but reported no details of his comparisons with other methods. He also appears, like everyone else, to have ignored the possible role of the instrument pump in floc breakage, as was found by Gibbs (1982).

McCave et al. (1986) evaluated the Malvern Instruments 3600E laser particle size analyzer using both 63 mm and 100 mm focal length lenses, and compared the instrument with data obtained from the same samples by an ESZ machine. Their principal finding was that the laser-based instrument seemed to be severely affected by light scattered by particles < 2 μm, which showed as modes in the cumulative particle size curves, irrespective of sample type and treatment. This did not occur in the curves obtained from ESZ measurements. The effect was most pronounced with the 63 mm lens, but it also occurred with the 100 mm lens, and varied in magnitude and, to some extent, with clay content. The effect is most marked in samples with clay contents (<2 μm ESD) of 35% or more. Konert and Vandenberghe (1997) reported that a laser-light scattering instrument "saw" clay particles as ca. 8 μm ESD, rather than the <2 μm ESD as determined by pipet. In contrast, Vitton and Sadler (1997) reported reasonable agreement between particle size distribution of soils measured by the hydrometer method and by laser-light scattering, although they noted that the agreement worsened as the mica content of the soils increased.

These findings reflect uncertainties found by others in the use of laser-light particle size analyzers for very small particles. Dodge (1984) reported discrepancies during calibration of such instruments, while Evers (1982) found that the Malvern and Microtrac instruments gave very different results for the same material.

In summary, light-scattering instruments offer the possibility of measuring particle size very rapidly with very small samples of material. However, the theories on which they are based are known rigorously only for simple particles (spheres, ellipsoids, disks, etc.), and the instruments clearly have problems in dealing with variation in this factor and with systems of particles of differing refractive index. Beuselinck et al. (1998) compared a Coulter LS-100 instrument with the pipet method. They concluded that as long as the clay mineralogy of samples was similar, then the results of particle size analysis of soils by the two methods could be compared through functions derived from principal components analysis. In order to do this, a database of analyses of soils by the two methods needs to be constructed, and this may be the way forward in eventually bringing the two approaches closer together. Laser-light scattering has been described in at least two national Standards (AFNOR, 1992; ASTM, 1998e).

ACKNOWLEDGMENTS

We thank Mrs. F. Cox (SRI) for word processing the manuscript. Silsoe Research Institute is grant-aided by the UK Biotechnology and Biological Sciences Research Council.

REFERENCES*

AFNOR (Association Française de Normalisation, Paris). 1979. *Granulométrie. Analyse granulométrique des poudres fines sur tamiseuse à dépression d'air.* Doc. NF-X-11-

* There is a very large number of Standards dealing with aspects of particle size analysis. No Standards organization publishes all its Standards in one volume. However, Standards are uniquely referenced by number and date. It is this information that is given here. All Standards for one organization are listed under a single heading for that organization. A full list of member organizations of the International Standards Organisation (ISO), as well as its publications, can be found at: http://www.iso.ch/. Other useful information and listings can be found at: http://catafnor.afnor.fr/; http://www.bsi.org.uk/; http://www.din.de/; http://www.aist.go.jp/jisc/. Standards can be replaced or updated as often as at five-year intervals. Users are advised to check the latest information at regular intervals, as this may have legal implications for the work of their laboratory.

640; and subsequent documents: 1981a: NF-X-11-683; 1981b: NF-X-11-630; 1982: NF-X-11-642; 1983a: NF-X-11-685; 1983b: NF-X-11-693; 1983c: NF-X-31-107; 1984: NF-X-11-682; 1990: NF-X-11-661; 1992: NF-X-11-667; 1997a: NF-X-11-671; 1997b: NF-X-11-636. Paris: AFNOR.

Allen, T., Davies, R., and Scarlett, B., eds. 1996. *Particle Size Measurement.* (2 vols.) London: Chapman and Hall.

Anon. 1997. *Macmillan Encyclopedia of Physics.* London: Macmillan Reference/Prentice Hall.

ASTM (American Society for Testing and Materials, Philadelphia). 1998a. *Annual Book of Standards.* Vol. 15.02. Designation C690–86 (Re-approved 1992); and additional sections: 1998b: Vol. 04.08. Designation D422–63 (Re-approved 1990); 1998c: Vol. 14.02. Designation E1617; 1998d: Vol. 15.02. Designation C1182–91 (Re-approved 1995); 1998e: Vol. 15.02. Designation C1070–86 (Re-approved 1992); 1998f: Vol. 15.02. Designation C690–86 (Re-approved 1992). Philadelphia: ASTM.

Atkinson, C. M. L., and Wilson, R. 1981. The mass integration method for the calibration of the electrical sensing zone technique used for the sizing and counting of fine particles. In: *Particle Size Analysis* (N. G. Stanley-Wood, and T. Allen, eds). New York: Wiley-Interscience, pp. 185–197.

Atterberg, A. 1916. Die Klassifikation der humusfreien und der humusarmen Mineralboden Schwedens nach den Konsistenzverhaltnissen derselben. *Int. Mitt. Bodenkd.* 6:27–37.

Avery, B. W., and Bascomb, C. L., eds. 1982. *Soil Survey Laboratory Methods.* Tech. Monogr. No. 6, Soil Survey. Harpenden, U.K., pp. 18–19.

Barak, P., Seybold, C. A., and McSweeney, K. 1996. Self-similitude and fractal dimension of sand grains. *Soil Sci. Soc. Am. J.* 60:72–76.

Barth, H. G., ed. 1984. *Modern Methods of Particle Size Analysis.* New York: John Wiley.

Barth, H. G., and Sun, S. T. 1985. Particle size analysis. *Anal. Chem.* 57:151R-175R.

Batch, B. A. 1964. The application of an electronic particle counter to size analysis of pulverized coal and fly-ash. *J. Inst. Fuel* 37:455–461.

Beuselinck, L., Govers, G., Poesen, J., and Dregaer, G. 1998. Grain-size analysis by laser diffractometry: Comparison with the sieve-pipette method. *Catena* 32:193–208.

Billi, P. 1984. Quick field measurement of gravel particle size. *J. Sedimentol. Petrol.* 54: 658–660.

BSI (British Standards Institution, London). 1981. *Code of Practice for Site Investigation,* Standard BS5930; and subsequent Standards: 1986: BS410; 1990: BS1377 (Part 2); 1993: BS3406 (Part 4); 1994a: BS3406 (Part 5); 1994b: BS 3406 (Part 6); 1998: BS7755–5.4. London: BSI.

Buchan, G. D., Grewal, K. S., Claydon, J. I., and McPherson, R. J. 1993. A comparison of Sedigraph and pipette methods for soil particle size analysis. *Aust. J. Soil Res.* 31: 407–417.

Buchter, B., Hinz, C., and Flühler, H. 1994. Sample size for the determination of coarse fragment content in a stony soil. *Geoderma.* 63:265–275.

Burt, R., Reinsch, T. G., and Miller, W. P. 1993. A micro-pipet method for water-dispersible clay. *Commun. Soil Sci. Plant Anal.* 24:2531–2544.

Caroni, E., and Maraga, F. 1983. Misure granulometriche in alvei naturali con un compasso registratore. *Geol. Appl. Idrogeol.* 18:19–31.

Carter, M. R., ed. 1993. *Soil Sampling and Methods of Analysis*. Boca Raton, FL: Lewis Publishers.

Christopher, T. B. S., and Mokhtaruddin, A. M. 1996. A computer program to determine the soil textural class in 1-2-3 for WINDOWS and EXCEL. *Commun. Soil Sci. Plant Anal.* 27:2315–2319.

Churchman, G. J., and Tate, K. R. 1987. Stability of aggregates of different size grades in allophanic soils from volcanic ash in New Zealand. *J. Soil Sci.* 38:19–28.

Coates, G. F., and Hulse, C. A. 1985. A comparison of four methods of size analysis of fine-grained sediments. *N.Z. J. Geol. Geophys.* 28:369–380.

Cooper, L. R., Haverland, R. L., Hendricks, D. M., and Knisel, W. G. 1984. Microtrac particle size analyzer: An alternative particle size determination method for sediment and soils. *Soil Sci.* 138:138–146.

Cornillault, J. 1986. HR850 Granulometer. *Spectra 2000* 111:27–29.

Coventry, R. J., and Fett, D. E. R. 1979. *A Pipette and Sieve Method of Particle-Size Analysis and Some Observations on Its Efficiency*. Div. Soils Divisional Rep. No. 38. Queensland, Australia: CSIRO.

Dahneke, B. E., ed. 1983. *Measurement of Suspended Particles by Quasi-Elastic Light Scattering*. New York: John Wiley.

Davis, C. N. 1947. Sedimentation of small suspended particles. *Trans. Ind. Chem. Eng.* S25:25–39.

DIN (Deutsches Institut für Normung, Berlin). 1983. *Partikelgrossenanalyse: Sedimentationanalyse in Schwerefeld; Pipette-Verfahren*. DIN 66115, Berlin: DIN.

DIN. 1996. *Bestimmung der Korngrößenverteilung,* DIN 18123. Berlin: DIN.

Dewell, P. 1967. A centrifugal sedimentation method for particle size analysis. In: *Particle Size Analysis*. London: Soc. Anal. Chem., pp. 268–280.

Dodge, L. G. 1984. Calibration of the Malvern particle sizer. *Appl. Opt.* 23:2415–2419.

Dudley, R. J. 1976. The particle size analysis of soils and its use in forensic science—The determination of particle size distribution within the silt and sand fractions. *J Forens. Sci. Soc.* 16:219–229.

Duke, M. B., Woo, C. C., Bird, M. L., Sellers, G. A., and Finkelman, R. B. 1970. Size distribution and mineralogical constituents. *Science* 167:648–650.

Evers, A. D. 1982. Methods for particle size analysis of flour: A collaborative test. *Lab. Pract.* 31:215–219.

Gee, G. W., and Bauder, J. W. 1986. Particle size analysis. In: *Methods of Soil Analysis*, Part I, 2d ed. (A. Klute, ed.). Madison, WI: Am. Soc. Agron., pp. 383–411.

Gibbs, R. J. 1982. Floc breakage during HIAC light blocking analysis. *Environ. Sci. Technol.* 16:298–299.

Griffiths, J. C. 1967. *Scientific Method in Analysis of Sediments*.New York: McGraw-Hill.

Grout, H., Tarquis, A. M. and Wiesner, M. R. 1998. Multifractal analysis of particle size distributions in soil. *Environ. Sci. Technol.* 32:1176–1182.

Head, K. H. 1992. *Manual of Soil Laboratory Testing, Vol. I. Soil Classification and Compaction Tests*. London: Pentech Press.

Hendrix, W. P., and Orr, C. 1971. Automatic sedimentation size analysis instrument. In: *Particle Size Analysis 1970* (M. J. Groves and J. L. Wyatt-Sargent, eds.). London: Soc. Anal. Chem., pp. 133–146.

Hodgson, J. M. 1978. *Soil Sampling and Soil Description*. London: Oxford Univ. Press.

Hodgson, J. M. 1997. *Soil Survey Field Handbook*. Technical Monograph No. 5. Silsoe, U.K.: Soil Survey and Land Research Centre, pp. 112–115.

Hunt, C. M., and Woolf, A. R. 1969. Comparison of some different methods for measuring particle size using microscopically calibrated glass beads. *Powder Technol.* 3:1–8.

Hyslip, J. P., and Vallejo, L. E. 1997. Fractal analysis of the roughness and size distribution of granular materials. *Eng. Geol.* 48:231–244.

ISO (International Standards Organisation, Geneva). 1973. *Guide to the Choice of Series of Preferred Numbers and Series Containing More Rounded Values of Preferred Numbers*. Doc. ISO-497-1973-(E) (1st ed.); and subsequent documents: 1977: ISO-387-(E) (1st ed.); 1981a: ISO-649-1-(E) (Part 1) (1st ed.); 1981b: ISO-649-2-(E) (Part 2) (1st ed.); 1988: ISO-2591-1-(E) (Part 1) (1st ed.); 1990a: ISO-565-(E) (3d ed.); 1990b: ISO-2395-(E) (2d ed.); 1990c: ISO-3310-1-(E) (Part 1) (3d ed.); 1990d: ISO-3310-2-(E) (Part 2) (3d ed.); 1990e: ISO-3310-3-(E) (Part 3) (1st ed.); 1995a: ISO-9276-1-(E); 1995b: ISO-9276-2-(E); 1996: ISO-13321-(E) (1st ed.); 1998: ISO-11277-1-(E) (1st ed.). Geneva: ISO.

ISSS (International Society of Soil Science). 1928. *The Study of Soil Mechanics and Physics*. Report of Commission I, Proc. 1st Int. Congr. Soil Sci, Part II, Washington, DC, pp. 359–404.

Kellerhals, R., Shaw, R., and Arora, V. K. 1975. On grain size from thin sections. *J. Geol.* 83:79–96.

Kennedy, S. K., Meloy, T. P., and Durney, T. E. 1985. Sieve data—Size and shape information. *J Sedimentol. Petrol.* 55:356–360.

Kiss, K., and Pease, R. N. 1982. Quantitative analysis of particle sizes: Estimation of the most efficient sampling scheme. *J. Microsc.* 126:173–178.

Klute, A., ed. 1986. *Methods of Soil Analysis, Part I. Physical and Mineralogical Methods*. 2d ed. Madison, WI: Am. Soc. Agron.

Konert, M., and Vandenberghe, J. 1997. Comparison of laser grain size analysis with pipette and sieve analysis: A solution for the underestimation of the clay fraction. *Sedimentology* 44:523–535.

Kosmas, C. S., Franzmeier, D. P., and Schulze, D. G. 1986. Relationship among derivative spectroscopy, color, crystallite dimensions, and Al-substitution of synthetic goethites and hematites. *Clays Clay Mineral.* 34:625–634.

Laxton, J. L. 1980. A method for estimating the grading of boulder and cobble grade material. In: *IGS Short Communications*. Report Inst. Geol. Sci. No 80/1. London: HMSO, pp. 31–35.

Lewis, G. C., Fosberg, M. A., Falen, A. L., and Miller, B. J. 1984. Identification of loess by particle size distribution using the Coulter counter TAII. *Soil Sci.* 137:172–176.

Lines, R. W. 1981. Particle counting by Coulter counter. *Anal. Proc.* 18:514–519.

Lloyd, P. J. 1982. Response of the electrical sensing zone method to non-spherical particles. In: *Particle Size Analysis 1981* (N. G. Stanley-Wood and T. Allen, eds.). New York: Wiley-Interscience, pp. 199–208.

Maeda, T., Takenaka, H., and Warkentin, B. H. 1977. Physical properties of allophane soils. *Adv. Agron.* 29:229–264.

McCave, I. N., and Jarvis, J. 1973. Use of the Coulter counter in size analysis of fine to coarse sand. *Sedimentology* 20:305–315.

McCave, I. N., Bryant, R. J., Cook, H. F., and Coughanowr, C. A. 1986. Evaluation of a laser-diffractions size analyzer for use with natural sediments. *J. Sedimentol. Petrol.* 56:561–564.

McTainsh, G. H., Lynch, A. W., and Hales, G. 1997. Particle-size analysis of aeolian dusts, soils and sediments in very small quantities using a Coulter "Multisizer." *Earth Surf. Proc. Landf.* 22:1207–1216.

Melik, D. H., and Fogier, H. S. 1983. Turbidimetric determination of particle size distributions of colloidal systems. *J. Colloid. Interface Sci.* 92:161–181.

Metz, R. 1985. The importance of maintaining horizontal sieve screens when using a Ro-Tap. *Sedimentology* 32:613–614.

Mohnot, S. M. 1985. Characterisation and control of fine particles involved in drilling. *J. Petrol. Technol.* 37:1622–1632.

Muller, R. N., and Tisne, G. T. 1977. Preparative-scale size fractionisation of soils and sediments and an application to studies of plutonium geochemistry. *Soil Sci.* 124: 191–198.

Mullins, C. E., and Hutchinson, B. J. 1982. The variability introduced by various subsampling techniques. *J. Soil Sci.* 33:547–561.

Nadeau, P. H. 1985. The physical dimensions of fundamental clay particles. *Clay Minerals* 20:449–514.

Nadeau, P. H., Wilson, M. J., McHardy, W. J., and Tait, J. M. 1984. Inter-particle diffraction: A new concept of interstratification of clay minerals. *Clay Mineral.* 19: 757–770.

Odén, S. 1915. Eine neue Methode zur mechanischen Bodenanalyse. *Int. Mitt. Bodenanal.* 5:257–311.

Olivier, J. P., Hickin, G. K., and Orr, C., Jr. 1971. Rapid, automatic particle size analysis in the sub-sieve range. *Powder Technol.* 4:257–263.

Oppenheimer, L. 1983. Interpretation of disk centrifuge data. *J. Colloid Interface Sci.* 92: 350–357.

Pennington, K. L., and Lewis, G. C. 1979. A comparison of electronic and pipet methods for mechanical analysis of soils. *Soil Sci.* 128:280–284.

Plantz, P. E. 1984. Particle size measurements from 0.1 to 1000 μm, based on light scattering and diffraction. In: *Modern Methods of Particle Size Analysis* (H. G. Barth, ed.). New York: John Wiley, pp. 173–209.

Pleijsier, L. K. 1986. The Laboratory Methods and Data Exchange Programme: Interim Report on the Exchange Round 86–1, Working Paper and Pre-print No. 86/4. Wageningen, The Netherlands: ISRIC.

Pritchard, D. T. 1974. A method for particle size analysis using ultrasonic disaggregation. *J Soil Sci.* 25:34–40.

Ringrose-Voase, A., and Bullock, P. 1984. The automatic recognition and measurement of soil pore types by image analysis and computer programs. *J. Soil Sci.* 35:673–684.

Ritter von Rittinger, P. 1867. *Lehrbuch der Aufbereitungskunde*. Berlin: Ernst & Korn Verlag.

Ross, D. A., Dhadwal, H. S., and Dyott, R. B. 1978. The determination of the mean and standard deviation of the size distribution of a colloidal suspension of sub-micron particles using the fiber-optic Doppler anemometer (FODA). *J. Colloid Interface Sci.* 64:533–542.

Rousseva, S. S. 1997. Data transformations between soil texture schemes. *Euro. J. Soil Sci.* 48:749–758.

Rybina, V. V. 1979. Use of conductimetry for the determination of the particle size composition of soils. *Sov. Soil Sci.* 11:482–486.

Sapetti, C. 1963. Misure granulometriche sul terreno. *Ann. Sper. Agrar.* 17:583–615.

Shiozawa, S., and Campbell, G. S. 1991. On the calculation of a mean particle diameter and standard deviation from sand, silt and clay fractions. *Soil Science* 152:427–431.

Stabinger, H., Leopold, H., and Kratky, O. 1967. Eine neue Präzisionsmethode zur Bestimmung der Dichte von Flüssigkeiten. *Monatsh. Chem.* 98:436–438.

Stein, R. 1985. Rapid grain-size analysis of clay and silt fraction by Sedigraph 5000-D: Comparison with Coulter counter and Atterberg methods. *J. Sedimentol. Petrol.* 55: 590–593.

Stokes, G. G. 1849. On the theories of the internal friction of fluids in motion, and of the equilibrium and motion of elastic solids. *Camb. Phil. Trans.* 8:287–319.

Sur, H. S., and Kukal, S. S. 1992. A modified hydrometer procedure for particle size analysis. *Soil Science* 153:1–4.

Svarovsky, L., and Svarovska, J. 1975. Centrifugal sedimentation data analysis by analogue deconvolution. *J. Phys. D: Appl. Phys.* 8:181–190.

Swithenbank, J., Beer, J. M., Taylor, D. S., and McCreath, G. C. 1977. A laser diagnostic technique for the measurement of droplet and particle size distribution. *Progr. Astron. Aeronaut.* 53:421–447.

Syvitski, J. P. M., ed. 1991. *Principles, Methods, and Applications of Particle Size Analysis.* Cambridge, U.K.: Cambridge Univ. Press.

Tanner, C. W., and Bourget, S. J. 1952. Particle shape discrimination of round and square holed sieves. *Soil Sci. Soc. Am. Proc.* 16:88.

Tanner, C. B., and Jackson, J. J. 1947. Nomographs of sedimentation times for soil particles and gravity or centrifugal acceleration. *Soil Sci. Soc. Am. Proc.* 12:60–65.

Tovey, N. K., and Smart, P. 1982. *Electron Microscopy of Soils and Sediments: Techniques.* Oxford: Clarendon Press.

Tyler, S. W., and Wheatcraft, S. W. 1992. Fractal scaling of soil particle size distributions—Analysis and limitations. *Soil Sci. Soc. Am. J.* 56:362–369.

USDA (US Dept. of Agriculture). 1996. *Soil Survey Laboratory Methods Manual.* Soil Survey Investigations Report No. 46, Version 3.0. Washington, DC.: Soil Conservation Service

Vitton, S. J., and Sadler, L. Y. 1997. Particle-size analysis of soils using laser light scattering and x-ray absorption techniques. *Geotech. Test. J.* 20:63–73.

Walker, P. H., and Hutka, J. 1971. *Use of the Coulter Counter (Model B) for Particle-Size Analysis of Soils.* Div. of Soils Tech. Paper No. 1. Melbourne, Australia: CSIRO.

Walker, P. H., Woodyer, K. D., and Hutka, J. 1974. Particle size measurements by Coulter counter of very small deposits and low suspended sediment concentrations in streams. *J. Sedimentol. Petrol.* 44:673–679.

Walton, E. K., Stephens, W. E., and Shawa, M. S. 1980. Reading segmented grain-size curves. *Geol. Mag.* 117:517–524.

Webster, R., and Oliver, M. A. 1990. *Statistical Methods in Soil and Land Resource Survey.* London: Oxford Univ. Press.

Wentworth, C. K. 1922. A scale of grade and class terms for clastic sediments. *J. Geol.* 30: 377–392.

Whalley, W. R. 1988. Theory and use of centrifugal photo-sedimentation for particle size analysis of clays. Ph.D. thesis, University of Aberdeen, U.K.

Whalley, W. R., and Mullins, C. E. 1991. Effect of saturating cation on tactoid size distribution in bentonite suspensions. *Clay Mineral.* 43:531–540.

Whalley, W. R., and Mullins, C. E. 1992. Oriented and random sedimentation of plate-like clay particles in high centrifugal fields. *J. Soil Sci.* 43:531–540.

Whalley, W. R., Mullins, C. E., and Livesey, N. T. 1993. Use of centrifugal photo-sedimentation to measure the particle size distribution of clays. *J. Soil Sci.* 44:221–229.

Zaneveld, J. R. V., Spinrad, R. W., and Bartz, R. 1982. An optical settling tube for the determination of particle size distributions. *Marine Geol.* 49:357–376.

8
Bulk Density

Donald J. Campbell and J. Kenneth Henshall
Scottish Agricultural College, Edinburgh, Scotland

I. INTRODUCTION

The wet bulk density of a soil, ρ, is its mass, including any water present, per unit volume in the field; its dry bulk density, ρ_s, is the mass per unit volume of field soil after oven-drying. These parameters are related to the soil gravimetric water content, W, as follows:

$$\rho_s = 100\left(\frac{\rho}{100 + W}\right) \tag{1}$$

where W is the mass of water expressed as a percentage of the mass of dry soil.

The methods available for the measurement of soil bulk density fall into two groups. In the first group are the long-established direct methods, which involve measurement of the sample mass and volume. The mass M_s of the oven-dried sample is obtained by weighing, and the total volume, V, of the soil including air and water is obtained by measurement or indirect estimation. The dry bulk density ρ_s is then given by

$$\rho_s = \frac{M_s}{V} \tag{2}$$

Such methods have been used by both agricultural soil scientists (Freitag, 1971) and civil engineers (DSIR, 1964), and many of them reduce essentially to the problem of the accurate determination of the sample volume. As these methods have not always proved entirely effective, a second group of methods has evolved in which the attenuation or scattering of nuclear radiation by soil is used to give an indirect measurement of bulk density. Radiation methods are capable of measuring more accurately and precisely than direct methods, but they too have limitations of their own.

Thus there is no single measurement method suitable for all circumstances. Sometimes a very crude but quick measurement is all that is required to characterize soil conditions, but in other circumstances it may well be appropriate to use a slower method involving expensive equipment, in order, for example, to detect detailed differences between experimental treatments.

II. RADIATION METHODS

A. Theory

Radiation methods involve measuring either the attenuation or the scattering of gamma radiation by the soil, both of which increase with density. Empirical calibration relationships are used to relate the magnitude of such effects to soil bulk density.

Gamma-ray photons are emitted by radioactive nuclei as they decay to form more stable nuclei of lower excitation. A specific source will emit gamma photons with the characteristic energy of one or more decay transitions. In passing through any medium, the probability that these photons will interact with the atoms of the medium is dependent on the density of the medium, as well as other factors such as the energy of the photon and the chemical composition of the medium. These interactions take the form either of complete absorption of the photon or of scattering, where the photon loses energy in relation to the angle of deflection. Since the photons interact principally with the electrons of the medium, the extent of the interaction depends on the electron density, which is related to the bulk density of the medium.

There are two main types of gamma-ray density equipment: backscatter gauges, which are designed to detect only scattered photons, and transmission gauges, which detect mainly unscattered photons. Depending on the level of energy discrimination, however, some simpler transmission systems also detect scattered photons to different extents.

B. Backscatter Gauges

In backscatter gauges, the gamma-ray source and detector are fixed relative to, and shielded from, each other in an assembly designed to prevent measurement of directly transmitted photons. This assembly either rests on the soil surface or, in some designs, is lowered into an access hole in the soil (Fig. 1). In either case, any photons incident upon the detector must have been deflected by one or more scattering interactions in the medium. Since there is only a low probability that a photon that has travelled an appreciable distance from the source will reach the

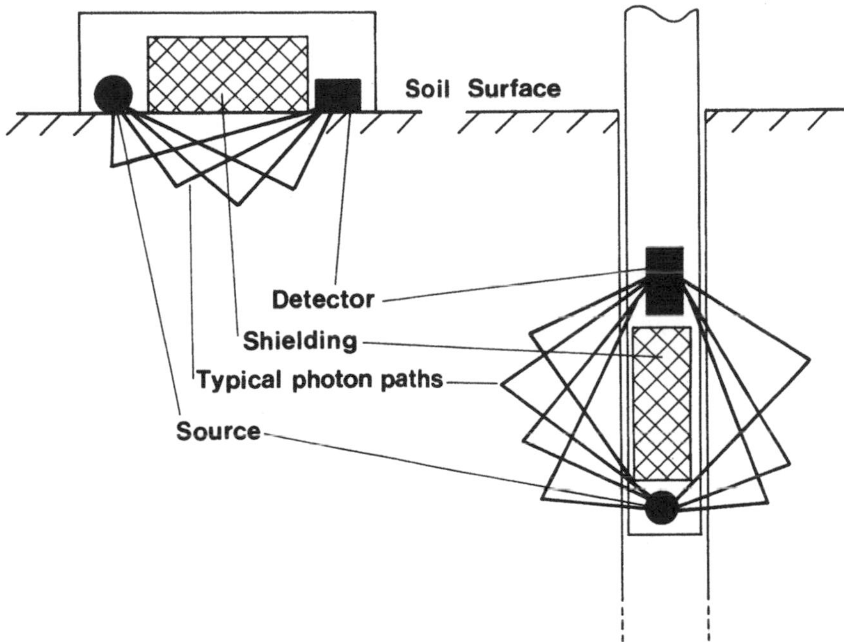

Fig. 1 Schematic diagrams of backscatter gamma-ray gauges in which the source and detector assembly either lies on the soil surface (left) or is lowered into an access hole in the soil (right).

detector, it follows that only a restricted volume of the medium close to the source/detector axis will influence the detected photon count rate. In practice, with a probe that is used in an access hole, it is found that the zone of influence does not extend more than about 75 mm from the source/detector axis and that 50% of the photons penetrate soil within only about 25 mm of this axis.

The relation between count rate and bulk density is complicated, since the degree of scattering increases with density, thereby increasing the count rate, but absorption of both scattered and unscattered photons also increases with density and so reduces count rate. Thus theoretical calibrations of backscatter gauges are impracticable, and empirical calibrations must be made.

Surface backscatter gauges require only that the surface of the soil be made perfectly level in order to exclude air gaps, but they yield little information, merely indicating the average density of the top 50–75 mm of the soil profile. Their main use is in civil engineering applications where bulk densities which are generally uniform with depth are to be measured. A typical level of accuracy for these gauges is ± 0.16 Mg m^{-3} (Carlton, 1961).

Single-probe backscatter gauges are normally lowered into lined access holes in a manner similar to neutron moisture probes (Chap. 1) and are available in combination with such probes. The major failing of these gauges results from the bias of their zone of influence close to the source/detector axis. This means that both the clearance gap of the probe in the liner tube and the tube itself influence the measurements unduly. The measurements are also very susceptible to any disturbance of the soil during installation of the liner tube.

C. Transmission Gauges

In transmission gauges (Fig. 2), the sample to be tested is located between the source and the detector of the gauge, and ideally only unattenuated photons passing directly from source to detector are counted. In this ideal case, where none of the photons has been degraded, the detected photon count rate, I, obeys Beer's law,

$$I = I_0 \exp[-\mu\rho x] \tag{3}$$

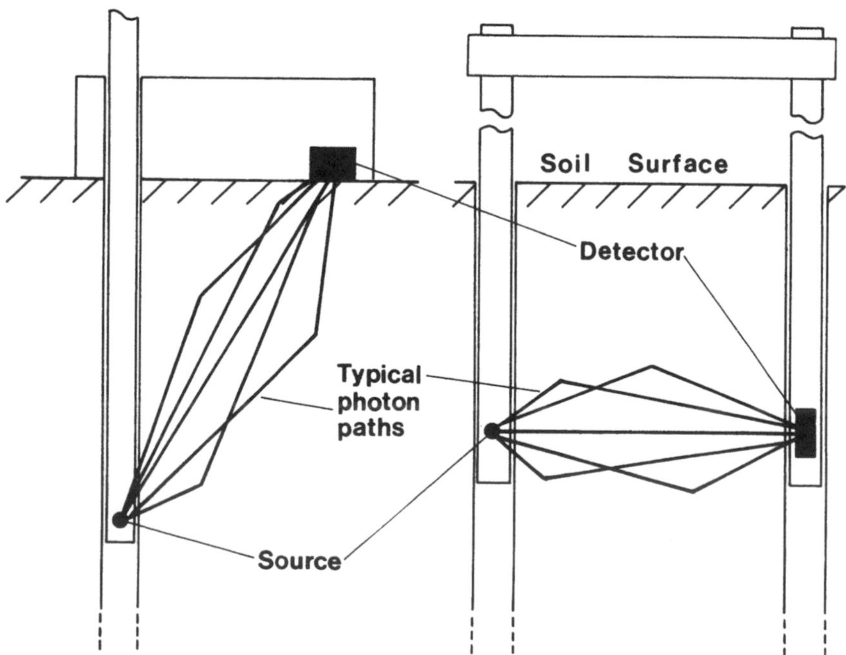

Fig. 2 Schematic diagrams of transmission gamma-ray gauges in which the detector either remains on the soil surface and the source is lowered into an access hole in the soil (left) or in which both the source and detector are lowered into separate access holes (right).

where I_0 is the photon count rate in the absence of a sample, μ is the mass attenuation coefficient for the specific photon energy and sample material concerned, ρ is the wet bulk density of the sample, and x is the sample length. The bulk density of the sample can then be calculated as

$$\rho = \frac{-1}{\mu x} \ln \left(\frac{I}{I_0} \right) \tag{4}$$

if values are available for μ, x and I_0.

In practice, several factors make such a theoretical calculation of density impracticable. The most important of these are

1. Inclusion in the count of some scattered photons
2. Determination of a single mass attenuation coefficient for soils of variable composition
3. Estimation of the photon count rate in the absence of a sample

1. Scattered Photons

With the exception of laboratory equipment in which a high degree of both collimation and energy discrimination is possible, scattered photons will always be included to some extent in the detected count rate. Scattered and unattenuated photons have different mass attenuation coefficients, and the presence of scattered photons therefore affects the linearity of the relationship between ρ and ln I/I_0. The reduced energy of these scattered photons also increases the dependence of the detected count rate on the chemical composition of the soil sample, as will be discussed later, and reduces the spatial resolution of the gauge by increasing the volume of soil, which influences the count rate.

While it is possible to reduce the number of scattered photons by collimation, limited space prevents this in field gauges. An alternative is to use an energy-discriminating detector, set to exclude photons with energies lower than the emission energy of the source. Gauges with this facility generally use a scintillation detector, such as a sodium iodide crystal, linked to a photomultiplier tube and pulse height analyzer. Energy-discriminating detectors need to be stabilized against temperature changes.

Simpler transmission gauges use Geiger–Müller detectors, which are not capable of energy discrimination and hence are susceptible to scattered photons. In effect, these gauges operate in both the transmission and backscatter modes simultaneously. Provided such a gauge is calibrated empirically, its only major disadvantage, other than a slight dependence on the chemical composition of the soil, is its low spatial resolution, which can affect measurements close to distinct boundaries such as the soil surface or a plow pan. For example, Henshall and Campbell (1983) found that a Geiger–Müller based gauge overestimated the density of water by 35% at a depth of 20 mm below an air/water interface and continued to overestimate the density by more than 5% to a depth of 90 mm.

Gauges employing energy discrimination can be adjusted to give high spatial resolution limited only by the dimensions of the detector, which can be as small as 10×10 mm cross-section. However, the need to ensure sufficiently high count rates forces lower resolution settings which, by including some scattered photons, results in the need for empirical calibration as with simpler gauges.

2. *Soil Composition*

As used in Eq. 3, the mass attenuation coefficient, μ, is an overall value for the bulk material examined. A theoretical value of μ would be the mean of the individual mass attenuation coefficients for each of the constituent elements, weighted according to the mass fraction of each element in the sample. Differences in the chemical composition of the soil can therefore affect the overall mass attenuation coefficient.

The mass attenuation coefficient of a chemical element varies with the atomic number of the element, Z, and the incident photon energy. Coppola and Reiniger (1974) showed that μ increased with increasing photon energy but that, for photon energies above about 0.3 MeV, there was little dependence of μ on Z below $Z = 30$, with the exception of hydrogen, which is discussed below. Caesium-137, which emits mono-energetic photons of 0.662 MeV, is the radioactive source most commonly employed in soil bulk density gauges. At this photon energy, calculations based on theoretical values of mass attenuation coefficient for nine different soils show that the error in estimated density due to the effect of composition is of the order of 0.5% in the most extreme case (Reginato, 1974). An energy-discriminating system set to exclude photons of energy lower than the caesium-137 emission energy would therefore not show a significant dependence on chemical composition of the soil. In contrast, Geiger–Müller detectors, which do not employ energy discrimination, are sensitive to photon energies as low as 0.04 MeV (Soane, 1976). Consequently, a significant proportion of the detected count rate will include scattered photons with energies that are below 0.3 MeV and so are susceptible to composition effects. Nevertheless, only a small proportion of the detected photons will have been scattered through angles large enough to result in such low energies so that the effect of composition on count rate is unlikely to be serious except in backscatter gauges, where it is only the less energetic scattered photons that are counted. Generally, transmission gauges, especially those with energy discrimination, are not susceptible to soil composition effects except in soils that have a large proportion of heavy elements, such as iron (Gameda et al., 1983).

3. *Photon Count Rate in the Absence of a Sample*

In order to apply Beer's equation (Eq. 3), it is necessary to know the photon intensity I_0 in the absence of a sample. The theoretical relation assumes an ideal situation where none of the detected photons in I or I_0 are attenuated or scattered. Although a measurement of I_0 directly, i.e., in the absence of any attenuation by the soil, would be very similar to this ideal situation, safety considerations make it impracticable. The normal method therefore is to make a reference measurement using a material of constant density such as a steel plate. The reference count rate, I_r, can be written as

$$I_r = I_0 \exp[\mu_r \rho_r x] \tag{5}$$

where ρ_r is the mean density, over the sample length, of the reference plate and air gap, and μ_r is the corresponding mass attenuation coefficient. This, combined with Eq. 3, gives

$$\frac{I}{I_r} = \exp[-x(\mu\rho - \mu_r \rho_r)] \tag{6}$$

thereby eliminating I_0. Relating test measurements to reference measurements in this way also allows for the gradual decrease with time in the activity of the source and any gradual change in the efficiency of the detection system.

D. Calibration

When a gauge is calibrated relative to a standard reference plate, Eq. 6 can be rearranged to give an expression for bulk density, namely

$$\rho = \frac{1}{\mu x}\left[\ln\left(\frac{I_r}{I}\right) - \mu_r \rho_r x\right] \tag{7}$$

or

$$\rho = A \ln\left(\frac{I_r}{I}\right) + B \tag{8}$$

where A and B are empirically determined constants. Since the gauge measures only wet bulk density, an independent measurement of gravimetric water content is required to give the dry bulk density ρ_s from Eq. 1.

Hydrogen, which in soil is most abundant in the water, does not conform with other elements in its attenuation of gamma photons, as it possesses only one nucleon per electron, whereas other atoms typically possess approximately two. While the gamma-ray attenuation system effectively measures the number of electrons per unit volume, bulk density is related to the number of more massive nucleons per unit volume, and so the density of hydrogen is overestimated by a factor of approximately two. Consequently, if the greater attenuation coefficient of hydrogen were not corrected for, the bulk density would be slightly overestimated. For samples with gravimetric water contents of 10, 25, and 100%, the theoretical overestimate would be 1, 2, and 5%, respectively. In many applications, this level of accuracy may be considered acceptable, but, if required, the error can be corrected for during calibration. Separating the effects of water and soil, Eq. 3 becomes

$$I = I_0 \exp[-x(\mu_s \rho_s + \mu_w \rho_w)] \tag{9}$$

where ρ_w is the mass of water per unit total sample volume, and μ_s and μ_w are the mass attenuation coefficients for soil and water, respectively. Expressing ρ_w as $(\rho_s W/100)$ and incorporating a reference standard as in Eq. 6, we have

$$\frac{I}{I_r} = \exp\left\{-x\left[\rho_s\left(\mu_s + \mu_w \frac{W}{100}\right) - \mu_r \rho_r\right]\right\} \tag{10}$$

which leads to

$$\rho_s = \frac{\ln(I_r/I) + \mu_r \rho_r x}{x(\mu_s + \mu_w W/100)} \tag{11}$$

which again can be simplified to

$$\rho_s = \frac{A \ln(I_r/I) + B}{100 + CW} \tag{12}$$

where constants A, B, and C are determined empirically.

E. Gauge Design

1. *Radioactive Source*

The primary requirements of a radioactive source for a soil density gauge are that it should have a single energy peak at an energy sufficiently high to reduce composition effects, that the emitted photons should have a suitable penetration range into the soil sample, and that the half-life should be long enough not to affect any series of experimental measurements and should preferably exceed the expected life of the gauge. Caesium-137, with a mono-energetic peak of 0.662 MeV and a half-life of 30 years, is the source most suited to these requirements. The optimum soil sample length for gamma photons of this energy has been suggested as 100 to 250 mm (Ferraz and Mansell, 1979).

The rate of emission of gamma photons from a radioactive source is not perfectly constant but subject to random fluctuations about a mean value. The resulting fractional error in count rate is inversely proportional to the square root of the total number of photons counted (Ferraz and Mansell, 1979), and so it is preferable to count as many photons as possible to achieve the highest level of precision. This can be achieved by counting for long periods of time and by using the highest possible activity of source. However, for portable field gauges, the practical limit of activity is set by safety considerations. The maximum source activity that can be shielded to give the statutory levels of safety without the gauge becoming unacceptably heavy for field use is of the order of 0.4 GBq (10 mCi). In laboratory gauges, larger shields allow much larger sources to be used, in which case the upper limit to source activity is determined by the dead time of the detection system. This results from the inability of the detector to respond within a fixed time after detecting a photon, thereby imposing a count rate limit irrespective of source strength. With gauges based on NaI(T1) detectors this limits source activity to about 7 GBq (200 mCi). Although it has been suggested (Herkelrath and Miller, 1976) that this could be increased to 70 GBq (2000 mCi) where plastic scintillators are used, this proposal has never been adopted.

2. *Probe Design*

Portable field transmission gauges are of either single or twin probe design (Fig. 2). In single-probe gauges, the radioactive source is lowered through the body of the gauge into a preformed access hole, normally to a depth of about 300 mm (Fig. 2). The detector, which is generally of the nondiscriminating type, is located on the base of the gauge body at a fixed distance from the source probe axis, so that it is in contact with the surface of the soil. The count rate at each depth then relates to the average bulk density between the source depth and the surface. Such a gauge avoids some operational problems common to twin-probe gauges but suffers from an inability to examine soil layers and also requires a separate calibration for each measurement depth. Commercial gauges are normally supplied with factory calibrations, but users generally find that recalibration is necessary (Gameda et al., 1983).

The probes of twin-probe gauges (Fig. 2) are normally clamped rigidly at a fixed separation of between 140 and 300 mm so that, after they have been lowered to any desired depth in the soil, horizontal layers of soil can be examined (Fig. 2). These gauges are more suited to the study of soils in the context of agriculture, forestry, and the natural environment, where considerable variation in bulk density with depth is usually found. Conversely, in civil engineering applications, the soil is likely to be more uniform with depth, since only subsoils, either in situ or excavated and subsequently compacted as fill material, are of concern. In such applications, single-probe gauges have proved more popular.

Because of the fixed probe separation in twin-probe gauges, a single calibration relationship is applicable to all depths, but it is essential either that the access holes remain parallel or that any deviation is corrected for. Most popular commercial gauges incorporate nondiscriminating detectors and are therefore susceptible to problems of lack of resolution close to either air/soil interfaces or abrupt soil density changes with depth. However, detectors that employ energy discrimination are available (Fig. 3).

Fig. 3 Gamma-ray transmission gauge developed at the former Scottish Centre of Agricultural Engineering, complete with transport box which incorporates material for making a reference measurement and a scaler in the lid.

F. Soil Water Content Determination

While water content data are normally obtained from soil samples that have been extracted by auger and oven-dried at 105°C, some gauges incorporate a facility that allows water content to be estimated by nucleonic methods. Some single-probe gamma transmission gauges incorporate a neutron backscatter apparatus either in the base of the gauge body or in the probe. In conditions of uniform water content, such systems give an adequate overall estimate, but where water content varies with depth, the neutron backscatter apparatus does not have sufficient spatial resolution to allow correction of individual density measurements, since it has a typical sphere of influence of 250 mm radius.

A much more sophisticated method of simultaneously measuring bulk density and water content involves the use of the double-energy gamma transmission gauge. By employing a low-energy source, usually ^{241}Am with an energy peak of 0.06 MeV, together with a ^{137}Cs source (0.662 MeV), this technique makes use of the effect of chemical composition, especially hydrogen content, on the attenuation of low-energy photons. By including the effects of both soil and water, as in Eq. 9, in separate calibrations for the two energies, the resulting simultaneous equations can be solved for both dry bulk density and water content. The major drawback to this method is that the dependence of the low-energy calibration on chemical composition may necessitate different calibrations for different soils or possibly even for different depths in the same soil. This limitation effectively restricts the usefulness of this method to repeated laboratory tests on a single soil where only a single set of calibrations would be needed. Because of their specialized nature, such gauges are not available commercially.

III. METHODS OF MEASURING BULK DENSITY

A. Direct Measurement of Sample Mass and Volume

1. Core Sampling

In this widely used method a cylindrical sampler is hammered or pressed into the soil. As the volume of the cylinder is known, trimming of the soil core flush with the ends of the cylinder allows the bulk density to be calculated (Lutz, 1947; Jamison et al., 1950). The method works best in soft, cohesive soils sampled at water contents in the region of field capacity. Sands and gravels cannot be sampled satisfactorily.

A possible source of error in the method, which is difficult to quantify, is soil disturbance, especially by compression, during insertion of the sampler. Baver et al. (1972) have suggested that insertion by hammering may cause shattering, while steady pressure may produce compression. In an extensive survey of core sampling for civil engineering purposes referred to by Freitag (1971), Hvorslev

(1949) considered sample distortion to be a minimum when the sampler was pressed steadily rather than hammered into the soil. He also built a core sampler in which a piston was used to reduce the air pressure acting on the upper surface of the sample in the cylinder. The diameter of the sample also influences the risk of compression, with small diameter samples being more susceptible. Constantini (1995) found that increasing the sample diameter beyond approximately 60 mm did not improve the accuracy of bulk density measurement. Baver et al. (1972) proposed a diameter of 75–100 mm as a satisfactory compromise for most work, while Freitag (1971) suggested that the diameter should be selected to give a sample of adequate size, and that the length should not be more than about three times the diameter. Generally, the cylinder wall should be as thin as possible consistent with being rigid (DSIR, 1964). Further aids to easy insertion of the sampler include relieving both the inner and outer diameter immediately behind the cutting edge (Buchele, 1961) and lightly greasing the inside of the sample cylinder (Veihmeyer, 1929).

In order to extend the range of soils from which core samples can be taken, rotary core samplers have been introduced for hard, brittle soils that may shatter during conventional core sampling (Buchele, 1961; Freitag, 1971).

2. *Rubber Balloon Method*

In this method a hole is excavated in the soil to the bottom of the layer being tested, and the removed soil is weighed and its water content determined. The volume of the sample is determined by inserting a thin rubber balloon into the excavated hole and filling it with water. For accurate results to be obtained, the excavated hole should have a regular shape so that the balloon can reasonably be expected to fill any irregularities which arise (DSIR, 1964; Blake, 1965; Freitag, 1971). To this end, apparatus has been developed in which the balloon is clamped to the base of a calibrated water container that includes a pump to force the water into the balloon (DSIR, 1964; Freitag, 1971). Generally, the method is considered to give unreliable results.

3. *Sand Replacement*

In the sand replacement method, the sample is excavated, weighed, and its water content determined as in the rubber balloon method. The hole produced is usually about 100 mm in diameter. A metal cylinder, usually referred to as a "sand bottle" (Fig. 4), containing dry sand is placed over the hole and a tap in the base of the cylinder is opened to allow the sand to fill the hole. The difference in weight of the cylinder, before and after filling the hole, is recorded. The bulk density of the sand is obtained from a calibration test in which sand from the bottle is used to fill a can of known volume, and this allows the volume of the excavated hole to be

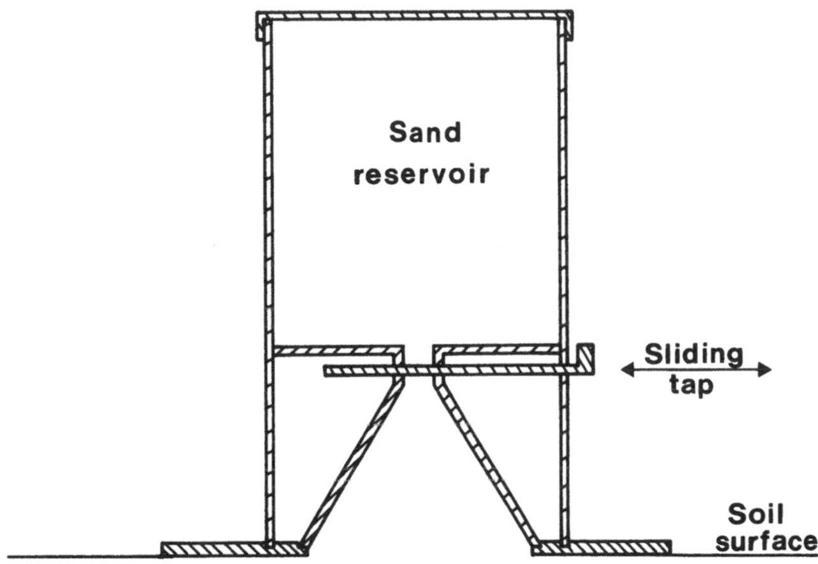

Fig. 4 Schematic section through a typical sand bottle used in the sand replacement method showing the sliding tap in the closed position.

calculated (DSIR, 1964; Blake, 1965). Allowance is made for the sand between the tap and the soil surface level by opening the tap while the equipment rests on a flat metal plate. In a variation of the method, which does not involve determination of the bulk density of the sand, a container for the sand is calibrated in terms of volume, as in a measuring cylinder, and the difference in volume before and after filling the hole gives the volume of the hole. The method is claimed to give smaller errors than the conventional sand replacement method (Cernica, 1980).

Several aspects of the test procedure require to be carefully controlled if reliable results are to be obtained. The volume of the calibration can should be similar to that of the excavated hole, since a 25 mm decrease in the depth of the can produces a decrease of about 1% in sand bulk density. A similar decrease in density is produced by a 50 mm reduction in the initial level of the sand in the cylinder (DSIR, 1964). The sand should be closely graded (typically, 0.2 to 2.0 mm material is used) to prevent segregation and hence variation in sand bulk density, and this is considered more important than the actual size range used.

The greatest care should be taken to ensure that the sand remains dry and uncontaminated by soil when it is recovered from the hole at the end of a test. Frequent checks on the calibration are the best way of checking whether this is occurring (Freitag, 1971). Although the sand replacement method is relatively slow, with a typical test time of 30 minutes, it has the advantage that it can be used on all soil types (Freitag, 1971).

4. Clod Method

In this method a clod is weighed and its volume is determined by coating it in paraffin wax and immersing it in a volumenometer. The volume of water displaced corresponds to that of the clod plus wax (DSIR, 1964). Alternatively, the waxed clod may be weighed in air and in water. In both versions of the method the wax coating must subsequently be removed and weighed. The wax coating is applied by suspending the clod from a fine wire and dipping the clod in paraffin wax at a temperature just above its melting point. Although the method gives satisfactory results, it is limited to cohesive soils and is a rather slow method when wax is used as the coating material. A useful summary of these techniques is given by Russell and Balcerek (1944).

Saran F-220 resin, dissolved in methyl ethyl ketone, was used as a substitute for wax by Brasher et al. (1966), who found that it was flexible, did not melt during oven drying at 105°C, and was permeable to water vapor but not to liquid water. It could therefore be used to study the drying and shrinkage characteristics of a clod. Rubber solution has also been used as the coating material, with claims of improved accuracy and convenience over the paraffin wax method (Abrol and Palta, 1968). A flotation technique has been used in which the clods were sprayed with a resin solution and then immersed sequentially in liquids of different relative density. The relative densities of the two liquids in which the clods just sank and just floated provided an upper and lower limit to the clod bulk density. As neither clod mass nor clod volume was determined, the technique was shown to be ten times as rapid as the wax coating method (Campbell, 1973).

It is possible to avoid coating the clod at all if the immersion fluid does not penetrate the soil pores. Although various viscous oils and mercury have been used, the technique is probably restricted to soils with very small pores. Thus one successful application was in a study of the density of puddled soils (Gill, 1959). Other published techniques for clod bulk density measurement include the use of x-rays (Greacen et al., 1967), elutriation in a vertical air stream (Chepil, 1950), and immersion in a bed of glass beads (Voorhees et al., 1966).

B. Radiation Methods

Several users have designed and built gamma-ray gauges to suit specific purposes. A selection of both backscatter and transmission gauges that are commercially available is given in Table 1.

Table 1 Details of Some Commercially Available Gamma-Ray Gauges

Supplier	Model	Configuration	Detector	Source and strength	Maximum measurement depth (m)	Data recording microprocessor	Comments
ELE Ltd Eastman Way Hemel Hempstead Hertfordshire HP2 7HB, UK	CPN Corp. MC-3 Portaprobe	Transmission (surface detector, single probe) or backscatter (surface source and detector)	Geiger—Müller	^{137}Cs, 10 mCi	0.2 or 0.3	Yes	Incorporates neutron backscatter gauge with source at surface
	CPN Corp. Strata gauge	Transmission (twin probe at approx. 300 mm separation)	Geiger—Müller	^{137}Cs, 10 mCi	0.6	Yes	Incorporates neutron backscatter gauge with source in probe
	CPN Corp. 501B Depthprobe	Backscatter (source and detector in single probe)	Geiger—Müller	^{137}Cs, 10 mCi	10.0	Yes	Incorporates neutron backscatter gauge with source in probe
Wykeham-Farrance Weston Road Slough, Berkshire SL1 4HW, UK	Humbolt Mfg. Co. 36530	Transmission (surface detector, single probe) or backscatter (surface source and detector)	Geiger—Müller	^{137}Cs, 10 mCi	0.2 or 0.3	Yes	Incorporates neutron backscatter gauge
Troxler Electronic Laboratories Inc PO Box 12057 North Carolina 27709	3430 Density gauge	Transmission (surface detector, single probe)	Geiger—Müller	^{137}Cs, 8 mCi	0.2 or 0.3	No	Incorporates neutron backscatter gauge with source at surface
Soils Department, SAC, Bush Estate Penicuik, Midlothian EH26 0PH, UK	SCAE density gauge	Transmission (twin probe at 220 mm separation)	Energy discrimination	^{137}Cs, 5 mCi	0.6	No	Detailed specification to order

1. Sample Preparation

For any type of nuclear density gauge it is important that the sample be always presented to the gauge in a consistent manner. In laboratory transmission gauges, each sample is placed in turn in a container located between the source and the detector. In field transmission gauges, either a single access hole or two parallel access holes must be made in the soil; equipment for this purpose is shown in Fig. 5. Access holes can be formed by hammering solid spikes through an alignment jig lying on the soil surface (Soane et al., 1971). Although a certain amount of disturbance takes place during this operation, this can be considered to be compensated for by providing access holes in calibration samples in exactly the same way, provided the soil is not fractured during spiking.

The provision of access holes by augering minimizes soil disturbance, but the procedure can be more difficult, particularly where parallel holes are required. Augering has several other advantages however, namely that the removed soil can be used for water content determination, calibration samples can be smaller, and it is easier to instal liner tubes in the access holes where they are required (Soane,

Fig. 5 Equipment used to provide two parallel access holes for transmission gamma-ray gauges either by hammering spikes through an alignment jig (left) or by augering (right). A liner tube has been inserted in the right-hand augered hole.

1968). In loose soil conditions, liners should be inserted progressively during augering to prevent soil entering the access hole.

2. *Calibration*

Except for laboratory gauges with high levels of collimation, for which it is possible to use theoretical values for mass attenuation coefficients, some form of empirical calibration is required. Some gauge manufacturers supply specimen calibrations with gauges, but most workers involved with agricultural soils have found it desirable to recalibrate their gauges. Some manufacturers also supply standard density blocks for calibration, which can be useful for periodic checks on calibration stability but are unlikely to be suitable for a full calibration, because both the mode of probe access to such blocks and their composition can be different from that in the field.

Calibrations with field soils can be made in situ either by comparison with a direct method, normally core sampling, or by repacking field soils into bins and determining their density independently from measurements of sample mass and volume (Henshall and Campbell, 1983; Soane et al., 1971). Both types of calibration are slow, and each has its merits. Comparison with core sampling has the advantage that soils of field structure are used, but core sampling, especially at depth, is time-consuming and unreliable. Such comparisons usually assume, without justification, that core sampling results are the more accurate. Unless minimal disturbance is ensured in the gauge method by using auger access, sampling at different positions for the two methods is required, with the resulting complication of accounting for the variability of field soils.

Calibration with remolded field samples packed in bins simplifies the direct measurement of bulk density (Henshall and Campbell, 1983; van Bavel et al., 1985), but where gauge access is by spiking, samples must be sufficiently large to ensure that the walls of the bin do not influence the soil disturbance during spiking, and tests have to be restricted to a single access position to avoid interaction between multiple spikings. Where insertion is by augering and only unattenuated photons are counted, samples that are only marginally larger than the probe spacing can be used, and multiple access positions will compensate for inconsistencies in the packing of the sample. It should be remembered that the zone of influence extends horizontally as well as vertically. Generally, samples must be carefully prepared in thin layers to achieve uniform packing (Fig. 6).

In calibration, the precision of both measurements made with the gauge and direct methods should be similar, and there is no advantage in making excessively long, precise measurements of count rate. If soil variability is high, short, less precise measurements should be made with the gauge, and the time saved spent in further sampling with both methods. In general, test counts normally

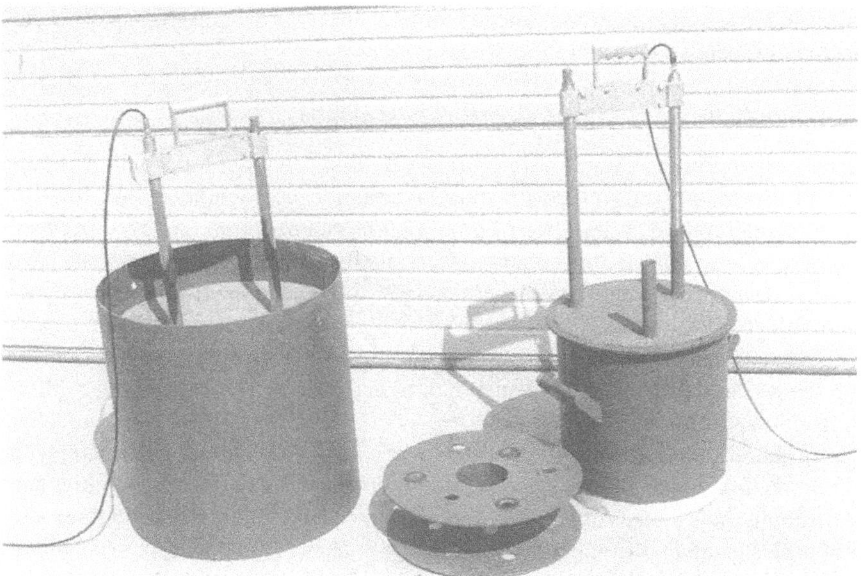

Fig. 6 Calibration samples for gamma-ray gauges in which access is provided by spiked holes (left) and by augered and lined holes (right). The alignment jig for the augered holes is also shown.

comprise between 2,000 and 10,000 counts, giving levels of precision of between 2.5 and 1%.

Standard reference counts should be made for each calibration sample, using the same reference plate as used with test measurements. Since the reference count is related to all measurements in a sample, and any errors could have a significant effect on the calibration, it is usually made over a longer period than that for test counts.

Finally, it should be stressed that it is essential that calibration samples be tested in exactly the same manner as the experimental samples to which the calibration is applied. This is particularly important with respect to the method of providing probe access.

3. *Experimental Considerations*

As with calibration, the decision between making a few highly detailed measurements or more replication in less detail is determined by sample variability. Since field soils tend to display large random variations in soil properties, it is generally more worthwhile to replicate measurements than to make very precise measurements in a few sampling positions. Typically, more than 5000 counts per measurement cannot be justified, and between 2000 and 3000 counts is adequate (Soane, 1976). In replicated field experiments, the number of measurement positions per treatment is typically two or three, giving coefficients of variation of about 10%, and is probably a good compromise (Soane et al.,1971). However, measurements of soil properties in sampling positions that are close together generally tend to be more similar than those made further apart (Burgess and Webster, 1980). When such spatial dependence is allowed for, the number of measurements required for a given level of precision can sometimes be reduced (McBratney and Webster, 1983).

Stones may present difficulties either by preventing the provision of access holes to the full depth or by deflecting the probes of a twin probe system and so altering the source/detector separation. Where access holes cannot be made, a new sampling position has to be tested instead, with the result that the mean bulk density may be biased in favor of those samples where stones lie between, rather than at, the positions of the two probes. Thus the bulk density of stony soil may be overestimated. The effect will depend on both the number and the size distribution of stones but appears not to have been investigated. The problem of possible probe deflection by stones can be overcome only by measuring, and correcting for, the actual source/detector separation at each depth (Soane, 1968). The statistical problems arising from soil variability and from stones have been examined in relation to the measurement of soil cone resistance; some of this information is relevant to the measurement of bulk density (O'Sullivan et al., 1987).

4. Operational Safety

All nuclear density gauges are potential health hazards. In the U.K., it is a legal requirement for radioactive sources to be registered with the Health and Safety Executive (Anon., 1985). A similar situation exists in the USA. In the U.K., a "System of Work" which describes an approved safe operating procedure for the gauge is normally incorporated in the registration. Most manufacturers supply an example of such a document with the gauge, but for nonstandard gauges or procedures, a system that minimizes the exposure of the operator to radiation must be devised, documented, and approved.

For field gauges, a safe operating procedure is one that ensures that the source is exposed for the minimum time possible. This can be achieved by lowering the probes through the base of the gauge so that the source is always shielded either by its shield or the soil, and by ensuring that, when not in use, the source is securely located in its shield. For laboratory gauges, interlocking devices on the shield are required to prevent accidental exposure, since much larger sources are generally used than in the field.

C. Comparison of Methods

The difficulty in extracting soil samples from the field without disturbance to both the sample and the wall of the remaining hole means that none of the direct methods of measuring bulk density can be relied upon to be totally accurate. Erbach (1983) described the sand replacement method as "good for use in gravelly soil," but for most soils the core sampling method is generally taken to be the standard method, despite its many forms of error. Raper and Erbach (1985) stated that "it is disturbing that a method with this many inherent errors is referred to as a standard." Many workers, when finding that density measurements recorded by gamma-ray gauges do not agree with direct measurements, have been inclined to dismiss the gamma gauge as inaccurate or unsatisfactory.

Several comparisons between direct and gamma-ray measurements have found general agreement between the two methods (King and Parsons, 1959; Blake, 1965; Soane et al., 1971; Gameda et al., 1983; Minaei et al., 1984; Schafer et al., 1984), with discrepancies in some soil types, which are normally attributed to inaccuracies in the gamma gauge. King and Parsons (1959) found reasonable agreement ($\pm 3\%$) between a single-probe gamma gauge and the sand replacement method in sandy and clay soils but unacceptably large differences of 11% in gravelly soils. Several explanations of the discrepancy were given, such as variation in gamma-ray absorption according to particle size, but no consideration was given to the more probable dependence of the sand replacement test on particle size (DSIR, 1964).

Gameda et al. (1983) compared single and twin-probe gamma gauges with the core sampling method on three soils to a depth of 0.6 m. They found a good correlation between the gamma and core measurements on sandy and clay soils but not on loamy soil. The poor correlation in loamy soil was attributed to the presence of stones in the soil and its high iron content. The data as presented suggests that the loam was very variable, perhaps due to stones, but a significant effect due to iron content seems unlikely. Although a good correlation was found between core density and the density values indicated by the factory calibrations for the gamma gauges, the test values for the gauges were significantly different from each other, confirming the need for calibration of gamma gauges in field soils.

Soane et al. (1971) found that, on three contrasting mineral soils, density measurements from a twin-probe gamma gauge agreed with corresponding core sample measurements within 3%, but that there was a discrepancy of 0.06 Mg/m^3 on low-density (0.28 Mg/m^3) organic peat samples. The coefficients of variation for both methods were found to be similar for a given soil. The gamma gauge was found to be faster in operation by a factor of 2 or 3; it also had the advantage that measurements could be made at close depth intervals in a soil profile with little disturbance. A single calibration relationship was applicable to all the soils tested. In a review of gamma-ray transmission systems, Soane (1976) reported that the accuracy of different laboratory measurement systems ranged from ±1.2% to ±3%.

A useful indication of the potential accuracy of gamma gauges was carried out by Schafer et al. (1984). Over a five-year period, core samples were removed from the field and tested in an empirically calibrated laboratory gamma gauge after direct measurement of their bulk density. For 80% of the 236 cores tested, the discrepancy was less than 1%, and the results for only two samples disagreed by more than 2%.

The gamma-ray transmission method is therefore potentially at least equal in accuracy to any of the direct methods of density determination and is simpler and quicker to use, especially where measurements at depth are required. The twin-probe gamma gauge is more accurate than the single-probe version, allowing much more detailed information on soil layers to be acquired, provided that the parallel access holes are carefully prepared or nonparallelism is allowed for.

The high cost of gamma-ray gauges compared with equipment for direct measurement and the requirement for compliance with radiation safety regulations (Anon., 1985) offsets the advantages of the gamma gauges where few measurements are required. In such cases, the core sampling method has proved to be the most popular alternative except in gravelly soils or where looseness of the soil prevents its retention within the core, in which case the sand replacement method is the best option.

Some comparisons have been made of the various direct methods available, in terms both of their practical advantages and disadvantages and of the errors associated with them (DSIR, 1964; Cernica, 1980). It might be expected that the clod method would give bulk densities greater than other measures of bulk density that include interclod spaces. Generally, however, core sampling and the clod method give similar results, while the sand replacement values are about 2% lower (DSIR, 1964). The rubber balloon method has proved relatively unreliable, with systematic errors of nearly 5% being found, in comparison with nearly 3% for the sand replacement method or 0.5% when sand volume rather than mass is measured (Cernica, 1980).

All methods of bulk density measurement may be hindered by the presence of stones, which may also create complications in the interpretation of treatment means from field experiments (O'Sullivan et al., 1987). Keisling and Smittle (1981) made measurements of the bulk density at which root growth was inhibited in a soil with between 5.8 and 11% of stones of 3 to 13 mm size. They found that bulk densities were between 0.097 and 0.12 Mg m^{-3} lower when the presence of stones was allowed for and that the corrected values corresponded to the limiting values for root growth in stone-free soil.

IV. APPLICATIONS OF BULK DENSITY MEASUREMENTS

Many of the direct methods of bulk density measurement have been widely used for civil engineering work, which generally results in little variation of bulk density within any sample. Here, the direct methods can be entirely appropriate. However, the limitations of all methods other than transmission methods employing energy discrimination can be very important in agricultural soils, in which large variations in bulk density can occur over very short horizontal and, especially, vertical distances as a result of the localized effects of tillage and traffic. Thus thin layers of soil of high bulk density, which may be very important in relation to such matters as root penetration or water infiltration, may pass undetected when making a mean bulk density measurement with such methods. Some examples of the use of bulk density methods will now be considered.

A. Soil Compaction by Wheels

Soil compaction by a wheel may be assessed by measuring bulk density at regular depth increments below the soil surface before the wheel runs over the soil and then making similar measurements under the center line of the wheel rut produced. The measurements may then be graphed as the variation of dry bulk density with depth both before and after the passage of the wheel. Figure 7 shows the results of such measurements made after the passage of an unladen tractor. Measurements were made with gamma-ray transmission equipment both with and without energy discrimination, and the data confirm that different results are produced by the two methods (Henshall, 1980). The depth interval between measurements can be varied so that measurements are more intensive in the region of any feature of interest, such as the top of a plow pan, but an interval of about 30 mm has been found to be an appropriate compromise for general purposes (Campbell and Dickson, 1984; Campbell and Henshall, 1984; Campbell et al., 1986).

Presentation of data at fixed depths in relation to the undisturbed soil surface as shown in Fig. 7 is satisfactory for many purposes, but difficulties can arise when comparisons are made of the effects of two or more vehicles, especially when they produce wheel ruts of different depths. Henshall and Smith (1989) developed a procedure in which the bulk density measurements are used to trace vertical movement of the soil mass arising from compaction. Consequently, comparisons between treatments can be made on soil elements that originated from the same depth in the undisturbed soil profile, irrespective of their depths in the compacted profiles (Fig. 8).

A further limitation to the value of the information provided by Fig. 7 is that it ignores the lateral distribution of compaction on either side of the center line of the wheel rut, which is of particular interest when soil compaction is being

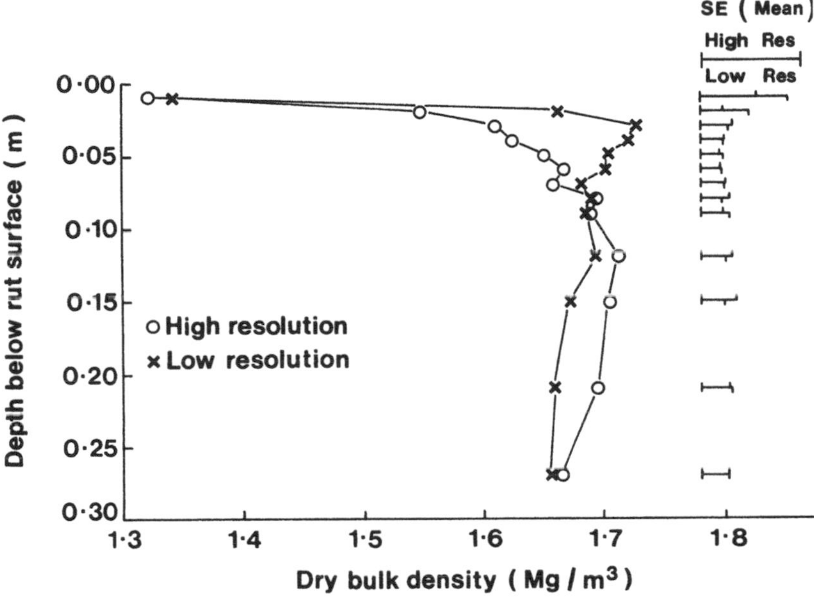

Fig. 7 Variation of dry bulk density with depth below a wheel rut produced in a sandy loam by an unladen tractor. Measurements were made with gamma-ray transmission equipment both with (high resolution) and without energy discrimination. (Based on data from Henshall and Campbell, 1983.)

studied in relation to crop growth. Such additional information can be obtained by making a series of measurements along a transect at right angles to the wheel rut. With such an arrangement, sampling positions can usually be no closer than about 100 mm before probe access disturbs adjacent positions (Dickson and Smith, 1986), but this limitation can be overcome with a two-dimensional scanning gamma-ray system, making measurements on a regular grid at right angles to the wheel rut. However, this requires the formation of carefully cut trenches on each side of the soil sample, which is time-consuming (Fig. 9). Nevertheless, the method can provide a detailed description of both the vertical and horizontal variation in bulk density across the wheel track (Fig. 10). Soane (1973) used an automated version of the method that employed energy discrimination and in which the source and detector probes were mounted on an electrically powered carriage. Readings were made on a 20 × 20 mm grid. The test sample was 1.4 m long at right angles to the wheel track, 0.3 m deep, and 0.3 m thick. This technique was used on a simulated seedbed in a sandy loam, to compare the distribution of compaction produced by a conventional tractor, the same tractor with the addition of cage wheels, and a crawler tractor.

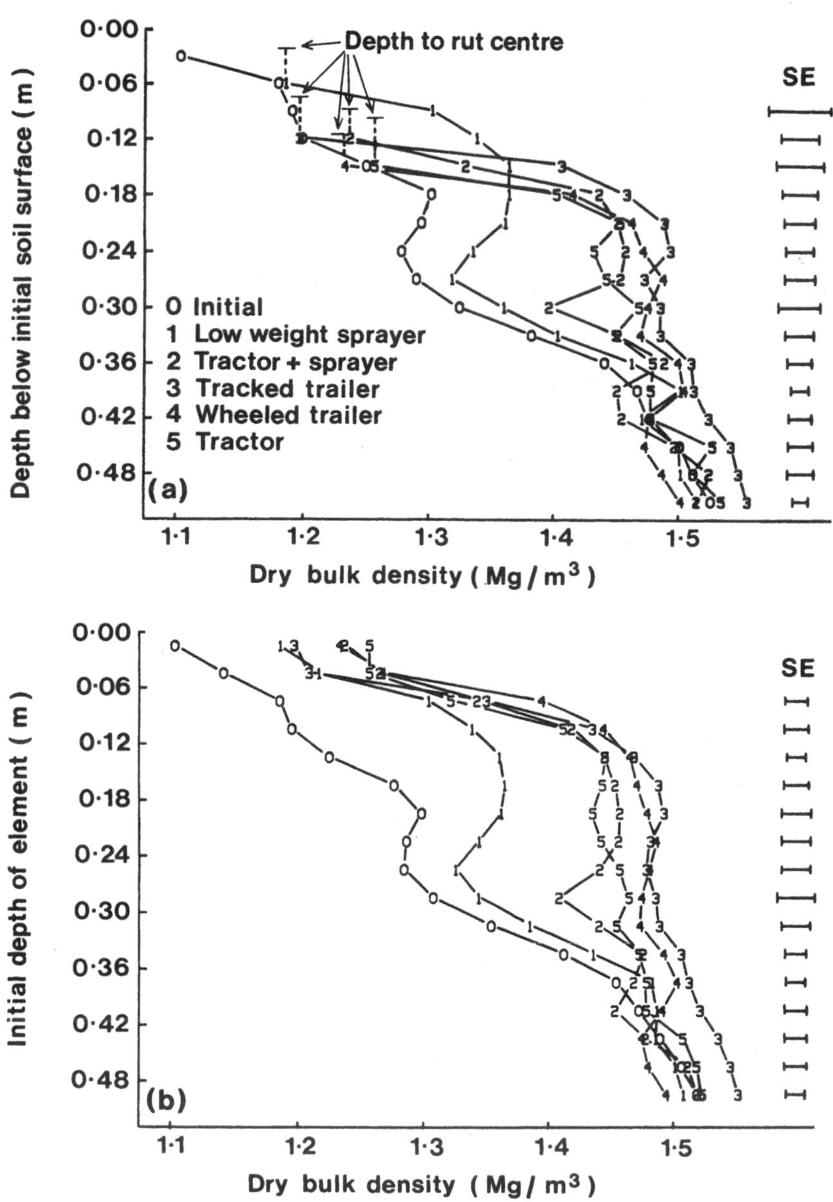

Fig. 8 Variation, for five treatments, of dry bulk density with (a) depth below the initial soil surface and (b) initial depth of each soil element. (Based on data from Henshall and Smith, 1989.)

Fig. 9 Gamma-ray transmission system designed and constructed at former Scottish Centre of Agricultural Engineering, which provides a two-dimensional scan of an undisturbed block of soil at right angles to a wheel rut.

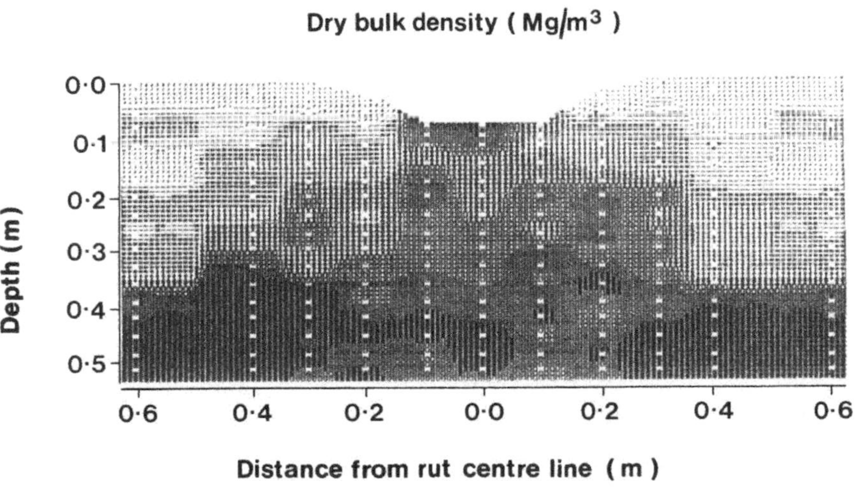

Fig. 10 The variation in bulk density produced in a sandy loam by a tyre with an inflation pressure of 84 kPa and a load of 2.47 t as measured with a scanning gamma-ray transmission system that employed energy discrimination. (D. J. Campbell and J. K. Henshall, unpublished data.)

B. Soil Tillage

There have been many attempts to determine the limiting bulk density for root growth for a variety of crops in a range of soils (Veihmeyer and Hendrickson, 1948; Zimmerman and Kardos, 1961; Edwards et al., 1964). Although good relationships have been found in the laboratory, such relationships are always much poorer in the field because of soil variability. Veihmeyer and Hendrickson (1948), who found that the limiting bulk density for the growth of sunflower roots in the laboratory ranged from 1.46 to 1.90 Mg m^{-3} depending on soil texture, demonstrated that the restriction to root growth was high bulk density and small pore size. Their conclusion was consistent with that of Wiersum (1957) who proved that the tip of a growing root will enter a pore only if that pore is larger than the root tip diameter. Wiersum (1957) also concluded that, for satisfactory root growth, the pore structure must not be too rigid, implying that both soil bulk density and soil strength are important in this context. Thus it is easily seen that with the inherent variability of soils in the field, any effect of bulk density on root growth will interact with the effects of soil strength, water status, aeration, and structure.

Many researchers have felt it worthwhile to measure soil bulk density in tillage experiments so that air-filled porosities may be derived. Where measurements of water release characteristics or permeability to air or water are required, the soil cores required for such measurements are often used for bulk density determination (Douglas et al., 1986). Typically, two or three cores per plot at each depth are considered sufficient in replicated experiments.

Bulk density measurements by the gamma-ray method are often used to measure the degree of loosening provided by tillage treatments or the extent of compaction following direct drilling (Pidegon and Soane, 1977; Ball et al., 1985). However, high soil variability both before and after the treatments can demand large numbers of measurements, if treatment differences are to be detected.

Soane (1970) used the scanning gamma-ray method in unreplicated measurements to illustrate the distribution of compacted soil in moldboard plowed land and in potato ridges and furrows. The two-dimensional scan possible with a cone penetrometer (see Chap. 10) (Bengough et al., 2000) is a more useful method of detecting compacted soil in such circumstances than is a scan of bulk density, because of the vastly greater speed of the cone penetrometer test, which in turn allows the replication required to overcome problems of soil variability.

Hand-held gamma-ray transmission equipment has been used successfully in a long-term experiment to compare three alternative plowing treatments with direct drilling on two different soils (Holmes and Lockhart, 1970; Soane et al., 1970; Pidgeon and Soane, 1977; O'Sullivan, 1985). Such measurements were made in two positions per plot in each of the four replications of the four treatments (Fig. 11). The work showed that the direct-drilled soil reached a bulk den-

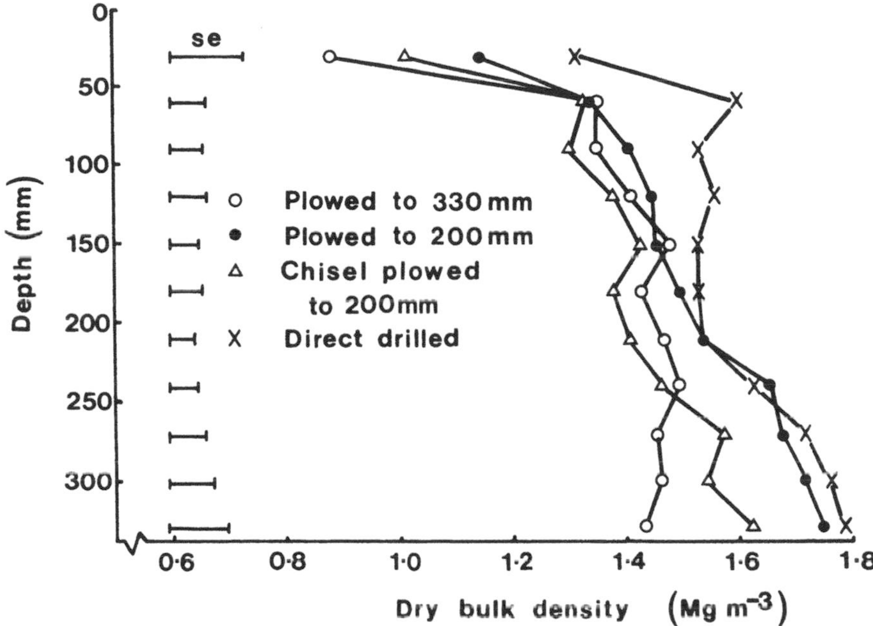

Fig. 11 Variation of soil bulk density with depth in a loam for four tillage treatments in the middle of a spring barley growing season. (Based on data from O'Sullivan, 1985.)

sity that was in equilibrium with the applied traffic after three years. Most of the soil that was loosened by the three plowing treatments had compacted to its original bulk density by the end of the growing season. Although each soil was compacted to a different bulk density in response to traffic, measurements of cone resistance showed no difference between soils. Thus although cone resistance depended only on tillage and traffic, bulk density was also influenced by soil compactibility and hence texture and water status. These results emphasize the potential dangers of assessing soil compaction in terms of changes in only one soil physical property. In this instance, measurements of cone resistance in isolation would not have detected the difference in response to the tillage treatments of the two soil textures (Pidgeon and Soane, 1977).

In addition to measurements of the density of the bulk soil, it is sometimes appropriate to measure the bulk density of the aggregates or clods within the soil mass. For example, in studies of the movement of fluids through bulk soil, both inter- and intra-aggregate porosities may be of interest, since the large inter-aggregate pores dominate fluid movement (Hillel, 1982). The bulk density of soil clods in potato ridges has been found relevant to problems in the harvesting of potatoes (Campbell, 1976). In measuring clod or aggregate bulk density, problems of variability associated with water status, bulk density gradients, and the range of clod sizes involved usually necessitate measurements on 50 to 100 clods per plot in replicated experiments. In such circumstances the older clod method (DSIR, 1964), in which the clod is coated in wax and weighed in air and in water, is unacceptably slow, and even the more recent flotation method (Campbell, 1973), which is ten times quicker, is still tedious to use (see Sec. III.A.4).

C. Soil Erosion

In a review of soil erosion in the U.K., Speirs and Frost (1987) noted that, although soil compaction had often been suggested as a cause of erosion, several cases had occurred where soil had eroded until a compact pan was reached that resisted further erosion. In the U.S.A., Jepsen et al. (1997) found that the rate of erosion from the end face of cores from river sediments decreased linearly with increasing bulk density for a given water flow rate. In contrast, Parker et al. (1995) did not find a direct correlation between bulk density and erodibility in laboratory studies using a 6.1 m long flume. At low bulk densities, ripples and dunes formed causing soil deposition, whereas at higher bulk densities, the soil surface remained flat, causing high water velocities close to the soil bed and hence higher erosion rates.

Both Jepsen et al. (1997) and Parker et al. (1995) determined bulk densities of of samples subjected to erosion by direct measurement of both sample mass and volume. While the method was appropriate for their laboratory studies, any field studies of the role of bulk density in the effect of, for example, tillage on erosion would require replicated measurements by a method with appropriate depth resolution for use in soils with pans or crusts. However, in erosion studies generally, many workers have found a satisfactory compromise in the use of core sampling (Comia et al., 1994; Ebeid et al., 1995; Sharratt, 1996).

Since erosion will not occur in the absence of runoff, soil infiltration rate is important in relation to erosion. Mbagwu (1997) related infiltration to land use and soil pore size distribution. He found that infiltration rate was strongly correlated with bulk density for 18 Nigerian soils when bulk densities were determined from the cores used to measure pore size distributions. Roth (1997) saturated air-dried soil crusts in low-viscosity oil and subsequently found their bulk densities from immersion in water. Systematic errors arose from clay shrinkage in some samples and from the surface of the uncrusted portion of some thicker crust samples being not well defined. Correction procedures were devised for both sources of error.

D. Soil Compaction Models

Empirical models relate inputs such as wheel load to outputs such as bulk density. On the other hand, mechanistic models attempt also to simulate the process relating inputs and outputs. O'Sullivan and Simota (1995) reviewed soil compaction models and their value in relation to environmental impact models. They considered that while empirical models are useful for integrating information for a specific site, mechanistic models are more useful for making predictions about unknown sites. However, mechanistic models usually have some empirical features.

Model inputs and outputs in terms of bulk density must be measured with the same considerations given to the selection of a measurement method as in any other application. For example, Smith (1985) used a gamma-ray transmission method to measure bulk density in the field at 30 mm depth intervals down to 0.51 m below a wheel track. Results generally compared favorably with those predicted by his mechanistic model but underestimated the compaction in loose soil overlying a dense layer, a situation commonly encountered in agricultural soils. Such underestimation is associated with the analytical method used to model the propagation of stresses through the soil under the applied wheel. A similar limitation applies when a finite element method is used to model stress propagation (Raper and Erbach, 1990). Because of these limitations, mechanistic models have more relevance to the comparison of compaction caused by different wheels (Smith, 1985) or to studies of the relative importance of soil or wheel characteristics to compaction (Kirby, 1989) than to the precise prediction of soil bulk density changes.

Further development of soil compaction models is little hindered by existing methods to measure soil bulk density. In contrast, areas in which progress is required include the use of stochastic models to take account of the high spatial variability in field soils (O'Sullivan and Simota, 1995) and the importance to compaction of both the shear forces produced by driven wheels (Kirby, 1989) and repeated wheel passes (Smith, 1985; Jakobsen and Dexter, 1989).

V. SUMMARY

Both direct and indirect measurements of soil bulk density are described. In the direct methods, the sample mass and volume are determined. In the indirect methods, the effect of the sample on gamma radiation is measured and related to bulk density by empirical calibration. The theory of the interaction of atoms of soil with gamma photons is discussed in relation to photon energy and intensity together with soil chemical composition and bulk density. Basically, photons from a gamma source are absorbed or scattered during interaction with the electrons of the soil atoms such that the number of photons incident on the detector in a given time is related to the bulk density of the soil sample.

Backscatter gauges detect only scattered photons, while transmission gauges are designed to detect unattenuated photons, provided the detector employs energy discrimination. Details of the construction of gamma gauges are given, together with calibration procedures and an assessment of the need for accurate water content measurements.

Both direct and indirect methods are detailed. Direct methods discussed include the core sampling, rubber balloon, sand replacement, and clod methods. Indirect methods include both the backscatter and transmission gamma methods which are described in relation to problems associated with sample preparation, calibration, operational safety, soil variability, and stones. Comparisons of methods are reviewed. Although there is general agreement between the results of direct and indirect methods, the latter tend to be more accurate, especially the gamma-ray transmission method, which is particularly suited to the layered soils usually found in agriculture, forestry, and the natural environment. Examples are given of the use of various methods to detect changes in bulk density associated with soil compaction by wheels, soil loosening by tillage implements and soil erosion, and in the development and application of soil compaction models.

REFERENCES

Abrol, I. P., and J. P. Palta. 1968. Bulk density determination of soil clod using rubber solution as a coating material. *Soil Sci.* 106:465–468.

Anon. 1985. *The Ionising Radiations Regulations 1985*. London: HMSO.

Ball, B. C., M. F. O'Sullivan, and R. W. Lang. 1985. Cultivation and nitrogen requirement for winter barley as assessed from a reduced-tillage experiment on a brown forest soil. *Soil Till. Res.* 6:95–109.

Baver, L. D., W. H. Gardener, and W. R. Gardener. 1972. *Soil Physics*. New York: John Wiley.

Bengough, G., D. J. Campbell, and M. F. O'Sullivan. 1998. Penetrometer techniques in relation to soil compaction and root growth. In: *Soil Analysis: Physical Methods* (K. A. Smith and C. Mullins, eds.), 2d ed. New York: Marcel Dekker.

Blake, G. R. 1965. Bulk density. In: *Methods of Soil Analysis, Part 1* (C. A. Black, ed. in chief). Madison, WI: Am. Soc. Agron., pp. 374–390.

Brasher, B. R., D. P. Franzmeir, V. Valassis, and S. E. Davidson. 1966. Use of Saran resin to coat natural soil clods for bulk density and moisture retention measurements. *Soil Sci.* 101:108.

Buchele, W. F. 1961. A power sampler of undisturbed soils. *Trans. Am. Soc. Agric. Eng.* 4: 185–187, 191.

Burgess, T. M., and R. Webster. 1980. Optimal interpolation and isarithmic mapping of soil properties: I. The semi-variogram and punctual kriging. *J. Soil Sci.* 31: 315–331.

Campbell, D. J. 1973. A flotation method for the rapid measurement of the wet bulk density of soil clods. *J. Soil Sci.* 24:239–243.

Campbell, D. J. 1976. The occurrence and prediction of clods in potato ridges in relation to soil physical properties. *J. Soil Sci.* 27:1–9.

Campbell, D. J., and J. W. Dickson. 1984. Effect of four alternative front tyres on seedbed compaction by a tractor fitted with a rear wheel designed to minimise compaction. *J. Agric. Eng. Res.* 29:83–91.

Campbell, D. J., and J. K. Henshall. 1984. Two new instruments to measure the strength and bulk density of soil in situ. *Proc. 6th Int. Conf. Mechanisation of Field Expts.*, Dublin, pp. 338–344.

Campbell, D. J., J. W. Dickson, B. C. Ball, and R. Hunter. 1986. Controlled seedbed traffic after ploughing or direct drilling under winter barley in Scotland, 1980–1984. *Soil Till. Res.* 8:3–28.

Carlton, P. F. 1961. Application of nuclear soil meters to compaction control for airfield pavement construction. In: Symposium on Nuclear Methods of Measuring Soil Density and Moisture, *Am. Soc. Testing Mater.*, Spec. Tech. Publ., 293:27–35.

Cernica, J. N. 1980. Proposed new method for the determination of density of soil in place. *Geotech. Testing J.* 3:120–123.

Chepil, W. S. 1950. Methods of estimating apparent density of discrete soil grains and aggregates. *Soil Sci.* 70:351–362.

Comia, R. A., E. P. Paningbatan, and I. Hakansson. 1994. Erosion and crop yield response to soil—conditions under alley cropping systems in the Philippines. *Soil Till. Res.* 31:249–261.

Constantini, A. 1995. Soil sampling bulk density in the coastal lowlands of south-east Queensland. *Aust. J. Soil Res.* 33:11–18.

Coppola, M., and P. Reiniger. 1974. Influence of the chemical composition on the gamma-ray attenuation by soils. *Soil Sci.* 117:331–335.

Dickson, J. W., and D. L. O. Smith. 1986. Compaction of a sandy loam by a single wheel supporting one of two masses each at two ground pressures. *Scot. Inst. Agric. Eng.* Unpubl. Dep. Note SIN/479.

Douglas, J. T., M. G. Jarvis, K. R. Howse, and M. J. Goss. 1986. Structure of a silty soil in relation to management. *J. Soil Sci.* 37:137–151.

DSIR (Department of Scientific and Industrial Research). 1964. Soil mechanics for road engineers. London: HMSO.

Ebeid, M. M., R. Lal, G. F. Hall, and E. Miller. 1995. Erosion effects on soil properties and soybean yield of a Miamian soil in western Ohio in a season with below normal rainfall. *Soil Tech.* 8:97–108.

Edwards, W. M., J. B. Fehrenbacher, and J. P. Vavra. 1964. The effect of discrete ped density on corn root penetration in a planosol. *Soil Sci. Soc. Am. Proc.* 28:560–564.

Erbach, D. C. 1983. Measurement of soil moisture and bulk density. Am. Soc. Agric. Eng., Paper No. 83-1553.

Ferraz, E. S. B., and R. S. Mansell. 1979. Determining water content and bulk density of soil by gamma-ray attenuation methods. Univ. of Florida Bull. No. 807.

Freitag, D. R. 1971. Methods of measuring soil compaction. In: *Compaction of Agricultural Soils* (K. K. Barnes, W. M. Carleton, H. M. Taylor, R. I. Throckmorton, and G. E. Vanden Berg, eds.). St. Joseph, MI: Am. Soc. Agric. Eng., pp. 47–103.

Gameda, S., G. S. V. Raghavan, E. McKyes, and R. Thériault. 1983. Single and dual probes for soil density measurement, Paper No. 83-1550. St. Joseph, MI: Am. Soc. Agric. Eng.

Gill, W. R. 1959. Soil bulk density changes due to moisture changes in soil. *Trans. Am. Soc. Agric. Eng.* 2:104–105.

Greacen, E. L., D. A. Farrel, and J. A. Forrest. 1967. Measurement of density patterns in soil. *J. Agric. Eng. Res.* 12:311–313.

Henshall, J. K. 1980. The calibration and field performance of a high resolution gamma-ray transmission system for measuring soil bulk density in situ. *Scot. Inst. Agric. Eng.*, Unpubl. Dep. Note SIN/299.

Henshall, J. K., and D. J. Campbell. 1983. The calibration of a high resolution gamma-ray transmission system for measuring soil bulk density and an assessment of its field performance. *J. Soil Sci.* 34:453–463.

Henshall, J. K., and D. L. O. Smith. 1989. An improved method for presenting comparisons of soil compaction effects below wheel ruts. *J. Agric. Eng. Res.* 42:1–13.

Herkelrath, W. N., and E. E. Miller. 1976. High performance gamma system for soil columns, *Soil Sci. Soc. Am. J.* 40:331–332.

Hillel, D. 1982. *Introduction to Soil Physics.* New York: Academic Press.

Holmes, J. C., and D. A. S. Lockhart. 1970. Cultivations in relation to continuous barley growing, I. Crop growth and development. *Proc. Int. Soil Tillage Conf.*, Silsoe, U.K., pp. 46–57.

Hvorslev, M. J. 1949. Subsurface exploration and sampling of soils for civil engineering purposes. Rep. on Res. Project of Am. Soc. Civ. Engrs., U.S. Army Engrs. Waterways Expt. Sta., Vicksburg, MS.

Jakobsen, B. F., A. R. Dexter, and I. Hakansson. 1989. Simulation of the response of cereal crops to soil compaction. *Swed. J. Agric. Res.* 19:203–212.

Jamison, V. C., H. H. Weaver, and I. F. Reed. 1950. A hammer-driven soil core sampler. *Soil Sci.* 69:487–496.

Jepsen, R., J. Roberts, and W. Lick. 1997. Effects of bulk density on sediment erosion rates, *Water, Air and Soil Pollution* 99:21–31.

Keisling, T. C., and D. A. Smittle. 1981. Soil bulk density corrections for providing a better relationship with root growth in gravelly soil. *Commun. Soil Sci. Plant Anal.* 12: 91–96.

King, F. G., and A. W. Parsons. 1959. Portable radioactive equipment for measuring soil density. Road Res. Lab., U.K., Res. Note RN/3628/FGK.AWP.

Kirby, J. M. 1998. Shear damage beneath agricultural tyres: A theoretical study. *J. Agric. Eng. Res.* 44:217–230.

Lutz, J. F. 1947. Apparatus for collecting undisturbed soil samples. *Soil Sci.* 64:399–401.

Mbagwu, J. S. C. 1997. Quasi-steady infiltration rates of highly permeable tropical moist savannah soils in relation to land use and pore size distribution. *Soil Tech.* 11: 185–195.

McBratney, A. B., and R. Webster. 1983. How many observations are needed for regional estimation of soil properties? *Soil Sci.* 135:177–183.

Minaei, K., J. V. Perumpral, J. A. Burger, and P. D. Ayers. 1984. Soil bulk density by core and densitometer procedures. Am. Soc. Agric. Eng. Paper No. 84-1041.

O'Sullivan, M. F. 1985. Soil responses to reduced cultivations and direct drilling for continuous barley at South Road 1979–1982. Scot. Inst. Agric. Eng., unpubl. dep. note SIN/430.

O'Sullivan, M. F., and C. Simota. 1995. Modelling the environmental consequences of soil compaction: A review. *Soil Till. Res.* 35:69–84.

O'Sullivan, M. F., J. W. Dickson, and D. J. Campbell. 1987. Interpretation and presentation of cone resistance data in tillage and traffic studies. *J. Soil Sci.* 38:137–148.

Parker, D. B., T. G. Michel, and J. L. Smith. 1995. Compaction and water velocity effects on soil erosion in shallow flow. *J. Irrig. and Drain. Eng.* 121:170–178.

Pidgeon, J. D., and B. D. Soane. 1977. Effects of tillage and direct drilling on soil properties during the growing season in a long-term barley mono-culture system. *J. Agric. Sci., Camb.* 88:431–442.

Raper, R. L., and D. C. Erbach. 1985. Accurate bulk density measurements using a core sampler. Am. Soc. Agric. Eng. Paper No. 85-1542.

Raper, R. L., and D. C. Erbach. 1990. Prediction of soil stresses using the finite element method. *Trans. Am. Soc. Agric. Eng.* 33:725–730.

Reginato, R. J. 1974. Gamma radiation measurements of bulk density changes in a soil pedon following irrigation. *Soil Sci. Soc. Am. Proc.* 38:24–29.

Roth, C. H. 1997. Bulk density of surface crusts: Depth functions and relationships to texture. *Catena* 29:223–237.

Russell, E. W., and W. Balcerek. 1944. The determination of the volume and airspace of soil clods. *J. Agric. Sci., Camb.* 34:123–132.

Schafer, G. J., P. R. Barker, and R. D. Northey. 1984. Density of undisturbed soil cores by gamma-ray attenuation. New Zealand Soil Bureau, Report 67.

Sharratt, B. S. 1996. Tillage and straw management for modifying physical-properties of a sub-arctic soil. *Soil Till. Res.* 38:239–250.

Smith, D. L. O. 1985. Compaction by wheels: A numerical model for agricultural soils. *J. Soil Sci.* 36:621–632.

Soane, B. D. 1968. A gamma-ray transmission method for the measurement of soil density in field tillage studies. *J. Agric. Eng. Res.* 13:340–349.

Soane, B. D. 1970. The effects of traffic and implements on soil compaction. *J. Proc. Inst. Agric. Engrs.* 25:115–126.

Soane, B. D. 1973. Techniques for measuring changes in the packing state and cone resistance of soil after the passage of wheels and tracks. *J. Soil Sci.* 24:311–323.

Soane, B. D. 1976. Gamma-ray transmission systems for the in situ measurement of soil packing state. In: Report for 1974–76, Scottish Inst. Agric. Eng., pp. 59–86.

Soane, B. D., D. J. Campbell, and S. M. Herkes. 1970. Cultivations in relation to continuous barley growing, II. Soil physical conditions. *Proc. Int. Soil Tillage Conf.*, Silsoe, U.K., pp. 58–76.

Soane, B. D., D. J. Campbell, and S. M. Herkes. 1971. Hand-held gamma-ray transmission equipment for the measurement of bulk density of field soils. *J. Agric. Eng. Res.* 16: 146–156.

Speirs, R. B., and C. A. Frost. 1987. Soil water erosion on arable land in the United Kingdom. *Res. and Dev. in Agric.* 4:1–11.

Van Bavel, C. H. M., R. J. Lascano, and J. M. Baker. 1985. Calibrating two-probe, gamma-gauge densitometers. *Soil Sci.* 140:393–395.

Veihmeyer, F. J. 1929. An improved soil-sampling tube. *Soil Sci.* 27:147–152.

Veihmeyer, F. J., and A. M. Hendrickson. 1948. Soil density and root penetration. *Soil Sci.* 65:487–493.

Voorhees, W. B., R. R. Allmaras, and W. E. Larson. 1966. Porosity of surface soil aggregates at various moisture contents. *Soil Sci. Soc. Am. Proc.* 30:163–167.

Wiersum, L. K. 1957. The relationship of the size and structural rigidity of pores to their penetration by roots. *Plant Soil* 9:75–85.

Zimmerman, R. P., and L. T. Kardos. 1961. Effect of bulk density on root growth. *Soil Sci.* 91:280–288.

9
Liquid and Plastic Limits

Donald J. Campbell
Scottish Agricultural College, Edinburgh, Scotland

I. INTRODUCTION

Plasticity is the property that allows a soil to be deformed without cracking in response to an applied stress. A soil may exhibit plasticity, and hence be remolded, over a range of water contents, first quantified by the Swedish scientist Atterberg (1911, 1912). Above this range, the soil behaves as a liquid, while below it, it behaves as a brittle solid and eventually fractures in response to increasing applied stress. The upper limit of plasticity, known as the liquid limit, is at the water content at which a small slope, forming part of a groove in a sample of the soil, just collapses under the action of a standardized shock force. The corresponding lower limit, the "plastic limit," is at the water content at which a sample of the soil, when rolled into a thread by the palm of the hand, splits and crumbles when the thread diameter reaches 3 mm. By convention, both water contents are expressed gravimetrically on a percentage basis. The numerical difference between the liquid and plastic limits is defined as the plasticity index. Remarkably, these simple empirical tests have been used, essentially unchanged, for nearly a century by soil engineers and soil scientists (BSI, 1990).

Engineers found the limits, particularly the plastic limit, to be useful in the design and control testing of earthworks and soil classification (Dumbleton, 1968) as a result of the development by Casagrande of apparatus to measure the limits (Casagrande, 1932). Although his apparatus was based on that of Atterberg, Casagrande appreciated the need, where empirical tests were concerned, to specify closely every detail of the test procedure so that both the repeatability of the test by one operator and the reproducibility between operators were optimized (Sherwood, 1970). Consequently, the Casagrande tests became widely adopted as the

Table 1 Relation Between Potato Harvesting Difficulty, as Indicated by the Number and Strength of Clods in Potato Ridges, and Plasticity Index of Soil

(A) Yield of 30–75 mm diameter clods (t/ha)	(B) Crushing resistance of 30–45 mm diameter clods (N)	(A) × (B)	Plasticity index
76.2	73.7	5615	12.8
95.0	17.6	1672	11.2
19.0	65.9	1252	10.3
60.5	40.4	2444	8.8
48.0	38.5	1848	8.1
29.2	26.8	782	6.2
26.8	19.4	519	5.1
1.4	52.2	73	3.6

official standard by engineers in the United Kingdom (BSI, 1990), the United States of America (Sowers et al., 1968), and elsewhere.

Soil scientists have made less use of the Atterberg limits, which do not feature in soil survey or land capability classification systems but have been used mainly as indicators of the likely mechanical behavior of soil (Baver et al., 1972; Archer, 1975; Campbell, 1976a). This has generally been done by establishing simple correlations between the plasticity limits or plasticity index and other properties considered important in determining soil behavior. An example is shown in Table 1. It has been suggested, however, that liquid and plastic limit values would be a useful addition to soil particle size distributions in the classification of soils in the laboratory (Soane et al., 1972). This is particularly relevant as the Atterberg limits are related to the field texture, as determined in the hand, a method often preferred by soil scientists concerned with practical problems of soil workability in the field (MAFF, 1984).

Two further index values can be derived from the Atterberg limits. The liquidity index, *LI*, is related to the percentage gravimetric soil water content, *w%*, the plastic limit, *PL*, and the plasticity index, *PI*, by

$$LI = \frac{w\% - PL}{PI} \tag{1}$$

The activity, *A*, is the ratio of the plasticity index to the percentage by weight of soil particles smaller than 2 μm, *C*, thus

$$A = \frac{PI}{C} \tag{2}$$

The activity of a soil depends on the mineralogy of the clay fraction, the nature of the exchangeable cations, and the concentration of the soil solution.

II. THEORIES OF PLASTICITY

In attempting to explain the mechanism behind the existence of the liquid and plastic limits, two basic approaches have been adopted. Traditionally, soil behavior is considered in terms of the cohesive and adhesive forces developed as a result of the presence of water between the soil particles (Baver et al., 1972). The critical state theory of soil mechanics that is used in the second approach has been detailed by Schofield and Wroth (1968) and is mathematically complicated. However, the basic concepts and their importance have been discussed by Kurtay and Reece (1970).

A. Water Film Theory

Cohesion within a soil mass is due to a variety of interparticle forces (Baver et al., 1972). Bonding forces include Van der Waals forces; electrostatic forces between the negative charges on clay particle surfaces and the positive charges on the particle edges; particle bonding by cationic bridges; cementation effects of substances such as iron oxides, aluminum, and organic matter; and the forces associated with the soil water. Taken together, these forces will determine whether a soil will, when stressed, undergo brittle failure, plastic flow, or viscous flow.

At low water contents, most of the soil water forms annuli around the interparticle contact (Haines, 1925; Norton, 1948; Schwartz, 1952; Kingery and Francl, 1954; Vomocil and Waldron, 1962). These annuli provide a tensile force that increases with decreasing particle size, through this relationship breaks down at higher water contents because the individual annuli of water start to coalesce (Haines, 1925). Just above the plastic limit, the soil becomes saturated, and, in a cohesive soil, the soil water tension and other bonding forces are in equilibrium with the repulsive forces due to the double layer swelling pressure. Nichols (1931) showed that, for laminar clay particles, the interparticle force F was related to the particle radius r, the surface tension of the pore water T, the angle of contact between the liquid and the particle α, and the distance between the particles d, by

$$F = \frac{4k\pi rT \cos \alpha}{d} \tag{3}$$

where k is a constant. He also showed that, for each of three soils, the product of the cohesive force and the water content was a constant at low water contents. At higher water contents, however, the cohesive force decreased rapidly with increasing water content.

Although the existence of a relationship between water content and cohesion, which exhibits a maximum, has been demonstrated experimentally (Nichols, 1932; Campbell et al., 1980), the relation is valid only for dry soils that have been rewetted. When puddled soil is allowed to dry, cohesion increases with decreasing water content and reaches a maximum when the soil becomes dry. This effect probably arises because, in puddled soils, the number of interparticle contacts are maximized, and hence cohesive forces other than those due to soil water are large.

Baver (1930) suggested that when a soil at the plastic limit is stressed, the laminar clay particles, which are each surrounded by a water film and which were previously randomly orientated in the friable state, are rearranged so that they slide over each other. Thus the cohesive forces associated with the tension effects in the water films are overcome, and the soil deforms. When the stress is removed, the particles remain in their new position under the action of the cohesive forces and there is no elastic recovery. The soil has undergone plastic deformation or flow. Before the soil reaches the liquid limit, the water films have completely coalesced, and the soil water tension has greatly decreased. Thus cohesion decreases and the soil is capable of viscous flow. As the water content and particle separation further increase, the liquid limit is reached, and the viscosity of the outermost layers of water is reduced to that of free water, allowing the soil to flow like a liquid (Grim, 1948; Sowers, 1965).

The liquid limit is related to clay content and its surface area for most types of clay mineral. Montmorillonite is an exception in that the liquid limit is controlled essentially by the thickness of the diffuse double layer, thereby giving a linear relation between the liquid limit and the amount of exchangeable sodium ions present (Sridharan et al., 1986).

Although the interparticle forces associated with soil water may not provide a comprehensive explanation of the mechanism of plasticity, it is clear the soil particle sizes, their specific surface, and the nature of the clay minerals are all important. This is consistent with the common experience that, generally, the liquid and plastic limits are both dependent on both the type and the amount of clay in a soil (DSIR, 1964).

B. Critical State Theory

If a relatively loose sample of soil is subjected to a progressively increasing uniaxial (deviatoric) stress while the confining stress (spherical pressure) is kept constant, then the soil volume will decrease. This will occur for both unsaturated soil and soil that is saturated but allowed to drain as it is compressed. Eventually, a point will be reached where the soil can be compressed no further. However, if the deviatoric stress is maintained and the soil continues to distort without any change in volume, then the soil is said to be in the critical state. In terms of the three-dimensional relationship of spherical stress, deviatoric stress, and specific volume, the point describing this critical state is one of the many possible critical state points that together form the critical state line. The critical state line is an extremely important concept in that it allows, within the confines of a single theory, the stress–strain behavior of a soil with any particle size distribution to be explained, be it wet or dry, dense or loose, confined or unconfined.

As the line describes all conditions under which a soil will undergo continuous remolding without a change in volume, it follows that soil being prepared for either the liquid or the plastic limit test must be described by a point on this line. Thus the liquid and plastic limit tests can give more than simple qualitative information about soil behavior.

During the liquid limit test, the soil water content, and hence the specific volume, is adjusted by adding water and remolding the soil until, in effect, the soil has a fixed undrained shear strength determined by the conditions of the test. Because the soil is continuously remolded as water is added, it is in the critical state and under the action of a negative pore water pressure.

When soil is prepared for the plastic limit test, it is continuously remolded and hence once again is in the critical state. However, since the soil is much drier than in the liquid limit test, the pore water pressure (matric potential) is even more negative. This negative pore water pressure acts in the same way as if the soil were subject to an additional externally applied stress and serves to increase the shear strength of the soil. It is reasonable to speculate that the plastic limit should, like the liquid limit, correspond to a state in which the soil has a fixed undrained shear strength. Atkinson and Bransby (1978) reported that the undrained shear strength data obtained for four clay soils by Skempton and Northey (1953) revealed that all four soils had very similar undrained shear strengths at the plastic limit. Perhaps more remarkably, the undrained shear strength of each soil at the plastic limit was almost exactly 100 times the undrained shear strength at the liquid limit.

Knowing the ratio of the shear strengths at the liquid and plastic limits, it is possible to define the slope of the critical state line on a plot of the logarithm of the spherical pressure versus the specific volume in terms of the plasticity index (Schofield and Wroth, 1968; Atkinson and Bransby, 1978). Thus the plasticity index can be used as a direct indicator of soil compressibility.

The description of soil behavior at the liquid and plastic limits offered by critical state theory is, at first sight, quite different from that given by the water film theory and may give the impression that soil water content is irrelevant. However, the water content is important in critical state theory, but only insofar as it affects the pore water pressures.

III. DETERMINATION OF THE LIQUID AND PLASTIC LIMITS

The methods initiated by Atterberg (1911, 1912) and subsequently developed by Casagrande (1932) were adopted by the British Standards Institution and the American Society for Testing and Materials as the standard tests in civil engineering. However, in 1975, a new test for the liquid limit, based on a procedure involving a drop-cone penetrometer, was introduced and is included in the current British Standard (BSI, 1990). The Casagrande tests were retained, but the cone penetrometer method was described as the preferred method for the determination of the liquid limit. Although various other methods of determining the liquid and plastic limits have been suggested, usually, but not always, based on correlation of the limits with other soil rheological properties, by far the most widely used methods are the Casagrande and, to a lesser extent, drop-cone tests.

A. Casagrande Tests

In the Casagrande liquid limit apparatus (BSI, 1990) (Fig. 1), the sample is contained in a cup that is free to pivot about a horizontal hinge and which rests on a rubber base of specified hardness. A rotating cam alternately raises the cup 10 mm above the base and allows it to drop freely onto the base. The test soil is mixed with distilled water to form a homogeneous paste, allowed to stand in an air-tight container for 24 hours and remixed, and then a portion is placed in the cup. The sample is divided in two by drawing a standard grooving tool through the sample at right angles to the hinge. The crank is then turned at two revolutions per second until the two parts of the soil come into contact at the bottom of the groove over a length of 13 mm. The number of blows to the cup required to do this is recorded and the test repeated. If consistent results are obtained, a subsample of the soil is taken from the region of the closed groove for the measurement of water content. More distilled water is added to the test sample and the procedure repeated. This is done several times at different water contents to give a range of results lying between 50 and 10 blows. The linear relation between the water content and the log of the number of blows is plotted, and the percentage water content corresponding to 25 blows is recorded, to the nearest integer, as the liquid limit of the soil.

A simplified test procedure for liquid limit determination using the Casagrande apparatus is that known as the "one point method." Essentially the method involves making up a soil paste such that the groove cut in the sample in the cup closes at a number of blows as close as possible to 25, and certainly between 15 and 35, blows. A correction factor, which varies with the actual number of blows, is applied to the water content of the soil to give the liquid limit (BSI, 1990). The method has the advantage of speed, but this is at the expense of reliability (Nagaraj and Jayadeva, 1981).

Fig. 1 The Casagrande grooving tool and liquid limit device, showing a soil sample divided by the tool prior to testing.

For the Casagrande plastic limit test (BSI, 1990), the sample is mixed with distilled water until it is sufficiently plastic to be molded into a ball. A subsample of approximately 10 g is formed into a thread of about 6 mm diameter, and the thread is then rolled between the tips of the fingers of one hand and a flat glass plate until it is 3 mm in diameter. The thread is then remolded in the hand to dry the sample and again rolled into a thread. The operation is repeated until the thread crumbles as it reaches a diameter of 3 mm. A second subsample is similarly tested, and the mean of the two water contents (expressed as percentages) at which the threads crumble on reaching a diameter of 3 mm is recorded, to the nearest integer, as the plastic limit of the soil. Where the plastic limit cannot be obtained or where it is equal to the liquid limit, the soil is described as nonplastic.

Both these tests are undertaken on air-dried material passing a 425 μm sieve, although it has been susggested that, when the bulk of the soil material passes 425 μm, it may be more convenient to test the whole soil (BSI, 1990). However, it is generally agreed that the results for soils tested in the natural condition may be different from tests conducted on material that has previously been air-dried, and this is certainly the case when soils are at above-ambient temperatures (Basma et al., 1994). This is particularly true of organic soils. Where an appreciable proportion of the soil is retained on the 425 μm sieve, removal of such material can influence the plasticity characteristics of the soil (Dumbleton and West, 1966). Because of these various aspects of the test procedures and because the tests are conducted on remolded soil, the results should be interpreted with caution in relation to the likely behavior of soil in the field.

B. Drop-Cone Tests

Most of the shortcomings of the Casagrande liquid limit test are related to its subjectivity and to the tendency for some soils to slide in the cup or liquefy from shock, rather than flow plastically (Casagrande, 1958). After reviewing five alternative cone penetrometer tests, Sherwood and Ryley (1968) concluded that a method developed by the Laboratoire Central des Ponts et Chaussées, 58 Boulevard Lefebre, F-75732 Paris Cedex 15, France (Anon., 1966) offered the possibility of a suitable method for liquid limit determination. The new method, which used apparatus already available in most materials testing laboratories, was shown to be easier to perform than the Casagrande method, to be less dependent on the design of the apparatus, to be applicable to a wider range of soils, and to be less susceptible to operator error. Largely as a result of the work of Sherwood and Ryley (1968), the drop-cone penetrometer test was adopted as the preferred method for liquid limit determination by the British Standards Institution (BSI, 1990) in the United Kingdom.

The apparatus used in the drop-cone penetrometer test is shown in Fig. 2. The mass of the cone plus shaft is 80 g, and the cone angle is 30°. The test soil, which is prepared to give a selection of water contents in exactly the same way as in the Casagrande test, is contained in a 55 mm diameter, 50 mm deep cup. At each water content, the soil is pushed into the cup with a spatula, so that air is not trapped, and then levelled off flush with the top of the cup. The cone is lowered until it just touches the soil surface, and the cone shaft is allowed to fall freely for 5 s before the shaft is again clamped and the cone penetration noted from the dial gauge. Usually, the 5 s release is automatically controlled via an electromagnetic solenoid clamp as shown in Fig. 2. A duplicate measurement is made, and the procedure is then repeated for a range of water contents. The linear relation between cone penetration and water content is plotted, and the percentage water content corresponding to a penetration of 20 mm is recorded, to the nearest inte-

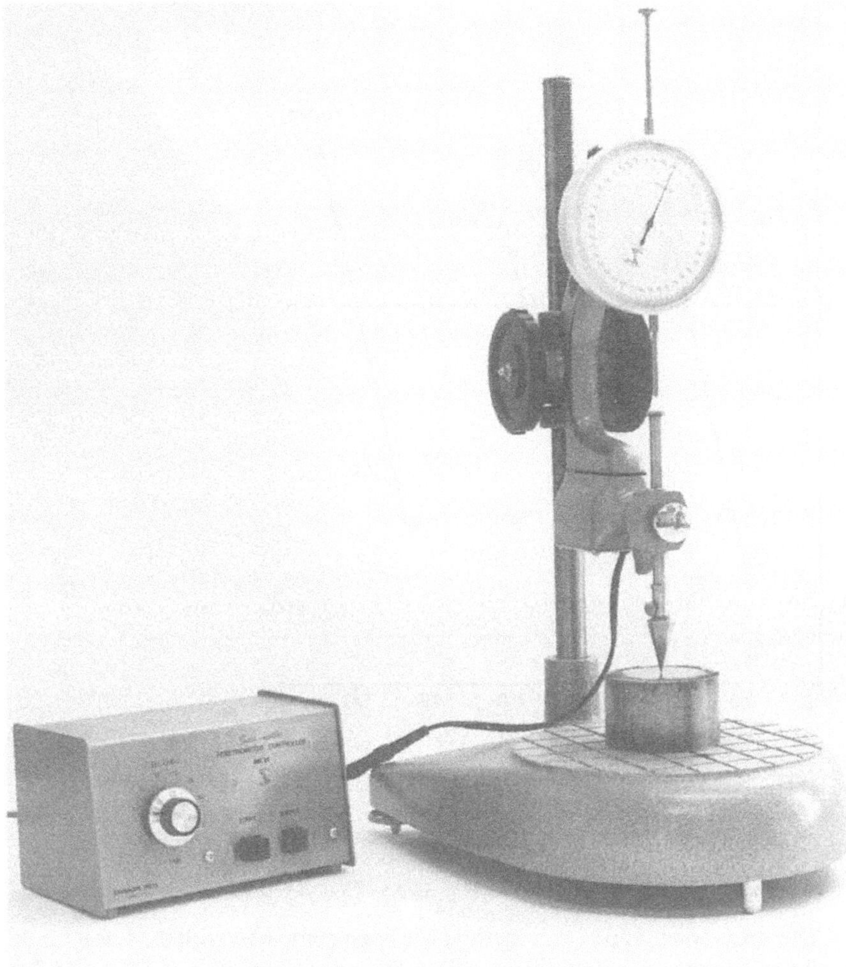

Fig. 2 The drop-cone penetrometer, showing the cone position at the start of a test.

ger, as the cone penetrometer liquid limit. Typical test results for four soils are shown in Fig. 3.

Attempts have been made to develop a one-point cone penetrometer liquid limit test analogous to the one-point Casagrande test. As with the latter, the method is a compromise between speed and accuracy but has been shown to be a satisfactory alternative (Clayton and Jukes, 1978). The one-point cone penetrometer test has been shown to be theoretically sound and not based simply on statistical correlations (Nagaraj and Jayadeva, 1981).

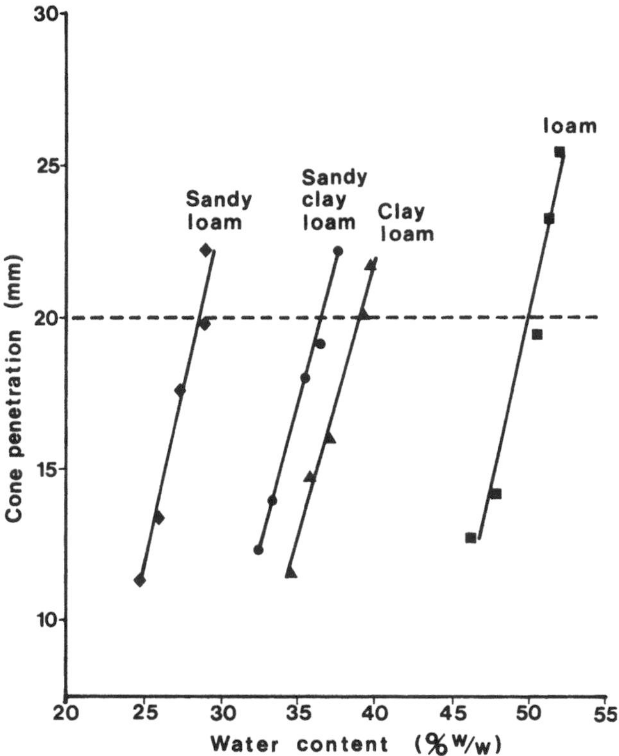

Fig. 3 The results of cone penetrometer liquid limit tests on four arable topsoils of contrasting texture. The horizontal broken line indicates the cone penetrometer liquid limit. (From Campbell, 1975.)

The drop-cone liquid limit method has been compared with the Casagrande method for a range of soils used in civil engineering (Stefanov, 1958; Karlsson, 1961; Scherrer, 1961; Sherwood and Ryley, 1968, 1970a, b) and agriculture (Towner, 1974; Campbell, 1975; Wires, 1984). Generally, the two tests give equivalent results (Littleton and Farmilo, 1977; Moon and White, 1985; Sivapullaiah and Sridharan, 1985; Queiroz de Carvalho, 1986). A comparison of the two methods is shown in Fig. 4, which also shows the reproducibility of the drop-cone method.

With the widespread adoption of the drop-cone method for measuring the liquid limit, there were obvious advantages in using the same apparatus to measure the plastic limit, if that were possible. Scherrer (1961) proposed a method of plastic limit determination that involved extrapolation of the linear relation between

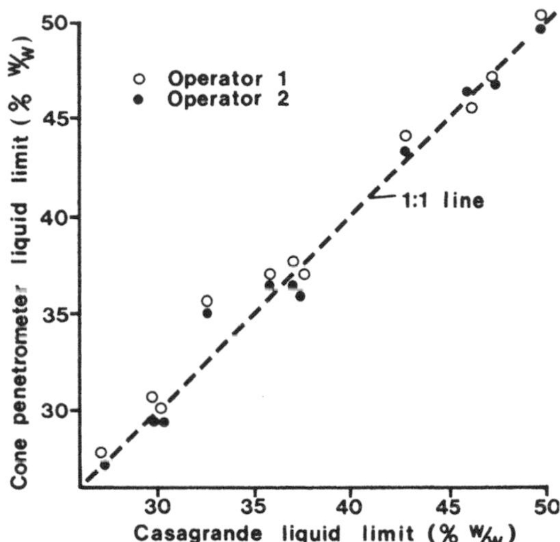

Fig. 4 The relation between the cone penetrometer liquid limit, as determined by two operators, and the Casagrande liquid limit determined by operator 1 for some arable topsoils. (From Campbell, 1975.)

water content and cone penetration found in the region of the liquid limit but conceded that the necessary extrapolation implied possible sources of inaccuracy in the method. In fact, Towner (1973) showed that, although the water content/cone penetration relation is linear in the region of the liquid limit, it becomes nonlinear at lower water contents, tending to show a minimum penetration. Campbell (1976b) made detailed measurements of the water content/cone penetration relations for 18 soils and found a pronounced minimum in the curve for each soil in the region of the Casagrande plastic limit. Results for three of the soils are shown in Fig. 5. The water content corresponding to the minimum of the curve was always numerically less than, but correlated closely with, the plastic limit. It was suggested that the plastic limit be redefined as the water content corresponding to the minimum of the curve and that it be referred to as the cone penetrometer plastic limit. The possibility of the establishment of a fixed penetration value corresponding to the plastic limit was considered (Towner, 1973; Campbell, 1976b; Allbrook, 1980) but was dismissed because variation in penetration between soils was unacceptably high (Campbell, 1976b). The cone penetrometer plastic limit was shown to offer reduced operator errors and to be a good indicator of soil behavior in an examination of the variation with water content of soil cohesion, soil–metal friction, susceptibility to compaction, implement draught, and the slope and intercept of the virgin compression line of critical state soil mechanics

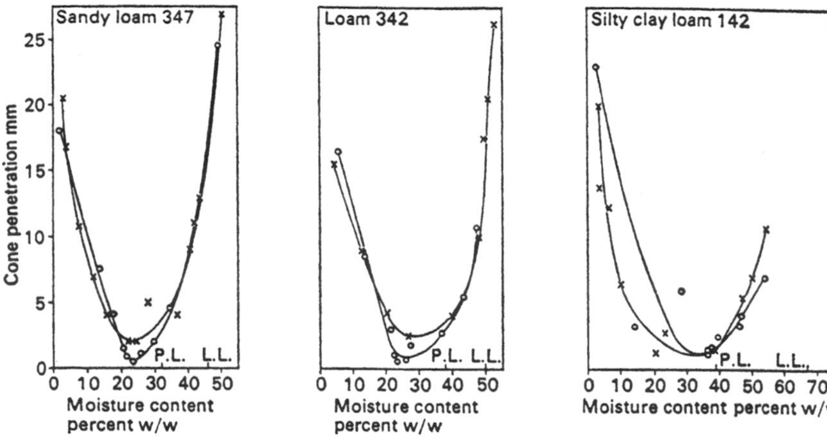

Fig. 5 Water content/cone penetration relations for three soils of contrasting texture in relation to the Casagrande liquid (LL) and plastic (PL) limits. Results obtained by two independent operators are shown. (From Campbell, 1976b.)

theory. For a given soil, all these relations were shown to exhibit turning points at a water content corresponding to the cone penetrometer plastic limit (Campbell et al., 1980).

A distinct approach to the use of the cone penetrometer to measure the plasticity index was made by Wood and Wroth (1978). They suggested that the plastic limit be redefined so that the undrained shear strength at the plastic limit is one hundred times that at the liquid limit. The proposal was based on the assumption that all soils have the same strength at their liquid limits, and this was shown to be reasonable. Further, it was shown that the proposal allowed a unique relation to be developed for remolded soil between strength and liquidity index and also between compression index and plasticity index (Wroth and Wood, 1978).

C. Other Methods

Several workers have devised methods of measuring liquid and plastic limits that depend either on correlation with other soil physical or mechanical properties or on a revision of the definition of the limits, which relates them more to changes in soil behavior. None of these methods has been widely adopted, but to a certain extent this is due to the difficulty of replacing long-established standard methods.

Faure (1981) related the liquid and plastic limits to turning points on the water content/dry bulk density relation of several soils, while Russell and Mickle (1970) attempted, with only limited success, to relate the limits to the water release characteristics. There have been attempts to relate the liquid and plastic limits to specific viscosities (Yasutomi and Sudo, 1967; Hajela and Bhatnagar, 1972), to the residual water content of a soil paste subjected to a standard stress (Vasilev, 1964; Skopek and Ter-Stephanian, 1975), and to various mechanical properties (Sherwood and Ryley, 1970a). However, none of these alternative methods has been widely adopted.

D. General Considerations

As both liquid and plastic limit tests are empirical, it is important that the test procedures be closely specified, if consistent results are to be obtained. Most test procedures specify that the soil should first be air-dried and then sieved through a 425 μm sieve (BSI, 1990), although wet sieving through a 425 μm sieve followed by air-drying has been proposed (Armstrong and Petry, 1986). However, it has been suggested that in some circumstances either air-drying (Allbrook, 1980; Pandian et al., 1993) or removal of any soil particle size fraction (Dumbleton and West, 1966; Sivapullaiah and Sridharan, 1985; BSI, 1990) can markedly affect the result obtained. The development of a practical in situ test might be desirable, but it is unlikely because of the difficulty in obtaining an appropriate sequence of test water contents without the complication of hysteresis effects as the soil alternately wets and dries in a random way (Campbell and Hunter, 1986). Such effects, probably together with cementation effects, have led to the need for samples prepared to a given water content to be thoroughly mixed (Sowers et al., 1968) and allowed to cure for 24 hours before being tested (BSI, 1990), although the latter is not universally agreed to be necessary (Gradwell and Birrel, 1954; Moon and White, 1985). In addition, sample preparation may be complicated by the fact that some soils undergo irreversible changes on drying (Allbrook, 1980), while other soils may give index values that depend on the number of times the test sample is remolded and cured prior to the test, especially where the liquid limit is concerned (Coleman et al., 1964; Davidson, 1983). The latter effect is thought to be due to particularly stable aggregates that break down only with prolonged remolding (Coleman et al., 1964; Sherwood, 1967; Pringle, 1975; Blackmore, 1976).

Although the standard test for the liquid limit using the drop-cone penetrometer includes a check on the sharpness of the cone used (BSI, 1990), Houlsby (1982) concluded that, in contrast to the work of Sherwood and Ryley (1970b), the effect of variations in cone sharpness was very small compared with the effect of the roughness of the cone surface. Both the cone angle (Budhu, 1985) and the cone mass (Budhu, 1985; Campbell and Hunter, 1986) affect the penetration obtained. Large variations in temperature affect the Casagrande liquid and plastic limits appreciably, due to variation in water viscosity (Youssef et al., 1961).

Table 2 Cone Penetrometer Liquid Limit and Proposed Cone Penetrometer Plastic Limit Determinations by Experienced and Totally Inexperienced Operators and the Corresponding Casagrande Limits for Some Arable Topsoils

			Liquid limit (% w/w)			Plastic limit (% w/w)		
				Cone penetrometer			Cone penetrometer	
Soil No.	USDA Field texture	Total organic matter (%)	Casagrande	Experienced operator	Inexperienced operator	Casagrande	Experienced operator	Inexperienced operator
1	SL	3.0	27	28	29	22	17	15
2	SL	3.9	30	31	31	26	17	18
3	SL	3.7	30	30	30	26	18	19
4	SCL	4.8	33	36	36	24	19	17
5	SCL	3.3	36	37	36	28	19	26
6	SCL	5.5	37	38	36	26	19	21
7	SL	7.4	37	37	38	31	25	22
8	SL	5.2	49	47	45	44	27	30

Source: Campbell, 1976b, 1975.

Lack of reproducibility between operators carrying out liquid (Dumbleton and West, 1966; Campbell, 1975; Wires, 1984) and plastic (Ballard and Weeks, 1963; Gay and Kaiser, 1973; Campbell, 1976b) limit tests led to the development of the drop-cone test for the liquid limit, but proposed improvements to the Casagrande plastic limit test (Gay and Kaiser, 1973) or alternative test procedures (Campbell, 1976b) have not been widely adopted. The reproducibility of the cone penetrometer liquid and plastic limit tests is shown for eight arable topsoils in Table 2.

When the Casagrande plastic limit either cannot be obtained or is greater than the liquid limit, the soil is described as nonplastic. However, it is common experience that such soils may indeed exhibit plastic behavior when subjected to the appropriate combination of stresses. In this respect, both the cone penetrometer plastic limit proposed by Campbell (1976b) and the plastic limit related to compactibility proposed by Faure (1981) have the advantage that a plastic limit can be determined for all soils.

IV. APPLICATIONS OF TEST RESULTS

The most widespread single application of the results of liquid and plastic limit tests is their use by engineers to classify soils (Anon., 1964), since the test results are related to properties such as compressibility, permeability (i.e., saturated hydraulic conductivity), and strength (Casagrande, 1947). Thus the test results can indicate the likely mechanical behavior of the soil in earthworks. The use of remolded soils in the tests is entirely appropriate in this context.

However, for soils used for plant growth, remolding of the soil prior to testing has always been considered a limitation to the value of the test result. Consequently, soil classification has always placed more emphasis on soil particle size distribution, although it has been suggested that liquid and plastic limit values could usefully be added to such classifications (Soane et al., 1972).

The following sections give some examples of the use of liquid and plastic limits in soil classification and describe some of the relations of the limits with other soil properties.

A. Soil Classification

Casagrande (1947) developed a system of classifying soils based on sieve analysis together with measurement of the liquid and plastic limits on the fraction smaller than 425 μm. Developments of this system now form the British Soil Classification System in the U.K. (Dumbleton, 1968) and the Unified Soil Classification System in the U.S.A. (ASTM, 1966). Casagrande plotted liquid limits against plasticity indices to give what he called the plasticity chart shown in Fig. 6. An empirical boundary known as the A-line on the chart separated the inorganic clays which lay above the line from the silty and organic soils which lay below. Both above and below the A-line, the liquid limit was used to divide solids into three classes of compressibility, namely low, intermediate, and high, corresponding to liquid limits <35, 35–50, and >50, respectively. In the British Soil Classification System, the chart was extended to include soils with very high (70–90) and extremely high (>90) liquid limits as shown in Fig. 6. Moreover, soils with liquid limits <20 were described as nonplastic, and it was recognized that organic soils could occur both above and below the A-line.

Much can be deduced about the mechanical properties of a soil from its position on the plasticity chart. For a given liquid limit, the greater the plasticity index of a soil, the greater is its clay content, toughness, and dry strength, and the lower is its permeability. For a given plasticity index, soil compressibility increases with increasing liquid limit. The liquid and plastic limits are both dependent on the amount and type of clay in a soil. Kaolinitic clays generally lie below the A-line and behave as silts, while montmorillonitic clays lie just above the A-line. Peats have very high liquid limits of several hundred percent but a small plasticity index.

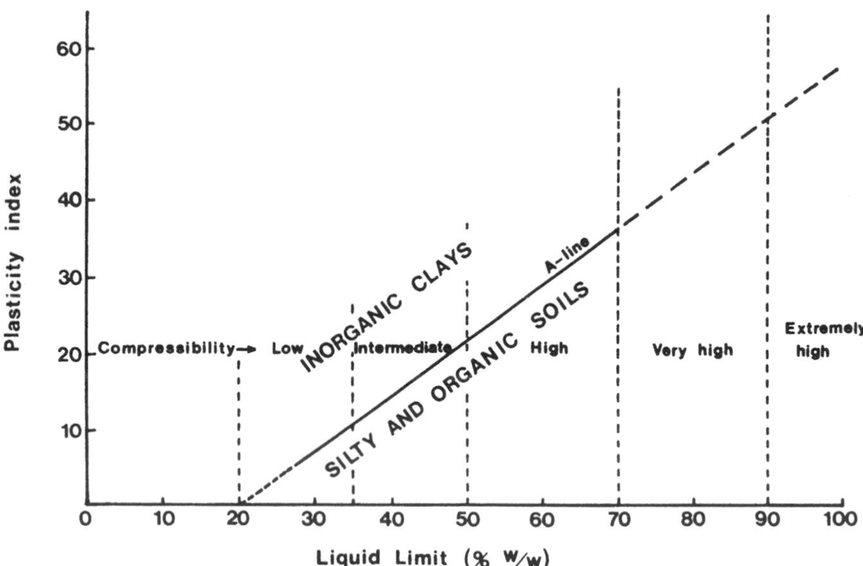

Fig. 6 The plasticity chart used in the British Soil Classification System. The original Casagrande system assigned all soils with liquid limits >50 to a single compressibility class.

B. Relations with Other Soil Properties

1. Texture and Organic Matter

Plasticity characteristics have been related to clay content by many authors (Odell et al., 1960; Archer, 1975; Humphreys, 1975; Yong and Warkentin, 1975; Mulqueen, 1976; de la Rosa, 1979). Several report a simple linear relation between plasticity index and clay content (Odell et al., 1960; Humphreys, 1975; Mulqueen, 1976), although a closer relationship was often found when other factors such as organic matter (Odell et al., 1960; de la Rosa, 1979) or silt content (Humphreys, 1975) were included. Odell et al. (1960) found a very close correlation for Illinois soils between plasticity index and a combination of clay percentage, clay percentage which is montmorillonite, and percentage organic carbon. Where the relation between plasticity index and clay content was weak, the effect may have been associated with particle sizes rather coarser than the clay fraction (Humphreys, 1975) or to the presence of strongly aggregated clay-sized particles (Coleman et al., 1964; Sherwood and Hollis, 1966). Baver (1928) found that swelling montmorillonite clay soils exhibit higher plasticity than nonswelling soils. Those with sodium-saturated exchange sites have a much greater plasticity index than those saturated with potassium, calcium, or magnesium.

Both particle shape and the percentage of organic material in the soil have an effect on the plasticity characteristics, and these factors usually interact. Farrar and Coleman (1967) found that the particle surface area, as indicated by adsorption of water, was strongly related to the liquid limit and rather less so to the plastic limit. Hammell et al. (1983) suggested that the liquid and plastic limits could be used as a less laborious method of measuring the surface area of soils. Although the liquid and plastic limits increase with particle surface area, they may not do so in simple proportion since the water involved in filling soil pores may be involved in addition to that increasing the thickness of the water layer between particles (Yong and Warkentin, 1975). Indeed, it has been suggested that soil specific surface determines the plasticity index and liquid limit only insofar as it determines the particle separation at the liquid and plastic limits (Nagaraj and Jayadeva, 1981).

Archer (1975) found that both liquid and plastic limits increased with organic matter content but that the plasticity index could either increase or decrease, depending on the soil texture. The data in Table 2 are generally consistent with his results. It has been suggested, however, that hydration of the organic matter in a soil must be fairly complete before water is available for film formation on the soil particles. Thus, although the plastic limit is increased, the quantity of water subsequently required to reach the liquid limit is unchanged and so the plasticity index remains the same (Baver et al., 1972). In general, organic matter influences the plasticity properties of a soil (Odell et al., 1960; Hendershot and Carson, 1978;

de la Rosa, 1979; McNabb, 1979; Hulugalle and Cooper, 1994; Emerson, 1995; Mbagwu and Abeh, 1998), but the role of organic matter in this context may vary with the nature of the organic material involved.

2. Workability in Relation to Tillage and Mole Drainage

The plastic limit has generally been taken to indicate the upper end of the range of water contents in which the soil is friable and most readily cultivated to produce a seedbed (Russell and Wehr, 1922). Although clod strength is low and breakage therefore relatively easy in the plastic range (Archer, 1975; Spoor, 1975), soils are also more susceptible to compaction and puddling and so clods are also easily formed (Smith, 1962; Spoor, 1975; Adam and Erbach, 1992). Moreover, both soil adhesion to metal and tine draught are at their maximum within the plastic range (Nichols, 1930), as is the angle of soil–metal friction (Spoor, 1975). Campbell et al. (1980) have shown that both the angle of soil–metal friction (Fig. 7) and the draught force on a tine are at a maximum at the cone penetrometer plastic limit.

Subsoiling is ineffective in loosening the subsoil unless it is drier than the plastic limit (O'Sullivan, 1992). Above the plastic limit, the soil will simply remold without shattering. In contrast, mole drainage channels can be satisfactorily established only when the soil at mole depth is above the plastic limit, although the soil immediately above the channel must remain friable enough to shatter and allow water access to the mole drain. Archer (1975) has suggested that the plasticity index should be at least 22 if a soil is to be considered suitable for mole drainage.

3. Compressibility

At water contents around the plastic limit, soil resistance to compaction drops sharply (Archer, 1975). Above the liquid limit, resistance to compaction can be very high, but relatively low compressive or shearing forces can easily destroy the pore structure of the soil, leaving it in a puddled state (Koenigs, 1963).

The optimum water content for compaction in the British Standard compaction test (2.5 kg rammer method) (BSI, 1990) has been shown to be correlated with the plastic limit (Weaver and Jamison, 1951; Soane et al., 1972; Campbell et al., 1980). However, it has been suggested that such a relationship is probably fortuitous, since the optimum water content for compaction decreases with increasing compactive effort (Campbell et al., 1980). Nevertheless, Bertilsson (1971) found that the soil water content associated with the maximum slope of the virgin compression lines, for two of the four soils he studied, corresponded to the optimum water content for compaction. Similarly, Campbell et al. (1980) found a maximum slope for the virgin compression lines of two soils at water contents lying between their Casagrande and cone penetrometer plastic limits. The water

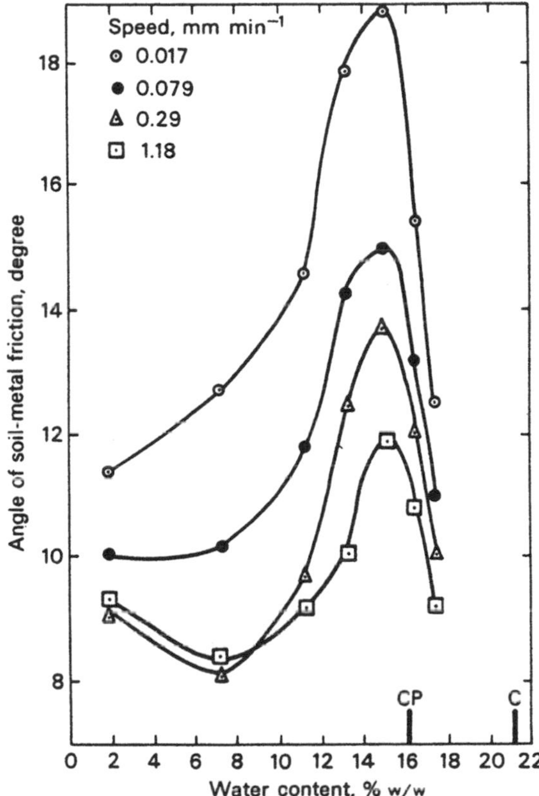

Fig. 7 The variation of soil–metal friction with water content at each of four sliding speeds for a sandy clay loam in relation to the cone penetrometer (CP) and Casagrande (C) plastic limits. (From Campbell et al., 1980.)

contents concerned were shown to correspond to the cone penetrometer "plastic limit" when this test was performed on intact aggregates of <10 mm diameter that had not been remolded. Since the maximum slope of the virgin compression line indicates the maximum susceptibility to compaction, they suggested that a soil is much more likely to compact if subjected to tillage and traffic at water contents close to the cone penetrometer "plastic limit," as determined on soil that has not been remolded but is in its natural state.

Compression characteristics have been related to the plasticity index either empirically (Carrier and Beckman, 1984) or with the aid of critical state theory, making the assumption that the strength at the plastic limit is one hundred times that at the liquid limit (Wroth and Wood, 1978). O'Sullivan et al. (1994) showed that both the normal consolidation and the critical state lines pivoted about a point as water content increased so that compactibility was greatest near the plastic limit.

4. *Water Regime*

Uppal (1966) found that for nine remolded soils with plastic limits ranging from 17 to 34% w/w, the plastic limit corresponded to a matric potential of −0.3 kPa on the wetting curve and −3 kPa on the drying curve. His work was extended by Livneh et al. (1970) to include a range of bulk densities and water contents, and they found the plastic limit to be in the range −6 to −60 kPa on the drying curve. Rather higher values of −13 to −100 kPa were found for the plastic limit on a drying curve by Stakman and Bishay (1976).

The value of field capacity relative to the plastic limit can affect the behavior of a soil during cultivation. Where the plastic limit is less than field capacity, the soil structure will be readily damaged when worked at water contents between the plastic limit and the field capacity. A soil for which the plastic limit is greater than the field capacity will have good workability. Similarly, susceptibility to slaking, which generally occurs above the liquid limit, depends on the relative values of field capacity and liquid limit (Boekel, 1963). Archer (1975) found that the field capacity was close to and generally slightly greater than the plastic limit for four contrasting soil textures (Fig. 8).

Benson et al. (1994) estimated the hydraulic conductivity of compacted clay liners by means of a multivariate regression equation involving the liquid limit, the plasticity index, and soil particle size fractions. Sewell and Mote (1969) made use of a relation between the logarithm of saturated hydraulic conductivity (permeability) and the liquid limit to determine the effectiveness of various chemicals for sealing ponds without the necessity of making large numbers of conductivity measurements. Similarly, Carrier and Beckman (1984) considered such simple correlations to be satisfactory for preliminary engineering design purposes. Using data from both the literature and their own experiments, Reddi and Poduri (1997) concluded that the liquid limit is a useful state to which the water release characteristic of a fine-grained soil at other states may be referred.

5. *Strength*

Many researchers have reported empirical relationships between the plasticity index and the shear strength (Nichols, 1932; Voight, 1973), the cohesion (Gibson, 1953), or the angle of internal friction of a soil (Gibson, 1953; Kanji, 1974; Humphreys, 1975). Wroth and Wood (1978) suggested that the plastic limit should be defined as that water content at which the soil has 100 times the strength it possesses at the liquid limit. On the assumption that all soils have the same strength

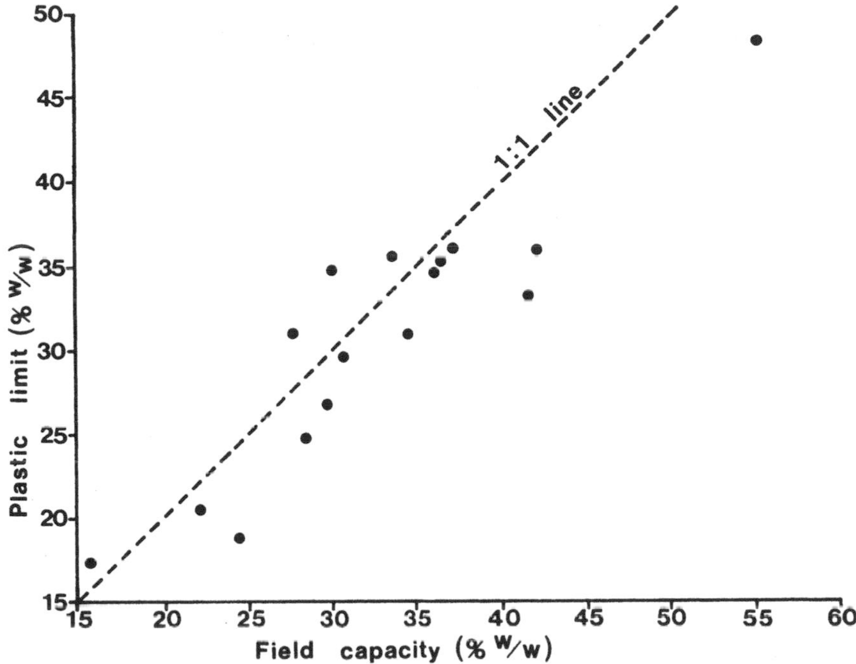

Fig. 8 The relation between plastic limit and field capacity for sixteen soils. (Based on data from Archer, 1975.)

at the liquid limit, they went on to use critical state soil mechanics theory to show that estimates of undrained shear strength depended only on the liquidity index of the soil.

V. SUMMARY

Plasticity is the property that allows a soil to be deformed without cracking in response to an applied stress. Such behavior can occur over a range of soil water contents, with the upper and lower limits of the range being referred to as the liquid and plastic limits, respectively.

The cohesive and adhesive forces associated with soil water and, especially, their variation with water content determine whether a soil will, when stressed, undergo brittle failure, plastic flow, or viscous flow. At the plastic limit, there is just sufficient water to surround each soil particle with a water layer so that the laminar particles can slide over each other under stress and remain in their new positions when the stress is removed. At the liquid limit, the water layers between particles are sufficiently thick for viscous flow to occur in response to an applied stress.

Dry soil to which water is added during continuous remolding to reach either the liquid or the plastic limit is said to be in the critical state in terms of the critical state theory of soil mechanics. This theory describes the stress–strain behavior of any soil in relation to the three-dimensional relationship of spherical pressure, deviatoric stress, and specific volume. All points on the critical state line within this relationship correspond to states in which the soil can be continuously remolded without any change in volume.

The liquid limit has traditionally been determined with the Casagrande apparatus, but more recently a drop-cone test has become the preferred British Standard method.

The plastic limit is defined in the traditional method, which is still the British Standard method, as the water content at which a thread of soil, rolled between the fingertips of the operator and a flat glass plate, just crumbles when the thread reaches a diameter of 3 mm. More recently there have been attempts to redefine the plastic limit using tests based on the drop-cone apparatus. One proposal is that the minimum of the penetration-water content relation corresponds to the plastic limit. It has also been suggested that the plastic limit be defined so that the undrained shear strength of the soil at the plastic limit is one hundred times that at the liquid limit.

Various other methods of measuring liquid and plastic limits have been proposed that depend either on a correlation with other soil properties or on a revision of the definitions of the limits so that they are more related to soil behavior.

The liquid and plastic limits have been widely used in soil engineering for soil classification because the limits are correlated with other important soil physical and mechanical properties. A possible objection to the tests so far as soils used in agriculture are concerned is that remolded soil is used. Nevertheless, the limits may provide a quicker, cheaper, or easier indication of other properties than their direct measurement where no great precision is required.

REFERENCES

Adam, K. M., and D. C. Erbach. 1992. Secondary tillage tool effect on soil aggregation. *Trans. Am. Soc. Agric. Eng.* 35:1771–1776.

Allbrook, R. F. 1980. The drop-cone penetrometer method for determining Atterberg limits. *N.Z. J. Sci.* 23:93–97.

Anon. 1966. Determination rapide des limites d'Atterberg à l'aide d'un pénétromètre et d'un picnomètre d'air. Dossier SGR/149. Paris: Laboratoire Central des Ponts et Chaussées.

Archer, J. R. 1975. Soil consistency. In: *Soil Physical Conditions and Crop Production*. Ministry of Agriculture, Fisheries and Food, Tech. Bull. 29. London: HMSO, pp. 289–297.

Armstrong, J. C., and T. M. Petry. 1986. Significance of specimen preparation upon soil plasticity. *Geotech. Testing J.* 9: 147–153.

ASTM (American Society for Testing and Materials). 1966. Tentative method for classification of soils for engineering purposes. In: *Book of Am. Soc. Testing and Mater. Standards*. ASTM, pp. 766–811.

Atkinson, J. H., and P. L. Bransby. 1978. *The Mechanics of Soils*. Maidenhead, U.K.: McGraw-Hill

Atterberg, A. 1911. *Die Plastizität der Tone. Int. Mitt. Bodenk.* 1: 10-43.

Atterberg, A. 1912. *Die Konsistenz und die Bindigkeit der Böden. Int. Mitt. Bodenk.* 2: 149–189.

Ballard, G. E. H., and W. F. Weeks. 1963. The human factor in determining the plastic limit of cohesive soils. *Mater. Res. Stand.* 3: 726–729.

Basma, A. A., A. S. Al-Homoud, and E. Y. Al-Tabari. 1994. Effects of methods of drying on the engineering behavior of clays. *Appl. Clay Sci.* 3: 151–164.

Baver, L. D. 1928. The relation of exchangeable cations to the physical properties of soils. *J. Am. Soc. Agron.* 20: 921–941.

Baver, L. D. 1930. The Atterberg consistency constants: Factors affecting their values and a new concept of their significance. *J. Am. Soc. Agron.* 22: 935–948.

Baver, L. D., W. H. Gardner, and W. R. Gardner. 1972. *Soil physics*. New York: John Wiley.

Benson, C. H., H. Zhai, and X. Wang. 1994. Estimating hydraulic conductivity of compacted clay liners. *J. Geotech. Eng.* 120: 366–387.

Bertilsson, G. 1971. Topsoil reaction to mechanical pressure. *Swed. J. Agric. Res.* 1: 179–189.

Blackmore, A. V. 1976. Subplasticity in Australian soils, IV. Plasticity and structure related to clay cementation. *Aust. J. Soil Res.* 14: 261–272.

Boekel, P. 1963. The effect of organic matter on the structure of clay soils. *Neth. J. Agric. Sci.* 11: 250–263.

BSI (British Standards Institution). 1990. British Standard methods of test for soils for civil engineering purposes. British Standard 1377. London: BSI.

Budhu, M. 1985. The effect of clay content on liquid limit from a fall cone and the British cup device. *Geotech. Testing J.* 8: 91–95.

Campbell, D. J. 1975. Liquid limit determination of arable topsoils using a drop-cone penetrometer. *J. Soil Sci.* 26: 234–240.

Campbell, D. J. 1976a. The occurrence and prediction of clods in potato ridges in relation to soil physical properties. *J. Soil Sci.* 27: 1–9.

Campbell, D. J. 1976b. Plastic limit determination using a drop-cone penetrometer. *J. Soil Sci.* 27: 295–300.

Campbell, D. J., and R. Hunter. 1986. Drop-cone penetration in situ and on minimally disturbed soil cores. *J. Soil Sci.* 37: 153–163.

Campbell, D. J., J. V. Stafford, and P. S. Blackwell. 1980. The plastic limit, as determined by the drop-cone test, in relation to the mechanical behaviour of soil. *J. Soil Sci.* 31: 11–24.

Carrier, W. D., and J. F. Beckman. 1984. Correlations between index tests and the properties of remoulded clays. *Géotechnique* 34:211–228.

Casagrande, A. 1932. Research on the Atterberg limits of soils. *Public Roads* 13:121–130.

Casagrande, A. 1947. Classification and identification of soils. *Proc. Am. Soc. Civ. Eng.* 73:783–810.

Casagrande, A. 1958. Notes on the design of the liquid limit device. *Géotechnique* 8: 84–91.

Clayton, C. I., and A. W. Jukes. 1978. A one point cone penetrometer liquid limit test? *Géotechnique* 28:469–472.

Coleman, J. D., D. M. Farrar, and A. D. Marsh. 1964. The moisture characteristics, composition and structural analysis of a red clay soil from Nyeri, Kenya. *Géotechnique* 14:262–276.

Davidson, D. A. 1983. Problems in the determination of plastic and liquid limits of remoulded soils using a drop-cone penetrometer. *Earth Surface Processes and Landforms* 8:171–175.

de la Rosa, D. 1979. Relation of several pedological characteristics to engineering qualities of soil. *J. Soil Sci.* 30:793–799.

DSIR (Department of Scientific and Industrial Research). 1964. Soil mechanics for road engineers. London: HMSO, pp. 541.

Dumbleton, M. J. 1968. The classification and description of soils for engineering purposes: a suggested revision of the British system. Report LR182. Crowthorne, U.K.: Transport and Road Res. Lab.

Dumbleton, M. J., and G. West. 1966. The influence of the coarse fraction on the plastic properties of clay soils. Report LR36. Crowthorne, U.K.: Transport and Road Res. Lab.

Emerson, W. W. 1995. The plastic limit of silty, surface soils in relation to their content of polysaccharide gel. *Aust. J. Soil Res.* 33:1–9.

Farrar, D. M., and J. D. Coleman. 1967. The correlation of surface area with other properties of nineteen British clay soils. *J. Soil Sci.* 18:118–124.

Faure, A. 1981. A new conception of the plastic and liquid limits of clay. *Soil Till. Res.* 1: 97–105.

Gay, C. W., and W. Kaiser. 1973. Mechanisation for remolding fine grained soils and for the plastic limit test. *J. Testing Eval.* 1:317–318.

Gibson, R. E. 1953. Experimental determination of the true cohesion and true angle of internal friction in clays. *Proc. 3rd Int. Conf. Soil Mech. and Found. Eng.*, Vol. 1, pp. 126–130.

Gradwell, M., and K. S. Birrell. 1954. Physical properties of certain volcanic clays. *N.Z. J. Sci. Tech.* B36:108–122.

Grim, R. E. 1948. Some fundmental factors influencing the properties of soil materials. *Proc. 2nd Int. Conf. Soil Mech. and Found. Eng.*, Vol. 3, pp. 8–12.

Haines, W. B. 1925. Studies in the physical properties of soils, II. A note on the cohesion developed by capillary forces in an ideal soil. *J. Agric. Sci., Camb.* 15:529–535.

Hajela, R. B., and J. M. Bhatnagar. 1972. Application of rheological measurements to determine liquid limit of soils. *Soil Sci.* 114:122–130.

Hammel, J. E., M. E. Sumner, and J. Burema. 1983. Atterberg limits as indices of external surface areas of soils. *Soil Sci. Soc. Am. J.* 47:1054–1056.

Hendershot, W. H., and M. A. Carson. 1978. Changes in the plasticity of a sample of Champlain clay after selective chemical dissolution to remove amorphous material. *Can. Geotech. J.* 15:609–616.

Houlsby, G. T. 1982. Theoretical analysis of the fall cone test. *Géotechnique* 32:111–118.

Hulugalle, N. R., and J. Cooper. 1994. Effect of crop rotation and residue management on properties of cracking clay soils under irrigated cotton-based farming systems of New South Wales. *Land Degrad. and Rehab.* 5:1–11.

Humphreys, J. D. 1975. Some empirical relationships between drained friction angles, mechanical analyses and Atterberg limits of natural soils at Kainji Dam, Nigeria. *Géotechnique* 25:581–585.

Kanji, M. A. 1974. The relationship between drained friction angles and Atterberg limits of natural soils. *Géotechnique* 24:671–674.

Karlsson, R. 1961. Suggested improvements in the liquid limit test, with reference to flow properties of remoulded clays. *Proc. 5th Int. Conf. Soil Mech. and Found Eng.*, Vol. 1, pp. 171–184.

Kingery, W. D., and J. Francl. 1954. Fundamental study of clay, XIII. Drying behavior and plastic properties. *J. Am. Ceram. Soc.* 37:596–602.

Koenigs, F. F. R. 1963. The puddling of clay soils. *Neth. J. Agric. Sci.* 11:145–156.

Kurtay, T., and A. R. Reece. 1970. Plasticity theory and critical state soil mechanics. *J. Terramech.* 7:23–56.

Littleton, I., and M. Farmilo. 1977. Some observations on liquid limit values with reference to penetration and Casagrande tests. *Ground Eng.* 10:39–40.

Livneh, M., J. Kinsky, and D. Zaslavsky. 1970. Correlation of suction curves with the plasticity index of soils. *J. Materials* 5:209–220.

MAFF (Ministry of Agriculture, Fisheries and Food). 1984. Soil textures. Leaflet 895. London: MAFF.

Mbagwu, J. S. C., and O. G. Abeh. 1998. Prediction of engineering properties of tropical soils using intrinsic pedological parameters. *Soil Sci.* 163:93–102.

McNabb, D. H. 1979. Correlation of soil plasticity with amorphous clay constituents. *Soil Sci. Soc. Am. J.* 43:613–616.

Moon, C. F., and K. B. White. 1985. A comparison of liquid limit test results. *Géotechnique* 35:59–60.

Mulqueen, J. 1976. Plasticity characteristics of some carboniferous clay soils in north central Ireland and their significance. *Ir. J. Agric. Res.* 15:129–135.

Nagaraj, T. S., and M. S. Jayadeva. 1981. Re-examination of one-point methods of liquid limit determination. *Géotechnique* 31:413–425.

Nichols, M. L. 1930. Dynamic properties of soil affecting implement design. *Agric. Eng.* 11:201–204.

Nichols, M. L. 1931. The dynamic properties of soil, I. An explanation of the dynamic properties of soils by means of colloidal films. *Agric. Eng.* 12:259–264.

Nichols, M. L. 1932. The dynamic properties of soils, III. Shear values of uncemented soils. *Agric. Eng.* 13:201–204.

Norton, F. H. 1948. Fundamental study of clay, VIII. A new theory for the plasticity of clay–water masses. *J. Am. Ceram. Soc.* 31:236–241.

Odell, R. T., T. H. Thornburn, and L. J. McKenzie. 1960. Relationships of Atterberg limits to some other properties of Illinois soils. *Soil Sci. Soc. Am. Proc.* 24:297–300.

O'Sullivan, M. F. 1992. Deep loosening of clay loam subsoil in a moist climate and some effects of traffic management. *Soil Use and Manage.* 8:60–67.

O'Sullivan, M. F., D. J. Campbell, and D. R. P. Hettiaratchi. 1994. Critical state parameters derived from constant cell volume triaxial tests. *Eur. J. Soil Sci.* 45:249–256.

Pandian, N. S., T. S. Nagaraj, and G. L. S. Babu. 1993. Tropical clays, I. Index properties and microstructural aspects. *J. Geotech. Eng* 119:826–839.

Pringle, J. 1975. The assessment and significance of aggregate stability in soil. In: *Soil Physical Conditions and Crop Production.* MAFF Tech. Bull. 29. London: HMSO, pp. 249–260.

Reddi, L. N., and R. Poduri. 1997. Use of liquid limit state to generalize water retention properties of fine-grained soils. *Géotechnique* 5:1043–1049.

Queiroz de Carvalho, J. B. 1986. The applicability of the cone penetrometer to determine the liquid limit of lateritic soils. *Géotechnique* 36:109–111.

Russell, E. R., and J. L. Mickle. 1970. Liquid limit values by soil moisture tension. *J. Soil Mech. Found. Eng. Am. Soc. Civ. Eng.* 96:967–989.

Russell. J. C., and F. M. Wehr. 1922. The Atterberg consistency constants. *J. Am. Soc. Agron.* 20:354–372.

Scherrer, H. U. 1961. Determination of liquid limit by the static cone penetration test. *Proc. 5th Int. Conf. Soil Mech. and Found. Eng.*, Vol. 1, pp. 319–322.

Schofield, A., and P. Wroth. 1968. *Critical State Soil Mechanics.* London: McGraw-Hill.

Schwartz, B. 1952. Fundamental study of clay, XII. A note on the effect of surface tension of water on the plasticity of clay. *J. Am. Ceram. Soc.* 35:41–43.

Sewell, J. I., and C. R. Mote. 1969. Liquid-limit determination for indicating effectiveness of chemicals in pond sealing. *Trans. Am. Soc. Agric. Eng.* 50:611–613.

Sherwood, P. T. 1967. Classification tests on African red clays and Keuper Marl. *Quart. J. Eng. Geol.* 1:47–55.

Sherwood, P. T. 1970. The reproducibility of the results of soil classification and compaction tests. Report LR339. Crowthorne, U.K.: Transport and Road Res. Lab.

Sherwood, P. T., and B. G. Hollis. 1966. Studies of Keuper Marl: Chemical properties and classification tests. Report LR41. Crowthorne, U.K.: Transport and Road Res. Lab.

Sherwood, P. T., and M. D. Ryley. 1968. An examination of cone-penetrometer methods for determining the liquid limit of soils. Report LR233. Crowthorne, U.K.: Transport and Road Res. Lab.

Sherwood, P. T., and M. D. Ryley. 1970a. An investigation of alternative methods of determining the plastic limit of soils. Tech. Note TN536. Crowthorne, U.K.: Transport and Road Res. Lab.

Sherwood, P. T., and M. D. Ryley. 1970b. An investigation of a cone penetrometer method for the determination of the liquid limit. *Géotechnique* 20:203–208.

Sivapullaiah, P. V., and A. Sridharan. 1985. Liquid limit of soil mixtures. *Geotech. Testing J.* 8:111–116.

Skempton, A. W., and R. D. Northey. 1953. The sensitivity of clays. *Géotechnique* 3: 30–53.

Skopek, J., and G. Ter-Stephanian. 1975. Comparison of liquid limit values determined according to Casagrande and Vasilev. *Géotechnique* 25:135–136.

Smith, N. 1962. Let's have traffic congestion in the potato field. *Fm. Mech.* 14:137.

Soane, B. D., D. J. Campbell, and S. M. Herkes. 1972. The characterization of some Scottish arable topsoils by agricultural and engineering methods. *J. Soil Sci.* 23:93–104.

Sowers, G. F. 1965. Consistency. In: *Methods of Soil Analysis, Part 1* (C. A. Black et al., eds.). Madison, WI: Am. Soc. Agron., pp. 391–399.

Sowers, G. F., A. Vesic, and M. Grandolfi. 1968. Penetration tests for liquid limit. Am. Soc. Testing and Materials, Spec. Tech. Publ. 254, pp. 216–226.

Spoor, G. 1975. Fundamental aspects of cultivation. In: *Soil Physical Conditions and Crop Production.* Ministry of Agriculture, Fisheries and Food Tech. Bull. 29. London: HMSO, pp. 128–144.

Sridharan, A., S. M. Rao, and N. S. Murthy. 1986. Liquid limit of montmorillonite soils. *Geotech. Testing J.* 9:156–159.

Stakman, W. P., and B. G. Bishay. 1976. Moisture retention and plasticity of highly calcareous soils in Egypt. *Neth. J. Agric. Sci.* 24:43–57.

Stefanov, G. 1958. Discussion on liquid limit. *Proc. 4th Int. Conf. Soil Mech. and Found. Eng.*, Vol. 1, p. 97.

Towner, G. D. 1973. An examination of the fall-cone method for the determination of the strength properties of remoulded agricultural soils. *J. Soil Sci.* 24:470–479.

Towner, G. D. 1974. A note on the plasticity limits of agricultural soils. *J. Soil Sci.* 25: 307–309.

Uppal, H. L. 1966. A scientific explanation of the plastic limit of soils. *J. Materials* 1: 164–178.

Vasilev, Y. M. 1964. Rapid determination of the limit of rolling out. *Pochvovedenie* 7: 105–106.

Voight, B. 1973. Correlation between Atterberg plasticity limits and residual shear strength of natural soils. *Géotechnique* 23:265–267.

Vomocil, J. A., and L. J. Waldron. 1962. The effect of moisture content on tensile strength of unsaturated glass bead systems. *Soil Sci. Soc. Am.* 26:409–412.

Weaver, H. A., and V. C. Jamison. 1951. Effects of moisture on tractor tire compaction of soil. *Soil Sci.* 71:15–23.

Wires, K. C. 1984. The Casagrande method versus the drop-cone penetrometer method for the determination of liquid limit. *Can. J. Soil Sci.* 64:297–300.

Wood, D. M., and C. P. Wroth. 1978. The use of the cone penetrometer to determine the plastic limit of soils. *Ground Eng.* 11:37.

Wroth, C. P., and D. M. Wood. 1978. The correlation of index properties with some basic engineering properties of soils. *Can. Geotech. J.* 15:137–145.

Yasutomi, R., and S. Sudo. 1967. A method of measuring some physical properties of soil with a forced oscillation viscometer. *Soil Sci.* 104:336–341.

Yong,. R. N., and B. P. Warkentin. 1975. In: *Developments in Geotechnical Engineering 5. Soil Properties and Behaviour.* Amsterdam: Elsevier, pp. 62–68.

Youssef, M. S., A. Sabry, and A. H. El Rami. 1961. Temperature changes and their effects on some physical properties of soils. *Proc. 5th Int. Conf. Soil Mech. and Found. Eng.*, Vol. 1, pp. 419–421.

10

Penetrometer Techniques in Relation to Soil Compaction and Root Growth

A. Glyn Bengough
Scottish Crop Research Institute, Dundee, Scotland

Donald J. Campbell and Michael F. O'Sullivan
Scottish Agricultural College, Edinburgh, Scotland

I. INTRODUCTION

Soil hardness is the resistance of the soil to deformation, be it by a plant root, the blade of a plow, or the tip of a penetrometer. Hard soils are a major problem in agriculture worldwide; they restrict root growth and seedling emergence, increase the energy costs of tillage, and impose restrictions on the soil management regimes that can be used.

Penetrometers are used commonly to measure soil strength. If a standard probe and testing procedure is used, penetrometers give an empirical measure of soil strength that enables comparisons between different soils. A penetrometer consists typically of a cylindrical shaft with a conical tip at one end, and a device for measuring force at the other (Fig. 1). Penetration resistance is the force required to push the cone into the soil divided by the cross-sectional area of its base (i.e., a pressure). The American Association of Agricultural Engineers specified a standard penetrometer design that gives a measurement called the cone index (ASAE, 1969). This standard has been adopted widely, but many nonstandard penetrometers are in use. Nonstandard penetrometers and testing procedures are more appropriate for some applications, as long as comparisons are made using the same procedure. The principles behind the testing procedure must be understood so that the results can be interpreted sensibly.

In this chapter we describe the theory behind the measurement of penetration resistance, and how penetration resistance is related to other soil properties.

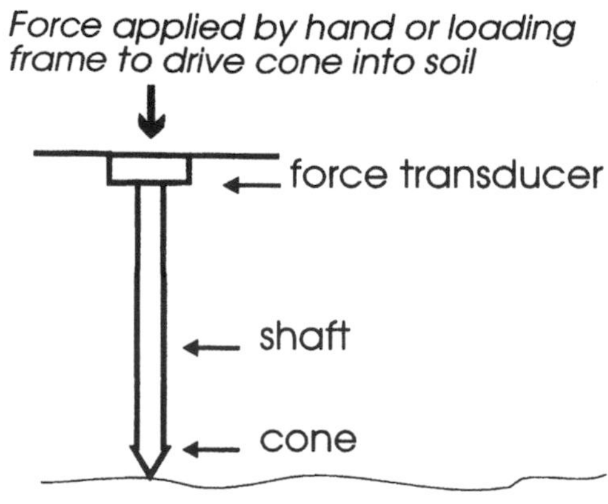

Fig. 1 Schematic diagram of a penetrometer showing cone, shaft, and force transducer.

We then consider the practical aspects of penetrometer measurements, including the design of the apparatus, the availability of equipment, the measurement procedure, and the interpretation of data. In the final section we discuss how to apply the technique to studies of trafficability, tillage, compaction, and root growth.

II. THEORY

A. Soil Penetration by Cones

Penetration resistance can, in principle, be estimated from the bulk mechanical properties of the soil. Farrell and Greacen (1966) developed a model of soil penetration in which penetration resistance consisted of two components: the pressure required to expand a cavity in the soil, and the frictional resistance to the probe. Penetrometer resistance, Q, is given by Eq. 1 (Farrell and Greacen, 1966), including the effects of adhesion (Bengough, 1992):

$$Q = \sigma(1 + \cot\alpha \tan\delta) + c_a \cot\alpha \tag{1}$$

where σ is the stress normal to the cone surface, α is the cone semiangle, δ is the angle of soil–metal friction, and c_a is the soil–metal adhesion. This equation assumes that the soil is homogeneous and isotropic, that the frictional resistance between the penetrometer shaft and the soil is negligible, that the cone angle of the penetrometer is sufficiently small so that no soil-body accumulates in front of the cone, and that the stress is distributed uniformly on the cone surface.

The normal stress, σ, was equated with the pressure required to expand a cylindrical or spherical cavity in the soil. Expansion of the cavity occurred through compression of the soil surrounding the probe. Two distinct zones were identified: a zone of compression with plastic failure surrounding the probe, with a zone of elastic compression immediately outside it (Farrell and Greacen, 1966). Calculating σ required measurements of many soil mechanical properties. The value of σ was predicted for three soils at different bulk densities and matric potentials. For cylindrical soil deformation, σ was only 0.25–0.45 of that for spherical deformation. Greacen et al. (1968) suggested that roots and penetrometers with narrow cone angles cause cylindrical soil deformation, while penetrometers with larger cone angles cause spherical deformation.

The detailed measurements and calculations required to predict σ show that it is much easier to measure penetration resistance than to predict it. One of the major findings of this work was the large contribution of friction to penetration resistance. Friction on a 5° semiangle probe accounts for more than 80% of the total penetration resistance (Eq. 1). This has been tested using a penetrometer with a rotating tip (Bengough et al., 1991, 1997). Rotation of the penetrometer tip decreased the resultant component of friction directed along the penetrometer shaft. The measured penetration resistance agreed closely with the predicted resistance in a range of soils.

When the cone angle exceeds 90° $-$ ϕ, where ϕ is the angle of internal friction of the soil, a cone of soil builds up on the probe tip (Koolen and Kuipers, 1983). This body of soil moves with the probe, so that friction occurs between the soil body and the surrounding soil, instead of between the metal and soil surfaces. Equation 1 can therefore be applied only to probes with relatively narrow cone angles. Penetrometer design, testing procedure, and the effects on penetration resistance are considered in Sec. III.

B. Effects of Soil Properties on Penetration Resistance

Penetration resistance depends on soil type—the distribution of particle sizes and shapes, the clay mineralogy, the amorphous oxide content, the organic matter content, and the chemistry of the soil solution (Gerard, 1965; Byrd and Cassel, 1980; Stitt et al., 1982; Horn, 1984). Within a given soil type, the penetration resistance depends on the bulk density, water content, and structure of the soil. Penetration resistance can be affected by the pretreatment of the soil prior to testing. Hence the penetration of samples that have been dried, sieved, rewetted, and remolded will probably be very different from the penetration resistance of the soil in the field. The purpose of the experiment must therefore be considered carefully before the soil is sampled or penetration resistance is measured.

Penetration resistance decreases with increasing soil water content, and it increases with increasing bulk density. Gravimetric water content is a useful measure of water status, as matric potential and volumetric water content may change as soil is compressed during penetration (Koolen and Kuipers, 1983). Matric potential, however, is the mechanistic link to effective stress and hence to soil strength, via the surface tension of water-films holding the soil particles together (Marshall et al., 1996). Water content has little effect on cone resistance in loose soil, but its effect increases with bulk density. The influence of bulk density on cone resistance is greater in dry than in wet soil. Different functions have been proposed to describe these relations (Perumpral, 1983). For a given soil, the simplest suitable function is

$$Q = k_1 + k_2\theta_m + k_3\rho + k_4\rho\theta_m \tag{2}$$

where θ_m is gravimetric water content, ρ is dry bulk density, and $k_1 \ldots k_4$ are empirical constants (Ehlers et al., 1983). This relation is applicable widely and is illustrated in Fig. 2, using values of the constants for a loess soil. In some soils, however, the changes in cone resistance with bulk density and water content are not linear: cone resistance changes most rapidly at high bulk densities and low water contents. The linear model (Eq. 2) may still be appropriate if the ranges of bulk density and water content are small or soil variability is high, but other models may be valid more generally (Perumpral, 1983).

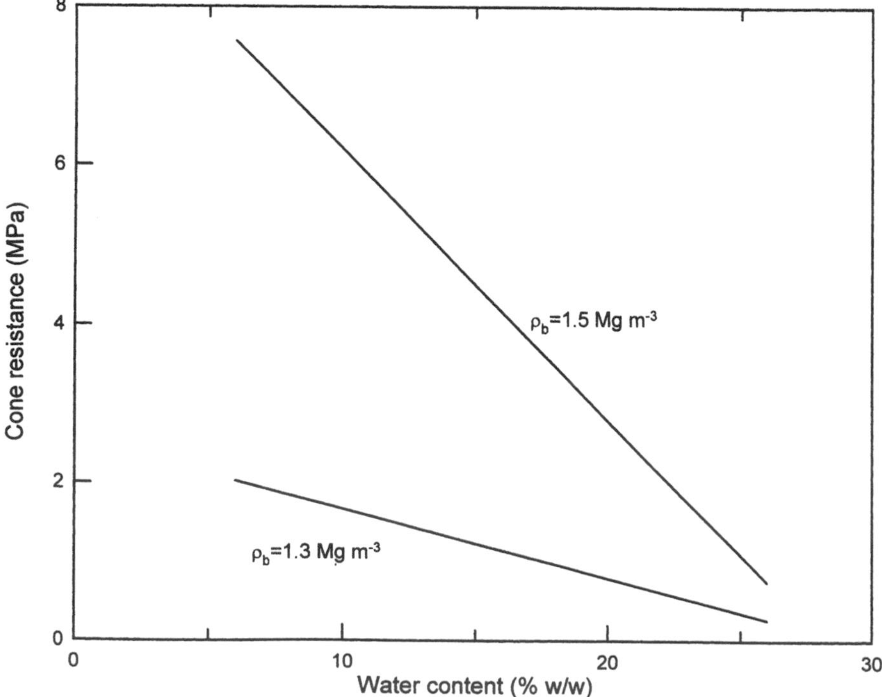

Fig. 2 Variation of penetrometer resistance with water content at different bulk densities. (Based on data from Ehlers et al., 1983.)

The relation between soil strength (in this case measured as penetration resistance) and matric potential is known as the soil strength characteristic. The main problem in deriving and applying such empirical relations is that soil strength changes with time, even if bulk density and water content remain constant (Davies, 1985). Soil management practices affect soil structure, changing the constants in these empirical relations.

At constant water content and bulk density, cone resistance tends to increase with decreasing particle size (Ball and O'Sullivan, 1982; Horn, 1984). Thus a clay will have a larger penetration resistance for a given gravimetric water content than a sand. This is due to the greater effective stress associated with the lower matric potential in the finer textured soil. In general, the decrease in organic matter associated with the intensive cultivation or deforestation of soils is associated with an increase in the gradient of the soil strength characteristic (Mullins et al., 1987).

III. PENETROMETER DESIGN

Details of a selection of commercially available penetrometers are given in Table 1. Penetrometers can be classified broadly as "needle" type if they have a diameter smaller than about 5 mm. Most needle penetrometers are used for testing of soils in the laboratory, though some have been used in the field. Penetrometers that are used in the field often have a diameter greater than 10 mm. Many penetrometers have also been designed for specific purposes. Needle penetrometer measurements can be made in the laboratory by attaching a suitable probe to the force transducer of a loading frame designed for material testing. In the following sections, the effects of penetrometer design and testing procedure on penetration resistance measurements are considered.

A. Cone Angle and Surface Properties

Penetrometer tips are generally cones, although flat-ended cylinders (Groenevelt et al., 1984) and shapes resembling the tips of plant roots (Eavis, 1967) have been used. The shape of the tip determines both the mode of soil deformation and the amount of frictional resistance on the tip. Penetrometer resistance is a minimum at a cone angle of 30° (Fig. 3; Gill, 1968; Voorhees et al., 1975; Koolen and Vaandrager, 1984). Increased cone resistance is associated at small cone angles with the increased component of soil–metal friction and, at large cone angles, with soil compaction in front of the cone (Gill, 1968; Mulqueen et al., 1977). Figure 3, which was derived from measurements made in 67 agricultural fields (Koolen and Vaandrager, 1984) shows the relationship between cone resistance and cone angle for a fixed cone base area. Soil tends to be displaced laterally at small cone angles, whereas the direction of displacement becomes more vertical with increasing cone angles (Gill, 1968; Tollner and Verma, 1984). Lateral soil displacement relates more closely to the mechanics of root growth than does the more axial displace-

Table 1 Suppliers of Some Penetrometers, Force Transducers, and Load Frames Available Commercially

Supplier	Address	Equipment	Approximate cost (US$)
ELE International Ltd.	In the UK: Eastman Way, Hemel Hempstead, Hertfordshire, HP2 7HB In the USA: 86 Albrecht Drive, P.O. Box 8004, Lake Bluff, Illinois 60044-8004	Field penetrometer with data logger, hand-held.	7500
Soil Test Inc.	2250 Lee Street, Evanston, Illinois 60202, USA	Proving ring penetrometer	
Eijkelkamp	P.O. Box 4, 6987ZG Giesbeek, The Netherlands	Field penetrometer with data logger, hand-held	8800
Leonard Farnell & Co. Ltd.	North Mymms, Hatfield, Hertfordshire AL9 7SR, UK	Simple hand-held penetrometer with dial gauge.	1000
Ametek	Mansfield & Green Division, 8600 Somerset Drive, Largo, Fl 34643, USA	Wide range of loading frames and force transducers. Agents also in UK.	
Pioden Controls Ltd.	Graham Bell House, Roper Close, Roper Road, Canterbury, Kent CT2 7EP, UK	Force transducers suitable ranges for needle penetrometers.	From about 270
Applied Measurements Ltd.	3 Titan House, Calleva Park, Aldermaston, Reading, Berkshire, RG7 4QW, UK	Force transducers suitable ranges for needle penetrometers	From about 225

Inclusion in this list does not constitute any recommendation of the product.

ment produced by probes with larger cone angles (Greacen et al., 1968). Conversely, the load-bearing characteristics of the soil are more closely related to the resistance encountered by larger cone angles. Penetrometers that are available commercially are generally fitted with 30° or 60° cones, but these can be easily interchanged.

The surface roughness of the cone is not an important factor in penetrometer design, as abrasion by soil particles quickly removes any minor irregularities. Lubrication of the cone decreases penetration resistance by decreasing soil–cone friction and the movement of soil in the axial direction (Gill, 1968; Tollner and Verma, 1984). Use of such a lubricated penetrometer is of questionable advantage, as the mechanics of penetration of a lubricated cone is poorly understood, and the lubricating technology may be difficult to standardize.

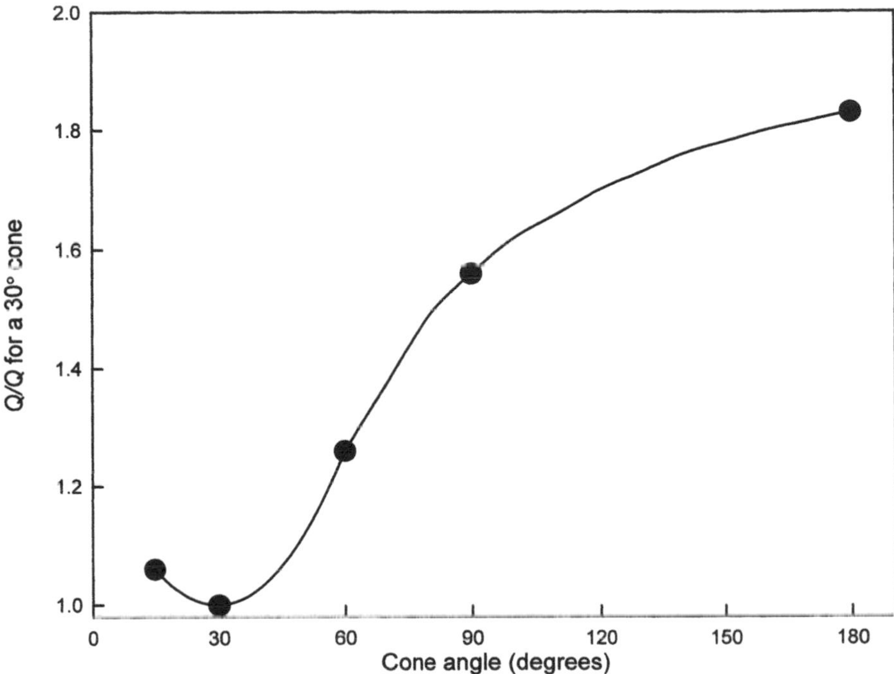

Fig. 3 Variation of penetrometer resistance with cone angle for a fixed cone base area. (From Koolen and Vaandrager, 1984. Reproduced with permission from *the Journal of Agricultural Engineering Research.*)

B. Cone Base Diameter

In general, the diameter of needle penetrometers is important and must be taken into account when comparing results from different instruments. Diameter is less important when comparing field penetrometers.

The diameter of the cone bases range from large field penetrometers (>10 mm) (Ehlers et al., 1983) to small needle penetrometers (<0.2 mm) (Groenevelt et al., 1984). Although cone resistance is expressed as a force per unit base area, it tends to increase with decreasing base area (Freitag, 1968). For field penetrometers, the standard of the American Society of Agricultural Engineers (ASAE, 1969) allows cone base areas of 320 mm^2 and 130 mm^2, both with a 30° cone angle. A 3% decrease in diameter is allowed for cone wear. In Europe, cones of 100 mm^2 base area are common, but cones with base areas of up to 500 mm^2 have been used.

Even in homogeneous soil, penetration resistance can depend on probe diameter as soil particles of finite size must be displaced. Diameter dependence is most noticeable for very small probes, which may have to displace particles of comparable size. The effect of probe diameter on penetration resistance depends on the soil type, water content, and structure (Whiteley and Dexter, 1981). In remolded soil cores with textures ranging from clay to sand, resistance to a 1 mm probe was typically 45–55% greater than to a 2 mm diameter probe (Whiteley and Dexter, 1981). Other studies found no significant effect of diameter among 1, 2, and 3 mm diameter probes in remolded sandy loam (Barley et al., 1965), between 3.8 and 5.1 mm probes in undisturbed cores (Bradford, 1980), and between 1 and 2 mm probes in both undisturbed clods and remolded soils (Whiteley and Dexter, 1981). There is need for a comprehensive study over a wide range of penetrometer diameters and soil textures.

In soils with well-developed structural units, the mechanism of penetration may differ between cones of different sizes. A cone with a small diameter, relative to the size of structural units, may penetrate aggregates or planes of weakness between aggregates, whereas a large cone will tend to deform aggregates (Jamieson et al., 1988).

C. Shaft Diameter

The surface area of a penetrometer shaft is directly proportional to its diameter, whereas the force on the penetrometer tip is proportional to the square of the tip diameter. Thus shaft friction is relatively more important for smaller probes, and this has been confirmed by experiment (Barley et al., 1965). To decrease soil–metal shaft friction, a relieved shaft (i.e., a shaft with a diameter 20% smaller than the probe tip) is used commonly.

Shaft friction can significantly increase the resistance even to a standard ASAE penetrometer, especially in wet clay (Freitag, 1968; Mulqueen et al., 1977). Freitag (1968) found that increasing the shaft diameter from 9.5 mm to 15.9 mm (the ASAE standard) increased the resistance threefold at 0.3 m depth on a standard 20.3 mm diameter cone. Similarly, Reece and Peca (1981) used a shaft 8 mm in diameter to eliminate the clay–shaft friction on the standard 20.3 mm diameter cone.

IV. PENETROMETER INSERTION AND MEASUREMENT

A. Force Measurement

The commonest and most easily interpreted penetrometer results are from measuring the resistance to a probe driven into soil at a constant speed. Other designs measure the magnitude or the rate of probe penetration under different constant loads (van Wijk, 1980). In this chapter only penetrometers designed to be used at a constant rate are considered.

1. *Laboratory Needle Penetrometers*

To obtain a constant rate of penetration in the laboratory, it is necessary either to drive the probe downward into the soil with some sort of motor (Barley et al., 1965) or to raise the soil sample on a moving platform toward a stationary probe (Eavis, 1967). The movable crosshead of a strength testing machine has a convenient drive capable of a wide range of speeds, and can accept force transducers to measure the force resisting penetration (Fig. 4; Callebaut et al., 1985; Bengough et al., 1991). Proving rings, strain gauges, and electronic balances have all been used to measure the force resisting penetration (Barley et al., 1965; Eavis, 1967;

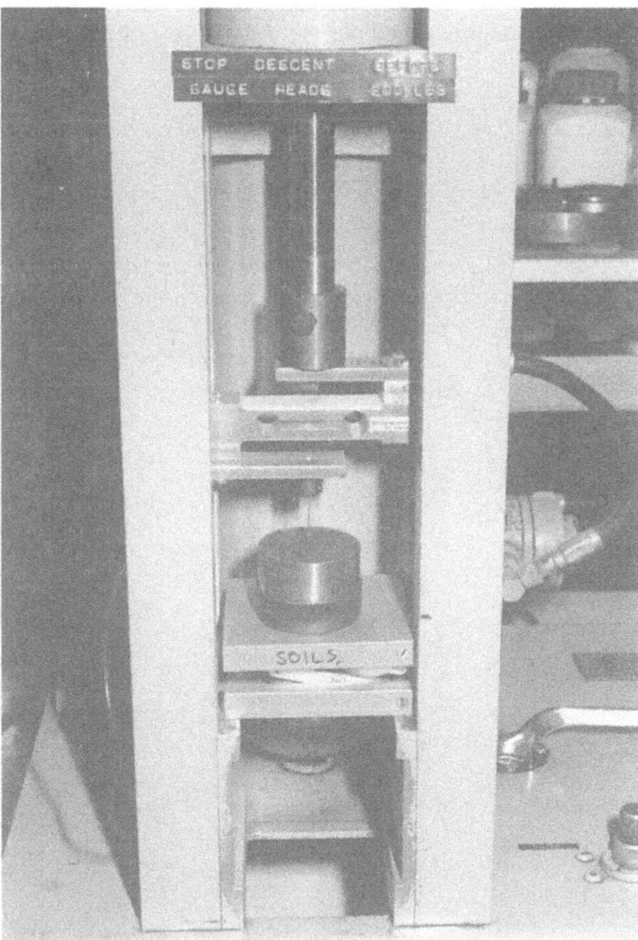

Fig. 4 Needle penetrometer attached to a force transducer on a loading frame.

Misra et al., 1986a). The advantage of an electronic balance or force transducer is that the output can be logged using the analog-to-digital converter of a datalogger or personal computer. Proving rings that are too flexible can result in small voids going undetected, as the proving ring expands when unloaded.

2. Field Penetrometers

A field penetrometer may be mounted on a rack to allow easy and precise location (Soane, 1973; Billot, 1982). This facilitates measurements on a regular, closely spaced grid. Hand-held penetrometers are more portable, are cheaper, and can be used in inaccessible field sites (Fig. 5).

Automatic logging of force is very advantageous, as it is difficult for the operator to record measurements at predefined depths. Analog recording using a

Fig. 5 Field penetrometer with data storage unit.

chart recorder records even rapid changes with depth. However, the graphical output must then be digitized for statistical analysis, which can be laborious.

Digital recording has the disadvantage that maxima and minima may be not be identified. This loss of information can be important when depth increments are large, especially if cone resistance changes abruptly with depth or if the depth of a cultivation pan varies between penetrations. Averaging data at predetermined depths can disguise such features.

B. Rate of Penetration

1. Laboratory Needle Penetrometers

Needle penetrometers are used most commonly to estimate the penetration resistance of the soil to roots. Roots elongate typically at a rate of 1 mm/h or less, which is an inconveniently slow rate at which to conduct penetrometer tests. Most needle penetrometer measurements are performed at rates of penetration between one and three orders of magnitude faster than root growth rates (Whiteley et al., 1981). Eavis (1967) found no effect of rate of penetration on the penetrometer resistance of a silty clay loam at rates between 5 and 0.1 mm/min. At slower rates of penetration, however, the resistance decreased, but only by 13% at a penetration rate 20 times slower. A small decrease in the penetrometer resistance of sandy loam and clay was noted at rates below 0.02 mm/min (Voorhees et al., 1975). In saturated clay, penetrometer resistance increases with penetration rate because water must be displaced as the probe compresses the soil (Barley et al., 1965). In such a saturated system, the penetration resistance depends on the saturated hydraulic conductivity in the soil surrounding the probe. Penetrometer resistance is relatively weakly dependent on penetration rate in unsaturated sandy soils at typical rates of testing. Given the large difference in penetration rate between roots and penetrometers, it is still an important factor that must be evaluated if estimating the penetration resistance to roots.

2. Field Penetrometers

Increasing penetration speed increases cone resistance in fine-textured soils (Freitag, 1968), in which strength depends on strain rate (Yong et al., 1972). In most soils, however, cone resistance is relatively insensitive to penetration rate within the range expected from operators of manual penetrometers aiming for the ASAE standard rate of 30 mm/min (Carter, 1967; van Wijk and Beuving, 1978; Anderson et al., 1980). The constant penetration rate possible with mechanically driven penetrometers is not a significant advantage. Exceptions are saturated clay (Turnage, 1973) and soils with a strong layer overlying a weak layer. The large force required to penetrate the strong layer may cause an excessive penetration rate in the underlying layer.

C. Variability

Penetration resistance readings can be very variable, even when penetrations are made close together (O'Sullivan et al., 1987). The coefficient of variation is typically between 20 and 50%, though it may be more than 70% near the soil surface (Voorhees et al., 1978; Cassel and Nelson, 1979; Gerrard, 1982; Kogure et al., 1985). Small cones give more variable results than large cones (Bradford, 1980). The resistance readings may have a skewed distribution, so that a logarithmic (McIntyre and Tanner, 1959; Cassel and Nelson, 1979) or square root (Mitchell et al., 1979) transformation is necessary to normalize the data. Data at individual depths may be normally distributed (Cassel and Nelson, 1979; Gerrard, 1982; O'Sullivan and Ball, 1982), but a logarithmic transformation may be necessary if depth is included as a factor in analyzing results.

The number of measurements, N, required can be predicted using the equation

$$N = \left[\frac{2CV}{L}\right]^2 \tag{3}$$

where L is the 95% confidence interval, expressed as a percentage of the mean, and CV is the coefficient of variation (%) (Snedecor and Cochran, 1967). This relation assumes that the data is normally distributed and is illustrated in Fig. 6 for values of CV that represent the normal range encountered. A fourfold increase in the number of replicates is required to double the expected degree of precision. The ASAE recommends seven measurements, giving a 95% confidence interval between about 15 and 38% of the mean. This is a very large error compared with the maximum 5% error they allow for cone wear, though such wear is a source of systematic error (ASAE, 1969).

Our estimates of the number of penetrations required assume that all measurements are independent. O'Sullivan et al. (1987) found that measurements made more than about 1 m apart were independent, but Moolman and Van Huyssteen (1989) found evidence of spatial dependence that extended to about 9 m.

The penetrometer is ideal for investigating the uniformity of a site because the measurements can be made cheaply, quickly, and easily. Furthermore, cone resistance is related to many other soil properties. Hartge et al. (1985) used the penetrometer to identify areas within a field experiment for more detailed investigation. Schrey (1991) showed that cone resistance data could be used to identify areas of shallow or compact soil or plow pans.

D. Problems in Use

1. *Laboratory Needle Penetrometers*

Most penetrometers designed for small cones are unsuitable for field use (Bradford, 1980). Large field penetrometers have been used successfully in root growth

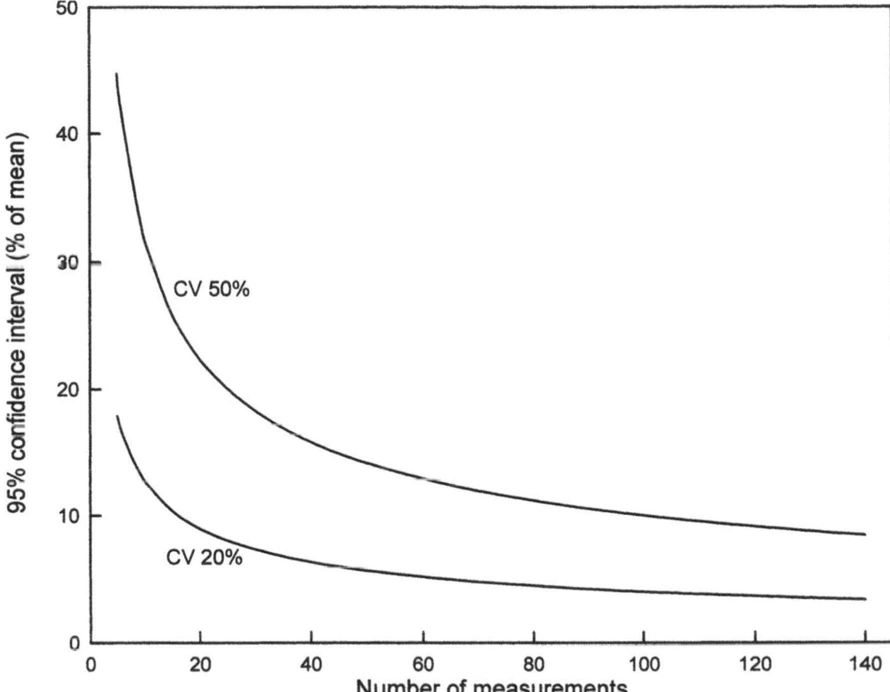

Fig. 6 Variation of the 95% confidence interval about the mean with the number of cone resistance observations, for two coefficients of variation.

studies (Ehlers et al., 1983; Barraclough and Weir, 1988; Jamieson et al., 1988), but these are very different from growing roots, in terms of diameter and penetration rate.

Care must be taken, when sampling soils for needle-penetrometer measurements, that the soil is compressed as little as possible during coring. Soil is compacted if cores are sampled too close together, or if soil is trampled by the fieldworker. Such compaction increases the penetrometer resistance.

Lateral confinement of the soil core may increase penetrometer resistance if the core diameter is less than about 20 times that of the probe (Greacen et al., 1969). Tensile failure of the core may occur if the core is unconfined laterally, decreasing the penetrometer resistance as the core cracks. Penetrometer resistance may also be affected if more than one penetration is performed on each core—cracks of tensile failure may form between the penetration holes (Greacen et al., 1969) though, under other circumstances, penetration resistance could be increased by compaction around the neighboring penetration hole.

Stones cause rapid increases in penetration resistance that can damage sensitive force transducers. Overload cutoffs should be included, if possible, to protect against such damage in motor-driven penetrometers. Force readings corresponding to stones should be specially identified in a data set. Roots can grow around stones and other localized regions of large resistance, and so it may be appropriate to remove these readings from the data set if the aim is to relate resistance to root growth. Penetrometer readings taken after a stone has been pushed aside may also have to be discarded in case the stone rubs against the penetrometer shaft, creating larger frictional resistance.

Penetrometer readings obtained as the probe is entering the surface layer of the soil (i.e., depths less than three times the probe diameter) should be discarded: the values of resistance are anomalously small because the soil failure mechanism near the soil surface is different from that in the bulk soil (Gill, 1968).

2. *Field Penetrometers*

The operator of a penetrometer that is driven by hand can often sense a sudden change in the force transmitted from the penetrometer cone when a stone is hit. The presence of stones increases the mean and standard deviation of the penetrometer resistance data, may introduce unrepresentative large values, and increase the shaft friction. Stone encounters may be identified as outliers, for example, more than three standard deviations from the mean. Such outliers should be eliminated from penetrometer data as they may bias treatment comparisons, though they are unlikely to affect treatment rankings (O'Sullivan et al., 1987). In very stony soils, however, all penetrations are affected to some extent by stones. Penetrations may fail to reach the required depth because they are obstructed by stones. When this happens, the penetration should not be abandoned. Discarding such data could bias the results, because stones are more likely to prevent penetration in strong than in weak soil. Missing observations can be replaced by their expected values (Glasbey and O'Sullivan, 1988). There are a number of less sophisticated techniques that can also be used to avoid bias, such as replacing the first missing value in each penetration by the maximum measurable value (Glasbey and O'Sullivan, 1988). The number of interrupted penetrations can also give an indication of soil stone content (Wairiu et al., 1993).

Measurements at adjacent depths in a penetration are generally not independent. O'Sullivan et al. (1987) showed that measurements made at depths closer than 0.25 m were correlated. A significant treatment effect at one depth is likely to be accompanied by significant effects at adjoining depths. Soil overburden pressure increases with depth, increasing penetration resistance (Bradford et al., 1971). Shaft friction increases with depth and may be increased further by bending of the shaft when high-strength layers or stones are encountered. The interpretation of cone resistance values therefore depends on the depth of measurement. Simple averaging of cone resistance over a number of depths may be misleading, and the geometric mean may be more appropriate than the arithmetic mean. Statistical methods such as covariance analysis and time series analysis can be used to correct for water content, bulk density, and depth effects and so increase the validity of treatment comparisons (Christensen et al., 1989).

Compaction and tillage treatments that cause large changes in the height of the soil surface create problems for interpreting penetrometer data. High resolution bulk density measurements beneath a wheel rut may establish the original depth of each layer in the compacted soil. This calculation cannot be made when only cone resistance is recorded, but a good approximation is to assume that each layer moves vertically by the same amount (Henshall and Smith, 1971). An example of this depth correction in a tillage experiment is given in Fig. 7. The average bulk density of the plowed soil was 1.2 Mg m^{-3} and that of the direct drilled soil was 1.5 Mg m^{-3}, with a plowing depth of 0.25 m. Thus the equivalent depth of direct-drilled soil was $0.25 \times 1.2/1.5 = 0.2$ m, and the scale factor to convert the actual depth in plowed soil to the equivalent depth in direct-drilled soil was 0.8 (= 0.2/0.25). Figure 7 shows that an apparent cultivation effect below the

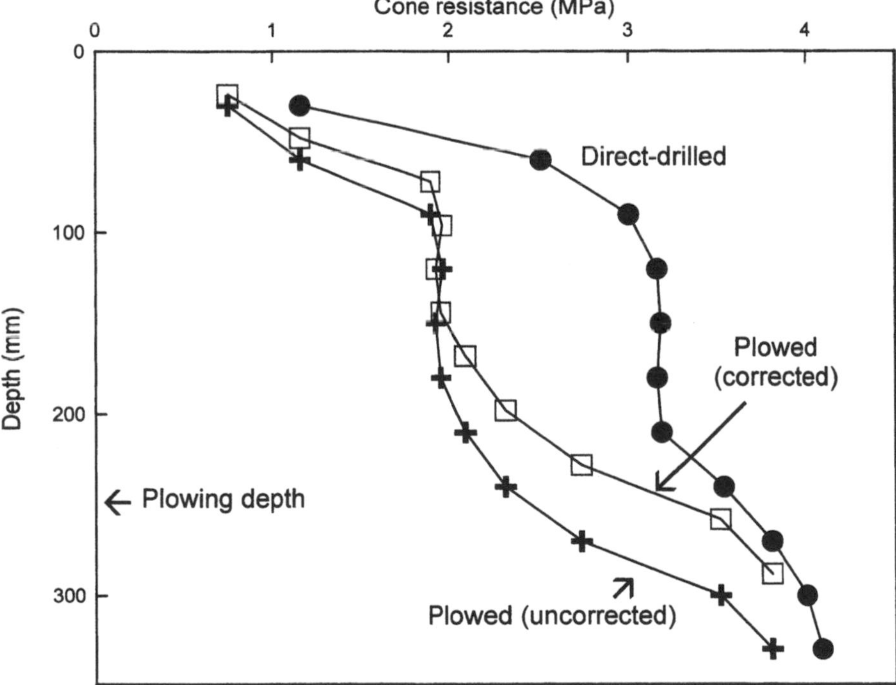

Fig. 7 Variation of soil cone resistance with depth for plowed and direct-drilled soils, before and after correction for the difference in surface level between treatments, due to compaction.

depth of plowing was merely a consequence of the greater depth of topsoil in the plowed than in the direct-drilled land. Such depth corrections are essential when differences in surface level between treatments are large and the investigation is concerned with the mechanism or processes that led to the measured values.

V. APPLICATIONS

A. Trafficability

Trafficability refers to the ability of the soil to allow traffic without excessive structural damage, and the term is also used to indicate its potential to provide adequate traction for vehicles. The cone penetrometer has been used widely for assessing soil trafficability (Knight and Freitag, 1962; Freitag, 1965; Turnage, 1972) and for predicting the performance of tires (Turnage, 1972; Wismer and Luth, 1973) and cultivation implements (Wismer and Luth, 1973). The main objections to the prediction of tire performance from cone resistance are that cone resistance alone is insufficient to characterize the strength of soils (Mulqueen et al., 1977), and that a penetrometer and a wheel induce markedly different strains in the soil (Yong et al., 1972). The calibration data also limit the accuracy of predictions, and the effects of soil compaction on cone resistance are not yet predictable. In common with all other empirical methods, results cannot be extrapolated to soils that have not been included in the calibration, and the method gives no insight into the processes involved. The advantages of penetrometers are that they are simple and fast to use, and that simple useful relations can be developed between cone resistance and wheel performance.

Predictions of whether a soil is trafficable (Knight and Freitag, 1962; Paul and de Vries, 1979) may be adequate for the limited range of vehicles and soils used in deriving empirical relations. Predictions of the effects of varying soil and wheel parameters on properties such as trafficability should be used only to rank treatments or make approximate comparisons.

Engineers of the U.S. Army developed a trafficability assessment system for fine-grained soils (Knight and Freitag, 1962). The "rating cone index" was measured as the average cone index of a critical layer, after an empirical correction for the softening of the soil under the action of the wheels. This critical layer was between 0.15 and 0.3 m thick for most military vehicles. The "vehicle cone index," required to allow 50 passes of a given vehicle, was estimated empirically from factors including the vehicle weight, tire–soil contact stress, engine power, and transmission type.

A dimensional analysis of tire–soil and cone–soil interaction led to the development of dimensionless mobility numbers for dry, cohesionless sands, and saturated, frictionless clays (Freitag, 1965). The clay and sand mobility numbers N_c and N_s are given by

$$N_c = Q\frac{bd}{W}\left(\frac{\Delta}{h}\right)^{1/2}\left[\frac{1}{1 + b/2d}\right] \tag{4}$$

$$N_S = G\left(\frac{bd}{W}\right)^{3/2}\frac{\Delta}{h} \tag{5}$$

where b, d, and h are the unloaded tire width, diameter, and section height, Δ is the tire deflection under load, W is the vertical load on the tire, Q is the cone index, and G is the gradient of cone index with depth. These mobility numbers were used as independent variables in empirical predictions of tire sinkage and torque, and hence drawbar pull (Turnage, 1972). The clay and sand mobility numbers required refining to reflect the variation in compactibility and strength between sands (Reece and Peca, 1981; Turnage, 1984).

Wismer and Luth (1973) recognized that wheel behavior differed between the unsaturated, cohesive–frictional soils, usual in agriculture, and the saturated clays for which Eq. 4 was developed. They proposed empirical equations to predict the towing force on an undriven wheel, the pull generated by a driven wheel, and tractive efficiency for agricultural soils from the "wheel numeric," C_n,

$$C_n = Q\frac{bd}{W} \tag{6}$$

They suggested that the average cone resistance of the top 150 mm should be used for Q if the tire sinkage was shallower than 75 mm. If the sinkage was greater, the average cone resistance of the 150 mm layer, which included the maximum sinkage of the tire, should be used. No guidance was given, however, for predicting tire sinkage. Another difficulty with this procedure is the tendency of agricultural soils to compact, with a large, but unpredictable, change in strength, during the passage of a wheel. Traction is therefore more closely related to the properties of the compacted than the uncompacted soil. Consequently, the cone resistance measured after compaction gives a better prediction than that measured before compaction (Wismer and Luth, 1973). The method is therefore of limited use in loose agricultural soils.

Paul and de Vries (1979) plotted cone resistance against the subsequent wheelslip of a tractor pulling a manure spreader and used the cone resistance at 20% wheel slip as a criterion of trafficability. They combined this value with empirical relations between cone resistance and water table depth (Paul and De Vries, 1979) and a numerical simulation of the drainage process (Paul and de Vries, 1983) to investigate the effects of drain spacing on soil trafficability. Good agreement was found between model output and farmers' assessments of trafficability.

B. Compaction and Tillage

Soane et al. (1981) and O'Sullivan et al. (1987) reviewed the use of the cone penetrometer in studies of traffic and tillage. The penetrometer is a useful rapid method for detecting compact layers; assessing the relative depth, intensity, and persistence of loosening or compaction between treatments; detecting changes in strength with time; and assessing whether soil strength will limit root growth (see Sec. V.C). Compaction and tillage have much greater effects proportionally on penetration resistance than on bulk density. Differences between treatments are greatest generally when the soil is dry.

Comparisons between traffic and tillage treatments are often complicated by differences in water content. Measurements made at field capacity decrease the effect of water content but also minimize treatment effects. Furthermore, the penetration resistance of the soil under dry conditions is often of greater interest. The soil water content should be measured at the same time as the penetration resistance, so that a soil strength characteristic can be constructed (Young et al., 1993). This allows penetration resistances to be compared at any given water content. The cone penetrometer is useful for making empirical comparisons between traffic and tillage treatments on the same soil type. Comparisons between soils are confounded because of the complex effects of soil type on cone resistance.

Measurements at field water content should be made as soon as possible after the passage of wheels, because changes in matric potential and hydraulic conductivity associated with compaction will eventually lead to changes in water content below the wheel track. Differences in cone resistance between treatments may be small if the average bulk density is low. Depth effects, as discussed earlier, may also complicate comparisons between treatments, even when a depth correction is made. Dickson and Smith (1986) measured both cone resistance and bulk density below the ruts of a wheel supporting one of two loads at each of two ground pressures. After depth corrections were made, bulk density results confirmed the theoretical predictions that ground pressure is important to compaction at shallow depth, while wheel load is more important at greater depths. In contrast, although cone resistance data were consistent with bulk density data at shallow depths, no treatment effects were detected at greater depths.

Penetrometers can be used to study the spatial effects of tillage implements (Cassel et al., 1978; Threadgill, 1982; Billot, 1985; O'Sullivan et al., 1987) and wheel traffic. Figure 8 shows penetration resistance profiles across the direction of travel of a slant-leg subsoiler, and below wheel tracks (O'Sullivan et al., 1987). In both of these diagrams, the arrangement of the loose and compacted regions of soil can be seen clearly.

In addition to its use for empirical comparisons of compaction, cone resistance has been related to compactive effort (O'Sullivan et al., 1987). The penetrometer has been used to estimate stresses and their distribution under wheels and other loads (Blackwell and Soane, 1981; Koolen and Kuipers, 1983; Bolling,

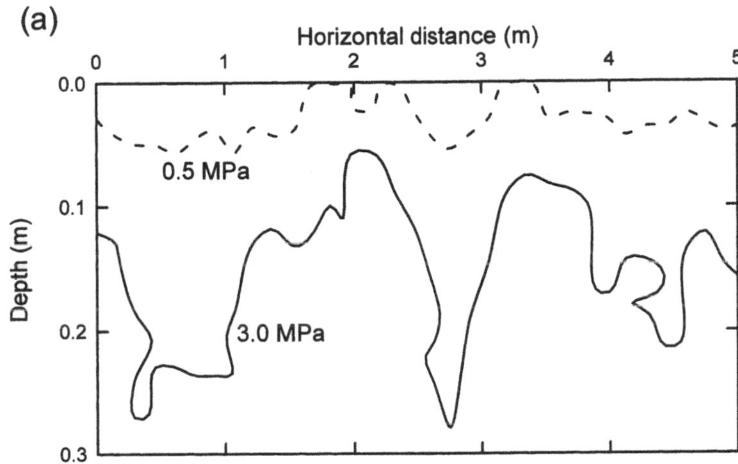

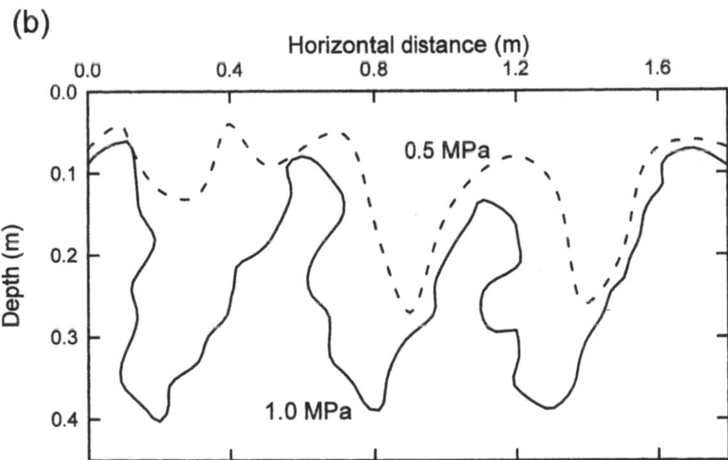

Fig. 8 Variation of cone resistance with depth: (a) across a field of conventionally grown winter barley. Large penetration resistances lie below the wheel tracks; (b) across the direction of travel of a slant leg subsoiler, showing the 0.5 and 1.0 MPa contours.

1985). Penetrometer resistance has also been used to predict plow draft (Wismer and Luth, 1973) and the performance of cultivator tines (Gill, 1968). However, soil deformation around a cone differs from that around a tine, and therefore the cone is not a good analog of cultivator performance (Freitag et al., 1970; Johnston et al., 1980).

C. Root Growth

1. *Comparisons Between Penetrometer Resistance and Root Resistance*

Few studies have compared directly root penetration resistance and penetrometer resistance, because of the experimental difficulties involved with the root measurements. Such comparisons are made by measuring the force exerted by a root as it penetrates a soil sample (Whiteley et al., 1981; Bengough and Mullins, 1991). The technique involves anchoring a root with plaster of Paris a few mm behind its apex. The root is allowed to grow into the surface of a soil core until the root has extended at least three times its diameter into the surface of a soil core, but before the tip becomes anchored by root hairs. The force exerted on the soil by the penetrating root tip is recorded using a balance or force transducer. To calculate the root penetration resistance, the root force must be divided by the cross-sectional area of the root. Roots often swell in response to mechanical impedance and, as a continuous record of root force and diameter cannot normally be obtained, it is not clear whether it is most relevant to measure the initial or the final root diameter. Indeed, because root tips are tapered, the distance behind the root tip at which diameter is measured can be of considerable importance. The best solution is to measure root diameter at 1 mm intervals behind the root tip. The diameter used in the calculation should be measured at the distance behind the root tip that is level with the soil surface when the force measurement is made. The root resistance then calculated should correspond to the normal stress on the surface of the root, if the stress is distributed uniformly on the root surface.

Direct comparisons have shown that penetrometers experience a resistance between two and eight times greater than roots (Table 2). Further indirect evidence of this difference comes from comparing studies of root elongation rate and penetrometer resistance with measurements of the maximum pressures that roots can exert. Critical values of penetrometer resistance at which root elongation ceases are in the 0.8–5.0 MPa range, depending on the soil and the crop (Greacen et al., 1969). The maximum axial pressures a root can exert vary between about 0.24 and 1.45 MPa, depending on species (Misra et al., 1986b). Such maximum pressure is limited by the cell turgor pressure in the elongation zone. Thus root elongation is halted in soil with a penetrometer resistance much greater than the maximum pressure the root can exert. The reason why penetrometers experience much greater resistance than roots is largely because they encounter much more frictional resistance (Bengough and Mullins, 1991). The relative importance of other factors is unclear, but the faster penetration rate of the penetrometer will certainly account for some of the difference, especially in finer-textured soils.

Root elongation rate decreases, approximately inversely, with increasing penetrometer resistance (Taylor and Ratliff, 1969; Ehlers et al., 1983). This is illustrated for two crop species in Fig. 9. A similar form of relation between ap-

Table 2 Comparisons of Penetrometer Resistance with Root Penetration Resistance Measured Directly

Soil	Probe diameter (mm)	Cone semi-angle (°)	Penetration rate (mm min^{-1})	Ratio, probe resistance/ root resistance	No. of replicates	Reference
Remolded sandy loam	1	Parabolic probe	1	4–8	12	Eavis (1967)
Remolded sandy loam	3	30	0.17	4.5–6	2	Stolzy and Barley (1968)
Sandy loam, remolded cores and undisturbed clods	1 to 2	30	3	2.6–5.3	120	Whiteley et al. (1981)
Clay loam aggregates	1	30	3	1.8–3.8	324	Misra et al. (1986b)
Sandy loam, undisturbed cores	1	30	4	4.5–9	14	Bengough and Mullins (1991)
Sandy loam, remolded cores	1	7.5	2	2.5–4.8	19	Bengough and McKenzie (1997)

Updated from Bengough and Mullins, 1990.

plied pressure and root growth has been obtained in studies using pressurized cells filled with ballotini (Abdalla et al., 1969; Goss, 1977). Voorhees et al. (1975) found that root elongation rate correlated better with the resistance to a 5° semi-angle probe after the frictional component of resistance (estimated by measuring the angle of soil–metal friction) had been subtracted.

2. *Small-Scale Variations in Soil Strength*

Penetrometers, unlike roots, follow a linear path through the soil and are unable to follow biopores, cracks, or planes of weakness in the way that roots have been observed to do (Russell, 1977). This limits the utility of penetrometers in structured soil, where the average resistance measured by large penetrometers will overestimate the resistance to root growth. Soil structure exists as a hierarchy (Dexter, 1988), so that even soils that are macroscopically homogeneous contain spatial variations in strength on a much smaller scale, which a root may be able to exploit. Ehlers et al. (1983) found that roots grew through untilled soil with a large penetration resistance, whereas root growth was halted in tilled soil with the same penetration resistance. The untilled soil contained more cracks and biopores that

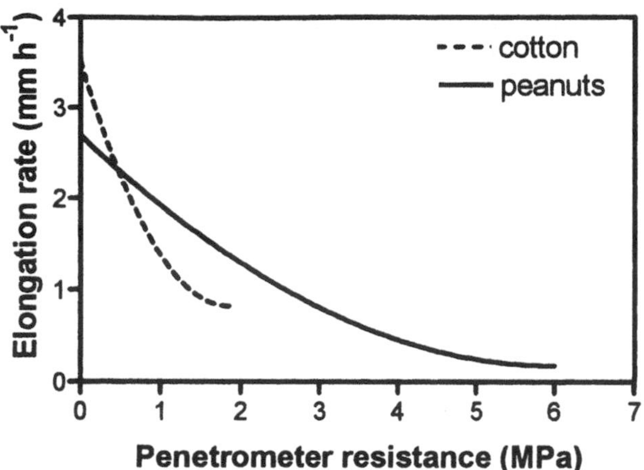

Fig. 9 Root elongation rate for peanuts and cotton versus soil penetrometer resistance. (Reproduced from H. M. Taylor and L. F. Ratliff, Root elongation rates of cotton and peanuts as a function of soil strength and water content. *Soil Science* 108:113–119 (1969). © by Williams and Wilkins, Baltimore, MD.)

were available for root growth, but were not detected by the field penetrometer with an 11 mm diameter cone.

Individual soil peds can be considered continuous in some soils, even though the soil itself is structured on a larger scale (Greacen et al., 1969). Dexter (1978) used this idea, together with the probability of roots penetrating peds, to model root growth through a bed of aggregates. The variability of penetrometer readings may increase with decreasing penetrometer diameter, even though the average resistance is unchanged (Bradford, 1980). Very small penetrometers may be used to determine the fraction of the soil that is penetrable by roots (Groenevelt et al., 1984). The "percentage linear penetrability" decreases with increasing soil bulk density. Spectral analysis of penetrometer data has been attempted (Grant et al., 1985), but not yet applied to root growth.

VI. SUMMARY

Soil strength can be measured using a penetrometer. Penetration resistance is expressed as penetration force per unit cross-sectional area of the cone base. Penetrometer resistance measurements are used widely, are relatively quick and easy to make, and can provide data that are valuable if interpreted carefully. Penetration resistance depends on many factors, but the dry bulk density and water content of the soil are important especially. Penetration resistance measurements are useful in studies of trafficability, compaction, tillage, and root growth. The probe shape and testing procedure must be chosen appropriately, so that the results are of maximum relevance to the application. The American Society of Agricultural Engineers has adopted a standardized penetrometer design and testing procedure to be used for field studies of trafficability, compaction, and tillage. A very different probe design and testing procedure should be used in laboratory studies of root growth. Root elongation rate and root penetration resistance are related to penetrometer resistance in soils that do not contain many continuous pores or channels available for root growth. The best estimates of root penetration resistance are obtained by subtracting the large frictional component of resistance from the total penetration resistance.

Acknowledgment

The SCRI receives grant-in-aid from the Scottish Executive Rural Affairs Department.

REFERENCES

Abdalla, A. M., D. R. P. Hettiaratchi, and A. R. Reece. 1969. The mechanics of root growth in granular media. *J. Agric. Eng. Res.* 14:263–268.

American Society of Agricultural Engineers (ASAE). 1969. Soil cone penetrometer. In: Recommendation ASAE R313, *Agricultural Engineering Yearbook.* St. Joseph, MI: Am. Soc. Agric. Eng., pp. 296–297.

Anderson, G., J. D. Pidgeon, H. B. Spencer, and R. Parks. 1980. A new hand-held recording penetrometer for soil studies. *J. Soil Sci.* 31:279–296.

Ball, B. C., and M. F. O'Sullivan. 1982. Soil strength and crop emergence in direct drilled and ploughed cereal seedbeds in seven field experiments. *J. Soil Sci.* 33:609–622.

Barley, K. P., E. L. Greacen, and D. A. Farrell. 1965. The influence of soil strength on the penetration of a loam by plant roots. *Aust. J. Soil Res.* 3:69–79.

Barraclough, P. B., and A. H. Weir. 1988. Effects of a compacted subsoil layer on root and shoot growth, water use and nutrient uptake of winter wheat. *J. Agric. Sci.* 110: 207–216.

Bengough, A. G. 1992. Penetrometer resistance equation—Its derivation and the effect of soil adhesion. *J. Agric. Eng. Res.* 53:163–168.

Bengough, A. G., and B. M. McKenzie. 1997. Sloughing of root cap cells decreases the frictional resistance to maize (Zea mays L.) root growth. *J. Exp. Bot.* 48:885–893.

Bengough, A. G., and C. E. Mullins. 1990. Mechanical impedance to root growth—A review of experimental techniques and root growth responses. *J. Soil Sci.* 41: 341–358.

Bengough, A. G., and C. E. Mullins. 1991. Penetrometer resistance, root penetration resistance and root elongation rate in 2 sandy loam soils. *Plant Soil* 131:59–66.

Bengough, A. G., C. E. Mullins, and G. Wilson. 1997. Estimating soil frictional resistance to metal probes and its relevance to the penetration of soil by roots. *Eur. J. Soil Sci.* 48:603–612.

Bengough, A. G., C. E. Mullins, G. Wilson, and J. Wallace. 1991. The design, construction and use of a rotating-tip penetrometer. *J. Agric. Eng. Res.* 48:223–227.

Billot, J. F. 1982. Use of penetrometer for showing soil structure heterogeneity application to study tillage implement impact and compaction effects. In: *Proc. 9th Conf. Int. Soil Tillage Research Organization,* Osijek, Yugoslavia, pp. 177–182.

Billot, J. F. 1985. Use of penetrometry in tillage studies. In: *Proc. Int. Conf. Soil Dynamics,* Vol. 2, Auburn, AL, pp. 213–218.

Blackwell, P. S., and B. D. Soane. 1981. A method of predicting bulk density changes in field soils resulting from compaction by agricultural traffic. *J. Soil Sci.* 32:51–65.

Bolling, I. H. 1985. How to predict the soil compaction of agricultural rites. In: *Proc. Int. Conf. Soil Dynamics.* Vol. 5. Auburn, AL, pp. 936–952.

Bradford, J. M. 1980. The penetration resistance in a soil with well-defined structural units. *Soil Sci. Soc. Am. J.* 44:601–606.

Bradford, J. M., D. A. Farrell, and W. E. Larson. 1971. Effect of soil overburden pressure on penetration of fine metal probes. *Soil Sci. Soc. Am. Proc.* 35:12–15.

Byrd, C. W., and D. K. Cassel. 1980. The effect of sand content upon cone index and selected physical properties. *Soil Sci.* 129:197–204.

Callebaut, F., D. Gabriels, W. Minjauw, and M. De Boodt. 1985. Determination of soil surface strength with a needle-type penetrometer. *Soil Tillage Res.* 5:227–245.

Carter, L. M. 1967. Portable recording penetrometer measures soil strength profiles. *Agric. Eng.* 48:348–349.

Cassel, D. K., and L. A. Nelson. 1979. Variability of mechanical impedance in a tilled one-hectare field of Norfolk sandy loam. *Soil Sci. Soc. Am. J.* 43:450–455.

Cassel, D. K., H. D. Bowen, and L. A. Nelson. 1978. An evaluation of mechanical impedance for three tillage treatments on Norfolk sandy loam. *Soil Sci. Soc. Am. J.* 42: 116–120.

Christensen, N. B., Sisson, J. B., and P. L. Barnes. 1989. A method for analysing penetration resistance data. *Soil Tillage Res.* 13:83–89.

Davies, P. 1985. Influence of organic matter content, soil moisture status and time after reworking on soil shear strength. *J. Soil Sci.* 36:299–306.

Dexter, A. R. 1978. A stochastic model for the growth of roots in tilled soils. *J. Soil Sci.* 29:102–116.

Dexter, A. R. 1988. Advances in characterization of soil structure. *Soil Tillage Res.* 11: 199–238.

Dickson, J. W., and D. L. O. Smith. 1986. *Compaction of a Sandy Loam by a Single Wheel Supporting One of Two Masses Each at Two Ground Pressures,* Unpubl. Dep. Note No. SIN/479, Scot. Inst. Agric. Eng.

Eavis, B. W. 1967. Mechanical impedance to root growth, Paper No. 4/F/39. In: *Agricultural Engineering Symp.* Silsoe, U.K., 1967, pp. 1–11.

Ehlers, W., U. Kopke, F. Hesse, and W. Bohm. 1983. Penetration resistance and root growth of oats in tilled and untilled loess soil. *Soil Tillage Res.* 3:261–275.

Farrell, D. A., and E. L. Greacen. 1966. Resistance to penetration of fine probes in compressible soil. *Aust. J. Soil Res.* 4:1–17.

Freitag, D. R. 1965. *A Dimensional Analysis of the Performance of Pneumatic Tires on Soft Soils.* U.S. Army Waterways Exp. Stn. Rep. No. 3-688.

Freitag, D. R. 1968. Penetration tests for soil measurements. *Trans. Am. Soc. Agric. Eng.* 11:750–753.

Freitag, D. R., R. L. Schafer, and R. D. Wismer. 1970. Similitude studies of soil-machine systems. *Trans. Am. Soc. Agric. Eng.* 13:201–213.

Gerard, C. J. 1965. The influence of soil moisture, soil texture, drying conditions and exchangeable cations on soil strength. *Soil Sci. Soc. Am. Proc.* 29:641–645.

Gerrard, A. J. 1982. The use of hand-operated soil penetrometers. *Area* 14:227–234.

Gill, W. R. 1968. Influence of compaction hardening of soil on penetration resistance. *Trans. Am. Soc. Agric. Eng.* 11:741–745.

Glasbey, C. A., and M. F. O'Sullivan. 1988. Analysis of cone resistance data with missing observations below stones. *J. Soil Sci.* 39:587–592.

Goss, M. J. 1977. Effects of mechanical impedance on root growth in barley (*Hordeum vulgare* L.): I. Effects on elongation and branching of seminal roots. *J. Exp. Bot.* 28: 96–111.

Grant, C. D., B. D. Kay, P. H. Groenevelt, G. E. Kidd, and G. W. Thurtell. 1985. Spectral analysis of micropenetrometer data to characterize soil structure. *Can. J. Soil Sci.* 65:789–804.

Greacen, E. L., D. A. Farrell, and B. Cockroft. 1968. Soil resistance to metal probes and plant roots. In: *Trans. 9th Int. Congr. Soil Sci.*, Adelaide, Vol. 1, pp. 769–779.

Greacen, E. L., K. P. Barley, and D. A. Farrell. 1969. The mechanics of root growth in soils with particular reference to the implications for root distribution. In: *Root Growth* (W. H. Whittington, ed.). London: Butterworths, pp. 256–268.

Groenevelt, P. H., B. D. Kay, and C. D. Grant. 1984. Physical assessment of a soil with respect to rooting potential. *Geoderma* 34:101–114.

Hartge, K. H., H. Bohne, H. P. Schrey, and H. Extra. 1985. Penetrometer measurements for screening soil physical variability. *Soil Tillage Research* 5:343–350.

Henshall, J. K., and D. L. O. Smith. 1971. An improved method for presenting comparisons of soil compaction effects below wheel ruts. *J. Agric. Eng. Res.* 42:1–13.

Horn, R. 1984. Die Vorhersage des Eindringwiderstandes von Böden anhand von multiplen Regressionsanalysen (The prediction of the penetration resistance of soils by multiple regression analysis). *Z. Kulturtechn. Flurbereinig.* 25:377–380.

Jamieson, J. E., R. J. Morris, and C. E. Mullins. 1988. Effect of subsoiling on physical properties and crop growth on a sandy soil with a naturally compact subsoil. In: *Proc. 11th Int. Conf. Int. Soil Till. Res. Organization*, Vol. 2, pp. 499–503.

Johnston, C. E., R. L. Jensen, R. L. Schafer, and A. C. Bailey. 1980. Some soil–tool analogs. *Trans. Am. Soc. Agric. Eng.* 23:9–13.

Knight, S. J., and D. R. Freitag. 1962. Measurement of soil trafficability characteristics. *Trans. Am. Soc. Agric. Eng.* 5:121–132.

Kogure, K., Y. Ohira, and H. Yamaguchi. 1985. Basic study of probabilistic approach to prediction and soil trafficability—Statistical characteristics of cone index. *J. Terramech.* 22:147–156.

Koolen, A. J., and H. Kuipers. 1983. *Agricultural Soil Mechanics.* Heidelberg: Springer-Verlag.

Koolen, A. J., and P. Vaandrager. 1984. Relationships between soil mechanical properties. *J. Agric. Eng. Res.* 29:313–319.

Marshall, T. J., J. W. Holmes, and C. W. Rose. 1996. *Soil Physics.* 3d ed. Cambridge: Cambridge University Press.

McIntyre, D. S., and C. B. Tanner. 1959. Anormally distributed soil physical measurements and non-parametric statistics. *Soil Sci.* 88:133–137.

Misra, R. K., A. R. Dexter, and A. M. Alston. 1986a. Penetration of soil aggregates of finite size: I. Blunt penetrometer probes. *Plant Soil* 94:43–58.

Misra, R. K., A. R. Dexter, and A. M. Alston. 1986b. Penetration of soil aggregates of finite size: II. Plant roots. *Plant Soil* 94:59–85.

Mitchell, C. W., R. Webster, P. H. T. Beckett, and B. Clifford. 1979. An analysis of terrain classification for long-range prediction of conditions in deserts. *Geog. J.* 145: 72–85.

Moolman, J. H., and L. A. Van Huyssteen. 1989. A geostatistical analysis of the penetrometer soil strength of a deep ploughed soil. *Soil Till. Res.* 15:11–24.

Mullins, C. E., I. M. Young, A. G. Bengough, and G. J. Ley. 1987. Hard-setting soils. *Soil Use Manag.* 3:79–83.

Mulqueen, J., J. V. Stafford, and D. W. Tanner. 1977. Evaluating penetrometers for measuring soil strength. *J. Terramech.* 14:137–151.

O'Sullivan, M. F., and B. C. Ball. 1982. A comparison of five instruments for measuring soil strength in cultivated and uncultivated cereal seedbeds. *J. Soil Sci.* 33:597–608.

O'Sullivan, M. F., J. W. Dickson, and D. J. Campbell. 1987. Interpretation and presentation of cone resistance data in tillage and traffic studies. *J. Soil Sci.* 38:137–148.

Paul, C. L., and J. de Vries. 1979. Effect of soil water status and strength on trafficability. *Can. J. Soil Sci.* 59:313–324.

Paul, C. L., and J. de Vries. 1983. Soil trafficability in spring: 2. Prediction and the effect of subsurface drainage. *Can. J. Soil Sci.* 63:27–35.

Perumpral, J. V. 1983. *Cone penetrometer application—A review.* Paper No. 83-1549. St. Joseph, MI: Am. Soc. Agric. Eng.

Reece, A. R., and J. D. Peca. 1981. An assessment of the value of the cone penetrometer in mobility prediction. In: *Proc. 7th Int. Conf. Int. Soc. Terrain-Vehicle Systems,* Vol. 3, Calgary, p. A1.

Russell, R. S. 1977. *Plant root systems: Their function and interaction with the soil.* London: McGraw-Hill.

Schrey, H. P. 1981. Die Interpretation des Eindringwiderstands zur flächenhaften Darstellung physikalischiede in Böden (The interpretation of penetration resistance in use of spatial discrimination of physical differences in soils). *Z. Pflanzenern. u. Bodenkunde* 154:33–39.

Snedecor, G. W., and W. G. Cochran. 1967. *Statistical Methods.* Ames, IA: Iowa State University Press.

Soane, B. D. 1973. Techniques for measuring changes in the packing state and cone resistance of soil after the passage of wheels and tracks. *J. Soil Sci.* 24:311–323.

Soane, B. D., P. S. Blackwell, J. W. Dickson, and D. J. Painter. 1981. Compaction by agricultural vehicles: A review: I. Soil and wheel characteristics. *Soil Till. Res.* 1: 207–237.

Stitt, R. E., D. K. Cassel, S. B. Weed, and L. A. Nelson. 1982. Mechanical impedance of tillage pans in Atlantic coastal plains soils and relationships with soil physical, chemical and mineralogical properties. *Soil Sci. Soc. Am. J.* 46:100–106.

Stolzy, L. H., and K. P. Barley. 1968. Mechanical resistance encountered by roots entering compact soils. *Soil Sci.* 105:297–310.

Taylor, H. M., and L. F. Ratliff. 1969. Root elongation rates of cotton and peanuts as a function of soil strength and water content. *Soil Sci.* 108:113–119.

Threadgill, E. D. 1982. Residual tillage effects as determined by cone index. *Trans. Am. Soc. Agric. Eng.* 25:859–867.

Tollner, E. W., and B. P. Verma. 1984. Modified cone penetrometer for measuring soil mechanical impedance. *Trans. Am. Soc. Agric. Eng.* 27:331–336.

Turnage, G. W. 1972. Tire selection and performance prediction for off-road wheeled vehicle operations. In: *Proc. 4th Int. Conf. Int. Soc. Terrain-Vehicle Systems.* Vol. 1. Stockholm, pp. 61–82.

Turnage, G. W. 1973. Influence of viscous-type and inertial forces on the penetration of saturated, fine-grained soils. *J. Terramech.* 10:63–76.

Turnage, G. W. 1984. Prediction of in-sand tire and wheel vehicle drawbar performance. In: *Proc. 8th Int. Conf. Int. Soc. Terrain-Vehicle Systems.* Vol. 1. Cambridge, pp. 121–150.

van Wijk, A. L. M. 1980. Soil water conditions and playability of grass sportsfields: I. Influence of tile drainage and sandy drainage layers. *Z. Vegetationstechnik* 3: 16–22.

van Wijk, A. L. M., and J. Beuving. 1978. Relation between soil strength, bulk density and soil water pressure head of sandy top-layers of grass sportsfields. *Z. Vegetationstechnik* 1:53–58.

Voorhees, W. B., D. A. Farrell, and W. E. Larson. 1975. Soil strength and aeration effects on root elongation. *Soil Sci. Soc. Am. Proc.* 39:948–953.

Voorhees, W. B., C. G. Senst, and W. W. Nelson. 1978. Compaction and soil structure modification by wheel traffic in the northern corn belt. *Soil Sci. Soc. Am. J.* 42: 344–349.

Wairiu, M., C. E. Mullins, and C. D. Campbell. 1993. Soil physical factors affecting the growth of sycamore (*Acer pseudoplatanus* L.) in a silvopastoral system on a stony upland soil in north-east Scotland. *Agroforestry Systems* 24:295–306.

Whiteley, G. M., and A. R. Dexter. 1981. The dependence of soil penetrometer pressure on penetrometer size. *J. Agric. Eng. Res.* 26:467–476.

Whiteley, G. M., W. H. Utomo, and A. R. Dexter. 1981. A comparison of penetrometer pressures and the pressures exerted by roots. *Plant Soil* 61:351–364.

Wismer, R. D., and H. J. Luth. 1973. Off-road traction prediction for wheeled vehicles. *J. Terramech.* 10:49–61.

Yong, R. N., C. K. Chen, and R. Sylvestre-Williams. 1972. A study of the mechanisms of cone indentation and its relation to soil-wheel interaction. *J. Terramech.* 9:19–36.

Young, I. M., A. G. Bengough, C. J. Mackenzie, and J. W. Dickson. 1993. Differences in potato development (*Solanum tuberosum* cv Maris Piper) in zero and conventional traffic treatments are related to soil physical conditions and radiation interception. *Soil Till. Res.* 26:341–359.

11
Tensile Strength and Friability

A. R. Dexter and Chris W. Watts
Silsoe Research Institute, Silsoe, Bedfordshire, England

I. INTRODUCTION

Tensile strength is defined as the stress, or force per unit area, required to cause soil to fail in tension, that is, to pull it apart. Tensile strength is remarkably sensitive to the soil microstructure, and this makes it a valuable parameter to measure in research into the structure and behavior of soil.

The tensile strength of a soil is of little interest in civil engineering, where it is usually assumed to be zero, as soils are maintained under compressive loads and are not meant to fail anyway. However, when soils are considered in agricultural and environmental contexts, this is not the case, and tensile strength is important. For example, the cracking and crumbling of soil that occurs during soil wetting and drying or during tillage operations are strongly dependent on the tensile strength characteristics.

Soil friability may be defined as the tendency of a mass of soil to crumble into a certain size range of smaller fragments under an applied stress. This property is crucial for the production of seedbeds during tillage operations. It is often observed that the results of tillage depend more on the soil conditions than on the details of the tillage implement. Intuitively, one can imagine that this crumbling property depends on the pre-existing micro-structure of the soil. Later, we show that it can be quantified through the variability of the tensile strength.

II. TYPES OF TENSILE STRENGTH TESTS

A. Indirect Tension Tests

Indirect tests of tensile strength are so called because the stress is not applied directly. Instead, a compressive force is applied across the diameter of a cylindri-

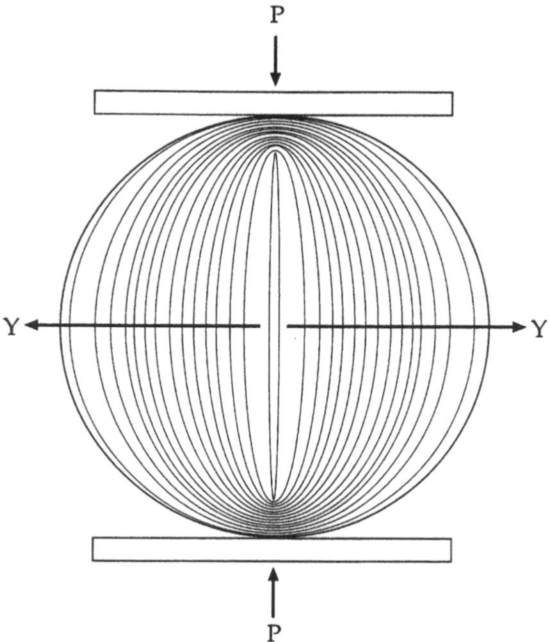

Fig. 1 Contours of equal tensile stress in a cylindrical sample loaded across a diameter by a force, *P*. Maximum tensile stress, *Y*, occurs at the center of the sample and is given by Eq. 1. The first two contours from the center have values of 0.96 and 0.89 of the maximum value, respectively.

cal, spherical, or quasi-spherical sample, and this gives rise to a tensile stress within the sample at right angles to the direction of the applied force.

Figure 1 shows contours of equal tensile stress within such a loaded cylindrical sample. Maximum tensile stress occurs on a vertical plane through the center of the sample. It can be seen from the stress contours in Fig. 1 that quite a large volume in the center of the sample is subjected to a fairly uniform level of tensile stress in this test. The maximum value of tensile stress, Y_{max}, within a cylindrical sample is given by

$$Y_{max} = \frac{2P}{\pi DL} \tag{1}$$

where P is the applied force and D and L are the diameter and length of the sample, respectively.

The corresponding equation for spherical samples is

$$Y_{max} = 0.576\frac{P}{D^2} \tag{2}$$

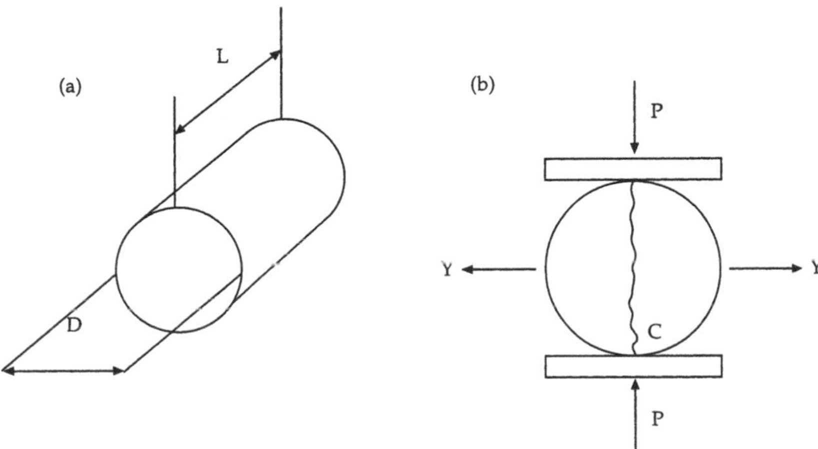

Fig. 2 A cylindrical sample (a) used in the Brazilian test has a length, *L*, and diameter, *D*. When it fails in tension (b), a crack *C*, is formed between the points of loading due to the tensile stress, *Y*, which acts at the center of the sample.

In tensile strength testing, the load *P* is increased steadily until the sample fails. This is apparent by the formation of a crack that runs through the sample from top to bottom, as shown in Fig. 2. The tensile strength is equal to the value of the tensile stress in the sample at failure, as given by Eqs. 1 and 2 and is denoted by *Y*. It has the usual units of mechanical stress, kPa or MPa.

The indirect test on cylindrical samples was first developed as a test for concrete by Akazawa (1943). However, it is often called the Brazilian test following its subsequent and independent development by Carneiro & Barcellos (1953). It has been analyzed by many people including Peltier (1954) and Wright (1955). The Brazilian test has been applied to soil cores (Kirkham et al., 1959, Frydman, 1964; Kemper and Rosenau, 1984).

The crushing test for soil aggregates was first described by Vilensky (1949) and by Martinson and Olmstead (1949). Originally, it was used as an arbitrary measurement of strength and was not related to tensile strength. This step required the work of Rogowski et al. (1964, 1966, 1976) and of Dexter (1975). These researchers used the photoelasticity measurements of Frocht and Guernsey (1952) to obtain the value of the coefficient in Eq. 2.

Different values of the coefficient in Eq. 2 have been used by different researchers with values ranging from 0.576 (Dexter, 1975; Braunack et al., 1979; Utomo and Dexter, 1981; Dexter and Kroesbergen, 1985; Macks et al., 1996), to 0.711 (Hadas and Lennard, 1988), 0.821 (Rogowski and Kirkham, 1976), 0.9 (Hiramatsu and Oka, 1996), 0.964, and 1.86 (Dexter, 1988a). Furthermore, there is some evidence that this coefficient is not a constant but may vary with soil water content (Vomocil and Chancellor, 1969), bulk density (Hadas and Lennard,

1988), and aggregate shape (Dexter, 1988a). However, for most studies, the exact value of this coefficient is not important, and we use the value given in Eq. 2 as standard.

B. Direct Tension Tests

In direct tension tests, the sample is pulled into two parts by a tensile force, P (Fig. 3). The tensile strength is given by

$$Y = \frac{P}{A} \tag{3}$$

where P is the value of the tensile force when the sample fails and A is the cross-sectional area of the failure surface. In this test, the sample fails with a crack that is perpendicular to the applied force, P.

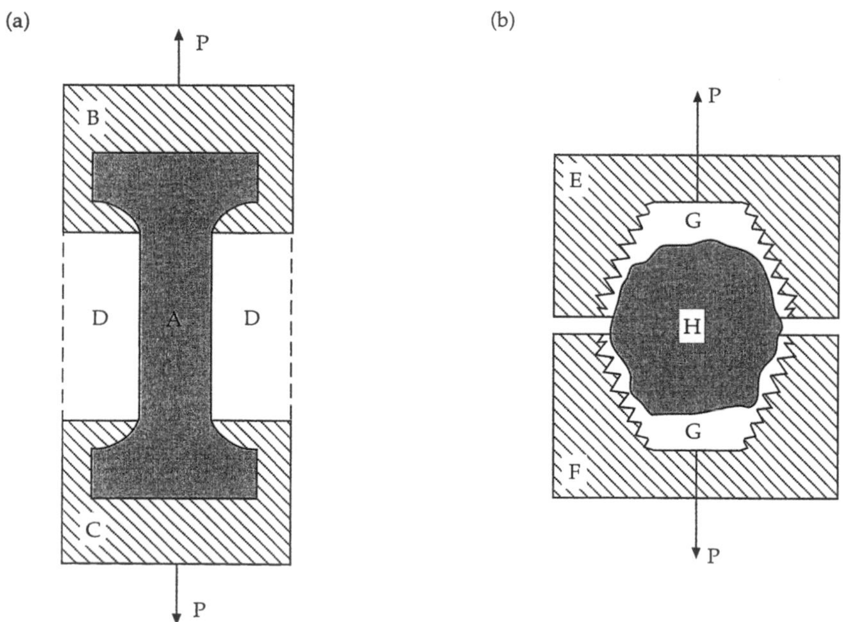

Fig. 3 Direct tension tests on soil samples. In (a), a remolded dog bone sample, A, is made in a mold comprising parts B, C, and D, which are initally clamped together. For the test, the mold is unclamped and parts D are removed. Parts B and C then form grips that are used to pull the sample apart in tension with a force, P. In (b), a natural soil aggregate or clod, H, is bonded into two grooved cups, E and F, with plaster of Paris, G. The sample is then pulled apart in tension by the force, P.

Direct tension tests are difficult to perform. Particular difficulties arise in dog bone tests (Fig. 3a) where it is difficult to prepare undamaged samples. In tests on natural aggregate samples (Fig. 3b), care must be taken to bond the samples to the end cups while maintaining them in alignment so that a straight pull is achieved. Usually, an aggregate sample is bonded into one cup first and then inverted and bonded into the second cup. Plaster of Paris is a convenient bonding agent because it has the advantage of hardening quickly. However, it hardens through crystallization, and when it becomes dry, water will flow into it from any moist soil sample, thereby changing the water content of the soil sample under test.

C. Difference Between Direct and Indirect Tension Tests in Moist Soils

There is an interesting and important difference between direct and indirect tension tests that can affect the measured strength of moist samples. Whereas the two types of test should give similar results for dry samples, this is not true for moist samples. The difference is caused by differences in the mean stresses in the sample. In a direct test, the mean stress is negative (as is a tensile stress). However, in an indirect test, both compressive and tensile stresses occur within the sample, and the mean stress is positive.

The effects of this can be measured in unsaturated soil by inserting microtensiometers, e.g., of 1 mm diameter (Gunselmann et al., 1987), into the center of the samples during the tests. Results show that in direct tests, the pore water pressure becomes more negative (i.e., the "suction" increases), whereas in indirect tests, the pore water pressure becomes less negative (i.e., the "suction" decreases) (Hallett, 1996). The different values of pore water pressure at the point of failure give rise to different values of effective stress (Aitchinson, 1961; Dexter, 1997) at failure and hence different apparent values of tensile strength.

Direct tension tests are not discussed further in this chapter, and attention is concentrated on the easier and more rapid indirect tension tests.

III. STATISTICAL THEORY OF BRITTLE FRACTURE

A. Basic Theory

The following analysis is drawn partly from the comprehensive work and review of the subject by Freudenthal (1968). The presentation follows Braunack et al. (1979) with some additions and corrections.

Several important assumptions are made in the theoretical analysis, which are summarized below. Flaws of various magnitudes are distributed throughout the solid being considered. The volume of the solid is considered to be composed of a number of equal volume elements, each of which is sufficiently large to contain a large number of flaws of various sizes. No interaction between flaws exists, that is, the stress fields surrounding each flaw are mutually independent. The strength of each volume element is determined by the stress at which the most severe flaw it contains propagates. The strength of the total volume is determined by the strength of the weakest volume element. The fracture of the total specimen is therefore determined by the unstable propagation of the most severe crack. This is the "weakest link" concept: the strength of the total solid is determined by the local strength of the weakest volume element, in the same way as the strength of a chain is determined by its weakest link.

It is useful to consider a volume element of soil containing a substantial number of cracks or other type flaws, each of which has a critical tensile stress, s, required to propagate it. There is a statistical distribution of the critical stresses associated with the cracks in the volume element. The statistical distribution of critical stresses will have a nonnegative left-hand bound, which is taken here as zero. That is, it is possible, although improbable for small enough stresses, for fracture to occur with any positive applied stress. However, the theory is not altered in essence if a nonzero minimum critical stress is considered.

Although we cannot know the actual distribution of critical stresses of the flaws in a volume element, interest lies not in this but in the distribution of the smallest critical stress. The distributions of extremes, largest or smallest, of large samples are in fact quite limited, as shown by Gumbel (1958), despite the initial distribution of the sample population. In this case, the distribution of smallest values of a large enough sample population, bounded by zero on the left, will be

$$H(s) = 1 - \exp\left[-\left(\frac{s}{s_0}\right)^{\alpha}\right] \tag{4}$$

where H(s) is the probability that the smallest critical stress random variable S is equal to or less than s. This distribution is known as the third asymptotic distribution of smallest values. A derivation and analysis of the three asymptotic distributions of extreme values is to be found in Gumbel (1958). The parameters α and s_0 are constants of the material, s_0 being the strength of the solid for which

$$H(s) = 1 - \exp(-1) \tag{5}$$

and $1/\alpha$ is proportional to the scatter of local flaw strengths.

The Eq. 4 for the distribution of smallest critical stress is for a volume element. The effect of the total volume of the sample is incorporated as follows. Suppose that there are n equal volume elements in the total volume. Then the probability that the minimum critical strength S is greater than s for one volume element is $1 - H$(s), and for n volume elements it is

$$[1 - H(s)]^n = \exp\left[-n\left(\frac{s}{s_0}\right)^{\alpha}\right] \tag{6}$$

Hence if $H_T(s)$ is the probability that S is equal to or less than s in the total volume, then

$$H_T(s) = 1 - \exp\left[-n\left(\frac{s}{s_0}\right)^{\alpha}\right] \tag{7}$$

If the volume element is V_0, and the total volume is V, then

$$nV_0 = V \tag{8}$$

and

$$H_T(s) = 1 - \exp\left[-\frac{V}{V_0}\left(\frac{s}{s_0}\right)^{\alpha}\right] \tag{9}$$

The mean critical stress $\bar{s}$ and the variance $(\sigma_s)^2$ are found from the moments of the extreme value distribution, namely

$$\bar{s} = \int_0^{\infty} s\,dH_T(s) \tag{10}$$

and

$$(\sigma_s)^2 = \int_0^{\infty} (s - \bar{s})^2\,dH_T(s) \tag{11}$$

From Eq. 9, these are

$$\bar{s} = s_0\left(\frac{V}{V_0}\right)^{-1/\alpha} \Gamma\left(1 + \frac{1}{\alpha}\right) \tag{12}$$

and

$$(\sigma_s)^2 = s_0^2\left(\frac{V}{V_0}\right)^{-2/\alpha}\left[\Gamma\left(1 + \frac{2}{\alpha}\right) - \Gamma^2\left(1 + \frac{1}{\alpha}\right)\right] \tag{13}$$

where Γ is the well known and tabulated gamma function.

The coefficient of variation of strength values is given from Eqs. 12 and 13 as

$$\frac{\sigma_s}{\bar{s}} = \frac{[\Gamma(1 + 2/\alpha) - \Gamma^2(1 + 1/\alpha)]^{1/2}}{\Gamma(1 + 1/\alpha)} \tag{14}$$

This equation enables the parameter $1/\alpha$ of the brittle fracture theory to be obtained from measurements of the coefficient of variation of strength, S, or as used later in Eq. 19, the coefficient of variation in aggregate strength, $\sigma_Y/\bar{Y}$. When logarithms have been taken twice, Eq. 9 becomes

$$\log_e\{-\log_e[1 - H_T(s)]\} = \log_e\left(\frac{V}{V_0}\right) + \alpha\log_e\left(\frac{s}{s_0}\right) \tag{15}$$

If a set of m observations of fracture strengths of aggregates of the same volume, ranked in ascending order, are taken to represent the distribution of tensile strengths, then the kth value can be given a cumulative frequency of

$$H_T(s_k) = \frac{k}{m + 1} \tag{16}$$

The denominator is taken as $m + 1$ primarily so that the first and last observation may be used (Gumbel, 1958). Then the tensile yield strength distribution can be found by fitting $y = \log_e[-\log_e[1 - (k/(m + 1))]]$ to $x = \log_e s$, i.e.,

$$x = \frac{1}{\alpha} y - \frac{1}{\alpha} \log_e \left(\frac{V}{V s_0^{\alpha}} \right) \tag{17}$$

and the material parameters α and $s_0 V_0^{1/\alpha}$ are obtained. If sets of observations of tensile yield strengths for different volumes of the same material are taken, then a set of parallel straight lines of the form of Eq. 17, shifted in the negative x-direction by an amount $1/\alpha \log(V_2/V_1)$, for volume V_2 greater than V_1, will be produced. Alternatively, volume effects can be considered by taking logarithms of Eq. 12, so that

$$\log_e \bar{s} = -\frac{1}{\alpha} \log_e V + \log_e \left[s_0 V_0^{1/\alpha} \Gamma \left(1 + \frac{1}{\alpha} \right) \right] \tag{18}$$

The material parameters α and $s_0 V_0^{1/\alpha}$ can now be found by a best straight line fit of $\log_e s$ to $\log_e V$, which will have a slope of $-1/\alpha$. Alternatively, a fit of $\log_e s$ against $\log_e D$, where D is the aggregate diameter, will have a slope of $-3/\alpha$.

B. Application to Friability Measurement

Soil is friable not because of its strength but because of the distribution of flaws or microcracks within it. The heterogeneity of strength resulting from these flaws controls the way in which soil crumbles. The distribution of flaw strengths is represented by $1/\alpha$ in the preceding equations (Freudenthal, 1958) and has been identified with the friability (Utomo and Dexter, 1981; Macks et al., 1996; Watts and Dexter, 1998).

The preceding equations give rise to three different methods for the determination of friability. Due to deficiencies in the theory and problems associated with sampling and measurement, which are discussed elsewhere as they occur, the different methods give rise to different estimates of friability, which are therefore denoted separately by F_1, F_2, and F_3.

The first method is based upon Eq. 14, from which it may be shown that the coefficient of variation $(\sigma_Y/\overline{Y})$ differs from $1/\alpha$ by less than 15% over the range of interest $(0 < (1/\alpha) < 1.2)$. Accordingly, we may define

$$F_1 = \frac{\sigma_Y}{\overline{Y}} \pm \frac{\sigma_Y}{\overline{Y}\sqrt{2n}} \tag{19}$$

where σ_Y is the standard deviation of measured values of tensile strength. $\overline{Y}$ is the mean of the tensile strength measurements of n replicates. The second term is the standard error of the coefficient of variation. F_1 may be related to the principal parameter of brittle fracture theory, $1/\alpha$, through Eq. 14. Because this equation is not easy to compute, Watts and Dexter (1998) developed a simpler empirical, approximate relationship

$$\log F_1 = 0.929 \log\left(\frac{1}{\alpha}\right) \tag{20}$$

which is accurate to within 2% over the range of interest, well within the experimental error. An example of results obtained using Eq. 19 is given in Fig. 4. This method has the advantages that only one size of soil sample is needed and that Eq. 19 is easy to compute and to think about.

The second method is based on the use of Eq. 17. As with the first method, only one size of soil sample is needed. In this method, a function of the ranking

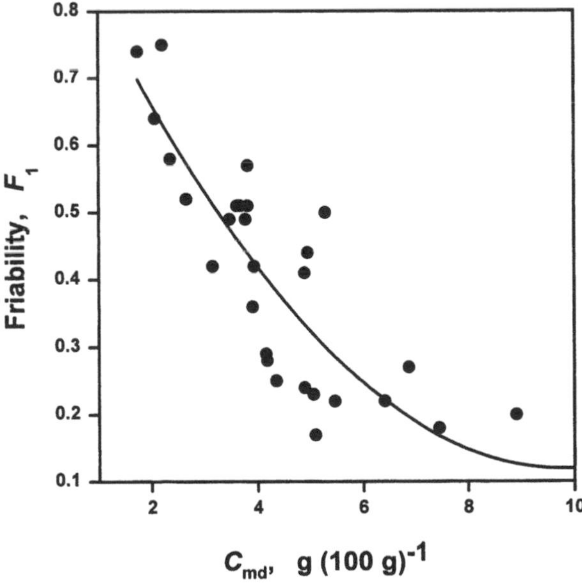

Fig. 4 An example of results obtained using the first method for determining soil friability, F_1. Here, F_1 is determined using Eq. 19. The example shows results from By-pass Field, Silsoe, where the friability is related to the amount of mechanically dispersible clay, C_{md}, in the soil. The total clay content of the soil is 35 g $(100g)^{-1}$.

order, with the sample strengths ranked, is plotted against the logarithm of tensile strength. The reciprocal of the slope gives the friability

$$F_2 = \frac{1}{\alpha} \tag{21}$$

Figure 5 shows an example of results obtained by this method.

The third method is based upon Eq. 18. A graph of $\log_e Y$ against $\log_e V$, where V is the sample volume, has a slope of $-1/\alpha$. An example is given in Fig. 6. Again,

$$F_3 = \frac{1}{\beta} \tag{22}$$

(The symbol α is replaced by β in Eq. 22 to indicate different methods of determination.)

Except for soils with no microstructure, which have zero friability, the strength of soil samples is always size dependent. Larger aggregates, for example, are always weaker than smaller aggregates because they contain larger flaws or

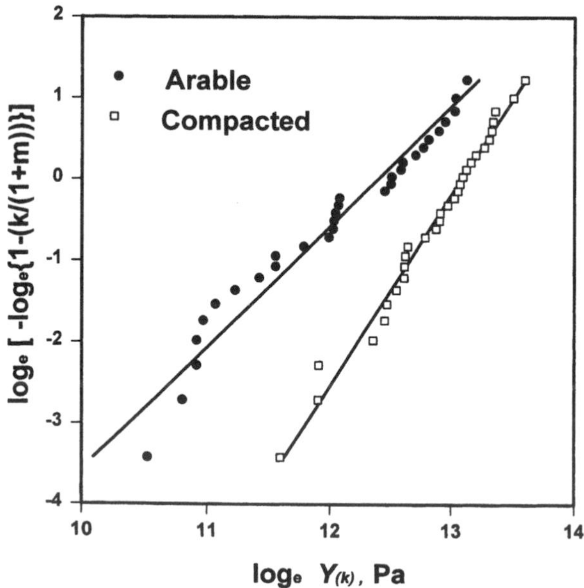

Fig. 5 An example of results obtained using the second method for determining friability, F_2. Here, F_2 is given by Eq. 21. The example shows results from Boot Field, Silsoe, where ● represents an arable plot where $F_2 = 0.70$, and □ represents compacted wheelways where $F_2 = 0.43$. (Data from Watts and Dexter, 1998.)

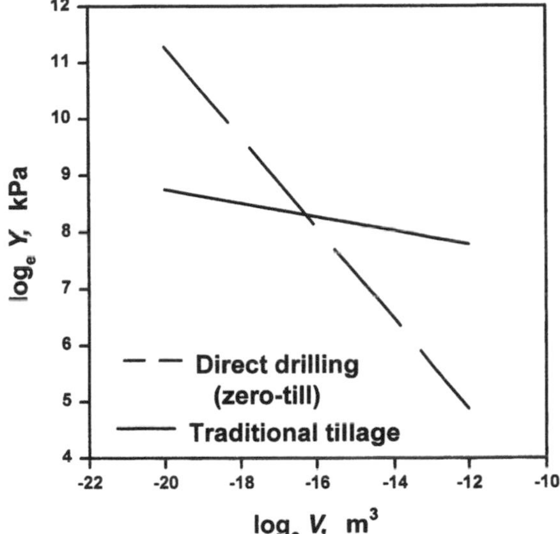

Fig. 6 An example of results obtained using the third method for determining friability, F_3. Here, F_3 is given by Eq. 22. The example shows results from a direct-drilled plot where $F_3 = 0.80 \pm 0.01$ and from a plot with traditional tillage where $F_3 = 0.12 \pm 0.01$. (Data from Macks et al., 1996.)

microcracks. Larger aggregates from the same population have a higher porosity than smaller aggregates for the same reason (Currie, 1966; Dexter, 1988b).

At least two factors contribute to the finding that F_2 is always larger than F_3 by a factor usually in the range 2 to 4 (Braunack, 1979). The first is that some variability of the force, P, for sample failure (Eqs. 1 and 2) is due to differences in the shape of individual soil aggregates. This factor influences F_2 but not F_3. The second is that, for natural aggregates, some of the flaws or microcracks are not very small compared with the sample size, and this negates one of the main assumptions of the weakest link theory of soil strength.

Watts and Dexter (1998) found, using experimental data, that values of F_1 obtained from Eq. 19 were very close to values obtained by method 2 and Eq. 20, i.e.,

$$\log_e F_1 \approx 0.929 \log_e F_2 \tag{23}$$

The first method, Eq. 14, is recommended as the standard method for measuring soil friability because it is easy to calculate and to think about, because it can be related to the principal parameter, $1/\alpha$, of brittle fracture theory, Eqs. 14 and 20, and because it requires fewer measurements than the third method.

IV. EXPERIMENTAL METHODOLOGY

A. Sample Collection, Storage, and Preparation

Samples should be collected from the field using a randomized sampling pattern. All samples must be collected in the same way and from the required, predetermined depth. The different treatments and plots of a given experiment should all be sampled within half a day to prevent subsequent changes in soil properties with aging or natural wetting or drying processes from influencing the sample properties.

At water contents above the plastic limit, the soil becomes increasingly sensitive to mechanical damage, and this has been shown to influence dry aggregate strength and soil friability (Watts and Dexter, 1998). It is therefore good practice, at all water contents, to minimize mechanical damage during sampling and transport of samples from the field to the laboratory.

To obtain samples of the desired size, it is often necessary to break up a larger soil mass or clod into its constituent aggregates. This is best done by carefully teasing the large sample apart in the hands. Scissors are useful for cutting enmeshing roots, particularly when collecting samples from under grassland. The desired size range is most easily obtained with the aid of sieves. However, mechanical sieve shaking should be avoided because of the risk of unnecessary additional damage associated with it.

Aging after mechanical disturbance such as tillage can result in an increase in strength, commonly by factors exceeding 2 (Utomo and Dexter, 1981; Dexter et al., 1988). Soil strength is also very sensitive to water content. Rapid wetting can generate microstructure in samples (Grant and Dexter, 1989; Kay and Dexter, 1992), and slow drying can cause large increases in strength (e.g., a factor of 2 for a decrease in water content of 2.5 g 100 g^{-1}). These potential problems illustrate the importance of controlling or taking these factors into account if confusing results are not to be obtained.

In the laboratory, the samples may be stored in sealed plastic bags for a few days before measurement, but this time should be minimized. Care must be taken to avoid condensation occurring within the sample bags. The heterogeneous drying by evaporation and wetting by water drops associated with this can modify the sample properties. Storage in a constant temperature room can reduce condensation. If the temperature is low (e.g., 4°C), then biological activity will also be minimal.

If it is required to measure the soil at a given water potential, then it will be necessary first to wet the samples slowly or under vacuum to a low suction (e.g., a potential of -5 kPa) and then to drain them on a pressure plate apparatus to the required potential.

If the samples are to be measured dry, then it is best to let them air dry slowly first and then to oven dry them at 105°C. They can then be allowed to cool in a vacuum desiccator at low humidity (e.g., over silica gel). They should be taken individually from the desiccator when required for measurement because they will rapidly absorb water vapor from the air.

B. Measurement of Sample Size

Aggregate diameter, D, has to be known before aggregate tensile strength can be calculated using Eq. 2. Because of the irregular shape of soil aggregates, exact determination of an "effective spherical diameter" is not possible, but several methods are available for its estimation. Five different methods for estimating the diameter, D, of soil aggregates were described by Dexter and Kroesbergen (1985). These are outlined below.

Method 1

The soil is sieved and aggregates are collected that pass through a sieve with an opening size of d_1 but not through a sieve with an opening size of d_2. The mean aggregate diameter is estimated from

$$D_1 = \frac{d_1 + d_2}{2} \tag{24}$$

This value is then assumed to be the diameter of all the individual aggregates in the sample. This method is useful for small ($D_1 < 3$ mm) aggregates, which are difficult to measure directly in other ways. The size range, $(d_1 - d_2)/d_2$, must be kept as small as possible.

Method 2

In this method, aggregates are measured individually with calipers or some other suitable measuring device. Calipers with a digital, electronic display (R.S. Components, P.O. Box 99, Corby, Northants, U.K.) are particularly suitable. The idea is that the use of individual aggregate diameters D with individual crushing forces P in Eq. 2 will reduce significantly the variance of the resulting values of Y. For aggregates larger than about 5 mm, it is possible to measure the longest (D_x), intermediate (D_y), and smallest (D_z) diameters of each aggregate.

In method 2, the arithmetic mean diameter is calculated by

$$D_2 = \frac{D_x + D_y + D_z}{3} \tag{25}$$

and the value of D_2 for each aggregate is used in Eq. 2.

Method 3

Individual aggregates are measured as in method 2, but the geometric mean diameter is calculated

$$D_3 = (D_x D_y D_z)^{1/3} \tag{26}$$

The diameter D_3 is the diameter of a sphere that has the same volume as an ellipsoid with principal diameters D_x, D_y, and D_z.

Method 4

The mean sieving diameter D_1 from method 1 is used for all aggregates, but the effective diameters are adjusted according to their individual masses M. The adjustment is done on the assumption that all aggregates have equal density, ρ. If the mean mass of a batch of aggregates is M_0, then

$$\rho = \frac{6M_0}{\pi D_1^3} = \frac{6M}{\pi D_4^3} \tag{27}$$

whence

$$D_4 = D_1 \left(\frac{M}{M_0}\right)^{1/3} \tag{28}$$

This method is particularly effective because D_1 is known and the masses can be obtained quickly and accurately by weighing.

Method 5

In this method, all aggregates are assumed to have equal density, as in method 4, and individual aggregate diameters are adjusted according to their individual masses. In this case, however, the mean aggregate density ρ is known. Therefore

$$D_5 = \left(\frac{6M}{\pi\rho}\right)^{1/3} \tag{29}$$

is the diameter of a sphere having the same volume as the irregular aggregate being measured. This is a good measure because during loading of an aggregate in the crushing test, elastic strain energy is distributed through the whole volume of the aggregate.

The density of the aggregates can be determined by the kerosene saturation method of McIntyre and Stirk (1954). However, because the aggregate density is measured when the aggregates are oven dry, the method can be applied only to aggregates of nonswelling soils or to aggregates that are to be crushed in the dry condition.

In most of our work on soil aggregates, we have chosen to use method 4, because we have found it to be quick, easy, and reliable.

C. Indirect Tension Tests

1. Methods for Strong or Large Samples

For soil cores, aggregates, or clods in the size range 4–100 mm, a loading frame as shown in Fig. 7 is commonly used. Loading frames vary considerably in their sophistication but consist essentially of two parallel plates between which the

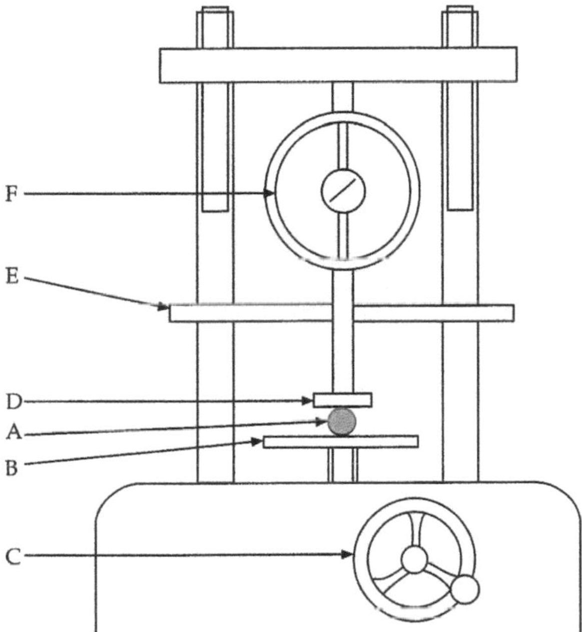

Fig. 7 Schematic diagram of a simple loading frame. Turning the handle, *C*, raises the lower platform, *B*, and applies a force across the soil sample, *A*, against the top plate, *D*. The force acting across the soil sample is measured by a load ring, *F*.

sample, *A*, is crushed. The lower plate, *B*, is raised at a constant rate either through a motor drive or manually by turning the handle, *C*. With some soils, the strain rate may influence sample strength, particularly with moist samples. It is therefore important that the same strain rate be used throughout any experiment. We have routinely used a strain rate of 0.07 mm s^{-1}. Figure 8 is a photograph of this apparatus being used to crush an aggregate.

The force applied to the sample is measured with a load-measuring device that is placed between the upper plate, *D*, and the cross beam of the loading frame. This may be either a proving ring (load ring), *F*, in which the deflection measured with a dial gauge is proportional to the applied load, or alternatively an electronic load cell. Load cells provide an electrical output that is proportional to the applied load and that can easily be recorded by data logging and computer systems. Load cells are relatively inexpensive. For oven-dried aggregates up to 20 mm diameter, sensors with a 0–2 kN range are usually suitable. However, for well-structured and weaker soils in this size range, a 0–500 N load cell provides an adequate range and better resolution.

Prior to crushing, the size of each aggregate is measured as described in Subsec. B, above. To provide a standard sample orientation, the test aggregate is

Fig. 8 Configurations used for measuring aggregate tensile strength using a loading frame. The force is measured in this case using a loading ring.

then placed flattest side downwards on the lower plate so that it will be crushed across its shortest axis.

When the sample is loaded, the force measured increases and, at the point of failure, a vertical crack appears in the sample and a rapid drop in the force is measured. The peak force, P, at failure is recorded and used in calculating the sample strength, Eqs. 1 and 2. Well-structured, friable soils, high in organic matter, tend to crumble progressively under applied load, making the determination of P rather difficult. Such soils produce a number of minor peaks in the force trace before a sample finally fails. By contrast, oven-dried samples of remolded soil fail abruptly, leaving no doubt about the point of failure.

Loading frames in conjunction with more sophisticated signal processing equipment can provide a picture in real time of the force against strain characteristics (Fig. 9). However, this should not distract the operator from observing the sample under load, as this often provides the best indication of the point of failure. Integration of the force against strain curve up to failure can give the energy used to fail the sample, if this is required.

The loading frame and its sensors can readily be adapted for direct tension tests as shown in Figs. 3a and b. The sample grips are then attached by bolts to the plates on the loading frame, and the lower plate is lowered rather than raised. The sensors then measure tensile force rather than crushing force.

2. *Methods for Weak or Small Aggregates*

For small aggregates in the 1–10 mm size range and having crushing forces in the 0.1–40 N range, it is possible to insert a digital balance into the loading frame to act as the load sensor, as shown in Figs. 10 and 11. If a digital electronic balance is used, then two additional refinements are required:

1. Firstly, the output from the balance must be logged, as it is impossible to follow the rapidly changing numbers by eye. Most modern balances have provision for the output to be logged externally by, for example, a lap top computer.
2. Secondly, the high stiffness of modern digital balances means that very small deflections can result in an excessively rapid increase in the force reading. Additional resilience can be added to the system with the aid of a spring. Dexter and Kroesbergen (1985) have suggested a spring rate that gives a 10 mm deflection at maximum balance load. We have found that a 0–500 N proving ring provides adequate resilience and has the added advantage of being readily incorporated into the loading frame.

The force applied to the sample is the product of the balance output in kg, and $g = 9.807\ \mathrm{ms}^{-2}$, which is the acceleration of gravity.

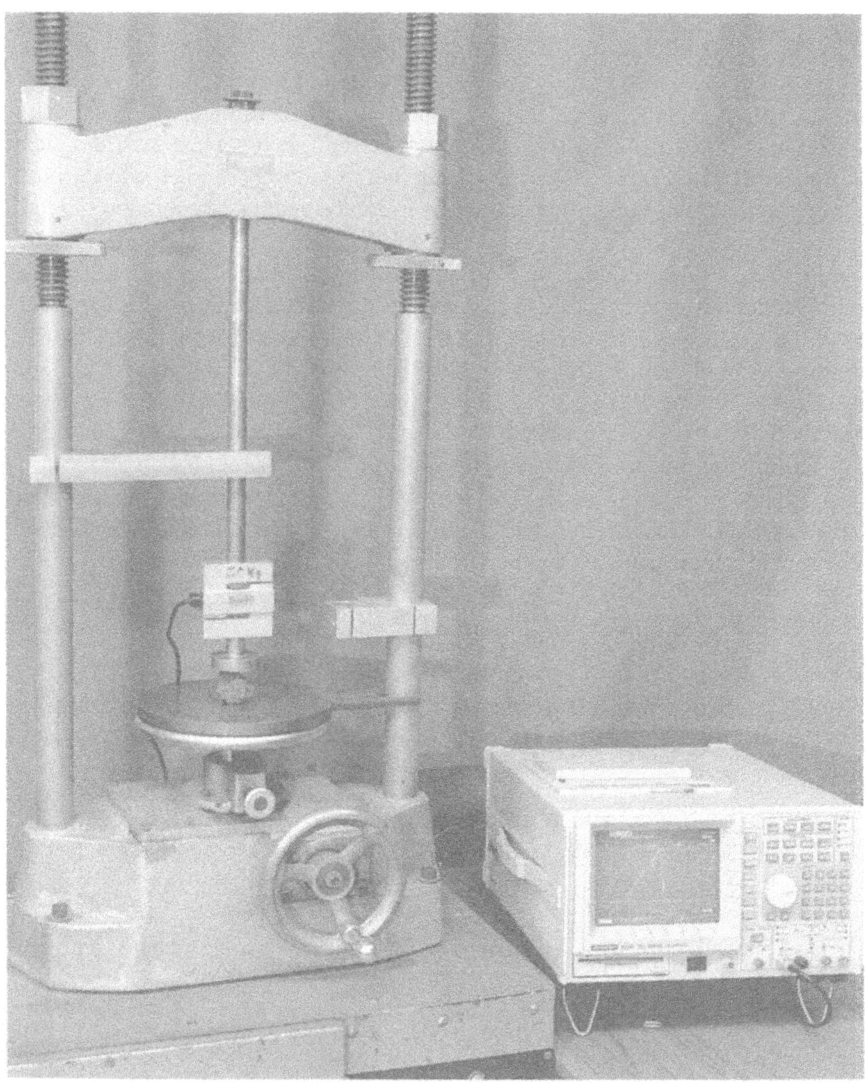

Fig. 9 Aggregate strength measurement. The load is measured using a load cell, the output of which is displayed on a signal analyzer.

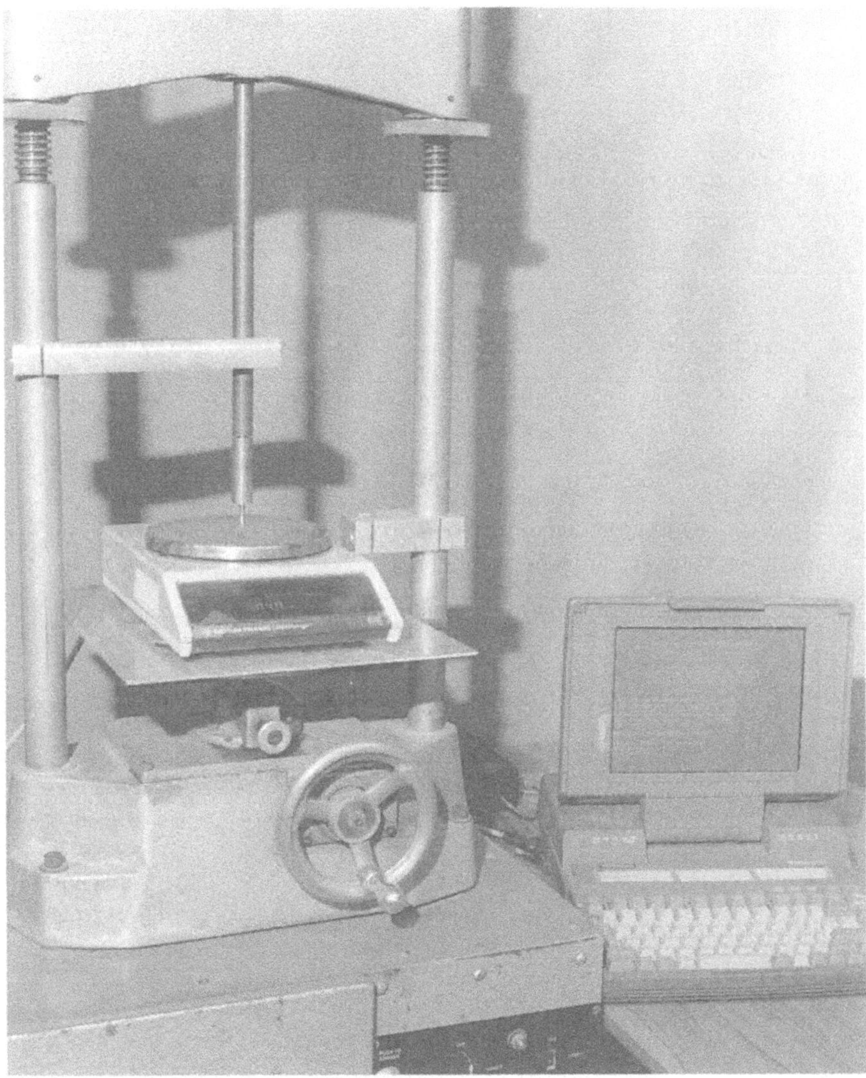

Fig. 10 Technique used for measuring the strength of small soil aggregates. The crushing force across the sample is being measured using a digital balance with the output recorded on a laptop computer.

Fig. 11 A close-up of the configuration used for loading small aggregates.

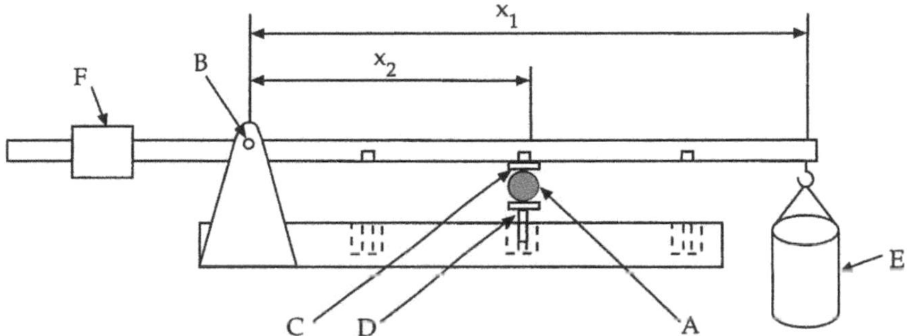

Fig. 12 A simple apparatus for measuring the force required to crush a soil aggregate. The component parts and the method of use are described in the text.

3. A Simple Apparatus

Loading frames, as described above, are routinely found in soil mechanics laboratories. However, most of the experiments described in this chapter can be done with very much simpler apparatus (Figs. 12 and 13). The equipment described here is based on a design by Horn and Dexter (1989). The equipment consists basically of two parallel arms that are hinged together. The upper arm is made from aluminum channel or box-section for lightness. The length of the lower arm is typically 500 mm, while that of the upper arm is 750 mm. The upper arm is free to rotate about a pivot, *B*, which is held on supports fixed to the lower arm. As in the preceding tests, an aggregate, *A*, is crushed between two parallel plates. The upper plate, *C*, is flush with the upper arm, while the lower plate, *D*, is adjusted so that the arms are parallel when the aggregate is in place. There are several positions for the plates *C* and *D* along the upper and lower arms. This allows the length x_2 from the pivot to be varied to give different lever ratios for different strengths of samples.

The apparatus is positioned overhanging the edge of a bench so that a plastic bucket, *E*, can be hung from a hook at the end of the upper arm. The bucket is then a distance x_1 from the pivot. A counterweight, *F*, is adjusted so that the upper arm just balances and there is no net force on aggregate *A*. Water is then run slowly into the bucket until the aggregate cracks. The weight of water, *W*, is then measured and the crushing force is calculated from

$$P = Wg\left(\frac{x_1}{x_2}\right) \tag{30}$$

where g is the acceleration of gravity. Alternatively, but less accurately, the amount of water in the bucket can be determined using a large measuring cylinder.

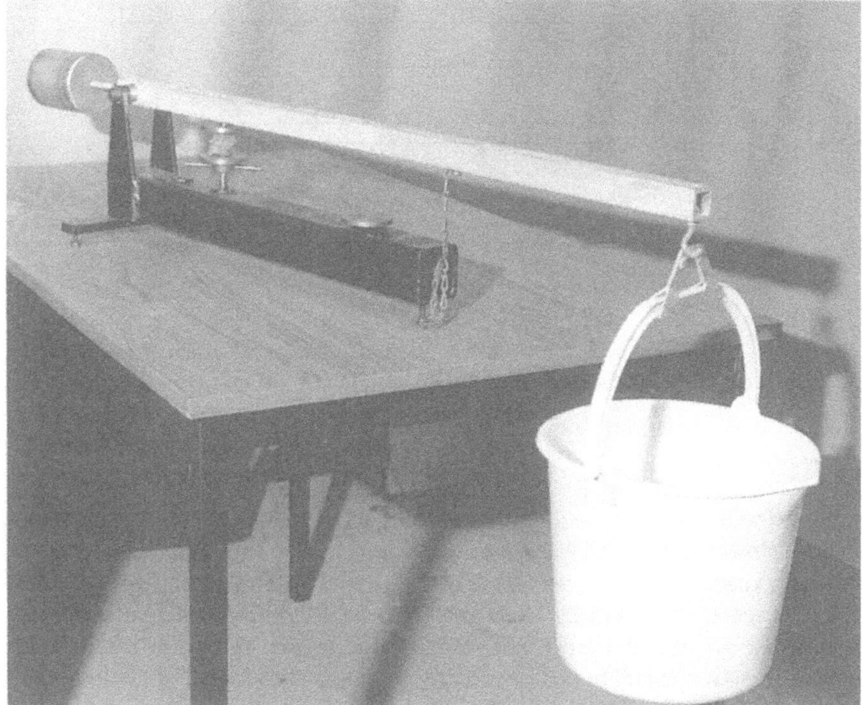

Fig. 13 Photograph of the simple apparatus shown in Fig. 12.

Larger or stronger aggregates may require a smaller value of x_2. We have found that a 10 L bucket and an x_1/x_2 ratio of 10 has been appropriate for most soil aggregates of 20 mm diameter. Water is usually run into the bucket at around 2 L min^{-1} through a rubber hose of 9 mm bore. When the aggregate fails, the flow can be stopped rapidly using a spring-loaded hose clamp.

With all the methods described above, an experienced operator can crush between 25 and 40 aggregates per hour.

D. Problems with Moist Samples and Sample Flattening

When soils are moist, samples may deform plastically to some extent before tensile failure. This usually takes the form of some flattening at the points of loading (the top and bottom of the cylindrical sample in Fig. 1).

Frydman (1964) studied this effect both theoretically and experimentally, and concluded that if the width of the flattened portion is smaller than 0.27 of the cylinder diameter, D, at failure, then the error involved in using Eq. 1 does not exceed 10%. It is likely that limited flattening of the poles of loaded spheres or soil aggregates will have a similar small effect on results obtained using Eq. 2.

More significant is the effect of soil water content on the type of failure. Provided that samples fail with sudden brittle failure, then Eqs. 1 and 2 are meaningful. However, if a sample fails with ductile failure (plastic deformation on the failure surface), then the method is not valid and the results must be rejected.

E. Levels of Replication

To distinguish between two populations of samples having mean values of tensile strength, Y_i and Y_j, a number, n, of replicate measurements are required. The replication, n, depends on the difference between Y_i and Y_j, on the coefficient of variation, COV, of the Y values, and on the level of statistical significance required. It can be shown that the minimum number of replicates required can be estimated from

$$n = \left[\frac{U \cdot COV}{\Delta}\right]^2 \tag{31}$$

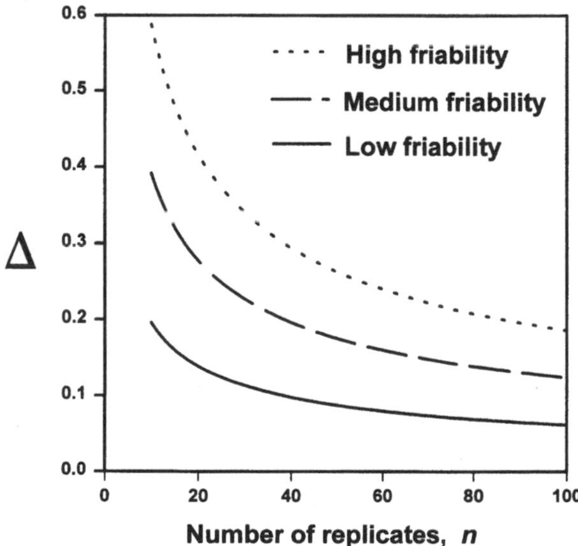

Fig. 14 Relationship between the proportional difference, Δ, between two values of tensile strength, and the number, n, of replicates required to distinguish between them at the $P = 0.05$ level of significance. Results are shown for soils of high, medium, and low friability, where the coefficients of variation are 0.6, 0.4, and 0.2, respectively.

Here, Δ is the proportional difference between Y_i and Y_j (such that $\Delta = 0.1$ for a 10% difference, etc.). U takes the approximate values 2.5, 3.1, 3.9, and 4.3 for levels of significance of $P = 0.10$, 0.05, 0.02, and 0.01, respectively. Figure 14 shows graphs of minimum values of replication, n, for the $P = 0.05$ level of significance.

To distinguish between two populations of samples having values of friability, F_i and F_j, the minimum number of replicates required is

$$n = V\left(\frac{F_i + F_j}{F_i - F_j}\right)^2 \tag{32}$$

where V takes values of 0.8, 1.2, 1.85, and 2.26 for the $P = 0.1$, 0.05, 0.02, and 0.01 levels of significance, respectively.

Irrespective of Eqs. 31 and 32, we would not recommend the use of fewer than $n = 10$ replicates in any experimental investigation.

V. APPLICATIONS

A. Effects of Dispersible Clay

In Fig. 4 we can see that friability of dry soil aggregates is greatly decreased when there is a greater content of mechanically dispersible clay in the soil. Similarly, the tensile strength is increased by the presence of mechanically dispersible clay. This effect is attributed to the deposition of the clay at the ends of microcracks as the soil dries. This strengthens the cracks and prevents them elongating under stress and contributing to aggregate failure or crumbling.

Factors that increase the quantity of mechanically dispersible clay in soil and that will therefore increase the tensile strength and reduce the friability of the dry soil include greater times for which the soil has been wet (Watts and Dexter, 1998; Currie, 1966), mechanical disturbance by tillage or wheel traffic (Watts and Dexter, 1998), and sodicity (Dexter and Chan, 1982). Low values of mechanically dispersible clay with consequent low tensile strength and high friability of the dry soil are associated with high contents of organic matter (Watts and Dexter, 1998), with calcareous soil (Dexter and Chan, 1991), and with other physicochemical factors that reduce clay dispersibility and that promote flocculation (Shanmuganathan and Oades, 1982).

B. Effects of Wetting/Drying Cycles on Soil Structure Generation

The effects of weathering in generating or ameliorating soil structure are well known. Whereas drying of soil tends to produce widely spaced cracks, rapid wetting can create closely spaced microstructure that makes the soil friable, among other things. The effects of rapid wetting have been studied on remolded soils in the laboratory (Grant and Dexter, 1989; Kay and Dexter, 1992; McKenzie and Dexter, 1985). It has been shown that the soil must be drier than a water potential of −1 MPa before there can be structure generation upon rapid wetting (Grant and Dexter, 1989; Sato, 1969). Natural wetting and drying cycles in the field have also been found to reduce the tensile strength of natural soil aggregates (Kay and Dexter, 1992).

The role of wetting and drying cycles in the formation and stabilization of soil aggregates adjacent to plant roots has been studied by Horn and Dexter (1989). They showed that the tensile strength of aggregates that were adjacent to roots was increased by the intense and periodic drying of the soil there, which was caused by evapotranspiration from the plant leaves. This was in contrast with aggregates not adjacent to roots or in soils that were kept permanently moist.

C. Effects of Soil Organic Matter Content

Using soils with a range of organic matter contents, Watts and Dexter (1998) found a very strong positive correlation between friability and organic matter content. In these soils, the tensile strength did not vary much with organic matter content, but the variability did [σ_Y in Eq. 19].

Kay and Dexter (1992) found that the tensile strength of natural soil aggregates in the field was not reduced as much by natural wetting and drying cycles at an organic matter content of 3.4 g $(100\ g)^{-1}$ as it was at an organic matter content of 2.2 g $(100g)^{-1}$.

D. Tillage Research

Macks et al. (1996) found that soils with a low friability were unsuitable for direct drilling (no-till). They also showed that direct drilling maintained a significantly higher value of soil friability than traditional tillage, as shown in Fig. 6. However, they were not sure if changing from traditional tillage to direct drilling would be possible for some of the soils that they examined because soil degradation had made direct drilling unfeasible, and the hard-setting, clod-forming, degraded soils seemed to require more rather than less tillage if plants were to be established. Clearly, methods other than tillage are required to ameliorate such severely degraded soils, and it is likely that friability tests in the laboratory would be an efficient way of rapidly screening a range of amelioration options.

Wheel traffic also reduces friability (Fig. 5) because the mechanical energy input associated with it breaks the bonds between soil particles and thereby increases the amount of mechanically dispersible clay. Tillage of soil that is too wet has a similar detrimental effect (Watts and Dexter, 1998). If soil is tilled when it is wetter than the lower plastic limit, then it is susceptible to mechanical damage. The amount of damage increases with the intensity of the tillage, as measured by specific energy input.

Friability is a function of water content and goes through a maximum just below the lower plastic limit (Utomo and Dexter, 1981; Watts and Dexter, 1998; Shanmuganathan and Oades, 1982). This agrees well with the optimum water content for tillage, which is the water content at which the crumbling effect is greatest (Ojeniyi and Dexter, 1979). Interestingly, for the clay soil that they studied, Watts and Dexter (1998) found that after drying it retained a "memory" of the structure that it had when moist in the field. This enabled them to use friability measurements on dry aggregate samples in the laboratory to determine the optimum water content for tillage.

The soil structure produced by tillage usually depends more on the soil properties than on the details of the tillage implement. The soil properties are themselves dynamic and vary with many factors including antecedent climatic conditions (Kay and Dexter, 1992). It seems highly desirable, therefore, in tillage trials to measure and record soil friability values in view of their dominant importance.

E. Index of Soil Structural Quality

Friability provides a valuable index of soil structural quality. It provides a numerical scale that can be applied to all soils. Although they do not tell us why a soil is bad, friability measurements can be used in simple laboratory tests to screen rapidly for causes and then for optimum ameliorative treatments to alleviate a problem. Sufficient comparisons have been made between the behavior of soils in the field, and the results of simple laboratory tests, to enable us to rely on the results of the latter with considerable confidence.

VI. SUMMARY AND CONCLUSIONS

The tensile strength of soil is sensitive to soil structure, and this makes it a valuable quantity in soil structure research. This is in contrast with other measures of soil strength, which fail soil in compression and which are insensitive to structure.

The theory of brittle fracture shows how the dispersion of strengths of soil structural features, flaws or microcracks, gives rise to a parameter that can be identified with soil friability. This parameter is also consistent with the observed behavior of soil in the field. The theory shows how the friability can be measured either from the variability of strength within a population of samples of a single size, or from the increase in strength of samples with decreasing sample size. However, the theory has some obvious deficiencies resulting from the limited validity of some of its assumptions for soils. As a result of these limitations, estimates of friability from estimates of the variability of strength of one size of sample (methods F_1 amd F_2) give significantly larger numerical values than those from the size-dependence method (method F_3). Nevertheless, friability measurements provide a powerful index for soil behavior and soil structure.

Measurement of the size of soil aggregates is particularly important, and five different methods for doing this have been described.

Techniques for measurement of sample crushing force are available that cover the range of sample sizes from 1 to 100 mm. In principle, there are no technical obstacles to prevent this range being widened by a factor of at least 2 in each direction.

A simple apparatus for measurement of crushing force (Figs. 12, 13) can be built at low cost. This is satisfactory for samples of medium size (5–25 mm) and can give results with accuracies similar to those from much more expensive equipment.

Examples of results are given that illustrate the applicability of measurements of soil tensile strength and friability to a wide range of problems associated with soil structure and soil physical quality.

Our favorite method at the moment is to determine friability, F_1 (the coefficient of variation of tensile strength values) using aggregate diameters determined by method 4, and measuring the aggregate crushing force, P, using a loading frame with a load cell and a signal analyser (Fig. 9) for weak and strong aggregates, and a digital balance and laptop computer (Fig. 10) for small and weak aggregates. However, we use all the methods described above depending on the nature and constraints of each particular experiment. All the methods described provide results that can give us valuable new insights into soil structure and behavior.

REFERENCES

Aitchinson, G. D. 1961. Relationships of moisture stress and effective stress functions in unsaturated soils. In: *Pore Pressure and Suction in Soils*. London: Butterworths, pp. 47–52.

Akazawa, T. A. 1943. A new method to find the tensile strength of concrete. *J. Jpn. Soc. Civil Eng*. 29:777–787.

Braunack, M. V., J. S. Hewitt, and A. R. Dexter. 1979. Brittle fracture of soil aggregates and the compaction of aggregate beds. *J. Soil Sci*. 30:653–667.

Carneiro, F. L. L. B., and A. Barcellos. 1953. Tensile strength of concretes. *Bull. R.I.L.E.M.* 13:103–125.

Currie, J. A. 1966. The volume and porosity of soil crumbs. *J. Soil. Sci*. 17:24–35.

Dexter, A. R. 1975. Uniaxial compression of ideal brittle tilths. *J. Terramech*. 12:3–14.

Dexter, A. R. 1988a. Strength of soil aggregates and of aggregate beds. *Catena Supplement* 11:35–52.

Dexter, A. R. 1988b. Advances in characterization of soil structure. *Soil Tillage Res.* 11: 199–238.

Dexter, A. R. 1997. Physical properties of tilled soils. *Soil Tillage Res.* 43:41–63.

Dexter, A. R., and. B. Kroesbergen. 1985. Methodology for determination of tensile strength of soil aggregates. *J. Agric. Eng. Res.* 31:139–147.

Dexter, A. R., R. Horn, and W. D. Kemper. 1988. Two mechanisms for age-hardening of soil. *J. Soil Sci.* 39:163–175.

Dexter, A. R., and K. Y. Chan. 1991. Soil mechanical properties as influenced by exchangeable cations. *J. Soil Sci.* 42:219–226.

Freudenthal, A. M. 1968. Statistical approach to brittle fracture. In: *Fracture: An Advanced Treatise. Vol. II. Mathematical Fundamentals* (H. Liebowitz, ed.). New York and London: Academic Press, pp. 591–619.

Frocht, M. M., and R. Guernsey. 1952. A special investigation to develop a general method for three-dimensional photoelastic stress analysis. Contract No. NAw—5959: Natnl. Adv. Comm. Aeronautics Tech., Note 2822.

Frydman, S. 1964. The applicability of the Brazilian (indirect tension) test to soils. *Aust. J. Appl. Sci.* 15:335–343.

Grant, C. D., and A. R. Dexter. 1989. Generation of microcracks in moulded soils by rapid wetting. *Aust. J. Soil Res.* 27:169–182.

Gumbel, E. J. 1958. *Statistics of Extremes.* New York and London: Columbia Univ. Press.

Gunselmann, M., U. Hell, and R. Horn. 1987. Die Bestimmung der Wasserspannungs/wasserleitfähigkeits Beziehung von Bodenaggregaten. *Z. Pflanzenernähr. Bodenk.* 150:400–402.

Hadas, A., and G. Lennard. 1988. Dependence of tensile strength of soil aggregates on soil constituents, density and load history. *J. Soil Sci.* 39:577–586.

Hallett, P. D. 1996. Fracture mechanics of soil and agglomerated solids in relation to microstructure. Ph.D. thesis, Univ. Birmingham, U.K.

Hiramatsu, Y., and Y. Oka. 1996. Determination of the tensile strength of rock by a compression test of an irregular test piece. *Int. J. Rock Mech. and Mining Sciences* 3: 89–99.

Horn, R., and A. R. Dexter. 1989. Dynamics of soil aggregation in an irrigated desert loess. *Soil Tillage Res.* 13:253–266.

Kay, B. D., and A. R. Dexter. 1990. Influence of aggregate diameter, surface area and antecedent water content on the dispersibility of clay. *Can. J. Soil Sci.* 70:655–671.

Kay, B. D., and A. R. Dexter. 1992. The influence of dispersible clay and wetting/drying cycles on the tensile strength of a red-brown earth. *Aust. J. Soil Res.* 30:297–310.

Kemper, W. D., and R. C. Rosenau. 1984. Soil cohesion as affected by time and water content. *Soil Sci. Soc. Am. J.* 48:1001–1006.

Kirkham, D., M. de Boodt, and L. de Leenheer. 1959. Modulus of rupture determination on undisturbed core samples. *Soil Sci.* 87:141–144.

Macks, S. P., B. W. Murphy, H. P. Cresswell, and T. B. Koen. 1996. Soil friability in relation to management history and suitability for direct drilling. *Aust. J. Soil Res.* 34: 343–360.

Martinson, D. C., and L. B. Olmstead. 1949. Crushing strength of aggregated soil materials. *Soil Sci. Soc. Am. Proc.* 14:34–38

McIntyre, D. S., and G. B. Stirk. 1954. A method for determination of apparent density of soil aggregates. *Aust. J. Appl. Sci.* 5:291–296.

McKenzie, B. M., and A. R. Dexter. 1985. Mellowing and anisotropy induced by wetting of moulded soil samples. *Aust. J. Soil Res.* 23:37–47.

Ojeniyi, S. O., and A. R. Dexter. 1979. Soil factors affecting the macrostructures produced by tillage. *Trans. Am. Soc. Agric. Eng.* 22:339–343.

Peltier, R. 1954. Theoretical investigation of the Brazilian test. *Bull. R.I.L.E.M.* 19:26–69.

Rogowski, A. S. 1964. Strength of soil aggregates. Ph.D. thesis, Iowa State Univ. Sci. Tech., Ames, Iowa.

Rogowski, A. S., and D. Kirkham. 1976. Strength of soil aggregates: Influence of size, density, and clay and organic matter content. *Meded. Fac. Landbouww.* (Rijksuniv, Gent) 41:85–100.

Rogowski, A. S., W. C. Moldenhauer, and D. Kirkham. 1968. Rupture parameters of soil aggregates. *Soil Sci. Soc. Am. Proc.* 32:720–724.

Sato, K. 1969. Changes in the shrinkage and slaking properties of clayey soils as an effect of repeated drying and wetting. *Trans. Jpn. Soc. Irrig. Drainage. Reclam. Eng.* 28:12–16.

Shanmuganathan, R. T., and J. M. Oades. 1982. Effect of dispersible clay on the physical properties of the B horizon of a red-brown earth. *Aust. J. Soil Res.* 20:315–324.

Utomo, W. H., and A. R. Dexter. 1981a. Soil friability. *J. Soil Sci.* 32:203–213.

Utomo, W. H., and A. R. Dexter. 1981b. Age-hardening of agricultural top soils. *J. Soil Sci.* 32:335–350.

Vilensky, D. G. 1949. *Aggregation of Soil: Its Theory and Application.* Moscow: U.S.S.R. Academy of Sciences. Transl. from the Russian by A. Howard, C.S.I.R.O., Melbourne.

Vomocil, J. A., and W. J. Chancellor. 1969. Energy requirements for breaking soil samples. *Trans. Am. Soc. Agric. Engrs.* 12:375–388.

Watts, C. W., and A. R. Dexter. 1998. Soil friability: Theory, measurement and the effects of management practices and organic carbon content. *Eur. J. Soil Sci.* 49:73–84.

Wright, P. J. F. 1955. Comments on an indirect tensile test on concrete cyclinders. *Mag. Concrete Res.* 7:87–96.

12
Root Growth: Methods of Measurement

David Atkinson
Scottish Agricultural College, Edinburgh, Scotland

Lorna Anne Dawson
Macaulay Land Use Research Institute, Aberdeen, Scotland

I. INTRODUCTION

The morphology of a plant root system is a function of its genetics and the environment in which it grows (Smucker, 1993; Aiken and Smucker, 1996). Morphology is also affected by interaction with soil microorganisms, e.g., arbuscular mycorrhizal fungi (Hooker et al., 1992). Both individual plant roots and whole root systems can and do show substantial variation within the potential range of their characteristics. Soil physical factors, particularly temperature, aeration, water potential, and mechanical impedance, are frequently the cause of limits to the expression of genetic potential. The morphology of the root system can thus be regarded as representing the integrated effects of three factors. This chapter first reviews those root properties that are most likely to be influenced by soil physical conditions and then describes methods that can be used in the field or laboratory to describe particular attributes of root systems. It illustrates some of the uses to which particular methods have been put and some of the limitations of their use. Bohm (1979) has given a more complete description of methods for measuring roots, and Atkinson (1981) has reviewed those methods relevant to tree crops. The impact of soil biological and chemical factors, and of the growth of the aerial parts of the plant, on root growth have been reviewed in general terms by Russell (1977).

In both field and laboratory, many of the methods give information on a range of parameters. For example, when a root system is observed directly (e.g., through an observation panel), measurements can be made of length, diameter, longevity, and branching. It therefore seems more logical to divide studies on root systems by type of method rather than by root system property. Consequently, in this chapter, the major groups of available methods and the significance of the measurements they facilitate are discussed together, but in the context of the need to determine how plant function can be influenced by soil physical conditions.

A. Root System Properties

Root systems are branched structures composed of a number of individual roots with normally up to four orders of branching. Individual roots are themselves made up of large numbers of cells. The processes relating to root development have been characterized at both cellular (e.g., Scheres et al., 1996) and molecular levels (e.g., Chriqui et al., 1996). The size, shape, and form of these cells, the numbers in a particular tissue (e.g., xylem or cortex), and their function (Smucker 1993) may be altered by the growing environment. Major soil physical factors, such as soil water potential and soil mechanical resistance, can affect root properties such as cell wall extensibility and wall pressure in a number of ways. Cell wall pressure is closely related to the rate of root growth, while osmoregulation is closely related to changes in soil water potential but less completely related to mechanical resistance. As a consequence of these effects, the length of individual roots, their rate of extension, and increases in root diameter can be changed by soil physical conditions, thus affecting the overall volume of soil exploited by roots, via effects on horizontal spread and the depth of penetration, which in turn influence the resources of water and nutrients available to the plant. Other parameters that can vary include the angle at which roots grow through soil (e.g., their susceptibility to geotropism).

The longevity of roots varies between species (Atkinson, 1985) and between root types in a species (Hooker et al., 1995). The rate of production, and the longevity, determine the total root length and average root length density, i.e., the length of root (LA) under an area of soil surface or the length (LV) in a volume of soil. These factors are important to the ability of the root system to obtain nutrients for plant growth. In addition to possible effects of soil factors on morphology, root function (e.g., nutrient uptake per unit root length, surface area, or volume) may be altered as a consequence of effects on the types and ages of root present. However, the exact effects of physical conditions on the above parameters are incompletely described or understood, and considerable plasticity clearly exists in respect of most properties (e.g., Reynolds and D'Antonio, 1996). Roots are normally considered in relation to their ability to supply water and nutrients to the plant, but they are also required to anchor the plant (Coutts, 1983) and to produce hormones, which may regulate the growth and performance of both root and shoot (Aiken and Smucker, 1996). The root system of most plants exists in nature in a symbiotic association with fungi (mycorrhizas), and so assessments of effects of

Table 1 Root System Characteristics That Can Be Affected by Soil Physical Conditions

Characteristic	Parameter
Anatomy	Cell size, cortex width, balance of xylem cell types, epidermal wall form, root shape
Individual root features	Diameter, growth rate, angle, length, mass, longevity, root hair length and density, penetration pressure
Branching pattern	Amount, density, number of orders, position, distance between branches
Whole root system properties	Horizontal distribution, vertical distribution, length, mass, absolute and relative distribution
Function	Absorption of nutrients and water, anchorage, production of biologically active molecules (e.g., enzymes, phenolics)

soil conditions on roots should also consider effects on mycorrhizas. Effects of physical factors on root characteristics are summarized in Table 1.

B. Potential Effects of Physical Properties

Many root system parameters can be influenced by a change in soil physical conditions (Table 1). Root effects depend upon many factors, including the nature of the changed soil variable, the species under investigation, and conditions in other parts of the soil. General principles were reviewed by Greacen and Oh (1972).

1. Case Study

In a study of the effect of zones of contrasting bulk density on root system development in oats, the effect of a given value of bulk density varied according to its relation to the density of other areas in the soil column (Schuurman, 1965). Compaction did not reduce branching, although it did influence root survival. The length of branch roots, which was normally highest in the surface, was affected (Fig. 1). Where the elongation of the main axis was reduced by a dense subsoil, its diameter increased and branching was stimulated.

2. Types of Response

In addition to changes in overall root system length, mass, or volume, there can be alterations in the partitioning of dry matter within the root system (e.g., by increasing root branching or root number: Goss, 1977). In soil, root elongation was reduced by 60% by a mechanical resistance of 1–8 MPa in ryegrass and by 2–6 MPa in pea (Gooderham, 1977). Goss (1977), using a ballotini (glass sphere) system, showed that the effect of increasing pressure on root growth inhibition varied between species, with barley being the most sensitive of the species tested.

Fig. 1 Effect of density of the topsoil (0–25 cm) and subsoil (>25 cm depth) on development of oat root systems. Bulk densities above and below the boundary are given in megagrams per cubic meter. (From Schuurman, 1965.)

When roots are prevented from elongation, a resultant increase in the diameter is found, mainly due to an increase in the cross-sectional area of the cortex (Wiersum, 1957). Goss showed that even when root system mass is unchanged, length can be reduced by 65% by mechanical impedance, while Logsdon et al. (1987) demonstrated that an increase in root diameter can compensate, in part, for a reduction in total length. Appropriate measurements are clearly needed to establish such effects, although their physiological significance is still poorly understood. The use of penetrometers in such studies has been discussed by Bengough et al. (1994). An increasing concentration of roots at the "soil surface" in the laboratory had no effect on nutrient supply (Goss, 1977), but in the field, in the absence of irrigation, it might be expected to have adverse consequences because the surface would quickly dry out. This exemplifies why results cannot be directly transferred from laboratory to field. There can also be changes in the internal root turgor pressure, which has been primarily studied in relation to water potential effects. A novel approach, using a force transducer, allows the turgor in an impeded root to be measured without the need to remove the root from the impeding environment (Clark et al., 1996).

C. Purpose of Measurements

The optimum method for the assessment of any root system will depend both on the characteristics of the root system itself and on the reason for making the measurement. The following are among the commonest reasons for measuring roots:

1. To assess the significance of changes in soil physical conditions on the plant

2. To help to interpret a plant response to a particular soil treatment by understanding effects on water or nutrient supply
3. To improve the use of inputs (e.g., irrigation water or fertilizers) or to optimize the effects of tillage and other soil management practices
4. To allow the development of better plant root systems by conventional breeding or genetic enhancement

D. Significance of Root Features

Criteria for the selection of methods of assessing root performance are limited by our current understanding of the significance of particular root characteristics and of the consequences of changes to them. Information exists on the importance of some characteristics, and so we can identify circumstances in which particular measurements will be useful.

1. Characteristics Influencing Water Supply

Soil water flux is the product of the hydraulic potential gradient between the soil and the root and the unsaturated hydraulic conductivity. Typical maximum flux rates seem to be around 2.5 mm d^{-1} (Russell, 1977). This should allow a rate of root water uptake of around 160 μL d^{-1} cm^{-1}. On this basis a root density of 2 cm of root per cm^2 of soil surface area should be able to supply transpirational needs in Northern Europe. The uniformity of root distribution and the mean distance between roots both influence water supply. Thus if an average root length density is clustered in one soil region, the flux to the root surface is likely to be inadequate. As soil water content and consequently unsaturated conductivity decrease, a greater root density will be needed to supply the same total flux to the shoot.

The total volume of soil exploited by the root system directly affects the total amount of water and nutrients available to that plant. This volume can be represented by the horizontal spread times the maximum rooting depth (or the depth containing 95% of roots), although the effective volume will be increased by capillary rise. Average root length density and root distribution within the soil volume exploited are both important. This, however, gives only a static picture. In the field, the transpiration rate will change during the season, as a consequence of weather and plant needs. It is thus important to be able to assess the root system as it grows, and to characterize factors such as root death; root longevity in some species is of the order of days, not weeks (Atkinson, 1985).

E. Commonly Measured Characteristics

Despite the long list of root parameters that might be measured (Table 1), researchers have commonly determined a smaller set. They are summarized in Table 2.

Table 2 Functional Significance of Major Root Systems Parameters and the Principal Means of their Estimation

Root parameter	Functional significance	Method
Length	Total system size Potential for the absorption of nutrients or water Soil microbial activity (Temporal variation: frequent estimation needed)	Monolith methods (soil coring, etc.) Rhizotron/minirhizotron methods (with assumptions) Profile wall methods (with assumptions and where limited precision is needed)
Mass	Total root system size	Monolith methods (especially soil coring) Excavation (woody perennials)
Number	Growth regulator production	Counts on soil cores Profile walls Rhizotron/minirhizotron
Root : shoot ratio	Relative allocation strategy	Calculation from root and shoot weights
Length density	Limitations to soil nutrient and water exploitation	Calculation from root length and soil volume
Specific length	Within-root-system allocation strategy Relative importance of soil exploitation	Calculation from root length and weight
Diameter	Potential for mycorrhizal development Regulation of water stress Potential for apoplast/symplast exchange Growth potential Response to soil physical conditions	Direct measurement or calculation from root length and volume
Amount of secondary thickened root	Investment in system infrastructure	Measurement of weight or length
Branching pattern	Intensity of soil exploitation	Measurement of lengths or number of roots of different orders
Longevity	Potential for rapid adjustment to root length Plasticity Flux of carbon to rhizosphere	Cohort analysis of rhizotron or minirhizotron images
Production	Overall potential for soil exploitation Ability to increase system length	Sequential soil core estimates or rhizotron/minirhizotron measurements
Mycorrhizal infection	Carbon allocation strategy Surface for nutrient uptake	Estimation from stained root samples
Vertical distribution	Physical stability Depth of soil exploited Potential for resource use	Soil core sampling Profile walls
Horizontal distribution	Stability Interaction with other species	Profile walls Excavation

II. METHODS OF STUDYING ROOT SYSTEMS IN THE LABORATORY

Laboratory studies are normally concerned with assessing the effect of either a single or a limited range of soil physical properties on plant and root performance. Single-factor studies usually relate to soil temperature, water potential, osmotic potential, or aeration, whereas studies of the effect of soil impedance often involve simultaneous changes in other factors. For example, when bulk density was changed from 1.24 to 1.52 Mg m^{-3} (Fig. 1), there was also a reduction in the volume of pores that were filled with air at field capacity and an increase in the volume of pores from which water would be unavailable to plants (Schuurman, 1965). Because the results of such experiments are likely to be influenced by the types of containers and media used, these are also briefly reviewed.

A. Containers

1. General Factors

Container methods permit the isolation of individual environmental factors that will normally interact with other characteristics and influence root growth in the field. Replication and "management" are easier than in the field, although container effects on root growth may be unnatural because of the restricted space and absence of soil organisms (bacteria, fungi, soil arthropods, etc.). Container methods are best suited for studying plants with small root systems or for investigating the early stages of plant development.

2. Container Types

The *size* of a container determines the total volume of soil available to a plant. Conventional plastic or clay plant pots, Mitscherlich pots, glass pots, petri dishes, tubes made from glass or plastic, and cardboard cartons have all been successfully used as containers, but their limited volume frequently results in roots concentrating near the walls of the vessel and around its bottom. As a consequence of moisture and temperature differentials, the concentration of roots between the wall of a pot and the soil tends to be greater in a porous clay pot than in a plastic pot.

In container experiments designed principally to study root growth and distribution, the depth of the container needs to be large because of the root's tendency to grow downward when restricted by the walls of a container (Bohm, 1979). Boxes 80 cm high, made from metal, wood, or plastic, have been used in this type of study, but cylindrical tubes are more common. Iron, clay, asbestos, plastic, acrylic, and glass have all been used. Where tubes are of a transparent material, or where boxes have glass windows set into their sides, roots can be directly observed and measured (Fig. 2). Containers of all dimensions can be

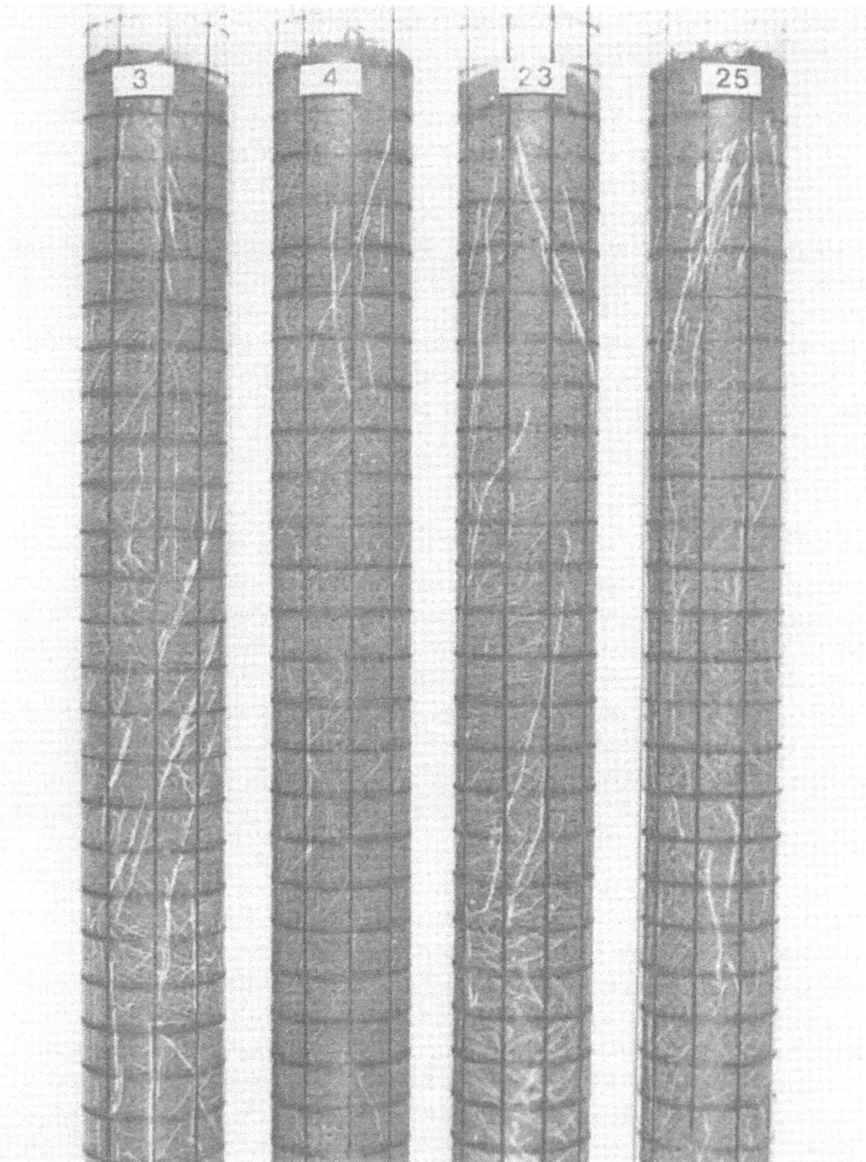

Fig. 2 Use of acrylic cylinders to observe differences in root form, density, and distribution in spring barley. To calculate information, such as total or average root lengths, the relationship between length at the observation surface and in the whole soil volume must be known. (From Atkinson, 1987.)

modified and constructed to have a viewing window or sides to permit repeated observations to be made of root growth (Mackie-Dawson et al., 1995a). Tubes may be buried in the soil or in insulated boxes to prevent the establishment of unrealistic temperature differentials (Mackie-Dawson et al., 1995a) and video recording equipment can be used to record root growth.

Root boxes vary in size in relation to the type of plant being investigated. For studies of M.1 apple rootstocks, boxes 60 × 17.5 × 42.5 cm high were used (Rogers, 1939a), while for studies of maize, Walker and Barber (1961) used smaller boxes. Such boxes with windows may be used simply as a means of observing the response of roots receiving particular treatments or as a means of assessing the uptake of radioisotopes incorporated into the soil adjacent to the soil-observation interface. The latter method has been used to observe the uptake of ^{86}Rb from the soil around individual roots (Walker and Barber, 1961).

Such boxes have been used to assess the effect of soil moisture and soil temperature on the root growth of grass and clover species (Garwood, 1968). Using the observation windows, it was possible to assess treatment effects on root system length, the elongation of individual roots, root diameter, and root number. All these parameters were affected by soil temperature.

Where plants are grown in containers that allow observation of the root system, the possible effects of light on root growth must be considered, although few studies have addressed this issue. A comparison of the effect of a range of different light exposures, varying from total darkness to total light, on the growth of apple rootstocks showed that while continuous illumination severely checked growth, increased suberization, and reduced the development of lateral roots, the short periods of 3 × 20 minutes per week or 2 hours per 2 weeks needed for observation had little obvious effect. Effects were greatest during periods of maximum root growth, with length being more affected than root number (Rogers, 1939a). Given the paucity of data on the effects of light, it is prudent to reduce unnecessary exposure to a minimum.

3. *Filling Containers*

Care must be taken when filling containers (both tubes and boxes), particularly when physical conditions (e.g., bulk density) are being controlled (Schuurman, 1965). Soil should be sieved when nearly air-dry, the appropriate fertilizer added, and the soil then moistened to a friable condition and mixed. It should then be put in the containers and compressed, layer by layer, several cm at a time, with the top few mm of each layer loosened before adding the next layer, to avoid stratification. All containers should normally be watered and allowed to stand for a period from a week to a month before the experiment begins, to allow the soil to settle. Where short-lived radioisotopes are being used, however, this may not be possible (Walker and Barber, 1961). These procedures are needed both for tubes and boxes.

Where an observation surface is to be viewed, it is essential that care be taken to prevent smearing and the formation of voids.

B. Media

For experimental purposes, plants can be grown in a solid, liquid, or gaseous rooting medium; the type used depends on the scope and aims of the experiment.

1. Solid Media

Soil. Soil is the most realistic growing medium for terrestrial plants and for long-term experiments. However, the extraction of whole root systems from soils other than very sandy ones is difficult, as is the complete removal of soil particles from recovered roots (Atkinson, 1987; McCully, 1987). Although exact nutrient compositions cannot be produced easily, soil temperature, water content, and compaction levels can all be manipulated. For example, containers of soil have been produced in which one layer is varied in bulk density, thickness, and depth from the surface (Baligar et al., 1981). Maintenance of a given matric potential is usually made using tensiometers or by weighing. A variation of this technique in which roots are grown in soil within porous membrane envelopes has been used successfully (Brown and ul Haq, 1984). Here the root system was confined within the "envelope," and water and nutrients were able to move across the membrane.

Undisturbed soil columns have the advantage that the structure, texture, and water availability, and also the complex structural and mycorrhizal network, remain relatively undisturbed. Columns allow the experimenter to control certain soil conditions (e.g., water content) and plant growing conditions (e.g., temperature), but properties such as pore size distribution, structure, and bulk density cannot be precisely determined before experimentation. Undisturbed columns may be very large, e.g., 1 m^3 (Belford, 1979), and can be collected by hand coring, power coring, or hydraulic sampling machines. The columns can be preserved with a coating of paraffin wax, plastic, foil, or liquid plastic material. Care needs to be taken during transportation so that artificial voids are not created.

Sand. Sand is often used in nutrient experiments because of its low buffering capacity and the ease with which it can be manipulated. It is also relatively easy to wash from the roots, although some may still firmly adhere (Atkinson, 1987). The physical properties of the medium can be altered by varying the particle size. Fine sand gives a higher water-holding capacity, while coarse sand is more freely draining and better aerated. Root systems obtained from experimental media of this type are generally similar to those grown in solution culture (Bohm, 1979).

Other Solid Media. Perlite, which is composed of expanded volcanic rock fragments, is uniform and inert and so is suitable for studies of germination or seedling development. However, it is less suitable for long-term growth experiments and for studies of physical effects on roots. Perlite seems to result in root system development similar to that in solution culture (Bohm, 1979). It is a good medium where roots are to be used for studies of ultrastructure. Particles of perlite embedded in roots cause less damage to knives used in sectioning than do sand grains. However, roots can penetrate perlite, from which they are difficult to extract. The differing penetrating abilities of roots have been studied using agar, paraffin, and wax materials of differing hardness (Taylor and Gardner, 1960a, b; Yu et al., 1995). A wax mixture layer has been used to assess the root penetration ability among rice roots (Yu et al., 1995). Vermiculite has also been used as a growing medium and seems to give growth comparable to that of roots grown in soil (Bohm, 1979).

2. Solution Culture

The major advantages of solution culture are that the ionic composition of the root environment can be defined, measured, and manipulated with precision and that the entire medium can be held under standard conditions (e.g., of temperature and aeration) (Atkinson, 1986). It can either be applied as a solution or as a mist application. Because of ion uptake/efflux by plant roots, however, nutrient solutions are liable to rapid changes in composition, and so require more routine maintenance than is needed for soil-based systems.

Solution culture is only of limited use in soil physical studies. The uniform medium, lack of physical resistance, and the absence of soil flora and fauna make it difficult to compare root growth in solution culture with that in soil. However, this approach has applications in studies of impeded aeration, water stress, and temperature. Roots can be maintained at precise temperatures by flowing the solutions through a refrigeration or heating unit before entry into the plant growth containers (Bhat, 1980), or by immersing plant-growth containers in thermostatically controlled water baths. Mist chambers have been used in the study of water uptake, using an applied dye, sulphorhodamine (Varney and Canny, 1993).

3. Special Techniques

Split-Root Techniques. Spatial variability in nutrient supply can be controlled using split-root containers, in which isolated parts of the root system receive different nutrient supplies, either in solution culture or in solid media. Individual roots can also be separated out to study specific effects.

Water Stress Control. Osmotic control of plant water stress can be obtained in solution culture using sodium chloride, polyethylene glycol 4000 (PEG), or a range of other chemicals. This method allows plant water stress to be accurately maintained and more easily reproduced than is possible in soil. However, the stresses brought about by the two methods have different physiological bases. The water stress to which roots are normally exposed in soils is primarily due to a substantial negative matric potential, while in a PEG-modified solution the stress results from a substantial negative osmotic potential. Although stresses due to both matric and osmotic potential have been shown to produce similar effects on plant growth, it must be remembered that the matric potential at the root surface can be considerably less than that measured in bulk soil.

Pressure Control. Many workers have studied effects of mechanical stress by growing roots in pressure cells through which aerated nutrient solution is circulated. The cell walls consist of flexible impervious polyester membranes. A known hydrostatic pressure is applied by suspending the cells in water-filled vessels, to which an external pressure is applied (Barley, 1962; Abdalla et al., 1969). Details of construction, use, and the types of measurements that can be made are given by Goss (1977). A technique that combines a pressurized wall with time-lapse video analysis has been used to study pea lateral root emergence (Gordon et al., 1992). Pressure has also been applied to roots grown behind thin rubber diaphragms forced against the root by gas under controlled pressure (Gill and Miller, 1956). However, the actual pressure experienced by the root is uncertain.

C. Measurement of Roots in Laboratory Media

1. Measurement in Soil

Impregnated Sections. Roots can be studied in undisturbed samples by impregnating the soil with wax or resin, using samples collected from pot experiments or field plots. The method involves removal of soil water in exchange for a solvent in which the concentration of resin is gradually raised. After addition of an accelerator or hardener, the cured soil blocks are sectioned for examination (Altemuller, 1986). Several combinations of fixative and impregnating resin have been used, including a mixture of acetic acid–formaldehyde and ethanol and a polyester resin for impregnation (Lund and Beals, 1965), acetone instead of ethanol in the preceding procedure (Altemuller, 1986), and glutaraldehyde–acetone, with a resin for impregnation (Darbyshire et al., 1985). The best methods can preserve the form of delicate biological materials, such as root tissues and protozoa cells (Altemuller, 1986). Staining roots in the blocks with methylene blue and basic fuchsin followed by sectioning can lead to good identification of the detailed structure of preserved materials. Fluorochromes, such as acidine orange (Darbyshire et al., 1985), can be used to increase the natural root fluorescence. Glutaraldehyde impregnation has been used to study the soil pore network available to protozoa and roots. Although the method is expensive and labor-intensive, it allows detailed examination of the soil-root interface. Root–soil interactions, at a more detailed level, can be assessed using the scanning electron microscope (Fig. 3)

Fig. 3 A root of spring barley and the soil attached to it as seen with the Stereoscan electron microscope.

Nuclear Magnetic Resonance Imaging. Nuclear magnetic resonance (NMR) has been used as a noninvasive tool for studying roots in situ (Bottomley et al., 1986). However, because the images are based on H detection, soil moisture levels have to be kept low. The technique has been used to obtain images of root systems grown in a range of soil types, vermiculite, sand, perlite, fritted clay, potting soil, and "peatlite," but the clarity of image varies according to the magnetic properties of the medium examined. To observe relatively dry soil, hence to optimize the root image, measurements were made at the end of the watering cycles. Recent developments have used both 2-D and 3-D images and have been able to distinguish plant vasculature from surrounding parenchymal tissue (MacFall and Johnson, 1994). Images can now be measured for root surface area, volume, and orientation (MacFall and Johnson, 1996). However, there are still limits to the resolution at which it can operate.

X-Ray Computer-Aided Tomography. This is a non-destructive x-ray technique that can separate out features such as roots, soil pores, and cracks, due to their low absorbing properties (Tollner et al., 1989). Currently, its resolution is approximately 1 mm, which limits its application for detailed root studies. Also, as the x-ray absorption of each pixel is a function of the water content, the dry bulk density, and chemical properties of the soil (Tollner et al., 1991), it has limitations for use in soil physical investigations. However, it is possible to calibrate CT measurements with root length density measurements from core samples, which allows simultaneous nondestructive measurement of both root length density and water removal from the same soil core.

Neutron Radiography. Neutron radiography has been used to produce two-dimensional images of plant root systems (Willat et al., 1978). Plant roots grown in narrow (2.5–5.0 cm wide) boxes, with neutron-sensitive back plates, were irradiated with thermal neutrons, and photographic images were obtained from these plates. Roots were identified because of preferential neutron scattering by the roots. In this way the elongation rates of soybean and maize roots were obtained from sequential radiographs. Lateral roots (< 0.33 mm) were poorly visible. There is a need for improved resolution before the method can be regarded as a practical means of producing quantitative data. Neutron radiography, CT, and NMR are rapidly developing techniques that are constantly improving in image resolution quality and have the advantage of being nondestructive and usable with soils under relatively natural conditions. However, they have the disadvantage of not being available to the majority of researchers.

Autoradiography. Radioisotopes have been used in a variety of ways to observe roots or give a measure of root activity in soil (Walker and Barber, 1961). If two species grown in mixed culture are injected, one with ^{32}P and one with ^{33}P, the roots of the two species can be subsequently identified in a section of a soil block containing the cut ends of the roots of both species. Mixing radioisotopes into the soil and assessing depletion around roots has been used by a number of workers as a means of assessing root activity in soil boxes (see also Sec. III.D.3). The technique could also be used to assess the effects of soil impedance on uptake.

2. Measurements of Root Parameters

Number. The number of roots can be counted in samples of washed roots or in situ (e.g., glass-faced columns). The number of root tips per unit volume of soil has been used as an indicator of root distribution in soil (Weller, 1971). It has been suggested that root number is closely related to leaf number (Richards and Rowe, 1976). Image analysis methods now make this easier to determine (Smucker, 1993).

Mass. For the determination of root mass, roots are washed free of soil, then oven-dried (usually at about 80°C) for 24–48 hours. Mass can be measured in all methods that permit roots to be destructively sampled. Mass characterizes the total amount of root but is not a good indicator of absorbing potential. It is less sensitive to soil factors than root length. Treatments with a major effect on root length may have no effect on mass. Measurements of mass are important to the production of carbon budgets (Gansert, 1994; Thomas et al., 1996).

Surface Area. Surface area can be related to water and nutrient uptake. An estimate of root surface area can be obtained from root length and diameter, if root hairs and the extramatricular hyphae of mycorrhizas are ignored. Direct methods used to estimate the root surface of roots washed free of soil include photoelectric attenuation (Morrison and Armson, 1968), dye adsorption, and the retention of calcium on the external surface of the root, following a brief immersion in a concentrated solution of calcium nitrate and centrifugation (Carley and Watson, 1966). Surface area is now most commonly estimated by image analysis (Smucker, 1993).

Strength. The tensile strength of single washed roots has been obtained from the force required to break 5–10 cm lengths of root of known diameter (Parlychenko, 1942). The buckling stress of clamped, excised roots may be used to characterize root elasticity. Weights are hung from one end of a 10 mm length of root clamped at its other end. Elasticity is related to the deflection caused by a known weight (Goss et al., 1987).

Diameter. The diameter of newly washed root samples from soil cores or solution culture, or from roots in impregnated soil blocks, can be measured directly and used to estimate either root surface area or length where volume is known (Bhat, 1983). Large numbers of measurements are needed to characterize diameter accurately. The effects of external pressure on root diameter have been studied in beds of ballotini (Goss, 1977). By varying the size of the glass beads, interactions between pore size distribution and root diameter can be assessed. Root diameter is usually measured using a microscope micrometer eyepiece. Diameter can also be measured on images captured using a minirhizotron system. By using pixel counts, the direct measurement of the average root diameter can be made (Lebowitz, 1988).

Length. Other than mass, this is probably the most common single measurement made.

BASIC CALCULATIONS. Length can be measured directly using calipers and samples of wet root in a water-filled dish, or by placing roots on graph paper and counting squares. This method is time-consuming, but for large roots (> 5 mm diameter) it is often the most practical (Atkinson et al., 1976). For samples of greater root length, measurement requires some type of sampling method, such as counting the number of intersections between roots and a random or regular pattern of lines. Total root length R can be estimated by

$$R = \pi \frac{AN}{2H} \tag{1}$$

where R is the total length of roots, A is the area of the field of view, and N is the number of intersections between the roots and a set of randomly oriented straight lines whose total length is H (Newman, 1966).

Newman (1966) applied this principle to a system where a number of fields of view were examined using a microscope with a hairline in the eyepiece, which was randomly reoriented before each new examination. Using this method, the time for root length measurement was reduced to less than half that required by direct measurement (e.g., 24 min to measure 3.4 m of root with a CV of 4.3% versus 67 min by direct measurement).

Since Eq. 1 requires only that the orientation between roots and a set of lines be random, this equation can be used equally well for regular arrangements of lines such as parallel straight lines or a grid, provided there is no preferred orientation of the roots in relation to these lines. Furthermore, where the line spacing or the spacing of a square grid is equal to the unit in which root length is measured, it is easy to show that A/H is 1 (for lines), and thus

$$R = \pi\frac{Nd}{2} \tag{2}$$

where d is the distance between grid lines, and $\frac{1}{2}$ (for a grid), or in general for a square grid

$$R = \pi\frac{Nk}{4} \tag{3}$$

where k is the grid spacing. The theory has been used by Marsh (1971) and Tennant (1975). A similar result was obtained empirically by Head (1966). With this modification of Newman's (1966) method, there is no necessity for the roots to be uniformly spread out over the counting area. A procedure that enables a single set of measurements to be obtained within 6 minutes, and with a coefficient of variation of 5% or less, has been given by Tennant (1975). It is appropriate to match grid size and the length of root to be measured and to keep N between about 100 and 500. Time can also be saved because all organic debris does not have to be carefully removed from the roots. This method was used in the study of Oosterhuis and Zhao (1994), who calculated root length as intercepts (with a 1×1 cm grid) $\times$ 0.7857.

AUTOMATED METHODS. The theory presented above underpins systems where the scanning of the root sample or the counting of the number of times roots intersect a regular pattern are automated. In one system (Richards et al., 1979; commercially available from Commonwealth Aircraft Corp. Ltd., Port Melbourne, Victoria 3207, Australia), roots are spread out on a transparent rotating turntable, which is traversed by a light beam and detector. The number of times that roots interrupt the beam is converted to give a direct readout in meters. This machine works best with samples of 20–40 m total root length. Greater root lengths tend to be underestimated because of overlapping, while lengths less than 20 m tend to be overestimated. The measurement time is about 10 min per sample, with CVs of about 3% for root lengths > 10 m.

Length can also be measured using a high-resolution scanning camera (Harris, 1986; Smucker, 1993) and an image-analyzing computer (such as the Quantimet). A computerized scanning system is commercially available from Delta-T-Devices Ltd. (Burwell, Cambridge CB5 0EJ, UK). When intersect systems have been used to process samples scanned with computer scanners (Kirchoff, 1992), video cameras (Harris and Campbell, 1989), or flatbed scanners (Lebowitz, 1988), care has to be taken with line intercept methods due to root overlap, and correction factors have been added to compensate for this (Sackville-Hamilton et al., 1991b; Kirchoff, 1992). Also, roots have to be well spread out to ensure random orientation. Alternatively, pixel-count methods can be used on a root skeleton and have been shown to be more accurate and more precise than the line intercept method (Lebowitz, 1988; Ewing and Kaspar, 1995).

Limitations to the size of the field of view mean that only small samples can be measured, and organic debris must be carefully separated from the root sample. Staining roots with dyes such as methyl violet can aid detection with some instruments. Use of fluorescent dyes permits roots to be distinguished from debris when illuminated with ultraviolet light (McGowan et al., 1983). Farrell et al. (1993) have shown that mean CVs between repeated measurements are normally greater for manual than for digitized methods.

INDIRECT METHODS. Root length can be calculated from counts of root numbers found in sectioned impregnated blocks of soil. Formulas are available to convert root number to root length for both random (Melhuish and Lang, 1968) and strongly anisotropic (Lang and Melhuish, 1970; Melhuish and Lang, 1971) root distributions. For the former, provided a reasonably large sample of randomly orientated sections are taken, the assumptions inherent in the calculations are met. The equation for random root distributions is

$$L_{\mathrm{T}} = 2N \qquad (4)$$

where L_{T} is the total length of root per unit volume of soil (cm cm^{-3} and N is the number of roots intersecting a plane of unit surface area (per cm^2).

Elongation Rate. Root elongation rates have been studied using pressurized ballotini-filled cells. Elongation is most commonly recorded as the difference in length between successive measurements made directly on still film shots or using time-lapse cinematography (Atkinson and Lewis, 1979). Time-lapse photography can also be used on observation units. It allows the quantification of detailed reactions between root and soil. The method has been used to study root nutation (Head, 1965), variation in root–soil contact, and variation in root diameter (Huck et al., 1970; Atkinson and Lewis, 1979).

Root Age. Root color and morphology are the commonest criteria for identifying root age. For apple, anatomical changes with age have been described in detail (MacKenzie, 1979). The fluorescence of roots has been shown to decrease with root age, disappearing when suberization begins. A positive correlation has been found between the intensity of fluorescence and rate of new root growth (Dyer and Brown, 1983). However, studies have shown that UV fluorescence cannot be used as a universal indicator of root age or functionality, but in some species it can be used to separate roots from the background, using image analysis techniques (Smit and Zuin, 1996). Root activity can also be assessed using stains such as tetrazolium blue, fluorescine diacetate, acridine orange, or pH indicators like bromocresol purple. Aging, by visual identification in situ, is normally performed when roots are visible through a glass face (Head, 1966, 1968).

Species Identification. It is possible to identify anatomical differences between plant roots (Schwaar, 1971). In observation units, different species can be distinguished on the basis of characteristics such as color (which can vary from translucent white to pale brown), diameter, branching pattern, root hair development, and UV fluorescence. In general, grass species show the highest levels of fluorescence (Smit and Zuin, 1996).

Distribution. Root distribution can be studied in situ (Schumacher et al., 1969) or in impregnated blocks of soil (Melhuish and Lang, 1968). Nuclear magnetic resonance techniques can also be used for qualitative assessments in situ of root and water distribution in relation to soil physical and chemical factors.

Branching. In solution culture, or with an easily washed-off potting medium, the entire root system of a plant can be extracted and the main and lateral roots measured. Hackett and Bartlett (1971) have given a detailed description of the density of branching and of changes in lateral length along axes, for plants grown in solution culture. Hooker et al. (1992) have described the impact of mycorrhizal infection on similar developments for a perennial species. Describing the branching of root systems and characterizing the amount of branching, the pattern of branching, and the different orders of branching has proved to be difficult. Fitter (1985, 1991) and Berntson (1995) have approached the problem by using a topological ("tree") system of analysis. This type of analysis provides a way of assessing the morphological effects of differences in soil physical properties. However, it remains unclear as to how well the system works on root systems extracted from soil. Root branching can also be studied in situ using observation chambers or a soil impregnation technique (e.g., Pages and Serra, 1994).

Volume. Root system volume can be calculated from measurements of length and mean root diameter. Asamoah (1984) measured root volume directly to an accuracy of ± 0.025 mL using a meteorological micrometer gauge. This fast, nondestructive method can be used to make nondestructive time sequence measurements on roots grown in solution culture.

III. FIELD METHODS

A. Introduction

There are no field methods that allow roots to be observed directly without either their removal from the soil or the establishment in the soil of an in situ observation surface. Where soil sampling is used, because observations cannot be repeated for the same volume of soil, temporal and spatial variation become confused in data from sequential samplings. Where spatial variation is very high, as is normal (Waddington, 1971), it prevents the detection of temporal variation. The techniques that can be used to assess root growth in the field divide into three main groups.

Root System Removal. The complete root system (excavation) or part of the root system (soil monolith, soil cores, needleboards) are either removed from the soil or assessed in situ (profile wall).

Observation Methods. A viewing surface is inserted into the soil, either a small observation window (Asamoah, 1984), a minirhizotron (Hendrick and Pregitzer, 1992a), or a large walk-in facility (Rogers, 1939b, 1969; Rogers and Head, 1963; Fogel and Lussenhop, 1991).

Indirect Methods. The presence of the root system is inferred from activity (e.g., the removal of soil water or the uptake of radioisotopes) (Newbould et al., 1971; McGowan, 1974). Indirect methods do not, however, predict what may happen under other conditions. Low root activity may result from few roots being present or from root inactivity.

This section deals with each of these groups of techniques in turn. Methods are illustrated by reference to a small number of selected papers, often the earliest published. Listing of all published variants of the basic methods is not possible. Most methods used in the field are simple in concept, and so emphasis has been placed on the interpretation of results, the situations in which the methods have been used, and the factors that may limit their use in studies of the effects of soil physical conditions.

B. Root System Removal

1. Excavation

For large plants, excavation involves the removal of more soil than is the case for any other sampling method (Atkinson, 1972, 1981; Tamasi, 1986). Total excavations are useful for determining the mass and distribution of large roots (Fig. 4). The large loss of fine roots during excavation means that excavation is unsuitable as a means of estimating root length. The method can be useful for studying the effects of soil type or soil features, such as impeded drainage or depth of induration, on the development of a whole root system (Rogers and Vyvyan, 1934).

Fig. 4 Root system of 26-year-old apple tree (Fortune/M9) excavated by the skeleton method. (From Atkinson, 1972.)

Total Excavation Methods. A major study of apple root systems has involved the excavation of 26 mature apple trees grown on either a light sand, a sandy loam, or a heavy clay loam with about 1000 Mg of soil having to be moved by hand (Rogers and Vyvyan, 1934). An entry trench was dug beyond the rooting volume of the tree under investigation. Beginning from this trench, soil was removed in 50 cm sections, moving systematically across the ground occupied by the root system under study. Soil was brushed away from the side of the trench with a small hand fork, leaving the root system exposed. Following such an excavation, the root system is kept as entire as possible to allow later reconstruction (Fig. 4).

Using an excavation method, it is possible to compare horizontal and vertical root distribution, the total amount of root, and the uniformity of distribution for different soil types (Table 3). In this study, it was found that a single quadrant (25% of the soil volume) could contain as little as 6% or as much as 51% of total root weight. This has major implications for the choice of sampling methods and the accuracy of the data they will produce. Although the time needed to sample a single individual tree will tend to limit the use of excavation, given the spatial variation inherent in tree root systems and the limited number of large roots which represent most of the root weight, excavation is probably the only method that will give a reliable estimate of total biomass. This method is easier to apply to small trees (Atkinson et al., 1976).

Table 3 The Effect of Soil Type on an Apple Root System (Lanes Prince Albert/M1)

Parameter	Soil texture (series)		
	Sandy (Wisley)	Sandy loam (Malling)	Clay loam (Wisborough Green)
Root system weight (kg)	4.2	16.3	8.5
Percentage of root weight in 1 m^2 of soil around trunk	48	34	49
Lowest percentage of root in 25% of soil volume	—	16	6
Percentage of root weight at >50 cm depth	20	21	22
Ratio of stem to root weight	0.92	2.36	2.18
Root system spread (m)	4.4	6.0	5.1

Source: Rogers and Vyvyan, 1934.

Partial Excavation. Partial excavations have been used for trees (Coker, 1958). This involves either the excavation of a section (usually one-quarter of the rooting volume) or of a combination of stump-pulling and root excavation. It has been estimated that this latter procedure removes about 38% of root weight (Atkinson et al., 1976). Partial excavation has been used to compare the effect of soil type on root weight and distribution (Coker, 1958; Tamasi, 1986).

2. *Profile Wall Method*

General Considerations. In this technique, a trench or pit is dug to expose a vertical soil profile from which records of partially exposed roots can be made. Horizontal areas, at different soil depths, can be prepared in the same way. Unlike pinboard and soil coring methods, this technique can be used on stony soils. The method is, however, labor-intensive and time-consuming, and it leads to extensive soil disturbance; moreover it can be difficult to obtain a statistically meaningful number of replicates. Nevertheless, it has been suggested that this method gives among the most favorable ratios of information gained to labor expended (Ward et al., 1978). The trench, initially dug at a distance from the crop, can be cut back serially toward the plant if information on lateral distribution is needed. For a row crop, the trench is usually dug across the rows. The trench can be dug by hand or using a mechanical digger, and it should be positioned so that a further layer of soil (~30–50 mm) can later be removed from the trench face so as to avoid damaging roots. The size of the trench required will depend on factors such as crop type.

The Spiral Trench Method. Special considerations apply to tree crops, for which the use of a logarithmic spiral trench has been suggested (Huguet, 1973). This attempts to weight the intensity of sampling relative to the amounts of soil at a distance from the tree trunk by using a trench in the form of a part of a spiral (Fig. 5a). At the periphery of the root system, root density tends to be at a minimum (Fig. 5b), but soil at this distance, nevertheless, contributes a very large proportion of the total soil volume. With the spiral trench method, such soil is

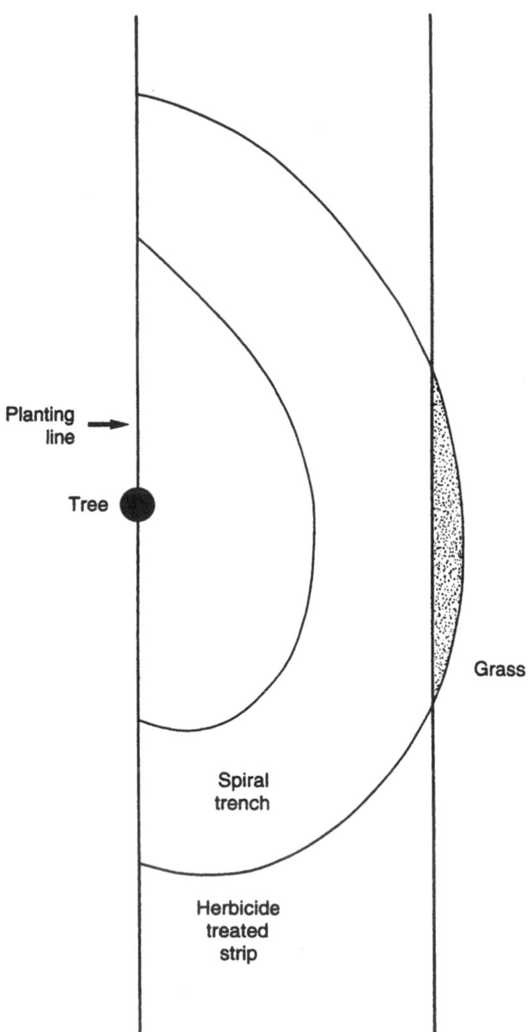

Fig. 5a A spiral trench normal to the plantation line, in a wide herbicide strip. ● tree trunk.

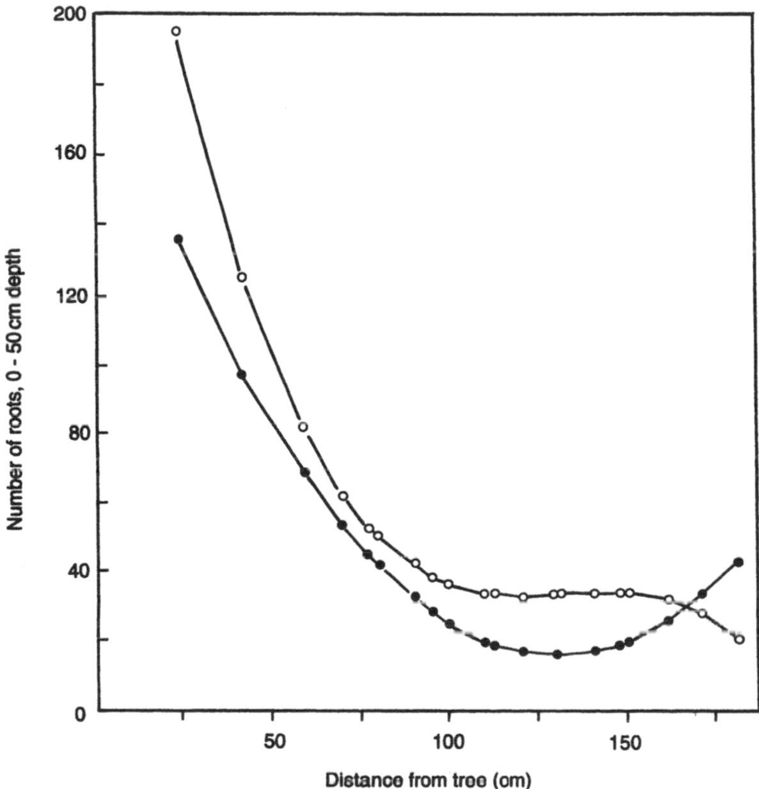

Fig. 5b The number of roots per 500 cm^2 of trench wall with distance from the trunk of trees grown under overall herbicide (○) or herbicide strip management (●). (Reproduced from Gurung, 1979.)

relatively heavily sampled. Comparisons between estimates derived from the logarithmic spiral trench and more conventional straight trenches give a higher average root density (0.017 roots cm^{-2} soil face area) for the spiral than for the straight trench (0.011 cm^{-2}) (Gurung, 1979). Where the soil cover is not uniform, for example, where trees are grown in herbicide-treated strips in grass, the orientation of the spiral trench will influence the density of roots detected. Where the orientation of the trench is such that it samples soil between tree and grass alley (normal to row), the estimated root density is higher than when the sampling is principally of soil between trees in the row.

Face Preparation and Measurement. When the trench, of whatever form, has been dug, the working face of the soil profile is prepared using a profile knife to remove a layer of around 10 to 20 mm. In stony soils, the preparation is best done with a spade, trowel, and knives. The roots exposed against the wall are cut off with scissors. Starting at this surface, the face is cut back by a further ~3–5 mm, but usually 10 mm for trees, so as to expose the root system. Water and air under pressure have been used to remove this soil layer. A frame containing a grid (the mesh size depending on root size and sampling strategy) is positioned over the prepared face and the root system recorded.

Measurements. Root number, length, diameter, and distribution may all be obtained from profile wall measurements.

ROOT NUMBER. The number of exposed roots visible in every square of the grid is recorded, either as a count or onto a prepared sheet of foil or graph paper, containing a matching array of boxes. A direct record of this type can be used to derive the average number of roots per unit area of profile wall, from which estimates of variation may be calculated. Pens of different colors can be used to distinguish between new and old roots. A direct count is faster than the mapping of individual roots (Fig. 6) directly onto either graph paper or transparent sheets. However, counts do not show individual roots in their natural position relative to the profile wall. A visual representation of root counting can be used to show the effects of treatments that result in differences in soil physical conditions. Counts can be plotted versus depth (Fig. 7). This method has been used to study the effects of localized irrigation (Levin et al., 1979).

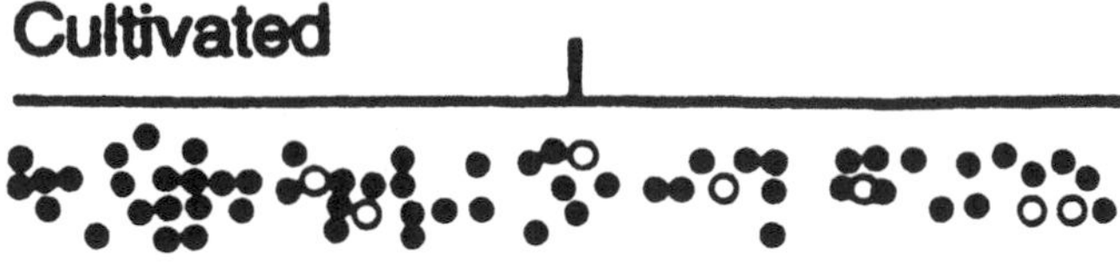

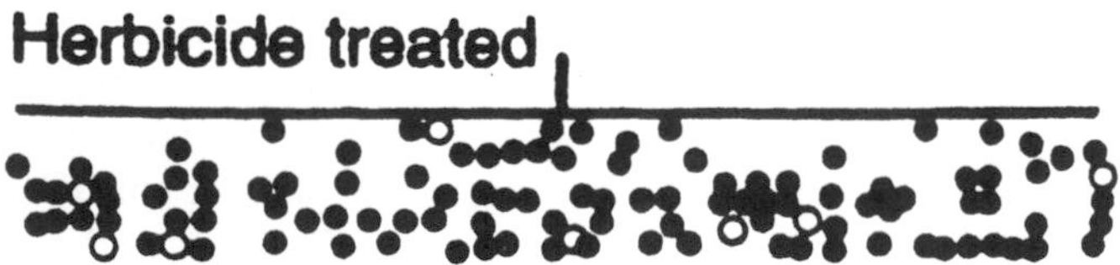

Fig. 6 Root distribution map obtained using a profile wall method and illustrating the difference in the root systems of mature apple trees, between 0 and 30 cm depth, when grown under cultivation or herbicide management: •, roots < 2 mm diameter; o, roots > 2 mm diameter. (From Gurung, 1979.)

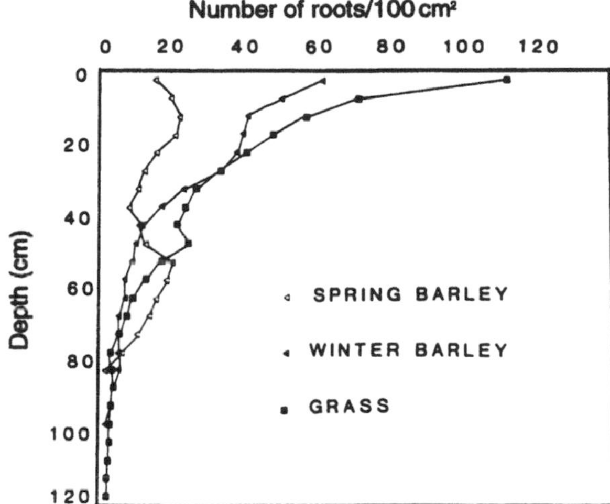

Fig. 7 Root distribution with depth, for three crops, determined by the profile wall method. (From Mackie-Dawson et al., 1988.)

The number of roots recorded using two variants of the profile wall method, counting and drawing onto foil, have been compared (Bohm and Kopkc, 1977). In general, counts obtained with the foil method were higher than those obtained from direct soil counts. Counts in densely rooted areas tended to be less than those obtained by other field methods.

ROOT LENGTH. Estimates of root length are based on the assumption that any root present in the soil will go back into the profile for at least the depth to which it has been exposed. In one variant of this method, one root unit is set equal to a 5 mm root length for a profile that has been cut back by 5 mm. Roots 10 mm long are counted as two root length units (Bohm, 1976). If root distribution is assumed to be uniform, root length per unit volume can be calculated. Estimates of root length obtained in this way are much lower than those obtained by washing roots from an undisturbed block of soil (Table 4). The cause of the underestimation is not clear, although the removal of the soil from the surface of the profile wall to expose the roots may result in the loss of some of the fine roots as well as roots growing parallel to the profile face. However, the method gives a good representation of root distribution. It seems to be most reliable for plants like trees, in which most roots are horizontally distributed. In grasses, where a large proportion of roots are vertical, many roots would be lost leading to an underestimation.

Table 4 Comparison of Estimates of Root Length Density of Maize (*Zea mays*)[a]

Method	Root length density (cm cm^{-2})			
	0–20	20–40	40–60	60–100
Monolith	15.3	5.3	2.7	1.0
Profile wall	7.1	3.1	1.2	0.6

[a] Obtained from profile walls and soil monoliths to 100 cm depth.
Source: Bohm, W. In situ estimation of root length at natural soil profiles. *J. Agric. Sci. Camb.* 87:365–368 (1976).

ROOT DIAMETER AND VOLUME. Using a small hand lens, a micrometer screw, or calipers, the diameter of exposed roots can be measured directly, in situ. The exposed roots can be distinguished by diameter on a root map (Fig. 6), or the number in each diameter class can be recorded directly. Root volume can be calculated from diameter and length (Bhaskaran and Chakrabarty, 1965).

Monolith Methods

A monolith is an undisturbed block of soil isolated from the surrounding soil by excavation or by inserting a metal frame. Monolith methods involve the removal of a sample of soil to represent the whole or part of the rooting volume of a plant (depending on the size of the plant and the volume of soil removed). Monolith samples can be washed to remove the roots from the whole or a part of the soil volume, or the roots can be held in something resembling their original position by a series of pins. This latter method is necessary where it is important to know how the root system distribution is spatially related.

Pinboard. The basic pinboard or needleboard method has been described in detail by Schuurman and Goedewaagen (1971). Monoliths 152 × 41 × 91 cm deep were removed using a root extraction frame (Nelson and Allmares, 1969) that could contain the root systems of four maize (*Zea mays* L.) plants. After its removal, 6 mm diameter brass pins were driven through the monolith into a board on a 5 cm grid pattern, using a compressed air gun. The monolith was then soaked, the side opposite from the pins removed, and the soil washed away. The root system was photographed under water, divided, dried, and weighed. This method has been used to assess the effects of treatments such as straw mulching on a total root mass and on horizontal and vertical distribution (Nelson and Allmares, 1969).

Pinboards (35 cm long × 20 cm deep) have been used to assess effects of soil physical condition on the root system of winter wheat in a comparison between plowing and direct drilling (Finney and Knight, 1973). Board sampling a volume 30 × 5 × 30 cm deep has been employed to study the cabbage root system (Goodman and Greenwood, 1976). In this study, roots and soil were removed from the board as 36 samples of 5 cm^3 each, and the roots were washed free of soil, and their length determined. Photographic records can be used to indicate the effects of treatments on branching. However, such records are difficult to quantify.

Modified Pinboard Methods. A method combining the relative ease of sampling given by soil coring but also providing the spatial information of the pinboard method has been developed (Gooderham, 1969). Samples obtained by soil coring were encased in a perforated acrylic cylinder and the roots held in place using nylon fishing line sewn through the holes in the cylinder with a needle. Soil was washed from the core and the remaining root system resuspended in 5% w/v gelatin. This technique allows the root system's geometry to be related to soil physical characteristics. The method is more rapid and flexible than the more traditional pinboard method. However, both these methods are likely to be superseded by more advanced spatial techniques, such as scanning NMR.

Soil Coring. Soil coring is the most frequently used method of root sampling. Coring is often used for sequential sampling of an experimental plot to give estimates of temporal change in root length or weight, although spatial variation may confound such estimates. When samples are taken in relation to the planting geometry of the crops, information on the spatial distribution of their roots can be obtained. Soil core samples can be obtained from points immediately adjacent to the sites of soil physical measurements, or in some cases, the same cores can be used for measurements of bulk density and pore size distribution as well as root length. The published literature on this subject is very large (Bohm, 1979, contains an extensive listing), and the papers quoted have been selected only to illustrate some of the variations in technique that have been used and to indicate the types of information that can be obtained.

Welbank et al. (1974) described the use of a powered soil coring system to assess the growth and development of cereal root systems; samples were taken using coring tubes fitted with hardened cutting tips (Fig. 8). Variations of this method have been used by many workers subsequently.

Measurement of Root Growth. To facilitate removal of the soil cores, the tubes are fitted with split liners. To avoid compaction of the soil core within the liners, the internal diameter of the cutting tip is manufactured slightly narrower than that of the liners. The coring tube is driven into the ground using a portable gasoline-powered motor hammer (Fig. 9). A depth of 1 m can be reached in moist, relatively stone-free sandy loam soil in around 30 seconds. In dry soils penetration to depth takes much longer. The presence of small stones does not cause a problem, but the corer will not penetrate large stones. Tubes are removed from the soil using a tripod and chain hoist. After extraction, the soil cores are divided into sections corresponding to different soil layers or soil depths.

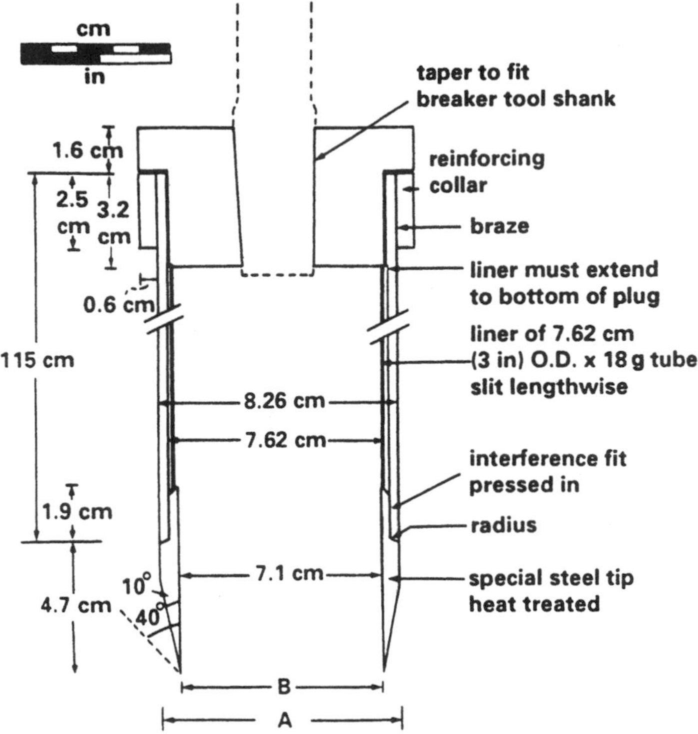

Fig. 8 A design for a soil coring tube. (From Welbank et al., 1974.)

In spite of using the narrower cutting tip, some compaction of the soil core inevitably occurs. Welbank et al. (1974) found that the bulk density of core samples usually showed increases of 1–5% compared with undisturbed samples for a range of depths, but in some soils, especially when wet, the compaction in a 1 m core could be as high as 25%. In this situation, allowance must be made for compaction before the cores are cut up. Injections of paint to known depths have been used to assess the distribution of soil compaction within the core and to allow for its correction. A simplified, low-cost version of the soil coring system has been developed (Prior and Rogers, 1994) using styrofoam plugs to allow the collection of multiple core samples within a plastic liner. This can reduce the time spent on individual sample collection.

After coring, soil can be separated from the roots by washing on sieves manually or automatically (Cahoon and Morton, 1961; Bohm, 1979). A range of washing and flotation techniques and the use of chemical dispersal agents, e.g., sodium pyrophosphate (Schuurman and Goedewaagen, 1971), have been

Fig. 9 Soil coring, using motor-driven hammer.

reviewed by Bohm (1979). The use of dispersal agents is complicated if samples are later to be used for chemical analysis. A modification of the basic method has been described (Smucker et al., 1982) as a hydropneumatic elutriation system. This combines the kinetic energy of pressurized spray jets and the low energy of air flotation. Air and water are used to isolate and deposit roots on a submerged sieve. Washing times vary from 3 to 10 min per sample and are a function of soil type, plant species, the concentration of dispersing agent, sieve size, and soaking time. Using this equipment, nearly 100% root recovery was achieved in around 2 min for a sandy soil, 6 min for a loam, and 10 min for clay (Smucker et al., 1982; Smucker, 1993). These units are available commercially from Gillison (Benzonia, MI 49616, U.S.A.) at a cost in 1997 of $6600 for an eight-chamber unit and $4250 for a four-chamber unit. An alternative four-bucket model, based on the same principle, was available from Delta T, (Burwell, Cambridge CB5 0EJ, UK) for £1815 in 1997.

Welbank et al. (1974) used root-washing cans (Fig. 10) coupled with the intersection method for measuring root length, to compare the effects of plant type and nutrition on root length, root weight, and specific root length (length per unit weight; see Table 5). Because specific root length can vary with crop variety, age, nutrition, depth, and soil physical conditions, use of a given value of specific root length to convert root mass to length is liable to systematic error (Table 5) and is inadvisable except where these values are obtained from representative subsamples taken from the actual samples being assessed. Using the extreme values given by Welbank et al. (1974), 1 g dry weight of roots could have a length as low as 33 m or as high as 199 m.

Core sampling can result in large errors in the assessment of living/functional roots. To overcome this problem, Ward et al. (1978) placed soil cores in bags made from 100 mesh (0.15 mm opening) cloth, which allowed clay, silt, and fine sand to be washed out but retained roots, organic residues, and coarse sand when agitated in water. Washed roots and organic materials were separated from sand by flotation, stained with 1% Congo red, and saturated with 95% ethanol. Under these conditions, living but not dead roots stained dark pink to bright red and were then measured by the grid intersect method (Sec. II.C.2).

These workers found that after staining by the method above, the percentage length of a sample, measured using a grid intersection method, increased by 18% compared with an assessment made by laying the root system out on a dark background. The increase was assumed to be due to an increase in the visibility of small rootlets. A comparison of the staining of different species indicated that monocotyledons stained more than dicotyledons. Some dicot species (e.g., sugar beet) stained very poorly. Storage of root samples can affect the quality of staining. For example, storage at 15°C for 35 days reduced the stainable root length of wheat roots by 65%, whereas at 5°C the decrease was only 13% (Ward et al., 1978). Other dyes have also been used. Ottman and Timm (1984) used trypan

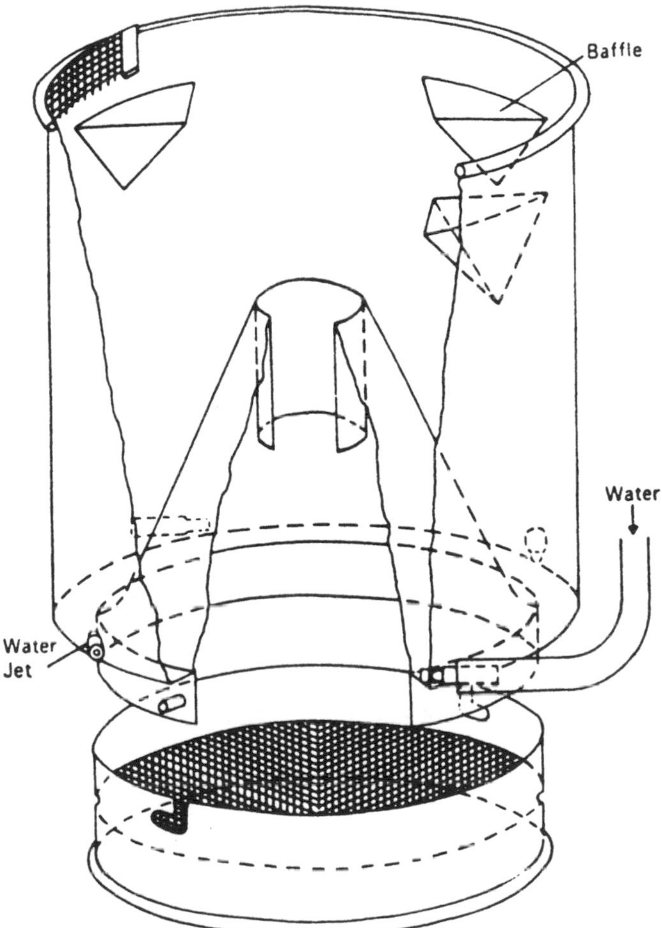

Fig. 10 Root washing can for separating roots from soil. (From Welbank et al., 1974.)

blue to identify the length of viable roots using an image analysis system. Color and texture can also be used to discriminate between living and dead roots.

Root data obtained through field sampling normally has a high spatial variability associated with the soil physical, chemical, and biological variability. A knowledge of the variability can help both in the design of subsequent experiments and in the interpretation of results. The coefficient of variation is often used to describe the variability and is useful provided it is based on a large enough set of individual measurements.

Table 5 Specific Root Length of Winter Wheat (Mean of CV Capelle–Desprez and Maris Ranger) in Two Separate Years, at a Range of Depths

	Specific root length (m g^{-1} dry wt)			
	1969–1970		1970–1971	
Depth (cm)	9 Dec	14 Apr	15 Dec	28 Apr
0–15	155	198	226	217
15–25	101	187	128	200
25–35	40	105	62	82
35–45	33	49	84	84
45–55		52		89

Source: Welbank et al., 1974.

After washing, roots may lose part of their weight due to respiration, leakage out of cells, or abrasion of tissue. In studies where root length is of critical importance and root weights have been taken for comparative purposes only, post-washing root losses are often disregarded. However, particularly in studies of the carbon balance in plant or soil, these losses should be critically assessed (Van Noordwijk and Floris, 1979). Soil remaining attached to roots can also be a problem. To avoid this, the ash content of root samples can be determined and subtracted from the dry mass, which allows root samples to be expressed on an ash-free organic matter basis.

Estimation of Root Length. The importance of root length as a measurement and the principal means available for its estimation have been discussed in Sec. II.C.2. When roots have been removed from soil, they can be treated in the same way as roots obtained from laboratory studies except for a greater need for the removal of contaminant materials (more a feature of field studies) and the need for greater care in the separation of live and dead roots.

Special Factors in Relation to Trees. With graminaceous crops, the horizontal component of the root system is relatively small, and planting densities normally result in relatively complete soil exploitation. The ratio of core volume to the soil volume exploited by an individual plant is also relatively high. However, for tree crops, rooting density varies greatly with depth, with distance from the tree, and between samples taken at comparable positions and also includes an extra woody category. In one study, at a distance of 150 cm from the trunk of apple trees and to 45 cm depth, a high proportion, around 70%, of cores contained no roots. Even close to the tree and near the soil surface, the amount of root in a core could be highly variable, e.g., 0–40 mg of roots < 2 mm in diameter (Atkinson and Wilson, 1980). High spatial variability is a common feature of the root distribution of trees. Data on Douglas fir have been discussed by Reynolds (1970).

A count of the number of root tips recovered from 40 mm diameter soil cores has been advocated as a means of assessing the distribution of fruit tree roots (Weller, 1971). In this study as in other studies (Reynolds, 1970; Atkinson and Wilson, 1980), spatial variation was high. Differences appeared to be smaller in good soils than in impeded soils. This type of effect needs to be considered when comparing soils with different physical characteristics.

Overall, the large spatial variability in root distribution makes it difficult to draw conclusions about the effect of treatments on tree root density or distribution.

Ingrowth Bags. The ingrowth or mesh bag is a distinctive modification of the monolith method: the removed soil core is replaced with a cylindrical nylon mesh bag filled with sieved soil (Steen, 1991). The bag is left in the soil for a period during which time roots from surrounding plants grow into the bag, so allowing estimation of root production during that period. As the soil is free of roots at the beginning of the period, any roots found in the bag after incubation can safely be assumed to have been produced during that period. Roots within the bag can be sorted into live and dead categories. The assessment will not, of course, estimate roots that are produced, die, and disappear during the incubation period. Direct observation studies, e.g., Atkinson (1985), have shown that roots can die and disappear in a few days. The need to use sieved soil in the ingrowth bag will limit the usefulness of this method for studies of soil physical problems, although Steen and Hakansson (1987) used the method to assess the effects of soil compaction on oilseed rape and spring wheat.

C. Observation Methods

1. Introduction

The methods described previously allow the root system to be quantified at a single moment in time and consequently do not allow the easy estimation of temporal variation. The development of root systems in situ can be observed by creating a window into the soil. This approach allows a sample of roots to be viewed but raises a number of questions about the representativeness of the sample, hence of the significance of deductions drawn from such observations. These questions represent the principal limitations to this group of methods. There are three main types of observation facility: large permanent facilities, smaller semipermanent rhizotrons, and minirhizotrons.

The earliest permanent root observation facilities were simple pits with glass-lined walls (Rogers, 1939b). These were ultimately developed into large permanent root observation laboratories (Rogers, 1969). General aspects of the methodology for this type of facility have been reviewed (Huck and Taylor, 1982). The alternative approach is to use an observation tube or minirhizotron methods (Waddington, 1972). Both developments ultimately depend on the observation of a sample of roots, and so common criteria apply to evaluations of the significance of results.

The minirhizotron approach offers the possibility of flexibility in location and greater replication. If resolution is adequate, then some of the information obtained through the use of rhizotrons can be obtained using minirhizotrons, at a lower cost.

2. Basic Criteria

The conditions that must be satisfied before data recorded from an observation surface can be converted into a root length per unit soil surface or root length per unit soil volume, are as follows (Atkinson, 1985; Mackie-Dawson and Atkinson, 1991):

1. The presence of the window must not result in an atypical root system.
2. Root density adjacent to the window should either be typical of that in a comparable volume away from the glass or should differ by a predictable amount.
3. The position of the observation panel in relation to the horizontal spread of the root system should give an acceptable representation (with respect to root density, distribution with depth, the timing of new growth and root turnover) of the whole soil volume exploited. This condition tends to be more important for perennial species (e.g., in trees, where there is extensive horizontal development of the root system and variation in production with distance from the stem, Fig. 5).
4. The sample of the root system observed through the observation window should be sufficiently large.

The extent to which the roots being observed represent a valid sample of the growth, behavior, distribution, or density of the population as a whole will vary for different plant species and perhaps for different ages of plant material. An understanding of this relationship is a key element in the use of data from these methods (Hendrick and Pregitzer, 1992a, 1996).

Calculation of Root Density. Calculations of root length density depend on the use of a conversion factor (Taylor et al., 1970; Atkinson, 1985; Atkinson and Fogel, 1997). This factor varies for roots of different diameter, is usually a function of species or variety, and will be influenced by the soil physical condition at the observation interface. In a long-term study of apple, there was no difference in root density adjacent to the glass of a root laboratory and at a distance from it (Atkinson, 1985; Mackie-Dawson and Atkinson, 1991).

In a study with peas (*Pisum sativum* L.), the rate of root elongation at the interface was lower than in the center of the container. Differences decreased at higher soil bulk densities. In another investigation with maize and tomato (*Lycopersicon esculentum* Mill.), there was no clear difference between root weight per unit soil volume at the glass–soil interface and that in the bulk soil (Taylor et al., 1970). In the latter investigation, it was found that root density varied widely over the observation surface, from 0.5 to 5 cm cm^{-1}. The frequency of high values increased with age. In 104 day-old plants, 60% of the sample area had a density between 0.5 and 1.9 cm cm^{-1}. In tomato, where total root length was one-third of that in maize, the range of root length density was similar, but a high proportion of area had a density of less than 0.5.

There has been concern that length of root could be influenced by the minirhizotron face itself. It has been suggested that number of roots (Upchurch, 1987) or first point of intersection of roots (Mackie-Dawson and Atkinson, 1991) could show a better relationship with the rooting intensity in the bulk soil.

3. *Minirhizotrons*

Waddington (1971) used a square 5 × 5 cm glass or acrylic tube inserted into the soil at 45°, with a fiber-optics probe, both to illuminate the soil and to observe the growth of wheat roots (Fig. 11a). Bohm (1974) used a similar system involving 64 mm diameter round glass tubes but with the soil adjacent to the tube illuminated by a bulb and a magnifying glass at the top of the observation tube (Fig. 11b). He found it necessary to pack air-dried soil around the tube to get good contact between glass and soil. Observations were made of the effects of cultivation on spring barley and oilseed rape.

Sanders and Brown (1978) used a medical duodenoscope to view root growth adjacent to 7.5 cm diameter tubes inserted at 45° to the soil surface and compared this method with soil coring for soybean. Root length was recorded by photographing the image of the root system, and length was estimated on the basis of an intersection method. Length at the tube–soil interface was converted into a length per unit volume, using the assumption that roots were seen up to 3 mm from the surface of the tube.

A number of materials have been used for minirhizotron tubes. Glass gives good visibility but tends to fracture under cold conditions or if stones are present in the soil profile; acrylic tubes tend to be easier to insert and to last longer, but can scratch where angular sand grains are present. Taylor and Bohm (1976) suggested that poor soil interface contact was more of a problem with plastic than with glass. Some of the problems inherent in rigid minirhizotron tubes can be alleviated by using inflatable ones (Merrill et al., 1987; Lopez et al., 1996). Such a system allows an endoscope to record root growth directly on the surface of the

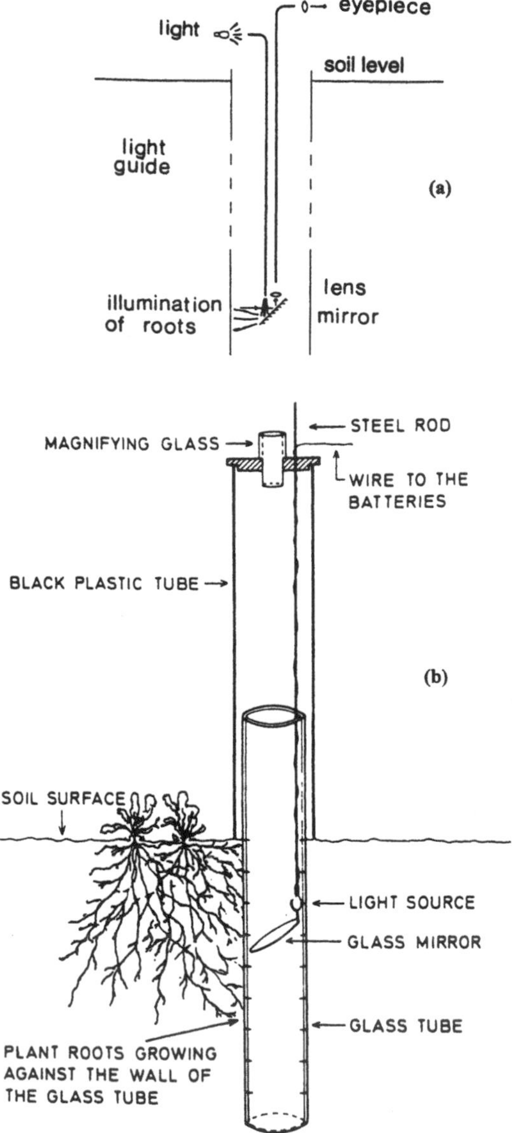

Fig. 11 Two minirhizotron methods using different equipment to measure root growth adjacent to the observation tube: (a) Waddington, 1971; (b) Bohm, 1974.

soil cavity. The method has utility on sandy soils and obviates the need for backfilling. Care is usually taken to minimize exposure of roots to light penetrating into the tubes, as light has been shown to alter root growth. Some minirhizotron systems include both visible and ultraviolet (UV) light, with live roots fluorescing more strongly than dead ones. For assessing the effects of a treatment on root death, it is possible to use either of the light sources. In a comparative study, Wang et al. (1995) concluded that either method gave good levels of accuracy for estimating live root proportions. However, for field studies, it was concluded that the visible light method was much easier to use and less costly. Smit and Zuin (1996) assessed the use of UV-induced fluorescence as an index of root functionality. They found variation in intensity between leek and Brussels sprouts, but that over time, i.e., with aging, fluorescence increased in leek roots but decreased in Brussels sprouts. They concluded that UV-induced fluorescence could not be used as a universal indicator of either age or functionality.

Tube Insertion. It has been suggested that the angle of tube insertion has a significant effect on the interpretation of the results (Merrill et al., 1994). In a comparison of vertical and 45° angle minirhizotrons with direct measurements from washed-out cores of oat roots, better agreement was found with the tubes at 45° (Bragg et al., 1983). However, vertical tubes are still used (De Ruijtar et al., 1996), particularly when conditions prohibit angled tubes being used. It has also been suggested that a combination of horizontal and vertical might minimize the effects of root anisotropy (Horgan et al., 1993). Minirhizotron tubes have been used as access tubes for the neutron probe, so allowing comparisons of root density and soil water depletion (Upchurch and Ritchie, 1983). In this study, comparison of minirhizotron measurements and soil core sampling showed poor correlation in the surface 20 cm.

Viewing Systems. The minirhizotron method has been combined with miniaturized television cameras; this combination allows the soil adjacent to the tube to be seen on a television screen, transferred to video tape, and later subjected to image analysis. This method has been used by a number of workers (e.g., Belford and Henderson, 1984; Upchurch and Ritchie, 1984; Taylor, 1987; Hendrick and Pregitzer, 1992a, b; Hooker et al., 1995; Goins and Russelle, 1996). A TV system increases the number of tubes that can be assessed, which is vital because of the number of tubes required to give good correlations with other assessments of root length density. A color video camera made it possible to differentiate between old and new maize roots and to assess the effects of water stress (Upchurch and Ritchie, 1984). Equipment of this type can be obtained from Bartz Technology Co., Santa Barbara, California, U.S.A. Recent advances in image analysis have facilitated direct measurement of length, diameter, and development of individual roots. Several programs are now available to analyze the sequential images, e.g., a PC-based system (MSU-ROOTS, Michigan State University, E. Lansing, MI,

U.S.A.) and a Macintosh-based system (RooTracker, Duke University, Durham, NC, U.S.A.).

The images are digitized, and computer processing for separating roots from the background as well as measuring root properties has become possible (Smucker et al., 1987; Smucker, 1993). The programmes can assign color composites to root images, which helps in the automated discrimination of roots (Heeraman et al., 1993). However, automated discrimination of viable roots from a variable soil background is still difficult (Andren et al., 1996), and manual intervention and efforts are still considerable, with care needed in the light selection and exposure levels for data collection and in the processing conditions. A system that allows the direct digitization of such images has been developed by Smucker (1993). This processes up to 11,800 video images of roots per week, without human intervention. The system uses computer algorithms for ridge detection to identify roots. The center lines of these segmented root images are identified by skeletonization algorithms. In known situations, this system is able to identify over 90% of roots. The difficulties likely to be encountered in applying image analysis to minirhizotron images captured on video tape have been detailed and discussed by Hendrick and Pregitzer (1996).

The major value of the minirhizotron method is, however, its ability to provide information on root longevity. The development of a cohort approach, where a population (cohort) of roots initiated within a discrete and short time period is followed, so as to document the survival of individual members of the cohort, has greatly improved knowledge of total productivity, root dynamics, and nutrient flows (Hendrick and Pregitzer, 1992b; Hooker et al., 1995; Goins and Russelle, 1996).

An alternative to the use of a TV system is to use a borescope, a rigid fiberoptic endoscopic system for remote visual inspection. This can be recorded directly or linked to a 35 mm camera or a video camera, with the resulting film analyzed as above. A borescope allows direct observations to be made in color (Mackie-Dawson et al., 1995a, b). Equipment of this type has been used to assess root production under the difficult conditions found in a salt marsh (Steinke et al., 1996). Equipment of this type can be obtained from KeyMed (Southend-on-Sea, Essex, SS2 5QH, U.K.) and from ITI (Westfield, MA 01086, U.S.A.). A telescopic lens can also be used instead of an endoscope, which provides photographs of a high quality and can easily produce a single image of the entire tube length (Poelman et al., 1996).

Comparison with Other Methods. Majdi (1996) discussed the ways in which minirhizotron data could be related to that obtained using soil coring and ingrowth bags. He concluded that the minirhizotron method was capable of quantifying root dynamics. Table 6 shows a comparison between soybean root length and density estimates obtained by coring and by the minirhizotron (Sanders and Brown,

Table 6 Comparison of Root Length and Density in Soybeans Determined by Soil Coring (SC) and Minirhizotron (MR) Methods

	Root length (m)		Root distribution (%)		Root length density (cm cm^{-3})	
Depth (cm)	SC	MR	SC	MR	SC	MR
0–18	38	25	68	46	10.6	6.8
18–36	9.5	13	17	24	2.6	3.6
36–54	2.8	4.9	5	11	0.6	1.4
54–72	5.5	10	10	19	1.5	2.8
Total	56	53				
CV%	53	14				

Source: Sanders and Brown, 1978.

1978). Estimates of root length were higher at all depths, other than 0–18 cm, using the minirhizotron method. Total estimated root length was similar for the two methods. The coefficient of variation for the core determinations was 53%, similar to the range quoted in the other publications reviewed: 61–95%. For the minirhizotron method, CVs were much lower. In contrast, CVs of around 95% have been found for minirhizotron observations of young apple trees.

In earlier work, Bohm et al. (1977) compared minirhizotron results with those from soil water depletion, a pinboard, core sampling, and the profile wall method for soybean on a loess soil. In this case, the minirhizotron method gave a higher percentage of roots in the surface 15 cm (85% vs. 63% for soil coring), in contrast to the report of Sanders and Brown (1978). However, in the Bohm et al. (1977) study, the soil adjacent to the tube was back-filled and the concentration of roots adjacent to the tube was higher than in the bulk soil. Gregory (1979) used a minirhizotron in a study of wheat and millet and estimated root length by counting intersections with a 0.5 cm grid in the periscope eyepiece. Root length, R, was calculated by

$$R = \frac{\pi}{4} N \times 0.5 \times 1.18 \tag{5}$$

where N is the number of intersections, 0.5 the grid size, and 1.18 a scaling factor accounting for the difference between the field size (40 mm) and the observed image (34 mm). As with most other studies, the minirhizotron method seemed to underestimate root length at the surface compared with auger sampling. For wheat sampled at 0–60 cm depth between May and June, rhizotron results indicated a constant rate of root growth, whereas coring showed a decrease. Total length estimates for the two methods on a number of dates were highly correlated

($r = 0.83$). However, the relationships between root length assessed using the two methods were different for the two species under study (Gregory, 1979). Similar studies were carried out by Kopke (1981).

Problems with relationships between minirhizotron estimates and root coring estimates have arisen when studying the effects of soil compaction on rooting (De Ruijter et al., 1996). The higher estimates of rooting at depth recorded in this study were probably due to preferential root tracking along the vertically orientated tubes. They conclude that for studying effects of biotic and most abiotic factors on total root systems, both soil coring and minirhizotrons could be used, but for bulk density effects, soil coring was the preferable technique. However, if minirhizotrons are installed at a 45° orientation, and light effects are minimized, estimates can be much better (Bragg et al., 1983; Horgan et al., 1993). A flexible-sided minirhizotron has also been used to study effects of bulk density on rooting and can give closer correspondence to core estimates than do rigid-walled tubes. At high soil bulk densities, the latter overestimated rooting in both wheat and beans (Volkmar, 1993).

Hendrick and Pregitzer (1996) documented a number of published correlation coefficients relating minirhizotron estimates to soil core data. They concluded that "not all minirhizotron soil core comparisons have proved to be poor" and that "too much attention has often been given to discussing the pros and cons of particular methods and too little to understanding if a particular tool is appropriate to answer the question at hand." Measurements made using the minirhizotron method are more rapid, taking around one-tenth of the time needed for other methods (Sanders and Brown, 1978; Gregory, 1979). The minirhizotron technique has limitations, particularly at the soil surface, for providing estimates of root biomass or length distributions, and is no more informative in this respect than the traditional coring method. However, because individual roots can be measured repeatedly throughout their life, the minirhizotron data can be used to quantify rates of root production and mortality independently.

By combining soil core data with minirhizotron data, biomass production and nutrient input into the soil by root mortality and decomposition can be estimated. Information on sampling systems to be used with minirhizotrons has been presented by Hendrick and Pregitzer (1992a). Also, information on root orientation can be obtained using the minirhizotron method (Merrill et al., 1994; De Ruijter et al., 1996).

Miniature Windows. The basic root laboratory method has been modified to use small observation windows set into the soil adjacent to trees. Asamoah (1984) used a sheet of plate glass held adjacent to the soil with a wooden frame. Soil was repacked beside the glass. Root growth was estimated directly without the need for a viewing device, which greatly reduced the time needed for measurements. Sword et al. (1996) used a similar approach in a study of *Pinus taeda* L. Images were captured on acetate sheets, which were analyzed using GSRoot software (PP System Inc., Bradford, MA, U.S.A.). In contrast to minirhizotrons, the viewing area with miniature windows is much larger, and access to roots or soil for sampling is possible. Installation costs tend to be low. Modifications to the basic method of installation are described in Haussling et al. (1985).

4. Root Observation Laboratories

Introduction. In contrast to minirhizotron methods, in which the observation system is taken to the field experiment, root laboratories are permanent or semi-permanent facilities around which experiments must be constructed. However, they allow a high degree of sophistication in measurement and have been used to assess the impact of soil physical conditions on plant growth.

Simple Laboratories. Early observation laboratories were simply holes in the ground with glass walls and roof (Rogers, 1939b). Simple designs of this type are still in use. The original Bangor rhizotron was 1.7 × 1.7 m square and 1.35 m deep, constructed largely of cement blocks with 12.5 mm thick glass (Carpenter et al., 1985). As one of the interests of this laboratory was the study of the soil surface litter layer, the windows extend above ground level. This design necessitates a range of measures to prevent light piping from the surface down through the glass to deeper layers. The Brooms Barn observation pits (Durant et al., 1973) were 1.8 m × 1.2 m and 1.8 m deep, with observation panels at the ends of the pits. A comparison of root growth in repacked and unrepacked soil showed that where soil was not repacked, severe soil slippage obscured root development. Comparisons were made of the root growth of sugar beet, potato, and barley and of the effects of drought.

Large Permanent Laboratories

THE EAST MALLING LABORATORY. The first of the large root observation facilities was that built in 1961 in England at East Malling (now part of Horticulture Research International) (Rogers and Head, 1963). A second facility to an improved design was completed in 1966 (Rogers, 1969). Both these laboratories provided 48 observation windows, each approximately 1 × 1.2 m in size. The windows were made from either four panes of glass, approximately 50–60 cm, or from up to 24 smaller removable panes of glass. All windows used 6 mm unwired plate glass, which was engraved with 12.5 mm squares to allow root length to be assessed using an intersection method (Head, 1966). The soil adjacent to the windows was removed before construction, in 10 cm layers. It was air-dried and later repacked layer by layer to give the same approximate densities. Repacking the windows in this way gave a good smear-free contact between glass and soil.

OTHER FACILITIES. Similar facilities, largely based on the East Malling design, have been constructed in a number of countries (Glover, 1967; Ovington and Murray, 1968; Taylor, 1969; Hilton et al., 1969; Soileau et al., 1974; Freeman and Smart, 1976; Karnok and Kucharski, 1982). The most recent are at Pellston, Michigan, U.S.A. (Fogel and Lussenhop, 1991) and Bangor, Wales (Sackville-Hamilton et al., 1991a). Most facilities have been used for studies of crop plants, but those in Michigan and Cambridge were built to study forest species. The Michigan Biotron (Fogel and Lussenhop, 1991) is set into an area of established hardwood forest. It has 34 1.2 × 1.2 m observation panels, each of which is comprised of 16 smaller removable panes. The facility allows root growth to be related to variables such as soil temperature and moisture. The removable panels allow the localized modification of the soil environment. The natural setting of the biotron allows root density and distribution for individual species and root types to be compared and for temporal variation in rates of root production and death to be quantified. The balance between these varies within and between seasons (Fig. 12).

WINDOW MATERIALS. In a comparison of the use of glass and acrylics in studies of soybean grown on a silt-loam soil, acrylic windows seemed to result in a greater concentration of roots at the observation surface, a poorer adhesion of soil to the surface, and shrinkage at high soil water potentials (Taylor and Bohm, 1976). Glass therefore seems to be the best material for observation studies.

ADVANTAGES AND DISADVANTAGES OF THE METHOD. The capital cost of observation laboratories is higher than for any other method. It is important, therefore, to characterize their main advantages and disadvantages. The principal advantages are as follows:

1. The ability to study the same volume of soil in a nondestructive manner, and so to document root longevity, turnover, and the times during the year when active growth occurs (Fig. 12).
2. The opportunity to assess changes in soil and the roots of perennial crops over very long periods of time, for example, to assess the stability of channels in the soil (resulting from worm or root activity) over periods of a decade or more (Atkinson and White, 1981).
3. The ability, by detailed sampling behinds windows, to relate root activity to soil conditions in the same place or to previously observed soil events.
4. The provision of a viewing area large enough to distinguish roots of different species in a mixed culture and to assess the amounts of woody roots.
5. The provision of a physically comfortable working environment which allows the selection of a range of recording methods.
6. The ability to use microscopes of sufficient resolution to permit repeated measurements on soil microorganisms, such as the external mycelium of arbuscular mycorrhizal fungi (Table 7). Current minirhizotron cameras have insufficient resolution for this purpose.

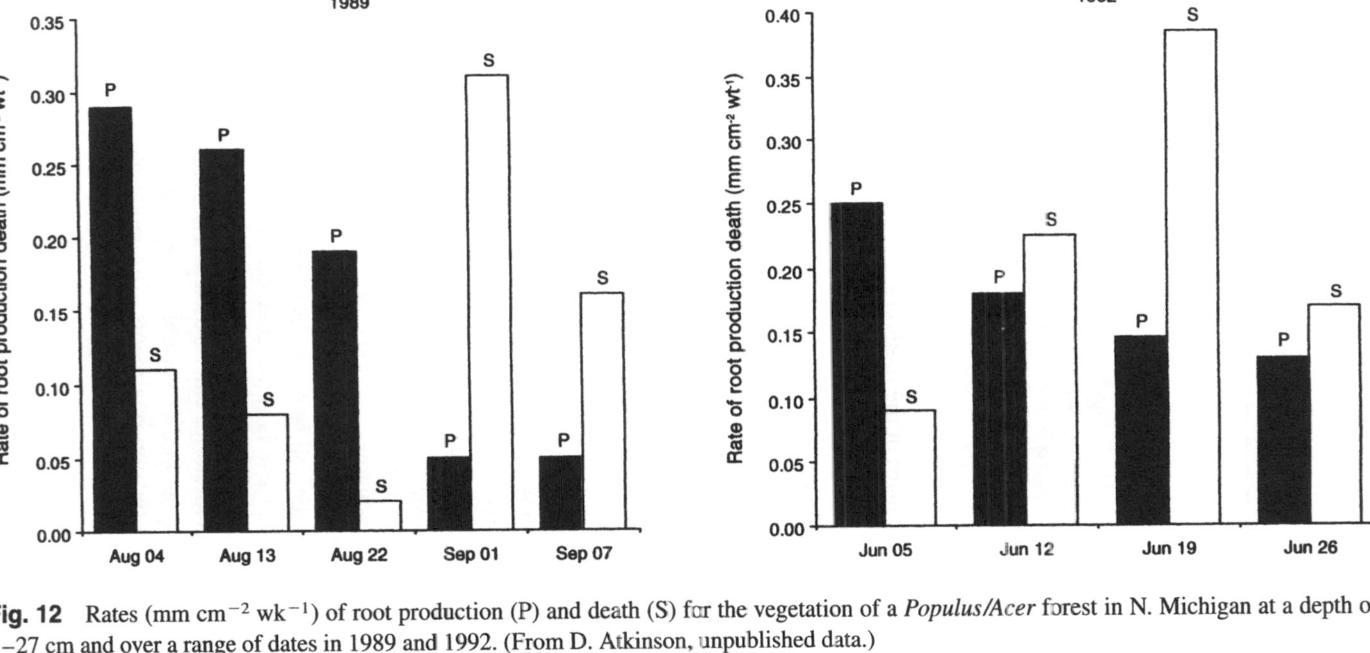

Fig. 12 Rates (mm cm^{-2} wk^{-1}) of root production (P) and death (S) for the vegetation of a *Populus/Acer* forest in N. Michigan at a depth of 0–27 cm and over a range of dates in 1989 and 1992. (From D. Atkinson, unpublished data.)

Table 7 Survival (%) of a Cohort of Fungal Hyphae at Different Times After Initiation Estimated on Both a Length (mm) and a Number Basis

	Survival (%)					
Parameter	0	3	7	15	21	24
Number	100	100	44	29	4	4
Length	100	100	42	28	5	5

The hyphae were adjacent to the roots of the under-story species, known to have arbuscular mycorrhizal infection of 52–94%, in a mature hardwood forest in N. Michigan.
Source: D. Atkinson, unpublished data.

The principal disadvantages of the method are as follows:

1. For good soil–glass contact, the soil adjacent to the glass must be replaced. In soils with a high sand content and in unstructured soils, this will have little effect, but the use of such materials in soil structural studies is limited.
2. Any experimental treatments that need to be investigated must be installed adjacent to the facility. The facility cannot be used to study physical problems in random field sites (the minirhizotron method can be used in this way). With a limit on the number of windows or compartments, as well as a need for buffer areas between treatments and the necessity for replication, the number of treatments that can be assessed at a given time is limited. This can be circumvented by running root laboratory experiments and conventional well-replicated field trials in parallel. With this strategy, the observation laboratories can be used to explain root-related effects.
3. The facilities are expensive to build and need continuous maintenance to keep the windows in good condition.
4. To prevent light ingress, the top of the observation window is usually sited below the soil surface. As a consequence, as with the minirhizotron method, there is some evidence that growth in the soil surface is underestimated (Atkinson, 1985).
5. While the laboratories can be used to assess the effects of differences in water supply and other variables, they are harder to use to assess the effect of cultivation treatments. The vibration occurring during cultivation may damage the windows.

Recording Methods. The commonest measurements made in observation facilities are of root length per unit area of window, the duration of root survival, and the relationship between individual roots and the soil. Root length is normally assessed by counting the numbers of intersections between roots and the grid system engraved on the windows and then converting to root length. For the 12.5 mm grid, which has often been used (Head, 1966), the relationship is

$$1 \text{ intersection} \equiv 1 \text{ cm length}$$

Taylor et al. (1970) found the relationship

$$y = 6.29 + 87.354x - 2.04x^2 \tag{6}$$

between the number of roots of maize and tomato crossing 100 cm horizontal transects at 15 cm depth intervals (y) and root length/cm^2 of viewing area (x). The viewing area was sampled at random in only 7% of squares at the same date and depth. The correlation between the two methods of assessment was $R = 0.97$, suggesting that a transect method, which saves time, is adequate. A high correlation was also found between the number of intersections with selected horizontal lines and counts in the whole area for timothy (*Phleum nodosum*) (Atkinson, 1977).

In a study with apple, Gurung (1979) found that results obtained using the laboratory window method and those from minirhizotron tubes adjacent to the window agreed in 10 of the 12 recorded weeks. During the period of most rapid root growth, data from the two methods differed. Where root growth is spatially variable, the larger area of the root laboratory windows is more likely to reveal areas of high growth.

Root survival, turnover, and longevity can be assessed by marking new roots at their initiation on strips of paper placed along particular horizontal or vertical grid lines or by tracings of whole or parts of windows. These records can then be used to establish a cohort that can be assessed by comparing root presence over a series of weeks to chart the survival of particular roots (Rogers, 1969; Atkinson, 1985). For perennial roots, changes in root diameter can be measured using a microscope fitted with a micrometer eyepiece and using the grid system to reidentify roots (Head, 1968). Relationships between individual root performance and soil features (e.g., root growth adjacent to an indurated layer) can be assessed either by direct measurement or by the use of time-lapse cinematography, which allows detailed analysis of root development (Fig. 13) (Atkinson and Lewis, 1979).

Use to Assess Effects of Soil Physical Properties. The number of studies relevant to the effects of soil physical properties on root development are too numerous to document fully. Examples will indicate a range of what is possible. Root laboratories have been used to assess the effect of varying rates of soil moisture depletion, induced by different planting densities, on root growth and distribution (Atkinson, 1978). Root distribution was deeper where soil water deficits were higher. Similar results were obtained in studies where different soil moisture deficits arose from growing trees under bare or grassed soil (Atkinson, 1977, 1983). In these studies, it was possible to relate root development at various depths in the soil to

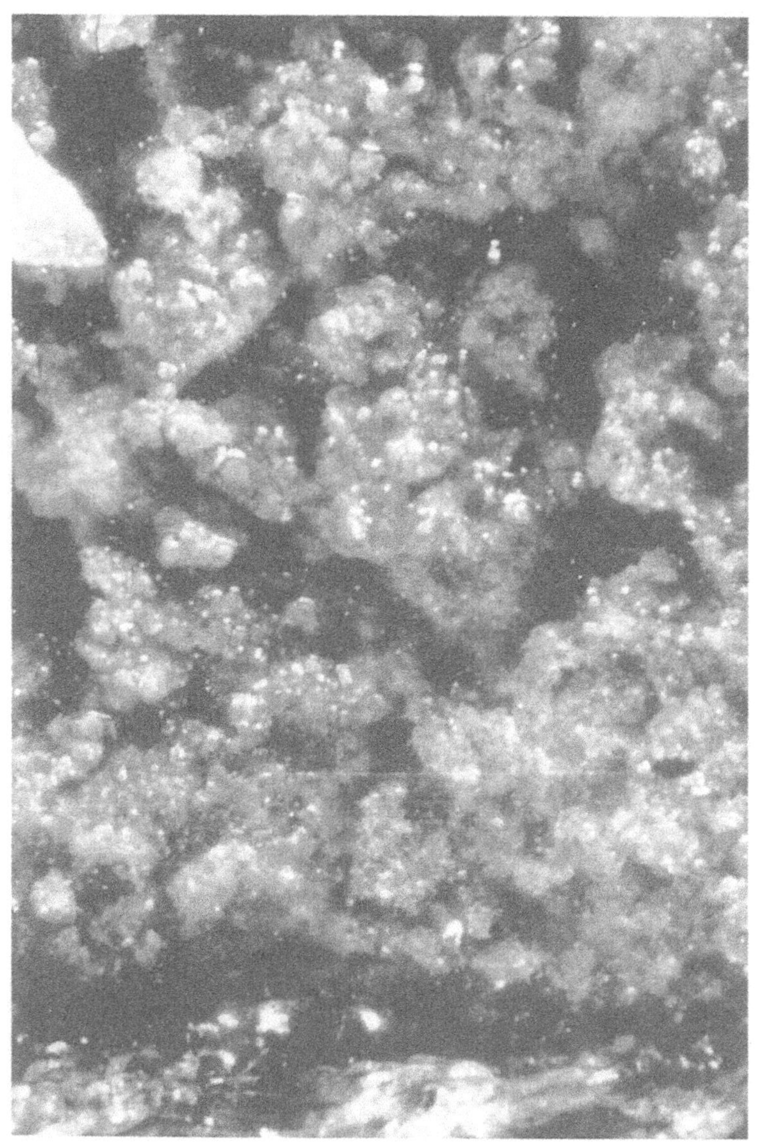

(a)

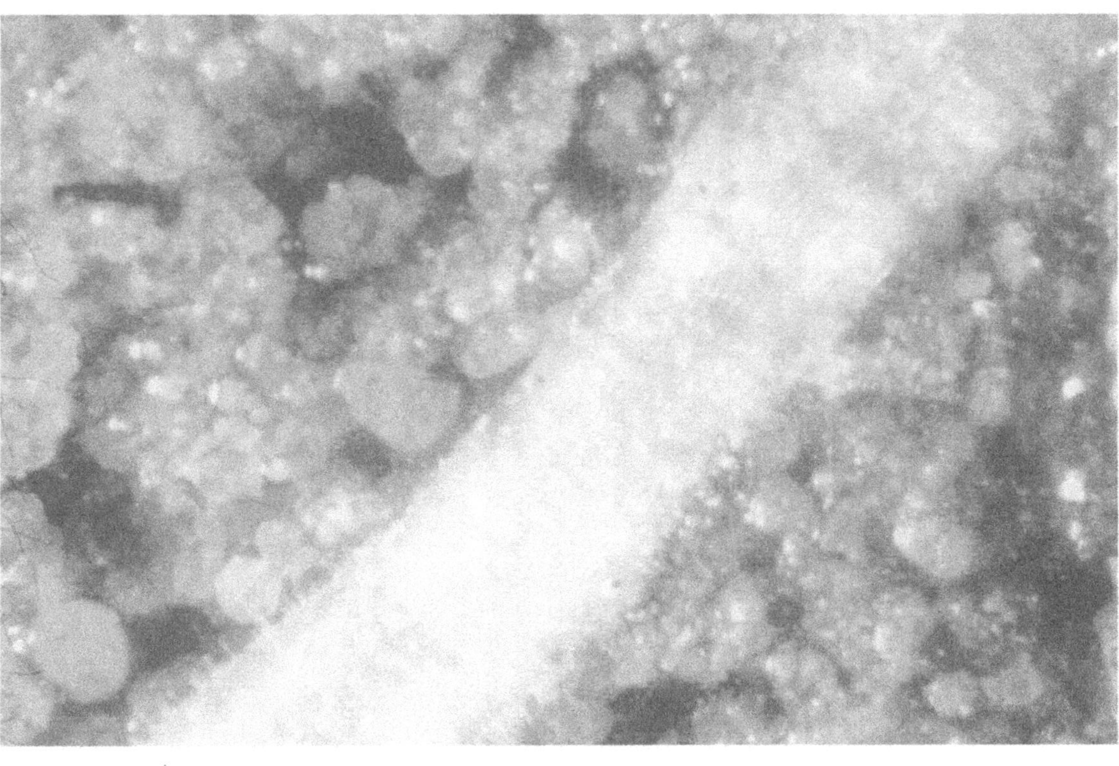

(b)

Fig. 13 Soil as seen in a root laboratory (a) before and (b) after the growth through it of an apple root.

Table 8 The Effect of Soil Water Supply on Root Growth, Assessed Using Rhizotrons

Crop	Root measurement	Effect of drought/poor water supply	Source
Maize	Rooting density	Inflow highest at low root density	Taylor and Klepper, 1973
Cotton	Elongation rate	Stopped at water content below 0.06 cm^3 cm^{-3}	Taylor and Klepper, 1974
Cotton	Rooting depth	Reduced new root growth	Browning et al., 1974
Cotton	Root distribution	Deepened	Davis and Huck, 1978
White oak	Elongation rate and number	Reduced below −0.3 MPa	Teskey and Hinkley, 1981
Soybean	Depth	Increased water depletion depth	Kaspar et al., 1984
Soybean	Cumulative growth	Increased length, number, depth	Huck et al., 1986

soil factors in the same layers. With sugar beet, potato, and barley, the root growth was usually 10–15 cm deeper than the maximum depth of soil moisture extraction (Durant et al., 1973). A large number of studies have assessed the effects of soil water supply on root development (Table 8).

Clearly, observation methods can be used to assess the effects of soil physical conditions on root growth, although care must be taken with installation, and there may be problems in accurately assessing growth in the surface soil.

D. Indirect Methods of Assessing the Activity of Root Systems

1. *Introduction*

The major functions of roots in soil are to absorb water and nutrients and to pass elements like carbon to soil microorganisms. By assessing the release of carbon, the depletion of water, or the removal of nutrients from soil, the distribution of roots in soil or their activity may be inferred. Correlations may, however, be variable, and root distribution is only imperfectly related to activity.

2. *Estimation from Soil Water Use*

The rate of water depletion from the soil depends on both soil water potential, which influences the rate of water movement to the root surface, and root density (for the depth-distribution of most root systems, axial resistance can be assumed to be small). When a root system begins to use water from moist soil, the pattern of water extraction and the rates of water use at various depths will reflect root density at those depths. However, as soon as roots begin to deplete soil water—and this will occur most rapidly in the zones with the highest root length density—the rate of depletion in these zones will begin to fall and the relationship between water depletion and root distribution will change. This relationship will be complicated by the redistribution of water in the soil profile, by rainfall, and by new root growth. The influence of some of these changes on root development have been detailed by Smucker et al. (1991) and Smucker and Aiken (1992). Drought can cause the death of lateral roots and consequentially reduce the absorptive potential of the system.

Notwithstanding these considerations, a number of studies have attempted to relate roots and water. Water depletion, measured with a neutron probe, was used as a means of estimating root distribution and density in a citrus orchard by Cahoon and Stolzy (1959). They concluded that (a) there was a good relationship between the cumulative moisture depletion, with depth, over the season and root distribution with depth, (b) a higher percentage of root weight in the soil surface resulted in a higher depletion of soil moisture, (c) the rate of moisture depletion indicated rather well differences in root density with depth among the soil profiles, and (d) in a soil that is not rewetted at depth during the growing season, the water content at any depth will, in the short term, remain fairly constant after drainage has ceased until there is water extraction by roots. Thus, by monitoring profile water content at a range of depths with a neutron probe, it has been possible to infer the progressive downward growth of the root system of annual crops from the dates on which there is a sudden drop in the water content at each depth (McGowan, 1974). It has also been suggested that the progress of the drying front down a soil profile allows the maximum depth from which quantities of water are extracted by crop roots, the "effective rooting depth" of the crop, to be estimated.

Water depletion can act as a general guide to rooting in annual crops and can yield good relationships between root counts and soil water content (Andren et al., 1991). However, as other factors interact independently, it is probably not suitable as a detailed indicator of rooting activity.

3. *Estimation from the Uptake of Isotopes*

Basic Principles. Estimates of root activity can be based on the uptake of a radioisotope, most usually ^{32}P, from a range of depths and horizontal positions. The basis of the method is that within a block of similar plants, treatment plots can be established and radioactive material injected to a range of soil depths. For any one plant, all injections are made to a single depth (or a combination of relatively similar depths). At time intervals, after the isotope injection, samples of foliage are taken and activity determined. After correction of values for dilution in the soil by unlabeled material, relative root activity at a range of depths can be established.

The use of tracers to measure nutrient uptake depends on isotopic depletion, that is,

$$N_p = \frac{T_p \cdot N_s}{T_s} \qquad (7)$$

where T_p and N_p are the quantities of tracer and nutrient absorbed by the plant, T_s is the quantity of tracer added to soil, and N_s is the quantity of labile nutrient (Newbould and Taylor, 1964).

For N_p to be a valid estimate of nutrient absorption, the following conditions need to be met:

1. The application of tracer must not alter absorption from the zone to which tracer is added, so the concentration of labile ions must not be appreciably altered. This is satisfied by using carrier-free tracer.
2. Labile nutrients in the soil must be uniformly labeled with added tracer. This is difficult to achieve in field experiments.

The requirements for satisfactory field experiments in both uniform and nonuniform soils have been detailed by Newbould and Taylor (1964). This method seems to have been fairly effective for annual crops, especially those with a high root length density, but rather less effective for perennial crops or plants with low root densities (Atkinson, 1974).

The same principles that relate to radioisotopes can also be applied to stable isotopes, whether injected (Atkinson, 1977) or present in amounts different from those in the atmosphere through natural processes, e.g., ^{15}N (Atkinson et al., 1985). Current concerns in relation to both worker and environmental safety have limited the use of radioisotopes in field studies and make the use of natural abundance levels of isotopes such as ^{15}N attractive.

Radioisotope Placement in Soil. In a study of cotton root growth (Bassett et al., 1970), a radioisotope was placed at 15 cm intervals over a distance of 120 cm along the planting row. Carrier-free ^{32}P (4 mL) was placed in 7 mL gelatin capsules, which were dropped into each hole. Each capsule was dissolved in 100 mL of warm water poured down the access hole, which was then covered, and uptake was assessed by periodic sampling of young main stem leaves. Injections, 960 μCi (35 MBq) of ^{32}P per plot were made at depths of 30, 61, 91, 122, 152, and 183 cm, 2 weeks after emergence. All placements were replicated three times. A similar method was used to assess the effect of cultivation and soil physical conditions on root activity in barley (Ellis et al., 1977). Tomar et al. (1981) used a similar method in a study of the effect of subsoil compaction on wheat with 500 μCi (18.5 MBq) of ^{32}P placed at depth. In common with a number of other studies (Broeshart and Nethsinghe, 1972; Patel and Kabaara, 1975), variation between plots prevented definitive conclusions about the effects of treatments being drawn.

In studies of soil effects on fruit trees, similar problems relating to high variability were found (Atkinson, 1974). Here coefficients of variation were often in excess of 50%. The distribution of activity within the tree canopy was variable, with up to 43% of total activity being found in one-quarter of the canopy. Despite this variation, a good relationship between new root growth and the uptake of ^{32}P injected into the soil at a number of depths was found. The variation detected was considered to be a consequence of the limited movement of ^{32}P in soils and the low density of roots found on fruit trees. Where ^{15}N, which has a higher mobility in soil, was used as a tracer, sample-to-sample variability decreased greatly (Broeshart and Nethsinghe, 1972; Atkinson, 1977). The use of ^{15}N as a tracer is complicated by the effects of the release of N by mineralization, giving variable dilution of added tracer. However, at times of the year when mineralization is likely to be small and for comparisons of similar soil depths, the method is effective. In a study where the CV for ^{32}P was 53%, it was only 27% for ^{15}N (Broeshart and Nethsinghe, 1972).

Clearly this method can be used to assess whether a plant obtains added tracer from a particular soil zone and so can be used to delimit the size of the soil volume being exploited. Where variation is not excessive, differences between soil physical treatments can be assessed.

Radioisotope Injection into Plants. In the method described above, the radioisotope is placed in the soil and its uptake assessed by counting foliage. In the converse to this approach, a radionuclide is injected into a plant and its presence in the root system is detected by counting soil plus roots. Ellis and Barnes (1973) used this method to assess root activity. They injected 5–10 μCi of ^{86}Rb (185–370 kBq) per plant into the leaf sheath at the base of a ryegrass or barley plant. Above-ground material was removed 24 h after injection, and 5 cm diameter cores of soil were taken, divided into 7.5 cm depth sections and bulked into 3 kg lots for γ-counting. Repacking of samples normally affected results by no more than 3%. Calibration of this method against the root mass in core samples gave close agreement for both ryegrass and barley. Agreement was better 9 weeks after sowing than at maturity. Comparison of results obtained with the ^{86}Rb technique with soil placement of ^{32}P showed a degree of similarity except that root activity indicated using ^{32}P was relatively higher at 12.5–22.5 cm depth.

Soil Carbon Metabolism. Developing interest in climate change and in the dynamics of CO_2 as a greenhouse gas have increased the availability of field data on soil carbon dynamics. Roots are the major pathway by which carbon is moved to the soil biota. Just as root presence/activity can be inferred from water depletion, it may also be inferred from CO_2 release from the soil. Thomas et al. (1996) related root distribution and soil surface carbon fluxes for *Pinus radiata*. Increasing the concentration of CO_2 in which the trees were grown increased root growth (assessed using minirhizotrons) by 120%, root carbon density by 61%, and soil respiration. Root respiration varied between root types and was related to mycorrhizal biomass at one of two field sites. Improved methods to measure CO_2 are likely to increase the use of this summative method in the future.

Root System Models. Many production models still assume a single linear relationship between root and shoot biomass despite significant studies showing the ability of roots to vary independently of shoots (Atkinson and Fogel, 1997). However, a number of valuable models of the root system have been produced that allow some root properties to be estimated from other values. Van Noordwijk and Van de Geijn (1996) assessed the root parameters required by process-orientated models of crop growth limited by resource supply. Using root length density (cm cm^{-3}), root diameter, and root position, plus some canopy and soil parameters, they were able to predict water and nutrient use. Adiku et al. (1996) modeled root growth and distribution with depth as affected by varying soil water supply. They found lateral root growth to be influenced by current root length density and soil water potential. New root mass was partitioned to the wetter areas of the soil. Pellerin and Pages (1996) developed a model of three-dimensional root growth in maize. Using information on root number, branching, root elongation, and angles of growth, they were able to predict soil colonization. Robinson (1996) modeled the effects of spatial heterogeneity of soil resources using information on relative nutrient uptake rate for a unit of root and root elongation. In general, it is important to assess why there is a need to measure root properties rather than to estimate them either from known relationships or from models such as those detailed above (Atkinson, 1998).

IV. CONCLUSIONS

There is no single method of root measurement applicable to all situations calling for assessments of the effects of soil physical conditions. The principal factors influencing the choice of methods are likely to be the availability and cost of equipment and facilities and the crop and soil under investigation. These considerations will essentially select the methods to be used and are detailed in Table 2. New techniques, mainly of an indirect nature, continue to be developed. By being aware of the limitations and advantages of indirect techniques, and if necessary by using a combination of techniques appropriate to the nature of the investigation, studies will continue to yield important new insights into below-ground responses. Just as a range of measurements is often made in dealing with the above-ground part of the plant, so with the below-ground portion a range of measurements is often needed to be able fully to characterize important factors such as the rate and type of growth, the standing crop of roots, and their activity.

REFERENCES

Abdalla, A. M., D. R. P. Hettiaratchi, and A. R. Reece. 1969. The mechanics of root growth in granular media. *J. Agric. Eng. Res.* 14:236–248.

Adiku, S. G. K., R. D. Braddock, and C. W. Rose. 1996. Modelling the effect of varying soil water on root growth dynamics of annual crops. *Plant Soil* 185:125–135.

Aiken, R. M., and A. J. M. Smucker. 1996. Root system regulation of whole plant growth. *Ann. Rev. Phytopathol.* 32:325–346.

Altemuller, H. J. 1986. Fluorescent light microscopy of soil root interactions. In: *Trans. 13th Int. Congr. Soil Sci.* Hamburg, pp. 1546–1547.

Andren, O., K. Rajkai, and T. Katterer. 1991. A non-destructive technique for studies of root distribution in relation to soil moisture. *Agric. Eco. Environ.* 34:269–278.

Andren, O., H. Elmquist, and A.-C. Hansson. 1996. Recording, processing and analysis of grass root images from a rhizotron. *Plant Soil* 185:259–264.

Asamoah, T. E. O. 1984. Fruit tree root system: Effects of nursery and orchard management and some consequences for growth, nutrient and water uptake. Ph.D. thesis, Univ. of London.

Atkinson, D. 1972. The root system of Fortune/M9. *Rept. E. Malling Res. Stn. 1971*, pp. 72–78.

Atkinson, D. 1974. Some observations on the distribution of root activity in apple trees. *Plant Soil* 40:333–342.

Atkinson, D. 1977. Some observations on the root growth of young apple trees and their uptake of nutrients when grown in herbicided strips in grassed orchards. *Plant Soil* 49:459–471.

Atkinson, D. 1978. The use of soil resources in high density planting systems. *Acta Hortic.* 65:79–89.

Atkinson, D. 1981. The distribution and effectiveness of the roots of tree crops. *Hortic. Rev.* 2:424–490.

Atkinson, D. 1983. The growth, activity and distribution of the fruit tree root system. *Plant Soil* 71:23–36.

Atkinson, D. 1985. Spatial and temporal aspects of root distribution as indicated by the use of a root observation laboratory. In: *Ecological Interactions in Soil* (A. H. Fitter, D. Atkinson, D. Read, and M. B. Usher, eds.). Oxford: Blackwell, pp. 43–65.

Atkinson, D. 1986. The nutrient requirements of fruit trees: Some current considerations. *Adv. Plant Nutr.* 2:93–128.

Atkinson, D. 1987. Variation in root distribution in spring barley. *Rept. Macaulay Inst. Soil Res. 1986*, pp. 174–181.

Atkinson, D. 1998. Root characteristics; Why and what to measure. In: *Root Research: A Handbook of Methods*. Dordrecht, The Netherlands: Kluwer (in press).

Atkinson, D., and R. Fogel. 1997. Roots: Measurement, function and dry matter budgets. In: *Scaling-Up from Cell to Landscape* (P. R. van Gardingen, G. M. Foody, and P. J. Curran, eds.). Cambridge: Cambridge Univ. Press, pp. 151–172.

Atkinson, D., and J. K. Lewis. 1979. Time-lapse cinematographic studies of fruit tree root growth. *J. Photogr. Sci.* 27:255–257.

Atkinson, D., and G. C. White. 1981. The effects of weeds and weed control on temperate fruit orchards and their environment. In: *Pests, Pathogens and Vegetation* (J. M. Thresh, ed.). London: Pitman, pp. 415–428.

Atkinson, D., and S. A. Wilson. 1980. The growth and distribution of fruit tree roots: Some consequences for nutrient uptake. In: *Mineral Nutrition of Fruit Trees* (D. Atkinson, J. E. Jackson, R. O. Sharples, and W. M. Waller, eds.). London: Butterworth, pp. 137–150.

Atkinson, D., D. Naylor, and G. Coldrick. 1976. The effect of tree spacing on the apple root system. *Hort. Res.* 16:89–105.

Baldwin, J. P., and P. B. Tinker. 1972. A method for measuring the lengths and uptake patterns of two interpenetrating root systems. *Plant Soil* 37:209–213.

Baligar, V. C., V. E. Nash, F. D. Whisler, and D. L. Myhre. 1981. Sorghum and soybean growth as influenced by synthetic pans. *Commun. Soil Sci. Plant Anal.* 12:97–107.

Barley, K. P. 1962. The effect of mechanical stress on the growth of roots. *J. Exp. Bot.* 13: 95–110.

Bassett, D. M., J. R. Stockton, and W. L. Dickens. 1970. Root growth of cotton as measured by ^{32}P uptake. *Agron. J.* 62:200–203.

Belford, R. K. 1979. Collection and evaluation of large soil monoliths for soil and crop studies. *J. Soil Sci.* 30:363–373.

Belford, R. K., and F. K. G. Henderson. 1984. Measurement of the growth of wheat roots using a TV camera system in the field. In: *Wheat Growth and Modelling* (W. Day and R. K. Aiken, eds.). New York: Plenum Press, pp. 99–105.

Bengough, A. G., C. J. Mackenzie, and H. E. Elangwe. 1994. Biophysics of the growth responses of pea roots to change in penetration resistance. *Plant Soil* 167:135–141.

Berntson, G. M. 1995. The characterisation of topology: A comparison of four topological indices for rooted binary trees. *J. Theor. Biol.* 177:271–281.

Bhaskaran, A. R., and D. C. Chakrabarty. 1965. A preliminary study on the variations in the soil binding capacity of some grass roots. *Indian J. Agron.* 10:326–330.

Bhat, K. K. S. 1980. A low cost, easy to install flow culture system, suitable for use in a constant environment cabinet. *J. Exp. Bot.* 31:1435–1440.

Bhat, K. K. S. 1983. Nutrient inflows into apple roots. *Plant Soil* 71:371–380.

Bohm, W. 1974. Mini-rhizotrons for root observations under field conditions. *Z. Acker. Pflanzerbau.* 140:282–287.

Bohm, W. 1976. In situ estimation of root length in natural soil profiles. *J. Agric. Sci. Camb.* 87:365–368.

Bohm, W. 1979. *Methods of Studying Root Systems.* Berlin: Springer-Verlag.

Bohm, W., and U. Kopke. 1977. Comparative root investigations with two profile wall methods. *Z. Acker. Pflanzenbau.* 144:297–303.

Bohm, W., H. Maduakor, and H. M. Taylor. 1977. Comparison of five methods for characterizing soybean rooting density and development. *Agron. J.* 69:415–419.

Bottomley, P. A., H. H. Rogers, and T. H. Foster. 1986. NMR imaging shows water distribution and transport in plant root systems in situ. *Proc. Nat. Acad. Sci. U.S.A.* 83: 87–89.

Bragg, P. L., G. Gobi, and R. Q. Cannell. 1983. A comparison of methods, including angled and vertical minirhizotrons for studying root growth and distribution in a spring oat crop. *Plant Soil* 73:435–440.

Broeshart, H., and D. A. Nethsinghe. 1972. Studies on the pattern of root activity of tree crops using isotope techniques. In: *Isotopes and Radiation in Soil-Plant Relationships, Including Forestry*. Vienna: IAEA, pp. 453–463.

Brown, D. A., and A. ul-Haq. 1984. A porous membrane-root culture technique for growing plants under controlled soil conditions. *Soil Sci. Soc. Am. J.* 48:692–695.

Browning, V. D., H. M. Taylor, M. G. Huck, and B. Klepper. 1974. Water relations of cotton: A rhizotron study. *Bull. Auburn Univ. Agric. Exp. Stn.* No. 467.

Cahoon, G. A., and E. S. Morton. 1961. An apparatus for the quantitative separation of plant roots from soil. *Proc. Am. Soc. Hortic. Sci.* 78:593–596.

Cahoon, G. A., and L. H. Stolzy. 1959. Estimating root density and contributions in citrus orchards by the neutron moderation method. *Proc. Am. Soc. Hortic. Sci.* 74: 322–327.

Carley, H. E., and R. D. Watson. 1966. A new gravimetric method for estimating root surface area. *Soil Sci.* 102:289–291.

Carpenter, A., J. M. Cherrett, J. B. Ford, M. Thomas, and E. Evans. 1985. An inexpensive rhizotron for research on soil and litter-living organisms. In: *Ecological Interactions in Soil* (A. H. Fitter, D. Atkinson, D. J. Read, and M. B. Usher, eds.). Oxford: Blackwell, pp. 67–71.

Chriqui, D., A. Guivarc'h, W. Dewitte, E. Prinsen, and H. van Onkelen. 1996. Rol genes and root initiation and development. *Plant Soil* 187:47–55.

Clark, L. J., W. R. Walley, A. R. Dexter, P. B. Barraclough, and R. A. Leigh. 1996. Complete mechanical impedance increases the turgor of cells in the apex of pea roots. *Plant, Cell Environ.* 19:1099–1102.

Coker, E. G. 1958. Root studies: XII. Root systems of apple on Malling rootstocks on five soil series. *J. Hortic. Sci.* 33:71–79.

Coutts, M. P. 1983. Root architecture and tree stability. *Plant Soil* 71:171–188.

Darbyshire, J. F., L. Robertson, and L. A. Mackie. 1985. A comparison of two methods of estimating the soil pore network available to protozoa. *Soil Biol. Biochem.* 17: 619–624.

Davis, J. M., and M. G. Huck. 1978. Identifying turgor responses of water-stressed cotton to rapid changes in net radiation using spectral analysis. *Crop Sci.* 18:605–612.

De Ruijter, R. J., B. Veen, and M. A. Van Oijen. 1996. A comparison of soil core sampling and minirhizotrons to quantify root development of field-grown potatoes. *Plant Soil* 182:301–312.

Drew, M. C. 1987. Function of root tissues in nutrient and water transport. In: *Root Development and Function* (P. J. Gregory, J. V. Lake, and D. A. Rose, eds.). Cambridge: Cambridge Univ. Press, pp. 71–102.

Durant, M. J., B. J. G. Love, A. B. Missem, and A. P. Draycott. 1973. Growth of crop roots in relation to soil moisture extraction. *Ann. Appl. Biol.* 74:387–394.

Dyer, D. J., and D. A. Brown. 1983. Relationship of fluorescent intensity to ion uptake and elongation rates of soybean roots. *Plant Soil* 72:127–134.

Ellis, F. B., and B. T. Barnes. 1973. Estimation of the distribution of living roots of plants under field conditions. *Plant Soil* 39:81–91.

Ellis, F. B., J. G. Elliot, B. T. Barnes, and K. R. Howse. 1977. Comparison of direct drilling, reduced cultivation and ploughing on the growth of cereals: 2. Spring barley on a sandy loam soil: Soil physical condition and root growth. *J. Agric. Sci. Camb.* 89: 631–642.

Ewing, R. P., and T. C. Kaspar. 1995. Accurate perimeter and length measurement using an edge chord algorithm. *Comp. Assist. Microscopy* 7:99–100.

Farrell, R. E., F. L. Walley, A. P. Kukey, and J. J. Germida. 1993. Manual and digital line-intercept methods for measuring root length—A comparison. *Agron. J.* 85: 1233–1237.

Finney, J. R., and B. A. G. Knight. 1973. The effect of soil physical conditions produced by various cultivation systems on the root development of winter wheat. *J. Agric. Sci. Camb.* 80:435–442.

Fitter, A. H. 1985. Functional significance of root morphology and root system architecture. In: *Ecological Interactions in Soil* (A. H. Fitter, D. Atkinson, D. J. Read, and M. B. Usher, eds.). Oxford: Blackwell, pp. 87–106.

Fitter, A. H. 1991. The ecological significance of root system architecture: An economic approach. In: *Plant Root Growth: An Ecological Perspective* (D. Atkinson, ed.). Oxford: Blackwell, pp. 229–243.

Fogel, R., and J. Lussenhop. 1991. The University of Michigan Soil Biotron: A platform for soil biology research in a natural forest. In: *Plant Root Growth: An Ecological Perspective* (D. Atkinson, ed.). Oxford: Blackwell, pp. 61–68.

Freeman, B. M., and R. E. Smart. 1976. A root observation laboratory for studies with grapevines. *Am. J. Enol. Vitic.* 27:36–39.

Gansert, D. 1994. Root respiration and its importance for the carbon balance of beech saplings (*Fagus sylvatica* L.) in a montane beech forest. *Plant Soil* 167:109–119.

Garwood, E. A. 1968. Some effects of soil water conditions and soil temperature on the roots of grasses and clover: 2. Effects of variation in the soil water content and in soil temperature on root growth. *J. Brit. Grassl. Soc.* 23:117–128.

Gill, W. R., and R. D. Miller. 1956. A method for study of the influence of mechanical impedance and aeration on the growth of seedling roots. *Soil Sci. Soc. Am. Proc.* 20: 154–157.

Glover, J. 1967. The simultaneous growth of sugar cane roots and tops in relation to soil and climate. *Proc. S. Afr. Sugar Technol. Assoc.* 1–16.

Goins, G. D., and M. P. Russelle. 1996. Fine root demography in alfalfa (*Medicago sativa* L.). *Plant Soil* 185:218–291.

Gooderham, P. T. 1969. A simple method for the extraction and preservation of an undisturbed root system from a soil. *Plant Soil* 31:201–204.

Gooderham, P. 1977. Some aspects of soil compaction, root growth and crop yield. *Agric. Prog.* 52:33–44.

Goodman, D., and D. J. Greenwood. 1976. Distribution of roots, water and nutrients beneath cabbage grown in the field. *J. Sci. Food Agric.* 27:28–36.

Gordon, D. C., D. R. P. Hettiaratchi, A. G. Bengough, and I. M. Young. 1992. Non-destructive analysis of root growth in porous media. *Plant Cell Environ.* 15:123–128.

Goss, M. J. 1977. Effects of mechanical impedance on root growth in barley (*Hordeum vulgare* L.): 1. Effects on the elongation and branching of seminal root axis. *J. Exp. Bot.* 28:96–111.

Goss, M. J., A. R. Dexter, and M. Evans. 1987. Mechanics of root elongation and the effects of 3,5-di-iodo-4-hydroxybenzoic acid (DIHB). *Plant Soil* 99:211–218.

Greacen, E. L., and J. S. Oh. 1972. Physics of root growth. *Nature (New Biol.)* 235: 24–25.

Gregory, P. J. 1979. A periscope method for observing root growth and distribution in field soil. *J. Exp. Bot.* 30:205–214.

Gurung, H. P. 1979. The influence of soil management on root growth and activity in apple trees. M.Phil. thesis, University of London.

Hackett, C., and B. O. Bartlett. 1971. A study of the root system of barley: III. Branching pattern. *New Phytol.* 70:409–413.

Harris, G. A. 1986. Root length measurements with a modified Delta-T area meter. Delta-T Ltd., Cambridge, U.K.

Harris, G. A., and G. S. Campbell. 1989. Automated quantification of roots using a simple image analyser. *Agron. J.* 81:935–938.

Haussling, M., E. Lei Sen, H. Marschner, and V. Romheld. 1985. An improved method for non-destructive measurements of pH at the root soil interface. *J. Plant. Physiol.* 117: 371–375.

Head, G. C. 1965. Studies of diurnal changes in cherry root growth and nutational movements of apple root tips by time lapse cinematography. *Ann. Bot.* 29:219–224.

Head, G. C. 1966. Estimating seasonal changes in the quantity of white unsuberized root on fruit trees. *J. Hortic. Sci.* 41:197–206.

Head, G. C. 1968. Seasonal changes in the diameter of secondarily thickened roots of fruit trees in relation to the growth of other parts of the tree. *J. Hortic. Sci.* 43:275–282.

Heeraman, D. A., P. H. Crown, and N. G. Juma. 1993. A color composite technique for detecting root dynamics of barley (*Hordeum vulgare* L.) from minirhizotron images. *Plant Soil* 157:275–287.

Hendrick, L. R., and S. K. Pregitzer. 1992a. Spatial variation in tree root distribution and growth associated with minirhizotrons. *Plant Soil* 143:283–288.

Hendrick, L. R., and S. K. Pregitzer. 1992b. The demography of fine roots in a northern hardwood forest. *Ecology* 73:1094–1104.

Hendrick, L. R., and S. K. Pregitzer. 1996. Applications of minirhizotrons to understand root function in forests and other natural ecosystems. *Plant Soil* 185:293–304.

Hilton, R. J., D. S. Bhar, and G. F. Mason. 1969. A rhizotron for in situ root growth studies. *Can. J. Plant Sci.* 49:101–104.

Hooker, J. E., M. Munro, and D. Atkinson. 1992. Vesicular-arbuscular mycorrhizal fungi induced alteration in poplar root system morphology. *Plant Soil* 145:207–214.

Hooker, J. E., K. E. Black, R. L. Perry, and D. Atkinson. 1995. Arbuscular mycorrhizal fungi induced alteration to root longevity of poplar. *Plant Soil* 172:327–329.

Horgan, G. W., S. T. Buckland, and L. A. Mackie-Dawson. 1993. Estimating three-dimensional line process densities from tube counts. *Biometrics* 49:899–906.

Huck, M. G., and H. M. Taylor. 1982. The rhizotron as a tool for root research. *Adv. Agron.* 35:1–35.

Huck, M. G., B. Klepper, and H. M. Taylor. 1970. Diurnal variation in root diameter. *Plant Physiol.* 45:529–530.

Huck, M. G., C. M. Peterson, G. Hoogenboom, and C. D. Burch. 1986. Distribution of dry matter between shoots and roots of irrigated and non-irrigated determinate soybeans. *Agron. J.* 78:807–813.

Huguet, J. G. 1973. A new method of studying the rooting of perennial plants by means of a spiral trench. *Ann. Agron.* 24:707–773.

Karnok, K. J., and R. T. Kucharski. 1982. Design and construction of a rhizotron-lysimeter facility at the Ohio State University. *Agron. J.* 74:152–156.

Kaspar, T. C., H. M. Taylor, and R. M. Shibles. 1984. Top-root elongation rates of soybean cultivars in the glasshouse and their relations to field rooting depth. *Crop Sci.* 24: 916–920.

Kirchoff, G. 1992. Measurement of root length and thickness using a hand-held computer scanner. *Field Crop Res.* 29:79–88.

Kopke, U. 1981. A comparison of methods for measuring root growth of field crops. *Z. Acker. Pflanzenbau.* 150:39–49.

Lang, A. R. G., and F. M. Melhuish. 1970. Lengths and diameters of plant roots in non-random populations by analysis of plane surfaces. *Biometrics* 26:421–431.

Lebowitz, R. J. 1988. Digital image analysis measurement of root length and diameter. *Environ. Exp. Bot.* 28:267–273.

Levin, I., R. Assaf, and B. Bravdo. 1979. Soil moisture and root distribution in an apple orchard irrigated by tricklers. *Plant Soil* 52:31–40.

Logsdon, S. D., J. C. Parker, and R. B. Renau. 1987. Root growth as indicated by aggregate size. *Plant Soil* 99:267–275.

Lopez, B., S. Sabate, and C. Gracia. 1996. An inflatable minirhizotron system for stony soils. *Plant Soil* 179:255–260.

Lund, Z. F., and H. O. Beals. 1965. A technique for making thin sections of soil with roots in place. *Soil Sci. Soc. Am. Proc.* 29:633–635.

MacFall, J. S., and G. A. Johnson. 1994. The architecture of plant vasculature and transport as seen with magnetic resonance microscopy. *Can. J. Bot.* 72:1561–1573.

MacFall, J. S., and G. A. Johnson. 1996. Root growth, turnover and transport as seen with three-dimensional magnetic resonance microscopy. In: *Proc. 5th Symp., Intern. Soc. Root Res.* S. Carolina, USA.

MacKenzie, K. A. D. 1979. The development of the endodermic and phi layer of apple roots. *Protoplasma* 100:22–32.

Mackie-Dawson, L. A., and D. Atkinson. 1991. Methodology for the study of roots in field experiments and the interpretation of results. In: *Plant Root Growth: An Ecological Perspective* (D. Atkinson, ed.). Oxford: Blackwell, pp. 25–47.

Mackie-Dawson, L. A., A. D. Walker, D. Atkinson, and J. S. Bibby. 1988. Water abstraction from the River Spey for domestic and agricultural purposes and its effect on agriculture. *Scot. Geog. Mag.* 104,2:91–96.

Mackie-Dawson, L. A., J. F. Darbyshire, and G. D. Wimaladasa. 1995a. Video enhanced photography of lateral roots of perennial ryegrass, *Lolium perenne* L., with and without potassium fertilizer amendments. *Eur. J. Soil Biol.* 31:81–86.

Mackie-Dawson, L. A., S. M. Pratt, S. T. Buckland, and E. I. Duff. 1995b. The effect of nitrogen on fine white root persistence in cherry (*Prunus avium*). *Plant Soil* 173: 349–353.

Majdi, H. 1996. Root sampling methods—Applications and limitations of the minirhizotron technique. *Plant Soil* 185:225–258.

Marsh, B. A'B. 1971. Measurement of length in random arrangements of lines. *J. Appl. Ecol.* 8:265–267.

McCully, M. E. 1987. Selected aspects of the structure and development of field-grown roots with special reference to maize. In: *Root Development and Function* (P. J. Gregory, J. V. Lake, and D. A. Rose, eds.). Cambridge: Cambridge Univ. Press, pp. 53–70.

McGowan, M. 1974. Depths of water extraction by roots. In: *Isotopes and Radiation Techniques in Soil Physics and Irrigation Studies.* Vienna: IAEA, pp. 435–445.

McGowan, M., M. J. Armstrong, and J. A. Corrie. 1983. A rapid fluorescent dye technique for measuring root length. *Exp. Agric.* 19:209–216.

Melhuish, F. M., and A. R. G. Lang. 1968. Quantitative studies of roots in soil: 1. Length and diameter of cotton roots in a clay-loam soil by analysis of surface ground blocks of resin impregnated soil. *Soil Sci.* 106:16–22.

Melhuish, F. M., and A. R. G. Lang. 1971. Quantitative studies of roots in soil: 11. Analysis of non-random populations. *Soil Sci.* 112:161–166.

Merrill, S. D., E. J. Doering, and G. A. Reichman. 1987. Application of a minirhizotron with flexible pressurised walls to a study of corn rot growth. In: *Minirhizotron Observation Tubes—Methods and Application for Measuring Rhizosphere Dynamics.* Special Publ. No. 50. Madison, WI: Am. Soc. Agron., pp. 131–143.

Merrill, S. D., D. R. Upchurch, A. L. Black, and A. Bauer. 1994. Theory of minirhizotron root directionality observation and application to wheat and corn. *Soil Sci. Soc. Am. J.* 58:664–671.

Morrison, I. K., and K. A. Armson. 1968. The rhizometer-A device for measuring the roots of tree seedlings. *For. Chron.* 44:21–23.

Nelson, W. W., and R. R. Allmaras. 1969. An improved monolith method for excavating and describing roots. *Agron. J.* 61:751–754.

Newbould, P., and R. Taylor. 1964. Uptake of nutrients from different depths in soil by plants. *Trans. 8th Int. Congr. Soil Sci.* 4:731–743.

Newbould, P., R. Taylor, and K. R. Howse. 1971. The absorption of phosphate and calcium from different depths in soil by swards of perennial ryegrass. *J. Brit. Grassl. Soc.* 26:201–208.

Newman, E. I. 1966. A method of estimating the total length of root in a sample. *J. Appl. Ecol.* 3:139–145.

Oosterhuis, D. M., and D. Zhao. 1994. Increased root length and branching in cotton by soil application of the plant growth regulator PGR-IV. *Plant Soil* 167:51–56.

Ottman, M. J., and H. Timm. 1984. Measurement of viable plant roots with an image analysis computer. *Agron. J.* 77:1018–1020.

Ovington, J. D., and G. Murray. 1968. Seasonal periodicity of root growth of birch trees. In: *Methods of Productivity Studies in Root Systems and Rhizosphere Organisms.* Leningrad: Nauka, pp. 146–153.

Pages, L., and V. Serra. 1994. Growth and branching of the taproot of young oak trees—A dynamic study. *J. Exp. Bot.* 45:1327–1334.

Parlychenko, I. K. 1942. Root systems of certain forage crops in relation to the management of agricultural soils. *Natl. Res. Comb. Can. Dominion, NRC* 1088. Ottawa: Dept. Agric.

Patel, R. Z., and A. M. Kabaara. 1975. Isotope studies on the efficient use of P fertilizers by *Coffea arabica* in Kenya: 1. Uptake and distribution of ^{32}P from labelled KH_2PO_4. *Exp. Agric.* 11:1–11.

Pellerin, S., and L. Pages. 1996. Evaluation in field conditions of a three-dimensional architectural model of the maize root system: Comparison of simulated and observed horizontal root maps. *Plant Soil* 178:101–112.

Poelman, G., J. van de Koppel, and G. Brouwer. 1996. A telescopic method for photographing within 8 × 8 cm minirhizotrons. *Plant Soil* 185:163–167.

Prior, S. A., and H. H. Rogers. 1994. A manual soil coring system for soil-root studies. *Comm. Soil Sci. Plant Anal.* 25:517–522.

Reynolds, E. R. C. 1970. Root distribution and the cause of its spatial variability in *Pseudotsuga taxifolia* (Poir) Britt. *Plant Soil* 32:501–517.

Reynolds, H. L., and C. D'Antonio. 1996. The ecological significance of plasticity in root weight ratio in response to nitrogen: Opinion. *Plant Soil* 185:75–97.

Richards, D., and R. N. Rowe. 1976. Root-shoot interactions in peach: The function of the root. *Ann. Bot.* 41:1211–1216.

Richards, D., F. M. Foubran, G. N. Garwali, and M. W. Daly. 1979. A machine for determining root length. *Plant Soil* 52:69–76.

Robinson, D. 1996. Variation, co-ordination and compensation in root systems in relation to soil variability. *Plant Soil* 187:57–66.

Rogers, W. S. 1939a. Root studies: VIII. Apple root growth in relation to rootstock, soil, seasonal and climatic factors. *J. Pomol. Hortic. Sci.* 17:99–130.

Rogers, W. S. 1939b. Root studies: IX. The effect of light on growing apple roots: A trial with root observation boxes. *J. Pomol. Hortic. Sci.* 17:131–140.

Rogers, W. S. 1969. The East Malling root observation laboratories. In: *Root Growth* (W. J. Whittington, ed.). London: Butterworths, pp. 361–376.

Rogers, W. S., and G. C. Head. 1963. A new root-observation laboratory. *Rept. E. Malling Res. Stn. 1962*, pp. 55–57.

Rogers, W. S., and M. C. Vyvyan. 1934. Root studies: V. Rootstock and soil effect on apple root systems. *J. Pomol. Hortic. Sci.* 12:110–150.

Russell, R. S. 1977. *Plant Root Systems—Their Function and Interaction with the Soil.* London: McGraw-Hill.

Sackville-Hamilton, C. A. G., J. M. Cherrett, G. R. Sagar, and R. Whitbread. 1991a. A molecular rhizotron for studying soil organisms: Construction and establishment. In: *Plant Root Growth: An Ecological Perspective* (D. Atkinson, ed.). Oxford: Blackwell, pp. 49–61.

Sackville-Hamilton, N. R., M. Jones, and J. L. Harper. 1991b. Automated analysis of roots in bulk soil. In: *Plant Root Growth: An Ecological Perspective* (D. Atkinson, ed.). Oxford: Blackwell, pp. 69–73.

Sanders, J. L., and D. A. Brown. 1978. A new fibre optic technique for measuring root growth of soybeans under field conditions. *Agron. J.* 70:1073–1076.

Scheres, B., H. McKhann, C. van den Berg, V. Willemsen, H. Wolkenfelt, G. de Vrieze, and P. Weisbeek. 1996. Experimental and genetic analysis of root development in *Arabidopsis thaliana. Plant Soil* 187:97–105.

Schumacher, R., F. Fankhouser, and E. Schlapfer. 1969. Development of apple roots; influence of the growth retardant Alar (2,2-dimethylhydrazide of suceinic acid) on root development. *Schweiz. Z. Obst. Wein* 107:409–452.

Schuurman, J. J. 1965. Influence of soil density on root development and growth of oats. *Plant Soil* 22:352–374.

Schuurman, J. J., and M. A. J. Goedewaagen. 1971. *Methods for the Examination of Root Systems and Roots*. Wageningen, The Netherlands: Pudoc.

Schwaar, J. 1971. Wurzeluntersuchungen aus Niedermooren. *Ber. Dtsch. Bot. Ges.* 84: 745–757.

Smit, A. L., and A. Zuin. 1996. Root growth dynamics of brussels sprouts (*Brassica olearacea* var. gemmifera) and leeks (*Allium porrum* L.) as reflected by root length, root color and UV fluorescence. *Plant Soil* 185:271–280.

Smucker, A. J. M. 1993. Soil environmental modification of root dynamics and measurement. *Ann. Rev. Phytopathol.* 31:191–216.

Smucker, A. J. M., and R. M. Aiken. 1992. Dynamic root responses to water deficits. *Soil Sci.* 154:281–289.

Smucker, A. J. M., S. L. McBurney, and A. K. Srivastava. 1982. Quantitative separation of roots from compacted soil profiles by the hydropneumatic elutriation system. *Agron. J.* 74:500–503.

Smucker, A. J. M., J. C. Ferguson, W. P. De Bruyn, R. K. Belford, and J. T. Ritchie. 1987. Image analysis of video-recorded plant root systems. In: *Minirhizotron Observation Tubes: Methods and Applications for Measuring Rhizosphere Dynamics* (H. M. Taylor, ed.). Spec. Publ. No. 50. Madison, WI: Am. Soc. Agron., pp. 67–80.

Soileau, J. M., D. A. Mays, F. E. Khasauneh, and V. J. Kilmer. 1974. The rhizotron-lysimeter research facility at TVA Muscle Shoals, Alabama. *Agron. J.* 66:828–832.

Steen, E. 1991. Usefulness of the mesh bag method in quantitative root studies. In: *Plant Root Growth: An Ecological Perspective* (D. Atkinson, ed.). Oxford: Blackwell, pp. 75–86.

Steen, E., and I. Hakansson. 1987. Use of in-growth soil cores in mesh bags for studies of relations between soil compaction and root growth. *Soil Tillage Res.* 10:363–371.

Steinke, W., D. J. von Willert, and F. A. Austenfield. 1996. Root dynamics in a salt marsh over three consecutive years. *Plant Soil* 185:265–269.

Sword, M., A. Gravatt, P. L. Faulkner, and J. L. Chambers. 1996. Seasonal branch and fine root growth of juvenile loblolly pine five growing seasons after fertilization. *Tree Physiol.* 16:899–904.

Tamasi, J. 1986. *Root Location of Fruit Trees and Its Agrotechnical Consequences.* Budapest: Akademiai Kiado.

Taylor, H. M. 1969. The rhizotron at Auburn, Alabama—A plant root observation laboratory, Circular No. 171, Auburn University.

Taylor, H. M., ed. 1987. *Minirhizotron Observation Tubes: Methods and Applications for Measuring Rhizosphere Dynamics.* Spec. Publ. No. 50. Madison, WI: Am. Soc. Agron.

Taylor, H. M., and Bohm, W. 1976. Use of acrylic plastic as rhizotron windows. *Agron. J.* 68:693–694.

Taylor, H. M., and H. R. Gardner. 1960a. Relative penetrating ability of different plant roots. *Agron. J.* 52:479–581.

Taylor, H. M., and H. R. Gardner. 1960b. Use of wax substrates in root penetration studies. *Soil Sci. Soc. Am. Proc.* 24:79–81.

Taylor, H. M., and B. Klepper. 1973. Rooting density and water extraction patterns for corn (*Zea mays* L.). *Agron. J.* 65:965–968.

Taylor, H. M., and B. Klepper. 1974. Water relations of cotton: I. Root growth and water use as related to top growth and soil water content. *Agron. J.* 66:584–588.

Taylor, H. M., M. G. Huck, B. Klepper, and Z. R. Lund. 1970. Measurement of soil grown roots in a rhizotron. *Agron. J.* 62:807–810.

Tennant, D. 1975. A test of a modified line intersect method of estimating root length. *J. Ecol.* 63:995–1002.

Teskey, R. O., and T. M. Hinkley. 1981. Influence of temperature and water potential on root growth of white oak. *Physiol. Plant.* 52:363–369.

Thomas, S. M., D. Whitehead, J. A. Adams, J. B. Reid, R. R. Sherlock, and A. C. Leckie. 1996. Seasonal root distribution and soil surface carbon fluxes for one-year-old Pinus radiata trees growing at ambient and elevated carbon dioxide concentration. *Tree Physiol.* 16:1015–1021.

Tollner, E. W., J. W. David, and B. P. Verma. 1989. Managing errors with x-ray Ct when measuring physical properties. *Trans. Am. Soc. Agric. Eng.* 32:1090–1096.

Tollner, E. W., R. Harrison, and C. Murphy. 1991. Interpreting the pixel standard deviation statistics from an x-ray tomographic scanner. *Trans. Am. Soc. Agric. Eng.* 34: 1054–1059.

Tomar, S. S., M. B. Russell, and A. S. Tomar. 1981. Effect of subsurface compaction on root distribution and growth of wheat. *Z. Acker. Pflanzenbau.* 150:62–70.

Upchurch, D. R. 1987. Conversion of minirhizotron root intersections to root length density. In: *Minirhizotron Observation Tubes: Methods and Applications for Measuring Rhizosphere Dynamics.* Spec. Publ. No. 50. Madison, WI: Am. Soc. Agron., pp. 51–65.

Upchurch, D. R., and J. T. Ritchie. 1983. Root observations using a video recording system in minirhizotrons. *Agron. J.* 75:1009–1015.

Upchurch, D. R., and J. T. Ritchie. 1984. Battery operated color video camera for root observations in mini-rhizotrons. *Agron. J.* 76:1015–1017.

Van Noordwijk, M., and J. Floris. 1979. Loss of dry weight during washing and storage of root samples. *Plant Soil* 53:239–243.

Van Noordwijk, M., and S. C. Van de Geijn. 1996. Root shoot and soil parameters required for process-oriented model of crop growth limited by water on nutrients. *Plant Soil* 183:1–25.

Varney, G. T., and M. J. Canny. 1993. Rates of water uptake into the mature root system of maize plants. *New Phytol.* 123:775–786.

Volkmar, K. M. 1993. A comparison of minirhizotron techniques for estimating root length density in soils of different bulk densities. *Plant Soil* 157:239–245.

Voorhees, W. B. 1976. Root elongation along a soil–plastic container interface. *Agron. J.* 68:143.

Waddington, J. 1971. Observation of plant roots in situ. *Can. J. Bot.* 49:1850–1852.

Walker, J. M., and S. A. Barber. 1961. Ion uptake by living plant roots. *Science* 133: 881–882.

Wang, Z., W. H. Burch, P. Mou, R. H. Jones, and R. J. Mitchell. 1995. Accuracy of visible and ultraviolet light for estimating live root proportions with minirhizotrons. *Ecology* 76:2330–2334.

Ward, K. J., B. Klepper, R. W. Rickman, and R. R. Allmaras. 1978. Quantitative estimation of living wheat root lengths in soil cores. *Agron. J.* 70:675–677.

Welbank, P. J., M. J. Gibb, P. J. Taylor, and E. D. Williams. 1974. Root growth of cereal crops. *Rept. Rothamsted Exp. Stn. 1973*, Vol. 2, pp. 26–66.

Weller, F. 1971. A method for studying the distribution of absorbing roots of fruit trees. *Exp. Agric.* 7:351–361.

Wiersum, L. K. 1957. The relationship of the size and the structural rigidity of pores to their penetration by roots. *Plant Soil* 9:75–85.

Willat, S. T., R. G. Struss, and H. M. Taylor. 1978. In situ root studies using neutron radiography. *Agron. J.* 70:581–586.

Yu, L. X., J. D. Ray, J. C. Otoole, and H. T. Nguyen. 1995. Use of wax-petrolatum layers for screening rice root penetration. *Crop Sci.* 35:648–687.

13
Gas Movement and Air-Filled Porosity

Bruce C. Ball
Scottish Agricultural College, Edinburgh, Scotland

Keith A. Smith
University of Edinburgh, Edinburgh, Scotland

I. INTRODUCTION

The movement of gases through the pore space of the soil is important in several respects. It plays a vital role in soil biological processes, in which the supply of oxygen, O_2, to respiring roots and microorganisms from the atmosphere above the soil is balanced by the outward flow of carbon dioxide, CO_2. Impedance of this gas exchange is frequently damaging to plant growth, due to deficiency in O_2 supply to the roots. Such conditions also give rise to emissions to the atmosphere of the microbially produced gases methane, CH_4, and nitrous oxide, N_2O, which, like CO_2, contribute to the greenhouse effect (Houghton et al., 1996); conversely, part of the methane in the atmosphere is removed by diffusion into well aerated soils, where it is oxidized by microorganisms. Soil fumigation to control diseases of horticultural crops depends on movement of the fumigant in the vapor phase; emissions of methyl bromide, the most widely used fumigant, contribute to stratospheric ozone depletion (as does N_2O). In a very different context, emissions of the radioactive gas radon into buildings, following the decay of radium present in underlying soils, may be sufficient to constitute a health hazard in some localities.

The mechanisms responsible for the transport of all these gases are *diffusion*, resulting in a net movement of gas from a zone of higher concentration to one of lower concentration, and *mass flow*, where the whole gas mixture moves in response to a pressure gradient. Most gas movement is by diffusion; mass flow is important only when pressure differences develop because of changes in barometric pressure, temperature, or soil water content. The movement occurs overwhelmingly in the air-filled pores, because diffusion in the gas phase is about four orders of magnitude greater than through water. As air-filled porosity varies with soil water content and soil structure, these factors have a major effect on the rate of gas movement in soils.

To measure this movement we need to identify the boundary conditions (i.e., soil depth, compactness, and water content), and take into account factors that might cause errors, such as temperature, matric potential gradients, soil respiration, and changes in absolute pressure at the soil surface. Establishing boundary conditions in the field is difficult; soil structure, bulk density, water content, and temperature can vary over only a few cm, and in particular may differ markedly between soil horizons. We consider here techniques for measuring diffusion and flow of gases and air-filled porosity in the laboratory and in the field, and the relationships of diffusion and flow to air-filled porosity. These methods include both direct measurements and indirect assessments from models. We also consider the applications of these techniques to the characterization of soil aeration and the impact on it of tillage and traffic, the study of trace gas exchange, and the investigation of the movement of radon and fumigants.

II. BASIC CONCEPTS

A. Air-Filled Porosity

Since gases move almost exclusively in the air-filled pores, measurement of porosity is vital to the understanding of gas movement in soil. Air-filled porosity is often used as an indicator of the likely aeration status of the soil and its ability to conduct and store gases.

Air-filled porosity (ϵ_A) is that fraction of the total soil volume that is occupied by air. Total porosity ϵ_T is the percentage of soil volume not occupied by solids. ϵ_A and ϵ_T are equal only in dry soils. ϵ_A is less than ϵ_T in moist soils because a fraction of the total porosity is occupied by soil water, this fraction being called the volumetric water content (θ). Thus air-filled porosity is

$$\epsilon_A = \epsilon_T - \theta \tag{1}$$

ϵ_T may be calculated from the dry bulk density of the sample, ρ_b, and from the particle density of the soil, ρ_p, as follows (Hall et al., 1975):

$$\epsilon_T = \left(1 - \frac{\rho_b}{\rho_p}\right) \tag{2}$$

ϵ_T in undisturbed soil cores may also be estimated as the volumetric water content at saturation (θ_S). Air-filled porosity is most useful when determined at a given water potential. This can be readily achieved by equilibrating on tension tables as used for the determination of pore size distribution (see Chap. 3). Alternatively, samples may be taken at field moisture content. This is best taken as field capacity when ϵ_A is the "air capacity" and corresponds to the soil drainable porosity or macroporosity (Hall et al., 1975).

Air-filled porosity may also be determined by use of an air pycnometer. This apparatus uses the principle of Boyle's Law. The volume of air in a soil sample is measured by observing the resulting pressure when a gas at a measured volume and pressure expands into a larger volume, which includes the sample. This method excludes pores whose entrances are blocked by water films unless they are compressed by the change in pressure, when part of the volume is measured and is used to calculate this volume of trapped air (Stonestrom and Rubin, 1989).

Air-filled porosity can be divided into three functional categories (arterial, marginal, and remote), using a simple model to interpret the results of a series of gas diffusion measurements in soils (Arah and Ball, 1994). Diffusion along the axis of a sample occurs through arterial pores, marginal pores do not contribute to axial diffusion, and remote pores are isolated from gas transport. Estimates of the three functional pore fractions were made by optimizing the fit between real and simulated data collected, using the technique described in Sec. IV.B.

B. Gas Diffusion

Gaseous molecules exhibit random movement as a result of their thermal energy. Where a gradient of partial pressure or concentration of a gas occurs, this random movement results in a net transfer, or *flux*, of gas along this gradient. This is the process of *gas diffusion*. In soils, gas diffusion is "counter-current"; that is, a flux of one gas is matched by a flux of another gas in the opposite direction. Oxygen is required by root cells for the metabolism involved in root growth and nutrient and water uptake, and also by microorganisms. The resulting consumption of oxygen causes a fall in concentration and the consequent creation of a concentration gradient between the soil and the atmosphere above. This is responsible for a net diffusive flux of oxygen into the soil. Carbon dioxide respired by the roots and microorganisms increases the soil concentration above that of the atmosphere, and, correspondingly, an outward flux occurs. Diffusion is also involved in the transfer of water vapor and soil gases (e.g., methane, nitrous oxide) produced under anaerobic conditions.

The diffusion coefficient of a particular gas is usually determined in the presence of another gas, commonly air. Kirkham and Powers (1972) cited a method for measuring the countercurrent diffusion coefficients of oxygen and nitrogen. Pritchard and Currie (1982) described a method to measure the countercurrent diffusion coefficients (D_0) of soil gases in air and gave values for carbon dioxide, nitrous oxide, ethylene, and ethane. The coefficient D_0 depends on absolute temperature and pressure, and can be calculated at the required values using the Boltzmann equation (see, e.g., Pritchard and Currie, 1982).

In soil, gas diffusion coefficients are considerably less than in free air because of obstruction by soil particles and water. Water effectively blocks gas diffusion, since the diffusion coefficients of the two gases of main interest in soils, oxygen and carbon dioxide, are nearly 10,000 times greater in air than in water (Grable, 1966). One-dimensional steady-state diffusion in soil is generally described by use of Fick's first law:

$$q_x = -D_S \frac{\delta C}{\delta x} \tag{3}$$

where q_x is the mass transfer rate of gas per unit area ($ML^{-2}T^{-1}$), D_s is the effective diffusion coefficient (L^2T^{-1}), C is the gas concentration (ML^{-3}), and x is the distance along the line of transfer (L). D_s is related to the diffusion coefficient in free air.

Currie (1960) proposed the relationship

$$D_s = a\epsilon D_0 \tag{4}$$

where ϵ is the air-filled cross-sectional area of soil (equal to the air-filled porosity) and a is a factor to account for the reduction in the effectiveness of ϵ for diffusion because of deviations in pore direction from the overall direction of gas movement (tortuosity) and roughness of the pore surfaces. This aspect is considered in greater detail in Sec. V. Field soil is generally aggregated and contains roots; as pointed out by Currie (1961), it cannot be regarded as homogeneous with randomly distributed pores. Currie suggested that soil contains two pore phases, the large pores between structural units (the intercrumb pores) and the small pores within the units (intracrumb pores). The diffusion coefficient within crumbs is considerably smaller than that between crumbs because of the greater complexity of the pore space within crumbs. Thus diffusion in the soil profile as a whole consists of contributions from diffusion in crumbs, between crumbs, through the water films surrounding roots, and through the plant roots themselves (Glinski and Stepniewski, 1985).

Uncertainty of boundary conditions makes the choice of appropriate diffusion solutions to Fick's law uncertain. However, field methods can give useful indications of gaseous exchange. Laboratory methods of measurement of gas diffusion offer the advantages that boundary conditions can be chosen, controlled, and specified and that sample size and volume can be chosen to represent the soil layer(s) of interest. The main disadvantages are soil disturbance during sampling and the problems associated with relating measurements to field conditions.

Both laboratory and field methods rely on solving Fick's first law. However, the application of this law to gases is empirical (Jaynes and Rogowski, 1983), and only under special circumstances is the diffusion coefficient contained in Fick's law a constant, independent of the mole fraction and the diffusive fluxes of other gases. In an atmosphere composed of O_2, CO_2, and N_2, and where the concentration of N_2 is constant (a system similar to the soil atmosphere), variations of about 10% from the tracer value of the diffusion coefficient of O_2 and CO_2 are possible with variations in the mole fraction (Jaynes and Rogowski, 1983). Nevertheless, Fick's law is almost universally used, and several solutions of it for different conditions were presented by Kirkham and Powers (1972).

Techniques for measurement of diffusion in the gaseous state are discussed in Sec. IV.B. Methods of measurement of oxygen diffusion rate (ODR) that use platinum electrodes (Stolzy and Letey, 1964) and relate to the rate of supply of oxygen through water films, such as those that occur at a root surface, are also dealt with in Sec. IV.B.

C. Mass Flow

Mass flow is the movement of molecules in response to a pressure gradient. Thus mass flow can cause gas exchange between the atmosphere and the soil air only if there are changes in temperature, barometric pressure, wind, or water content (Henderson and Patrick, 1982). Mass flow is a less important mechanism than diffusion (Evans, 1965) and accounts for only a few percent of the normal gas exchange (Henderson and Patrick, 1982). However, air permeability has been shown recently to be more relevant to gas exchange through its contribution to nitrous oxide flux (Ball et al., 1997a), to soil venting (Stylianon and De Vantier, 1995), and to soil vapor extraction (Poulsen et al., 1998). Since mass flow, unlike diffusion, is sensitive to the size of individual pores, measurements of air permeability are more relevant to the characterization of soil structure. They provide an alternative to hydraulic conductivity measurements to describe soil structure, since air flow at low pressure differences causes negligible sample disturbance (Janse and Bolt, 1960).

The flow of gases through soil is comparable to that of water, with certain restrictions. Darcy's law applies if flow is laminar or viscous, as it is when the flow rates are relatively small (Janse and Bolt, 1960):

$$q = \left(\frac{K}{\eta}\right)\left(\frac{dp}{dx}\right) \tag{5}$$

where q is flow rate, p is pressure, x is distance, K is gas permeability, and η is viscosity. In a tube of radius r and length L, flow rate can be calculated from Poiseuille's law:

$$q = \frac{\pi r^4 \Delta p}{8\eta L} \tag{6}$$

where Δp is the difference in pressure between the ends of the tube. It follows that flow rate, hence permeability in soils, depends on the fourth power of the pore radius, whereas diffusion depends only on the square of the radius. Flow is less subject than diffusion to changes due to small temperature differences, though ambient temperature, pressure, and humidity affect flow by their influence on gas viscosity (Grover, 1955). Deviations from laminar or viscous flow occur when large pressure differences are applied to samples containing pores of large enough radius to give flow velocities sufficiently high for a Reynolds number of about 2000 or greater; under these conditions, flow becomes turbulent. (For an explanation of the Reynolds number concept, see a fluid mechanics or physics text, e.g., Denny, 1993). Alternatively, gas slippage (i.e., gas moving along pore surfaces) may occur in very small pores. As for gas diffusion, air flow is blocked by water-filled pores, so that air permeability decreases as soil water content increases.

Field and laboratory techniques are available involving either steady-state or non-steady-state flow. Steady-state measurements of gas permeability are more generally applied than the non-steady-state variety; this is the reverse of the situation relating to measurements of diffusion. Field techniques are often rather inconclusive, because the variability of soil structure in the upper layers is large and is nonnormally distributed. Thus laboratory methods are preferable and are discussed here.

III. SAMPLING AND DIRECT MEASUREMENT

A. Sampling

The choice of soil sample size, the degree of replication, and the extent of pretreatment depend on the objective of the experiment and will guide the choice of measurement technique. Where it is appropriate to use disturbed soil for a diffusion measurement, sufficient is needed to fill a small cell (say 20–30 mm diam. and 20–30 mm long). However, when minimally disturbed samples are required that are representative of field conditions, the choice of sample size and dimensions is difficult. Ideally, the sample volume should be equal to or greater than the representative elementary volume (REV), i.e., the smallest volume that contains a representative packing of particles that is repeated throughout the porous region (Youngs, 1983). Bouma (1983) recommended that a representative sample should contain at least 20 peds and that the REV should be increased as the texture becomes finer and the structure becomes coarser. The REV classes suggested by Bouma for the sphere of influence relevant to individual plants are given in Table 1.

Sampling techniques relevant to the laboratory determination of ϵ_A using minimally disturbed cores are discussed in Chap. 3 (see also McIntyre, 1974; Hall et al., 1975; Hodgson, 1976). Sample size and sampling intensity can be less for ϵ_A than for assessment of gas movement, because porosity and bulk density vary less than flow and diffusion properties within a soil horizon, since porosity does

Table 1 Four Hypothetical Classes of Representative Elementary Volumes of Samples Relative to Soil Texture and Structure

Class	Texture	Structure	Hypothetical REV (cm^3)
1	Sandy	No peds	10^2
2	Loamy, silty	Small peds	10^3
3	Clayey	Medium peds, continuous macropores	10^4
4	Clayey	Large peds, continuous macropores	10^5

Source: Bouma (1983).

not depend on pore continuity. The Soil Survey of England and Wales recommended triplicate sampling of individual horizons (Hodgson, 1976).

Guidance for the construction of sampling equipment and for collection and preparation of minimally disturbed samples is given by McIntyre (1974). Solutions for the diffusion coefficient and equations for the calculation of air permeability generally include sample volume and length. Thus these variables should be kept as constant as possible, to minimize error.

The sample size for most reported diffusion measurements (100–300 cm^3) is smaller than the typical REV of 10^3 cm^3 assumed for a soil structure made up of small peds (Table 1). Thus the use of larger samples, as reported by de Jong et al. (1983), is desirable, even though changing or measuring the temperature or water content of such large samples is difficult. In such cases, a field method of diffusion measurement (discussed below) may be more suitable.

Many reported measurements of diffusion in minimally disturbed samples (Table 2) relate to the description of tillage treatments on specific soil layers, for example, those around a germinating seed. Where these layers are narrow and well-defined, samples can be relatively small, e.g., 35–75 mm deep (Bakker and Hidding, 1970; Ball et al., 1981). However, the great sensitivity of air permeability to pore diameter means that sample disturbance such as cracking or shrinking from the sides of the holder has a greater effect on this parameter than on measurement of diffusion. This sensitivity to pore and crack size also demands a greater requirement than for diffusion measurements for samples to be as large as the representative elementary volume (Bouma, 1983). The use of smaller samples can be justified if the largest channels, such as those produced by earthworms and cracks, are avoided (Ball, 1982; Groenevelt et al., 1984), provided that these are not required in the assessment.

The variation of air permeability among samples is large, with standard errors of replicated data often greater than the means. Ball (1982) attributed this

Table 2 Relationships Between Relative Diffusivity and Air-Filled Porosity in Minimally Disturbed Soil Cores

Soil texture	Soil origin/ management/depth	Soil moisture content	Range of ϵ_A	Relationship between D_S/D_0 and ϵ_A	Source
Silt loam	Compacted layer 4 cm deep at the surface of grass field	Field content or equilibrated to 50 m bar tension	0–0.4	$0.27\ \epsilon_A$	Bruckler et al. (1989)
Range	Range of distinct structures	Varied by changing soil water tension	0.05–0.35	$\sim 0.5\ \epsilon_A$	Ayres et al. (1972)
Sandy loams and silt loams	Arable topsoils	Field content	0.04–0.3	$0.85\ \epsilon_A^2$ (non-puddled) $2.0\ \epsilon_A^2$ (puddled)	Bakker and Hidding (1970)
Range	Range: 5–140 cm depth	Varied by changing soil water tension	0–0.55	$\sim\epsilon_A^{1.5-3}$	Flühler (1972)
Clay loam	Arable cultivation experiment, top 18 cm	Field content	0.02–0.1	$\sim\epsilon_A^{2.5}$ (scattered)	Boone et al. (1976)
Sand, silt, clay loams	Arable cultivation experiment, top 20 cm	Field content	0–0.3	$0.17\ \epsilon_A$ or ϵ_A^2	Richter and Grossgebauer (1978)
Silt loam, clay loam	Arable cultivation experiment, top 20 cm	Varied by changing soil water tension	0–0.5	$\epsilon_A^{2-2.5}$ (for ZL) $\epsilon_A^{2.1-4.5}$ (for CL)	Ball (1982)
Loam	Compaction experiment	Varied by changing soil water tension	0–0.28 0.28–0.5	$3\ \epsilon_A^{2.75}$ $0.34\ \epsilon_A$	Glinski and Stepniewski (1985)
Sandy clay loam	Compaction and tillage experiments	Field capacity	0.05–0.3	$0.3–2.1\ \epsilon_A^{1.7-2.6}$	Ball et al. (1988)
Sandy loams	Arable and woodland topsoils	Field content	0.05–0.5	$0.69\ \epsilon_A^{1.9}$	Ball et al. (1997b)
Sand	Forest	Varied by changing soil water tension	0.05–0.35	$(\epsilon_A/\epsilon_{AM})^{2.3}$ and $\epsilon_T^{1.33}(\epsilon_A/\epsilon_T)^{2.3}$	Moldrup et al. (1996)
Sand, loamy sand, sandy loam	Forest and arable	Field content and varied by changing soil water tension	0–0.5	$0.45\ \epsilon_A^3/\epsilon_T^2$	Poulsen et al. (1998)

ZL: silt loam; CL: clay loam; ϵ_{AM} is maximum measured air-filled porosity, corresponding to maximum measured D_S. ϵ_A, ϵ_T are defined in the text.

variability to the great variation among replicates of the radius, length, and continuity of the largest air-filled pores. Thus a relatively large number of samples, usually 15 to 30 per treatment, is required for adequate assessment of air permeability (Kirkham et al., 1958; Janse and Bolt, 1960; Ball, 1982). Significant scale dependence is also found with air permeability. Garberi et al. (1996) found that air permeability increased dramatically with sampling scale and that standard methods of air permeability assessment could underestimate advective transport of gas phase contaminants in soils.

Sampling distributions of relative diffusivities and air permeabilities may be skewed rather than normal. In such cases, conventional parametric statistics do not strictly apply. Coefficients of variation of replicated relative diffusivities can be up to twice as great as those of the air-filled porosities measured on the same sample using conventional water release calculations (Ball, 1982). Thus a greater number of samples may be required than for, say, assessment of soil water release.

Laboratory treatment of samples may also influence choice of size. If the matric potentials of samples have to be adjusted (e.g., if the intracrumb pores have to be blocked by water by wetting to field capacity to assess diffusion in the intercrumb pores), the time for attainment of equilibrium throughout the sample increases with sample length. We commonly use samples 50 mm long for such experiments (Ball, 1982).

If a soil core in a sample holder is dried in stages, and diffusion or air permeability measurements are made at each stage (see, for example, Ball, 1982), then shrinkage may occur from the walls of the holder. In such cases the gap can be filled with paraffin wax and the sample diameter remeasured; or, as suggested by de Jong et al. (1983), samples can be cut from larger blocks and the nondiffusing surfaces coated with wax.

B. Measurement of Air-Filled Porosity

To measure air-filled porosity at a specific water potential, the samples require equilibration on tension tables, as discussed in Chap. 3 (see also Ball and Hunter, 1988). Samples can also be used for subsequent measurement of diffusion and air permeability. To minimize equilibration times, sample lengths no greater than 50 mm are recommended (Hall et al., 1975). Methods of measurement of ρ_p and ρ_b, necessary for assessment of ϵ_T and thence ϵ_A, are given by McIntyre (1974) and Vomocil (1965). In cores of known volume, ρ_b is easily calculated from the weight of the soil core. In soils containing significant quantities of organic matter, the estimation of ρ_b by liquid pycnometry may overestimate the soil particle density because organic matter is destroyed in this technique. In such cases, a better estimate of ϵ_T may be the volumetric water content at saturation, θ_S. This may be determined after saturation either by capillary wetting and immersion or under vacuum (McIntyre, 1974). The first method may leave air trapped in the sample, thereby underestimating θ_S, and the second method may give structural breakdown and slaking in soils that are structurally unstable, as trapped air is rapidly released from aggregates. Ball and Hunter (1988) found that in the laboratory θ_S agreed best with ϵ_T after saturation by capillary wetting and subsequent estimation of θ_S by weighing the sample immersed in water. The calculation steps for air-filled porosity are described by Carter and Ball (1993).

Field assessment of air-filled porosity is best achieved by using the gamma probe to measure bulk density and then making one or more assessments of water content by time domain reflectometry, neutron moisture meter, or gravimetric measurement on samples taken with an auger (see Chaps. 1, 8). Separate measurement of particle density is required.

C. Measurement of Gas Diffusion

1. Laboratory Methods

Methods in current use involve non-steady-state diffusion where the concentration gradient and the flux of molecules change with time. The method recommended by Rolston (1986) involves measurement of the mutual diffusivity of argon (Ar) and nitrogen (N_2) but also applies to other gases of interest. Argon is used because it is relatively unreactive and has approximately the same values for gas diffusivity and solubility in water as O_2. In this method (Fig. 1), Ar is used to displace most of the air from a diffusion vessel that is initially isolated. The initial concentration of N_2 in the vessel is C_0. The diffusion vessel is slid under the soil sample and lines up with its open lower face, so that nitrogen in the air above the sample and the argon in the diffusion vessel can counterdiffuse through the soil. The change in N_2 concentration, C, in the diffusion vessel is monitored regularly by taking samples and analyzing them in a gas chromatograph. In this method, diffusion is in the unsteady state and is described by Fick's second law as

$$\epsilon \frac{\delta c}{\delta t} = \frac{\delta}{\delta x}\left[D_s \frac{\delta c}{\delta x}\right] \tag{7}$$

Rolston (1986) solved this equation for ϵ remaining constant in space and time, and for soil that was uniform with respect to diffusivity. His solution (with slight amendment to the symbols originally used) was

$$\frac{C - C_0}{C_0 - C_s} = \frac{2h \exp(-D_s\, \alpha_1^2\, t/\epsilon)}{\lambda(\alpha_1^2 + h^2) + h} \tag{8}$$

where $h = \epsilon/(a\epsilon_c)$, ϵ_c is the air content of the chamber, a is the chamber height, λ is the length of the soil sample, α_1 is the first root of $\alpha_1\lambda \tan \alpha_1 = h\lambda$ (values are tabulated by Rolston, 1986), and C_s is the concentration of N_2 in the atmosphere. To calculate D_s, $\ln[(C - C_s)/(C_0 - C_s)]$ is plotted vs. time, t. This is a straight

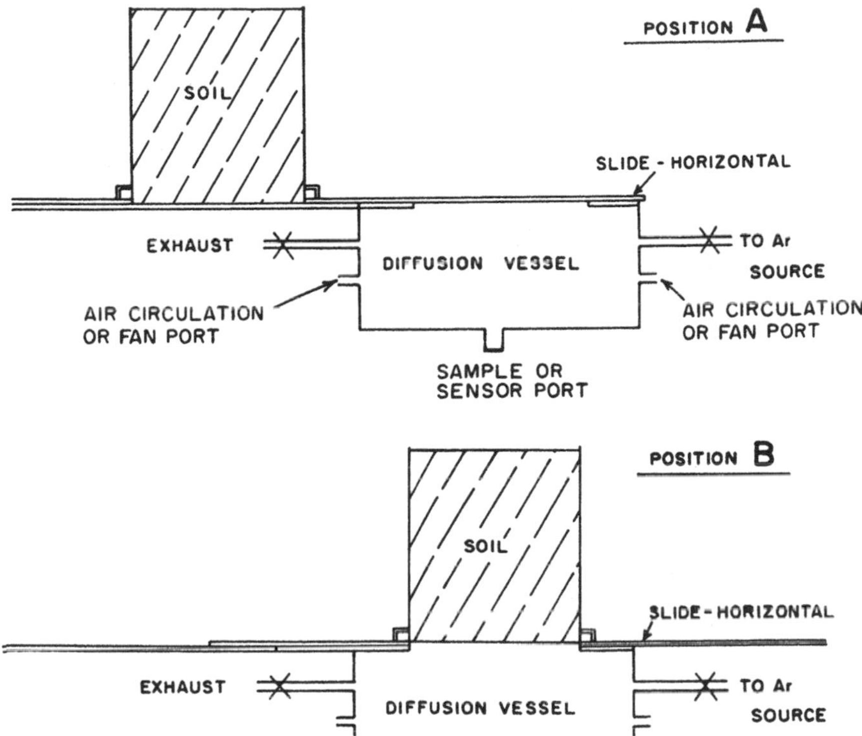

Fig. 1 Apparatus for measurement of gas diffusion in soil, using argon and nitrogen as counterdiffusing gases. (From Rolston, 1986.)

line with slope $-D_s\alpha_1^2/\epsilon$ for sufficiently large t. A problem with this system is the need to disturb and change the diffusion system by the withdrawal of samples for gas analysis.

Several variants on this method have been produced, monitoring nondestructively the changes in gas concentration (commonly O_2) in the diffusion vessel. In that of Schjønning (1985a), the chamber is initially filled with N_2 and contains an electrode to monitor O_2 concentration. Up to 12 samples are run simultaneously, with automatic data-logging. In undisturbed core samples the error in the determination of gas diffusivity due to consumption of O_2 is generally <0.5% but may be greater if the soil is recently disturbed or amended with organic matter subject to rapid microbial turnover (P. Schjønning, pers. comm., 1999).

One problem with such a system, identified by Rust et al. (1956), is that early measurements fail to take into account mass flow when the diffusivities of the two counterdiffusing gases differ significantly. In addition, such systems have one face of the sample open to the atmosphere, so that uniform boundary conditions of concentration, temperature, and pressure are difficult to maintain. Other methods overcome these problems by enclosing the sample between two gas-filled chambers and by using gases at trace concentrations as the diffusing species, to overcome the problem of mass flow. Such systems allow precise control of experimental conditions and can give accurate measurements of diffusion coefficients in soils of very low air-filled porosities (e.g., those that are nearly saturated and in which soil aeration is likely to limit plant growth). Three such methods used ^{85}Kr (Ball et al., 1981), sulfur hexafluoride (SF_6) (Reible and Shair, 1982), and freon-12 (CCl_2F_2) (Sallam et al., 1984; Jin and Jury, 1996), respectively, as tracers. All these gases have low solubility in water and are neither strongly adsorbed on soil surfaces nor consumed by soil microorganisms. In two of these methods (Ball et al., 1981; Reible and Shair, 1982), pressure differences between the end faces of the sample, which could cause mass flow, are monitored by a micromanometer capable of detecting differences as small as 0.01 Pa. In the method of Reible and Shair (1982), syringe samples of a SF_6–air mixture are taken from each chamber at regular intervals and analyzed for their SF_6 concentration, using an electron capture gas chromatograph. Samples of relatively small (2.54 cm) diameter are tested. The air–freon mixture is sampled at the beginning and at the end of the diffusion measurement. In the method of Sallam et al. (1984), the size of the chambers enclosing the sample varies according to the expected sample porosity.

The method of Ball et al. (1981) assesses trace gas concentration nondestructively and was designed for the use of minimally disturbed field samples held in their sampling cylinders. Samples 76 mm diameter and 50 mm long, or 150 mm diameter and 100 mm long, can be inserted directly into the apparatus in the field-moist condition or after equilibration to a given matric potential. This method, with its self-contained apparatus, which is relatively quick and easy to use, is briefly described below; fuller details are given in Ball et al. (1981).

In the apparatus (Fig. 2), two cylindrical gas chambers with scintillator disks and photomultipliers attached are sealed on to the ends of the stainless steel sample holder. A mixture of air and radioactive ^{85}Kr with an activity of the order of 400 GBq m^{-3} ($\sim$1 Ci m^{-3}) is injected into one gas cell and diffuses through the soil until an equilibrium concentration is achieved throughout the apparatus. The concentration of tracer is measured by regular counting of β radiation detected at each photomultiplier. In the latest version of this equipment, the counting data are recorded and analyzed by a PC.

It is assumed that after a short initial period (<5 min), the relationship between the count rates in the two gas chambers is given by

$$C_I - C_R = 2C_e e^{-kt} \tag{9}$$

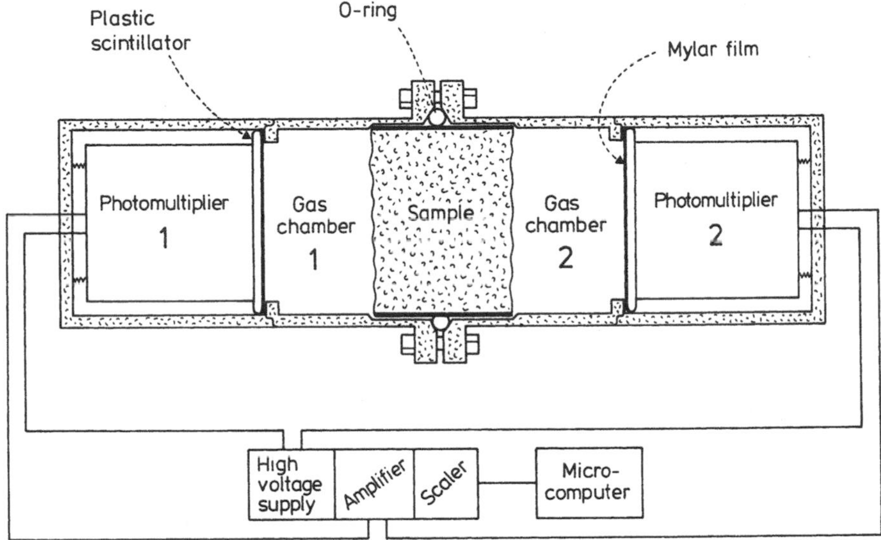

Fig. 2 Apparatus for measurement of gas diffusion and permeability, using ^{85}Kr as tracer gas. (From Ball et al., 1981, with slight adaptation.)

where C_I and C_R are the concentrations of gas in the injection and receiving gas chambers, respectively, t is time (s), C_e is the concentration in each gas chamber at equilibrium, and k is given by

$$k = \frac{2D_s A_s}{VL_s} \tag{10}$$

whence D_s may be found. A_s and L_s are the area and length of the sample, and V is the volume of the gas cell.

In practice, to speed up the measurements, diffusion is usually monitored only halfway to equilibrium. Values of $C_I - C_R$ and t, excluding those detected in the first 5 minutes, are fitted to Eq. 7, and k and C_e are estimated by exponential regression. This modification allows the making of diffusion measurement on samples at or below field capacity, typically in under an hour. Samples wetter than field capacity, particularly if they are compact or fine-textured, may require up to 15 hours for significant diffusion to occur and are best measured overnight.

Interest has recently increased in the movement of volatile organic compounds in soils. Batterman et al. (1996) reviewed the theory and methods for measurement of the diffusion of volatile organic components in the laboratory. They presented a novel one-flow sorbent-based technique. The system maintains a constant concentration gradient across a soil column using a test gas flow at one side and a high-capacity sorbent at the other. The diffusion coefficient of trichloroethylene was estimated using the difference between the inlet and outlet concentrations. The measurement of the transport of reactive gases which hydrolyze in water, such as SO_2 and CO_2, is problematic, and dependent on soil structure and soil solution pH (Rasmuson et al., 1990). The assessment of gas diffusivity in such soils may best be assessed using modeling (see Sec. IV.A).

2. *Field Methods*

a. Large Scale

Field methods of diffusion measurement overcome some of the soil disturbance problems of sampling, but have their own complexities. Methods involve withdrawal and analysis of gases injected into the soil or nondestructive sampling of trace gases.

McIntyre and Philip (1964) pointed out that early methods allowed no rigorous analysis because the geometry of the diffusion path was irregular and the boundary conditions not known. They developed a technique that measured soil surface gas exchange. A thin-walled brass cylinder was driven into the soil, the soil in the cylinder was flushed with air to give a known concentration initially, and then oxygen from a chamber placed on the cylinder was allowed to diffuse through the soil. The oxygen concentration in the chamber was measured with a membrane-covered oxygen cathode. This method takes into account errors due to temperature, relative humidity, and changes in soil porosity, but it suffers from problems of oxygen storage and consumption. However, the approach of measurement of the diffusion of gases across the soil water interface is of great relevance to water evaporation, soil aeration, loss of nitrogen and volatile organic compounds. Rolston et al. (1991) modified the technique to use freon-13 ($CClF_3$) as a tracer and proposed an analytical solution for the diffusion coefficient and a thorough appraisal of the boundary conditions, including comparison with core values of diffusivity.

A smaller-scale technique was proposed by Lai et al. (1976). The method is based on the theory of radial diffusion of a finite quantity of a gas into a semi-infinite medium. Oxygen is injected through a needle inserted into the soil, small aliquots of soil air are withdrawn at regular intervals, and oxygen and nitrogen concentrations are measured using a portable gas chromatograph. Two principal advantages were claimed for their method: there is no removal or alteration of the soil from its natural state, and minimal instrumentation is required at the site of measurement. In a modification of the method of Lai et al. (1976), Jellick and Schnabel (1986) used a numerical finite-difference model to allow the initial concentration profile within the sphere of injected gas to vary, based on experimental

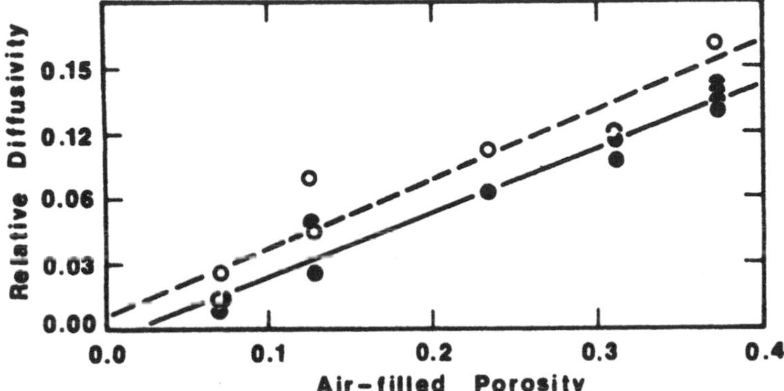

Fig. 3 Relationship between D_s/D_0 and volumetric air content for lawn soils. •: diffusion measurements with the injection method using the numerical model; ○: measurements with the core method. (From Jellick and Schnabel, 1986.)

data. The diffusion coefficients compared favorably with those determined on minimally disturbed core samples (Fig. 3).

These techniques suffer from the need to take samples from the site of diffusion, thereby changing the concentration and pressure of the tracer solution. Also the concentration of samples of the gas may change before analysis due to leakages. Further, such tests give no indication of the likely magnitude of the diffusion coefficient to help determine the frequency and duration of sampling. Ball et al. (1994) developed an apparatus that can be used either as a buried reservoir capable of sampling several soil layers in succession or as a surface chamber (Fig. 4). The method initially used ^{85}Kr as the diffusing gas, which was monitored continuously and nondestructively as it diffused from a cell surrounding a Geiger–Müller tube at the base of a probe (Ball et al., 1994). Later, in response to the inconvenience of satisfying radiological protection procedures, freon-22 ($CHClF_2$) was used as the diffusing gas and was monitored nondestructively at an electrical sensor (Ball et al., 1997b). In the buried reservoir mode, the probe is inserted within a hole augered in the soil and is used to measure diffusion at soil depths below about 150 mm. In the surface chamber mode the probe is located within a chamber enclosing the soil surface, and the system measures the rate of diffusion into the surface. In both modes the gas cell containing the detector is isolated from the probe above it by lightly inflating the rubber membrane above the gas cell. In use, ^{85}Kr or freon-22 (or freon-23) is injected into the gas cell so as to form a cylindrical source, and the decrease in concentration is monitored regularly (usually at intervals of 15 s) until it has decreased to $\frac{1}{2}$–$\frac{1}{3}$ of its original value. In order to calculate diffusivity, it is necessary first to simulate diffusion

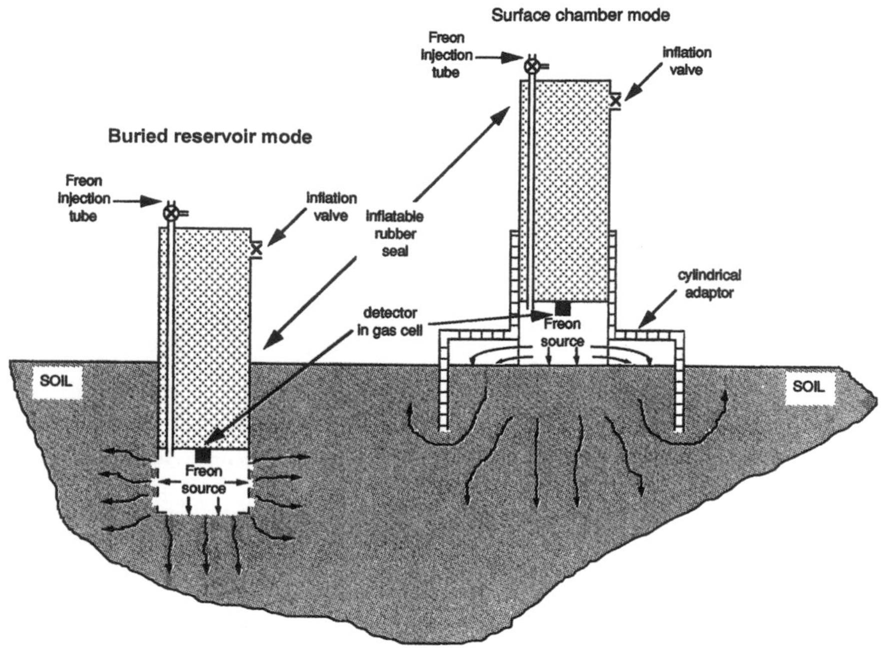

Fig. 4. Equipment to measure gas diffusion in soils in situ. Diagram shows use in either buried reservoir mode or surface chamber mode. (From Ball et al., 1997b.)

numerically using Fick's equation. The time axis of the simulation is expanded or contracted until it matches the observed decrease in concentration (Ball et al., 1994). The advantage of this numerical system is that an exact volume of tracer need not be injected; the initial concentration can vary, provided it is known.

In both laboratory and field measurements of diffusivity, the temperature should be stated when the results are reported. The following equation, cited by Rolston (1986) allows the gas diffusivity measured at any temperature to be calculated for any other temperature:

$$D_{T_2} = D_{T_1} \left(\frac{T_2}{T_1} \right)^{1.72} \tag{11}$$

where D_{T_2} and D_{T_1} are the diffusivities at temperatures T_2 and T_1 (in degrees Kelvin), respectively.

b. Small Scale

Dissolved oxygen can be determined in soil water samples using a polarographic oxygen electrode controlled by a purpose-built electronic analyzer (Blackwell, 1983). Ray et al. (1987) showed that the same method can be used successfully to

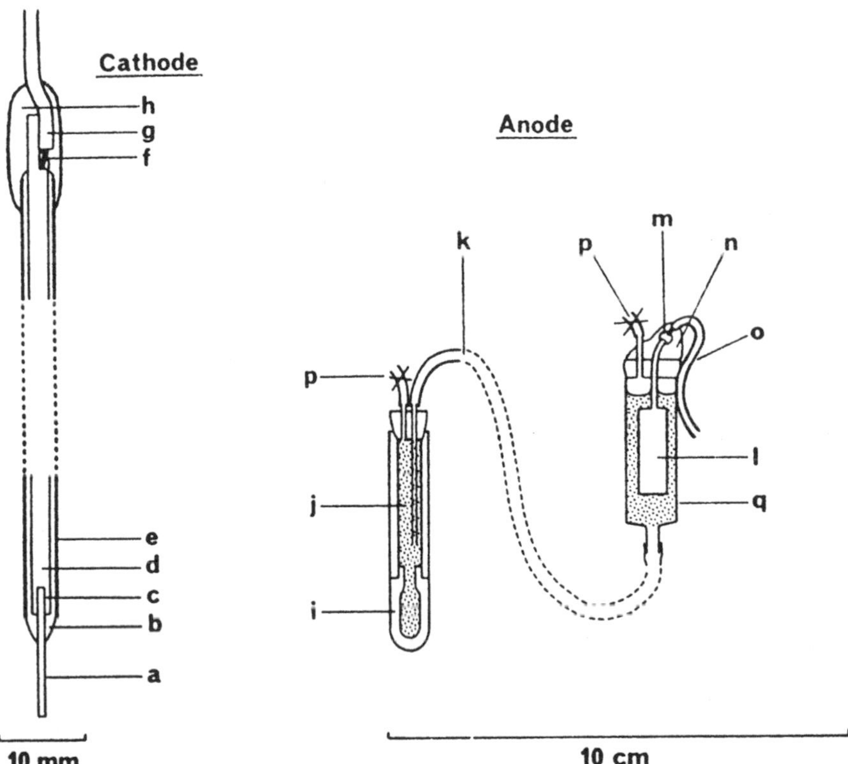

Fig. 5 Structure of anodes and cathodes used for measurements of oxygen flux with bare platinum electrodes. Key: a, Platinum wire; b, epoxy resin; c, crimped and soldered joint (Pt wire in hole at end of conductor); d, mild steel conductor; e, heat-shrinkable insulating sleeve; f, soldered connection; g, wire connected to cathode plug; h, self-amalgamating insulation tape; i, porous pot (air entry pressure approximately 100 kPa); j, saturated KCl solution; k, flexible connecting tube filled with solution; l, prepolarized silver sheet, surface area approximately 25 cm^2; m, crimped and soldered joint; n, silicone rubber cement; o, wire connected to anode plug; p, tap for bleeding air from solution; q, syringe body. (From Blackwell, 1983.)

determine the oxygen content of gaseous samples and that the analysis was more rapid and required less expensive equipment than for gas chromatography.

Soil oxygen flux and redox potential can be measured in waterlogged soil using polarographic techniques. These methods are based on the reduction of oxygen at a platinum wire cathode buried in the soil. This is linked to a calomel or silver–silver chloride anode placed in electrical contact with the soil (Fig. 5). Redox potential is a measure of the intensity of reduction in soils containing no molecular oxygen. Oxygen flux (also termed oxygen diffusion rate, ODR) to a cathode is a measure of the rate of supply of oxygen from the air through the surrounding soil through a film of soil water. This flux is comparable to the maximum required by roots respiring in moist soil (Blackwell, 1983). In measurement of ODR, Armstrong and Wright (1976) recommended that the relationship between current and voltage be established (polarogram). Where a plateau is reached on the polarogram, current is related to the flux of oxygen to the electrode.

Both redox potential and ODR can be measured with the same pair of electrodes. Blackwell (1983) showed that platinum cathodes can be left in the soil and remain functional for several months without removal for cleaning. These techniques work best in wet soil. In unsaturated soil, variations in pH and aeration status alter the shape of polarograms, and it may be necessary to measure soil electrical resistance before ODR can be calculated (Callebaut et al., 1982). Reviews of the principles and the conditions under which this equipment can be used, and detailed descriptions of electrodes and electronic instrumentation required for multiple assessments of both measurements for a lysimeter installation and for field use, can be found in Armstrong and Wright (1976), Callebaut et al. (1982), and Blackwell (1983).

D. Measurement of Mass Flow

Several methods are based on the steady-state method proposed by Grover (1955). Grover devised a permeameter with a float, a thin-walled cylinder that can be suspended to keep it centered (Janse and Bolt, 1960). The float is open only at the bottom and forms an air chamber that fits over an annular water reservoir (Fig. 6). The air pressure can be increased by adding weights to the reservoir. The air is displaced directly into field soil (Grover, 1955) or through a core sample sealed on the bottom (Janse and Bolt, 1960). Bowen (1985) proposed improvements to this apparatus, to incorporate a sensitive flowmeter and manometer, the latter reading to a maximum of 0.5 kPa. The direct reading of flow and pressure considerably speeds up measurements, since it is otherwise necessary to time the fall of the float for a given distance to be able to calculate permeability. The main advantage of this technique is that constant low pressures (0.03–1 kPa) can be applied. Kirkham (1946) discussed in some detail the errors and the assumptions involved in air permeability measurement, particularly that of neglecting gas compressibility. He integrated Eq. 6 into a form applicable to most air permeability measurement techniques:

$$q_v = \frac{K \, \Delta P \, A_s}{\eta L_s} \tag{12}$$

where q_v is the volumetric flow rate [$L^3 \, T^{-1}$], K is air permeability [L^2], ΔP is the pressure difference across the sample [$M \, L^{-1} \, T^{-2}$], A_s is the cross-sectional

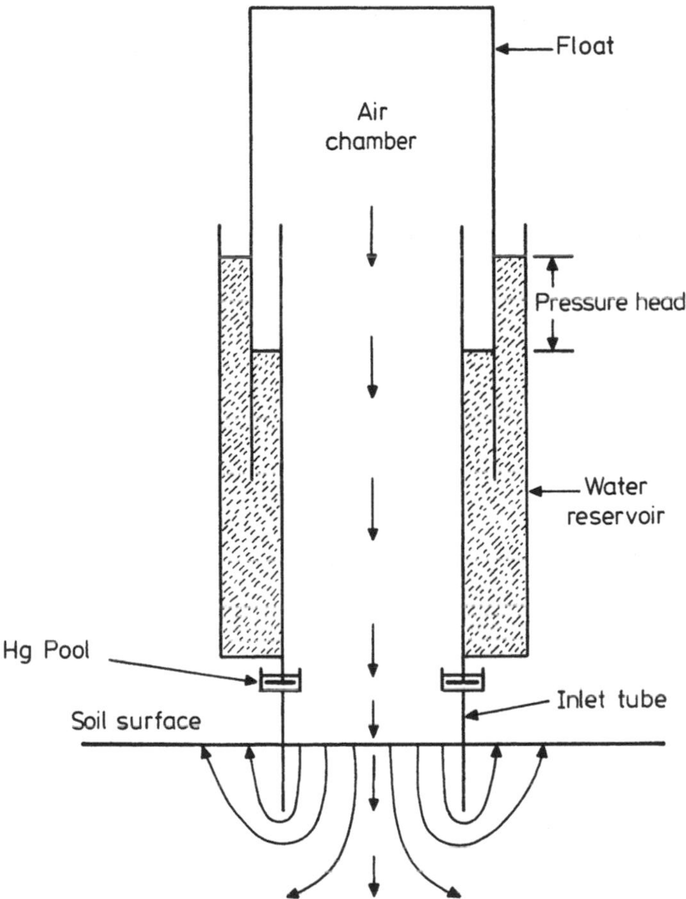

Fig. 6 Schematic diagram of air permeameter with mercury seal to allow quick attachment and good seal between permeameter and soil insert. See Grover (1955) for construction details.

area $[L^2]$, L_s is the length of the sample [L], and η is the dynamic gas viscosity $[M\,L^{-1}\,T^{-1}]$ corrected for temperature.

Permeability in topsoil layers is likely to be anisotropic. Janse and Bolt (1960) measured air permeability on undisturbed cores using a Grover-type permeameter and found that "vertical" samples were about twice as permeable as "horizontal" samples, part of the effect being attributed to greater compression during horizontal sampling than during vertical sampling. Conversely, McCarthy and Brown (1992) found the horizontal air permeabilities were greater than the vertical, an effect they attributed to the alluvial origins of their soils.

Due to the increasing interest in soil structure, steady state methods involving soil cores have developed further. The method of Corey (1986) allows for control of matric potential and volumetric water content by hydraulic contact around the circumference of the soil sample. This method is only applicable to disturbed, repacked samples. However, Roseberg and McCoy (1990) modified the method to allow the use of intact cores 77 mm long and 70 mm diameter. Their equipment allowed simultaneous measurement of air permeability and the soil water potential in the 0 to −6 kPa matric potential range. This was accomplished by counteracting the effects of the gravitational potential gradient within the sample using controlled air pressure. This enabled air flow measurements at or near saturation, where only macropores conduct air, and assessment of macropore continuity.

Ball et al. (1981) devised a method for intact samples that requires the same two-chamber apparatus as that used for the measurement of diffusion (Fig. 2). This method applies a constant pressure difference and measures the resultant flow. A differential micromanometer sensitive to pressure differences as small as 0.01 Pa is connected across the two gas chambers, and compressed air from a bottle is fed via a regulator and a flow controller to one gas chamber. Exhaust air is piped from the other gas chamber into a soap-film bubble meter or suspended-ball flowmeter. To preserve laminar flow, the pressure differences applied are kept small (0.15–300 Pa), as are the resultant rates of flow (0.15–6 $cm^3 s^{-1}$). For each sample, flow is measured at two or more pressure differences. This permits investigators to check their proportionality, giving two or more permeabilities, using Eq. 12. An advantage of this technique is that air permeability can be measured immediately after a diffusion measurement in 2–3 min without disturbing the sample. In addition, flow and pressure differences are measured with high accuracy. Other similar methods, using constant flow and measuring the resultant pressure difference with a water manometer, were proposed by McCarthy and Brown (1992) and Grant and Groenevelt (1993).

The statistical distribution of air permeabilities from a given depth and treatment within replicated field experiments is generally nonnormal. The distributions are skewed and usually log-normal (Kirkham et al., 1958; Ball, 1982; Groenevelt et al., 1984). In such cases either nonparametric tests should be applied to reduce the data and get an indicator of statistical degree of spread, such as statistical rank analysis (Kirkham et al., 1958) or the Mann–Whitney U Test (Groenevelt et al., 1984), or parametric tests should be applied to log-transformed data (Ball, 1982).

E. Soil–Atmosphere Trace Gas Exchange

Much attention has been given in recent years to the measurement of fluxes of trace gases between soils and the atmosphere, particularly the greenhouse gases CO_2, CH_4, and N_2O, because of their role in global climate change. The principal methods used are the *enclosure* (or *chamber*) methods outlined below, and *micrometeorological* methods, which are beyond the scope of this chapter but are described by Fowler and Duyzer (1989) and Lenschow (1995).

Enclosure methods normally involve covering an area of soil surface (<0.1–1 m^2) with a chamber, usually consisting of a plastic or metal cylinder or bottomless box, 20 cm to 1 m high (Fig. 7). The height should exceed that of any herbaceous vegetation cover. The bottom of the chamber is normally inserted a few cm into the soil to make a seal, and the top is covered by a lid sealed with a rubber gasket (Fig. 7a) or by a water-filled channel (Fig. 7b). A variant on this arrangement is to insert a "collar" into the soil so that it protrudes a few cm above the surface, and then seal the main chamber onto the collar when making a measurement (Fig. 7c).

There are two main modes of operation: "dynamic" or "open," and "static" or "closed." In the former mode, a steady stream of air is pumped through the chamber, and the gas of interest emitted from the soil is measured directly in the air stream (or sometimes adsorbed on a suitable trapping material for subsequent release and analysis) (e.g., Christensen, 1983; Skiba et al., 1992; Fang and Moncrieff, 1996). The latter mode simply involves closing the chamber with a gas-tight lid (or sealing a one-piece chamber onto a collar), typically for periods of 20–60 min, and taking gas samples at intervals for analysis (Hutchinson and Mosier, 1981), or circulating the chamber air through a nondestructive infrared gas analyzer (IRGA) (Norman et al., 1992).

When measuring gas emissions with static chambers, it is often recommended to measure the gas concentration three or more times during closure to check that the concentration increases linearly with time (e.g., IAEA, 1992). This provides a check against leaks, or against emissions increasing during the closure period. Closure times should be kept as short as is consistent with analytical sensitivity, to minimize such effects. The argument has been made that increases are also nonlinear because of the diminishing concentration gradient between the source in the soil and the chamber headspace (e.g., Healy et al., 1996). The model on which this is based is invalid, as it assumes that the concentration of the target gas in the soil remains constant, whereas if the gas is produced at a constant rate the soil concentration (which is the result of the balance between production and escape) increases as the chamber concentration increases (Conen and Smith, 2000); there is other direct evidence of linear concentration increases in closed chambers, e.g., for ^{222}Rn (Dörr and Münnich, 1990) and N_2O (Matthias et al., 1980).

It has also been argued (e.g., Hutchinson and Mosier, 1981) that to minimize flux measurement errors, "closed" chambers actually require an open vent

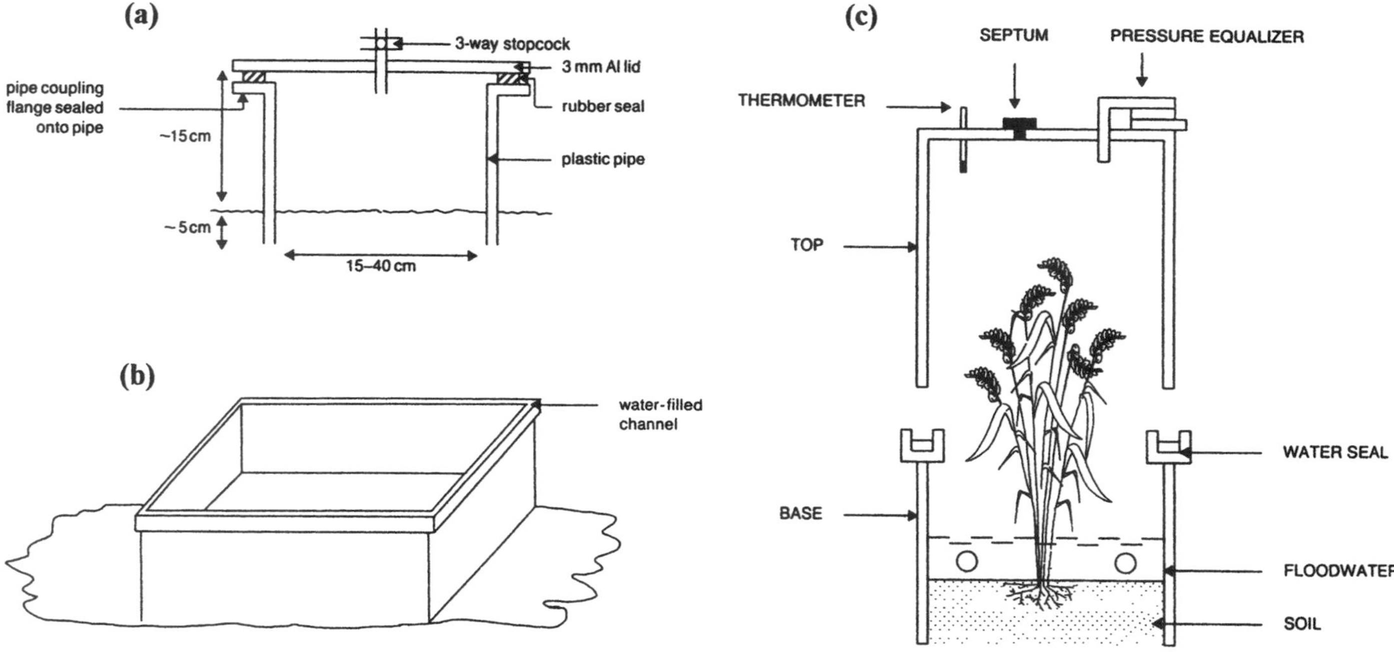

Fig. 7 Different versions of closed flux chambers used for measurement of gas fluxes between soil surfaces and the atmosphere: (a) cylindrical chamber with lid sealed with rubber gasket; (b) square chamber with water-filled channel in which flange on lid is immersed to make gas seal; (c) removable chamber located on permanent base (or "collar") in flooded rice field, also using water seal. (From IAEA, 1992.)

(typically 10 cm of 4–5 mm i.d. tubing) to equilibrate internal air pressure with fluctuating ambient conditions without causing significant loss of gas by mass flow. However, Conen and Smith (1998) have shown that in windy conditions there can be a substantial flux of air out of the chamber through such a vent, brought about by a reduction in pressure at its outer end: the Venturi effect. This can cause larger errors than those arising from use of a fully sealed chamber, as a result of increased emission by mass flow of air from the soil, where the trace gas concentration may be orders of magnitude higher than in the air above the surface. This effect can be reduced by using a tube venting close to the soil surface, where wind speeds are low (Norman et al., 1992). The potential errors resulting from chamber depressurization have also been recognized in relation to dynamic chambers, where blowing the air through the chamber by a fan placed upstream is preferred to drawing it through by a fan downstream (Kanemasu et al., 1974). A new open-top design for a dynamic chamber has been described that minimizes the effect of pressure differences on CO_2 efflux from the soil (Fig. 8).

The chamber method can also be used to measure radon (^{222}Rn) fluxes (Ussler et al., 1994). When such measurements are combined with measurements of ^{222}Rn concentrations at different depths in the profile (taking gas samples via sampling tubes), the diffusivity of the soil can be calculated from Fick's law; and if the concentration profile of another gas (CO_2 or CH_4) is also measured, the diffusivity value may then be applied to this gas also, and its flux calculated (Dörr and Münnich, 1990).

IV. INDIRECT AND MODELING TECHNIQUES

A. Gas Diffusion

Gas diffusivity can be related to air-filled porosity either empirically or by modeling. Relative diffusivity, being dimensionless, is used; the majority of the relationships can be grouped under three generalized forms. The first is

$$\frac{D_s}{D_0} = a(\epsilon_A - b) \tag{13}$$

This is a straight line corresponding to a slope of unity, i.e., $a = 1$, for straight tubes aligned with the concentration gradient. If the tubes twist along the direction of the concentration gradient (i.e., are tortuous), then $a < 1$. A value of 0.66 for a was suggested for soils by Penman (1940). Positive values of b are measures of air-filled porosity blocked against diffusing gas; a ranges between approximately 0.25 and 0.5, with most values near 0.3, and b ranges up to about 0.1 v/v for minimally disturbed samples (Gradwell, 1961; Ayres et al., 1972; Ball et al., 1988). Authors reporting the above relationship may have data covering a restricted range; see Troeh et al. (1982).

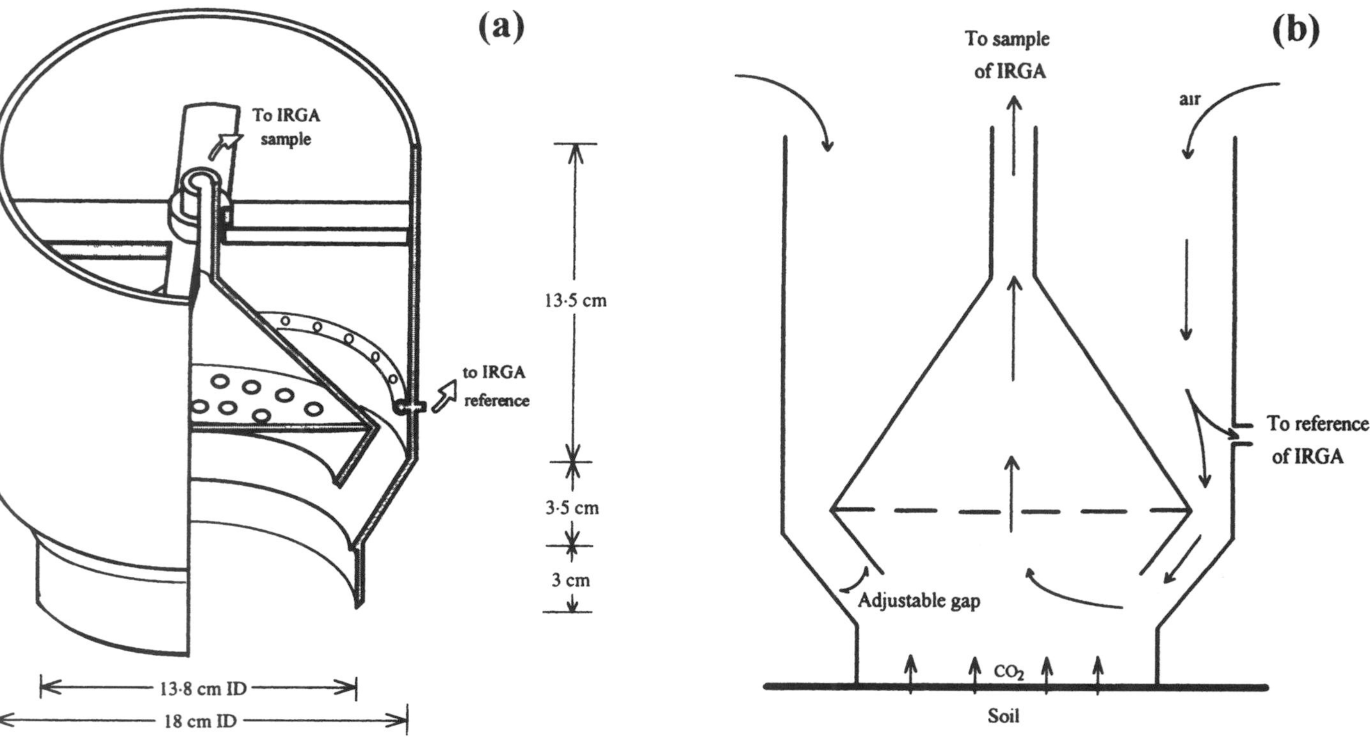

Fig. 8 Open-top dynamic chamber: (a) construction in cut-away section; (b) pattern of airflow through system. (From Fang and Moncrieff, 1998.)

The second relationship, to which many published results on minimally disturbed cores have been fitted (Currie, 1960), is curvilinear:

$$\frac{D_s}{D_0} = \gamma \epsilon \frac{\mu}{A} \tag{14}$$

where $\gamma \leq 1$ and $\mu \geq 1$ are taken as measures of pore shape in dry materials (Currie, 1960). In wet materials, γ and μ are interpreted as measures of pore continuity. μ has also been considered to equal the effective area for flow (Millington and Quirk, 1960). This relationship is similar to that predicted from a model of pore space comprising a series–parallel arrangement of pores arising from random apposition of planes within the porous material.

Millington and Quirk (1960) assumed greater pore interconnection (smaller μ value) and proposed the equation

$$\frac{D_s}{D_0} = \frac{\epsilon_A^{10/3}}{\epsilon_T^2} \tag{15}$$

where ϵ_T is the total porosity.

Many authors express their data using Eq. 12; Troeh et al. (1982) reported that γ ranged from 1 to 5 and μ from 1.6 to 3.4. They proposed a combination of Eqs. 11 and 12 that allows for both blocked porosity and a power relationship between D_s/D_0 and ϵ_A:

$$\frac{D_s}{D_0} = \frac{(\epsilon_A - c^d)}{1 - c} \tag{16}$$

where c has a similar physical significance to b of Eq. 11 and is a measure of the blocked air-filled porosity, and the exponent d controls the degree of curvature of the relationship.

A summary of the relationship between D_s/D_0 and ϵ_A for minimally disturbed samples is given in Table 2. Most of the data were collected in order to describe tillage or compaction treatments. The relationships were mostly curvilinear and were similar whether ϵ_A changed with field variation or with soil water potential on individual samples. More of these relationships correspond to Eq. 15 rather than to Eq. 16, where greater pore interconnection is assumed. Curved relationships were also reported by De Jong et al. (1983), Schjønning (1985b), and Boone et al. (1986). Approximate relationships derived from diffusion measurements in the field were $D_s/D_0 = \epsilon_A{}^{1.7}$ (Raney, 1949) and $D_s/D_0 = 0.37\,\epsilon_A$ (Jellick and Schnabel, 1986).

Jin and Jury (1996) evaluated these relationships and found that, although the Troeh model fitted experimental data for disturbed and intact samples when both model parameters were varied simultaneously, no obvious correlation was found between soil properties and the parameters. However, in disturbed soils of a range of textures the Millington–Quirk relationship (Eq. 15) fitted well.

The Millington–Quirk model was adapted by Moldrup et al. (1996) for predicting D_s/D_0 from the Campbell (1974) soil water retention model when no soil water release characteristic data are available. Their version, suitable for intact samples, is

$$\frac{D_s}{D_0} = \left(\frac{\epsilon_A}{\epsilon_T}\right)^{(1.5-3/b)} \epsilon_T^{1.33} \tag{17}$$

where b is estimated from soil texture and bulk density:

$$b = (0.303 - 0.093 \ln BD) - 0.0565(\ln CL + 0.00003\,FS^2)^{-1} \tag{18}$$

and where CL is clay content (% < 0.002 mm), FS is fine sand content (% between 0.02 and 0.2 mm), and BD is bulk density (g cm^{-3}). If water release characteristic data are available, then b is the Campbell (1974) water retention model parameter equal to the slope of the soil water retention curve in a $\log(\theta) - \log(\psi)$ coordinate system. θ is the volumetric soil water content [$L^3\ L^{-3}$], and ψ is the soil water potential [L].

If water release characteristic data and a measured gas diffusivity at high soil air content are available, Moldrup et al. (1996) provide empirical equations analogous to capillary tube models for unsaturated hydraulic conductivity to predict gas diffusivity as a function of soil air content. The measured gas diffusivity is used as a matching value.

Network modeling can be used successfully to simulate the relationship between diffusion coefficient and air-filled porosity, including hysteresis, but use of measured pore size distributions was unsuitable for calibration due to the limitation of the assumptions of the capillary model (Steele and Nieber, 1994).

B. Air Permeability

Air permeability and air-filled porosity at and near the soil surface can be determined acoustically in situ (Moore and Attenborough, 1992). The techniques use sound propagated near to and through the soil surface. The measured difference in acoustic spectra received by two vertically separated microphones above the ground surface and by probe microphones beneath the surface were matched theoretically to deduce the porosity of the continuous air-filled pores and an effective air permeability. These techniques were successfully used to monitor surface sealing and near-surface layering.

The relationship between air-filled porosity and air permeability is poorer than with relative gas diffusivity because of the greater dependence of air permeability on pore size. Ball et al. (1988) proposed a log–log relationship using an empirical form of the Kozeny–Carman relationship for air permeability:

$$\log K = \log M + N \log \epsilon_A \tag{19}$$

where M and N are empirical constants determined by regression and related to pore continuity. This relationship varied between soil types and tillage treatments according to macropore continuity, a feature also observed by Roseberg and McCoy (1990). McCarthy and Brown (1992) found approximately linear relationships between air permeability and air-filled porosity for soils containing less than 20% clay. However, the linear equation was unique for each soil and was related to soil structure and texture.

The relationship between diffusivity and air permeability is rather better than that of K and ϵ_A. Washington et al. (1994) found a linear relationship between log oxygen diffusivity and log air permeability for eight different soil types. They claim that the relationship is useful for allowing estimation of the bulk diffusion coefficient for any gas from the more easily measured air permeability. Moldrup et al. (1998) proposed a predictive gas permeability model based on that of Ball et al. (1981) for flow of gas in a porous medium consisting of unconnected tortuous tubes of uniform radius r_t, which they combined with an equation for relative diffusivity to yield an expression for predicting gas permeability as a function of air-filled porosity. This expression (with modified notation) is

$$K = 0.66\left(\frac{r_t^2\epsilon_A^3}{8\epsilon_T^2}\right) \tag{20}$$

where a value of $r_t = 71\ \mu m$ for sandy soils was suggested. Moldrup et al. (1998) also presented an equation for air permeability relative to reference-point measurements at given ϵ_A. Poulsen et al. (1998) used this in combination with Eq. 20 to yield an improved overall relationship for determining gas permeability as a function of ϵ_T, ϵ_A, and soil type. Soil type is determined by values of r_t, β (a pore interconnection factor), and b, the water retention parameter from Campbell (1974). The equation (with modified notation) is

$$K_A = \left(0.45\frac{r_t^2(\epsilon_A^*)^3}{8\epsilon_T^2}\right)^\beta\left(\frac{\epsilon_A}{\epsilon_A^*}\right)^{(1+0.25b)} \tag{21}$$

where ϵ_A^* corresponds to the value at -10 kPa. For a range of sandy soils, β and r_t values of 1.45 and 50 μm, respectively, were used, and b ranged from 3 to 46.

V. APPLICATIONS

A. Investigation of Soil and Root Aeration

The relationships discussed in Sec. IV.A between relative diffusivity D_s/D_0 and air-filled porosity ϵ_A have been widely used to estimate (1) the minimum porosity at which aeration limits plant growth, (2) the continuity or tortuosity of the air-filled pore system, (3) the content of air-filled pores blocked to entry of diffusing gas, (4) the extent of soil aggregation, and (5) gas exchange dynamics.

The frequency of occurrence of blocked air-filled pores is important in describing the aeration status of a soil layer. Grable and Siemer (1968) and the present authors have detected air-filled porosities of up to 10% when gas diffusion through samples was zero. In contrast, more recent data indicate significant D_s/D_0 in many soils at air-filled porosities of less than 10%, particularly in well-structured soils (Bakker and Hidding, 1970; Ball, 1982; Schjønning, 1985b; Ball et al., 1988). Lower limits of aeration status might be better specified as relative diffusivities (Stepniewski, 1981, suggested D_s/D_0 of 0.005–0.01) or as oxygen diffusion rates (Callebaut et al., 1982). However, such limits depend on soil structure, water distribution, soil respiration rate, and crop. Glinski and Stepniewski (1985) presented a detailed discussion of the influence of these factors on soil aeration.

Grable and Siemer (1968) demonstrated that plant growth was controlled by soil aeration only over a narrow range of relative diffusivities (<0.005–0.009). These diffusivities corresponded to conditions wetter than a matric potential of −2 kPa (the air-entry value). In conditions drier than a potential of approximately −100 kPa, soil strength limited root growth more than soil aeration. They suggested that the gas diffusivity limits corresponded to ϵ_A limits of less than 0.10–0.12 v/v. Similarly, Boone et al. (1986) defined lower and upper critical aeration limits (LCAL, UCAL) for root growth calculated from relative diffusivities, respiration rates, and water release characteristics. LCAL ranged between −2 and −8 kPa, corresponding to ϵ_A's of 0.05–0.08 v/v. UCAL ranged between −6 and −20 kPa, corresponding to ϵ_A's of 0.15–0.20 v/v. D_s/D_0 and ϵ_A were related differently according to soil type.

1. Tillage and Compaction Effects

Soil disturbance by tillage generally increases air-filled porosity and the rate of gas diffusion in soil (Bruce and Webber, 1953), while compaction decreases them (Grable, 1971). However, Grabert (1968) showed that tillage may stimulate respiration, thus decreasing oxygen concentration. The value of measurements of gas movement in assessing tillage and compaction effects lies in their sensitivity to the continuity of air-filled pores. These effects may not be detected from measurement of air-filled porosity or bulk density. Relationships between D_s/D_0 and ϵ_A are extensively used.

The importance of maintaining a nonpuddled and unsmeared soil surface has been shown from measurements of diffusion (Taylor, 1949; Domby and Kohnke, 1956), which reveal that wet, puddled, or crusted surfaces, where ϵ_A is less than 0.2, can have a D_s/D_0 of only one-tenth that in nonpuddled soils (Bakker and Hidding, 1970), though dry crusts may actually improve gas exchange (Domby and Kohnke, 1956). Air permeability measurements also are sensitive indicators of surface sealing (Grover, 1955), crusting (Evans and Kirkham, 1949), or capping (Green and Fordham, 1975), though Evans and Kirkham (1949) stressed that care must be taken to avoid cracks or wormholes, since these can give very high permeabilities.

Diffusion coefficients have been compared in zero-tilled and in plowed soils, and it has been shown that at similar air-filled porosities the values can be higher in zero-tilled conditions (Boone et al., 1976; Richter and Grossgebauer, 1978). The depth of sampling is important; Douglas and Goss (1987) found that at -1 kPa matric potential, D_s/D_0 and K_A were greater in plowed than in zero-tilled topsoil, but both were greater at the boundaries between topsoil and subsoil and in the upper subsoil after zero-tillage (20–35 cm depth, Fig. 9). The latter effect was attributed to compaction below plow depth and to the disruption of the continuity of channel-type macropores.

Air permeabilities have been found to be greater in general in plowed than in zero-tilled topsoil but similar or less after plowing than in zero-tilled soil below plowing depth (Ball, 1982; Janse and Bolt, 1960; Douglas and Goss, 1987).

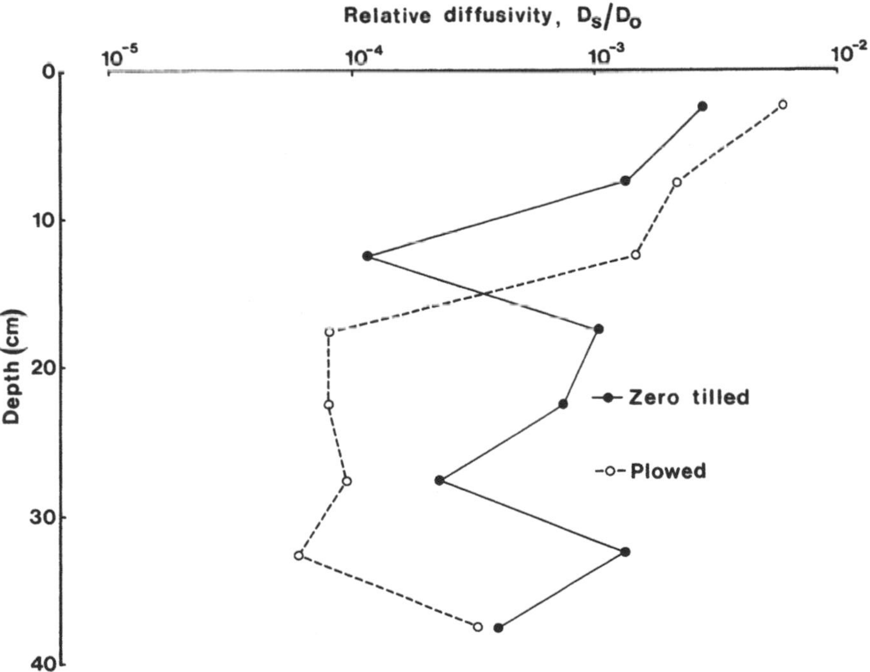

Fig. 9 Relative diffusivity at -1 kPa water potential in an English clay loam, after 6 years of contrasting cultivation treatment. (Reproduced with slight modification from Douglas and Goss, 1987.)

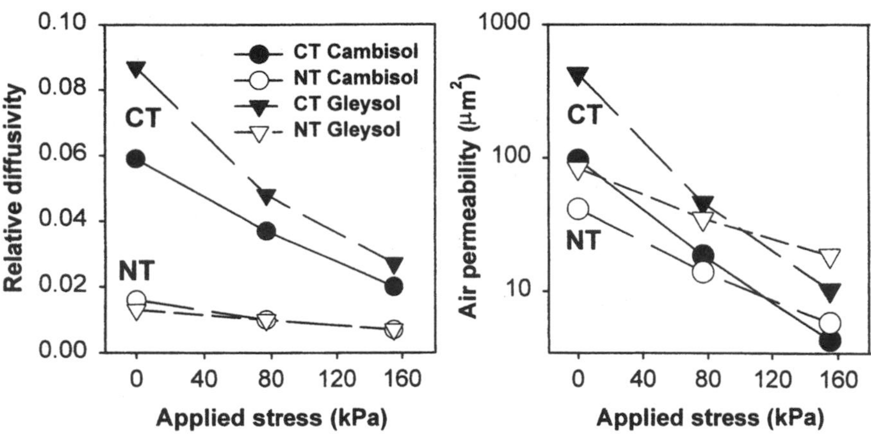

Fig. 10 Effect of uniaxial stress on relative gas diffusivity and air permeability on the same intact cores. CT: conventional tillage; NT: no-tillage.

The progressive increase in weight and frequency of use of tractors and wheeled implements has brought a new demand for techniques to reveal the influence of compaction on soil conditions for cropping. Gas movement measurements are suitable to define not only the influence of traffic on seedbed aeration but also the extent of recovery of soil from compaction (Campbell et al., 1986). Measurements of relative diffusivity and air permeability on intact cores in response to uniaxial stress can reveal differences in bulk soil compactibility attributable to tillage treatment or organic matter content (Fig. 10) (Ball and Robertson, 1994).

In the study of soil deformation, air permeability was shown by Kirby (1991) to be a useful indicator of pore continuity, and the results can be interpreted in terms of critical state soil mechanics to construct lines of critical state for permeability.

B. Exchange of Greenhouse Gases

The question of whether terrestrial ecosystems are net sinks or net sources of atmospheric CO_2 is an important one for predicting future climate change. Chamber methods have been widely used to measure soil CO_2 emissions due to respiration, in the course of quantifying carbon balances. The design shown in Fig. 8 has been used for this purpose by Fang and Moncrieff (1998), in a slash pine ecosystem in Florida. Other typical systems studied with various chamber designs include grassland (e.g., Norman et al., 1992), a maize crop (Desjardins, 1985), peatland (Shurpali et al., 1995) and an oak forest (Hanson et al., 1993).

Closed chamber methods have been the method of choice for measuring methane fluxes, and estimating the net balance between emission and soil microbial oxidation, in a wide range of ecosystems (Smith et al., 1999). These studies have often been accompanied by studies of the effect on the balance of soil water content and/or bulk density, which affect gas diffusivity in the soil. Examples include temperate forest and agricultural land (Lessard et al., 1994; Ball et al., 1997b); humid tropical forest and pasture soils (Keller and Reiners, 1994); a temperate wetland (Melloh and Crill, 1996); and a tropical soil permeated by termite galleries (MacDonald et al., 1998). Fluxes of CH_4 from flooded (paddy) rice are also widely determined by closed chamber methods, in which both the CH_4 transported by diffusion through the plant stems, and that emitted from the water surface after transport from the soil by ebullition and diffusion, is collected during the closure period (e.g., Butterbach-Bahl et al., 1997).

Chamber-based measurements of CH_4 emissions through landfill cover soils have been made in several countries. Bogner et al. (1995) showed that at one site with a dry coarse-textured soil cover the emission was reduced by two orders of magnitude after installation of a gas recovery system. Nozhevnikova et al. (1993) and Czepiel et al. (1996) measured fluxes on a regular grid pattern and showed large spatial variations over the landfill surface.

Soils are the major source of atmospheric N_2O. Emissions are governed by a complex combination of soil physical, biological, and chemical properties. Several studies (e.g., Keller and Reiners, 1994; Velthof and Oenema, 1995; Clayton et al., 1997) have examined the relationship between fluxes and soil water-filled porosity, which by virtue of controlling aeration determines the development of anaerobic zones and therefore of N_2O production by the denitrification pathway. Ball et al. (1997a) found that N_2O emission from N-fertilized grassland related better to air permeability than to gas diffusivity, indicating the importance of soil structure in regulating production, consumption, and transport of this gas. This study indicated the importance of the surface soil layer. Gas transport conditions in this layer are also important in regulating the emission of nitric oxide, NO, and carbon monoxide, CO (Sanhueza et al., 1994).

C. Radon Emissions

The release of the alpha-radioactive gas radon (^{222}Rn) from soils is a major concern for health and safety in many countries. In the U.K., for example, the accumulation of this nuclide (produced naturally by the decay of uranium in underlying strata) in domestic buildings, following emission from the underlying soils, is regarded as the most significant hazard associated with environmental radioactivity, even though it is concentrated in a few areas. Because of the problem, considerable effort has been devoted to predicting the extent of ^{222}Rn diffusion. Nielson et al. (1984) produced a mathematical model for calculating radon diffusion coefficients from soil water content and pore size distribution. The model considers pores to be composed of serial combinations of the size increments from the measured pore size distribution. The model uses diffusion in these pores, both air- and water-filled, and includes radon solubility in water. Riley et al. (1996) concluded from modeling work that wind-generated ground-surface pressures play a significant role in increasing radon entry rates into houses. Another model by Mowris and Fisk (1988) predicted that the flow of air induced by exhaust ventilation fans used in some houses could substantially increase indoor ^{222}Rn concentrations when soil permeabilities were between 10^{-12} and 10^{-10} m^2 and the ^{222}Rn concentrations in soil air were above average.

Rogers and Nielson (1991) and Washington et al. (1994) have related observed ^{222}Rn profiles in the soil to models relating diffusivity and air-filled porosity, and have indicated the importance of air permeability in governing the rate of escape of radon from soil. Rogers and Nielson (1991) developed simple correlations for predicting the ^{222}Rn diffusion coefficient and air permeability based on degree of water saturation, total porosity, and mean particle diameter.

D. Soil Fumigants

Fumigants are frequently used to control nematodes and fungi before planting horticultural crops. They are applied as liquids or gases via injection chisels drawn through the soil. Transport of the gas away from the injection point is primarily by diffusion, and low soil moisture contents result in rapid evaporative loss (Smelt et al., 1974). However, prediction of the distribution of fumigant in the soil needs to take into account not only the factors that influence diffusion but also adsorption and dissolution (reversible processes) and decomposition and chemical bonding (Hemwall, 1960; Rolston et al., 1982).

Methyl bromide has been one of the most widely used fumigants, but it is now implicated in stratospheric ozone depletion and is due to be phased out as an agrochemical in the next few years. Measurements of emissions from agricultural fields include flux chamber-based studies (e.g., Yates et al., 1996).

VI. SUMMARY AND CONCLUSIONS

Many methods for measurement of diffusion, both laboratory and field, have been developed in the past twenty years. Diffusion rates are expressed commonly in the dimensionless form of relative diffusivity.

Laboratory techniques increasingly use minimally disturbed samples and inert tracer gases, which may be radioactive. Such gases are not consumed or stored in the sample, as is possible when CO_2 or O_2 are used to trace diffusion.

Methods where the sample end faces are enclosed by two chambers offer the advantage of precise control of boundary conditions, particularly pressure and temperature, and the possibility of additional measurement of effective porosity. Methods for measurement of mass flow of air, expressed commonly as air permeability, have been less extensively developed than methods for measurement of diffusion. The same techniques are generally applicable in the field and in the laboratory. Steady-state laboratory techniques offer the advantage of preserving laminar flow of gases by application of constant low flow rates in controlled conditions.

All techniques have benefited from technological developments such as improved gas chromatographs, photomultipliers, electronic differential manometers, and digital flowmeters. Data may be collected and processed by microprocessors that can scan detectors rapidly. Finite element analysis is increasingly used to process data. Portable gas detectors are proving valuable for field measurements of diffusion.

Air permeability measurements are used mainly to determine soil structure, whereas gas diffusion measurements may be used to determine both soil structure and aeration. Field diffusion measurements reveal aeration status more directly than laboratory measurements, particularly where soil layers are uniform. However, boundary conditions are hard to specify. Air permeability and air-filled porosity are not related well because the former is very sensitive to large pores and cracks.

The application of gas movement measurements to tillage and compaction studies allows assessment both of short term effects, such as on seedbed aeration, and of long term effects such as buildup of soil structures under zero tillage or during recovery from compaction. Gas movement measurements also help to quantify the improvement of aeration by drainage which results directly from lowering of the water table and indirectly from amelioration of soil structure.

Chamber methods have been widely used to measure fluxes of greenhouse gases and methyl bromide between soils and the atmosphere, and the emission of radon into dwellings from the underlying soil. Such methods are often the only ones that are readily available, but they need to be applied with care, to avoid inducing effects such as changes in gas pressure, which can greatly affect the flux measurements.

REFERENCES

Arah, J. R. M., and B. C. Ball. 1994. A functional model of soil porosity used to interpret measurements of gas diffusion. *Eur. J. Soil Sci.* 45:135–144.

Armstrong, W., and E. J. Wright. 1976. A polarographic assembly for multiple sampling of soil oxygen flux in the field. *J. Appl. Ecol.* 13:849–856.

Ayres, K. W., R. G. Button, and E. de Jong. 1972. Soil morphology and soil physical properties. I. Soil aeration. *Can. J. Soil Sci.* 52:311–321.

Bakker, J. W., and A. P. Hidding. 1970. The influence of soil structure and air content on gas diffusion in soils. *Neth. J. Agric. Sci.* 18:37–48.

Ball, B. C. 1982. Pore characteristics of soils from two cultivation experiments as shown by gas diffusivities and permeabilities and air-filled porosities. *J. Soil Sci.* 32: 483–498.

Ball, B. C., and R. Hunter. 1988. The determination of water release characteristics of soil cores at low suctions. *Geoderma* 43:195–212.

Ball, B. C., and E. A. G. Robertson. 1994. Effects of uniaxial compaction on aeration and structure of ploughed or direct drilled soils. *Soil Till. Res.* 31:135–148.

Ball, B. C., K. E. Dobbie, J. P. Parker, and K. A. Smith. 1997b. The influence of gas transport and porosity on methane oxidation in soils. *J. Geophys. Res.* 102:23,301–23,308.

Ball, B. C., C. A. Glasbey, and E. A. G. Robertson. 1994. Measurement of soil gas diffusivity in situ. *Eur. J. Soil Sci.* 45:3–13.

Ball, B. C., W. Harris, and J. R. Burford. 1981. A laboratory method to measure gas diffusion in soil and other porous materials. *J. Soil Sci.* 32:323–333.

Ball, B. C., G. W. Horgan, H. Clayton, and J. P. Parker. 1997a. Spatial variability of nitrous oxide fluxes and controlling soil and topographic properties. *J. Environ. Qual.* 26: 1399–1409.

Ball, B. C., M. F. O'Sullivan, and R. Hunter. 1988. Gas diffusion, fluid flow and derived pore continuity indices in relation to vehicle traffic and tillage. *J. Soil Sci.* 39: 327–339.

Batterman, S., I. Padmanabham, and P. Milne. 1996. Effective gas phase diffusion coefficients in soils at varying water content measured using a one-flow sorbent-based technique. *Env. Sci. Technol.* 30:770–778.

Blackwell, P. S. 1983. Measurements of aeration in waterlogged soils: Some improvements of techniques and their application to experiments using lysimeters. *J. Soil Sci.* 34: 271–285.

Bogner, J., K. Spokas, E. Burton, R. Sweeney, and V. Corona. 1995. Landfills as atmospheric methane sources and sinks. *Chemosphere* 31:4119–4130.

Boone, F. R., S. Slager, R. Miedema, and R. Eleveld. 1976. Some influences of zero-tillage on the structure and stability of a fine-textured river levee soil. *Neth. J. Agric. Sci.* 24:105–119.

Boone, F. R., H. M. G. van der Werf, B. Kroesbergen, B. A. ten Hag, and A. Boers. 1986. The effect of compaction of the arable layer in sandy soils on the growth of maize for silage. I. Critical matric water potentials in relation to soil aeration and mechanical impedance. *Neth. J. Agric. Sci.* 34:155–171.

Bouma, J. 1983. Use of soil survey data to select measurement techniques for hydraulic conductivity. *Agric. Water Manage.* 6:177–190.

Bowen, H. D. 1985. Air permeability measurement. Proc. Int. Conf. Soil Dynamics, Auburn, Ala. Vol. 3, pp. 481–489.

Bruce, R. R., and L. R. Webber. 1953. The use of a diffusion chamber as a measure of the rate of oxygen supplied by a soil. *Can. J. Agric.* 33:430–436.

Bruckler, L., B. C. Ball, and P. Renault. 1989. Gaseous diffusion coefficient and porosity effective for diffusion using krypton-85 tracer and a finite element calculation method. *Soil Sci.* 147: 1–10.

Butterbach-Bahl, K., H. Papen, and H. Rennenberg. 1997. Impact of gas transport through rice cultivars on methane emission from rice paddy fields. *Plant Cell Environ.* 20: 1175–1183.

Callebaut, F., D. Gabriels, W. Minjauw, and M. De Boodt. 1982. Redox potential, oxygen diffusion rate, and soil gas composition in relation to water table level in two soils. *Soil Sci.* 134: 149–156.

Campbell, G. S. 1974. A simple method for determining unsaturated conductivity from moisture retention data. *Soil Sci.* 117:311–314.

Campbell, D. J., J. W. Dickson, B. C. Ball, and R. Hunter. 1986. Controlled seedbed traffic after ploughing or direct drilling under winter barley in Scotland. *Soil Till. Res.* 8: 3–28.

Carter, M. R., and B. C. Ball. 1993. Soil porosity. In: *Soil Sampling and Methods of Analysis* (M. R. Carter, ed.). Boca Raton, FL: Lewis, pp. 581–588.

Christensen, S. 1983. Nitrous oxide emission from a soil under permanent grass: Seasonal and diurnal fluctuations as influenced by manuring and fertilisation. *Soil Biol. Biochem.* 15:531–536.

Clayton, H., I. P. McTaggart, J. Parker, L. Swan, and K. A. Smith. 1997. Nitrous oxide emissions from fertilised grassland: A two-year study of the effects of N fertiliser form and environmental conditions. *Biol. Fertil. Soils* 25: 252–260.

Conen, F., and K. A. Smith. 1998. A re-examination of closed flux chamber methods for the measurement of trace gas emissions from soils to the atmosphere. *Eur. J. Soil Sci.* 49:701–707.

Conen, F., and K. A. Smith. 2000. An explanation of linear increases in gas concentration under closed chambers used to measure gas exchange between soil and the atmosphere. *Eur. J. Soil Sci.* 51: 111–117.

Corey, A. T. 1986. Air permeability. In: *Methods of Soil Analysis*, Part 1 (A. Klute, ed.). Madison, WI: Am. Soc. Agron., pp. 1121–1136.

Currie, J. A. 1960. Gaseous diffusion in porous media. Part 2. Dry granular materials. *Brit. J. Appl. Phys.* 11:318–324.

Currie, J. A. 1961. Gaseous diffusion in the aeration of aggregated soils. *Soil Sci.* 92: 40–45.

Czepiel, P. M., B. Mosher, R. C. Harriss, J. H. Shorter, J. B. McManus, C. E. Kolb, E. Allwine, and B. K. Lamb. 1996. Landfill methane emissions measured by enclosure and atmospheric tracer methods. *J. Geophys. Res.* 101: 16,711–16,719.

de Jong, E., J. T. Douglas, and M. J. Goss. 1983. Gaseous diffusion in shrinking soils. *Soil Sci.* 136: 10–18.

Denny, M. W. 1993. *Air and Water*. Princeton, NJ: Princeton Univ. Press.

Desjardins, R. L. 1985. Carbon dioxide budget of maize. *Agric. For. Meteor.* 36:29–41.

Domby, C. W., and H. Kohnke. 1956. The influence of soil crusts on gaseous diffusion. *Soil Sci. Soc. Am. Proc.* 20: 1–5.

Dörr, H., and K. O. Münnich. 1990. ^{222}Rn flux and soil air concentration profiles in west Germany. Soil ^{222}Rn as tracer for gas transport in the unsaturated soil zone. *Tellus,* 42B: 20–28.

Douglas, J. T., and M. J. Goss. 1987. Modification of pore space by tillage in two stagnogley soils with contrasting management histories. *Soil Till. Res.* 10:303–317.

Evans, D. D. 1965. Gas movement. In: *Methods of Soil Analysis* (C. A. Black et al., eds.). Madison, WI: Am. Soc. Agron., pp. 319–330.

Evans, D. D., and D. Kirkham. 1949. Measurement of the air permeability of soil in situ. *Soil Sci. Soc. Am. Proc.* 14:65–73.

Fang C., and J. B. Moncrieff. 1996. An improved dynamic chamber technique for measuring CO_2 efflux from the surface of soil. *Funct. Ecol.* 10:297–305.

Fang C., and J. B. Moncrieff. 1998. An open-top chamber for measuring soil respiration and the influence of pressure difference on CO_2 efflux measurement. *Funct. Ecol.* 12:319–325.

Flühler, J. 1972. *Oxygen Diffusion in Soils*. Zurich: Beer and Co.

Fowler, D., and J. Duyzer. 1989. Micrometeorological techniques for the measurement of trace gas exchange. In: *Exchange of Trace Gases Between Terrestrial Ecosystems and the Atmosphere* (M. O. Andreae and D. S. Schimel, eds.). Chichester, UK: John Wiley, pp. 189–207.

Garberi, K., R. G. Sextro, A. L. Robinson, J. D. Wooley, J. A. Owens, and W. W. Nazaroff. 1996. Scale dependence of soil permeability to air measurement method and field investigation. *Water Resources Res.* 32:547–560.

Glinski, J., and W. Stepniewski. 1985. *Soil Aeration and its Role for Plants.* Boca Raton, FL: CRC Press.

Grabert, D. 1968. Measurements of soil respiration in model experiment on deepening the mould. *Albrecht-Thaer-Arch., Muncheberg* 12:681–689.

Grable, A. R. 1966. Soil aeration and plant growth. *Adv. Agron.* 18:58–106.

Grable, A. R. 1971. Effects of compaction on content and transmission of air in soils. In: *Compaction of Agricultural Soils.* St. Joseph, MI: Am. Soc. Agric. Eng., pp. 154–164.

Grable, A. R., and E. G. Siemer. 1968. Effects of bulk density, aggregate size and soil water suction on oxygen diffusion, redox potentials and elongation of corn roots. *Soil Sci. Soc. Am. Proc.* 32:180–186.

Gradwell, M. W. 1961. A laboratory study of the diffusion of oxygen through pasture topsoils. *N.Z. J. Sci.* 4:250–270.

Grant, C. D., and P. H. Groenevelt. 1993. Air permeability. In: *Soil Sampling and Methods of Analysis* (M. R. Carter, ed.). Boca Raton, FL: Lewis, pp. 645–650.

Green, R. D., and S. J. Fordham. 1975. A field method for determining air permeability in soil. In: *Soil Physical Conditions and Crop Production.* Tech. Bull. No. 29. London: HMSO, pp. 273–288.

Groenvelt, P. H., B. D. Kay, and C. D. Grant. 1984. Physical assessment of soil with respect to rooting potential. *Geoderma* 34:101–114.

Grover, B. L. 1955. Simplified air permeameters for soil in place. *Soil Sci. Soc. Am. Proc.* 19:414–418.

Hall, D. G. M., M. J. Reeve, A. J. Thomasson, and V. F. Wright. 1975. *Water Retention, Porosity and Density of Field Soils*. Tech. Monogr. No. 9. Soil Survey, Harpenden, U.K.

Hanson, P. J., S. D. Wullschleger, S. A. Bohlman, and D. E. Dodd. 1993. Seasonal and topographic patterns of forest floor CO_2 efflux from an upland oak forest. *Tree Physiol.* 13:1–15.

Healy, R. W., R. G. Striegl, T. F. Russell, G. L. Hutchinson, and G. P. Livingston. 1996. Numerical evaluation of static-chamber measurements of soil-atmosphere gas exchanges: Identification of physical processes. *Soil Sci. Soc. Am. J.* 60:740–747.

Hemwall, J. B. 1960. Theoretical considerations of several factors influencing the effectivity of soil fumigants under field conditions. *Soil Sci.* 90:157–168.

Henderson, R. E., and W. H. Patrick. 1982. Soil aeration and plant productivity. In: *CRC Handbook of Agricultural Productivity*, Vol. 1. Boca Raton, FL: CRC Press, pp. 51–69.

Hodgson, J. H., ed. 1976. *Soil Survey Field Handbook.* Tech. Monogr. No. 5. Soil Survey, Harpenden, UK.

Houghton, J. T., L. G. Meiro Filho, B. A. Callander, N. Harris, A. Kattenberg, and K. Maskell, eds. 1996. *Climate Change 1995—The Science of Climate Change.* Cambridge: Cambridge Univ. Press.

Hutchinson, G. L., and A. R. Mosier. 1981. Improved soil cover method for field measurement of nitrous oxide fluxes. *Soil Sci. Soc. Am. J.* 45:311–316.

IAEA (International Atomic Energy Agency). 1992. *Manual on Measurement of Methane and Nitrous Oxide Emissions from Agriculture.* IAEA-TECDOC-674. Vienna: IAEA.

Janse, A. R. P., and G. H. Bolt. 1960. The determination of the air permeability of soils. *Neth. J. Agric. Sci.* 8:124–131.

Jaynes, D. B., and A. S. Rogowski. 1983. Applicability of Fick's law to gas diffusion. *Soil Sci. Soc. Am. J.* 47:425–430.

Jellick, G. J., and R. R. Schnabel. 1986. Evaluation of a field method for determining the gas diffusion coefficient of soils. *Soil Sci. Soc. Am. J.* 50:18–23.

Jin, Y., and W. A. Jury. 1996. Characterizing the dependence of gas diffusion coefficient on soil properties. *Soil Sci. Soc. Am. J.* 60:66–71.

Kanemasu, E. T., W. L. Powers, and J. W. Sij. 1974. Field chamber measurements of CO_2 flux from soil surface. *Soil Sci.* 118:233–237.

Keller, M., and W. A. Reiners. 1994. Soil-atmosphere exchange of nitrous oxide, nitric oxide, and methane under secondary succession of pasture to forest in the Atlantic lowlands of Costa Rica. *Glob. Biogeochem. Cycl.* 8:399–409.

Kirby, J. M. 1991. The influence of soil deformations on the permeability to air. *J. Soil Sci.* 42:227–235.

Kirkham, D. 1946. Field method for determination of air permeability of soil in its undisturbed state. *Soil Sci. Soc. Am. Proc.* 11:93–99.

Kirkham, D., and W. L. Powers. 1972. *Advanced Soil Physics.* New York: Wiley-Interscience.

Kirkham, D., M. De Boodt, and L. De Leenheer. 1958. Air permeability at the field capacity as related to soil structure and yields. Proc. Int. Symp. on Soil Structure, Ghent, Belgium, pp. 377–391.

Lai, S. H., J. M. Tiedje, and A. E. Erickson. 1976. In situ measurement of gas diffusion coefficient in soils. *Soil Sci. Soc. Am. Proc.* 40:3–6.

Lenschow, D. H. 1995. Micrometeorological techniques for measuring biosphere-atmosphere trace gas exchange. In: *Biogenic Trace Gases: Measuring Emissions from Soil and Water* (P. A. Matson and R. C. Harris, eds.). Oxford: Blackwell, pp. 126–163.

Lessard, R., P. Rochette, E. Topp, E. Pattey, R. L. Desjardins, and G. Beaumont. 1994. Methane and carbon dioxide fluxes from poorly drained adjacent cultivated and forest sites. *Can. J. Soil Sci.* 74:139–136.

MacDonald, J. A., P. Eggleton, D. E. Bignell, F. Forzi, and D. Fowler. 1998. Methane emission by termites and oxidation by soils, across a forest disturbance gradient in the Mbalmayo Forest Reserve, Cameroon. *Glob. Change Biol.* 4:409–418.

Matthias, A. D., A. M. Blackmer, and J. M. Bremner. 1980. A simple chamber technique for field measurement of emissions of nitrous oxide from soils. *J. Environ. Qual.* 9: 251–256.

McCarthy, K. P., and K. W. Brown. 1992. Soil gas permeability as influenced by soil gas-filled porosity. *Soil Sci. Soc. Am. J.* 56:997–1003.

McIntyre, D. S. 1974. *Methods for Analysis of Irrigated Soils* (J. Loveday, ed.). Tech. Comm. 54. Commonw. Agric. Bur., Australia, pp. 12, 21.

McIntyre, D. S., and J. R. Philip. 1964. A field method for measurement of gaseous diffusion into soils. *Aust. J. Soil Res.* 2:133–145.

Melloh, R. A., and P. M. Crill. 1996. Winter methane dynamics in a temperate peatland. *Glob. Biogeochem. Cycl.* 10:247–254.

Millington, R. J., and J. P. Quirk. 1960. Transport in porous media. Trans. 7th Int. Congr. Soil Sci. Madison, WI, U.S.A. 1:97–106.

Moldrup, P., C. W. Kruse, D. E. Rolston, and T. Yamaguchi. 1996. Modelling diffusion and reaction in soils. III. Predicting gas diffusivity from the Campbell soil-water retention model. *Soil Sci.* 161:366–375.

Moldrup, P., T. G. Poulsen, P. Schjønning, T. Olesen, and T. Yamaguchi. 1998. Gas permeability in undisturbed soils: Measurements and predictive models. *Soil Sci.* 163: 180–189.

Moore, H. M., and K. Attenborough. 1992. Acoustic determination of air-filled porosity and relative air permeability of soils. *J. Soil Sci.* 43:211–228.

Mowris, R. J., and W. J. Fisk. 1988. Modeling the effects of exhaust ventilation on ^{222}Rn entry rates and indoor ^{222}Rn concentrations. *Health Phys.* 54:491–501.

Nielson, K. K., V. C. Rogers, and G. W. Gee. 1984. Diffusion of radon through soils: A pore distribution model. *Soil Sci. Soc. Am. J.* 48:482–487.

Norman, J. M., R. Garcia, and S. B. Verma. 1992. Soil surface CO_2 fluxes and the carbon budget of a grassland. *J. Geophys. Res.* 97:18,845–18,853.

Nozhevnikova, A. N., A. B. Lifshitz, V. S. Lebedev, and G. A. Zavarin. 1993. Emission of methane into the atmosphere from landfills in the former U.S.S.R. *Chemosphere* 26: 401–417.

Penman, H. L. 1940. Gas and vapour movements in the soil. I. The diffusion of vapour through porous solids. *J. Agric. Sci. Camb.* 30:437–462.

Poulsen, T. G., P. Moldrup, P. Schjønning, J. W. Massman, and J. A. Hansen. 1998. Gas permeability and diffusivity in undisturbed soils: SVE implications. *J. Environ. Eng.* 124:979–986.

Pritchard, D. T., and J. A. Currie. 1982. Diffusion coefficients of carbon dioxide, nitrous oxide, ethylene and ethane in air and their measurement. *J. Soil Sci.* 33:175–184.

Raney, W. A. 1949. Field measurement of oxygen diffusion through soil. *Soil Sci. Soc. Am. Proc.* 14:61–65.

Rasmuson, A., T. Gimmi, and M. Flühler. 1990. Modelling reactive gas uptake, transport and transformation in aggregated soils. *Soil Sci Soc. Am. J.* 54:1206–1213.

Ray, D., D. G. Pyatt, and I. M. S. White. 1987. The effect of the frequency of sampling on the observed concentration of oxygen in an afforested peat soil. *J. Soil Sci.* 38: 115–122.

Reible, D. D., and F. H. Shair. 1982. A technique for the measurement of gaseous diffusion in porous media. *J. Soil Sci.* 33:165–174.

Richter, J., and A. Grossgebauer. 1978. Investigation of the soil gas regime in a tillage experiment. II. Apparent diffusion coefficients as a measure of soil structure. *Z. Pflanzener. Bodenk.* 141:181–202.

Riley, W. J., A. J. Gadgil, Y. C. Bonnefous, and W. W. Nazaroff. 1996. The effect of steady winds on radon-222 entry from soil into houses. *Atmos. Environ.* 30:1167–1176.

Rogers, V. C., and K. K. Nielson. 1991. Correlations for predicting air permeabilities and ^{222}Rn diffusion coefficients of soils. *Health Phys.* 61:225–230.

Rolston, D. E. 1986. Gas diffusivity. In: *Methods of Soil Analysis*, Part 1 (A. Klute, ed.). Madison, WI: Am. Soc. Agron., pp. 1089–1102.

Rolston, D. E., R. D. Glauz, and B. D. Brown. 1982. Comparisons of simulated and measured transport and transformation of methyl bromide gas in soils. *Pestic. Sci.* 13: 653–664.

Rolston, D. E., R. D. Glauz, G. L. Grundmann, and D. T. Louie. 1991. Evaluation of an in situ method for measurement of gas diffusivity in surface soils. *Soil Sci. Soc. Am. J.* 55:1536–1542.

Roseberg, R. J., and E. L. McCoy. 1990. Measurement of soil macropore air permeability. *Soil Sci. Soc. Am. J.* 54:969–974.

Rust, R. H., A. Klute, and J. E. Gieseking. 1956. Diffusion-porosity measurements using a non-steady state system. *Soil Sci.* 84:453–463.

Sallam, A., W. A. Jury, and J. Letey. 1984. Measurement of gas diffusion coefficient under relatively low air-filled porosity. *Soil Sci. Soc. Am. J.* 48:3–6.

Sanhueza, E., L. Cárdenas, L. Donoso, and M. Santana. 1994. Effect of plowing on CO_2, CO, CH_4, N_2O, and NO fluxes from tropical savannah soils. *J. Geophys. Res.* 99: 16,429–16,434.

Schjønning, P. 1985a. A laboratory method for determination of gas diffusion in soil. *Tidsskr. Planteavl. Specialserie*, Beret. nr. S1773, Statens Planteavlsforsog, Denmark.

Schjønning, P. 1985b. Soil pore characteristics. I. Models and soil type differences. *Tidsskr. Planteavl.* 89:411–423.

Shurpali, N. J., S. B. Verma, J. Kim, and T. J. Arkebauer. 1995. Carbon dioxide exchange in a peatland ecosystem. *J. Geophys. Res.* 100:14,319–14,326.

Skiba, U., K. J. Hargreaves, D. Fowler, and K. A. Smith. 1992. Fluxes of nitric and nitrous oxides from agricultural soils in a cool temperate climate. *Atmos. Environ.* 26A: 2477–2488.

Smelt, J. H., M. Leistra, M. C. Sprong, and H. M. Nollen. 1974. Soil fumigation with dichloropropene and metham-sodium: Effect of soil cultivations on dose pattern. *Pestic. Sci.* 5:419–428.

Smith, K. A., K. E. Dobbie, B. C. Ball, L. R. Bakken, B. K. Sitaula, S. Hansen, R. Brumme, W. Borken, S. Christensen, A. Priemé, D. Fowler, J. A. MacDonald, U. Skiba, L. Klemedtsson, Å. Kasimir-Klemedtsson, A. Degórska, and P. Orlanski. 2000. Oxidation of atmospheric methane in Northern European soils, comparison with other ecosystems, and uncertainties in the global terrestrial sink. *Glob. Change Biol.* (in press).

Steele, D. D., and J. L. Nieber. 1994. Network modelling of diffusion coefficients for porous media. II. Simulation. *Soil Sci. Soc. Am. J.* 58:1346–1354.

Stepniewski, W. 1981. Oxygen diffusion and strength as related to soil compaction. II. Oxygen diffusion coefficient. *Polish J. Soil Sci.* 14:3–13.

Stolzy, L. H., and J. Letey. 1964. Characterizing soil oxygen conditions with a platinum microelectrode. *Adv. Agron.* 16:249–276.

Stonestrom, D. A., and J. Rubin. 1989. Water content dependence of trapped air in two soils. *Water Resources Res.* 25:1947–1958.

Stylianon, C., and B. A. De Vantier. 1995. Relative air permeability as a function of saturation in soil venting. *J. Environ. Eng.* 121:337–347.

Taylor, S. A. 1949. Oxygen diffusion in porous media as a measure of soil aeration. *Soil Sci. Soc. Am. Proc.* 14:55–61.

Troeh, F. R., J. D. Jabro, and D. Kirkham. 1982. Gaseous diffusion equations for porous materials. *Geoderma* 27:239–253.

Ussler, W., J. P. Chanton, C. A. Kelley, and C. S. Martens. 1994. Radon 222 tracing of soil and forest canopy trace gas exchange in an open canopy boreal forest. *J. Geophys. Res.* 99:1953–1963.

Velthof, G. L., and O. Oenema. 1995. Nitrous oxide fluxes from grassland in the Netherlands: II. Effects of soil type, nitrogen fertilizer application and grazing. *Eur. J. Soil Sci.* 46:541–549.

Vomocil, J. A. 1965. Porosity. In: *Methods of Soil Analysis*, Part I (C. A. Black et al., eds.). Madison, WI: Am. Soc. Agron., pp. 299–314.

Washington, J. W., A. W. Rose, E. J. Ciolkosz, and R. R. Dobas. 1994. Gaseous diffusion and permeability in four soil profiles in central Pennsylvania. *Soil Sci.* 157:65–76.

Yates, S. A. R., J. Gan, F. F. Ernst, and D. Wang. 1996. Methyl bromide emissions from a covered field: III. Correcting chamber flux for temperature. *J. Environ. Qual.* 25:892–898.

Youngs, E. G. 1983. Soil physical theory and heterogeneity. *Agric. Water Manage.* 6:145–159.

14
Soil Temperature Regime

Graeme D. Buchan
Lincoln University, Canterbury, New Zealand

I. INTRODUCTION

Temperature has a fundamental control on almost all processes in the environment. In cool climates, it demarcates growing and "nongrowing" seasons. Storage and release of heat in soil control the temperature of both the soil and the lower atmosphere, thus affecting the whole terrestrial biosphere. Yet soil temperature and its effects were traditionally poorly researched, greater attention being given to water, mainly because, with adequate temperature established within the growing season, it becomes the major and often erratic determinant of growth, while being more controllable via irrigation or drainage. More recently, a wider need has arisen to either measure or model the soil temperature regime, defined here to include the depth and time variations of both temperature and heat flux. Thus the literature shows increased attention to effects of soil temperature on soil biological processes, nutrient and fertilizer transformations, physical processes including solute transport, and environmental issues such as soil–atmosphere gas exchanges, the global carbon budget, and the transformations and transport of contaminants. Also, crop growth and evapotranspiration models require improved submodels or measurements of soil temperature regime. Climate modeling and remote sensing require more accurate data, for both heat flow and soil (especially surface) temperature.

Recent decades have seen significant advances in (1) theory: the analysis of coupled flows of heat and water, and of flow and phase-change processes in freezing soils; (2) applications, including (a) more realistic modeling of heat flow, or simultaneous heat and water flows, by inclusion of the surface energy balance as the governing boundary condition; (b) measurement and recording techniques for temperature, heat flux, and thermal properties; (c) engineering applications, e.g. ground heat pumps, and particularly (d) more intensive investigation of soil temperature as a key controller of biosphere processes including soil–atmosphere gas exchanges, transport and reactivity of solutes, and the fate of contaminants.

The basic mechanisms of coupled heat and water flows in soil were first described by Philip and de Vries (1957). Despite this, the potentially large impact of this coupling is not yet fully appreciated. While models of simultaneous flows in field soils have correctly incorporated the coupled flow equations, in the design of experimental techniques and interpretation of field measurements, the assumption is often made that the heat flow equation can be viewed as "uncoupled" from the moisture flow equation (i.e., that heat flow in soils is "conductive," and equal to a thermal conductivity λ times a temperature gradient, where λ implicitly contains the *thermal* vapor flux driven by the temperature gradient). While this assumption is valid in a uniformly moist soil, it can fail badly in the presence of a strong moisture (i.e., water potential) gradient, which drives an *isothermal* vapor flux. This both contributes to the total soil heat flux and implies latent heat demand at the sites of vaporization. This occurs in drying soils, where much of the total *soil* evaporation can derive from "subsurface evaporation," which exerts a strong influence on heat flux and the temperature profile. Neglecting such effects can lead to large errors in measurements of heat flux and thermal properties (de Vries and Philip, 1986).

This chapter therefore has a dual role. First, it reviews underlying theory and experimental methods. Second, as many of these methods assume that heat flow is purely conductive, it clarifies the potentially large effects of coupled flows on field measurements. The vital concept is the correct interpretation of the soil heat flux, including its surface value G_0 appearing in the energy balance equation.

A review of solutions of the uncoupled conduction equation includes periodic solutions and Fourier methods; basic characteristics of the diurnal and annual waves, and noncyclic effects; "transient" solutions from Laplace transform and other methods; and numerical methods. The calculation of thermal properties from physical composition is described. A brief section reviews theories of freezing soil. The measurement section reviews (a) techniques of measuring temperature, heat flux, and thermal properties, and (b) sampling criteria and data smoothing.

There is a remarkable dearth of works on soil temperature regime, with a few exceptions (Gilman, 1977; Farouki, 1986), notably in the Soviet literature (Chudnovskii, 1962; Shul'gin, 1965), though several texts devote sections to basic aspects (e.g., Hillel, 1980; Jury et al., 1991). This chapter should help to remedy this deficiency and to correct some prevalent misconceptions.

Because the theory and measurement are so intimately related, Sec. II below concerns the theory underlying measurements, and its extension to modeling of soil temperature regime. Thus the reader concerned solely with field measurements may go straight to Sec. III. However, to understand the principles and potential pitfalls of measuring soil heat flux and thermal properties, as well as the use of measurements in modeling, the theory of Section II is necessary.

II. THEORY

A. Surface Energy Balance

The most powerful models of soil heat flow incorporate its fundamental driving mechanism, the energy balance at the soil surface. The net radiation R_n received per unit area of the soil surface is

$$R_n = (1 - \alpha)R_s + \epsilon L_d - L_u \tag{1a}$$

where R_s and L_d are incident solar and longwave radiation, and α and ϵ are the shortwave reflection coefficient and longwave emissivity of the soil surface, respectively. L_u (the longwave emission) $= \epsilon\sigma T_0^4$, where σ is the Stefan–Boltzmann constant. This longwave emission is detected during infrared thermometry of the surface temperature T_0 (Huband, 1985). R_n is partitioned at the soil surface according to the energy balance equation

$$R_n = H + L_v E + G_0 \tag{1b}$$

where H is the sensible heat flux from soil to air, $L_v E$ is the latent (evaporative) heat flux (L_v = the latent heat of vaporization), and G_0 is the heat flux into the soil. For vegetated soil, L_d "seen" by the surface will include plant as well as sky emissions, H will include a small stem heat conduction term as well as convection, and E, the *soil* evaporation, will be only a portion of total evapotranspiration (Main, 1996). Note that "sensible" implies heat flow causing a local change of temperature. Thus most of G_0 produces sensible heat (i.e., temperature) change, but in a drying soil some supplies the latent heat required for evaporation within the bulk of the soil.

The dominant solar term R_s in Eq. 1a, with its diurnal and annual cycles, drives similar cycles in surface temperature T_0 and air temperature T_a, while $L_v E$, H, and L_d are controlled by atmospheric temperature and vapor pressure. Thus Eq. 1b mechanistically relates soil temperature to meteorological variables and could help explain empirical relationships, e.g., between soil and air temperature (e.g., Hasfurther and Burman, 1973; Gupta et al., 1984), though under vegetation complex modeling of intracanopy exchanges would be required. Equation 1b also enables mechanistic understanding of practical alteration of temperature regime, e.g., by mulching.

1. *Components of the Total Soil Heat Flux, G_{tot}*

In practice the "surface" for the energy exchanges in Eqs. 1a and 1b will be a thin layer, with thickness controlled by the surface microprofile, but typically several

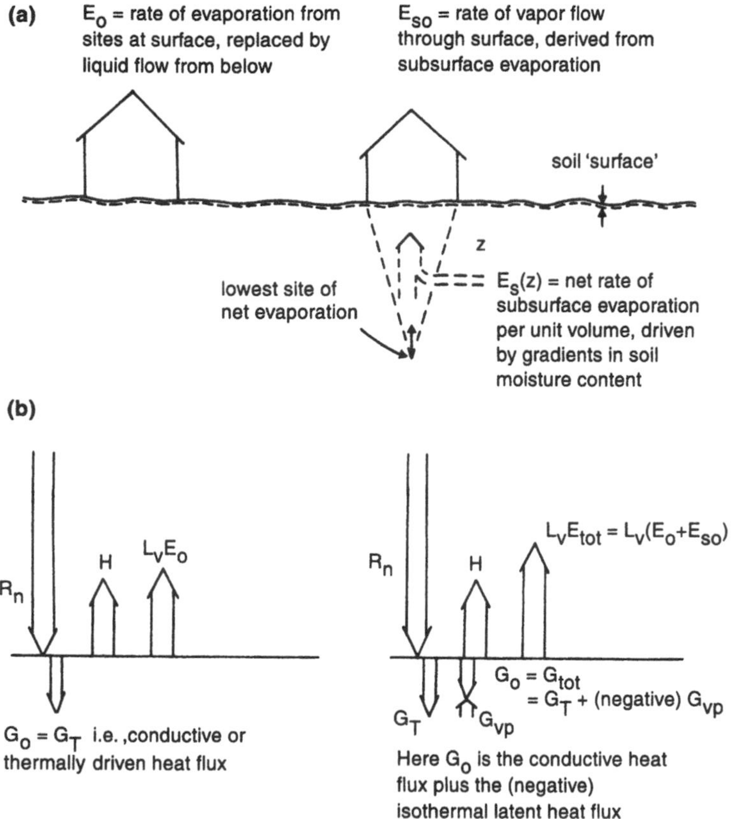

Fig. 1 (a) Partitioning of total evaporation $E_{tot} = E_0 + E_{s0}$ at the surface of a drying soil. (b) The two possible interpretations of the terms in Eq. 1b, shown under typical daytime conditions. At night the direction of G_T will usually reverse.

mm for a crumb-structured surface. However this layer is not necessarily the site of total soil evaporation, E_{tot}. In drying soils the evaporation sites retreat, at least partially, into subsurface layers (de Vries and Philip, 1986). This is critical for interpretation of both Eq. 1b and the soil heat flux $G(z, t)$, a function of soil depth z, with surface value G_0. As shown in Fig. 1a, E_{tot} is partitioned as

$$E_{tot} = E_0 + E_{s0} \tag{2}$$

$$E_{s0} = \int_0^\infty E_s(z)\, dz \tag{3}$$

Here E_0 is the evaporation sourced at the surface (replaced by liquid flow from below), and E_{s0} derives from subsurface evaporation. $E_s(z)$ (kg m^{-3} s^{-1}) is the vapor source strength per unit volume at depth z, contributing to upwards vapor flow driven by the *moisture* gradient. (Vapor distillation induced by the temperature gradient is included in the effective thermal conductivity; see Sec. II.C). E_{s0} will be dominant in a soil with a dry surface. Equation 1b may then be interpreted in two ways; see Fig. 1b. First, if $E = E_0$, then $G_0 = G_T(0, t)$ (i.e., the surface value of the conductive or thermally driven heat flux, $G_T(z, t)$; see Sec. II.C). Divergence in $G_T(z, t)$ (i.e., variation of G_T with depth) within the soil will then result from both changes in temperature and the subsurface phase change $E_s(z)$, corresponding to evaporation or condensation at depth z. Second, if, as is normally assumed, $E = E_{tot}$, then G_0 must be reduced by an amount $L_v E_{s0}$, corresponding to the subsurface evaporative energy demand. Then G_0 becomes the surface value of the *total* soil heat flux G_{tot} (see Sec. II.C) given by

$$G_{tot} = G_T + G_{vp} \tag{4}$$

The term "isothermal latent heat flux" is introduced here for G_{vp} $(= -L_v E_{s0})$ i.e., the latent heat carried from evaporating subsurface layers by the isothermal vapor flux (i.e., driven by a moisture gradient). For example, during daytime heating of a drying soil, G_T at the surface will be positive (into the soil), but $G_{tot} = G_T + G_{vp}$ will be reduced by the negative G_{vp}. Then divergence in G_{tot} is required to fuel only changes in soil temperature. Thus in the customary use of Eq. 1 to calculate *total* soil evaporation E_{tot}, it is vital to identify G_0 with G_{tot}. However, G_0 is often erroneously identified with the "thermal soil heat flux" G_T, which (Sec. III.C) is the heat flux obtained by methods detecting the temperature gradient (e.g., the heat flux plate).

B. Heat Conduction: Uncoupled Equations

Conduction of heat down a temperature gradient dT/dz is governed by the Fourier equation

$$G_T = -\lambda \frac{dT}{dz} \tag{5}$$

where the thermal conductivity λ (W $m^{-1}K^{-1}$) includes a vapor distillation term (Sec. II.D). Divergence in G_T causes heat changes, both sensible and latent, and so obeys energy conservation:

$$C\frac{\partial T}{\partial t} = -\frac{\partial G_T}{\partial z} + S(z, t) \tag{6}$$

where C (J m^{-3} K^{-1}) is the volumetric heat capacity. S (W m^{-3}) represents local heat sinks or sources, i.e., usually phase changes of water (Secs. II.C, II.F). Neglecting S (considered below) and spatial variations in λ, Eqs. 5 and 6 give the simple uncoupled heat diffusion equation

$$\kappa^{-1}\frac{\partial T}{\partial t} = \frac{\partial^2 T}{\partial z^2} \tag{7a}$$

$$= \frac{\partial^2 T}{\partial r^2} + r^{-1}\frac{\partial T}{\partial r} \tag{7b}$$

where $\kappa = \lambda/C$ (m^2 s^{-1}) is the thermal diffusivity. Equation 7b in cylindrical coordinates applies to the use of cylindrically symmetric probes (Sec. II.E). Equation 7 is uncoupled in the sense that, with the thermal vapor flux implicit in λ, it can be solved independently of the moisture flow equation. Its use implies a no-coupling assumption, invalid in soil undergoing aqueous phase changes, in particular subsurface evaporation.

The thermal properties λ, C, and κ are (a) functions of physical composition and hence both position and time, so that analytic solutions require simplifying assumptions (Sec. II.E), and (b) relatively weak functions of T itself, so that Eqs. 7a and 7b are, strictly, weakly nonlinear. Equation 7 in three-dimensional form has $\partial^2 T/\partial z^2$ replaced by $\nabla^2 T$.

C. Heat Flow: Moisture Coupling

Heat and water flows can interact strongly in soil. This interaction is small in soil close to absolute dryness or saturation, but important at intermediate states of wetness. The main coupling of flows is by two mechanisms: (a) the influence of gradients of temperature on water flow, in the liquid phase by its effect on surface tension, and more importantly in the vapor phase by its much stronger effect on vapor pressure (i.e., thermally driven water flow); and conversely (b) the influence of gradients of water potential, driving liquid and vapor flow, on the flow of heat (i.e., water potential driven heat flow). The interaction of heat and *liquid* water flow is often negligible (de Vries, 1975), with a few important exceptions. Examples corresponding to mechanisms (a) and (b) are the often rapid migration of liquid water under temperature gradients towards a freezing front, possibly leading to frost heave or formation of "ice lenses"; and heat convection by intense infiltration of water.

By contrast, heat and *vapor* flows may be strongly coupled, so conduction may be accompanied by a large latent heat flux. The source of this coupling is apparent in the one-dimensional (vertical) vapor flux J_v (Bristow et al., 1986), the sum of the thermal (J_{vT}) and isothermal (J_{vp}) vapor fluxes.

$$J_v = -D_v \frac{de}{dz} = J_{vT} + J_{vp} \tag{8}$$

$$J_{vT} = -\eta D_v hs \left[\frac{dT}{dz}\right] \tag{9}$$

$$J_{vp} = -D_v e_s(T) \left[\frac{dh}{dz}\right] \tag{10}$$

Here, e is the actual vapor pressure in the air phase, $e_s(T)$ is the saturation vapor pressure (svp), $s = de_s/dT$ is the slope of the svp curve, and $h = e/e_s$ is the relative humidity. $D_v = \alpha\theta_a\nu D_{va}$ is the apparent vapor diffusivity (kg m^{-1} s^{-1} Pa^{-1}) in soil air, where D_{va} is the diffusivity in bulk, still air, θ_a is air-filled porosity, and α is a pore space tortuosity factor. The mass flow factor $\nu = p/(p - e) \approx 1$ (where p is the total air pressure in soil) accounts for a small mass flow contribution to vapor transfer (Philip and de Vries, 1957). In Eq. 9, the added enhancement factor η is required to give the effective thermal vapor diffusivity ηD_v (Philip and de Vries, 1957; Cass et al., 1984; Bristow et al., 1986).

Thus the vapor flux, Eq. 8, has two components. The *thermal vapor flux* J_{vT} (Kimball et al., 1976) represents thermally driven vapor transfer. This carries latent heat from hotter (higher e_s) to cooler (lower e_s) regions, contributing to the effective thermal conductivity, λ. Conversely, the *isothermal vapor flux* J_{vp} represents a water-potential-driven latent heat transfer, $L_v J_{vp}$. Thus, neglecting osmotic effects, a moisture gradient controls humidity h in Eq. 10 according to

$$h = \exp\left[\frac{\psi_m M_w}{RT}\right] \tag{11}$$

where ψ_m (J kg^{-1}) is the matric potential and $M_w = 18.016 \times 10^{-3}$ kg mol^{-1} is the molecular weight of water. Equation 11 implies $h > 0.99$ for $\psi_m > -13$ bar. Thus J_{vp} will typically be relatively small in soils wetter than the wilting point. Then only J_{vT} (already inherent in λ) need be considered. However J_{vp} is significant under strong moisture gradients, e.g., in the upper layers of drying soils.

Following Eq. 8, we may define a *total soil heat flux* G_{tot}

$$G_{tot} = G_c + G_{vT} + G_{vp} = G_c + L_v J_{vT} + L_v J_{vp} \tag{12}$$

containing a "pure" conduction component G_c, a "thermal latent heat flux" G_{vT}, and an "isothermal latent heat flux" G_{vp}. In reality, pure conduction and thermal distillation (G_{vT}) are intertwined as complex series–parallel processes, and so are not strictly additive. However, both processes are proportional to $-dT/dz$, and may be combined into a single "thermal soil heat flux"

$$\begin{aligned} G_T &= G_c + G_{vT} \\ &= -\lambda \frac{dT}{dz} \end{aligned} \tag{13}$$

where λ is the apparent thermal conductivity (i.e., as calculated by the Philip–de Vries model discussed below).

The uncoupled heat diffusion Eq. 6 then becomes the coupled equation (Philip and de Vries, 1957)

$$C\frac{\partial T}{\partial t} = -\frac{\partial G_{tot}}{\partial z} = \frac{\partial(\lambda \partial T/\partial z)}{\partial z} - L_v\frac{\partial J_{vp}}{\partial z} \tag{14}$$

where the last term accounts for phase change induced by a moisture gradient. Divergence in J_{vp} represents a heat sink (a site of net evaporation) or source (a site of net condensation). In field soils undergoing subsurface evaporation, the heat sink effect will tend to increase divergence in G_T, and hence the curvature of the temperature profile. We will return to the practical impact of this on heat flux measurement in Sec. III.C.

The concept of an effective thermal conductivity, enhanced by thermal vapor distillation, can be treated theoretically in two distinct ways. The first method solves simultaneously the coupled flow equations (e.g., Milly, 1982; Bristow et al., 1986). Thus Eq. 14 is the heat transfer equation. However this method, while more comprehensive and accurate, requires complex numerical modeling.

The second method (Philip and de Vries, 1957) essentially builds the thermal vapor flux, Eq. 9, into the de Vries (1963) thermal conductivity model, which calculates λ from the conductivities of individual soil components (see next section). As vapor transfer occurs in the air filled pores, with net distillation from warm to cold ends, the air phase conductivity becomes

$$\lambda_{av} = \lambda_a + h\lambda_{vs} \tag{15}$$

Here λ_a is the conductivity of still air and

$$\lambda_{vs} = L_v \nu D_{va}\frac{de_s}{dT} \tag{16}$$

is the vapor distillation term for saturated air, ν is the mass flow factor discussed below Eq. 10. Eq. 16 is essentially the same thermal vapor flux effect as Eq. 9 but contains the simple bulk air diffusion coefficient rather than an effective one for a complex pore space. The latent heat term $h\lambda_{vs}$ can be "very effective in increasing the thermal conductivity of soils, since it multiplies the conductivity of the air-filled pores by a factor ranging from 2 at 0°C to 20 near 60°C" (de Vries, 1975). The advantage of this second method, albeit more approximate, is that it incorporates thermal vapor transfer into a single macroscopic conductivity, λ, effectively decoupling the heat and moisture flow equations. It does not, of course, account for heat transfer induced by a moisture gradient.

The theory of coupled flows in porous media can be approached more abstractly using irreversible thermodynamics (de Vries, 1975; Raats, 1975; Sidiropoulos and Tzimopoulos, 1983). Essentially this provides only an overlying formalism for the above coupled-flow approach. Phenomenological transport coefficients are introduced, but they still need to be derived using the mechanistic ideas of that approach.

Flow coupling can accumulate to visible level under prolonged steady-state heat flow. This can lead to marked thermally induced redistribution of moisture (e.g., around underground cables or pipes, or in laboratory determination of λ (Sect. III.D).

D. Calculation of Thermal Properties

Soil thermal conductivity and heat capacity depend on physical composition, especially moisture content, so single measurements are of limited use. Theory to predict the variation with moisture content is thus required.

1. *Volumetric Heat Capacity, C*

The heat capacity C of a unit volume of soil is, simply and exactly, the sum of the heat capacities of its phases (de Vries, 1975):

$$\begin{aligned} C &= x_m C_m + x_o C_o + x_w C_w \\ &= 4.18 \times 10^6 (0.46 x_m + 0.60 x_o + x_w) \text{ J m}^{-3} \text{ K}^{-1} \end{aligned} \tag{17}$$

where x denotes the volume fraction and C the volumetric heat capacity of a phase, with subscripts m, o, and w indicating mineral solids, organic matter, and liquid water, respectively. Air (moist) makes a negligible contribution. Table 1 shows thermal properties.

Table 1 Thermal Properties of the Principal Soil Phases (Solids at 10°C, Ice at 0°C)

Material	Volumetric heat capacity, C (MJ m^{-3} K^{-1})	Thermal conductivity (W m^{-1} K^{-1})
Quartz	2.0	8.8
Clay minerals	2.0	2.9
Organic matter	2.5	0.25
Water	4.2	$0.552 + 2.34 \times 10^{-3}T - 1.10 \times 10^{-5}T^2$
Ice	1.9	2.2
Air	1.25×10^{-3}	$0.0237 + 0.000064T$[a]

[a] T in degrees Celsius.
Source: de Vries (1975); Hopmans and Dane (1986a).

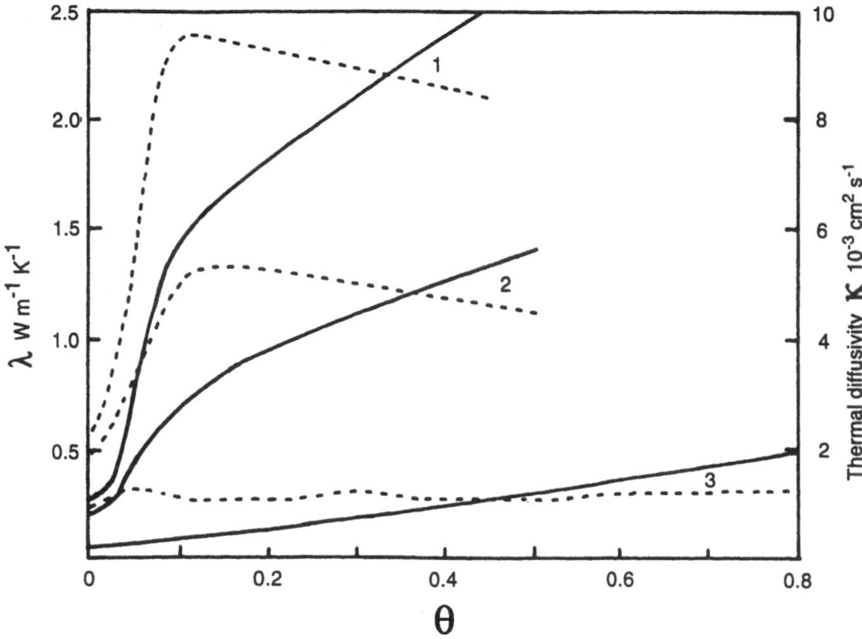

Fig. 2 Variation of soil thermal conductivity (solid curves) and diffusivity (broken curves) with volumetric water content θ for (1) quartz sand ($x_m = 0.55$); (2) loam ($x_m + x_0 = 0.50$); (3) peat ($x_0 = 0.20$). (From de Vries, 1975, Courtesy of Hemisphere Publ. Corp.)

2. *Thermal Conductivity, λ*

The macroscopic conductivity λ of Eq. 5 summarizes a heat flow that is spatially averaged over microscopically complex paths and so cannot be calculated exactly. An approximate "dielectric analog" model was developed by de Vries (1963), by application to a granular medium of "potential theory," which treats systems in which an induced response (here, a flow of heat) at any point is proportional to the local gradient of a potential (here temperature). Figure 2 shows typical variations of λ with water content for sand, loam, and peat soils.

The model views soil as a continuous medium (subscript c, either liquid water in moist soil, or air in drier soil), with volume fraction x_c and conductivity λ_c, in which are dispersed regularly shaped "granules" of the other four components (either air or water, plus quartz, clay, and organic matter). The overall conductivity is then a weighted mean of the component conductivities (Table 1),

$$\lambda = \frac{x_c \lambda_c + \Sigma k_j x_j \lambda_j}{x_c + \Sigma k_j x_j} \tag{18}$$

Each weighting factor k_j is the ratio of the average temperature gradient in a granule of phase j to that in the background phase. Assuming spheroidal granules, potential theory gives, to a good approximation,

$$k_j = \frac{2}{3}\left[1 + \left(\frac{\lambda_j}{\lambda_c} - 1\right)g_1\right]^{-1} + \frac{1}{3}\left[1 + \left(\frac{\lambda_j}{\lambda_c} - 1\right)(1 - 2g_1)\right]^{-1} \tag{19}$$

where g_1 is a shape factor for phase j. The assumption that all granules of phase j, though varying in scale, are geometrically similar spheroids, with principal axes in the ratio $a_1 = a_2 = na_3$, allows use of a *single* factor g_1. A single k_j factor (along with the factors of $\frac{1}{3}$ and $\frac{2}{3}$ in Eq. 19) emerges from averaging over random granule orientation.

For both sand and clay soils, de Vries (1963) deduced representative averages $n = 5$ and $g_1 = 0.125$ for the soil particles. The model, summarized as follows, subdivides the entire moisture range into four regions (Hopmans and Dane, 1986a).

a. Dry Soil

Here air is the continuous medium, and large ratios λ_j/λ_c (Table 1) require λ from Eq. 18 to be multiplied by an empirical factor of 1.25. Table 2 shows k_j from Eq. 19 with $g_1 = 0.125$ and data of Table 1.

b. Moist Soil Between Saturation and PWP, $x_{PWP} < x_w < x_{sat}$

Water is now the continuous medium, so $x_c = x_w$, and above the permanent wilting point (PWP) $h \cong 1$ in Eq. 15. With progressive drying, the air spheroids become increasingly elongated, and de Vries (1963) suggested a linear interpolation for the air shape factor, $g_a = 0.035 + (x_w/x_{sat})(0.333-0.035)$, between 0.333 for spherical bubbles close to saturation and 0.035 for dry soil. This formula, along with temperature-dependent λ_{av} in Eq. 15, gives k_j for air in Eq. 19. Table 2 shows k_j for the other, solid phases, again using $g_1 = 0.125$ and Table 1.

c. Moist Soil Below PWP, $x_{crit} < x_w < x_{PWP}$

With progressive drying below PWP, both the air shape factor g_a and humidity h decrease, the latter from ~ 1 to 0 at absolute dryness. de Vries suggested a linear

Table 2 Weighting Factors k_j for Thermal Conductivity: Eq. 19

	k_j			
Continuous medium	Quartz	Clay	Organic matter	Air
Water (moist soil)	0.267	0.523	1.30	See text
Air (dry soil, $x_w < x_{crit}$)	0.0161	0.047	0.36	2.0

interpolation for g_a between 0.013 at $x_w = 0$, and the value at PWP derived from above, and a linear approximation $\lambda_v = (x_w/x_{PWP})\lambda_{vs}$ to the vapor term $h\lambda_{vs}$ in Eq. 15.

d. *Soil Below a Critical Water Content, $x_w < x_{crit}$*

de Vries suggested the transition from water to air as the continuous medium occurs at a critical water content x_{crit} of about 0.03 for coarse-textured and 0.05 to 0.10 for fine-textured soils. Below this he recommended a linear interpolation of λ versus x_w, between its dry value (subsection a above) and the value at x_{crit} (Subsec. c). The model predicts λ values "with an accuracy of usually better than 5%, except in the interpolation range, where the error becomes of the order of 10%" (de Vries, 1975).

The air shape factor is determined in a "somewhat ad hoc manner" (de Vries and Philip, 1986). However the errors should be small as follows. First, there is a partial cancellation of error in calculating k_a from g_a via Eq. 19, and in turn λ from k_a via Eq. 18. In essence, the relative conductivity of a phase matters much more to the overall conductivity than small variations in the shape of its granules, particularly when their orientations are randomized. Second, the air phase contribution to λ is in any case small, except in two cases: (a) in very dry soil, when results rely more on calculation of Subsec. a, for which no g_a is required, and (b) at higher temperatures ($T >$ about 30°C), when λ_{av} is large. (In fact $\lambda_{av} = \lambda_w$ at $T = 59$°C; de Vries, 1963.) However, the reduced contrast between λ_{av} and λ_w will then reduce the sensitivity to shape factor. Hence fastidious computation of g_a is unwarranted. The model's greatest limitations are its use of (a) the assumption that intergranule spacing is sufficient to avoid disturbance of intragranule temperatures in potential theory; and (b) idealized spheroidal granules for pore-occupying phases.

In summary, the model accounts well for the strong moisture dependence of conductivity and also for its density dependence. It has also been applied successfully to swelling soils, with soil solids as the continuous medium (Ross and Bridge, 1987). Temperature dependence, due almost entirely to vapor distillation, may be considered weak over restricted ranges of temperature, particularly below 30°C.

A curve found empirically to represent the moisture dependence of conductivity has the equation (McInnes, 1981; Campbell, 1985)

$$\lambda(x_w) = A + B\left[\frac{x_w}{x_{sat}}\right] - (A - D)\exp\left[-C\left(\frac{x_w}{x_{sat}}\right)^E\right] \tag{20}$$

where A, B, C, D, and E are parameters determined by curve fitting, to values from either measurement or the de Vries model (D is the dry soil conductivity). Alternatively, for use in numerical models of nonuniform soils, approximate relationships of these parameters to composition and density have been developed (Campbell, 1985). Earlier empirical formulae for estimation of λ from density and water content were developed by Kersten (1949).

A computer package has been developed (Tarnawski et al., 2000), and calculates soil thermal properties (c and λ) for the user over a wide range of temperatures, suitable for agronomic, environmental, or engineering applications.

E. Solutions of the Conduction Equation

This section deals with solutions of the uncoupled conduction equations of Sec. II.B, primarily Eq. 7. These solutions have practical application, both in the field measurement of thermal properties and in the extrapolation of soil temperature regime from a restricted set of field measurements (e.g., Buchan, 1982a, b, c). In the field, complex variations of both soil thermal properties and surface weather, and hence of $T_0(t)$, require numerical simulation for greatest accuracy. Figure 3 illustrates complexity in $T_0(t)$ measured over a 3-day period. However, simplifying assumptions enable analytical solutions. These include neglect of the weak T-variation of thermal properties, uniformity or analytic variation of thermal properties with depth, and analytic boundary and initial conditions.

1. Analytical Methods

Analytical theory deals with two main types of time variation: periodic variations; or simple nonperiodic variations, i.e., transient or short-term heat flow. The two main methods are Fourier transform (FT) and Laplace transform (LT), respectively. Via integral transforms, both methods remove the time dependence in $T(\mathbf{r}, t)$, so that the partial differential Eq. 7 becomes an ordinary differential equation in the space ($\mathbf{r}$) coordinates only. We consider only one-dimensional solutions, for vertical (z) variations: and also the radial (r) solution for the cylindrical probe (Sec. III.D).

a. Periodic Variations

The Fourier method analyzes temperature variation into a set of harmonics of the dominant diurnal or annual waves. An irregular, *continuous* signal of finite duration can be broken down into an infinite sum of harmonics (Bloomfield, 1976). However, temperature data usually form a *discrete* sequence of N points in time, called a time series (e.g., $N = 24$ for hourly data over one day). Then the infinite sum becomes a finite sum of $M = N/2$ harmonics (assuming N is even), the so-called discrete Fourier transform (DFT); for example, a periodic N-point surface variation can be transformed to

$$T_0(t) = \overline{T}_0 + \sum_{n=1}^{M} A_n \sin(n\omega_1 t + \phi_n) \tag{21}$$

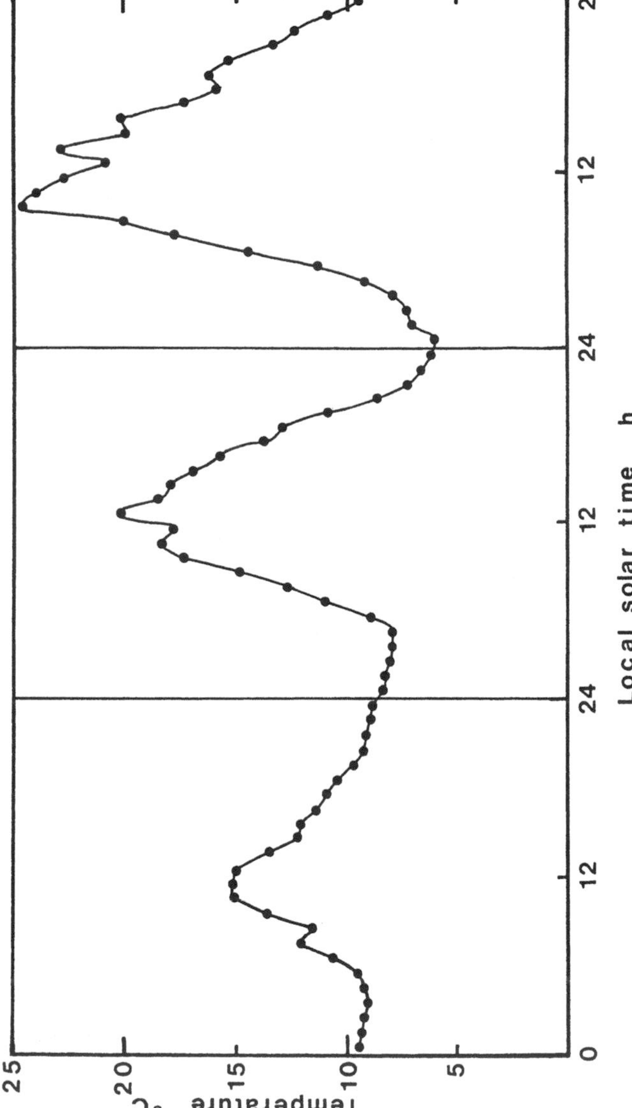

Fig. 3 Soil surface weather: hourly-measured bare soil surface temperature T_0 over a 3-day period, 8–10 June, 1979, at Aberdeen, Scotland. Note noncyclic changes.

where $\omega_1 = 2\pi/\tau$ is the fundamental angular frequency, with period $\tau = 24$ h or 12 months for the diurnal or annual wave. The N parameters, i.e., $\overline{T}_0$ plus amplitudes, A_n, and phases, ϕ_n, are determined from the N measured data (Bloomfield, 1976; Buchan, 1982a). Assuming Eq. 7 is linear, the depth penetration of $T_0(t)$ is simply the sum of the penetrations of each harmonic (van Wijk and de Vries, 1963; Carslaw and Jaeger, 1967):

$$T(z, t) = \overline{T}_0 + \sum A_n \exp\left(-\frac{z\sqrt{n}}{D_1}\right)\sin\left(n\omega_1 t + \phi_n - \frac{z\sqrt{n}}{D_1}\right) \tag{22}$$

Three implicit assumptions should be satisfied, at least approximately, for Eq. 22 to apply in the field:

1. The uniform soil assumption, that thermal properties are constant with depth.
2. An initial condition assumption, that the actual initial T-profile equals $T(z, 0)$ given by Eq. 22. This implies an isothermal assumption, that temperatures at all depths vary around the same average, $\overline{T}_0$.
3. $T(z, t)$ is approximately periodic, i.e., the noncyclic change, defined as the difference between successive midnights (or between a given month in successive years for the annual wave) is close to zero.

Conditions 2 and 3 can be satisfied using a superposition trick, i.e., by exploiting the linearity of Eq. 7 to subtract out, and solve separately for, the difference between the measured T-variation and that required by the condition. For example, periodicity in a noncyclic diurnal variation (e.g., Fig. 3) can be achieved by subtracting a linear ramp variation from single-day data (Buchan, 1982c). Also, by climatically averaging the diurnal variation over several days, a smoother periodic variation is achieved (Fig. 4) (Buchan, 1982a, b).

Equation 22 represents a damped, phase-delayed penetration of each harmonic (see Fig. 5). $D_1 = \sqrt{2\kappa/\omega_1}$ is the "damping depth" of the fundamental ($n = 1$), with values between about 8 and 16 cm for the diurnal wave ($\omega_1 = 2\pi/86400\ \mathrm{s}^{-1}$) in mineral soils (de Vries, 1975). Higher harmonics are more rapidly damped, with damping depth decreasing as $D_n = D_1/\sqrt{n}$. The amplitude is attenuated to 5% of A_n at depth D_n; and 0.7% at 5 D_n, representing an approximate limit of penetration. For the annual wave, the $\sqrt{n}$ rule implies a damping depth $\sqrt{365} = 19$ times the diurnal value. Thus a typical diurnal damping depth $D_d = D_1 = 0.12$ m gives an annual value $D_a = 2.29$ m.

From Eq. 22, the conductive soil heat flux $G_T = -\lambda\partial T/\partial z$ is

$$G_T(z, t) = \sum_{n=1}^{M} A_n\sqrt{\lambda C n\omega_1}\exp\left(-\frac{z}{D_n}\right)\sin\left(n\omega_1 t + \phi_n - \frac{z}{D_n} + \frac{\pi}{4}\right) \tag{23}$$

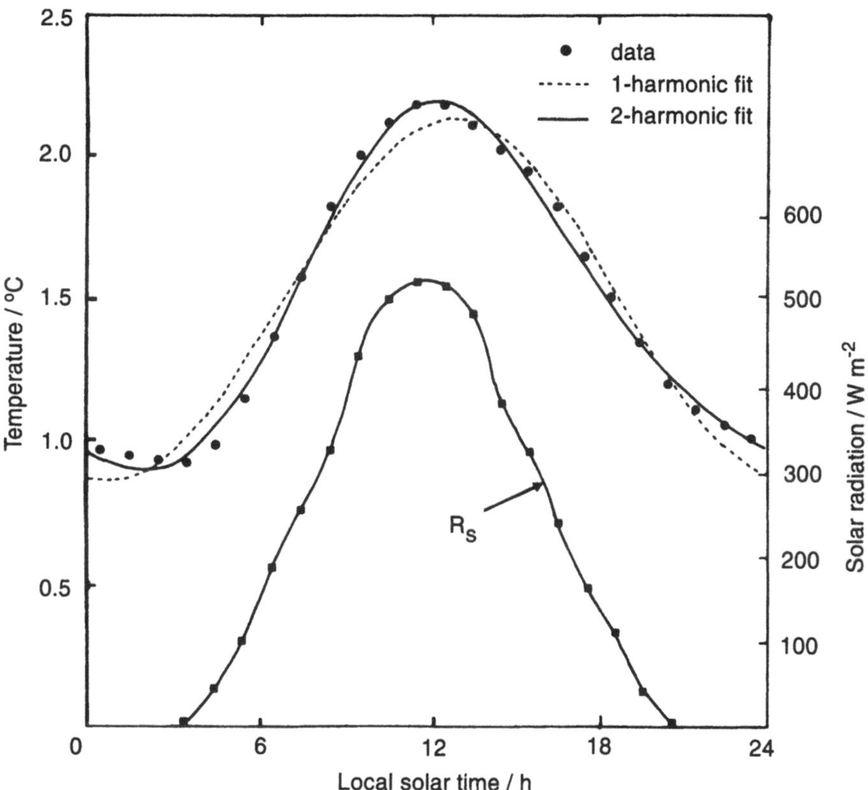

Fig. 4 Soil surface climate: 15-day average diurnal variations of bare soil surface temperature, T_0, showing measured data and one- and two-harmonic fits to data, and solar radiation, R_s. Note: Period (6–20 June, 1979) includes days of Fig. 3.

At the surface

$$G_T(0, t) = \sum_{n=1}^{M} A_n \sqrt{\lambda C n \omega_1} \sin\left(n\omega_1 t + \phi_n + \frac{\pi}{4} \right) \tag{24}$$

Thus for each harmonic the temperature variation lags the heat flux by phase $\pi/4$, i.e., a time lag of $\pi/4n\omega_1 = \tau/8n$. For the fundamental, this is 3 h for the diurnal and 1.5 months for the annual variation. However, this is not the lag of extrema in T_0 behind extrema in solar irradiation, because (a) higher harmonics contribute to $T_0(t)$ and (b) extrema in G_0 are determined by the *total* surface energy balance (see Figs. 3, 4, and 6). For a typical diurnal wave in moist bare soil, T_0 peaks at about 1300 h local solar time (van Wijk and de Vries, 1963; Buchan, 1982a), and

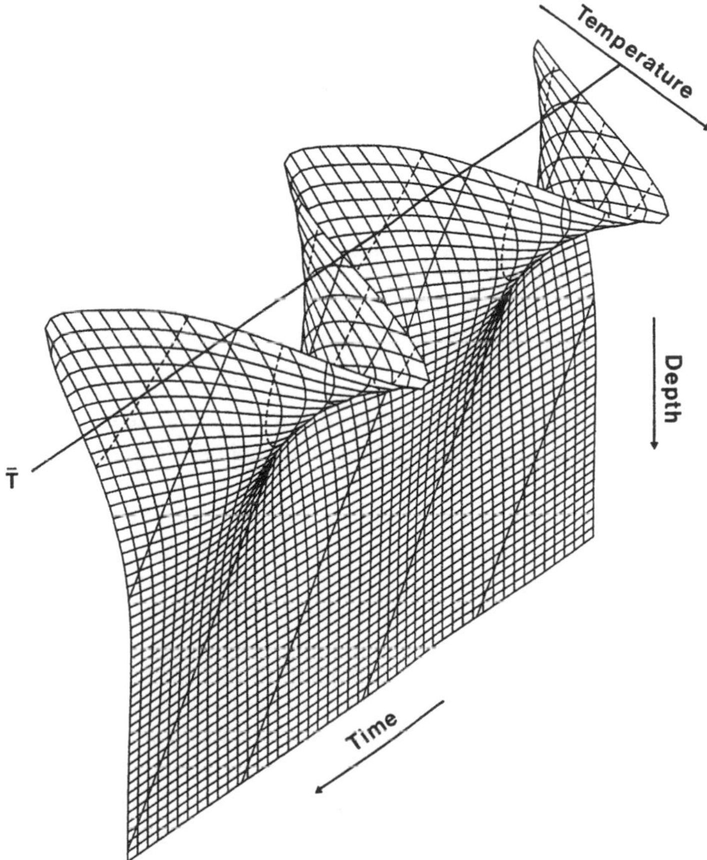

Fig. 5 Three-dimensional plot of soil temperature, showing decay of amplitude and increasing phase-lag with depth. Plot shows a two-harmonic springtime wave at Aberdeen, Scotland, with $A_1 = 4.8$ K, $A_2 = 1.1$ K, $\phi_1 = -17°$, $\phi_2 = -89°$ in Eq. 22. Note wave asymmetry due to second harmonic. (After G. S. Campbell, *An Introduction to Environmental Biophysics*, Springer-Verlag, Berlin, 1977.)

minimum T_0 is around sunrise. There are additional lags under vegetation, typically about 0.5 h for short grass and 1 h for cereal crops.

A simple model of the diurnal or annual wave (subscripts a and d) assumes a single harmonic for each. Their combination is

$$T_0(t) = T_0 + A_a \sin(\omega_a t + \phi_a) + A_d \sin(\omega_d t + \phi_d) \quad (25)$$

where ω_a and ω_d are the fundamentals and $\omega_a = \omega_d/365$. Hence a small noncyclic change is an integral feature of the diurnal wave, with a net 24-h heat gain (or loss) by the soil in the warming (cooling) half of the year. Averaged over each semiannual period, the noncyclic change in heat storage, drawn from the annual wave every day, is (de Vries and Philip, 1986)

$$\Delta_a S = \frac{2\lambda A_a}{\omega_d D_a} \, (\mathrm{Jm^{-2}d^{-1}}) \tag{26}$$

For the diurnal cycle, the net flow into (out of) the soil during the warming (cooling) semidiurnal period is

$$\Delta_d S = \frac{2\lambda A_d}{\omega_d D_d} \, (\mathrm{Jm^{-2}d^{-1}}) \tag{27}$$

For a cool-temperate bare soil, the annually averaged diurnal amplitude A_d is typically about 5 K, and the annual amplitude A_s about 9 K (author's data), implying $\Delta_a S/\Delta_d S = 0.19$.

Consider noncyclic change in surface temperature $T_0(t)$. Its semiannual average is $\Delta_a T = \pm 2\omega_a A_a/\pi = \pm 4A_a/365$ (K per day). Assuming $A_a = 9$ K gives an average of only 0.1 K per day. Thus while vagaries of weather may produce large (e.g., 5 K or more) single-day noncyclic changes, the average over many days is usually negligible (Buchan, 1982a).

However, a single 24-h harmonic is inadequate to represent the diurnal wave. For an irregular wave, at least 6 harmonics are required (Kimball et al., 1976; Buchan, 1982c). For multiday average variations, two harmonics are often adequate (Buchan, 1982b; Gupta et al., 1984), with amplitude ratio A_2/A_1 typically around one quarter or less in the summer months (Carson, 1963; Buchan, 1982b), but may approach 0.8 in winter (Carson, 1963). Figure 4 shows a 15-day average $T_0(t)$ for bare soil. The typical asymmetry contrasts with the nearly symmetrical solar radiation curve, $R_s(t)$. Three stages are identified: (a) steep morning rise; (b) slower afternoon decline; (c) even slower nocturnal cooling. The asymmetry of stages a and b is due to heat storage in soil and atmosphere partly offsetting afternoon heat losses. Stage c is due to the dominant control of nighttime microclimate by, first, net longwave exchange (the difference between surface and effective sky radiation temperatures being less than in daytime), and, second, the upwelling soil heat flux. The pronounced second harmonic reflects (a) a strong second harmonic in the driving solar radiation, $R_s(t)$ (Buchan, 1982b), imposed mainly by abrupt nighttime zeroing of the R_s curve (Fig. 4), and (b) soil and atmosphere heat storage. While the storage effect produces asymmetry, it in fact weakens the second harmonic in $T_0(t)$ compared to $R_s(t)$. Thus in Fig. 4, A_2/A_1 is 0.14 for T_0 but 0.24 for R_s.

For the smoother annual wave (Fig. 6), a two-harmonic fit is adequate for both soil (van Wijk and de Vries, 1963; Persaud and Chang, 1985) and air (Tabony, 1984) temperature. In soil, A_2/A_1 (typically 0.12 to 0.15; van Wijk and de

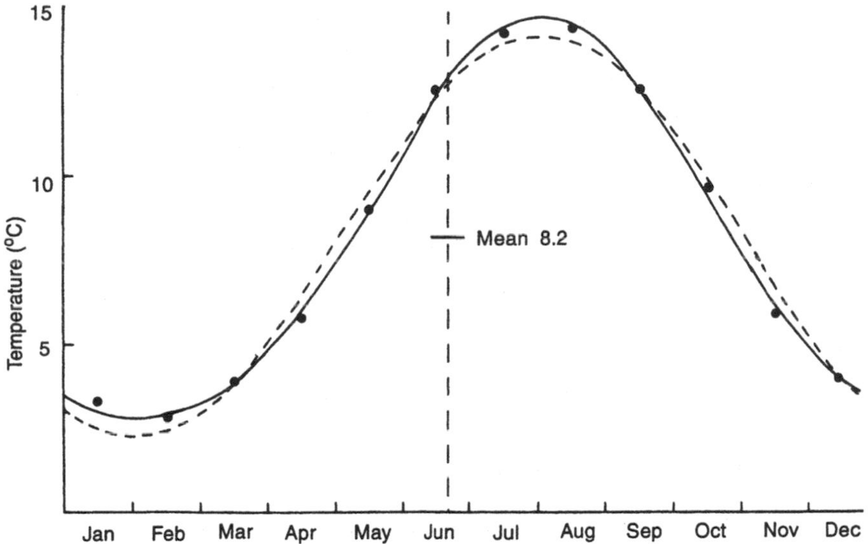

Fig. 6 Annual wave of soil temperature at 30 cm depth, Aberdeen, Scotland (1966–1975 10-year mean). Symbols: observed data. Dashed and solid curves: one- and two-harmonic fits to data, respectively. Center vertical line marks midsummer day.

Vries, 1963; Persaud and Chang, 1985) is less than for the diurnal wave. This reflects the smoother annual progression of R_s, with no analog of abrupt nighttime darkening, except at very high latitudes. Also, the asymmetry in the annual wave is less (Miller, 1981).

The rate at which heat is absorbed into the soil under given surface conditions will clearly increase with both λ and C and is measured by the term $\sqrt{\lambda C}$ in Eq. 24. This term has various names, including thermal admittance, from the analogy with electrical theory (Menenti, 1984; Novak, 1986). It controls daytime heat absorption and nighttime heat release. The strong control of the latter over nighttime microclimate explains why soils with lower λ and C, e.g., peats, can exacerbate frosts (de Vries, 1975). The insulation effect of plant cover has similar effects. Admittance, a measure of the rate of surface heat absorption, contrasts with the thermal diffusivity (λ/C), a measure of the rate at which soil attempts to equalize its temperature by internal diffusion of heat.

b. Nonperiodic Variations

The Laplace time-transform of $T(z, t)$ is given by

$$L <T(z, t)> = \int_0^\infty T(z, t)e^{-st}\, dt \tag{28}$$

(van Wijk, 1963) and is a function of z and s only, where s is the dimensionless Laplace parameter. Thus while the FT method decomposes $T(z, t)$ into a set of harmonics and their parameters, the LT employs only one parameter and so is more useful for analyzing simple transient (e.g., rising or decaying) variations. The LT of the heat diffusion Eq. 7 is the ordinary differential equation (van Wijk, 1963)

$$\frac{\kappa d^2 L < T(z, t) >}{dz^2} - sL < T(z, t) > + T(z, 0) = 0 \tag{29}$$

There are two distinct uses of the LT in soil:

1. The conventional or "analytical" use, i.e., solution of Eq. 29 for $L(z, s)$, then inversion L^{-1} of the transform, to obtain an explicit solution for $T(z, t)$. Here s plays a purely algebraic role: no numerical value is assigned. The LT is rarely used in this way. One example is solution of the cylindrical heat flow equation (Eq. 7), with r replacing z in Eq. 28, for the case of a heated hollow cylindrical probe used for conductivity measurement (Moench and Evans, 1970).
2. The predominant "numerical" use, used to analyze the propagation of a transient heat perturbation as a means of deriving thermal properties (λ or κ), without detailed solutions for $T(z, t)$. This requires only the forward numerical transform of measured data: in essence, $L\{T\}$ is used in lieu of T itself (van Wijk, 1963). The precise value of s is now important, as $\exp(-st)$ "weights" the temperature record in Eq. 28. The choice $s \geq 5.0/t_{max}$, where t_{max} is the duration of the record, ensures $\exp(-st) < 0.007$ beyond t_{max} (Asrar and Kanemasu, 1983).

Assuming initially isothermal soil, $T'(z, 0) = 0$, where $T'(z, t)$ is the difference between $T(z, t)$ and the initial isothermal value. Then a solution to Eq. 29 for a semi-infinite soil subject to some surface boundary condition is

$$L < T'(z, t) > = \text{const} \exp\left[-z\sqrt{\frac{s}{\kappa}} \right] \tag{30}$$

where the constant is actually a function of s, depending on the boundary condition applied (van Wijk, 1963). However, given data for two depths z_1 and z_2, this drops out in the ratio

$$\frac{L_1}{L_2} = \exp\left[-(z_1 - z_2)\sqrt{\frac{s}{\kappa}} \right] \tag{31}$$

Thus κ can be determined from temperature records for two or more depths: see Sec. III.D.

This method can be applied even without initially uniform temperature, by using a superposition trick. Then $T(z, t) = T_b(z, t) + T'(z, t)$ is viewed as the superposition of the transient T' on the "background" course T_b that T would have taken in the absence of the transient (van Wijk, 1963). This requires interpolation on longer records to estimate T_b.

c. Variations Round a Heated Line Source: Soil Probes

Conduction of heat away from a heated line source (wire or needle) inserted in soil provides increasingly popular methods for measuring soil thermal properties. There are three modes of heating with corresponding radial solutions of the conduction equations: (a) a continuously heated and (b) an instantaneously heated line source; and (since in practice instantaneous heating is not possible) (c) a short-duration heat pulse.

The cylindrical probe for measuring λ is a continuously heated line source. The solution for this problem is simpler than for the finite-radius probe mentioned above (Moench and Evans, 1970). For a probe in initially isothermal soil, with constant heating rate per unit length Q (W m^{-1}) switched on at $t = 0$, solution of Eq. 7b gives for probe temperature rise (Sepaskah and Boersma, 1979)

$$T_2 - T_1 = \frac{Q}{4\pi\lambda}\ln\left(\frac{t_2}{t_1}\right) \tag{32}$$

Solutions for instantaneously and pulse-heated line sources are given in Bristow et al. (1994). The latter enables measurement of soil thermal properties with a dual probe, i.e., a pulse-heated wire or needle with a parallel needle containing a temperature sensor (see Sec. III.D).

Details of additional analytical techniques developed for homogeneous, inhomogeneous, and layered soils can be found elsewhere in the literature (Lettau, 1962; van Wijk and Derksen, 1963; Gilman, 1977).

2. Numerical Methods

The advantages of numerical methods include their ability to deal with nonuniform soils; with irregular boundary and initial conditions; with multidimensional flows; and with strong nonlinearities, for example in the moisture flow equation if this is solved simultaneously. The soil volume is discretized into a set of volume *elements*, separated by boundary interfaces or *nodes*. Fig. 7 shows the case of horizontal layering. Local average temperature and conductivity values, and heat storage (equivalently a heat capacity value, C_i) are attributed to either the elements or the nodes, indexed by i. The heat flow equation is then transformed into a set of algebraic equations, one for each i, including the upper (soil surface) and lower boundaries. Computer solution is by matrix algebra. The key to numerical meth-

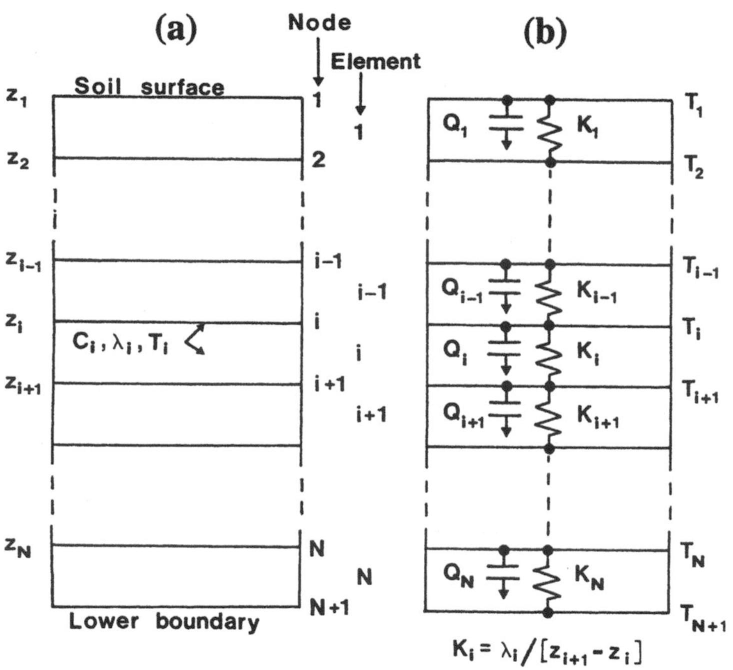

Fig. 7 Schematic layering of soil for numerical simulation of heat flow. (a) Finite difference and finite element methods. Values of T_i, λ_i, and centers of heat storage (with heat capacities C_i) are variously attributed to either nodes or elements, according to method used. (b) Network analysis method, showing equivalent resistors and capacitors. (From Campbell, 1985.)

ods is replacement of analytic time-integration by time-stepping from t_j to $t_{j+1} = t_j + \Delta t$. Temperatures are updated using

$$T_i^{j+1} = T_i^j + \frac{\Delta Q_i}{C_i} \tag{33}$$

where ΔQ_i is the net heat flow toward i from nodes (or elements) $i - 1$ and $i + 1$ over time step Δt. To obtain improved approximations to the true *average* ΔQ_i, various interpolation schemes for either the temperature or the heat content of i can be used, bridging both backward and forward in time. For temperature, a simple linear weighting can be used ($0 \leq \eta \leq 1$)

$$\overline{T} = \eta T_i^{j+1} + (1 - \eta)T_i^j \tag{34}$$

Thus $\eta = 0$ computes the net heat flow at the new time t_{j+1} from temperatures and their gradients at the previous time t_j, the so-called forward-difference scheme, which gives a direct or explicit expression for T_i^{j+1} in terms of the known T^j at $i - 1$, i, and $i + 1$. With $\eta > 0$ this simplicity is lost. Then T_i^{j+1} depends in an implicit way on spatially adjacent temperatures at t_{j+1} (Campbell, 1985). An assumed exponential decay or rise of $T_i(t)$ over the time step corresponds to $\eta = 0.57$ (Riha et al., 1980). More sophisticated interpolations exist (de Wit and van Keulen, 1972; Gerald and Wheatley, 1985).

There are three main numerical methods, differing in the ways they divide the space-time grid into discrete elements, attribute variables (to either nodes or elements), and refine the time-integration. They are

1. Finite difference, which assumes that node and time spacings are so small that parameters within them can be considered constant, and differentials may be replaced by their finite-difference forms (Carslaw and Jaeger, 1967; Mahrer, 1982).
2. Finite element, which uses elements of finite size and prescribes the variation of key parameters across the element, e.g., a constant heat flux, or a linear variation of temperature (Riha et al., 1980; Sidiropoulos and Tzimopoulos, 1983). This reduces the number of nodes and hence computational time.
3. Network analysis (Campbell, 1985; Bristow et al., 1986), which, developed for general flow processes in soil, also uses finite-sized elements, but with a physically based analysis of flow and storage analogous to resistance–capacitance networks in electrical circuit theory. To each element is attributed a conductivity K_i (the analog of a resistance), while a heat capacity and a temperature are ascribed to each node (the capacitance analog) (see Fig. 7). The method is recommended for its comparative simplicity, accuracy, and retention of physical insight (Campbell, 1985).

A fourth alternative is the use of ready-made computer simulation packages, obviating the need to write detailed numerical algorithms, e.g., CSMP (de Wit and van Keulen, 1972; Lascano and van Bavel, 1983) or ACSL.

For computational economy, grid spacings can be expanded in approximate inverse proportion to local rates of change of temperature. For example, node spacing can be progressively increased away from the soil surface (Wierenga and de Wit, 1970); or the algorithm can automatically increase Δt as simulation of a transient progresses. Algorithms are usually calibrated by comparing their output with exact analytical results for simpler problems. Element and time-step sizes are subject to two constraints: absolute values must be less than certain coarsest values, determined by trial variation, above which there is loss of accuracy (Milly, 1984); while their relative values may be constrained to ensure numerical stability (Campbell, 1985).

F. Freezing and Frozen Soil

Soil water freezes either as polycrystalline ice within the soil matrix or as separate ice lens inclusions that accrete when water migrates towards a slowly moving freezing front. Freezing brings large reduction in hydraulic conductivity and large increase in soil strength. Frost heave, which can lift soil, roots, and overlying structures, occurs only at or close to saturation, and usually only in frost-susceptible soils, i.e., those with texture dominated by silt or noncolloidal (>0.2 μm) clay fractions (Miller, 1980). On melting, holdup of surface water makes the thawed layers greatly susceptible to mechanical damage or erosion. The prediction of freezing temperature and frost and thaw penetration in soil is important for frost heave, and direct damage to roots, underground pipes, cables etc.

This section summarizes the theory of freezing point depression (ΔT), heat flow, and thermal properties. An approximate distinction can be made between *freezing* (or thawing) and *frozen* soil. In the former, phase change is an ongoing process, accompanied by freezing-induced redistribution of moisture, and by large effects on apparent thermal properties (Fuchs et al., 1978). In frozen soil, ice formation has effectively ceased and thermal properties have stabilized.

The depression of freezing point, a shift in the ice–water equilibrium, is due primarily to the lowering of the free energy (i.e., water potential) of soil water. It is given by (Miller, 1980)

$$\frac{L_f \Delta T}{273.15} = \frac{\psi_m + \pi}{\rho_l} - \frac{P_i}{\rho_i} \tag{35}$$

where $L_f = 3.33 \times 10^5$ J kg^{-1} is the latent heat of fusion of ice, ψ_m and π are the matric and osmotic components of the liquid water potential, ρ_l and ρ_i are the densities of liquid water and ice, and P_i is the ice pressure. For soil with low heave pressure, or unsaturated soil (Fuchs et al., 1978), $P_i = 0$, and then $\Delta T = 8.2 \times 10^{-7}(\psi_m + \pi)$. Thus with $\pi = 0$ and $\psi_m = -15$ bar (PWP), onset of freezing will occur at $T = -1.23$°C. As T is lowered beyond freezing onset, the ice phase grows progressively, initially in larger pores, possibly as water-drawing lenses, and later into surface-adsorbed layers. The persistence of liquid is explained mainly by the lower energy (hence ψ_m) of adsorbed water on particle surfaces, and partly by the tendency of water to freeze as pure ice, concentrating the solutes and lowering π in the remaining liquid. The former effect will clearly increase with clay content. Thus while most water freezes between 0 and −2°C in soils low in clay (Fuchs et al., 1978), the unfrozen water content in clay soils can be large at very low temperatures (e.g., as much as 10% by weight at −20°C; Penner, 1970; Yong and Warkentin, 1975; Jumikis, 1977).

The theory of heat flow in freezing soil exists at two levels. Earlier work, aimed at practical prediction of frost (or thaw) penetration, was dominated by the moving boundary approach, in which the freezing (thawing) zone is simplified to a sharp, moving front at depth $z_f(t)$, and the rate of latent heat production, proportional to $L_f\, dz_f/dt$, is balanced by net conduction away from the front (Yong and Warkentin, 1975; Jumikis, 1977; Bell, 1982; Hayhoe et al., 1983b). Later, more mechanistic models are based on simultaneous solution of heat and water transport equations, including phase change (Fuchs et al., 1978; Miller, 1980; Kung and Steenhuis, 1986). Striking features of these models include large thermally induced water flux and dramatic increases of thermal properties due to the phase change. Two major problematic quantities of the theory requiring more accurate description are the ice-formation characteristic dx_i/dT (Spaans, 1994) and the thermally driven water flux causing moisture redistribution.

Thermal properties of freezing soil exceed those of frozen soil, by up to several orders of magnitude, due to phase change effects. In freezing soil, continuing ice formation requires introduction of an apparent heat capacity (Fuchs et al., 1978; Miller, 1980):

$$C_{\mathrm{app}} = C - \rho_i L_f \frac{dx_i}{dT} \tag{36}$$

where C is the volumetric heat capacity of Eq. 17 with an added ice-fraction term, $x_i C_i$. The second, latent heat term causes C_{app} to "increase abruptly by several orders of magnitude as soon as ice is formed" (Fuchs et al., 1978), and, though diminishing as T decreases, dominates C_{app} down to a texture-dependent lower temperature where ice formation slows to a negligible level (about -2°C for the silt loam of Fuchs et al., 1978). The temperature range between onset of freezing and this lower limit defines a freeze–thaw zone of finite thickness, in contrast to the sharp front assumed in the simpler moving-boundary models. The apparent thermal conductivity, λ_{app}, of freezing soil is similarly increased, by the contribution of thermally driven water flow. This transports latent heat of fusion in a manner analogous to transfer of latent heat of vaporization by thermally driven vapor flow in ice-free soil (Fuchs et al., 1978). For *frozen* soil, C may be calculated using Eq. 17 with an ice term $x_i C_i$, and conductivity can be obtained from the theory of de Vries (see Sec. II.D), with about the same accuracy as for unfrozen soil (Penner, 1970; Jame and Norman, 1980).

III. MEASUREMENT TECHNIQUES

A. Temperature

1. Sensor Characteristics

An understanding of the general characteristics of temperature sensors is essential for proper choice and use of probe type. There is the question of the *type of output*, Q (e.g., voltage, current), and of the *temperature range*: likely near-surface extremes are -30 to $+50°$C, though bare surfaces can exceed 60°C in hotter regions (Miller, 1980).

There are two sources of measurement error.

1. *Precision* is a measure of a sensor's ability to reproduce a given value; it can be defined as the standard deviation of a set of repeated measurements of a fixed temperature.
2. *Accuracy* represents the deviation of the measured mean of the set from the true temperature on an established standard scale. It depends on care of calibration, including choice of interpolation formulae relating measured output to true temperature. Thus accuracy cannot be less than precision, but can be made close to it by careful calibration.

Stability refers to drift in accuracy with time. The *uniformity* of a sensor group or manufacturing method is the maximum expected difference in accuracy between sensors and determines their interchangeability. (*Tolerance* is also used, denoting typical or maximum deviation from a theoretical Q–T relationship.)

Table 3 summarizes typical maximum errors for various measurement objectives. The *resolution* of a device is the smallest difference in temperature it can detect. Thus precision cannot be less than resolution, though often the two are identical. Resolution is most commonly used to describe the readability of a total measurement system (e.g., electronic sensor plus meter or recorder). It is typically a fraction (e.g., one half) of a scale graduation, or one digit of a digital display.

Other priority characteristics include robustness, especially to exposure in soil, and, for electrical sensors, immunity to error signals (e.g., spurious connection emf's for thermocouple wires, or interference pickup).

Table 4 summarizes the principal sensor types. The *temperature coefficient* $Q^{-1}\, dQ/dT$ (or dQ/dT) expresses output sensitivity, important for choice of range and resolution of a connected meter or recorder. *Nonlinearity* is the maximum deviation from linear response over a chosen range (e.g., Fig. 8). It can be handled using linearizing bridge circuitry, e.g., for the thermocouple (Woodward and Sheehy, 1983) or the strongly nonlinear thermistor (Fritschen and Gay, 1979).

Table 3 Typical Error Requirements and Suitable Sensors

Objective	Allowable error (°C)	Most suitable sensors
Plant response and function (Woodward and Sheehy, 1983)	± 0.5	Any
Validation of soil T prediction models	± 0.2	Any electrical sensor
Temperature gradients, spatial variability, physical management contrasts	$\leq \pm 0.2$	Thermocouple, thermistor, resistance

Table 4 Comparison of Characteristics of Popular Sensors

Characteristic	Thermocouple (type K)	Resistance (Pt)	Thermistor	Semiconductor junction[b]	Mercury in glass
T-sensitive property	emf	Resistance	Resistance	I-V characteristic	Volume
Range (°C)	−50 to 400	−260 to 800	−50 to 150	−55 to 150	−10 to 55
T coefficient (°C^{-1})	~40 μV	0.39%	Variable, ~4%	1 μA	0.016%
Nonlinearity (°C)[a]	0.5	0.4	Large	0.8 max, 0.1 with trim	
Typical accuracy (°C)	± 0.1	± 0.1	± 0.2	± 0.2	≥ ± 0.2
Stability	Good	<0.05°C per year	<0.015°C per year	Long-term drift, <0.1°C	Good
Uniformity	Excellent for single spool of wire	± 0.3	± 0.2		
Time constant(s)[c]	0.2	1	0.8	>1	30
Dissipation constant (mW °C^{-1})		50	20		

[a] Nonlinearity over approximate range −10 to 50°C.

[b] Characteristics quoted refer to Analog Devices AD590 (Sec. III.A.2).

[c] Typical time constant in soil.

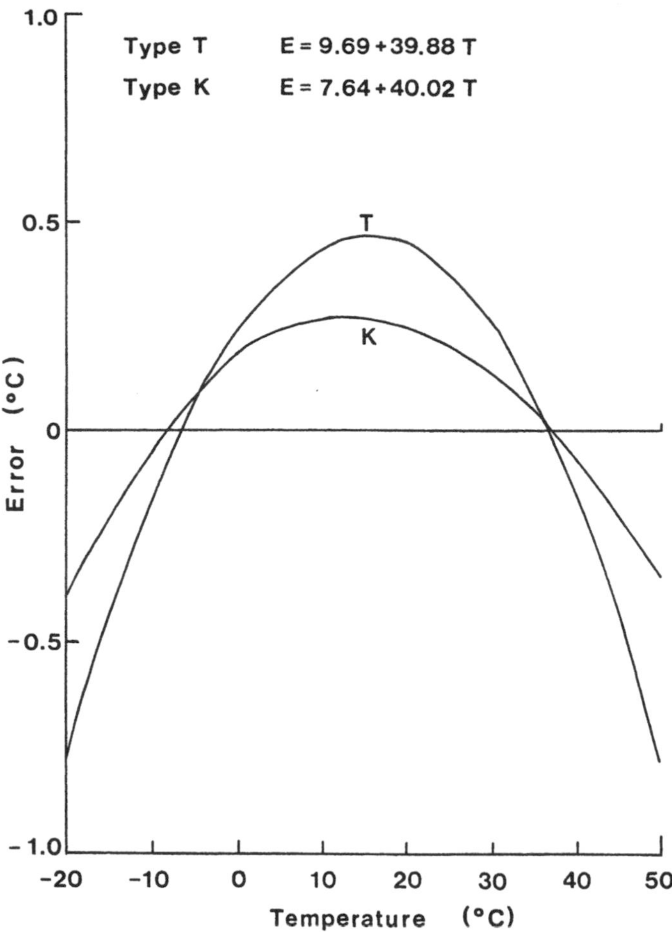

Fig. 8 Nonlinearity errors for popular type T and K thermocouples arising from the assumption of linear relationships between emf and temperature (see Table 4).

This was favored when direct readout instruments prevailed, but with modern logging and data processing, numerical conversion of unconditioned signal data is preferable and more accurate. The *time constant* (or response time) τ measures the delay in response to a step change in ambient *T*. It is the time taken for sensor *T* to reach $(1 - e^{-1}) = 63\%$ of the step change (Fritschen and Gay, 1979; Woodward and Sheehy, 1983) and determines sensor frequency response (Sec. III.B). Self-heating occurs in current-carrying sensors: the *dissipation constant*, *k*, is the power (mW) required to raise the sensor 1 C° above ambient. Both τ and *k* depend on the thermal properties of the sensor's environment. Values quoted in Table 4 are typical for a sensor in soil. In air, k may be up to 50 times smaller, and τ will be larger.

2. *Sensor Types*

Only the main features of sensors are discussed here. Further information can be found in reviews of temperature measurement (Sydenham, 1980; Meteorological Office, 1981; Woodward and Sheehy, 1983; Benedict, 1984; Bell and Rose, 1985) and of basic measurement circuitry (Woodward and Sheehy, 1983; Horowitz and Hill, 1989). Table 4 summarizes the most popular types. Some of the data quoted (e.g., range) are for a specific, good commercial sensor, and may vary for other probes.

a. Liquid-in-glass

Liquid-in-glass thermometers remain the standard soil probes in the meteorological services of many countries. Spirit-in-glass types placed just above the surface (e.g., grass, bare soil) measure minima (e.g., grass minimum). In-soil probes are invariably *mercury-in-glass*, with two main types (Meteorological Office, 1981). The World Meteorological Organisation (WMO) standard depths are 5, 10, 20, 50, and 100 cm (WMO, 1971); in the UK 30 cm (c.1 ft) is included. The first type is for depths to 20 cm, or even 30 cm. The stem has a right-angled bend with a graduated horizontal portion. The second, sheathed pattern type, for 30 cm and deeper, has a straight stem inside a sheath of stout glass tubing, suspended by a chain inside a hollow steel tube in the soil. The bulb is embedded in wax. This increases the time constant, enabling the thermometer to be lifted out of the steel tube for reading.

b. Electrical and Electronic

These are essential for intensive, automated measurements. Only the thermocouple is self-energized; others (i.e., resistance, thermistor, and semiconductor junction types) require an external power supply.

(1) *Thermocouples.* These are the most popular and are low in cost and easily constructed. They are differential devices, based on the Seebeck effect: a temperature-dependent contact emf is developed at the junction of two dissimilar metals. If two junctions at different temperatures are connected in a circuit (Fig. 9), the emf imbalance E increases as $T - T_r$ increases. Thermocouples can be used for accurate measurement of temperature differences, but more usually for absolute measurement of T (°C), either by separate measurement of T_r or by arranging $T_r = 0°C$ for the reference or cold junction, e.g., with an ice–water mix. More convenient is an ice point electronic reference junction (ERJ), which is basically a resistance bridge (with battery or AC power supply) containing a resistance that varies with ERJ temperature T_j. As T_j varies with ambient temperature, the bridge generates a compensation voltage equal and opposite to the thermo-

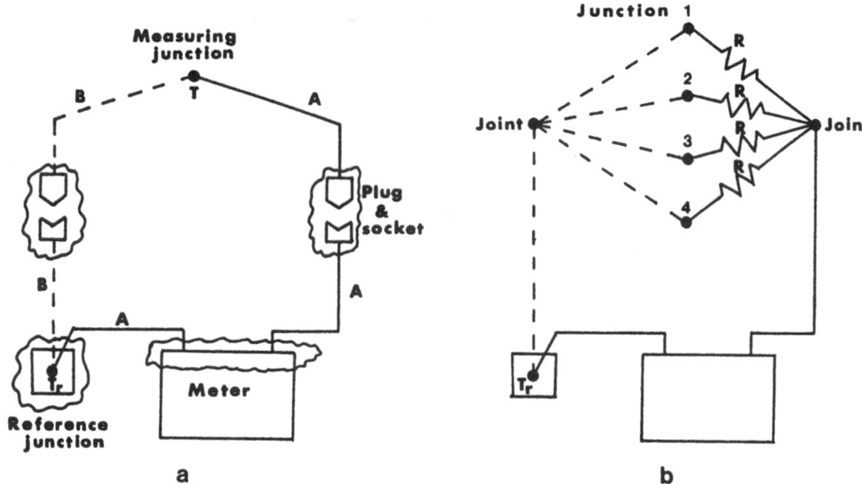

Fig. 9 Schematic thermocouple circuits. (a) Junction pair formed from metals A and B. Areas enclosed in wavy lines must form isothermal masses to avoid parasitic emf's. (b) Parallel arrangement for temperature averaging, with matched (e.g., 200 Ω ± 1%) swamping resistors R.

electric emf of an ordinary reference junction at T_j, thus forcing the ERJ to simulate the ice point (Woodward and Sheehy, 1983). The ERJ and its connections should form an isothermal mass to ensure correct compensation. Even then, imperfect compensation introduces an ERJ error, typically about ±0.1°C (Bell and Rose, 1985). With $T_r = 0°C$, output $E(T)$ is slightly nonlinear (Fig. 8), though the temperature coefficient is approximately constant over narrow ranges (e.g., about 40 $\mu V\ °C^{-1}$ for the most popular T or K types over 0–50°C). A good approximation is (Woodward and Sheehy, 1983)

$$E = a + bT + cT^2 \tag{37}$$

where a, b, and c are given in Table 5. Equation 37 is readily inverted to give $T(E)$. More accurate formulae and tables are available (Fritschen and Gay, 1979).

A major problem with thermocouple circuits is that all connections represent additional junctions, and so are potential temperature sensors producing error or parasitic emf's. These added junctions occur in pairs; each pair must separately be isothermal to avoid error emf's, e.g., in Fig. 9, each probe lead connection must be isothermal, as must paired meter connections J_1 and J_2 (Schimmelpfennig, 1976). Fortunately, temperature gradients *within* leads contribute no thermoelectric emf's.

The type T thermocouple, once the most popular, can help to minimize parasitic emf's if it has copper leads connected to a predominantly copper-conductor

Table 5 Characteristics of Popular Thermocouples[a]

Type	a (μV)	b (μV °C^{-1})	c (μV °C^{-2})
T Copper-constantan	−0.09	38.7	0.041
K Chromel-alumel	1.28	39.5	0.019
E Chromel-constantan	0.61	58.6	0.046
J Iron-constantan	1.02	50.4	0.026

[a] emf E is given by $E = a + bT + cT^2$, where T is the temperature (°C) and the reference junction is at 0°C.

circuit. Increasingly reliable, and popular with manufacturers, is Type K, which is more linear and resilient than Type T; and Type E, which offers greater uniformity, higher temperature coefficient, and potentially greater resolution.

Connection of N junctions in series gives a series thermopile: total emf is then the sum of the N emf's. This is used to amplify output in the soil heat flux plate (Sec. III.C). For spatial averaging of temperature, parallel connection is used (Radke et al., 1985). This maintains the signal level of a single device, requiring only one ERJ (Fig. 9). However, imbalance between thermocouple resistances will lead to error, so each junction should be series-connected to a matched (±1%) "swamping" resistor, with a value about 20 times the thermocouple resistance (Fritschen and Gay, 1979).

The main problems with thermocouples are the need for a reference junction; parasitic emf's; poor uniformity between wire batches; and avoiding electrical interference in the low-voltage (mV) DC output signal. Additional precautions include rigorous insulation of leads and junctions to eliminate connections to ground, especially in wet soil, and use of screening or twisted-pairs to minimize interference. High input impedance measurement circuitry, which minimizes thermocouple loading, allows very long leads but increases the risk of interference. Details of construction are given in Woodward and Sheehy (1983), Fritschen and Gay (1979), and Schimmelpfennig (1976).

(2) *Resistance Thermometers.* These are formed from metal conductors, in wire or film form. Most popular is the platinum resistance thermometer (PRT), described here, though other metals (e.g., nickel, tungsten) are used. The resistance $R(T)$ has very good linearity but more accurately follows a quadratic relation (Woodward and Sheehy, 1983). Though the temperature coefficient is low (Table 4), resistance can be measured with great accuracy. Most commercial PRT units have $R(0°\text{C}) = 100\ \Omega$, and an interchangeability of 0.1 Ω (= 0.3 K) at 0°C. A disadvantage is their low resistance. With two-wire measurement, increase in lead length beyond a few m will increase cable resistance error, which, though varying predictably with length, will change unpredictably with cable temperature. Compensating leads can then be used (Meteorological Office, 1981; Wood-

ward and Sheehy, 1983). Errors from (fixed polarity) thermal emfs in connectors and sensor junctions can be minimized by consecutive measurements with reversed excitation currents, or by using an AC bridge (Woodward and Sheehy, 1983). The large size of PRT's makes them unsuitable for 'point' measurements (e.g., at the soil surface) or for steep subsurface gradients.

(3) *Semiconductor Sensors.* The two basic types are the thermistor ("thermally sensitive resistor") and the junction device, based on the temperature sensitivity of a diode or transistor junction. The *thermistor* exploits the strong negative temperature coefficient of the resistance of semiconductor (metal oxide) material. The rate of promotion of electrons from the valence to the conduction band, across an effective energy gap E, and hence the electrical conductivity, is governed by the Boltzmann probability factor $\exp(-E/T)$. Thus resistance varies approximately as

$$R(T) = R(T_0) \exp \mathrm{B}\left(\frac{1}{T} - \frac{1}{T_0}\right) \tag{38}$$

where T is absolute temperature and B is a constant. The thermistor is often specified by its resistance $R(T_0)$ (usually several kΩ) at $T_0 = 298.15$ K $= 25°$C. From Eq. 37, a useful representation of calibration data is a linear regression of the form $\log R = mT^{-1} + c$, which is easily inverted to give $T(R)$. Small corrections can then be made for deviations from this linearity. Other calibration relationships are possible (e.g., a fifth-order polynomial for $T(R)$, Campbell Scientific Instruments, 1985).

Advantages of thermistors are their low cost, stability, robustness, and high temperature coefficient (ca. 10 times that of a PRT). They come in a variety of sizes suitable for soil use, down to <1 mm diameter catheter-types, excellent for point and surface contact measurement (Buchan, 1982b). Their large resistance minimizes interference and thermal emf errors, and it swamps connector and cable resistances, with cable errors typically only ca. 0.001°C m^{-1}. Self-heating effects are negligible with modern logging methods, which apply only a short-pulse excitation at sampling times.

Junction devices measure temperature via the the temperature sensitivity of the p–n junction (Woodward and Sheehy, 1983; Horowitz and Hill, 1989). The voltage across a forward-biased silicon diode carrying constant current decreases linearly by about 2 mV °C^{-1}. As the base-emitter junction of a bipolar transistor behaves essentially as a diode, both diodes and transistors can be used. More recent integrated circuit (IC) sensors use the temperature sensitivity of transistors fabricated on the chip and have integral amplifier and conversion circuitry. The Analog Devices AD590 (Table 4) is a two-terminal IC, is unfussy about supply voltages (+4 to +30 V), and has a nominal PTAT ("proportional to absolute temperature") current output of 1 μA K^{-1} (e.g., 298.2 μA at 25°C). With simple external trim circuitry and a two-point calibration, nonlinearity error can be reduced to about $\pm$ 0.2°C over the environmental range (R.S. Components, 1983).

Alternative voltage-output devices produced by National Semiconductor (LX and LM series) behave like the zener diode (a "constant" voltage device) with a temperature coefficient of 10 mV K^{-1} (e.g., output 2.982 V at 25°C) (Horowitz and Hill, 1989). These small IC devices remove the need for linearizing or resistance-measurement circuitry, cold-junction compensation, and, in the case of the AD590, for lead-wire compensation.

(4) *Infrared Thermometers* (IRT) (Huband, 1985; Woodward and Sheehy, 1983; Bell and Rose, 1985). The IRT measures surface temperature T_0 by remote sensing of the thermal radiation L_r transmitted to its detector, where

$$L_r = \tau[\epsilon\sigma T_0^4 + (1 - \epsilon)L_d] \tag{39}$$

Here ϵ is the emissivity of the surface, $1 - \epsilon$ is its reflectivity for ambient long-wave irradiance equal to the sky radiation L_d for a horizontal surface, and σ is the Stefan–Boltzmann constant. Radiation enters the detector via a sharp band-pass IR filter, which has transmissivity τ, usually over a waveband in the 8–14 μm range (i.e., the atmospheric window from which major H_2O and CO_2 absorption bands are absent). This window conveniently includes the blackbody radiation peak for terrestrial radiation, which varies according to Wien's law, $\lambda_{max} = 2900/T$ μm, between 10.6 μm (at 273 K) and 8.7 μm (at 333 K). Most commercial IRTs are calibrated for "grey-body" emission (i.e., $L_r = \tau\epsilon\sigma T_0^4$). The ϵ value is preset on a switch or dial. The reflected component in Eq. 39 is thus ignored, though smarter devices detect and correct for sky radiation. Unfortunately ϵ is very difficult to measure accurately, and literature values are usually used. For a typical bare loam soil, ϵ may vary between 0.90 (dry) and 0.95 (wet) (Brooks, 1960). Extreme values are 0.85 for pure dry quartz sand and 0.98 for water (Becker, 1980). Thus IRT errors will arise from uncertainty in ϵ and in L_d if corrections are made for reflected radiation. Both errors may be of the order of 1°C (Woodward and Sheehy, 1983). Care must be taken to distinguish between IRT *resolution*, which may be as small as 0.1°C, and IRT *accuracy* under the prevailing conditions. The latter may be ± 0.5°C or better using a high-quality IRT, plus an accurate ϵ value and reflection correction, but may be ± 1°C or poorer (Huband, 1985). However for comparative measurements under given radiative conditions (e.g., of spatial variability), resolution rather than accuracy is a better guide to discrimination ability. Devices, which may be handheld, are available with both narrow-angle (e.g., 3°) and wide-angle (e.g., 60°) fields of view.

Air- or satellite-borne scanning IRTs are used in remote sensing; they record spatial averages (e.g., the Skylab scanner with a 72 m resolution). The longer radiation-travel paths require either corrections for atmospheric absorption and emission or calibration with simultaneous ground-truth measurements (Curran, 1985).

(5) *Other Sensor Types.* The *quartz-crystal* thermometer produces a temperature-dependent oscillation frequency (Woodward and Sheehy, 1983) and is capable of very high accuracy (± 0.02°C) and resolution (10^{-4}°C). Despite its

great expense, it has been used in soils (Oliver et al., 1987). Several designs of *integrating thermometer* give either average temperature or the integral above some lower limit, e.g., "thermal time" measured in day-degrees (Hartley and MacLauchlan, 1969; Jones, 1972; Johnson and Thornley, 1985). Electronic integration of sensor output is now the obvious choice (e.g., Green et al., 1983).

3. Comparison of Sensors; Calibration

For field measurement, if error of ± 0.1°C is acceptable (Table 3), the author's preference is closely matched thermistors. They are low in cost, robust, reliable, and available in small sizes for good "point" (including surface) measurements. Their high resistance enables simple two-wire connection without lead compensation. They also enable measurement in the volt, rather than the millivolt, range of thermocouples, giving greater immunity to interference, including thermal emf's. Reference junctions are avoided. However, where smaller error is required (e.g., measuring localized gradients or spatial contrasts), thermocouples should be used. Temperature difference between two points can be measured more accurately by direct differential use of a thermocouple junction pair, rather than separately referenced junctions.

For surface thermometry, contact measurement with small, fast-response thermistors gives very good results. With careful installation (Buchan, 1982b), the error (about ± 0.3°C) is better than for typical use of an IRT, though the latter has clear advantage where rapid scanning or areal averaging are required.

Details of *calibration* can be found elsewhere (Benedict, 1984; Bell and Rose, 1985; Taylor and Jackson, 1986). If accuracy of about ± 0.2°C is acceptable, calibration against a secondary reference mercury-in-glass thermometer in a closely controlled temperature bath should suffice.

B. Sampling and Smoothing

Here we consider space and time sampling of soil temperature, and methods of data smoothing. Heat flux is considered in Sec. III.C. A discussion of electrical/electronic aspects of sampling (i.e., screening and grounding to reduce noise and interference, and sensor-recorder interfacing) can be found in Fritschen and Gay (1979), Weichert (1983), and Woodward and Sheehy (1983).

1. Spatial Variations

a. Vertical Sampling

The exponential decay $A_1 \exp(-z/D_1)$ of the dominant fundamental in the diurnal wave, Eq. 22, dictates closer spacing of sensors towards the soil surface. The optimum sequence will ideally be scaled to the damping depth D_1 (e.g., closer spacing in a peat than a loam). The author used the geometrical progression 0, 2, 4, 8,

16, 32, 64 cm in a loam, for which $D_1 = 12$ cm (Buchan, 1982b). However, closer spacing is required for accurate heat flux determination by gradiometric or calorimetric methods (Sec. III.C). One possible rationale is to space sensors at points marking constant decrement in amplitude, by a fraction f (e.g., 1/20th) of the surface value A_1. Defining a scaled depth z^* as z/D_1 generates the sequence

$$z^*_{n+1} = z^*_n - \ln(1 - f \exp z^*_n) \tag{40}$$

which can be initialized with $z^*_0 = 0$. This over-samples toward the surface compared to conventional practice, but with, say, $f = 1/20$th, only a partial sampling of the implied 20 depths need be used. Where seasonal progression is important (e.g., Sec. II.E), measurements to 1 m or deeper will be required. The WMO recommended depths are 5, 10, 20, 50, and 100 cm, plus discretionary "additional depths," under bare soil or grass (WMO, 1971).

Smoothing and interpolation of temperature profile data can be achieved with the cubic-spline method (Kimball, 1976). Measurements are grouped into successive sets of three, and curve segments (cubic polynomials) are fitted to each triplet. Imposing continuity of slope where segments meet determines the fourth parameter in each polynomial. This matches physical reality by ensuring continuity of dT/dz, and hence of soil heat flux.

b. *Horizontal Sampling*

Horizontal replication helps with two problems: spatial variability and rogue or faulty data. Both worsen toward the surface. Variability has both a deterministic and a stochastic character (Philip, 1980). Localized measurements involve mainly the latter, arising from variations in both surface condition (e.g., wetness, compaction), and in bulk thermal properties. Variability of T_0 at the surface is amplified with increased radiative forcing. Thus Buchan (1982b), sampling T_0 with three sensors (spacings ~ 10 cm), found a range averaging 0.4 K (daytime) and 0.2 K (night). Instantaneous values had a range up to ca. 1 K. This should attenuate rapidly with depth, due to the isothermalizing effect of horizontal heat flow. For surface contact thermometry, where faulty exposure can be an added problem, a minimum three-point replication enables averaging and possible "rogue-rejection." A two-point minimum is recommended at depths down to about D_1.

c. *Sensor Installation*

The main sources of installation error are probe size, soil disturbance, "stem" conduction, and, for a surface-placed probe, errors due to poor positioning, unrepresentative radiative characteristics, and "nonevaporation" from its surface. Near-surface measurement, including accurate gradiometry, requires small probes (i.e., thermocouples or bead thermistors). Horizontal installation of rod-shaped probes and leads will minimize conduction errors. At the surface, the author used a miniature (2 mm) bead thermistor, mimicking a small surface via a glued coating of fine soil (Buchan, 1982b).

2. *Temporal Variations*

Temperature fluctuations in soil are damped by its thermal mass, so the highest frequencies present are usually considerably lower than the cutoff frequency (related to the time constant) of a well-chosen sensor. The problem then is to find a suitable sampling frequency f_s, high enough to avoid information loss but low enough to avoid information excess, particularly with limited data storage. The sampling theorem (Woodward and Sheehy, 1983) states that faithful recording of the detail of a signal containing a maximum frequency f_{max} requires $f_s \geq 2f_{max}$ (the so-called Nyquist frequency). (This is suggested by the Fourier series method, Eq. 21.) Smaller f_s risks the aliasing effect: signal frequencies between f_s and f_{max} do not go unmeasured, but by intermittent sampling (at rate f_s), appear disguised as lower frequency variations (Fritschen and Gay, 1979). However this is an unlikely problem in measuring soil temperature. Thus f_s should be chosen to match the highest significant frequency component f_M required for adequate reconstruction (i.e., $f_s \geq 2f_M$). For the diurnal variation, recommendations are as follows. For the climatic (i.e., multi-day average) diurnal wave, $M = 2$ or 3 Fourier harmonics are adequate (Buchan, 1982b; Gupta et al., 1984). This implies a minimum of $2M$ (say 6) measurements per day. For the weather variation of a single day, a minimum of 6 harmonics, particularly on cloudy days, implies 12 or more samples per day (Buchan, 1982c; Horton et al., 1983). In practice, the interval $1/f_s = \Delta t = 0.5$ h or 1 h should be adequate, even at the surface. Smoothing can then be achieved by fitting a Fourier series (Eq. 22) to the data and truncating it at a cutoff frequency to suppress higher, noise-like harmonics. Cutoff at $M = 10$ (= 0.42 cycles/h) has been used (Kimball, 1974; Horton and Wierenga, 1983).

An incorrect procedure (Persaud and Chang, 1985), for both air and soil, is to use $(T_{min} + T_{max})/2$ for mean temperature $\overline{T}$ (i.e., effectively a two-point characterization). While correct for a simple sinusoidal wave, it is invalid for a multi-harmonic or more complex variation, and in practice overweights T_{max}. For the multi-day mean surface temperature wave, it has been found by the author to overestimate $\overline{T}$ by 0.5 to 1.0 K. For single-day variation the error could be greater. An improved procedure uses a weighted average, $\overline{T} = (\eta T_{min} + (1 - \eta)T_{max})/2$, with a weighting factor $\eta > 0.5$ (e.g., $\eta = 0.59$ at the soil surface (Parton, 1984)). Use of a daily averaged temperature, e.g., in calculating "thermal time" for biological processes (Johnson and Thornley, 1985), represents a single-datum characterization of the diurnal wave. This will be inadequate for a process with non-linear response to temperature.

C. Soil Heat Flux

The main use of measured heat flux G is to determine the surface flux ($G_0(t)$ in Eq. 1), usually for more accurate assessment of evaporation. Heat flux at depth

$G(z, t)$ is important for research into heat and coupled flows. Diurnal and annual cycles of the conductive component $G_T(z, t)$ may be decomposed into a set of harmonics (Eq. 23), each decaying according to $\exp(-z/D_n)$. The amplitude of the dominant fundamental decays to 10% of its surface value at $z = 2.3D_1$ (roughly 30 cm for the diurnal wave in a loam), so measurements may usually be confined to shallower depths. While the daily average of G_0 may be small (but see Sec. II.E on noncyclic heat storage), instantaneous values may reach up to about 250 W m^{-2} during peak daytime inflow (Tanner, 1963; Gupta et al., 1984). As a relative measure, the flux ratio G_0/R_n is often used and varies from about 0.5 for bare soils down to <0.1 under dense vegetation. Thus contrary to frequent assumption, G_0 may be a large fraction of the energy balance, and indeed dominant at night for sparsely vegetated surfaces.

As discussed in Secs. II.A and II.C, care must be taken with the heat flux concept where moisture gradients are large, as the *total* heat flux G_{tot} may contain a large isothermal latent heat component. This is not accounted for in the "conductive" component $G_T = -\lambda\, dT/dz$, which incorporates only the thermal latent heat flux. Thus in the common use of Eq. 1 to determine total soil evaporation E_{tot}, the frequent mispractice of using G_T instead of G_{tot} may give rise to substantial phase-change errors (Buchan, 1989; Mayocchi and Bristow, 1995). These errors arise from neglect of subsurface evaporation (or conversely condensation), or similarly from neglect of freeze/thaw effects.

Five methods of measurement or estimation of heat flux can be identified: transducer, the temperature gradient method, calorimetric methods, the Fourier analysis method, and the flux regression (i.e., G_0 versus R_n) method. The first two methods determine G_T via the temperature gradient, so they automatically exclude the isothermal latent heat component G_{vp} (Eq. 12).

1. *Transducers*

The heat flux plate (HFP) consists of a thermopile embedded in a thin, flat plate, usually a disk of glass or resin about 50 mm in diameter (d) and 5 mm in thickness (h). It impedes both liquid and vapor water flow, including evaporative supply to the surface, and so should not be placed where moisture gradients are large; 5–10 cm or deeper is recommended (Tanner, 1963; Horton and Wierenga, 1983). Despite this, depths of 1 cm or less have been used (Idso et al., 1975; Oliver et al., 1987). To obtain G_0, two methods can be used: the combination method (Tanner, 1963; Kimball and Jackson, 1975), i.e., the rate of change of heat storage above the HFP is determined by calorimetry, or a method proposed by Passerat de Silans et al. (1997). A major problem is the heat flux disturbance caused by difference between the plate conductivity λ_p and that of the soil, λ. Philip (1961), using the model of a thin spheroidal plate, derived an expression for the ratio of heat flux in the soil to that in the plate, G_p (the so-called heat flow disturbance factor) (Weaver and Campbell, 1985):

$$\frac{G}{G_p} = 1 - 1.7\left(\frac{h}{d}\right)\left(1 - \frac{\lambda}{\lambda_p}\right) \tag{41}$$

Plate thermopile output is a measure of ΔT, the temperature difference across the plate, and so of $G_p = \lambda_p \Delta T/h$. Hence use of a single calibration factor for the HFP presupposes a unique ratio G/G_p. However, as λ is variable, G/G_p should be close to 1 to minimize errors. This implies three design requirements: (a) The plate should be thin ($h/d \ll 1$); (b) λ/λ_p should be close to 1 (e.g., by ensuring plate conductivity is close to the soil value), and (c) since heat flow disturbance will still occur (i.e., $G/G_p \neq 1$), the HFP should be calibrated in a medium with conductivity close to the soil average. As an extreme example of the effect of calibration medium, Weaver and Campbell (1985) found that calibration of a plate with thickness $h = 4$ mm and conductivity $\lambda_p = 0.4$ W m^{-1} K^{-1}, first in dry and then wet sand ($\lambda = 0.4$ and 2.0 W m^{-1} K^{-1}, respectively), led to a doubling of the calibration constant—i.e., a potential for mismeasurement by a factor of two.

As an alternative to the thermopile HFP, commercial Peltier coolers have been used (Weaver and Campbell, 1985). These are thermopile-like devices, but with junctions of dissimilar (n- and p-type) semiconductor materials, rather than dissimilar metals. They are designed for cold-junction cooling under an applied electric current, i.e., exploiting the Peltier effect, the inverse of the Seebeck effect (Sec. III.A). They can, however, be used in Seebeck (thermoelectric) mode. The devices used by Weaver and Campbell had sensitivities about 70 mV kW^{-1} m^{-2}, i.e., about eight times that of a thermopile. However their temperature coefficient, at about 0.25% K^{-1}, was also greater, by about a factor of four. While cheaper and more sensitive than thermopiles, their conductivity (about 0.4 W m^{-1} K^{-1}) is lower than typical soil values, so heat flow disturbance may be large. Close plate–soil contact is essential. Air gaps, particularly in drying soil close to the surface, can lead to large errors. A minimum threefold replication is recommended for near-surface measurements. To increase sensitivity at deeper levels, where G_T is lower, series connection of plates (in fours) has been used (Fuchs and Hadas, 1972).

Details of flux plate calibration (Fuchs and Tanner, 1968; Biscoe et al., 1977; Woodward and Sheehy, 1983) and construction (Tanner, 1963; Fuchs and Tanner, 1968) are given elsewhere.

2. *Temperature Gradient Method*

Quite simply, this method computes the thermal heat flux $G_T(z) = -\lambda\, dT/dz$ from λ and dT/dz. Accurate gradiometry, especially close to the surface, requires accurate thermometry, with small sensors sufficiently close to avoid errors due to temperature profile curvature. Local conductivity may either be measured (Sec. III.D) or calculated (Sec. II.D).

3. Calorimetric Methods

These are based on the depth integral of the heat-conservation equation, Eq. 14, i.e.,

$$\begin{aligned} G_{\text{tot}}(z) &= G_{\text{tot}}(z_r) - \int_{z_r}^{z} C\left(\frac{\partial T}{\partial t}\right) dz \\ &= G_{\text{tot}}(z_r) - \sum C_i \Delta z_i \left(\frac{\Delta T_i}{\Delta t}\right) \end{aligned} \tag{42}$$

where z_r is a reference depth where G_{tot} is known. In practice, discrete sampling requires the second, finite-difference form of Eq. 41: a soil layer i, thickness Δz_i, is ascribed to each temperature sensor, recording change ΔT_i over interval Δt. Heat capacities C_i can be calculated from Eq. 17. Large errors may arise using short Δt (e.g., <30 min) as the ΔT_i may become too small (Kimball and Jackson, 1975).

Eq. 42 may be used or misused as follows. The correct use recognizes G_{tot} as the total heat flux, i.e., $G_T + G_{vp}$ (Eq. 12). However, most practitioners of calorimetry have neglected subsurface evaporation and assumed in effect that G_{tot} is the conductive flux, G_T (de Vries and Philip, 1986). This is the incorrect use. While this becomes correct when $G_{vp} = 0$ (i.e., when the heat summation is through soil layers effectively free from phase change), it can give rise to large phase-change errors where summation is through strong sites of net evaporation, e.g., drying near-surface layers (Mayocchi and Bristow, 1995). Following Eqs. 42 and 14, the correct equation for $G_T(z)$ is

$$G_T(z) = G_T(z_r) - \int_{z_r}^{z} C\left(\frac{\delta T}{\delta t}\right) dz - H(z, z_r) \tag{43}$$

where

$$H(z, z_r) = L_v[J_{vp}(z) - J_{vp}(z_r)] = L_v \int_{z_r}^{z} E_s(z)\, dz \tag{44}$$

Here H is the latent heat consumed by subsurface evaporation between z_r and z (but excluding that induced by the thermal vapor flux). Thus $H(z_1, 0) = L_v E_{s0}$, i.e., the total heat consumption by subsurface evaporation, where z_1 is a depth below which evaporation effectively ceases, i.e., "the lowest site of net phase change" (de Vries and Philip, 1986). This may be the dominant soil evaporative heat demand in a soil with a dry surface.

Thus the correct procedure for the conductive flux, G_T, requires, in the presence of subsurface evaporation, separate monitoring of the water (matric) potential profile to evaluate divergence in the vapor flux J_{vp} and hence Eq. 44. However, in the main practical use of calorimetry (i.e., to obtain G_0 at the surface for evaluation of *total* soil evaporation E_{tot}), it turns out (Sec. II.A) that $G_{tot}(z = 0)$ is required. Fortuitously, this is correctly delivered by the "simple" calorimetry of Eq. 42 if z_r is below the lowest site of evaporation, i.e., details of subsurface evaporation and the moisture profile are not required.

The key to calorimetry is knowledge of G at depth z_r. Several choices of z_r are possible. We look at each practice in turn, pointing out where subsoil evaporation can vitiate results. First, z_r may be chosen very deep (e.g., 100 cm), where $G_{tot} = G_T \approx 0$, and may either be neglected, or estimated from $\lambda\ dT/dz$.

Second, the null-point method, locates z_r at a null point, i.e., where $dT/dz = 0$. While this implies $G_T = 0$, error will result in Eq. 42 if z_r is within the evaporating layers. Unfortunately, weak curvature in $T(z)$ means that null points are poorly defined, except a few hours after sunrise or before sunset (Fig. 10). To enable calculation of $G(z)$ throughout the day, Kimball and Jackson (1975) introduced the so-called *null-alignment method.* Serious criticisms of this method have been made (de Vries and Philip, 1986), as its practitioners have usually implemented it in soils undergoing strong subsurface evaporation, while neglecting phase change by incorrect use of Eq. 42. This can accrue serious errors, both in $G(z)$ and in derived λ-values, which may be underestimated by 50% or more (de Vries and Philip, 1986).

The *combination method* determines the surface flux G_0 using calorimetry, with $G_T(z_r)$ measured by a heat flux plate, and $z = 0$ in Eq. 42. However if the

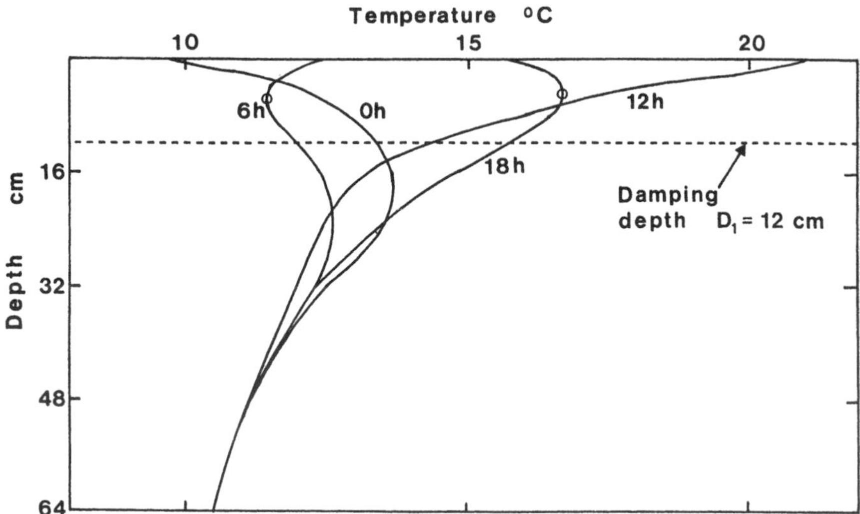

Fig. 10 Tautochrones (depth profiles) of soil temperature, showing null points (circled) 2.5 h after sunrise and before sunset. Data are for a 15-day average diurnal variation, as in Fig 4.

plate is too shallow (z_r above the lowest site of net phase change), this method will give the required total heat flux at the surface, *less* the isothermal latent heat flux at z_r not measured by the plate [i.e., $G_0 = G_{tot}(0) - L_v J_{vp}(z_r)$]. Passerat de Silans et al. (1997) proposed an alternative method to calculate G_0, from G and T measured simultaneously at a depth some cm below the surface.

In summary, calorimetry-based methods are attractive as they appear to require only monitoring of the temperature profile, plus knowledge of $C(z)$, and of G at one reference level. The null-alignment method appears to dispense with even the last requirement. However calorimetry, particularly the null-alignment method, can fail badly in soils with significant subsurface evaporation. Then correct procedure requires attention to (and some measurement of) vapor diffusion induced by water potential gradients. However, it turns out that for the main practical use of calorimetry (i.e., to determine $G_{tot}(0)$ for calculating evaporation) this is not required: simple calorimetry using a reference level below the lowest site of evaporation should give the correct value.

4. *Fourier Analysis Method*

This method fits a Fourier series, Eq. 22, to measured temperatures, then substitutes the derived amplitude and phase parameters in Eq. 23 to obtain $G_T(z, t)$. For accuracy the three assumptions following Eq. 22 should be approximately satisfied. Soil thermal properties (assumed uniform) must be known. If both λ and C are known, $G(z, t)$ may be calculated using Fourier parameters obtained from $T(z, t)$ at a single depth. Alternatively, if, say, only C is known, Fourier analysis of T at two or more depths can be used to evaluate κ (Sec. III.D), and then $\lambda = C\kappa$. The method has been used with a 10-harmonic fit to single-day data (Horton and Wierenga, 1983), giving good agreement with calorimetry, and with a two-harmonic fit to multi-day averaged diurnal variations (Gupta et al., 1984). Note that significant subsurface evaporation vitiates this method, since Eqs. 22 and 23 are solutions to the uncoupled heat flow equation.

5. *Flux Regression Method*

Several studies have sought approximate statistical relationships between the heat flux G and net radiation R_n, often as a simple ratio G/R_n. The main motivations are (a) the diurnal oscillations of G_0 and R_n are roughly sympathetic, and (b) $R_n - G_0$ is the available energy for partitioning between $L_v E$ and C in Eq. 1, so that relationships involving the *surface* flux, G_0, will enable better assessment of evaporation. However, relationships are complicated by their strong dependence on time of day; vegetative cover, and height (e.g., G_0/R_n ranges from about 0.5 for dry bare soil (Idso et al., 1975) to about 0.1 for grass or cropped soil (de Bruin and Holtslag, 1982; Clothier et al., 1986); the depth at which G is measured; and soil water content θ. The θ dependence appears to apply only to bare soils (Clothier et al., 1986): with G_T measured at 1 cm depth by heat flux plate, Idso et al. (1975) found the ratio G_T/R_n ranged from 0.5 for dry soil to 0.3 for wet soil. Comparison of studies is further complicated by the different methods used to measure G, giving different components of the heat flux.

D. Thermal Properties

Thermal conductivity and diffusivity are measured by inference, i.e., via observed temperature variations using the conductive heat equations. As these are uncoupled equations, the following methods require care in the presence of nonconductive heat flow, particularly subsurface evaporation.

1. Heat Capacity, C

This is usually calculated from Eq. 17, using volumetric sampling data, though calorimetric measurement is possible (Taylor and Jackson, 1986b). C may also be measured in situ (along with κ and λ) using the dual-probe heat-pulse method (Bristow et al., 1994). See Sec. 3 below.

2. Thermal Conductivity, λ

Methods can be classified as steady-state, transient, the diffusivity method, and the heat-flux method.

Steady-State Methods

These laboratory methods establish a steady, uniform temperature gradient dT/dx in a containerized sample by supplying heat at a rate Q (W m^{-2}) at one end. Then $\lambda = -Q/(dT/dx)$. Cylindrical containers have been used, with a planar electrical heat source at one end, and dT/dx measured close to the source (Hadas, 1974). The divided bar method is used for both rocks and soils (Williams, 1982). The sample is placed between plates of a uniform material of known conductivity, and a steady-state established across the sandwich, Q being calculated from dT/dx in the outer plates. A serious problem with steady-state methods is the migration of water from warmer to colder zones, leading to nonuniformity. Hadas (1974) overcame this by measuring both temperature and moisture gradients close to the heat source to obtain $\lambda(\theta)$. While stationary methods are acceptable for saturated or very dry soils, transient methods should be used for moist soils.

Transient Methods

The *cylindrical heat probe* has been used extensively in laboratory and field (de Vries and Peck, 1958; van Wijk and Derksen, 1966; Fritton et al., 1974; Sepaskah and Boersma, 1979). Its transient heat flow avoids major moisture redistribution. It consists of a continuously heated line heat source, usually a cylindrical probe containing an electrical heating wire or element. Soil temperature increase $T(t)$ is measured with a small sensor next to the heating wire. From Eq. 32, a plot of T against log t is a straight line whose slope gives λ. However, this is valid only for probes of small diameter. Large-diameter probes require a more complex solution to the conduction equation (van Wijk and Derksen, 1966; Jackson and Taylor, 1986). Imperfect probe–soil contact is a potential problem, particularly in drier soil: the extra thermal resistance involves an additional contact factor in the solution (de Vries and Peck, 1958; Hadas, 1974). A laboratory variant of the single probe is the twin transient-state cylindrical-probe method (Kasabuchi, 1984). One probe is placed axially in a cylinder of the soil, and its $T(t)$ is referenced to an identical probe placed in a cylinder of standard material (agar gel).

An alternative powerless probe uses a cylindrical glass or aluminum probe which is preheated (or cooled) and thrust into the soil (Riha *et al.*, 1980); then its temperature is monitored via a built-in thermocouple. λ is that value giving closest simulation of the measured temperature trend. Advantages of the method are its large, more durable probe, and the absence of heater circuitry.

Diffusivity Method

Here, conductivity is obtained as $\lambda = \kappa C$, where diffusivity κ can be measured using methods described below, including the dual-probe heat-pulse method.

Heat-Flux Method

This is the inverse of the temperature-gradient method for measuring the conductive heat flux G_T. Thus $\lambda = -G/(dT/dz)$ is obtained from simultaneous measurement of G and dT/dz. However G must be the conductive component G_T, e.g., from a heat flux plate. In field soils with strong subsurface evaporation, use of an incorrect G, e.g., from erroneous calorimetry, has been shown to give serious errors in measured λ-values (de Vries and Philip, 1986).

3. *Thermal Diffusivity, κ*

Diffusivity controls the dynamic redistribution of heat in soil and so is deduced from the observed propagation of temperature variations, either periodic or transient. Field methods are based on solutions (Sec. II.E) to the one-dimensional uncoupled heat conduction equation, Eq. 7. Hence they yield a so-called apparent diffusivity, κ_a, i.e., a κ that subsumes nonconductive (distillation) effects into the theoretically assumed pure conductive flow (Chen and Kling, 1996). Methods use either transient or periodic temperature variations.

Propagation of Transients

A recent successful innovation is the dual-probe heat-pulse method (Bristow et al., 1994). The probe consists of dual small-diameter (ca. 1 mm) needles, spaced ca. r = 5 to 10 mm apart. A short-duration (ca. 5 to 15 s) heat pulse is applied to a line heating element in one needle, and temperature change $\Delta T(r, t)$ is recorded from a sensor in the second needle. The properties extracted are the diffusivity κ and volumetric heat capacity C; and hence $\lambda = C\kappa$ is calculated (Bristow et al., 1995; Kluitenberg et al., 1995). This versatile probe can be made multipurpose, additionally yielding soil temperature and volumetric water content θ_v (deduced from C). Also, measurement of interneedle electrical resistance can give an estimate of soil electrical conductivity, and hence enable in situ monitoring of e.g. a solute breakthrough curve. A commercial version of the dual probe is available. Noborio et al. (1996) developed a three-rod time-domain reflectometry (TDR) probe with the dual probe incorporated into two of its rods, enabling simultaneous measurement of water content θ_v and thermal properties.

A different field method uses Laplace transform analysis of the shallow penetration of a transient change in soil surface heating, giving an effective κ for, typically, the upper 5–10 cm (van Wijk, 1963; van Wijk and Derksen, 1966; Asrar and Kanemasu, 1983). In the original technique, for field or laboratory use, an artificial heat pulse is imposed by heating the surface with a lamp for a short period (5–30 min) (van Wijk and Derksen, 1966). Conversely, soil shading can be applied on bright days. $T(z, t)$ is measured at two or more depths reached by the pulse. Following Eq. 30 et seq., a plot of $L\{T'\}$ versus depth should give a straight line of slope $-\sqrt{(s/\kappa)}$. Natural transients under broken cloud can also be used. Again, the method is based on the uncoupled conduction equation, Eq. 7, so that any nonconductive flows will be built into an apparent ("conductive") diffusivity.

An alternative mathematical approach analyzes a short-term portion of the diurnal variation of $T(z, t)$ in upper soil layers, starting at a time of isothermal conditions (Singh and Sinha, 1977). Two parameterized curves are fitted to observed temperature variations, (a) an analytic function to $T_0(t)$ (e.g., linear rise or fall, or sine wave) and (b) a cubic spline to the near-surface temperature profile at a fixed time $t_0 > 0$, to give dT/dz at the surface at t_0. Analytic expressions give κ in terms of the parameters of the fitted curves.

The diffusivity of soils packed in long cylinders can be determined using an unsteady-state method (Parikh et al., 1979; Hopmans and Dane, 1986a). The sealed column is equilibrated in a water bath at temperature T_1, then transferred to a bath at temperature T_2, the heating or cooling curve $T(t)$ at the column center being measured with a sensor. A plot of $\log[(T_2 - T)/(T_2 - T_1)]$ against t is a straight line whose slope gives κ.

Propagation of Periodic Variations

Methods are based on depth penetration of the diurnal or annual wave, represented in Fourier series form (Eq. 22). Thus, in addition to the assumption of uncoupled heat flow, the three assumptions following Eq. 22 are implicit. κ is determined from temperature measurements at two depths z_1 and z_2 by one of three methods: amplitude decay, phase lag, or matching of wave penetration. The first two methods yield κ via the damping depth $D_n = \sqrt{(2\kappa/n\omega_1)}$, usually for the fundamental ($n = 1$), using Eqs. 45 and 46. From Eq. 22, D_n is given by

$$D_n = \frac{[z_2 - z_1]}{\ln[A_n(z_1)/A_n(z_2)]} \tag{45}$$

$$D_n = \frac{[z_2 - z_1]}{[\gamma_n(z_1) - \gamma_n(z_2)]} \tag{46}$$

where $A_n(z)$ is the amplitude, and $\gamma_n(z) = \phi_n - z/D_n$ is the phase at depth z for the nth harmonic. In the third method, a Fourier series is fitted to measured T at depth z_1, and κ is the value in Eq. 22 that best reproduces measurements at z_2, e.g., by minimizing the sum of squared differences (Horton et al., 1983).

The first two methods have been found by the author and others (Horton et al., 1983) to give erratic results for the diurnal wave, chiefly due to large relative errors in the ratio and difference terms in Eqs. 45 and 46. However Horton et al. (1983) found that the third method gave consistent, reliable results, provided the number of observations per day at both depths was between 8 (on clear days) and 12 (on cloudy days, with more irregular variations). A simplified practice to be avoided uses Eq. 45 with the approximation $A_1(z) = (T_{max} - T_{min})/2$ for the fundamental. While correct for a simple sine wave, this is in error for a typical, complex diurnal variation and gives an unreliable κ value (Horton et al., 1983).

E. Frost and Thaw Penetration Depth

Five main methods of measuring the depth of the frozen–unfrozen interface in soil are summarized here. First, the *0° C isotherm* can be located by interpolation on a measured temperature profile (Hayhoe et al., 1983b). Depression of the freezing point (Sec. II.F) may make this an inaccurate indicator of the interface (Brach et al., 1985), particularly when temperature gradients are small and the freezing front diffuse. The *frost-tube* (Rickard and Brown, 1972; Caprio et al., 1977; Hayhoe et al., 1983b) consists of a bottom-sealed length of PVC pipe installed in the soil, containing a removable inner clear tube filled with sand saturated with fluorescein solution. On freezing, the dye changes color from green to pale yellow, indicating the depth of the frozen–unfrozen interface within the tube. The *time-domain reflectometry* (TDR) method (Chap. 1) responds to the different dielectric characteristics of water in the liquid and frozen state (Hayhoe et al., 1983a). The interface is detected by one of two techniques, a change in apparent *liquid* water content at the frost or thaw front, or a change in the reflection coefficient for electromagnetic waves at the interface, shown as a kink in the TDR trace. A comparison of the isotherm, frost-tube, and TDR methods gave comparable results for frost penetration depth but showed that in the thawing soil the TDR method was superior in detecting the presence and depth of the unfrozen–frozen interface (Hayhoe et al., 1983a). An electrical *capacitance probe* developed by Brach et al.

(1985) also exploits the large drop in dielectric constant (from about 80 to 3) as water freezes. *Gypsum blocks* placed in soil undergo an abrupt change in electrical conductivity as water freezes within them and can be used with a low-power DC supply (Burgess and Hanson, 1979). However, they lose discrimination when dry, i.e., in dry soils.

IV. APPLICATIONS

Soil thermal regime has application in five main areas: (a) its effects on biological processes; (b) its effects on chemical and physical processes, including soil weathering; (c) its role in determining above-ground climate; (d) remote sensing applications, and (e) engineering applications. Applications may be broadly classified into two categories, "passive" measurement and modeling and "active" manipulation.

A. Measurement and Modeling

In recent years, increasing environmental concerns have ushered in an upsurge in recognition of the crucial role of soil temperature as a key controller of processes in the soil mantle and the wider biosphere. Examples include soil temperature as a controller of soil–atmosphere exchange of greenhouse gases and soil carbon storage (Bouwman, 1989; Lal et al., 1995), solute transport and reaction processes (Hopmans and Dane 1986b; Barrow, 1992; Nassar and Horton, 1992), and transformations and transport of soil contaminants, including pesticides (Cheng, 1990; Lehmann et al., 1993), heavy metals (Selim and Amacher, 1997), and bacteria.

Many workers have characterized soil temperature regimes under contrasting types of ground cover, e.g., forest (Bocock et al., 1977; Halldin, 1979), grassland and field crops (Deardorff, 1978; Horton et al., 1984; Parton, 1984; Roodenburg, 1985; Main, 1996), bare soil (Buchan, 1982a; Schieldge et al., 1982); and in response to variations in environmental factors, e.g., altitude (Green and Harding, 1979), latitude and geographical location (Toy et al., 1978; Meikle and Gilchrist, 1983), and snow cover and freezing (Rieger, 1983). However, the theory of heat and water flow in the soil, coupled to energy exchanges at the surface (particularly below complex canopies) is insufficiently developed to provide a comprehensive framework capable of accurate prediction of these contrasting regimes.

B. Manipulation

Manipulation is directed at areas a, c, and e listed above (Sec. IV). Many reports and studies have assessed the effects on soil temperature regime of drainage (Scotter and Horne, 1985), mulching (Tanner, 1974; Davies, 1975; Rosenberg et al.,

1983; Bristow et al., 1986; Liakatis et al., 1986), tillage (Allmaras et al., 1977; Gupta et al., 1984; Radke et al., 1985), alteration of surface characteristics (Potter et al., 1987), plant cover and height (Rosenberg et al., 1983; Green et al., 1984), shading (Stigter, 1984), and artificial heating (Rosenberg et al., 1983). The mitigation of both ground and air frost is a prime example of manipulation (Tanner, 1974; Rosenberg et al., 1983). However, the multiprocess nature of the soil–atmosphere system often leads to a lack of clear, unequivocal results. Again, there is a distinct need for careful development of mechanistic models, capable of resolving the effects of changes (often subtle) in surface characteristics, soil thermal properties, and nonconductive heat flow.

Another classic example of manipulation is the elevation of soil temperature via soil *solarization* (Katan and DeVay, 1991), to control soil pathogens and weeds. "Engineering" applications include heat exchange with underground structures (e.g., cables, heat pumps); and deliberate manipulation, e.g., artificial ground freezing (Frémond, 1994), and the effects of ground covering (e.g., asphalt or concrete) on both surface temperature (e.g., road surfaces, Jacobs and Raatz, 1996) and above-ground microclimate (e.g., the urban heat island effect, Asaeda and Ca, 1993).

V. CONCLUSION

Practical interest in soil thermal regime focuses on two quantities. First, *temperature*, which controls temperature-sensitive soil, plant, and microclimate processes. Second, soil *heat flux*, which controls both within-soil and soil–atmosphere energy exchanges: its main utility is in improving estimates of soil surface evaporation, E, as a component of total evapotranspiration.

However, it is essential to realize that E is not sourced purely at the surface. Subsurface evaporation, often dominant in drying soils, can, via its heat sink effect, profoundly influence heat flux and temperature profiles. Unfortunately, in many experimental studies, this nonconductive heat flow is overlooked. This chapter clarifies the concepts essential to proper design and interpretation of experiments, in particular the concept of total soil heat flux, and its role in interpreting the surface energy balance equation.

Soil temperature regime is sometimes misconceived as one of the solved or inactive disciplines of soil physics. However it has several unsolved problems and deficiencies. The greatest problems are posed by soils undergoing aqueous phase change, requiring description of coupled heat and mass transfer processes. The processes in freezing soils (Fuchs et al., 1978) qualify as an outstanding unknown. In unfrozen field soil, fuller study of the extent of subsurface evaporation is required (de Vries and Philip, 1986; Mayocchi and Bristow, 1995).

The coupling of soil temperature models to the surface energy balance may explain hitherto empirical observations, e.g., of relations between soil and air temperatures (e.g., Gupta et al., 1984). Relationships between the harmonic composition of soil temperature and of its driving variables (solar radiation, air temperature (Tabony, 1984)) deserve further study (see Sec. II.E); e.g., in climatological analysis of geographical and long-term variations of the annual wave. The effects of *time-varying* temperature on biological processes are relatively poorly understood. Industrial thermographic imaging techniques offer methods to study variability in temperature and hence physical properties of soil. Finally, enormous scope for study of practical issues is offered by the intensifying attention to the interactions of soil temperature regime, and soil-based environmental processes, such as soil–atmosphere gas exchange, solute transport, and the fate of soil and groundwater contaminants.

ACKNOWLEDGMENTS

The author is grateful for support from colleagues at Lincoln University, and for cooperative research with Prof. V. Tarnawski; St Mary's University, Canada.

LIST OF SYMBOLS

A amplitude (d, diurnal, a, annual), °C
C volumetric heat capacity, J m^{-3} K^{-1}
D damping depth (D_n for nth harmonic), m
D_v apparent vapor diffusivity in soil air, kg m^{-1} s^{-1} Pa^{-1}
D_{va} vapor diffusivity in bulk air, kg m^{-1} s^{-1} Pa^{-1}
E generalized rate of evaporation to air, kg m^{-2} s^{-1}
E_{tot} $= E_0 + E_{s0}$ = total soil evaporation to air, kg m^{-2} s^{-1}
E_0 component sourced at soil "surface," kg m^{-2} s^{-1}
E_{s0} subsurface-sourced component, kg m^{-2} s^{-1}
$E_s(z)$ rate of net evaporation at depth z, driven by gradients in moisture content, kg m^{-3} s^{-1}
e vapor pressure, Pa
e_s saturation vapor pressure, Pa
G generalized soil heat flux, W m^{-2}
G_0 surface value of G, W m^{-2}
G_{tot} $= G_T + G_{vp}$ = total soil heat flux, W m^{-2}
G_T $= -\lambda\, dT/dz$ = thermal soil heat flux, W m^{-2}
G_{vp} $= L_v J_{vp}$ = isothermal latent heat flux induced by a moisture gradient, W m^{-2}
G_{vT} $= L_v J_{vT}$ = thermal latent heat flux induced by a temperature gradient, W m^{-2}
G_c "pure" conduction component of G, W m^{-2}
g granule shape factor
H sensible heat flux from soil surface to air, W m^{-2}
h relative humidity
k weighting factor for thermal conductivity
J_v total vapor flux in soil air phase, kg m^{-2} s^{-1}

J_{vT} thermal vapor flux, kg m^{-2} s^{-1}

J_{vp} isothermal vapor flux, kg m^{-2} s^{-1}

$L<f>$ Laplace transform of function f

L_d downward long-wave irradiance ("sky radiation"), W m^{-2}

L_u upward long-wave irradiance, W m^{-2}

L_f latent heat of fusion of ice, J kg^{-1}

L_v latent heat of vaporization, J kg^{-1}

M_w molecular weight of water, kg mol^{-1}

n harmonic number

N number of data points

N number of parameters in Fourier series

p total air pressure in soil, Pa

r radius, m

R gas constant, 8.314, J mol^{-1} K^{-1}

R_n net radiation, W m^{-2}

R_s solar radiation, W m^{-2}

s $= de_s/dT =$ slope of svp curve, Pa K^{-1}

s parameter in Laplace transform

t time, s

T temperature (subscripts: 0, soil surface; a, air) °C or K

x volume fraction of a soil component (subscripts: m, mineral solids; w, water; i, ice; c, continuous phase)

z soil depth, m

α short-wave reflection coefficient

α pore space tortuosity factor

ϵ long-wave emissivity

ΔS daily change in soil heat storage, J m^{-2} d^{-1}

η weighting factor

θ volumetric water content ($= x_w$)

θ_a air-filled porosity, m^2 s^{-1}

κ thermal diffusivity, m^2 s^{-1}

λ thermal conductivity, W m^{-1} K^{-1}

ν mass flow factor

ρ density (subscripts as for x), kg m^{-3}

σ Stefan–Boltzmann constant, W m^{-2} K^{-4}

τ period, time constant, transmissivity, s

ϕ phase angle

ψ_m matric potential, J kg^{-1}

π osmotic potential

ω angular frequency, s^{-1}

REFERENCES

Allmaras, R. R., Hallauer, E. A., Nelson, W. W., and Evans, S. D. 1977. *Surface energy balance and soil thermal property modifications by tillage-induced soil structure.* Univ. Minnesota, Agric. Exp. Station, Tech. Bull. 306.

Al-Nakshabande, G., and Kohnke, H. 1965. Thermal conductivity and diffusivity of soils as related to moisture tension and other physical properties. *Agric. Meteorol.* 2: 271–279.

Asaeda, T., and Ca, V. T. 1993. The subsurface transport of heat and moisture and its effect on the environment: A numerical model. *Boundary-Layer Meteorol.* 65:159–179.

Asrar, G., and Kanemasu, E. T. 1983. Estimating thermal diffusivity near the soil surface using Laplace transform: Uniform initial conditions. *Soil Sci. Soc. Am. J.* 47: 397–401.

Barrow, N. J. 1992. A brief discussion on the effect of temperature on the reaction of inorganic ions with soil. *J. Soil Sci.* 43:37–45.

Becker, F. 1980. Thermal infra-red remote sensing principles and applications. In: *Remote Sensing Application in Agriculture and Hydrology* (G. Fraysse, ed.). Rotterdam: A. A. Balkema, pp. 153–213.

Bell, C. J., and Rose, D. A. 1985. The measurement of temperature. In: *Instrumentation for Environmental Physiology* (B. Marshall and F. I. Woodward, eds.). Cambridge: Cambridge Univ. Press, pp. 79–99.

Bell, G. E. 1982. The prediction of frost penetration. *Int. J. Num. Anal. Meth. Geomech.* 6: 287–290.

Benedict, R. P. 1984. *Fundamentals of Temperature, Pressure and Flow Measurements.* 2d ed. New York: John Wiley.

Biscoe, P. V., Safell, R. A., and Smith, P. D. 1977. An apparatus for calibrating soil heat flux plates. *Agric. Meteorol.* 18:49–54.

Bloomfield, P. 1976. *Fourier Analysis of Time Series: An Introduction.* New York: John Wiley.

Bocock, K. L., Jeffers, J., Lindley, D. K., Adamson, J., and Gill, C. 1977. Estimating woodland soil temperature from air temperature and other climatic variables. *Agric. Meteorol.* 18.

Bouwman, A. F. 1989. *Soils and the Greenhouse Effect.* Chichester, UK: John Wiley.

Brach, E. J., Mack, A. R., Hayhoe, H., and Scobie, B. 1985. Electrical determination for frost depth in soil. *Agric. For. Meteorol.* 34:173–181.

Bristow, K. L., Campbell, G. S., Papendick, R. I., and Elliott, L. F. 1986. Simulation of heat and moisture transfer through a surface residue–soil system. *Agric. For. Meteorol.* 36:193–214.

Bristow, K. L., Kluitenberg, G. J., and Horton, R. 1994. Measurement of soil thermal properties with a dual-probe heat-pulse technique. *Soil Sci. Soc. Am. J.* 58:1288–94.

Bristow, K. L., Bilskie, J. R., Kluitenberg, G. J., and Horton, R. 1995. Comparison of techniques for extracting soil thermal properties from dual-probe heat-pulse data. *Soil Sci.* 160:1–7.

Brooks, F. A. 1960. *An introduction to physical microclimatology.* Syllabus No. 397, Univ. of California, Davis.

Buchan, G. D. 1982a. Predicting bare soil temperature. I. Theory and models for the multiday mean diurnal variation. *J. Soil Sci.* 33:185–197.

Buchan, G. D. 1982b. Predicting bare soil temperature. II. Experimental testing of multiday models. *J. Soil. Sci.* 33:199–209.

Buchan, G. D. 1982c. Predicting bare soil temperature. III. Extension to single-day variation. *J. Soil Sci.* 33:365–373.

Buchan, G. D. 1989. Soil heat flux and soil surface energy balance: A clarification of concepts. *Proc. 4th Australasian Conf. on Heat and Mass Transfer, Univ. of Canterbury, New Zealand*, pp. 627–634.

Burgess, M. D., and Hanson, C. L. 1979. Automatic soil-frost measuring systems. *Agric. Meteorol.* 20:313–318.

Campbell, G. S. 1985. *Soil Physics with BASIC*. Amsterdam: Elsevier.

Campbell Scientific Instruments 1985. *Model 107 Temperature Probe*. Logan, UT: Campbell Scientific.

Caprio, J. 1977. Soil freeze tube depths. *Abs. Nat. Conf. Agric. Forest Meteorol., Am. Meteorol. Soc., Tucson, Arizona, Oct. 1977.*

Carslaw, H. S., and Jaeger, J. C. 1967. *Conduction of Heat in Solids*. 2d ed. Oxford: Clarendon Press.

Carson, J. E. 1963. Analysis of soil and air temperatures by Fourier techniques. *J. Geophys. Res.* 68:2217–2232.

Cass, A., Campbell, G. S., and Jones, T. L. 1984. Enhancement of thermal water vapor diffusion in soil. *Soil Sci. Soc. Am. J.* 48:25–31.

Chen, D., and Kling, J. 1996. Apparent thermal diffusivity in soil: Estimation from thermal records and suggestions for numerical modeling. *Phys. Geog.* 17:419–430.

Cheng, H. H. 1990. *Pesticides in the Soil Environment: Processes, Impacts and Modeling*. Madison, WI: Soil Sci. Soc. Am. Book Series 2.

Chudnovskii, A. F. 1962. *Heat Transfer in the Soil*. Jerusalem: Israel Prog. Sci. Transl.

Clothier, B. E., Clawson, K. L., Pinter, P. J., Moran, M. S., Reginato, R. J., and Jackson, R. D. 1986. Estimation of soil heat flux from net radiation during the growth of alfalfa. *Agric. For. Meteorol.* 37:319–329.

Curran, P. J. 1985. *Principles of Remote Sensing*. London: Longman.

Davies, J. W. 1975. Mulching effects on plant climate and yield. Tech. Note No. 136, WMO, Geneva.

Deardorff, J. W. 1978. Efficient prediction of ground surface temperature and moisture with inclusion of a layer of vegetation. *J. Geophys. Res.* 83:1889–1903.

de Bruin, H. A. R., and Holtslag, A. 1982. A simple parameterization of the surface fluxes of sensible and latent heat during daytime compared with the Penman-Monteith concept. *J. Appl. Meteorol.* 21:1610–1621.

de Vries, D. A. 1963. Thermal properties of soils. In: *Physics of Plant Environment* (W. R. van Wijk, ed.). Amsterdam: North Holland, pp. 210–235.

de Vries, D. A. 1975. Heat transfer in soils. In: *Heat and Mass Transfer in the Biosphere* (D. A. de Vries and N. H. Afgan, eds.). New York: Scripta, pp. 5–28.

de Vries, D. A., and Peck, A. J. 1958. On the cylindrical probe method of measuring thermal conductivity with special reference to soil. I. Extension of theory and discussion of probe characteristics. II. Analysis of moisture effects. *Aust. J. Phys.* 11:255–271, 409–423.

de Vries, D. A., and Philip, J. R. 1986. Soil heat flux, thermal conductivity, and the null-alignment method. *Soil Sci. Soc. Am. J.* 50:12–18.

de Wit, C. T., and van Keulen, H. 1972. *Simulation of Transport Processes in Soils*. Wageningen, The Netherlands: Pudoc.

Farouki, O. T. 1986. *Thermal Properties of Soils*. Clausthal-Zellerfeld, Germany: Trans Tech Publications.

Frémond, M. 1994. Ground freezing. *Proc. 7th Int. Symp., Nancy, France, Oct. 1994.* Rotterdam: Balkema.

Fritschen, L. J., and Gay, L. W. 1979. *Environmental Instrumentation.* Berlin: Springer-Verlag.

Fritton, D. D., Busscher, W. J., and Alpert, J. E. 1974. An inexpensive but durable thermal conductivity probe for field use. *Soil Sci. Soc. Am. Proc.* 38:854–855.

Fuchs, M., and Hadas, A. 1972. The heat flux density in a non-homogeneous bare loessial soil. *Boundary-Layer Meteorol.* 3:191–200.

Fuchs, M., and Tanner, C. B. 1968. Calibration and field test of soil heat flux plates. *Soil Sci. Soc. Am. Proc.* 32:326–328.

Fuchs, M., Campbell, G. S., and Papendick, R. I. 1978. An analysis of sensible and latent heat flow in a partially frozen unsaturated soil. *Soil Sci. Soc. Am. J.* 42:379–385.

Gerald, C. F., and Wheatley, P. O. (1985). *Applied Numerical Analysis.* 3d ed. New York: Addison-Wesley.

Gilman, K. 1977. Movement of heat in soils. Rept. No. 44. Wallingford, Oxfordshire, UK: Inst. Hydrol.

Green, C. F., Schaare, P. N., and Bates, C. N. 1983. A temperature sensor for temperature integration in the field. *J. Exp. Bot.* 34:226–229.

Green, F. H. W., and Harding, R. J. 1979. The effects of altitude on soil temperature. *Meteorol. Mag.* 108:81–91.

Green, F., Harding, R. J., and Oliver, H. R. 1984. The relationship of soil temperature to vegetation height. *J. Climatol.* 4:229–240.

Gupta, S. C., Larson, W. E., and Allmaras, R. R. 1984. Predicting soil temperature and soil heat flux under different tillage-surface residue conditions. *Soil Sci. Soc. Am. J.* 48: 223–232.

Hadas, A. 1974. Problems involved in measuring the soil thermal conductivity and diffusivity in a moist soil. *Agric. Meteorol.* 13:105–113.

Halldin, S., ed. 1979. *Comparison of Forest Water and Energy Exchange Models.* Amsterdam: Elsevier.

Hartley, G. S., and MacLauchlan, J. 1969. A simple integrating thermometer for field use. *J. Ecol.* 57:151–154.

Hasfurther, V. R., and Burman, R. D. 1973. Soil temperature modeling using air temperature as a driving mechanism. *Trans. ASAE Structures Environ. Div.* 16:78–81.

Hayhoe, H. N., Topp, G. C., and Bailey, W. G. 1983a. Measurement of soil water contents and frozen soil depth during a thaw using time-domain reflectometry. *Atmos.-Ocean* 21:299–311.

Hayhoe, H. N., Topp, G. C., and Edey, S. N. 1983b. Analysis of measurement and numerical schemes to estimate frost and thaw penetration of a soil. *Can. J. Soil. Sci.* 63: 67–77.

Hillel, D. 1980. Fundamentals of Soil Physics. New York: Academic Press, pp. 287–317.

Hopmans, J. W., and Dane, J. H. 1986a. Thermal conductivity of two porous media as a function of water content, temperature and density. *Soil Sci.* 142:187–195.

Hopmans, J. W., and Dane, J. H. 1986b. Temperature dependence of soil hydraulic properties. *Soil Sci. Soc. Am. J.* 50:5–9.

Horowitz, P., and Hill, W. 1989. *The art of Electronics.* New York: Cambridge Univ. Press.

Horton, R., and Wierenga, P. J. 1983. Estimating the soil heat flux from observations of soil temperature near the surface. *Soil Sci. Soc. Am. J.* 47:14–20.

Horton, R., Wierenga, P. J., and Nielsen, D. R. 1983. Evaluation of methods for determining the apparent thermal diffusivity of soil near the surface. *Soil Sci. Soc. Am. J.* 47:25–32.

Horton, R., Aguirre-Luna, O., and Wierenga, P. J. 1984. Observed and predicted two-dimensional soil temperature distributions under a row crop. *Soil Sci. Soc. Am. J.* 48:1147–1151.

Huband, N. 1985. An infrared radiometer for measuring surface temperature in the field. I. Design and construction. II. Calibration and performance. *Agric. For. Meteorol.* 34:215–234.

Idso, S. B., Aase, J. K., and Jackson, R. D. 1975. Net radiation–soil heat flux relations as influenced by soil water content variations. *Boundary Layer Meteorol.* 9:113–122.

Jackson, R. D., and Taylor, S. A. 1986. Thermal conductivity and diffusivity. In: *Methods of Soil Analysis*, Part 1. 2d ed. (A. Klute, ed.). Wisconsin: Am. Soc. Agron., pp. 945–956.

Jacobs, W., and Raatz, W. E. 1996. Forecasting road-surface temperatures for different site characteristics. *Meteorol. Appl.* 3:243–256.

Jame, Y. W., and Norman, D. 1980. Heat and mass transfer in a freezing, unsaturated brown medium. *Water Res.* 16:811–819.

Johnson, I. R., and Thornley, J. 1985. Temperature dependence of plant and crop processes. *Ann. Bot.* 55:1–24.

Jones, R. 1972. The measurement of mean temperatures by the sucrose inversion method: A review. *Soils Fertil.* 35:615–619.

Jumikis, A. R. 1977. *Thermal Geotechnics*. New Brunswick, NJ: Rutgers Univ. Press.

Jury, W. A., Gardner, W. R., and Gardner, W. H. 1991. *Soil Physics*. 5th ed. New York: John Wiley.

Kasabuchi, T. 1984. Heat conduction model of saturated soil and estimation of thermal conductivity of soil solid phase. *Soil. Sci.* 138:240–247.

Katan, J., and DeVay, J. E. 1991. *Soil Solarization*. Boca Raton, FL: CRC Press.

Kersten, M. S. 1949. Thermal properties of soils. Univ. of Minnesota, Eng. Exp. Station Bull. 28.

Kimball, B. A. 1974. Smoothing data with Fourier transformations. *Agron. J.* 66:259–262.

Kimball, B. A. 1976. Smoothing data with cubic splines. *Agron. J.* 68:126–129.

Kimball, B. A., and Jackson, R. D. 1975. Soil heat flux determination: A null-alignment method. *Agric. Meteorol.* 15:1–9.

Kimball, B. A., Jackson, R. D., Reginato, R. J., Nakayama, F. S., and Idso, S. B. 1976. Comparison of field-measured and calculated soil heat fluxes. *Soil Sci. Soc. Am. J.* 40:18–25.

Kluitenberg, G. J., Bristow, K. L., and Das, B. 1995. Error analysis of the heat-pulse method for measuring soil heat capacity, diffusivity and conductivity. *Soil Sci. Soc. Am. J.* 50:719–26.

Kung, S. K., and Steenhuis, T. S. 1986. Heat and moisture transfer in a partly frozen non-heaving soil. *Soil Sci. Soc. Am. J.* 50:1114–1122.

Lal, R., Kimble, J., Levine, E., and Stewart, B. A. 1995. *Soil Management and the Greenhouse Effect*. Adv. Soil Sci., Boca Raton, FL: CRC Press.

Lascano, R. J., and van Bavel, C. 1983. Experimental verification of a model to predict soil moisture and temperature profiles. *Soil Sci. Soc. Am. J.* 47:441–448.

Lehmann, R. G., Fontaine, D. D., and Olberding, E. L. 1993. Soil degradation of flumetsulam at different temperatures in the laboratory and field. *Weed Res.* 33:187–195.

Lettau, H. H. 1962. A theoretical model of thermal diffusion in non-homogeneous conductors. *Gerlands Beitr. Geophys.* 71:257–271.

Liakatis, A., Clark, J. A., and Monteith, J. L. 1986. Measurements of the heat balance under plastic mulches. I. Radiation balance and soil heat flux. *Agric. For. Meteorol.* 36: 227–239.

Mahrer, Y. 1982. A theoretical study of the effect of soil surface shape upon the soil temperature profile. *Soil Sci.* 134:381–387.

Main, B. 1996. Development of mechanistic energy balance models for grassland. Ph.D. thesis, Lincoln Univ., Canterbury, New Zealand.

Mayocchi, C. L., and Bristow, K. L. 1995. Soil surface heat flux: Some general questions and comments on measurements. *Agric. For. Meteorol.* 75:43–50.

McInnes, K. J. 1981. Thermal conductivities of soils from dryland wheat regions of Eastern Washington. M.S. thesis, Washington State Univ., Pullman.

Meikle, R. W., and Gilchrist, A. J. 1983. A mathematical method for estimation of soil temperatures in England and Scotland. *Agric. Meteorol.* 30:221–226.

Menenti, M. 1984. Physical aspects and determination of evaporation in deserts applying remote sensing techniques. Rept. No. 10. Wageningen, The Netherlands: ICW.

Meteorological Office 1981. *Handbook of Meteorological Instruments. 2. Temperature.* 2d ed. London: HMSO.

Miller, D. H. 1981. *Energy at the Surface of the Earth.* New York: Academic Press.

Miller, R. D. 1980. Freezing phenomena in soils. In: *Applications of Soil Physics* (D. Hillel, ed.). New York: Academic Press, pp. 254–299.

Milly, P. C. 1982. Moisture and heat transport in hysteretic, inhomogeneous porous media: A matric head-based formulation and a numerical model. *Water Resour. Res.* 18: 489–498.

Milly, P. C. 1984. A simulation analysis of thermal effects on evaporation from soil. *Water Resour. Res.* 20:1087–1098.

Moench, A. F., and Evans, D. C. 1970. Thermal conductivity and diffusivity of soil using a cylindrical heat source. *Soil Sci. Soc. Am. Proc.* 34:377–381.

Nassar, I. N., and Horton, R. 1992. Simultaneous transfer of heat, water, and solute in porous media: I. Theoretical development. *Soil Sci. Soc. Am. J.* 56:1350–1356.

Noborio, K., McInnes, K. J., and Heilman, J. 1996. Measurements of soil water content, heat capacity and thermal conductivity with a single TDR probe. *Soil Sci.* 161: 22–28.

Novak, M. D. 1986. Theoretical values of daily atmospheric and soil thermal admittances. *Boundary Layer Meteorol.* 34:17–34.

Oliver, S. A., Oliver, H. R., Wallace, J. S., and Roberts, A. M. 1987. Soil heat flux and temperature variation with vegetation, soil type and climate. *Agric. For. Meteorol.* 39:257–269.

Parikh, R. J., Havens, J. A., and Scott, H. D. 1979. Thermal diffusivity and conductivity of moist porous media. *Soil Sci. Soc. Am. J.* 43:1050–1052.

Parton, W. J. 1984. Predicting soil temperatures in a shortgrass steppe. *Soil Sci.* 138: 93–101.

Passerat de Silans, A., Monteny, B., and Lhomme, J. P. 1997. The correction of soil heat flux measurements to derive an accurate surface energy balance by the Bowen ratio method. *J. Hydrol.* 188:453–465.

Penner, E. 1970. Thermal conductivity of frozen soils. *Can. J. Earth Sci.* 7:982–987.

Persaud, N., and Chang, A. C. 1985. Computing mean apparent soil thermal diffusivity from daily observations of soil temperature at two depths. *Soil Sci.* 139:297–304.

Philip, J. R. 1961. The theory of heat flux meters. *J. Geophys. Res.* 66:571–579.

Philip, J. R. 1980. Field heterogeneity: Some basic issues. *Water Resour. Res.* 16: 443–448.

Philip, J. R., and de Vries, D. A. 1957. Moisture movement in porous materials under temperature gradients. *Trans. Am. Geophys. Union* 38:222–232.

Potter, K. N., Horton, R., and Cruse, R. M. 1987. Soil surface roughness effects on radiation reflectance and soil heat flux. Soil *Sci. Soc. Am. J.* 51:855–860.

Raats, P. A. C. 1975. Transformations of fluxes and forces describing the simultaneous transport of water and heat in unsaturated porous media. *Water Resour. Res.* 11: 938–942.

Radke, J. K., Dexter, A. R., and Devine, O. J. 1985. Tillage effects on soil temperature, soil water, and wheat growth in South Australia. *Soil Sci. Soc. Am. J.* 49:1542–1547.

Rickard, W., and Brown, J. 1972. The performance of a frost tube for the determination of soil freezing and thawing depth. *Soil Sci.* 113:149–154.

Rieger, S. 1983. *The Genesis and Classification of Cold Soils*. New York: Academic Press.

Riha, S. J., McInnes, K. J., Childs, S. W., and Campbell, G. S. 1980. A finite element calculation for determining thermal conductivity. *Soil Sci. Soc. Am. J.* 44:1323–1325.

Roodenburg, J. 1985. Estimating 10-cm soil temperatures under grass. *Agric. For. Meteorol.* 34:41–52.

Rosenberg, N. J., Blad, B. L., and Verma, S. B. 1983. *Microclimate: The Biological Environment.* 2d ed. New York: John Wiley.

Ross, P. J., and Bridge, B. J. 1987. Thermal properties of swelling clay soils. *Aust. J. Soil Res.* 25:29–41.

RS Components. 1983. Semiconductor temperature sensors. Data Sheet 3992, RS Components, London, UK.

Schieldge, J. P., Kahle, A. B., and Alley, R. E. 1982. A numerical simulation of soil temperature and moisture variations for a bare field. *Soil Sci.* 133:197–207.

Schimmelpfennig, H. 1976. Basic thermocouple thermometry. Application Note No. W-15, Wescor, Inc., Logan, UT.

Scotter, D. R., and Horne, D. J. 1985. The effect of mole drainage on soil temperatures under pasture. *J. Soil Sci.* 36:319–327.

Selim, H. M., and Amacher, M. C. 1997. *Reactivity and transport of heavy metals in soils.* Boca Raton, FL: CRC Press.

Sepaskah, A. R., and Boersma, L. 1979. Thermal conductivity of soils as a function of temperature and water content. *Soil Sci. Soc. Am. J.* 43:439–444.

Shul'gin, A. M. 1965. *The Temperature Regime of Soils.* Jerusalem: Israel Prog. Sci. Transl.

Sidiropoulos, E., and Tzimopoulos, C. 1983. Sensitivity analysis of a coupled heat and mass transfer model in unsaturated porous media. *J. Hydrol.* 64:281–298.

Singh, S. R., and Sinha, B. K. 1977. Soil thermal diffusivity determination from over-specification of boundary data. *Soil Sci. Soc. Am. J.* 41:831–834.

Spaans, E. 1994. The soil freezing characteristic: Its measurement and similarity to the soil moisture characteristic. Ph.D. thesis, Univ. of Minnesota.

Stigter, C. J. 1984. Shading: A traditional method of microclimate manipulation. *Neth. J. Agric. Sci.* 32:81–86.

Sydenham, P. H. 1980. History and technique of temperature measurement. In: *Transducers in Measurement and Control.* Bristol, UK: Adam Hilger, pp. 36–44.

Tabony, R. C. 1984. Non-sinusoidal features of the seasonal variation of temperature in mid-latitudes. *Meteorol. Mag.* 113:64–71.

Tanner, C. B. 1963. Basic instrumentation and measurements for plant environment and micrometeorology. Soils Bull. 6. Madison, WI: Univ. of Wisconsin.

Tanner, C. B. 1974. Microclimatic modification: Basic concepts. *Hortic. Sci.* 9:555–560.

Tarnawski, V. R., Gori, F., Wagner, B. and Buchan, G. D. 2000. Modeling approaches to predicting thermal conductivity of soils at high temperatures. International J. of Energy Research 24:403–423

Taylor, S. A., and Jackson, R. D. 1986a. Temperature. In: *Methods of Soil Analysis*, Part 1. 2d ed. (A. Klute, ed.). Madison, WI: Am. Soc. Agron., pp. 927–940.

Taylor, S. A., and Jackson, R. D. 1986b. Heat capacity and specific heat. In: *Methods of Soil Analysis*, Part 1. 2d ed. (A. Klute, ed.). Madison, WI: Am. Soc. Agron., pp. 941–944.

Toy, T. J., Kuhaida, A. J., and Munson, B. E. 1978. The prediction of mean monthly soil temperature from mean monthly air temperature. *Soil Sci.* 126:181–189.

Van Wijk, W. R. 1963. General temperature variations in a homogeneous soil. In: *Physics of Plant Environment* (W. R. van Wijk, ed.). Amsterdam: North Holland, pp. 144–170.

Van Wijk, W. R., and Derksen, W. J. 1963. Sinusoidal temperature variation in a layered soil. In: *Physics of Plant Environment* (W. R. van Wijk, ed.). Amsterdam: North Holland, pp. 171–209.

Van Wijk, W. R., and Derksen, W. J. 1966. Thermal properties of a soil near the surface. *Agric. Meteorol.* 3:333–342.

Van Wijk, W. R., and de Vries, D. A. 1963. Periodic temperature variations in a homogeneous soil. In: *Physics of Plant Environment* (W. R. van Wijk, ed.). Amsterdam: North Holland, pp. 102–143.

Weaver, H. L., and Campbell, G. S. 1985. Use of Peltier coolers as soil heat flux transducers. *Soil Sci. Soc. Am. J.* 49:1065–1067.

Weichert, L. 1983. The avoidance of electrical interference in instruments. *J. Phys. E: Sci. Instrum.* 16:1003–1012.

Wierenga, P. J., and de Wit, C. T. 1970. Simulation of heat transfer in soils. *Soil Sci. Soc. Am. Proc.* 34:845–848.

Williams, P. J. 1982. *The Surface of the Earth: An Introduction to Geotechnical Science.* London: Longman.

Woodward, F. I., and Sheehy, J. E. 1983. *Principles and Measurements in Environmental Biology.* London: Butterworths.

World Meteorological Organisation (WMO) 1971. *Guide to Meteorological Instrument and Observing Practices.* Publication 8. TP. 3. Geneva: WMO.

Yong, R. N., and Warkentin, B. P. 1975. *Soil Properties and Behaviour.* Amsterdam: Elsevier.

15

Soil Profile Description and Evaluation

Tom Batey
University of Aberdeen, Aberdeen, Scotland

I. INTRODUCTION

The preceding chapters cover a wide range of soil physical measurements. In contrast, this chapter deals with the often neglected topic of the visual and tactile methods of assessment that can be made directly in the field. Both have their place.

Systematic examination of soil in the field should be a basic skill to evaluate its physical state. This was one of the conclusions of the international conference called Problems in Modern Soil Management (van Ouwerkerk et al., 1992). Information obtained in such a way can be used independently or can be used to complement and supplement measurements made by instruments in the field or the laboratory. Field examination should also precede the collection from the field of samples that are to be subject to other tests in the laboratory.

A. General Background

Expressions used to describe the field characteristics of soil go back to the beginnings of a settled agriculture. When manual work was required to till the soil and remove weeds, differences in particle size were readily detected by contact with the foot and hand. "Light" and "heavy," expressions still in use, did not refer to soil bulk density but to the stickiness of wet soil, which is texture related.

Despite the wide range of instruments available to measure physical properties of soils, there are many circumstances where such tests cannot be done. The equipment may not be available, the cost may be high, and the time taken to complete a test may be so long that the results cannot be available in time to deal with a practical problem. Unless the soil is examined first, samples taken for subsequent analysis may be taken from material that crosses physical boundaries and includes layers with dissimilar properties. There are also situations where the lateral distribution of a particular physical condition must be determined. Where any test is time-consuming or costly it may be possible to undertake it at only a few spots; examination of the soil is required to select a representative area.

Field techniques have been widely used in pedology and soil surveys, in land evaluation for crop growth, and in the use and management of soils. For these purposes, techniques have been developed with specific emphasis on particular properties.

1. Pedology and Soil Surveys

The identification of soil horizons and their sequence feature prominently in studies of soil genesis, soil distribution, and soil classification. For these purposes, there is an emphasis on criteria such as soil color and texture, which are relatively permanent, and on the examination of soils under "natural" conditions. A soil classification name may be given to the profile as a whole, based on the sequence of horizons that are identified. Although the names and nomenclature may differ between classification systems, they share a common core of diagnostic criteria to identify a particular horizon. The methods used for describing soils in the field, including any for diagnostic horizons, are described in detail in soil survey manuals or reports accompanying soil surveys. Although local or national systems of classification may reflect more accurately the circumstance of a particular territory (e.g., Glentworth and Muir, 1963; Taylor and Pohlen, 1976; Avery, 1990; Soil Survey Staff, 1993), there are two major soil classification systems that are used worldwide, U.S. Soil Taxonomy (Soil Survey Staff, 1975, 1998) and FAO (1998).

Some systems of soil classification rely on features that can be identified in the field (e.g., Avery, 1990); others require climatic data or laboratory analysis to supplement the field-based descriptions (e.g., Soil Survey Staff, 1993).

2. Land Evaluation

Key features that are required for the evaluation of land quality are related to the growth of crop plants and are climate specific. These include the amount of available water within the potential rooting zone (based on soil texture, aeration, and consistence), drainage class (based on color, texture, and porosity), and soil erodibility (based on soil texture) (Corbett and Tatler, 1970; FAO, 1976; Bibby et al., 1982; MAFF, 1988).

3. Soil Management

Where the physical properties of soil are altered, for example as the result of tillage or the application of mechanical pressure, it is often necessary to find out what changes have taken place. These could include soil compaction, surface crusting, erosion, structure degradation, or reduced permeability to air or water (Simpson, 1983; Davies and Payne, 1988; Batey, 1988, 1989; Daniells et al., 1996). Strictly speaking, permeability to air relates to the gaseous diffusivity (Chap. 13), and permeability to water to the saturated hydraulic conductivity of the soil (Chap. 4), although the earlier but less precise terminology has persisted in the literature on applications. With the advent of heavy machinery in agriculture and forestry, considerable emphasis is placed on the assessment of compaction and whether remedial deep tillage is required. There is also an accompanying need to evaluate the effects of a test run after soil has been loosened to confirm that landwork is effective. Such investigations must be done on the spot and the results evaluated immediately so that appropriate action can be taken.

For whatever purpose, properties that can be determined in the field by sight or by handling the soil have an important part to play in soil physical analysis. Some tests such as soil texture are of general applicability; others have been developed for situations where the physical properties have been changed by mankind's use of the soil. Such includes use as urban parks, playing fields, sports grounds, and paths and tracks as well as for crop production, grazing, or forestry. Profile examination is particularly appropriate for land that has been subject to high mechanical pressure under wet conditions, e.g., during harvesting of root crops, or to major disturbance such as extraction of minerals, renewal of landscapes, or installation of pipelines (e.g., Lowe, 1993), or after prolonged periods of industrial use.

B. Advantages of Direct Field Assessment of Soil Physical Conditions

The advantages of making assessments of soil physical conditions directly in the field are as follows:

1. The examination and evaluation can be done on the spot in a relatively short time, and the results are immediately available.
2. The examination can be comprehensive and thorough.
3. The methods are flexible and can deal with a wide range of situations. They can be done at any time of the year whether the land is bare, under crop, grassland, or forest.
4. Little equipment is required—simply a means to dig a hole in the ground, by spade or mechanical digger, followed by dissection of the profile with a knife or pointed trowel. For some properties, further information can be obtained from examination of the soil extracted by an auger.

5. Slight changes in physical conditions can often be detected that may be difficult to determine by other means.
6. Values for some key physical characteristics can be estimated by combining data on related properties determined in the field, for example saturated hydraulic conductivity, from field assessments of soil texture, structure, and porosity.

II. METHODS AND APPLICATIONS

A. Techniques of Field Examination and Evaluation

To be effective, examination of soils in the field requires access to a soil profile, the vertical face of which has been carefully prepared to expose both natural horizons and any features created as a result of the use and management of the land. The techniques described below are based on Batey (1975, 1988), Hodgson (1978), Simpson (1983), Pizer (1990), and McKenzie (1998).

1. When to Look

The techniques can be applied at any time of the year. If it is possible to choose the timing, examination should be made preferably when the maximum amount of information can be obtained. Under annual crops, this would be when the crops are close to their peak of vegetative growth and while the soil is moist. In many climates this would be in late spring. However, other factors may dictate the timing, such as access to the land. Postharvest examination is frequently made both because of easy access and because of the need to assess soil compaction, so that remedial deep tillage may be done timeously prior to the establishment of the next crop. Some of the information obtained may be limited by the conditions under which the examination is made. For example, if the soil is very dry, it is difficult to distinguish between layers that are hard because they are compact and those that are hard because they are dry. If the land is very wet, it may not be possible to prepare a hole without excessive damage to the soil in its vicinity, nor to make an adequate examination under soft and wet ground conditions.

In some circumstances it may be possible to use extremes of weather, such as drought or heavy rain, to supplement the information obtained from profile examination. The reaction of soil to heavy rain can be used to assess its hydraulic conductivity, its erodibility and the stability of soil structure. A wet and soft surface present after heavy rain may be caused by an impermeable compact layer below (Sec. II.E). If the surface of the land is bare, the degree of breakdown of structure and the degree of slumping can be determined (Sec. II.C).

2. *Where to Look*

This depends on the reason for the examination. Unless the diagnosis of a specific problem is the objective, care should be taken to avoid gateways, tracks, headlands, wheelings, and other abnormally disturbed ground. A representative area of land should be selected that is uniform in appearance.

When undertaking soil examination to determine the reason for a variation in crop growth, the pattern of growth can be a useful diagnostic feature and enables holes to be made in areas of good and poor growth. In times of drought, areas of shallow, rocky, or gravelly land (and archaeological foundations) may be shown up by pale or stunted vegetation. A similar appearance can be caused by soil compaction. Deeper soils may be shown up by darker, more vigorous plant growth. Photographs are recommended as a means of recording permanently the distribution of variations in soil color or of crop appearance, whether caused by inherent differences in soils or by the effects of management of the land. These may be taken at field level, from high ground or buildings, or from the air, and can be used subsequently to locate areas for detailed soil examination. Satellite imagery can be used to record variations in soil properties or plant growth. It can be very informative to dig a trench at right angles across the principal direction of tillage or harvest so that any compaction related to wheeltrack patterns can be more readily identified.

3. *How to Look*

a. *Digging a Hole and Preparation of the Profile Face*

A mechanical digger is recommended, provided that there is access to the location required without causing excessive crop damage. Alternatively, a hole can be dug with a clean sharp spade, supplemented if necessary with a pickaxe or crowbar. An auger maybe used to extract soil from depth. Details of augers and other equipment suited to soil examination are given in Sec. b below.

The dimensions of a hole depend on the question being asked and on how far the deepest zone of interest lies below the surface. Rarely would the depth be less than 50 cm, and it could be 1.2 meters or more. There should be enough space at the bottom of the hole to accept waste soil taken off the face during examination. While digging, two edges of the hole should be left untrampled, and the soil dug out should be kept well away from these sides. One or more vertical faces should then be cleared of any soil that was smeared or compressed while the hole was dug. The next objective is to highlight the physical characteristics of the soil. Using a small pointed trowel or penknife, the face should be gently probed, beginning at the surface then working down the face to restore natural features and to search for any human-induced changes. Where coarse blocky structure (Sec. II.C.1) is found, this can be levered out with a spade, beginning near the base of the hole.

Where a mechanical digger is used, a trench can be dug readily to a depth of 1.5 m or more, to provide a hole wide enough to walk along and to have top soil almost at eye level. Safety aspects must be considered and local regulations followed when working in trenches more than 1 m deep. Care must be taken when digging a hole in loose soil or where marked vertical fissures are present; the risk of injury as a result of wall collapse must be evaluated.

If compaction of the soil just below plow or cultivation depth is suspected, most of the loose soil above can be lifted off by spade or trowel and the last remnants lying on the upper surface of the suspect compact layer brushed or flicked off.

b. Equipment for Examining Soils in the Field

The suggestions made in this paragraph are based on the author's experience (see also ADAS, 1971). Catalogs of equipment for field use can be obtained from several of the suppliers listed in Table 1 of Chap. 10. These contain a much wider range of equipment than that described here, with some dedicated for specific purposes. Local suppliers may also be able to provide suitable equipment. Some examples of equipment are shown in Fig. 1.

Spades: A conventional rectangular spade, typically 20 × 25 cm, is often used. However, this may be difficult to push into firm or dry soil. It can be modified to penetrate more easily by cutting off the corners to make it U-shaped. A smaller spade 15 × 20 cm in size with a concave face is also often used.

Screw augers: These are usually modified wood-boring augers of 2 or 2.5 cm diameter with a screw length of 20 cm, to which a stem has been welded to increase its length to 1 m or possibly longer. If the original point is cut off, the auger can more easily penetrate soils which are slightly stony. Because a large pull is often required to extract the auger from the soil, care must be taken to avoid back strain or injury. Screw augers are suitable for taking samples for tests where structure is of no significance. The soil core retained on the screw can be examined for texture and color but not for structure.

Dutch augers: These are specially designed for soil examination and have an open twist tapered head about 20 cm long, typically of 3 or preferably of 5 cm diameter. The head is at the end of a stem 1.2 m long. Despite their larger diameter, they usually take less force to insert and pull out of the soil than screw augers. A core can be extracted that is partly intact, and about 15 cm long; this can be used to examine the texture, color, root numbers, and, to a certain extent, structure.

Crescent-shaped open corers (sometimes called cheese corers): These are semicircular in cross-section and some 2 to 2.5 cm in diameter. The length of the core may be limited to a specific distance of 15 or 30 cm for taking samples to that depth. Alternatively it may extend to 1.0 m, the whole length of the corer. After insertion into the soil and giving it a half rotation, an entire core can be held on the corer when it is pulled out. By cutting the exposed part off with a blade, an undisturbed soil profile can be retained on the corer for examination. Such corers

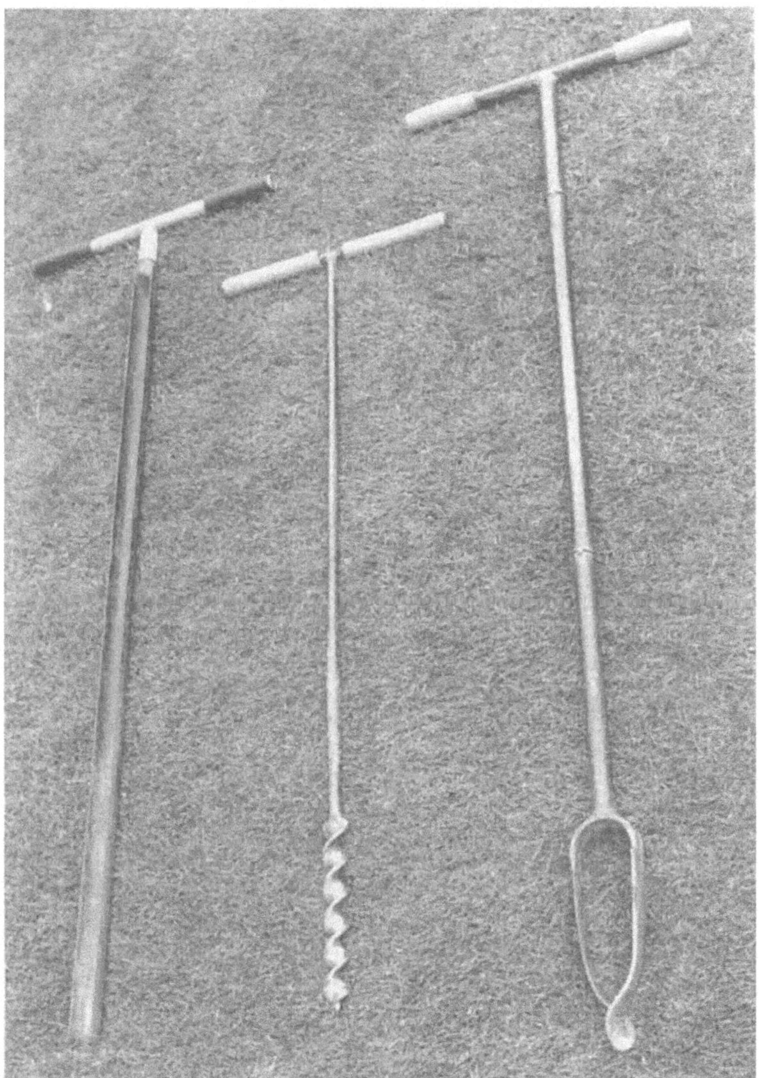

Fig. 1 Augers used for obtaining soil samples. From left to right: gouge, screw, Dutch auger.

cannot be used where stones are present. They can be used to extract deep cores in wet or soft soil such as peat, but in mineral or dry soils the force required to insert and extract long cores may be too great for manual use.

Mechanical corers and split samplers: Where cores are required of a size or depth that exceed human endeavor, mechanical equipment as used in geology or engineering can be used. Those used for extracting cores for root measurements are shown in Chap. 12.

4. What to Look For—Examination and Interpretation

Physical properties that can be determined by tactile and visual examination directly in the field are described in the following sections of this chapter. To assess their characteristics, it is convenient to divide the soil into four layers: the soil surface, the layer disturbed by normal cultivations, the soil just below the cultivated soil, and subsoil undisturbed by normal cultivations. Visual and tactile examination can also help to locate the optimum position for instrumental measurements to be made, or for samples to be taken for testing later in the laboratory [e.g., bulk density (Chap. 8) or gas movement (Chap. 13)].

a. The Soil Surface

If a bare soil has been exposed to rain, any disintegration of aggregates can be used as an indication of the stability of the structure. Individual aggregates may have partially collapsed, and if severe, a smooth surface can be the result (see Sec. II.B). The presence of such a layer can be confirmed by probing and levering up the surface with a pointed blade. Such a crust may act as a seal on the surface, which excludes air when it is wet; when dry, it may become hard and impenetrable to emerging seedlings. More stable aggregates and large mineral particles such as coarse sand or small stones can sometimes be seen firmly embedded in the crust and projecting above the otherwise smooth surface. Below a crust, aggregates can be firmly attached to the underside of the crusted portion. Soils with a high content of fine or very fine sand and silt, particularly where the organic matter content is low, are prone to show this feature (Davies, 1974). If rain is heavy and prolonged or the land is flooded for a while, a crust may develop into a layer 3–5 cm thick.

Compaction of the surface is widespread, caused by the treading of animals (including wildlife and human activity) and by the passage of wheeled or tracked vehicles. The surface is depressed by the pressure applied and the pattern is related to the movement of the animals or machines. The effects are worst when the land is soft. The primary effect is a decrease in porosity and infiltration that may lead to water flowing downslope and inducing erosion. In hot, dry regions of the world, hard and compact soil may be found extending from the surface throughout the topsoil and even deeper (see Sec. II.C.5). These are known as hardsetting soils and may be found occasionally in temperate regions where intensive management has reduced soil organic matter content (Mullins et al., 1987).

b. Within the Cultivated Layer

This refers to the layer disturbed by cultivation, usually to between 20 and 30 cm below the surface (i.e., the depth to which the deepest working implement operates). The term "cultivation" includes any operation done by a moldboard or disc plow, or by a rotary, tined or other implement. Multiple cultivations are common and may take place at a range of depths.

Most types of cultivation implement can form a thin zone of compressed soil, just below their operating depth (often called a cultivation pan). In this zone, a pan may be detected by the relative resistance to a probe pushed manually into the soil (a spade, auger, or stick can be used). Such pans occur not only just below plow depth (see the next section) but can also be found within the cultivated layer due to shallower secondary cultivations or the use of shallower implements in the later stages of seedbed preparation. A pan can often be seen from above as a smooth, slightly shiny smeared surface, which may be continuous or discontinuous, and may bear the imprint of the blade or implement responsible for its formation. In a thick panned layer, aggregates pack tightly together to form a slab of visibly dense soil, with reduced or no visible pores. When dry, this would be detected as a hard layer. Thick pans usually have a greater adverse effect than thin ones on water or root penetration, but the depth at which they occur is important. Shallow pans can have more severe effects (see Sec. II.D). Soils of all types, including sands and peats, may exhibit smeared or compacted layers.

On very sandy soils an unusual method to detect thin compact layers is to remove carefully an entire spadeful of dry soil and lay it on its side, tap the spade, and blow away any loose sand. If compact layers are present they may be seen as thin or thick layers separated by cleavage planes often lying parallel to the surface (Harrod, 1975).

In wet weather, water may build up above a compacted or smeared layer and can be seen seeping out and running down the side of an inspection pit. On sloping land, if water cannot drain through a pan, the risk of erosion is considerably enhanced. Other changes may also accompany soil compaction; for example, dark gray anaerobic pockets with a malodorous smell may be seen where recent crop residues have been incorporated into the compact soil (Sec. II.F).

c. *Just Below the Base of the Cultivated Layer*

This is the position of the classic plow-pan; it is one of the most critical for root and water penetration. Above it, the soil is loosened regularly by normal cultivations; within and below it, soil is rarely disturbed. However, it is not only plowing that may be responsible for compaction. Wheels of tractors, harvesters and loaded trailers running on the surface can transmit pressure to this depth (or even below) and can cause severe compaction (Soane and van Ouwerkerk, 1994; Hakansson and Petelkau, 1994; McKenzie, 1998).

The signs of compaction are high density as determined by probing, reduced hydraulic conductivity leading to an accumulation of water above the compact layer, a marked discontinuity in structural form often with horizontal laminated or platy units within the compact layer, which may have a smooth shiny upper surface, and the absence of pores, fissures, roots, or earthworm holes within it. Tortuous root paths with common horizontal segments provide a good indication of compaction (McCormack, 1986). The upper surface may also bear the imprint of cultivator tines or the lugs of tractor tires. Such imprints may even be found in prehistoric fields that have not been subsequently cultivated (Ashmore, 1996), which is an indication of the potential longevity of unrelieved compaction.

The pattern of roots can be used directly to assess the significance to the crop of any suspect compact layer. In a crop growing under unrestricted physical conditions, the root pattern would be related to the species and variety of the crop, to the soil water regime, to acidity, and sometimes to differences in soil nutrient status. The concentration of roots is usually greatest in the topsoil, with a relatively steady reduction in numbers with depth (Gregory, 1988). A compact or smeared layer can restrict the number of roots penetrating below it. A mat or an increased density of roots may be found on the upper surface of severely compacted soil. Roots that are able to grow a short way into compacted soil are often much thickened and distorted.

If roots have been unable to grow much into or below a compact soil, a sharply differentiated moisture profile may develop, with dry soil within and above the compact layer and moist below. This is caused by the lack of roots below the compaction to extract moisture. On the other hand, if the soil has dried to some depth in the subsoil below the compaction, this may be a useful indication that roots have been able to penetrate and extract moisture. However, the change in consistence at the base of the cultivated layer may be mistaken for the upper surface of a compacted layer, particularly in late summer when subsoils may be dry and hard. The unloosened subsoil is harder than the topsoil without necessarily having been compacted (discussed further in Sec. II. D.3).

Although the upper surface of a compacted layer may be readily located, it is more difficult to determine how far down the compaction persists. The most compacted soil is found on the upper surface. The severity of compaction then declines with depth until the layer merges with unaffected soil at some depth below. If possible, a comparison should be made between the physical properties of soil nearby that has not been compacted. The effects that compact layers may have on crops and on soil properties is discussed in Sec. II.D.

d. The Subsoil

This section deals mainly with the identification of natural soil features, as the physical properties of subsoil are not normally affected by grazing or cropping. However, there is increasing evidence that the continued use of tractors and harvesters of large mass transmit pressure deep into the subsoil (McKenzie et al., 1990; Hakansson and Petelkau, 1994; Sullivan and Montgomery, 1998). These effects have yet to be fully evaluated. The signs to look for are increased density, lack of porosity, and reduced penetration of water or roots.

The agronomic role of the subsoil is to provide entry and egress for water and to permit the entry and extension of crop roots to extract water and nutrients. Roots can grow into quite stiff soil by deforming it but tend to grow mainly down pores and cracks in structured soils if the peds are fairly dense. Pores and fissures may be up to several mm across and can be observed directly by eye, and by the presence within them of roots either living (white when young) or dead (brown). In some soils, fissures and pores may have been created many years ago when the land was in forest or marsh and when trees or other species were growing with roots much thicker than those of the present vegetation. The imprint of roots can persist on the faces of fissures for long periods, possibly centuries. Earthworm holes often contain roots and darker colored topsoil, and sometimes follow former root channels. Particularly where they are numerous, they may significantly improve root penetration and drainage.

In some subsoils dominated by sand, root penetration may be very poor, without obvious signs of compaction or hardness. In such soils roots may extend only 8–10 cm into the subsoil sand and also show a characteristic swollen appearance (Batey, 1988). This phenomenon is thought to be due to the close packing of the grains and their resistance to moving apart to create the space needed for roots to expand and grow normally (Hettiaratchi, 1987). Loamy subsoils in their natural state usually provide excellent conditions for root growth, unless affected by acidity or waterlogging.

In some clay soils distinctive and characteristic vertical cracks develop due to shrinkage of the clay when it dries (see also Sec. II.C.4). These cracks frequently re-form in the same position each year and roots therefore grow also in the same position. The degree and depth of fracturing is related to the magnitude of the soil water deficit, to the clay content, and to the type of clay present (Wilkinson, 1975). Topsoil often falls down cracks in summer, and whether this is a beneficial effect is equivocal. Chemical fertility may be enhanced in the subsoil, but the extra material may give rise to a tighter seal when the clay expands as it rewets (Smart, 1998; Batey and McKenzie, 1999). Because roots can be so readily seen on crack faces in the subsoil, their presence is a good guide to the absence of any major limiting feature higher up the soil profile.

e. *Cemented and Indurated Layers*

Hard layers may develop by natural processes. In northern latitudes, indurated layers occur in many sandy and loamy soils within 30–50 cm of the surface; these are relics of the Ice Age (Fitzpatrick, 1956; Glentworth and Muir, 1963). Their presence is rarely in doubt, as they are extremely hard even when wet; a strong blow with a spade may penetrate less than 1 cm. If shallow, they may adversely affect drainage, the growth of crops, and land capability. However, their direct significance for crop growth is often less than expected because of the cooler, wetter climate in which they are usually found (Batey, 1988). Dense layers can also occur due to pedogenic processes. These include the downward movement of clay, and cementing by iron and other oxides and oxyhydroxides (plinthite) (Soil Survey Staff, 1993).

B. Soil Texture

The expression "soil texture" is used to describe the feel and molding characteristics of moist soil. Words such as clay, sand, and loam have been used to distinguish soils with different properties since the beginnings of a settled agriculture. Hand texturing is one of the most important single tests that can be done in the field.

Four terms are used in varying combinations: *sand*, *loam*, *silt*, and *clay*, together with adjectives qualifying the size of the sand grains, to describe just over 20 different classes of texture. Texture must not be confused with soil structure, which describes the way the individual particles are assembled and bound into groups, usually called aggregates.

Soil texture gives a guide to many soil characteristics. The textural class provides an indication of soil water retention and the available water capacity (Chap. 3); particle size distribution (Chap. 7); the likely development and stability of soil structure; cation exchange capacity (and hence nutrient retention and availability, and the activity and retention of residual soil-acting herbicides); erodibility by wind or water; stickiness and ease of cultivation; drainage characteristics, saturated hydraulic conductivity and suitability for mole draining; cropping suitability; and thermal properties of soils (Chap. 14).

1. Soil Texture Classes and Particle Size

The size ranges of soil particles are classified into three groups, sand, silt and clay, with the upper limit of "soil" set at 2 mm. However, there is no general consensus regarding the size range of each group, as discussed in Chap. 7 and by Hodgson (1978). One system that is widely accepted classifies particles as follows (Hodgson, 1974):

Sand: between 2 mm and 60 μm
Silt: between 60 and 2 μm
Clay: less than 2 μm

Particles larger than 2 mm, i.e., stones (2 to 600 mm in size) and boulders (>600 mm), are important where present in a significant proportion. Stone sizes can be further subdivided into very small (2–6 mm), small (6–20 mm), medium (20–60 mm), large (60–200 mm) and very large (200–600 mm) (Hodgson, 1974).

There are two types of method available to determine soil particle size distribution, laboratory analysis and field assessment. To avoid confusion between the two, it is recommended that the term particle-size class be used to express the descriptive names applied to different mixtures of sand, silt, and clay size particles based on laboratory analysis. The term soil texture is then reserved for the estimation based on a field test as described in Sec. II.B.2 below.

Particle size analysis provides precise values for the proportions of particles in a number of size classes. The terms already used to describe soil texture classes are then used to describe soils with different proportions of sand, silt, and clay. The conversion is done using a triangular or orthogonal diagram (Chap. 7). Details of size classes and of the naming of various mixtures of these are discussed by Hodgson (1978). However, particle size analysis is conventionally done after removing cementing materials such as organic matter, carbonate, and iron and aluminum oxides and hydroxides. Laboratory results therefore cannot always be expected to relate accurately to the field behavior of soils. Furthermore, the textural diagrams commonly used take no account of the range of particle sizes within a class, so that important qualifying adjectives such as "coarse" or "fine" cannot be applied. Care must be taken when interpreting soil surveys where particle size analysis figures are used to verify field estimates of texture (Avery, 1990).

In some red tropical soils, estimating the texture by hand gives a result of silty loam, whereas a laboratory determination shows a high content of clay. This difference is due to the intense microaggregation of the clay particles, which masks some of the cohesive properties of the clay when manipulated by hand. In such soils, the results of hand texturing give a much better indication of the field behavior and capability of the soil (Trapnell and Webster, 1986).

When hand-based assessment of soil texture has been tested against laboratory analysis, the results showed that, for soils within a limited geographic region, those with experience of hand assessment could confidently estimate the particle size distribution of a wide range of samples (Hodgson et al., 1976; Pizer, 1990).

2. *Field Method*

Methods for assessment of soil texture are given in national and international soil survey manuals (e.g., Hodgson, 1974, 1978). A brief description of one method that has been used successfully over many years for field evaluation in England and Wales is given here (further details can be found in ADAS, 1971; Batey, 1988; and Pizer, 1990). The field properties associated with each class are described in the next section.

The procedure is as follows: Take about half a handful of soil, and if it is dry, add water gradually until the particles hold together to form a moist ball. No excess water should be present. The assessment is made by kneading the moist soil between fingers and thumb. It is important to work the soil thoroughly to eliminate any small lumps (aggregates) present. The assessment is then done by estimating the contribution that the different particles, sand, silt, and clay, make to the feel of the soil as a whole.

The physical properties of the individual fractions that determine the texture of a soil are as follows.

Sand consists of grains that feel gritty and are large enough to grate against each other; they may be detected individually by both touch and sight. Four subgrades can be distinguished, coarse, medium, fine, and very fine.

With silt, individual grains cannot be detected; silt feels smooth, soapy, or silky. It adheres readily to the fingers.

Clays are sticky; some dry clays require a great deal of moistening and working between the fingers before they develop their maximum stickiness. Clay coheres and can come away fairly cleanly from the fingers. A moist surface will take a slight polish when a finger or thumb is rubbed firmly across it.

Each class of soil texture has a characteristic feel (Table 1) and is best thought of as a single entity. For those unfamiliar with the technique of hand texturing, expert advice should be sought initially until experience is gained to make accurate assessments. Practicing on samples that have already been classified is a good way to gain experience.

3. Soil Texture Descriptions and Associated Physical Properties

In Table 1, one column describes the tactile characteristics of each class and the other the physical characteristics associated with each class. Although this is based on U.K. experience, the method can be applied with only minor modifications to the soils of most countries. In Table 1, particles between 20 and 60 μm in size are referred to as very fine sand. In some systems (Hodgson, 1974) this size range is called coarse silt.

4. Hand Assessment of Soil Texture

The characteristics of the different textural classes are set out below:

a. Sandy Soils

Those with a significant amount of grittiness. Test the binding and cohesion.

None	Sand
Slight	Loamy sand
Readily molded into a cohesive ball, does not form threads	Sandy loam
Moderately cohesive, sticky and plastic, forms threads, will take a polish on the surface	Sandy clay loam
Very cohesive, very sticky, forms long threads, will take a high polish when rubbed	Sandy clay

For each of the sand groups it is also important to identify the grade of sand, and the main classes should be prefixed accordingly, though for sandy clay loam and sandy clay it is less easy to identify the sand grades.

Coarse sand	Very harsh feel	2.0–0.6 mm
Medium sand	Moderately harsh feel (e.g., sea shore sand)	0.6–0.2 mm
Fine sand	Slightly gritty (e.g., dune sand)	0.2–0.06 mm
Very fine sand	Smooth and powdery, only just visible to the naked eye	0.06–0.02 mm

b. *Clayey Soils*

Those which are not gritty, but are strongly cohesive, form threads and rings easily and have a surface that readily takes a polish when rubbed with thumb or finger.

Moderately sticky, deforms readily when squeezed	Clay loam
Extremely sticky, moderately smooth, difficult to deform	Silty clay loam
Extremely cohesive, forms long threads and rings, high degree of polish when rubbed	Clay
Extremely cohesive, high polish, also smooth and silky	Silty clay

c. *Silty Soils*

Those dominated by a smooth, soapy slipperiness or silkiness, moderately cohesive. Silt adheres readily and fingers become very dirty; clay coheres, i.e., sticks to itself and fingers remain relatively clean.

Smoothness and silkiness dominant	Silty loam

d. *Loam*

Where none of the above fractions, sand, silt, or clay, imparts a dominant feel.

Moderately smooth and can be rolled into short threads; no polish can be obtained when rubbed	Loam

C. Soil Structure

In its broadest sense, soil structure refers to the physical organization of soil materials as expressed by the arrangement of solid particles and voids (Avery, 1990). Field descriptions place emphasis on the degree of development, and on the size, shape, and arrangement of naturally formed aggregates that are separated from each other by voids or planes of weakness (Hodgson, 1974; Avery,1990).

Soil structure has also been described as the architecture of the soil (Russell, 1961). Certainly it has to do with space, construction, stability, pathways, and

Table 1 Soil Textural Descriptions and Associated Physical Properties

Texture Description	Associated Physical Properties
SANDY SOILS: soils dominated by sands; divided into three groups (sands, loamy sands, and sandy loams), depending on the proportion of sand present. Each group is then subdivided into four (coarse, medium, fine, very fine), according to the dominant size of the sand grains	
SANDS feel gritty, lack any cohesion, loose when dry, not sticky at all when wet, do not stain the fingers	*Low retention of water and nutrients*
COARSE SAND (2–0.6 mm): harsh to the touch	Very droughty, fast draining, readily eroded by water
MEDIUM SAND (600–200 μm): sands of the seashore	Very droughty, erodible by wind and water, root entry difficult
FINE SAND (200–60 μm): dune sand	Very erodible by wind and water, root entry difficult
VERY FIND SAND (60–20 μm): loess, barely visible to the naked eye, powdery	Very erodible, root entry difficult
LOAMY SANDS feel gritty, slight cohesion—can be molded into a ball when sufficiently moist, do not stick to the fingers	*Low retention of nutrients and usually of water*
LOAMY COARSE SAND: harsh to the touch	Very droughty, fast draining, erodible by water
LOAMY MEDIUM SAND: as medium sand	Low water retention, very prone to erosion by wind, erodible by water
LOAMY FINE SAND: as fine sand	Reasonable water retention weak structure, liable to collapse in heavy rain, crusts and caps on surface, very erodible by wind and water
LOAMY VERY FINE SAND: very fine powder	Very weak structure, collapses readily, easily compacted, forms hard surface cap

Texture Description	Associated Physical Properties
SANDY LOAMS feel gritty, show a fair degree of cohesion, can be molded quite readily into a ball when just moist	*Free draining, easily tilled yet easily deformed*
COARSE SANDY LOAM: harsh and gritty	Very fast draining, free working, low water retention
MEDIUM SANDY LOAM: gritty, firmly molded	Fast draining, free working, reasonable water retention, stable structure, few physical problems
FINE SANDY LOAM: slight grittiness, firmly molded	Fast draining, free working, good retention of water—high proportion available, erodible by water, structure slightly weak, liable to cap
VERY FINE SANDY LOAM: grittiness barely detectable, firmly molded, fine and powdery when dry	Moderately porous, weak structure and liable to cap, surface ponding common, excellent retention of water, high available water capacity, erodible by water, very high value in dry areas
LOAMS	
LOAM: no fraction dominates the feel of the soil, readily molded into a ball although sand present, does not feel obviously gritty; insufficient silt to impart silky feel, insufficient clay to make it sticky or to take a polish	Good water retention, porous, easy working, stable structure
SILTY LOAM: smooth silky feel, sticky when wet, firmly cohesive	Good water retention, adhesive and difficult to work when wet, structure usually stable but may break down if overworked, high value in dry areas, less good in wet
SILT: as silty loam but smoothness, silkiness, and adhesion more distinct, surface takes a weak polish when rubbed with finger	As silty loam but more sticky, moderately slow draining
CLAY LOAM: sticky, binds together strongly when moist and resists deformation, takes a polish on surface	Good water retention, slow draining, high retention of nutrients, strongly developed stable structure, weathers into fine aggregates on surface, high draught requirement, readily smeared, may shrink on drying to form deep cracks

(*continued*)

Table 1 *Continued*

Texture Description	Associated Physical Properties
SANDY CLAY LOAM: as clay loam but also gritty	As clay loam but is extremely hard when dry, difficult to manage under tillage, readily smeared and compressed
SILTY CLAY LOAM: as clay loam but more sticky, adhesive and smooth	As clay loam but with a higher draught requirement
CLAYS	
CLAY: very sticky, binds together very strongly, ball of moist clay is very difficult to deform by hand, takes a high polish when moist clay is rubbed	As clay loam but clay characteristics more extreme, widespread as a subsoil below clay loam, forms deep and wide vertical cracks on drying
SANDY CLAY: as clay but also gritty	As sandy clay loam but clay characteristics more extreme, very difficult under tillage
SILTY CLAY: as clay but extremely sticky and adhesive	As clay but very difficult under tillage

Note: The physical properties of any of the clay textures may be altered if a few percent of chalk is present [e.g., in the Chalky Boulder Clays in eastern England; these soils develop a finer structure, drain faster, and are easier to till than equivalent clays that do not contain chalk (ADAS, 1971)].
Source: Batey, 1988; Pizer, 1990.

microhabitats. It is also an ephemeral property. Structure may change over a range of time scales from instantaneous disruption to slow modifications over decades. Consequently, it is often more appropriate to describe the structure of a soil than it is to measure it. There are no shortages of methods to measure structure; over 200 are listed in De Boodt et al. (1967).

1. *Description of Soil Structure*

Because of its compound and complex nature, soil structure can be described in several ways. The very act of breaking soil apart alters its properties, so that care is needed in deciding on the amount of effort applied before making a description. A number of pedological terms have been developed to describe soil structure on which there is more or less international agreement. Examples of these can be found in many soil survey manuals and textbooks (e.g., Payne, 1988; White, 1997).

Words such as *crumb*, *blocky*, and *angular* are used in their normal sense in soil descriptions; however others, such as *prismatic*, have specific definitions that deviate from the normal use of the word. Furthermore, it is important to realize that the terms used when describing structure have been precisely defined and

Table 2 Widely Used Terms to Describe Different Types of Soil Structure

Term	Description
Single grain	No perceptible bonding between particles to form compound units
Massive	An amorphous lump of soil devoid of fracture planes
Platy	Platelike with planes of weakness orientated horizontally
Crumb	Distinct rounded porous aggregates up to 1 cm in diameter
Blocky	Aggregates with similar dimensions in all three planes; they may have edges that are angular or subangular
Prismatic	Subsoil separated into vertically orientated units up to 1 m or more long, often pentagonal or hexagonal in cross-section and between 5 and 30 cm across
Columnar	As for prismatic but with the topmost part rounded

The terms can also be subdivided into size classes (Hodgson, 1974).

should not be confused with everyday descriptive terms that have not been defined in a particular system. Some terms to describe the structure of a soil are given in Table 2.

Structure can also be examined at a microscopic level (Brewer and Sleeman, 1988; Fitzpatrick, 1993). However, microstructures are not discussed further in this chapter.

2. Soil Structure Assessment

A comprehensive description of soil structure is time-consuming to complete. However carefully this is done, it is sometimes difficult to compare one description with another and to determine what are the essential differences between them. McKenzie and MacLeod (1989) found that conventional descriptions of structure, including grade, ped type, ped size, fabric, and macroporosity, were poor predictors of agronomically important soil properties on a broad range of irrigated soils. Gameda et al. (1994) found that soil profile assessment was a very useful tool for complementing conventional procedures and that parameters such as bulk density and penetration resistance, although they provided a reasonable indication of the degree of compaction and freedom from waterlogging, were inadequate to describe the overall suitability of soil structure for crop growth.

A numerical scale can be used to overcome some of the difficulties inherent in a purely descriptive method or to evaluate the result of a specific measurement. An early example by Peerlkamp (1967) assessed the structure of the cultivated layer as a whole. This test was originally designed for arable land and is to be done on moist soil in spring or autumn. However, it can be adapted for grassland, and for dry soil, and to be done at other times of the year.

The method is based on an examination of the structure in the cultivated layer and a ranking of its quality as a medium for root development. The assumptions made are that medium crumb structures, low cohesion, high porosity, and the absence of surface capping and dense clods are all beneficial qualities. The method involves visual examination and manipulation by hand of a spadeful of soil lifted up and laid on the ground. A number between St 1 (= poor structure) and St 10 (= very good structure) is assigned to the whole spadeful of soil, after consideration of the size, shape, and density of the aggregates; the porosity of the entire layer; and the ease with which surface aggregates break down (Table 3). To make adequate comparisons between different situations, a minimum of 10, or preferably 20, tests should be made on each area or treatment. The results can be statistically evaluated.

Experience has shown the method to be capable of detecting small changes in structure. The St value has been related to soil organic matter, to soil consistency, to the concentration of calcium in solution, to the residual effects of grass leys, and to crop yields (Eagle, 1975). The concept of using a numeric scale to assess a complex physical property has been further developed to assess the degree of compaction under irrigated cotton grown on vertisols in Australia (Daniells and Larsen, 1991; McKenzie, 1998) (discussed further in Sec. II.D). This SOILpak score has been tested against a range of physical measurements and has been found to be a successful predictor of soil conditions (Greenhalgh, 1995; McKenzie, 1997).

Scoring methods can be sensitive and flexible and can provide an accurate evaluation of structure. They help to focus on key properties and encourage a detailed evaluation of the whole soil environment; they can also be adapted and modified for specific situations wherever appropriate criteria can be established.

Table 3 Numerical Assessment of Soil Structure

St 1	Plow layer consists entirely of big clods, smooth dense crack faces, reducing conditions, roots only in cracks
St 3	Plow layer big dense aggregates, smooth crack faces, roots between aggregates. or Top 6 cm angular dense aggregates, very dense below
St 5	Plow layer big but porous aggregates, rather smooth crack faces or Top 7–8 cm small porous aggregates with denser layer below
St 7	Plow layer mostly porous crumbs partly combined as porous aggregates. Occasional denser clod
St 9	Plow layer all porous crumbs, very few dense aggregates

Source: Based on the method of Peerlkamp (1967). These are the criteria for clay and loam soils; those for sandy soils can be found in the original reference.

3. *Soil Structural Stability*

Structure is an ephemeral property. Its deterioration is almost always on or close to the soil surface and is associated with the exposure of bare soil to rain, with arable cropping, and with a decline in the concentration of soil organic matter.

There are many laboratory tests available to measure the stability of structure (e.g., De Boodt et al., 1967; Kemper and Rosenau, 1986). Most are based on the behavior of aggregates when immersed in water (and often shaken, too). There is no universal agreement on a standard method. Some have been developed for use in the field (e.g., McKenzie, 1998).

Signs of an unstable structure that may be seen in the field include

(a) Collapse or partial disintegration of aggregates when exposed to the direct impact of raindrops or irrigation, or when immersed under water. The reaction of a bare soil surface to such pressures ("verschlumping") can be assessed on a visual scale of 1 to 9 developed by De Boodt (pers. communication, 1971), with 9 representing stable aggregates unchanged by pressure and 1 representing soils showing total collapse of the aggregates after drop impact or immersion, to form a continuous crusted surface.

(b) In soils containing a significant content of fine or very fine sand, small pockets or thin layers of pale sand grains may be seen within the cultivated layer or concentrated at its base (De Leenheer, 1967; Batey, 1988). These are the consequences of the disintegration of aggregates into their component particles. From the same process, very thin skins of clay (cutans) can sometimes be seen within the soil profile and on the surface, often in minor depressions where water has lain temporarily (Davies, 1975).

4. *Structure and Clay Mineralogy*

The structure of subsoils that contain a significant content of clay is usually dominated by a pattern of vertical cracks. These open as a soil dries and close as it rewets. The degree to which soils shrink as they dry is determined by the content, type, and organization of the clay. The depth and width of the cracks are determined by the soil moisture deficit. Although it is not possible to identify specific clay minerals by examining the frequency and width of cracks found in a soil, an indication of the overall proportion of different groups can be obtained. Soils with a high proportion of smectitic (montmorillonite) clays form cracks that are both wide and deep, as exemplified by many vertisols. Soils dominated by illitic clays show only narrow cracks. When kaolinitic clays dominate, crack formation is likely to be less well developed.

The shrink–swell phenomenon has a marked effect on the overall characteristics of a clay soil. The cracks often are the principal pathway for root entry into the subsoil. In cracking clays, roots of cereal crops may reach depths of 2 m or more. In dry areas such soils may thus be of high value. On the other hand, a soil with a similar clay content in the same area but with few cracks would have a much lower value (Wilkinson, 1975).

5. Hardsetting Soils

The phenomenon of hardsetting has been described by Mullins et al. (1987, 1990). When dry, such soils are hard, compact, and apparently structureless, and the surface cannot be indented by the pressure of a forefinger (Northcote et al., 1975). They may be soft when wet. The hard condition may extend from the surface throughout the depth of cultivation, to a depth of 30 cm or more. Textures that may exhibit this phenomenon range from loamy sand to sandy clay with the clay fraction frequently dominated by kaolinite and/or illite. The particle size distribution is such that the individual particles will pack to a high density upon wetting alone. Hardsetting becomes worse as the amount of exchangeable sodium increases (McKenzie, 1998). The phenomenon is widespread in hot, dry regions of the world and may be found occasionally in temperate regions where intensive management has reduced soil organic matter content.

D. Soil Compaction

There is great concern worldwide over the adverse effects of soil compaction on crop production (Larsen et al., 1994). These include restricted root growth, reduced aeration, reduced availability of nutrients, reduced infiltration and reduced drainage. Any or all of these can lead to a reduction in plant growth (Batey, 1988, 1989; Larsen et al., 1994), although the manifestation of adverse effects are related to the weather. Soil compaction has been the subject of several international conferences (Anon., 1989; Herman, 1992; Nugis and Lehtveer, 1992) and books (Soane, 1983; Soane and van Ouwerkerk, 1994). However, there is not much guidance available on how to recognize soil compaction in the field. It is particularly important to establish such techniques because of the potential effects of compaction on land degradation and the need to enable appropriate remedies to be applied timeously. In some cropping systems, there may be only a matter of days between the harvest of one crop and the establishment of the next. Such an interval may preclude the use of complex or time-consuming tests for compaction.

The state of packing of a soil can be estimated in the field by careful visual and tactile examination. Such estimates are linked to the texture of the soil and will be of most use in making comparisons within a profile or between parts of the same land, which have been treated differently. The results can be used directly to make decisions on soil tillage, particularly subsoil loosening (Batey, 1988; McKenzie et al., 1990; McKenzie, 1998). They can also be used as an adjunct to measurements of saturated conductivity (Chap. 4), strength (Chap. 11), penetration resistance (Chap. 10), or bulk density (Chap. 8).

1. The Damaging Effects of Compaction and Their Identification

The physical characteristics that can be used to identify compaction are lack of visible pores, high density, high strength, and massive structure. The damaging effects may be caused by a reduction in macroporosity, which will reduce air porosity, saturated hydraulic conductivity, and root penetrability. Such a reduction may bring about changes in other properties, some of which can also be used as diagnostic features (Batey, 1988). These include

1. The formation of anoxic layers or pockets within a compact soil, detectable by a malodorous smell or by a field test to confirm the presence of ferrous ions (Batey and Childs, 1982). One of the consequences of anoxia is denitrification, which can reduce the nitrate content of the soil and therefore reduce plant growth.
2. The presence of wet soil above the compacted layer after rain or irrigation (see Sec. II.E).
3. Comparatively dry soil within and above the compacted layer, due to the greater uptake of water by plants from a shallower layer because of a restriction in root penetration.
4. Reduced nutrient uptake from the dry soil and consequent symptoms of deficiency (e.g., of N).
5. Tortuous root patterns often with a marked horizontal orientation.
6. A reduction in crop growth. This last effect is related to climate and to a reduction in crop root penetration and accessibility to available water. The pattern of crop growth may help diagnosis if also related to the degree of traffic passing over the land.

In many instances one can identify compact soil but cannot tell whether its effects will adversely affect crop yield or quality. Only when a problem is apparent can one deduce that compaction was the likely cause.

Compaction can also induce secondary effects. For example, the retention of water above a compact layer can create a soft zone rendering it susceptible to further compaction (see Sec. II.E). This is common where soils are puddled or poached (surface damage by hooves or feet) (Scholefield and Hall, 1985).

Based on a detailed and systematic examination of the soil, Daniells and Larsen (1991) developed a key with three ratings for soil structure for firm soil (shown in Table 4) and three for loose soil. In a further refinement of their SOILpak scoring procedure (McKenzie, 1998), eight separate factors are assessed and a weighting is given to each to provide a combined score between 0.0 (poor) and 2.0 (good). This system is used extensively by cotton agronomists in Australia to assess soil compaction and to provide options for soil loosening and soil management. It has been shown to correlate well with soil aeration and soil strength mea-

Table 4 Numerical System for Classifying Soil Structure

Firm, most soil: soil below the tilled layer or below the natural loose mulch; aggregates fit together along faces and it requires force to lever them apart
Firm 0 (F0, poor structure)
General: Difficult for spade or knife to penetrate; lumps of soil levered off made up of large tight fitting blocks. These fracture along the lines of forced applied in any direction into units with sharp right-angled corners. Finely grained and even internal surfaces with no pores visible or no subaggregates projecting from the fractured surface. Breaks like heavy dough or plasticine. Low number of new roots
Clod shape: massive, platy, or conchoidal
Clod or ped size: Usually >5 cm towards the surface, larger at depth
Clod faces: dull
Firm 1 (F1, moderate structure)
General: some natural planes of separation but distinct force needed to part the blocks, fracturing taking place mainly along the line of force applied to produce angular and mainly nonporous surfaces
Clod shape: Mixed shapes
Clod or ped size: 0.5–5 cm towards the soil surface, larger at depth
Clod faces: Occasionally shiny faces
Firm 2 (F2, good structure)
General: Parts readily into porous subunits along natural fracture planes that have a smooth and shiny face, or the fractured faces may be polyhedral with the exposed internal surfaces multifaced and with subangular units protruding. Good penetration by new roots
Clod shape: Polyhedral, subangular blocky or lenticular
Clod or ped size: Usually <5 cm towards the soil surface, larger at depth
Clod faces: Shiny

Note: The ratings can be subdivided, e.g., "Firm 1.5" for a structure that is not quite ideal.
Source: After Daniells and Larsen, 1991.

sured by several methods (McKenzie, 1997). A list of diagnostic features has also been made by Wildman (1980).

2. The Location of Compact Soil Layers

Compact soil can be present in several positions within the soil profile. It can be found on the surface as a hardsetting horizon or as a compact layer caused by the application of pressure from the wheels of tractors, trailers, and harvesters, or from the feet of livestock or other animals (including humans). It can occur within the layer of soil disturbed by regular tillage, where more than one denser layer may be found. Smearing as well as pressure may be involved in the formation of thin compact layers.

A particularly important location is just below the working depth of plows, where the plowshare or disk scrapes across the soil and where tractor wheels (or the feet of draught animals) gain traction in the open furrow. Dense layers can also occur due to pedogenic processes as discussed in Sec. II.A.4e. These include induration, in areas subject to past or present periglacial conditions (Fitzpatrick, 1956), and cementing by plinthite and other agents (Soil Survey Staff, 1993).

It is often straightforward to identify the upper surface of a compact layer but difficult to determine the depth to which it extends. Such information is required to make key decisions on deep tillage or subsoiling. The signs characteristic of compaction must be carefully assessed. Otherwise deep loosening may be ineffective or even harmful to the soil. It is also important to reassess the soil condition after a test run of deep tillage to determine whether the work has been effective.

3. Physical Discontinuities

On land that has been tilled and also on land in its natural state, the upper horizon usually has a relatively low bulk density. At the base of this horizon, there is often a sharp increase in bulk density. The looser consistence of the A horizon may be the result of mechanical loosening (tillage) and/or due to greater biotic activity. The lower layer may be firmer simply because it has not been disturbed, but it must be examined carefully to decide whether the hardness is due to compaction. Dry and hard layers can be found in the subsoil after harvest of cereal crops, when the land is rewetting from the surface downwards. In terms of root penetration it is important to realize that a sudden change in bulk density can result in the deflection of roots that would have been able to penetrate the subsoil from a more compact upper horizon that did not so readily permit root buckling (Whiteley et al., 1982).

E. Soil Bearing Capacity

This property is related principally to the texture, to water tension, and to bulk density of the soil. It is applied almost exclusively to surface soils and is linked to the response of soil to traffic by agricultural or military vehicles and to feet or the hooves of animals.

Although this property can be measured using instruments (Chap. 10), a simple field assessment of relative soil resistance can be obtained by pushing a spade, auger, or probe manually into the ground and noting its depth of penetration. An assessment of softness can be made using the squelch test. This was first used as an assessment of field drainage status (MAFF, 1969); a revised version is described below (Batey, 1988).

1. The Squelch Test

The softness of the surface can be ranked on a numerical scale either at a particular spot or on a grid pattern to provide information on the variation in this property over an area. This scale utilizes the reaction of the soil to the pressure applied underfoot by a normal boot:

- SQ 1: The surface is firm and no significant imprint is made by foot pressure.
- SQ 2: The surface is loose or slightly soft with penetration of up to 2 cm by foot pressure.
- SQ 3: The surface is soft underfoot, with imprints between 2 and 5 cm deep.
- SQ 4: The surface is distinctly soft with imprints greater than 5 cm deep.
- SQ 5: Very soft and wet with water standing on the surface and feet making an imprint deeper than 7 cm.
- SQ 6: Water standing on a firm surface.

F. Soil Color

Soil color is easily and accurately determined using a standard chart (Munsell Soil Color Charts, Munsell Color Company Inc., Baltimore, Maryland 21218, U.S.A.). It is a key characteristic in the classification of soil horizons and in the identification of world soil groups (Fitzpatrick, 1988; FAO, 1990; Avery, 1990; Soil Survey Staff, 1993, 1998). When recording the color, it is necessary to state whether the soil is wet or dry, as color is altered by the presence of a film of moisture on the surface of soil particles.

Soil color can also be described using general language. Standardized expressions should be used to describe the color and its intensity. Brown, yellow, orange, red, and gray are the most frequent basic colors; these terms are often combined, for example, yellowish brown. The intensity can be expressed using terms such as *pale* or *dark*. Adjectives with a less universal meaning such as "chocolate" or "warm" (brown) and "mousey" (gray) should be avoided.

1. Soil Color and Soil Profile Characterization

The color of an individual soil horizon, together with the sequence of horizons, are key diagnostic criteria in many systems of soil classification (e.g., FAO, 1990; Soil Survey Staff, 1998). Examples are the pale ash-colored eluviated horizon characteristic of podzols and the gray or mottled horizon of gleysols. The color changes are the result of soil-forming processes acting upon the parent material. A full discussion of these processes can be found in standard books on pedology (e.g., Avery, 1990).

2. Soil Color and Organic Matter

Organic matter tends to darken the color of a soil. This effect is widely used to distinguish the soil on the surface (topsoil, A horizon) from the paler subsoil beneath. The color difference between the two horizons can be somewhat crudely used to give an indication of the relative content of organic matter.

Soils with a high content of sand frequently have a low content of organic matter, and the color difference between topsoil and subsoil may be only slightly discernible. Exceptions to this are podzolic subsoils, which may be black in color due to surface coatings of organic matter.

3. Soil Color and Patterns of Mottling

The presence of mottled color patterns in a soil is an important criterion related to aeration and hence waterlogging. Mottling is caused by changes in the distribution, concentration, and state of oxidation of iron and manganese compounds, which are present in many minerals in the soil. The changes are caused by reducing conditions, which in turn result from microbial activity during anoxic conditions.

The factors governing the intensity of mottling and the distribution of the colors are those affecting the vigor of microbial activity. These include the presence of microorganisms, a substrate for their growth and temperature. Under aerobic conditions organic compounds are utilized as electron acceptors for microbial respiration, with carbon dioxide as the end product. When the supply of oxygen is unable to satisfy the respiratory demand, other electron acceptors are utilized. These include nitrate, ferric, manganic, and sulphate ions. As far as color changes are concerned, the key process is that of the chemical reduction of insoluble reddish ferric compounds to soluble grayish ferrous compounds. The latter may be redistributed within a soil horizon or removed as water moves through it. If and when air reenters the reduced soil, the process is reversed, and any ferrous compounds present are oxidized to the ferric state. The entry of air is often uneven and zones in the vicinity of aerated pathways become enriched with ferric compounds. These may be seen as rust-colored nodules or tubules with a harder crisp consistence. The redder zones can also be seen as diffuse pockets or bands.

In soils with marked vertical fissuring, the faces of the fissures are often uniform pale gray in color. Just behind the gray faces, orange or reddish colors are present. The gray colors are the result of the repeated development of reducing conditions along the fissure, which produces soluble ferrous compounds. Their subsequent leaching leads to the gradual depletion of ferric iron from the soil adjacent to the fissure. This is illustrated in Table 5. It can be seen that the sample taken within 2 mm of the uniformly colored face of a subsoil fissure had a much

Table 5 Patterns of Mottling: The Concentration of Total and Extractable Fe, Mn, Al, and Si Within 2 mm of a Vertical Fissure and from the Center of Adjacent Aggregates

		Face of aggregate	Center of aggregate
Distance of sample from fissure face		0–2 mm	3–8 mm
Color		Gray	Gray matrix, orange/brown mottles
Uniformity		Uniform	Mottled
Iron (% Fe)	Extractable	0.08	0.52
	Total	1.89	2.68
Manganese (% Mn)	Extractable	0.00	0.00
	Total	0.016	0.034
Aluminum (% Al)	Extractable	0.12	0.14
	Total	7.91	7.96
Silicon (% Si)	Extractable	0.07	0.08
	Total	31.2	31.2

Source: Batey, 1981, unpublished data; the samples were taken from a depth of 60–80 cm from a Temuka series profile, Lincoln University Farm, New Zealand.

lower content of both extractable iron and manganese than the brighter colored soil within the aggregates.

Although iron compounds usually dominate the changes in color, manganic compounds can be coprecipitated with the ferric compounds to give much darker colored mottles and nodules that may be almost black.

4. *Historic and Contemporary Mottles*

It is only possible to redistribute iron in the soil when it is in the soluble ferrous form. In situations where the depth or duration of saturation has decreased, mottles created during former anoxic conditions will persist because there is no process for redistribution of iron in aerated conditions. This situation might occur when there has been a decline in the regional water table or artificial drainage within a field. Mottled patterns can therefore be a relict feature, persisting long after the conditions responsible for their formation have gone. Thus, although mottled coloring can be used to assess the wetness class of a soil profile (e.g., Hodge et al., 1984) and in the design of field drainage schemes (Farr and Henderson, 1986), its use must be supported by other evidence of periodic saturation such as water table levels.

5. *Soil Color and Structure Degradation*

When soil aggregates disintegrate as a result of raindrop impact, tillage, or prolonged flooding, the individual particles tend to separate. Clay particles may migrate downwards into the soil or move away if suspended in surface flow. They can also be seen as a differently colored skin on the surface when the soil dries. Fine and very fine sand grains can also be seen as small pockets or thin layers of paler colored material within the topsoil or just beneath it. This phenomenon is used as an indication of structure degradation in soils that contain a significant content of very fine sand or silt (De Leenheer, 1967; Davies, 1975).

G. Soil Saturated Hydraulic Conductivity

Several tests are available to measure the hydraulic conductivity of a soil in either a saturated (Chap. 4) or an unsaturated state (Chap. 5). Because of the time and cost required for these tests, it may be necessary to obtain estimates based on other soil characteristics that can be readily obtained in the field. McKenzie and Jacquier (1997) found that useful predictions of saturated hydraulic conductivity were possible using soil texture, grade of structure, areal porosity, bulk density, dispersion index, and horizon type, all determined in the field. Information on the two properties most strongly related to hydraulic conductivity, soil texture and soil structure, can also be used to select a typical area within a field with relatively uniform properties so that any measurements made can be done at spots that can be regarded as representative of the land unit.

Table 6 shows some typical values of saturated hydraulic conductivity (Smedema and Rycroft, 1983). However, neither texture nor structure alone can be considered to be wholly reliable guides to hydraulic conductivity; the size and

Table 6 Typical Values of Saturated Hydraulic Conductivity Based on Texture and Other Soil Properties

Soil properties	Order of magnitude of saturated hydraulic conductivity (m day^{-1})
Coarse gravelly sand	10–50
Medium sand	1–5
Sandy loam/fine sand	1–3
Loam/clay loam/clay, well structured	0.5–2
Very fine sandy loam	0.2–0.5
Clay loam/clay, poorly structured	0.02–0.2
Dense clay, not cracked, no biopores	<0.002

The values will also depend on the type of clay mineral present in fine-textured soils and on the presence of root channels and holes made by earthworms.

Source: Smedema and Rycroft, 1983.

number of visible pores and the direction of the easiest fracture seem most correlated with conductivity. McKeague et al. (1982) determined eight classes of saturated hydraulic conductivity using visual estimates of structure and porosity. Hydraulic conductivity can be much reduced by compaction and induration (Farr and Henderson, 1986).

H. Sampling Soils for Physical Tests

Several measurements require samples to be taken in the field for subsequent analysis in the laboratory. The location of the sample and how it was taken may have a profound effect on the results. Furthermore, information about the exact position of the sample will be required in the interpretation of the results. For example, if a core contained two horizons with dissimilar properties, the result may be mainly determined by only one of the layers, or if the result is a mean of the two, it would be representative of neither horizon. The disparity in properties is even more striking when taking samples from a strongly fissured clay. Examination in the field can show clear and distinct differences in many physical properties related to the proximity of a major vertical fissure.

Careful examination in the field is therefore required before the location and depth of sampling can be chosen. This proviso also applies when inserting probes, such as tensiometers (Chap. 2), for monitoring soil physical properties. A "standard" sampling depth cannot be recommended for physical tests, nor should samples be taken blindly from below the surface with no knowledge of the properties of the zone from which they were extracted.

Some equipment suitable for examining and sampling soils is described in Sec. II.A.3b. During sampling, transport, and storage, care is required to ensure that no significant changes in physical properties occur prior to analysis. Where analysis is to be performed on soil that has been dried and ground, it is also important that a reliable method of subsampling such as chute splitting or rotary subsampling be used (Mullins and Hutchinson, 1982).

III. SUMMARY AND CONCLUSIONS

Visual and tactile methods can provide a unique, sensitive, detailed, and flexible means of assessing the physical condition of the soil directly in the field. These methods can locate narrow zones of significance and can also provide a picture of the soil as a whole. Such basic tests should not be displaced by, but should complement, those that can measure specific properties with greater accuracy. They can also be a key aid to the interpretation of data obtained by measurement.

Tests for soil examination and evaluation that can be done in the field are described in Sec. II; they include soil texture, structure, compaction, bearing capacity, color, and hydraulic conductivity.

The techniques of field examination and evaluation are important skills to be acquired, not only by soil scientists but also by anyone concerned with the use and management of soils. With training and experience they can be used successfully by farmers, farm workers, agronomists, foresters, ecologists, and others concerned with the protection and best use of the soil and conservation of the environment.

REFERENCES

Ashmore, P. J. 1996. *Neolithic and Bronze Age Scotland*. London: B. T. Batsford.

ADAS (Agricultural Development and Advisory Service). 1971. *Soil Field Handbook*. ADAS Advisory Papers No. 9. London: Ministry of Agriculture, Fisheries and Food.

Avery, B. W. 1990. *Soils of the British Isles*. Wallingford, England: CAB International.

Batey, T. 1975. Soil examination in the field. In: *Soil Physical Conditions and Crop Production*. MAFF Tech. Bull. No. 29. London: HMSO, pp. 207–217.

Batey, T. 1988. *Soil Husbandry*. Aberdeen, Scotland: Soil and Land Use Consultants Ltd.

Batey, T. 1989. Control of compaction on the farm. In: *Soil Compaction as a Factor Determining Plant Productivity*. Abs. Internat. Soil Sci. Soc. Conf., Lublin, Poland, pp. 24–25.

Batey, T. 1992. Visual assessment of the physical properties of arable and grassland soils. In: Herman, M., ed. *Problems in Modern Soil Management*. Proc. Internat. Conf. Soil Tillage Res. Org., Brno, Czechoslovakia, pp. 25–27.

Batey, T., and C. W. Childs. 1982. A qualitative field test for locating zones of anoxic soil. *J. Soil Sci.* 33:563–566.

Batey, T., and D. C. McKenzie. 1999. Deep subsoil compaction—Letter to the editor. *Soil Use Manage.* 15:136.

Bibby, J. S., H. A. Douglas, A. J. Thomasson, and J. S. Robertson. 1982. *Land Capability Classification for Agriculture*. Soil Survey of Scotland Monograph. Aberdeen, Scotland: Macaulay Land Use Res. Inst.

Brewer, R., and J. R. Sleeman. 1988. *Soil Structure and Fabric*. Melbourne, Australia: CSIRO Div. Soils.

Corbett, W. M., and W. Tatler. 1970. *Soils in Norfolk: (Beccles North)*. Survey Record No. 1. Harpenden, England: Soil Survey of England and Wales.

Daniells, I. G., and D. L. Larsen, eds. 1991. *SOILpak: A Soil Management Package for Cotton Production on Cracking Clays*. Narrabri, NSW, Australia: NSW Agriculture.

Daniells, I. G., D. L. Larsen, D. C. McKenzie, and D. T. W. Anthony. 1996. SOILpak: A successful decision support system for managing the structure of vertisols under irrigated cotton. *Aust. J. Soil Res.* 36:879–889.

Davies, D. B. 1974. Soil structure and crop production. In: Mackney, D., ed. *Soil Type and Land Capability*. Tech. Monograph No. 4. Harpenden, England: Soil Survey of England and Wales, pp. 117–124.

Davies, D. B. 1975. Field behaviour of medium textured and 'silty' soils. In: *Soil Physical Conditions and Crop Production*. MAFF Tech. Bull. No. 29. London: HMSO, pp. 52–75.

Davies, D. B., and D. Payne. 1988. Management of soil physical properties. In: Wild, A., ed. *Russell's Soil Conditions and Plant Growth*. 11th ed. London: Longman, pp. 412–448.

De Boodt, M., H. Frese, A. J. Low, and P. K. Peerlkamp, eds. 1967. *West European Methods for Soil Structure Determination*. Ghent, Belgium: State Faculty Agric. Sci.

De Leenheer, L. 1967. Determination in the field of soil structure deterioration. In: De Boodt, M., H. Frese, A. J. Low, and P. K. Peerlkamp, eds. *West European Methods for Soil Structure Determination*. Ghent, Belgium: State Faculty Agric. Sci., pp. 70–72.

Eagle, D. J. 1975. ADAS ley fertility experiments. In: *Soil Physical Conditions and Crop Production*. MAFF Tech. Bull. No. 29. London: HMSO, pp. 344–359.

Farr, E., and W. C. Henderson. 1986. *Land Drainage*. Longman Handbooks in Agriculture. London: Longmans.

FAO (Food and Agricultural Organisation of the United Nations). 1976. *A Framework for Land Evaluation*. Soils Bull. No. 32. Rome: FAO.

FAO. 1990. *Guidelines for Soil Profile Description*. 3d ed. rev. Soil Resources, Management and Conservation Service, Land and Water Development Div. Rome: FAO.

FAO. 1998. *World Reference Base for Soil Resources*. No. 84. World Soil Resource Reports. Rome: FAO.

Fitzpatrick, E. A. 1956. An indurated soil horizon formed by permafrost. *J. Soil Sci.* 7:248.

Fitzpatrick, E. A. 1988. *Soil Horizon Designation and Classification*. Tech. Paper No. 17. Wageningen, The Netherlands: Internat. Soil Reference and Information Centre.

Fitzpatrick, E. A. 1993. *Soil Microscopy and Micromorphology*. London: John Wiley.

Gameda, S., G. S. V. Raghaven, E. McKyes, G. R. Mehuys, A. Watson, and J. Duval. 1994. Profile characteristics of a clay soil under heavy axle load compaction. *Soil Till. Res.* 29:189–194.

Glentworth, R., and J. W. Muir. 1963. *The Soils Around Aberdeen, Inverurie and Fraserburgh*. Memoirs of Soil Survey of Scotland. Edinburgh: HMSO.

Greenhalgh, S. E. 1995. An evaluation of field soil structural assessment techniques for grey cracking clays under irrigated cotton production. M. Rural Sci. thesis, Univ. of New England, NSW, Australia.

Gregory, P. J. 1988. Growth and functioning of plant roots. In: Wild, A., ed. *Russell's Soil Conditions and Plant Growth*. 11th ed. London: Longman, pp. 113–167.

Hakansson, I., and H. Petelkau. 1994. Benefits of limited axle load. In Soane, B. D., and van Ouwerkerk, C., eds. *Soil Compaction in Crop Production*. Amsterdam: Elsevier, pp. 479–499.

Harrod, M. F. 1975. Field behaviour of light soils. In: *Soil Physical Conditions and Crop Production*. MAFF Tech. Bull. No. 29. London: HMSO, pp. 22–51.

Herman, M., ed. 1992. *Problems in Modern Soil Management*. Proc. Conf. Internat. Soil Tillage Res. Org., Brno, Czechoslovakia.

Hettiaratchi, D. R. P. 1987. A critical state soil mechanics model for agricultural soils. *Soil Use Manage.* 3:94–105.

Hodge, C. A. H., R. G. O. Burton, W. M. Corbett, R. Evans, and R. S. Seale. 1984. *Soils and Their Use in Eastern England*. Bull. No. 13. Harpenden, England: Soil Survey of England and Wales.

Hodgson, J. M., ed. 1974. *Soil Survey Field Handbook*. Tech. Monograph No. 5. Harpenden, England: Soil Survey of England and Wales.

Hodgson, J. M. 1978. *Soil Sampling and Soil Description*. Oxford: Oxford Univ. Press.

Hodgson, J. M., J. M. Hollis, R. J. H. Jones, and R. C. Palmer. 1976. A comparison of field estimates and laboratory analyses of the silt and clay contents of some West Midlands soils. *J. Soil Sci.* 27:411–419.

International Soil Science Conference. 1989. *Soil Compaction as a Factor Determining Plant Productivity*. Abs. Internat. Soil Sci. Soc. Conf., Lublin, Poland.

Kemper, W. D., and R. C. Rosenau. 1986. Aggregate stability and size distribution. In A. Klute, ed. *Methods of Soil Analysis, Part I*. 2d ed. Madison, WI: Soil Sci. Soc. Am., pp. 425–442.

Larsen, W. E., A. Eynard, A. Hadas, and J. Lipiec. 1994. Control and avoidance of soil compaction in practice. In Soane, B. D., and van Ouwerkerk, C., eds. *Soil Compaction in Crop Production*. Amsterdam: Elsevier.

Lowe, J. A. H. 1993. An investigation into soil compaction (and other soil problems) on land reinstated after pipeline laying. M.Sc. thesis, Univ. of Aberdeen, Scotland.

Mackney, D., ed. 1974. *Soil Type and Land Capability*. Tech. Monograph No. 4. Harpenden, England: Soil Survey of England and Wales.

MAFF (Ministry of Agriculture, Fisheries and Food). 1969. *Field Drainage Experimental Unit Annual Report 1969*. Cambridge, England: MAFF.

MAFF. 1975. *Soil Physical Conditions and Crop Production*. Tech. Bull. No. 29. London: HMSO.

MAFF. 1988. *Agricultural Land Classification of England and Wales*. London: HMSO.

McCormack, D. E. 1986. Land evaluations that consider soil compaction. *Soil Till. Res.* 10:21–27.

McKeague, J. A., C. Wang, and G. C. Topp. 1982. Estimating saturated hydraulic conductivity from soil morphology. *Soil Sci. Soc. Am. J.* 46:1239–1244.

McKenzie, D. C. 1997. Measurement and management of compaction damage on vertisols under irrigated cotton. Ph.D. thesis, Univ. of Sydney, Australia.

McKenzie, D. C. 1998. *SOILpak for Cotton Growers*. 3d ed. Orange, NSW, Australia: NSW Agriculture.

McKenzie, D. C., T. S. Abbott, D. T. W. Anthony, P. J. Hulme, D. A. MacLeod, and F. R. Higginson. 1990. Management of subsoil degradation in an Australian vertisol used for irrigated cotton production. *Trans. 14th Internat. Congr. Soil Sci., Kyoto*, Vol. 6, pp. 176–181.

McKenzie, N. J., and D. A. MacLeod. 1989. Relationships between soil morphology and soil properties relevant to irrigated and dryland agriculture. *Aust. J. Soil Res.* 27: 235–258.

McKenzie, N. J., and D. J. Jacquier. 1997. Improving the field estimation of saturated hydraulic conductivity in soil survey. *Aust. J. Soil Res.* 35:803–825.

Mullins, C. E., and B. J. Hutchinson. 1982. The variability introduced by various subsampling techniques. *J. Soil Sci.* 33:547–561.

Mullins, C. E., I. M. Young, A. G. Bengough, and J. G. Ley. 1987. Hard-setting soils. *Soil Use Manage.* 3:79–83.

Mullins, C. E., D. A. MacLeod, K. H. Northcote, J. H. Tisdall, and I. M. Young. 1990. Hardsetting soils; Behaviour, occurrence and management. *Adv. Soil Sci.* 2: 37–108.

Northcote, K. H., G. D. Hubble, R. F. Ishbell, C. F. Thompson, and E. Bettany. 1975. *A Description of Australian Soils*. Australia: CSIRO.

Nugis, E., and R. Lehtveer, eds. 1992. *Soil Compaction and Soil Management*. Proc. Conf. Internat. Soil Tillage Res. Org., Tallinn, Estonia.

Payne, D. 1988. Soil structure, tilth and mechanical behaviour. In: Wild, A., ed. *Russell's Soil Conditions and Plant Growth.* 11th ed. London: Longman, pp. 378–411.

Peerlkamp, P. K. 1967. Visual estimation of soil structure. In De Boodt, M., H. Frese, A. J. Low, and P. K. Peerlkamp, eds. *West European Methods for Soil Structure Determination.* Ghent, Belgium: State Faculty Agric. Sci., Vol. 2, p. 11.

Pizer, N. H. 1990. *Understanding Soils: The Experience of an Adviser.* Ashford, Kent, England: Wye College.

Russell, E. J. 1961. *The World of the Soil.* London: Collins.

Scholefield, D., and D. M. Hall. 1985. A method to measure the susceptibility of pasture soils to poaching by cattle. *Soil Use Manage.* 1:134–138.

Simpson, K. 1983. *Soil.* London: Longman.

Smart, P. 1998. Deep subsoil compaction—Letter to the editor. *Soil Use Manage.* 14:69.

Smedema, L. K., and D. W. Rycroft. 1983. *Land Drainage.* London: Batsford.

Soane, B. D., ed. 1983. *Compaction by Agricultural Vehicles: A Review.* Tech. Rep. No. 5. Penicuik, Scotland: Scottish Inst. Agric. Eng.

Soane, B. D., and C. van Ouwerkerk, eds. 1994. *Soil Compaction in Crop Production.* Amsterdam: Elsevier.

Soil Survey Staff. 1975. *Soil Taxonomy: A Basic System of Soil Classification for Making and Interpreting Soil Surveys.* Handbook No. 436. Washington, D.C.: U.S. Dept. of Agriculture.

Soil Survey Staff. 1993. *Soil Survey Manual.* Agric. Handbook No. 18. Washington, D.C.: U. S. Dept. of Agriculture.

Soil Survey Staff. 1998. *Keys to Soil Taxonomy.* 8th ed. Washington, D.C.: USDA Natural Resources Conservation Service.

Sullivan, L. A., and I. L. Montgomery. 1998. Deep subsoil compaction of two cracking clays used for irrigated cotton production in Australia. *Soil Use Manage.* 14: 56–57.

Taylor, N. H., and I. J. Pohlen. 1976. *Soil Survey Method: NZ Handbook for the Field Study of Soils.* Bull. No. 25. New Zealand Soil Bureau, Wellington.

Trapnell, C. G., and R. Webster. 1986. Microaggregates in red earths and related soils in East and Central Africa, their classification and occurrence. *J. Soil Sci.* 37: 109–123.

van Ouwerkerk, C., A. Canarache, T. Batey, and M. Herman. 1992. Conference Conclusions. In: Herman, M., ed. *Problems in Modern Soil Management.* Proc. Conf. Internat. Soil Tillage Res. Org., Brno, Czechoslovakia, pp. 302–303.

White, R. E. 1997. *Principles and Practice of Soil Science.* 3d ed. Oxford: Blackwell Science.

Whiteley, G. M., J. S. Hewitt, and A. R. Dexter. 1982. The buckling of plant roots. *Physiol. Plant.* 54:333–342.

Wild, A., ed. 1988. *Russell's Soil Conditions and Plant Growth.* 11th ed. London: Longman.

Wildman, W. E. 1980. *Diagnosing Soil Physical Problems.* Leaflet 2664. Univ. of California Div. Agric. Sci.

Wilkinson, B. 1975. Field experience on heavy soils. In: *Soil Physical Conditions and Crop Production.* MAFF Tech. Bull. No. 29. London: HMSO, pp. 76–93.

Index

Milton Keynes UK
Ingram Content Group UK Ltd.
UKHW031536071024
449327UK00024B/1882